MODERN METHODS OF PLANT ANALYSIS

FOUNDED BY

K. PAECH M. V. TRACEY

CONTINUED BY

H. F. LINSKENS M. V. TRACEY

VOLUME V

CONTRIBUTORS

K. BIEMANN · N. K. BOARDMAN · B. BREYER · S. P. BURG · W. L. BUTLER
D. J. DAVID · P. S. DAVIS · A. E. DIMOND · A. C. HILDEBRANDT
F. A. HOMMES · O. KRATKY · H. F. LINSKENS · H. MOOR
K. H. NORRIS · I. J. O'DONNELL · J. V. POSSINGHAM · H. PRAT
D. H. M. VAN SLOGTEREN · E. STAHL · J. A. VAN DER VEKEN
J. P. H. VAN DER WANT · E. F. WOODS

WITH 228 FIGURES

SPRINGER-VERLAG
BERLIN · GÖTTINGEN · HEIDELBERG
1962

MODERNE METHODEN DER PFLANZENANALYSE

BEGRÜNDET VON

K. PAECH M. V. TRACEY

FORTGEFÜHRT VON

H. F. LINSKENS M. V. TRACEY

5. BAND

BEARBEITET VON

K. BIEMANN · N. K. BOARDMAN · B. BREYER · S. P. BURG · W. L. BUTLER
D. J. DAVID · P. S. DAVIS · A. E. DIMOND · A. C. HILDEBRANDT
F. A. HOMMES · O. KRATKY · H. F. LINSKENS · H. MOOR
K. H. NORRIS · I. J. O'DONNELL · J. V. POSSINGHAM · H. PRAT
D. H. M. VAN SLOGTEREN · E. STAHL · J. A. VAN DER VEKEN
J. P. H. VAN DER WANT · E. F. WOODS

MIT 228 ABBILDUNGEN

SPRINGER-VERLAG
BERLIN · GÖTTINGEN · HEIDELBERG
1962

ISBN-13:978-3-642-45995-5
DOI: 10.1007/978-3-642-45993-1

e-ISBN-13:978-3-642-45993-1

Library of Congress Catalog Card Number A 55—6022

Softcover reprint of the hardcover 1st edition 1962

Druck der Brühlschen Universitätsdruckerei Gießen

Vorwort zu den Bänden V, VI und VII.

Die Pläne für die ersten vier Bände der „Modernen Methoden der Pflanzenanalyse" wurden vor einer Reihe von Jahren entworfen, der letzte Band erschien schließlich im Jahre 1956. Prof. Dr. KARL PAECH, der eigentliche Initiator des Gesamtwerkes, erlebte das vollständige Erscheinen nicht mehr. Sein allzu früher Tod war ein großer Verlust für die biochemische Pflanzenphysiologie. Das Hinscheiden wurde als besonders schmerzlich von jenen empfunden, die das Glück hatten, mit ihm auf seinen mannigfaltigen Interessengebieten zusammenarbeiten zu dürfen.

Prof. Dr. H. F. LINSKENS, Universität Nijmegen (Holland), hat als Mitherausgeber die Nachfolge von Prof. PAECH angetreten.

In den Jahren seit der Veröffentlichung der ersten vier Bände haben sich zahlreiche Methoden in der Pflanzenanalyse durchgesetzt, deren Entwicklung sich damals noch in den Anfangsstadien befand oder erst abzeichnete. Der Band V wurde daher als Ergänzung zum Band I geplant und umfaßt allgemeine Analysen-Methoden. Im Band VI sind weiterhin Stoffgruppen behandelt, so daß dieser eine Ergänzung zu den Bänden II—IV darstellt.

In der ursprünglichen Planung des Gesamtwerkes waren die Methoden der Enzymologie ausgeklammert. Der Rahmen wurde nunmehr erweitert: Im Band VI werden zunächst die allgemeinen Methoden der Enzymchemie behandelt, während Band VII die einzelnen Enzymgruppen bei Pflanzen umfaßt.

Die Herausgeber glauben, daß die vorgelegten Ergänzungsbände dazu beitragen, die „Modernen Methoden der Pflanzenanalyse" als ein nützliches Laboratoriumshilfsmittel auf dem neuesten Stand zu halten. Die Aufnahme der pflanzlichen Enzymologie wird für viele Untersucher, so hoffen wir, eine willkommene Bereicherung sein.

Wir waren besonders glücklich, für die Vorbereitung der enzymologischen Teile uns der tätigen Mitarbeit von Herrn Prof. SANWAL, Universität of Manitoba, Winnipeg, Man., (Canada) erfreuen zu können.

H. F. LINSKENS
M. V. TRACEY

Nijmegen und Sydney, 1962

Introduction to Volumes V, VI and VII.

The first four Volumes of "Modern Methods of Plant Analysis" were planned some years ago, the last volume appearing early in 1956. Professor PAECH, the initiator of the work, did not live to see publication completed; his early death was a great loss to plant biochemistry and one that was felt particularly sharply by those who had been fortunate enough to be associated with him in his many active fields of interest. Professor LINSKENS of the R. K. University, Nijmegen, has replaced Professor PAECH as co-editor.

During the years that have elapsed since the publication of the first four volumes many methods, then in the early stages of development or only foreshadowed, have become widely used. Volume V has been planned to supplement and bring up to date the original Vol. I which was concerned with analytical methods of general application. Volume VI is concerned in part with adding to the range of groups of compounds dealt with in Volumes II—IV. The work as originally planned did not include a consideration of methods for the detection and assay of enzymes. The scope of the work has now been enlarged and in Volume VI will also be found a treatment of general methods of enzyme chemistry while in Vol. VII individual groups of enzymes are covered.

It is hoped and believed that the publication of these three additional volumes will make "Modern Methods of Plant Analysis" even more useful than was indicated by the encouraging reception of the earlier volumes. A welcome feature to many will be the widening of its scope by the inclusion of material on enzyme methods.

We were fortunate in being able to secure the assistance of Professor SANWAL of the University of Manitoba in the preparation of those sections concerned with enzymes.

Sydney and Nijmegen, 1962.

M. V. TRACEY
H. F. LINSKENS

Aus der Einleitung zu den Bänden I—IV.

Viele bedeutende Errungenschaften der modernen Biologie sind erst nach Erfindung und Anwendung von leistungsfähigen Methoden möglich geworden. Man kann zwar nicht sagen, daß wir in einem „Zeitalter der Methoden" leben, aber jeder Schritt zur Lösung eines wichtigen biologischen Problems mußte und muß durch die Entwicklung einer geeigneten Methode vorbereitet werden. Es gibt nicht viele „klassische" Verfahren, die sich in der physiologisch-chemischen Forschung erhalten haben. Physik und Chemie entdecken immer wieder spezifische, genauere und bequemere analytische Möglichkeiten, die von Biochemikern übernommen und für die Anwendung auf Pflanzenmaterial abgewandelt werden. Wenn solche neue Methoden in rein analytischen Zeitschriften veröffentlicht werden, so sind sie zwischen analytischen Verfahren für alle möglichen Gebiete der Chemie, vom Petroleum bis zu Schwermetallegierungen, verborgen. Erscheinen sie aber als methodischer Teil in den Arbeiten über spezielle biochemische Probleme, so gibt oft weder der Titel der Arbeit noch die Zusammenfassung einen Hinweis auf die wertvolle Methode und ihre Anwendungsmöglichkeiten. Neue leistungsfähige Methoden erreichen deshalb oft nicht unmittelbar alle die Forscher, die bei ihren Arbeiten größten Nutzen daraus ziehen könnten.

Das waren die entscheidenden Überlegungen, die zur Planung dieses Handbuchs führten. Nach unseren Erfahrungen besteht ein Bedürfnis für eine moderne Sammlung zuverlässiger Methoden der Pflanzenanalyse vor allem auch in weiten Kreisen der angewandten Botanik, von landwirtschaftlichen und Gartenbau-Instituten bis zu pharmazeutischen und technischen Untersuchungsanstalten, die mit pflanzlichem Material zu tun haben.

Abgesehen von der Stoffwechselphysiologie, für die natürlich seit jeher chemisch-analytische Methoden unerläßlich waren, haben sich inzwischen auch andere Zweige der reinen Botanik so entwickelt, daß sie auf die genaue Kenntnis der chemischen Zusammensetzung der Pflanzen oder bestimmter Inhaltsstoffe angewiesen sind. Das gilt für die Genetik, die Wachstumsphysiologie und sogar für die Taxonomie. Einerseits die zerstreute Publikation der analytischen Methoden und andererseits die fehlende Übung in der Beurteilung der Leistungsfähigkeit chemischer Verfahren machen es für Neulinge auf dem Gebiet der Pflanzenanalyse manchmal schwierig, die für ihre Zwecke am besten geeignete Methode aufzufinden oder auszuwählen. Die in diesem Handbuch gebotene Sammlung kann deshalb wohl auch dazu dienen, den Weg zur Lösung von Problemen der reinen Botanik auf denjenigen Gebieten zu ebnen, wo die chemische Zusammensetzung der Pflanzen nur als Indicator anderer Funktionen als des Stoffwechsels von Bedeutung ist.

Es kann nicht die Absicht dieses Handbuches sein, ein biochemisches System der Pflanzenstoffe zu bieten oder die Physiologie des pflanzlichen Stoffwechsels darzustellen, sondern hier sollen rein analytische Gesichtspunkte vorherrschen. Unser Ziel war, eine möglichst umfassende, zuverlässige und zeitgemäße Hilfe für die Laboratoriumsarbeit zu schaffen. Wir sind den Autoren der einzelnen Kapitel zu größtem Dank verbunden, daß sie die Last auf sich genommen haben, die Methoden und Verfahren so zusammenzustellen, daß ihre persönliche Kenntnis und Erfahrung in der experimentellen Biochemie den Benutzern des Werkes in weitestem Umfange zugute kommen wird. In dankenswerter Weise haben sich die Autoren einer ausführlicheren, für sie viel verlockenderen Diskussion der dynamischen Seite der Pflanzenstoffe, ihrer Biogenese und ihres Umsatzes, enthalten.

Erfahrungsgemäß erfordert fast jedes Pflanzenmaterial eine gewisse Modifikation und Anpassung des analytischen Vorgehens. Diese Notwendigkeit scheint uns jedoch den Wert der hier gebotenen Sammlung von Methoden nicht wesentlich zu beeinträchtigen, denn die Abwandlungen sind im allgemeinen nicht prinzipiell, und sie bauen auf alle Fälle auf bestehende und erprobte Verfahren auf. Die in diesem Werk aufgenommenen Methoden sind kritisch gesichtet, ohne daß die dargebotene Auswahl zu stark beschnitten worden wäre. Erst wenige Methoden haben den Wert von Standard-Methoden für die betreffenden Substanzen erreicht. Dank der großen Erfahrung der Mitarbeiter konnten meist die Grenzen der Verfahren angedeutet und mögliche Fehlerquellen bei der Anwendung auf neues, noch unbekanntes Material aufgezeigt werden.

Die verschiedenen Beiträge zu diesem Handbuch sind weder in der Auffassung der Aufgabe noch in der Ausgestaltung ganz uniform. Eine gewisse Einheitlichkeit ist natürlich angestrebt und wohl auch erreicht worden, aber die Persönlichkeit des Autors gibt jedem Kapitel sein besonderes Gepräge, und die Herausgeber haben ihren Ehrgeiz nicht darein gesetzt, durch ein starres Schema die individuelle Gestaltung zu verwischen.

Tübingen und Rothamsted, Juli 1954.

K. Paech
M. V. Tracey

From the Introduction to Volumes I—IV.

Many outstanding advances in modern biology have been made as a result of the invention and application of efficient methods. This does not necessarily mean that we are living in an age of methods but every stride towards the solution of an important biological problem has been, and will have to be, prepared by the forging of the appropriate method. There are not very many 'classical' techniques that have become part of the standard equipment of biological research. Again and again chemistry and physics provide more specific, more exact, and more convenient analytical devices that are taken over by biochemists and modified to suit plant material. If published in purely analytical journals these recent methods are hidden in a welter of other methods applying to all fields of analysis from petroleum to heavy metal alloys. When they appear in the methods section of papers on specialised biochemical subjects they may be obscured by a title and discussion giving no hint of the presence of a valuable method or of its general applicability. As a result new methods very often do not immediately reach those manifold groups of research workers to whom they would be of the greatest use.

These are the considerations that led to the plan of this handbook. We are convinced that there is a real need for a collection of reliable up-to-date methods for plant analysis in large areas of applied biology ranging from agricultural and horticultural experiment stations to pharmaceutical and technical institutions concerned with raw materials of plant origin.

Apart from the study of plant metabolism in which analytical methods are essential many branches of pure botany, originally not concerned with the chemical composition of plants, have developed in such a way as to depend on accurate knowledge of the nature and amounts of plant constituents. This applies among others to genetics, the physiology of growth, and taxonomy. The scattered publication of analytical methods and lack of experience in judging chemical methods makes it hazardous for many of the newcomers in the field of plant analysis to find or select the best available methods for their purposes. The collection offered in this handbook may help to pave the way for tackling problems in those fields of pure botany where chemical components are useful as indicators of the varied activities of the living plant cell.

It is not in the scope of this work to produce a handbook of plant metabolism or a biochemical system of plant substances. Had it been, then the structure of the book and the arrangement of the contents would have been more logical than that which has been imposed by the purely analytical considerations that prevail in the present work, for example, mineral substances could not have been separated from many other sections. The ambition of the editors, however, was to produce a laboratory manual of the highest standard possible. We are greatly indebted to the authors of the various chapters for their willingness to undertake the task of discussing the detailed techniques of analysis in such a way that the user of these volumes will be able to make the greatest use of the authors' experience and knowledge of experimental biochemistry. The purpose of the work is furthered by the fact that the authors have avoided discussing problems of the biogenesis and metabolism of the substances with which they are concerned, tempting and fruitful though these physiological considerations may have appeared to them.

It is true that each particular plant material may require some modification of the usual methods and techniques. Nevertheless this necessity does not reduce the value of the collection given here, for the modifications will generally be slight and not fundamental, and be based on existing techniques. The methods set forth in the present work have been critically selected, some of them have reached the status of standard methods for the substances concerned. Owing to the wide experience of the contributors, proper emphasis has been laid on the limitations of the techniques described and reference has been made, where possible, to pitfalls likely to be met with in their application to unknown plant material.

The various contributions are not uniform in either style or approach. The author's personality gives each an individual form and attempts by the editors to impose a rigid mould and extinguish individual preferences have been avoided.

Rothamsted and Tübingen, July 1954.

M. V. Tracey
K. Paech

Inhaltsverzeichnis. — Contents.

Inhalt der übrigen Bände. — Contents of other Volumes.

1. Band.— Volume I.

2. Band. — Volume II.

3. Band. — Volume III.

4. Band. — Volume IV.

6. Band. — Volume VI

General Methods of Enzyme Chemistry.

7. Band — Volume VII.

Individual Groups of Enzymes.

Enzymes of Amino Acid Metabolism.

Mitarbeiter von Band V. — Contributors to Volume V.

KLAUS BIEMANN, Associate Professor of Chemistry, Massachusetts Institute of Technology, Cambridge 39, Massachusetts (USA).

N. K. BOARDMAN, Doctor, C.S.I.R.O., Division of Plant Industry, P.O. Box 109, Canberra City, A.C.T. (Australia).

B. BREYER, Professor D.Sc., M.D., Dr. phil., State Hospital, Lidcombe, N.S.W. (Australia).

STANLEY P. BURG, Assistant Professor, Department of Physiology, University of Miami, Medical School, and Visiting Investigator Howard Hughes Medical Research Institute, Miami, Florida (USA).

WARREN L. BUTLER, Biophysicist, Instrumentation Research Laboratory, AMS, USDA, Beltsville, Maryland (USA).

DAVID J. DAVID, B.Sc., M.Sc., C.S.I.R.O., Division of Plant Industry, P.O. Box 109, Canberra City, A.C.T. (Australia).

P. S. DAVIS, Australian Atomic Energy Commission, Research Establishment, Lucas Heights, N.S.W. (Australia).

A. E. DIMOND, Chief, Department of Plant Pathology and Botany, The Connecticut Agricultural Experiment Station, P.O. Box 1106, New Haven 4, Connecticut (USA).

ALBERT C. HILDEBRANDT, Ph.D., Professor of Plant Pathology, Department of Plant Pathology, University of Wisconsin, Madison 6, Wisconsin (USA).

F. A. HOMMES, Dr., Research Associate, Department of Biochemistry, School of Medicine, University of Nijmegen, Nijmegen (The Netherlands).

OTTO KRATKY, o. Professor Dr., Institut für physikalische Chemie der Universität, Graz, Halbärthgasse 5 (Austria).

H. F. LINSKENS, Dr. phil., o. Professor der Botanik, Botanisches Laboratorium der Universität, Driehuizerweg 200, Nijmegen (The Netherlands).

HANS MOOR, Dr., Laboratorium für Elektronenmikroskopie, Institut für allgemeine Botanik, Zürich 6 (Switzerland).

KARL H. NORRIS, Leader, Instrumentation Research Laboratory, AMS, USDA, Plant Industry Station, Beltsville, Maryland (USA).

I. J. O'DONNELL, M.Sc., C.S.I.R.O., Division of Protein Chemistry, Wool Research Laboratories, Parkville, Victoria (Australia).

J. V. POSSINGHAM, C.S.I.R.O., Division of Plant Industry, P.O. Box 109, Canberra City, A.C.T. (Australia). Present address: C.R.S., Merbein, Victoria (Australia).

HENRI PRAT, Professeur Dr., Faculté des Sciences, Place Victor Hugo, Marseille 3e (France).

D. H. M. VAN SLOGTEREN, Ir., Laboratorium voor Bloembollenonderzoek, Heereweg 345 a, Lisse (The Netherlands).

EGON STAHL, Professor, Dr. rer. nat., Institut für Pharmakognosie der Universität des Saarlandes, 66 Saarbrücken-15 (Germany).

J. A. VAN DER VEKEN, Ir., Laboratorium voor Bloembollenonderzoek, Heereweg 345 a, Lisse (The Netherlands).

J. P. H. VAN DER WANT, Professor Dr. Ir., Laboratorium voor Virologie, Landbouwhogeschool, Wageningen (The Netherlands).

E. F. WOODS, M.Sc., C.S.I.R.O., Division of Protein Chemistry, Wool Research Laboratories, Parkville, Victoria (Australia).

Emission and Atomic Absorption Spectrochemical Methods.

By

D. J. David.

With 7 Figures.

In these methods of analysis the sample is subjected to temperatures at which its constituent compounds are dissociated into free atoms or converted to temperature-stable molecules. In emission methods, use is made of the fact that a small proportion of the free ground-state atoms or molecules present in a flame, arc or spark are excited to energy levels above the ground-state by heat or electrical effects, then dissipate this energy, either directly or stepwise, by emission of light at wave-lengths characteristic of the atomic or molecular species involved; atomic absorption analysis is based on the fact that free ground-state atoms of an element are capable of absorbing light at the same wave-lengths as the light which they would emit if excited to energy levels above the ground-state.

Reference to sources of spectroscopic equipment is avoided in this article, except in the case of atomic absorption equipment, because so many alternatives are available for performance of a given operation that description and assessment of each could only be given at the expense of description of methods. The range and characteristics of available spectroscopic equipment is, in any case, freely available from the manufacturers' agents.

In using the methods described, it is essential that all reagents should be of the highest possible purity. Water and analytical-reagent grade acids should be redistilled from glass or silica. Ammonium hydroxide and ammonium salts should be prepared using gaseous ammonia.

A. Flame Emission Methods.

The earliest flame emission methods were qualitative or semi-quantitative and consisted in the manual or mechanical introduction of platinum wire or gauze, moistened with the sample solution, into a flame. These and other similar methods of sample introduction, as well as the use of fine droplets from an effervescing sample solution, have been superseded, for quantitative analysis, by the entrainment of fine droplets of the sample solution in the combustible gases by means of an atomiser operated on the air or oxygen supply to the flame. MITCHELL (1948) adequately reviews the development of flame emission methods of analysis.

The most commonly used flame gases are air-propane, air-acetylene, oxy-acetylene and oxy-hydrogen. Such flames excite only lines of low excitation potential and, for this reason, flame emission spectra are very much simpler than arc or spark emission spectra. Instruments of low resolving power, such as the medium spectrograph, or, for some elements, light filters can, therefore, be used for isolation of the analysis lines or bands.

In flame emission analysis there are three known possible sources of inaccuracy arising from variation in the physical and chemical properties of the solutions sprayed into the flame. The rate of delivery and mean droplet size of the spray entering the flame from the atomiser is dependent on the viscosity and surface tension of the solution sprayed; the presence of varying concentrations of ions in the solution which can form refractory compounds with the element being determined will cause variable depression of its emission from the flame; an extraneous element in the solution may emit a line or band which is coincident with that emitted by the analysis element or may increase the general background emission from the flame. These effects are alleviated either by so constituting the sample and standard solutions as to ensure uniformity in chemical and physical properties or, in the case of interference from radiation by extraneous elements, by suppression of the radiation or by correction for it.

With the recent rapid development of photoelectric detection, the Lundegardh (1928) photographic method is losing ground. A description of it is, however, given here because it has the attributes of permanently recording the concentrations of several elements in a sample simultaneously and of permitting more precise correction for flame background than is possible by flame photometric methods.

I. LUNDEGARDH Method (MITCHELL, 1948).

Apparatus. LUNDEGARDH atomiser-burner assembly; compressed air and acetylene supplies accurately controlled by diaphragm valves; pressure gauge and water manometer for air and acetylene supplies, respectively; quartz spectrograph of medium dispersion (2000—8000 Å over 12—25 cm spectrum length); Ilford Zenith photographic plates and photographic developing equipment; non-recording microphotometer; general laboratory glass- and silica-ware; balance; muffle furnace; steam bath.

Ten grams of oven dry, ground plant material in a silica dish is ashed by ignition at 450° C in a muffle furnace overnight. The ash is evaporated to dryness twice with concentrated hydrochloric acid on a steam bath, taken up in 25 ml of 2 N HCl, filtered, washed and the combined filtrate and washings made up to 500 ml with water. This solution can generally be used for the determination of magnesium, sodium and manganese. For potassium and calcium, five-fold or greater dilution is necessary. Dilutions should be carried out with hydrochloric acid solution of concentration equal to that in the undiluted sample solution.

Standards, as chlorides, in solutions of equal hydrochloric acid concentration to that of the sample solutions are made up to contain 0—M/4000 calcium, 0—M/200 magnesium, 0—M/200 potassium, 0—M/100 sodium and 0—M/4000 manganese at suitable concentration intervals.

The LUNDEGARDH burner is so positioned that its top is 1—2 cm below the optic axis of the spectrograph and so that its flame intersects the optic axis 5—10 cm from the slit. The air and acetylene pressures are adjusted to give a clear blue flame of height about 15 cm and having a clearly defined zone of unburnt gases about 1.5 mm high. They are held at these pressures throughout the analytical run. The slit width and length are set at 0.04 mm and 1.5 mm respectively, and the sliding mask on the plateholder opened.

The sample solutions and standards are atomised into the flame, which is exposed to the plate for two periods of 60 seconds for each solution atomised. The spectrograms should be so positioned on the plate that duplicates are well separated and so that neither the samples nor the standards are grouped together. The plate is tray-developed in mentol-hydroquinone developer for 6 minutes at 20° C, fixed, washed and dried at room temperature.

A microphotometer fitted with a linear scale is used for measurement of the analysis lines in the spectrograms. The wave-lengths of these lines are given in Table 1. The microphotometer scale is set on zero with the slit shutter closed and, with the slit shutter open, the pointer set to full scale deflection, by means of a variable resistance in the lamp circuit, for transmission through an unexposed portion of the plate. In all cases except that of magnesium, the analysis line deflection and two background deflections, which are averaged, near the line are read. For calcium and sodium, one background reading is taken on each side of the line. For potassium and manganese both background readings are taken to one side of the line. The weaker component of the potassium doublet and shortest wave-length component of the manganese triplet are read. The magnesium line is superimposed on an OH line, but another OH line of equal intensity exists slightly to the shorter wave-length side of it. This is read and taken as background for the magnesium line.

Table 1. *Analysis line wavelengths and their approximate sensitivities in the* LUNDEGARDH *flame emmission method.*

Element	Analysis line wavelength, Å	Lower limit of sensitivity
Calcium	4227	10^{-5} M
Magnesium . . .	2852	2×10^{-4} M
Sodium	3302	5×10^{-4} M
Potassium	4044	2×10^{-4} M
Manganese	4031	5×10^{-6} M

Calibration curves relating concentration to background/line deflection ratio are set up each element and the concentrations of the elements in the sample solutions read off.

High concentrations of sodium and iron can affect magnesium and potassium results, respectively, due to the close proximity of Na 2852.8 Å and Fe 4045.8 Å to the analysis lines of these elements. It has been shown that these effects can be disregarded if the concentrations of sodium and iron in the solutions analysed do not exceed M/25 and M/100, respectively. In the light of recent investigations into the flame photometric and atomic absorption analysis of plant material, it appears that the LUNDEGARDH method, as described, may suffer from interference by the formation of refractory compounds in the flame. An analyst should, therefore, examine the method for the limits of such interferences before adopting it. This is particularly important in the determination of calcium in the presence of the large concentrations of phosphorus found in plant material.

The results of OERTEL and STACE (1946) and CHOLAK and HUBBARD (1944) suggest that the coefficients of variation of single determinations to be expected in analysis by the LUNDEGARDH method are about $\pm 9\%$ for magnesium and $\pm 3-5\%$ for sodium, potassium, calcium and manganese. The poorer precision for magnesium is caused by its lower response of emitted intensity with concentration.

II. Flame Photometric Method for Sodium, Potassium and Calcium. (WILLIAMS and TWINE, 1960).

Apparatus. Flame photometer burning a propane-air mixture and having filters which will isolate the sodium doublet at 589.0/589.6 mμ, the potassium doublet at 766.5/769.9 mμ and the calcium oxide band at 620 mμ; general laboratory glass-ware; KJELDAHL flasks and digestion equipment.

The difficulties associated with plant analysis using filter photometers result from transmission of extraneous light by the filters and the formation of refractory compounds in the flame. In the method described, interference resulting from the sodium filter transmitting calcium light is overcome by the addition of aluminium chloride to the sample solution which suppresses calcium emission. That resulting from the formation of refractory calcium compounds is overcome by conditioning both sample and standard solutions with sodium, potassium, magnesium and sulphuric acid (c.f. DAVID, 1958). Since the quality of filters varies from one make of flame photometer to another, it is advisable to test them for transmission of light other than that of the element sought before use in analysis.

Dry, ground plant material, of weight between 0.3 and 1 g, is digested in 100 ml KJELDAHL flasks using, initially, 15 ml of concentrated nitric acid, 0.5 ml of 60% perchloric acid and 0.5 ml of concentrated sulphuric acid. If charring occurs when the nitric acid is driven off, the flask is removed from the heater, cooled, a few ml more of nitric acid added and the flask returned to the heater. Heating is continued until all nitric and perchloric acid is driven off leaving only the inorganic residue from the plant material and the sulphuric acid initially added. After cooling, 15 ml of water is added, the solution boiled and the contents of the flask filtered through a Whatman No. 42 filter paper into a 50 ml volumetric flask. The KJELDAHL flask and filter are then washed repeatedly with small volumes of cold water and the combined filtrate and washings made up to the mark with water.

For *sodium analysis*, sufficient aluminium chloride to give a concentration of about 200 p.p.m. Al is added to aliquots of the digest solutions or, if outside the range of the standards, to aliquots of suitable dilutions of them. These solutions are then measured flame photometrically against standards containing sulphuric acid in equal concentration to the sample solutions, 10 p.p.m. of calcium as $CaCl_2$, 200 p.p.m. aluminium as $AlCl_3$, 10 p.p.m. phosphorus as H_3PO_4, 25p.p.m. potassium as KCl and 0—5 p.p.m. sodium at suitable concentration intervals.

For *potassium analysis*, the plant digest solutions, or suitable dilutions of them, are measured flame photometrically against standards containing 0—10 p.p.m. potassium at suitable concentration intervals, 4 p.p.m. phosphorus as H_3PO_4 and sulphuric acid equal in concentration to that in the sample solutions.

For *calcium analysis*, 5 ml of a solution containing 4% magnesium as $MgSO_4$, 0.1% sodium as NaCl and 0.1% potassium as KCl is added to 15 ml aliquots of the digest solutions or, if this gives calcium concentration outside the range of the standards, to 15 ml aliquots of suitable 1% sulphuric acid dilutions of them. These solutions are measured flame photometrically against standards in 1% sulphuric acid containing 0—100 p.p.m. calcium at suitable concentration intervals, 1% magnesium as $MgSO_4$, 250 p.p.m. sodium as NaCl and 250 p.p.m. potassium as KCl. It has been found that these concentrations of magnesium sulphate, sodium chloride, potassium chloride and sulphuric acid in both sample and standard solutions prevent interference of phosphate, silicate, aluminium, sodium and

Table 2. *Comparison of results of analysis of several plant material samples for sodium, potassium and calcium by flame photometric and chemical methods.*

Sample	Sodium, %		Potassium, %		Calcium, %	
	Flame-photometric	Chemical	Flame-photometric	Chemical	Flame-photometric	Chemical
Sub-clover . .	0.73	0.75	1.95	1.84	0.92	0.92
Sub-clover . .	0.34	0.33	2.80	2.96	0.65	0.59
White clover. .	0.36	0.36	2.69	2.67	1.14	1.08
Oats	0.92	0.93	0.60	0.57	0.095	0.098
Tobacco . . .	0.18	0.15	4.41	4.44	1.32	1.35
Bromus . . .	0.27	0.29	4.85	4.88	0.50	0.55

potassium with calcium emission to concentrations well above those found in plant digest solutions prepared as described.

After reading the sodium, potassium and calcium concentrations in the solutions flamed from calibration curves prepared from the standards, the concentrations of these elements in the original plant material are calculated from the weights of plant material taken and the dilutions made on the digest solutions during preparation.

Satisfactory recoveries have been obtained of sodium, potassium and calcium added to digests of a variety of plant matter samples. Further evidence concerning accuracy is given in Table 2 in which results of the flame photometric methods described are compared with those of magnesium uranyl acetate, cobaltinitrite and oxalate titration methods for sodium, potassium and calcium, respectively.

III. Flame Spectrophotometric Method for Magnesium (Knutson, 1957).

Apparatus. Quartz spectrophotometer with Beckman oxyacetylene flame attachment, Kjeldahl flasks and digestion equipment, general laboratory glassware, balance, hotplate, oven.

About 1 g, accurately weighed, of dry, ground plant material is digested in a Kjeldahl flasks using a mixture of nitric and perchloric acids until all of the organic matter is decomposed leaving a residue of the mineral constituents of the sample and perchloric acid. This is diluted with water and filtered to remove silica. The filtrate is evaporated over a hot plate, then heated in an oven at 200° C to remove the water and perchloric acid, cooled, and the residue dissolved and made up to 100 ml with 0.02 M hydrochloric acid. The resulting solution is filtered through Pyrexglass-wool to remove any suspended matter before flame emission measurement is carried out on it.

Standard solutions in 0.02 M hydrochloric acid are made up to contain 1, 2.5, 5, 7.5, 10, 12.5, 25, 50, 125, 250, 500, and 1000 p.p.m. magnesium as $MgCl_2$. The stock magnesium chloride solution for these standards should be prepared by dissolving clean magnesium ribbon in hydrochloric acid. Its magnesium concentration should be checked gravimetrically before use.

After switching on the detecting and measuring equipment of the spectrophotometer, the flame is lit and the oxygen and acetylene flow rates to it adjusted to give a partly incandescent flame having a yellowish-white zone (due to incandescent carbon particles) of height 45 mm. A diaphragm, having an opening 10 mm high, is placed between the flame and entrance slit of the monochromator so that the slit is illuminated only by light from between 11 and 25 mm above the burner tip.

The slits of the monochromator are set at a width of 0.03 mm and the wavelength drum adjusted to give peak measurement of Mg 2852 Å on atomising a magnesium solution.

Since the atomising of magnesium-free solutions affects the flame conditions in such a manner as to cause an apparent negative magnesium reading, all measurements on standards and sample solutions are made against magnesium-free 0.02 M hydrochloric acid. The procedure used in making a measurement on a sample solution is to take 6 readings of it and 6 readings of the standard on each side of it in concentration. The average reading for each standard and the sample solution is then calculated and the magnesium concentration in the sample solution estimated by interpolation between the standards. The concentration in the original dry matter sample taken can then be estimated from the weight taken and final volume of the digest solution.

Aluminium, phosphate and sulphate in the sample solutions have been shown, under certain circumstances, to interfere with magnesium determination by this method. If aluminium in the sample solution exceeds a few parts per million, it must be removed before magnesium determination is carried out. No phosphate or sulphate interference with magnesium emission occurs if the calcium concentration in the sample solution is twice that of the phosphate or equal to that of the sulphate. Magnesium-free calcium chloride solution should be added to the digest solution if the quantity of plant calcium does not meet this requirement.

Recovery tests carried out using the method described on synthetically prepared solutions containing Na, K, Ca, Al, Mn, PO_4''' and SO_4'' concentrations in excess of, or equal to, those expected in plant digest solutions have given results differing from the known added amounts of magnesium by 4% or less. The coefficients of variation based on 10 analyses each of solutions containing 10 and 2 p.p.m. magnesium were $\pm 0.9\%$ and $\pm 0.8\%$, respectively. The detection limit is stated to be 0.06 p.p.m. in solution. The means of duplicate analyses for magnesium carried out by the method described on three birch leaf samples differed by less than 4% from the means of duplicate analyses by a gravimetric method.

IV. Flame Spectrophotometric Method for Iron, Manganese and Copper (Berneking and Schrenk, 1957).

Apparatus. Quartz spectrophotometer with Beckman oxyacetylene flame attachment, pH meter, general laboratory glass- and platinum-ware, muffle furnace, balance.

In this method of analysis, the iron, manganese and copper in plant material are separated by 8-quinolinol precipitation both to concentrate them and to prevent interference from other plant constituents in their determination.

Two-gram samples of dry, ground plant material are ashed in 50 ml platinum dishes, first by careful heating over a Bunsen flame to destroy most of the organic matter, then in a muffle furnace for 16 hours at 500° C to destroy the remainder. After cooling in a desiccator, 8 ml of concentrated hydrochloric acid is added to each and dissolution effected by gentle heating over a Bunsen flame. The solutions are then filtered into 400 ml beakers and the dishes and filters washed with small portions of cold water to a total volume of combined filtrate and washings of about 70 ml.

To separate the iron and copper from the manganese, 0.5 ml of a solution of aluminium chloride containing 4000 p.p.m. aluminium is added to each solution followed by 10 ml of a 5% solution of 8-quinolinol in 2 N acetic acid. Ammonium hydroxide (1 + 1) is added until the pH of the solution reaches 1.2 to 1.3, then 40 ml of 2 N ammonium acetate, with vigorous stirring, and 1 + 1 ammonium hydroxide until the pH reaches 5.1 as determined by the pH meter. The beakers are covered, left to stand overnight, then the precipitate separated under mild suction by filtration through a Whatman No. 52 filter paper. The beakers are washed with the filtrates to ensure complete transfer of the precipitates to the filters. Both precipitates and filtrates are retained.

The filter papers containing the 8-quinolinol precipitates are transferred to small silica crucibles and ashed by first carefully charring over a Bunsen flame, then igniting in a muffle furnace at 850° C for 16 hours. After cooling in a desiccator the ashes are washed into 30 ml beakers using, for each, 3 ml of concentrated hydrochloric acid, then several portions of water to a final volume of 12 ml. These solutions are evaporated just to dryness on a hot-plate and the residues taken up in 5.0 ml of 0.1 N hydrochloric acid for flame spectrophotometric analysis.

Manganese is separated from the filtrates from the copper and iron precipitations by adding to each filtrate 5 ml of 8-quinolinol in 2 N acetic acid followed by 1 + 1 ammonium hydroxide solution, with vigorous stirring, to a pH of 7.1 as determined on the pH meter. The beakers are covered and, after standing overnight, the 8-quinolinol precipitates are separated and treated in exactly the same manner as described for copper and iron determination to give, finally, solutions of the ashes in 5.0 ml of 0.1 N hydrochloric acid.

Separate sets of **standards** are prepared for copper, iron and manganese by suitably diluting aliquots of 2000 p.p.m. stock solutions of each element.

Flame spectrophotometric emission readings are made on distilled water, on the iron-copper and the manganese sample solutions, on the standards for each element and on blank solutions prepared with the samples. The distilled water reading is subtracted from each sample, standard and blank solution reading, calibration curves set up and the concentrations of the elements in the sample and blank solutions read from them. After correction for blanks, the copper, iron and manganese concentrations in the plant material can be calculated from the weights taken and final volume of the solutions flamed. Measurements are made using the lines Cu 3248 Å, Fe 3720 Å and Mn 4033 Å, the slits being set at widths of 0.04 mm for copper, 0.025 mm for iron and 0.06 mm for manganese. The sensitivity is appropriately adjusted to the concentration range of the samples and standards in each case.

It has been shown that if the procedure described is followed accurately, no interference of aluminium with the determination of any of these elements will occur. Manganese is precipitated separately from iron and copper because iron has been found to interfere with flame emission spectrophotometric determination of manganese.

Comparison of the results of analysis of a wide variety of plant materials by the method described with those of the A.O.A.C. (1950) methods for copper, iron and manganese has shown slight disagreement in a few cases, but it is uncertain which method causes the disagreement. The results of a test of the precision of the method, based on ten analyses of the same sample in each case, are 32.2 ± 3.6 p.p.m. for copper, 42.1 ± 3.4 p.p.m. for iron and 37.6 ± 1.6 p.p.m. for manganese, the variations stated being standard deviations of single determinations.

B. Arc Emission Analysis.

The variety of d.c. arc emission methods for mineral powders reported in the literature is so great that it is impossible to survey them here. Methods which have been used for plant analysis are similarly diverse. The reason for this is that the complicated relationship which the many variables existing in the d.c. arc excitation of a sample bear to each other dictates an empirical approach in devising a method.

The factors affecting the precision, accuracy and sensitivity of d.c. arc analysis are electrode shape and spacing, electrode material, the major element composition of the sample, volatility of the compounds composing the sample, arc current, arc stability, whether the sample is carried in the anode or cathode, self absorption of the analysis line, photographic and microphotometric considerations. Of these, major element composition and arc instability cause the analyst most difficulty in the production of precise and accurate results. These effects are minimised either by mixing the sample with a spectrographic buffer or diluent or by co-precipitating the element under analysis with a large excess of compounds of other elements. MITCHELL (1948) and AHRENS (1950) discuss, in detail, the variables associated with d.c. arc emission analysis of plant ash and non-conducting powders generally and describe methods of alleviating the effects of these variables.

The main advantage to be gained from d.c. arc analysis using photographic detection and evaluation is that, in employing it, it is possible with little effort to simultaneously detect, on a qualitative or semi-quantitative basis, some 70 chemical elements, mostly at concentrations of a few parts per million. If quantitative estimation of spectrographically detectable elements is required, it is necessary to replicate many times or to use an internal standard which behaves in the arc in a similar manner to the element under analysis. Both entail considerable microphotometric measurement and calculation.

Direct current arc analysis is generally confined to the analysis of plant ash for minor elements. It cannot compete in precision with conventional analytical methods in analysis for major components.

Three methods will be described, namely, two complementary internal standard, cathode layer methods and the "addition" method. In the first method described, the "matrix" effect associated with the wide variation in major element composition from one plant ash to another is overcome by chemically depositing a number of trace elements in a uniform $Fe_2O_3—Al_2O_3$ matrix and determining them against synthetically prepared standards. In the second method, which was devised to determine elements not quantitatively precipitated by the chemical treatment in the first, the matrix effect is overcome by the use of a large excess of a spectrographic buffer or diluent and analysing against similarly buffered standards prepared synthetically to simulate plant ash in major element composition. In the third method described the matrix effect is overcome by making known additions of the analysis elements to portions of the sample, thereby ensuring that sample and "standard" have identical matrices.

The use of a large excess of diluent prevents the detection of a number of elements at low concentration in plant ash which would otherwise be detectable and probably does not wholly correct the matrix effect. Although the addition method does not require a diluent, its accuracy is, to some extent, impaired by the fact that the assumption of linearity between concentration and emitted intensity of the analysis line, on which the method is based, is not always valid. The reason for this is that, at the low concentration of some elements in plant ash, there is no alternative to the use of resonance lines in analysis. These lines are always self-absorbed. The error arising from the use of a self-absorbed line in the addition method increases with increasing concentration of the element in the sample.

I. The Variable Internal Standard, Cathode Layer Method (MITCHELL and SCOTT, 1947; MITCHELL, 1948.)

Apparatus. General laboratory glass-, silica- and platinum-ware, silica-lined muffle furnace, steam bath, purified carbon electrodes, large quartz spectrograph (dispersion 2000—8000 Å in 70 cm), arc stand, suitable quartz and glass lenses, six-step 2:1 ratio stepped sector, suitably ballasted and variable d.c. power supply capable of delivering 10 amps, ILFORD Chromatic plates and tray developing equipment, non-recording microphotometer, torsion balance, agate mortar and pestle.

Twenty grams of dried and ground plant material is ashed in a platinum dish by ignition overnight at 450° C in a muffle furnace. The ash is transferred to a platinum crucible, fused with 4 g of sodium carbonate, the melt dissolved in 50 ml of 1 + 2 hydrochloric acid, evaporated to dryness on a steam bath and taken up again in 50 ml of 1 + 2 hydrochloric acid. The silica is removed by filtering through a Whatman No. 41 filter paper, washing several times with hot water, and the combined filtrate and washings made up to about 150 ml with water.

Aluminium chloride, equivalent to about 30 mg of Al_2O_3, and 10 ml of a solution containing 0.4 mg of cadmium as cadmium chloride are added to the solution, then 10 ml of 5% 8-quinolinol in 2 N acetic acid followed by 1 + 1 ammonium hydroxide dropwise, with stirring, until the colour of the solution changes from yellow to emerald green (pH 1.8—1.9). Precipitation is then effected by the addition of 30 ml of 2 N ammonium acetate followed by 2 ml of 10% tannic acid in 2 N ammonium acetate, 2 ml of 1% thionalide in glacial acetic acid and 20 ml of ammonia solution equivalent to the latter, stirring throughout.

After standing overnight, the precipitate is separated by filtration through an 11 cm Whatman No. 540 filter paper, washed with cold water and dried. The precipitate and filter paper are transferred to a silica crucible and ashed by ignition in a muffle furnace at 450° C. After allowing the ash to come to moisture equilibrium with the laboratory atmosphere, it is weighed on glazed paper using a torsion balance.

Ten milligrams of this ash is analysed colorimetrically for iron (for example, by the sodium salicylate method of SCOTT and MITCHELL, 1943) and the rest prepared in the following manner for spectrographic treatment: The ash is mixed with twice its weight of carbon powder by grinding in an agate mortar. Successive small portions of this mixture are pressed, by means of a ground-down steel needle, into a cavity of depth 8 mm and diameter 1 mm in the 15 mm-long, 2.8 mm-diameter turned down end of a 5 mm diameter carbon rod until the cavity is full.

Eight series of **standards** in Fe_2O_3—Al_2O_3 matrices containing, respectively, 2.0, 3.5, 5.0, 7.0, 10.0, 13.0, 18.0 and 25.0% Fe_2O_3 are prepared. Each should be a geometric dilution series, of interval $\sqrt{10}$, ranging from 1% to 1 p.p.m. of each element for which the samples are being analysed. These standards are prepared for spectrographic treatment in the same manner as were the sample ashes. For zinc determination, sufficient cadmium chloride should be present in each standard to give a known cadmium concentration approximately equal to those in the sample ashes.

The spectrograph is set up for cathode layer analysis by placing a spherical quartz lens of appropiate focal length close to the slit and so positioning the arc stand on the optical bench that an image of the arc will be focussed at the collimator of the spectrograph. The prism face is so masked that only the light from the extreme tip of the lower electrode and a small portion of the arc column near it will pass. The stepped sector is placed between the quartz condensing lens and slit and the slit length appropriately adjusted. The slit width is set at 0.02 mm. To ensure accurate positioning of the electrodes during arcing, a glass lens is placed behind the arc stand to focus the arc at a screen which is marked to indicate the size and position of the aperture of the spectrograph.

The electrodes containing the samples and standards are placed in the lower (cathode) jaw of the arc stand, the counter electrode in the upper jaw being a cleanly broken piece of 5 mm diameter carbon rod. The arc is started by touching the electrodes together with sufficient resistance in series to limit the current to

3 amps, then separating them to a gap of 10 mm, simultaneously increasing the current to 9 amps. The arc is run until the sample is completely consumed. Correct separation and positioning of the electrodes is maintained throughout the arcing by observation of the image of the arc on the screen behind the arc stand.

The plates are developed for $2^1/_2$ minutes at 18° C in Ilford I.D. 2 metol-hydroquinone developer, fixed, washed and dried at room temperature.

The spectrograms are evaluated by the constant density, variable internal standard method, the first step being to set up standard curves for each element. Microphotometer deflections of steps of optical density between 1.0 and 0.3 of each analysis and internal standard line (see Table 3) in the spectrograms of the synthetic standards are read and converted to optical densities. These densities are plotted, as is shown in Fig. 1, against their log relative exposures as determined from the step at which the density was read and a blackening curve drawn for each line. The logarithm of exposure of the weakest step is arbitrarily set at zero and, since the ratio between steps is 2.0, the logarithms of the exposures of successive steps occur at intervals of 0.3 on the relative exposure axis.

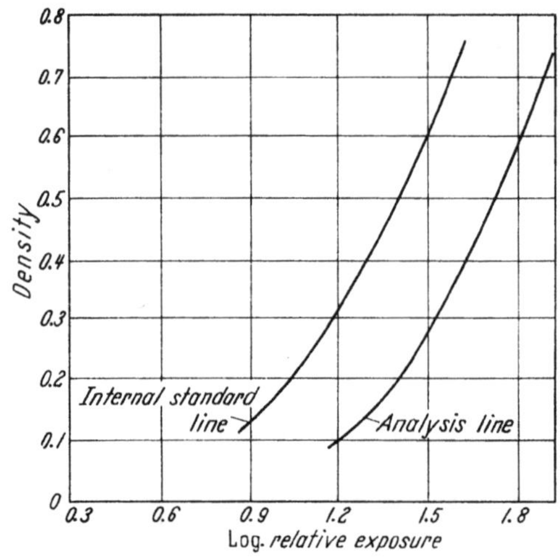

Fig. 1. Blackening curves of analysis and internal standard lines for use in the constant density method of platecalibration.

If the separation, measured parallel to the abscissa at a given density (usually 0.4), between the blackening curves of an analysis line and its internal standard line in the spectrograms of the standards is plotted against the log concentration of the analysis element, a family of curves similar to those shown in Fig. 2 is obtained. The working curves so obtained should be parallel to each other and displaced along the ordinate according to the concentration of the variable internal standard (Fe_2O_3 in this case). The curves in Fig. 2 being parallel, it is possible, for the purpose of correction for variation in internal standard concentration, to draw a curve relating working curve displacement to internal standard concentration which will be valid at all concentrations of the analysis element. Such a curve is shown in Fig. 3.

Having set up the standard curves illustrated in Figs. 1, 2 and 3, microphotometric density measurements are carried out on 3 steps of density between 0.3 and 1.0 of the analysis and internal standard lines in the spectrograms of the samples. Blackening curves, which will be similar to those shown in Fig. 1, are set up and the separations between curves for analysis and internal standard lines measured at the same density (0.4) as were the separations for the standards. These separations are corrected, according to the colorimetrically determined Fe_2O_3 concentrations in the samples, to an arbitrarily chosen datum internal standard concentration (usually 5% Fe_2O_3) using Fig. 3 and the concentrations of the analysis element read from the curve corresponding to that datum concentration in Fig. 2.

If analysis of samples for zinc is required, curves similar to those shown in Figs. 1, 2 and 3 are set up using the zinc analysis line and cadmium internal standard line. The concentration of cadmium in the ashed precipitate of a sample can be calculated from its weight and the original cadmium addition. Cadmium is used as internal standard for zinc because cadmium and zinc are more similar in physical properties, particularly in volatility, than are iron and zinc.

If correction for spectral background is deemed necessary, log relative intensity is plotted in place of log relative exposure in Fig. 1 and the log of the ratio of analysis to internal standard line intensity at constant density in place of blackening curve separation in Fig. 2. Background curves are drawn in the modified Fig. 1, extrapolated to constant density and subtracted from the corresponding log line intensities by use of subtraction logarithms. Subsequent treatment is similar to that described for the treatment of lines uncorrected for background. Reference should be made to SCOTT (1944, 1946) and MITCHELL, SCOTT and FARMER (1944, 1946) for a detailed description of the procedure and for methods of simplifying the calculations.

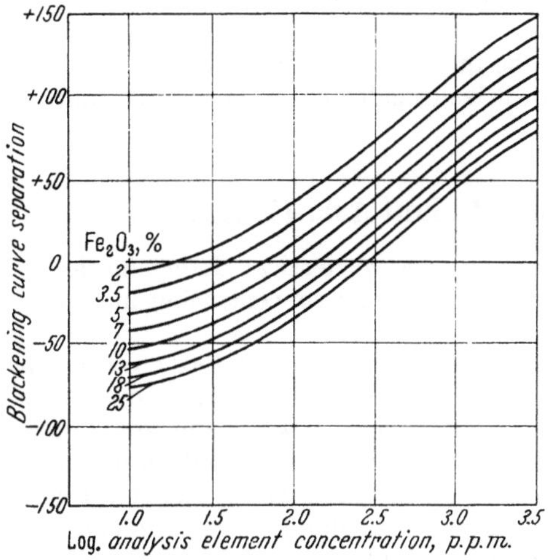

Fig. 2

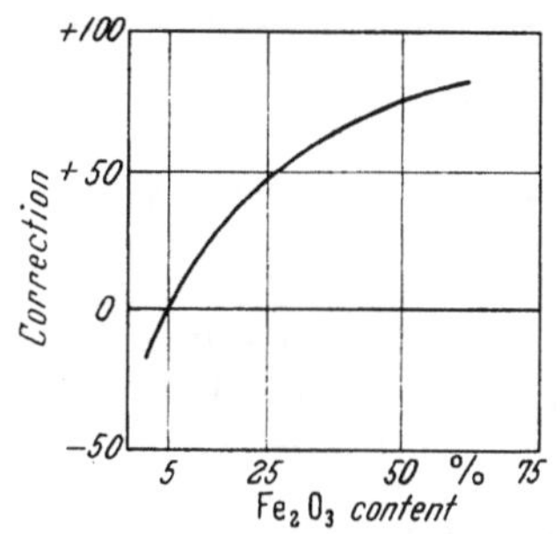

Fig. 3

Fig. 2. Working curves for variable internal standard method of arc emission analysis.

Fig. 3. Curve for correction of blackening curve separation to datum internal standard concentration.

Once working curves are established it is not necessary to run standards with each batch of samples analysed provided that care is exercised in ensuring that the chemical and spectrographic treatment of the samples is accurately reproduced each time. It is advisable to check working curves at regular time

Table 3. *The wavelengths and sensitive ranges of analysis lines used, together with their internal standard elements and line wavelengths, in the spectrochemical analysis of samples in an alumina matrix by the variable internal standard, cathode layer method.*

Element	Wavelength Å	Internal standard line Å	Useful range p.p.m.
Vanadium	4379	Fe 4251	10—3000
Chromium	4254	Fe 4251	3—3000
Cobalt	3454	Fe 3452	10—3000
Nickel	3415	Fe 3413	10—3000
Zinc	3345	Cd 3261	10^3—10^5
Silver	3281	Fe 3306	3—1000
Copper.	3274	Fe 3306	3—1000
Vanadium	3185	Fe 3197	10—3000
Tin	3175	Fe 3197	100—3000
Molybdenum . . .	3170	Fe 3197	10—3000
Beryllium	3130	Fe 3117	30—3000
Germanium . . .	3039	Fe 3117	100—3000
Gallium	2944	Fe 2929	10—3000
Tin	2840	Fe 2838	30—3000
Lead	2833	Fe 2838	100—3000
Thallium	2768	Fe 2768	100—3000

intervals (say, every 6 months) or when starting to use a new batch of photographic plates.

Table 3 gives a list of elements for which plant samples may be analysed by the chemical concentration method and variable internal standard, cathode layer spectrographic treatment described. The useful ranges given in the table refer to concentrations in the ashed 8-quinolinol-tannic acid-thionalide precipitate which is about 500-fold more concentrated in these elements than the original sample of dry plant material.

Although copper has been shown to be quantitatively recovered by the concentration method described, analysis for it is difficult due to contamination from reagents and other sources.

II. Direct Cathode Layer Analysis of Plant Ash. (FARMER, 1950).

This method was designed to cover the spectrochemical determination of elements not quantitatively recovered by the chemical concentration method of MITCHELL and SCOTT (1947).

One gram of oven-dried plant material is ashed by ignition at 450° C overnight, weighed, and 15 mg of the ash mixed thoroughly, by grinding in an agate mortar, with 15 mg of potassium sulphate and 30 mg of carbon powder containing about 250 p.p.m. silver and 150 p.p.m. chromium. This mixture is pressed, in duplicate, into 2.4 mm diameter by 4.5 mm deep cavities in the 4.0 mm-diameter turned down ends of 5 mm diameter pieces of carbon rod. **Standards** are prepared by mixing a synthetic base similar in composition to plant ash, and containing the analysis elements at varying known concentrations, with potassium sulphate and carbon powder containing silver and chromium in the same manner as were the sample ashes. The base is prepared by grinding together 1.8 g of KH_2PO_4, 2.0 g of K_2SO_4, 1.5 g of K_2CO_3, 0.5 g of MgO, 0.9 g of $Na_2CO_3 \cdot H_2O$, 0.9 g of $CaCO_3$ and 2.0 g of SiO_2, then allowing it to stand for 7 to 10 days to come to moisture equilibrium with the laboratory atmosphere. Barium and strontium carbonates and manganese, iron and copper oxides are mixed with the base to give a top standard containing 10,000 p.p.m. each of strontium, barium and manganese, 31,600 p.p.m. of iron and 3160 p.p.m. of copper. The magnesium, sodium and calcium standards are prepared by modifying the concentrations of the element or elements concerned in the synthetic base while keeping the concentrations of all other elements constant. The base of the top standard for magnesium

Table 4. *The useful ranges, precision and wavelengths of analysis and internal standard lines used in the cathode layer arc analysis of plant ash for a number of elements.*

Element	Analysis line, Å	Internal standard line, Å	Useful range, p.p.m. on ash	Coefficient of variation, %
Strontium	4607	Cr 4254[1]	30—1000	6.8
Strontium	3465[1]	Cr 3593	1000—10000	—
Barium	4554	Cr 4254[1]	30—1000	4.6
Manganese	4034[1]	Cr 4254[1]	30—1000	3.7
Iron	3441[1]	Cr 3593	100—30000	4.5
Copper	3247[1]	Ag 3281	10—3000	4.8
Sodium	3303[1]	Cr 3593	$2200—1.5\times10^5$	5.4
Magnesium	3337[1]	Cr 3593	$1800—1.2\times10^5$	3.9
Calcium	4579	Cr 4254	$7000—2.8\times10^5$	3.7

[1] Background correction necessary.

and sodium is made up to contain 20% each of MgO and Na_2O (as $Na_2CO_3 \cdot H_2O$) while that for calcium is made up to contain 40% CaO (as $CaCO_3$). The lower standards, in each case, are prepared by dilution of the top standard base with synthetic base material free from the element or elements for which the standards are being prepared.

The spectrographic treatment is the same as that described in the preceding section except that the samples are analysed in duplicate and ILFORD Thin Film Half tone plates are used.

Evaluation is by the blackening curve separation method already described. Table 4 gives the analysis and internal standard lines used together with an indication as to whether or not background correction is necessary, the useful analytical range and the precision of the mean of duplicate determinations for each element.

FARMER (1950) has compared the results of this method with those of other methods and finds the average differences to be 10% or less depending on the element sought and the nature of the sample analysed. This accuracy is stated to be adequate for many agricultural purposes.

III. The Method of Successive Additions (OERTEL and McKENZIE, 1954; OERTEL, 1950a, 1950b, 1955 and unpublished data; McKENZIE, 1956, 1959).

Apparatus. General laboratory glass- and silica-ware, silica-lined muffle furnace, oven, electric heaters, purified graphite electrodes, large quartz spectrograph (2000—8000 Å in 70 cm), arc stand, suitable quartz and glass lenses, front faced concave aluminium-on-glass mirror, stepped sectors, suitably ballasted and variable d.c. power supply capable of delivering 20 amps at 200 volts, ILFORD Thin Film Half-tone plates and tray developing equipment, non-recording microphotometer, balance, agate mortar and pestle.

The samples of oven-dry, ground plant material are arranged in batches according to the predictable approximate major element composition of their ash. For example, it is known that grass-like plants are high in silica and citrus leaves high in calcium.

Approximately 5 grams of plant dry matter is grab sampled from each sample in a batch. Additionally, grab samples are taken from each sample in the batch and contributed towards a composite sample from which the standards for the batch are prepared. This composite sample, which should weigh about 15 g, is thoroughly mixed to ensure homogeneity and divided into 3 composite sub-samples, each about 5 grams in weight. The plant dry matter samples taken and the composite sub-samples are each accurately weighed and placed in silica dishes.

To prepare the **standards,** solutions of volume between 1 and 5 ml containing the analysis elements are added to 2 of the composite sub-samples to give, in one, an addition equal to and, in the other, an addition of 4 times, the estimated average concentrations of the analysis elements in the batch of samples. Pure water is added to the third composite sub-sample in approximately equal volume to the solutions added to the other two. The composite sub-samples are then dried in an oven at 90° C.

Concentrated sulphuric acid, in the proportion of 1 ml of acid to 1 g of plant dry matter, is added to each sample and composite sub-sample and thoroughly mixed in using a small glass rod. (The rod is wiped clean after each mixing with a small piece of ashless filter paper which is placed in the dish with the sample). They are then heated on electric heaters, first gently until fumes of sulphuric acid appear, then stongly ntil the excess of sulphuric acid is driven off leaving aggregates of charred material. The charred aggregates are broken up with glass rods and the samples ashed by ignition in a well ventilated muffle furnace at 500—600° C for four hours.

These sulphated ashes are each ground in a clean agate mortar to ensure homogeneity and cavities in the 3 mm-diameter by 6 mm long turned down ends of 5 mm-diameter graphite spectroscopic electrode filled with them. The cavities should be of depth 3.2 mm and diameter 1.6 mm. They are filled by repeatedly pushing the opening of the cavity into the ash on a hard, smooth surface and pressing the ash into the cavity with a ground down steel needle until full, then smoothing the top with a spatula. Since the ash in the electrodes is not weighed, both the dimensions of the electrodes and the filling technique should be accurately reproduced from one electrode to another. Three replicates of the sample ashes and six replicates of the composite sub-sample ashes are loaded into electrodes.

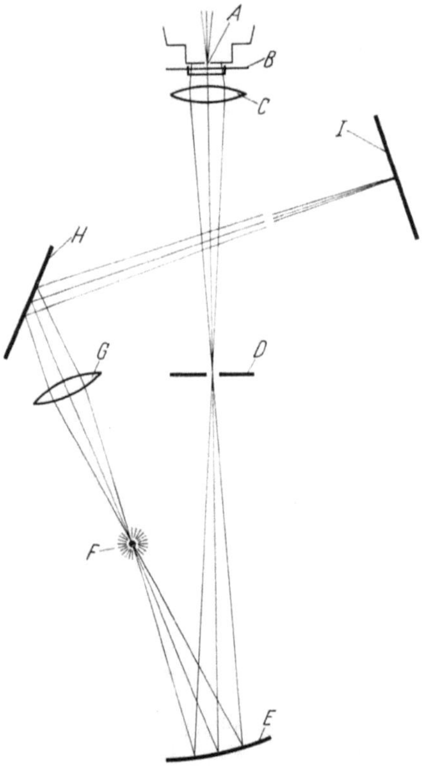

Fig. 4. Diagram of optical system for arc emission spectrochemical analysis of mineral powders. *A*, slit of spectrograph; *B*, stepped sector; *C*, quartz condensing lens; *D*, diaphragm; *E*, frontfaced, concave, aluminium-on-glass mirror; *F*, arc; *G*, glass lens; *H*, plane mirror; *I*, viewing screen.

The optical arrangement used in the spectrographic treatment of the samples is shown in Fig. 4. An image of the arc (*F*) is focused at the diaphragm (*D*) by the front-faced aluminium-on-glass mirror (*E*), at the collimator of the spectrograph by the quartz lens (*C*) and at the viewing screen (*I*) by the glass lens (*G*). The focal lengths and spacings of the optical components and the opening in the diaphragm should be such that an image of the full 4 mm arc column is focused at, and fills, the collimator of the spectrograph and that the light emitted from the incandescent electrodes during arcing, which would produce a continuous background if allowed to pass, is screened out at the diaphragm. The viewing screen is marked to show the position and size of the aperture of the spectrograph and the correct separation of the electrodes.

Electrodes carrying the samples and standards are placed in the lower (anode) jaw of the arc stand. The upper electrode, which is replaced with a fresh one after each arcing, is a neatly broken piece of 5 mm diameter graphite rod. The anode is adjusted to the correct position by viewing its image with respect to the image of the opening in the diaphragm (*D*), both of which can be observed by placing the eye behind the lens (*G*). The arc is started by touching the electrodes together with the rheostat in the d.c. supply circuit set to deliver 5 amps. The electrodes are moved apart to a gap of 4 mm and the current simultaneously increased to 15 amps over a period of 3—5 seconds. The arc column is kept within the aperture of the spectrograph and the arc gap held at 4 mm throughout the arcing by viewing the image of the arc cast on the screen and manually adjusting the arc stand accordingly. The arc is switched off 5 seconds after all of the sample has been consumed, this point being detected by a change in sound of the arc.

To minimise the effect of variation in emulsion characteristics over the surface of a plate and between plates, the spectrograms should be so arranged that two replicate spectrograms of a sample are situated adjacent to each other and the third replicate in any position on another plate. Each set of six replicate spectrograms of the standards is arranged so that 3 replicates are on one plate (two adjacent to each other and one separated), two, separated from each other, on a

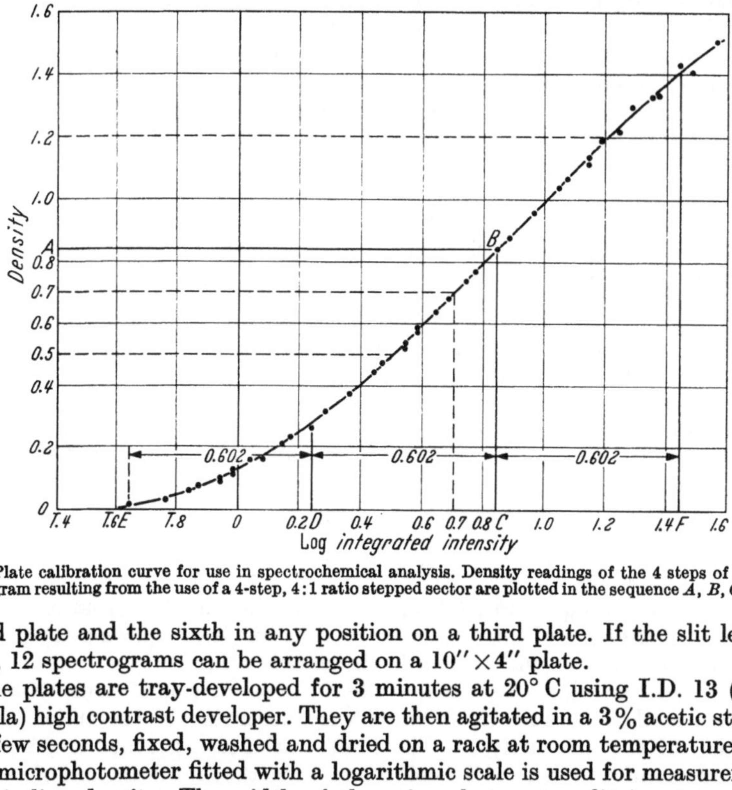

Fig. 5. Plate calibration curve for use in spectrochemical analysis. Density readings of the 4 steps of a line in a spectrogram resulting from the use of a 4-step, 4:1 ratio stepped sector are plotted in the sequence *A*, *B*, *C*, *D*, *E*, *F*.

second plate and the sixth in any position on a third plate. If the slit length is 6 mm, 12 spectrograms can be arranged on a 10″×4″ plate.

The plates are tray-developed for 3 minutes at 20° C using I.D. 13 (Ilford formula) high contrast developer. They are then agitated in a 3% acetic stop bath for a few seconds, fixed, washed and dried on a rack at room temperature.

A microphotometer fitted with a logarithmic scale is used for measurement of analysis line density. The width of the microphotometer slit is set to slightly less than the product of the spectrograph slit width and the magnification of the optics of the microphotometer. With the microphotometer slit shutter closed so that no light reaches the photocell, the scale is moved so that the infinity mark coincides with the index line and, with the slit shutter open, the reading is adjusted to zero for unexposed plate transmission by means of the variable resistor in the microphotometer lamp circuit. The image of the analysis line is moved slowly across the microphotometer slit and minimum background density on each side of the line, as well as the density maximum at the centre of the line, read. These three density measurements and the step in the spectrogram at which the measurements were made are noted.

The plates are calibrated by reading the peak density of the analysis line or a strong line in its vicinity in all steps of some 20 spectrograms evenly spread over them. A calibration curve, similar to that shown in Fig. 5, is prepared from these density readings in the following manner: The density differences between adjacent steps within the density range 0.6—1.6 in the spectrograms are averaged to determine the slope of the straight-line portion of the characteristic curve. A line of this slope, covering the range 0.5 to 1.2 on the density axis, is drawn through an arbitrarily chosen datum point (usually density 0.7, log integrated intensity 0.7). The log integrated intensity corresponding with a step density reading within the limits of this line is read and the densities of the other steps of the spectral line back plotted at intervals along the log integrated intensity axis corresponding with the stepped sector ratio.

To find the log integrated intensities of the analysis lines from the microphotometric density readings in the spectrograms of the samples and standards, the average of the background log integrated intensities, as read from the calibration curve, is subtracted from that of the line plus background using subtraction logarithms. These log integrated intensities are corrected to the first step by adding, if a 4:1 ratio stepped sector is used, 0.602 $(n - 1)$ where n is the number of the step in which the density reading was made.

In estimating the concentration of an analysis element in an ash sample by the addition method, it is assumed that the emitted integrated intensity of the analysis line of an element is related to the concentration of that element in the ash by the expression

$$\log C = A + \log It \ldots (1)$$

where C is the concentration of the element in the ash, A is a constant and It the integrated intensity of the analysis line of the element. If x and $I_1 t$ are the concentration and analysis line integrated intensity, respectively, of the element in the standard without addition and $(x + a)$ and $I_2 t$ the concentration and analysis line integrated intensity, respectively, of the element in a standard with addition, then

$$\log x = A + \log I_1 t$$

and

$$\log (x + a) = A + \log I_2 t \, .$$

Since, in this pair of equations, $I_1 t$, $I_2 t$ and a are known, the constant A can be calculated. Substitution of this value for the constant A in equation (1) gives the concentration of the analysis element in the sample ash.

It is not possible to give definite lower limits of estimation of elements in plant ash by this method because these limits vary according to the major element composition of the ash analysed. The limits for most elements are, however, claimed to be similar to those of the cathode layer method if the amount of material arced exceeds 5 mg. The sensitivity is adequate for the direct determination of the essential elements Cu, Mn, B, Fe and Mg in sulphated plant ash. The molybdenum concentration in some plant ash samples is below the detection limit of this method and in others it is necessary to measure the intensity of the molybdenum analysis line at levels close to the detection limit. Neither zinc nor cobalt are detectable if present at normal levels in the ash arced. Special boron-free electrodes are essential if boron is to be determined, in which case B 2497.7 Å is used. The same analysis lines as those listed for the cathode layer method (see Tables 3 and 4) are used in this method.

The precision of the method has been closely studied and the reader referred to the original papers for details. It is only necessary to state here that the coefficient of variation of a single line intensity determination to be expected is ± 10—15%, depending on the element sought and line used. Arc instability is the main contributor to the error.

C. Spark Emission Methods.

Although spark methods are preferred to arc methods, due to their greater precision, in the analysis of metallurgical samples, they have not, until recently, been extensively used in the analysis of mineral samples such as plant ash. Poorer sensitivity and the necessity to use electrically conducting samples have restricted the use of spark methods in plant analysis.

Two methods, both employing direct reading equipment, are described. The first, that for magnesium, was devised specifically as an adjunct to flame photometric analysis for Na, K and Ca because insensitivity and interference from OH bands makes the flame photometric analysis of solutions for magnesium difficult.

I. Porous Cup Solution Spark Method for Magnesium (Scott and Ure, 1958).

Apparatus. General laboratory glass- and silica-ware, silica-lined muffle furnace, steam bath, small quartz spectrograph (2000—8000 Å in 8 cm) fitted with a 2-channel integrating direct reading attachment, spark generator (15000 volts, 0.02 mH inductance, 0.001 μF capacitance and no added resistance), spark stand, 5.5 mm diameter carbon electrodes.

The electrodes used in this method are a pointed (70° included angle) lower electrode and an upper cup electrode of length 16 mm, internal diameter 3.2 mm and base thickness 0.60 ± 0.1 mm. The cups are heated in a muffle furnace at 450° C for 2 hours to improve porosity and should retain satisfactory porosity for at least a week after this treatment.

In preparation for an analytical run, the spark stand is adjusted to give an electrode gap of 2 mm situated on the optic axis 20 cm from the slit of the spectrograph. No condensing lens is used. Slits of the direct reader channels are set on the magnesium analysis line at 2802 Å and strontium internal standard line at 4077 Å. The entrance slit width is set at 0.02 mm, the magnesium exit slit at 0.05 mm and the strontium exit slit at 0.1 mm. If the emission of the strontium internal standard line is in excess of that which can be measured by the direct reader, it can be brought into range by supplying minimum voltage to the photomultiplier dynodes and placing a neutral filter behind the strontium exit slit.

Ten grams of ground plant dry material is ashed overnight at 450° C and the ash twice evaporated to dryness on a steam bath with concentrated hydrochloric acid to separate the silica. It is then taken up in 25 ml of (1 + 4 *v/v*) hydrochloric acid, filtered, washed and the filtrate plus washings made up to 500 ml with water. The bulk of this solution can be used for the estimation of elements other than magnesium. For magnesium determination, a 2 ml aliquot of this solution together with 5 ml of strontium chloride solution of strength 5000 p.p.m. Sr, is made up to 50 ml in a volumetric flask with 2.5% acetic acid solution.

The upper cup electrode on the spark stand is carefully filled with the sample solution using a fine-pointed pipette in such a manner as to prevent the adherence of air bubbles to the inner walls of the cup. After opening the slit shutter and setting the direct reader to the integrating position, the spark is started and run for 56 seconds. (The most precise sparking procedure of several tested is to allow no pre-sparking time and to effect no readjustment of the spark gap to its initial 2 mm during sparking. About 0.1 ml of solution is consumed under these sparking conditions). After completion of the exposure, the voltages on the magnesium analysis line condenser and the strontium internal standard line condenser are measured by applying to them an opposing e.m.f. to null point using a potentiometer. If the concentration of magnesium in the sample causes the voltage on the magnesium condenser to exceed that measurable by the potentiometer, an auxiliarly condenser can be inserted in parallel with it to bring the voltage into measurable range.

Standard solutions, in 2.25% acetic acid, are made up to contain potassium, calcium, sodium, phosphate and sulphate in the same proportions as in an average plant ash solution, 0—24 p.p.m. magnesium at suitable concentration intervals and 500 p.p.m. strontium. These solutions are sparked in exactly the same manner as the sample solutions and a standard curve prepared from the ratios of the magnesium and strontium condenser voltage readings. Ratios obtained from readings of the sample solution are applied to the standard curve and the magnesium concentrations in the plant material calculated.

This method has been operated using a laboratory-built direct reading attachment, the details of which are available in the original paper. If it were operated using a commercial direct reader and different spectrograph, appropriate adjustments to such details as sparking time and concentration range of standards would be necessary.

The method, as described, will cover the range 0.04 to 3% magnesium in plant material. The coefficient of variation of a single determination, based on 20 readings carried out on a solution containing 1.8 p.p.m. magnesium, is $\pm 1.86\%$ and is roughly independent of magnesium concentration. Errors due to slight variations in electrode dimensions, in position of the electrodes on the optic axis of the spectrograph and variation in major element composition from one plant sample to another have all been proved negligible. Comparison of the results of this method with those of a Titan Yellow colorimetric method in the analysis of 20 turnip samples gave mean values for magnesium concentration of 0.0687 and 0.0685% respectively, with coefficients of variation of 2.4 and 5.6% respectively.

II. The Pelleted Rotating Disc Spark Method (SCOTT, 1960).

Apparatus. Silica-lined muffle furnace, general laboratory equipment, rotating disc spark stand (see Fig. 6), triggered low-voltage a.c. condensed spark generator (13.4 μF capacitance, 0.13 mH inductance, 9 ohms resistance), pellet press and dies, medium quartz spectrograph (2000—8000 Å in 20 cm) with 11-channel integrating direct reader attachment, balance.

The ground plant dry matter sample, weighing 0.2 g, is ashed at 450° C in a silica-lined muffle furnace. The ash from this or 20 mg of ash, whichever is the lesser, is diluted with 50 mg of a mixture of equal weights of potassium and calcium sulphates. This mixture is ground with a 550 mg mixture consisting of 350 mg of boron-free graphite, 50 mg of potassium sulphate, 50 mg of calcium sulphate, 100 mg of cellulose powder and sufficient potassium dichromate to give 500 p.p.m. Cr. Inaccuracy, which would otherwise result from variation in major element composition from one plant ash to another, is avoided by the large excesses of potassium and calcium sulphates involved in this method of preparation.

An 8-ton load is applied to this mixture by means of a hydraulic press designed to produce a pellet $^1/_2''$ in diameter with a 4 mm-wide conical depression of 90° included angle in the centre of one side. The quantity of material stated produces a pellet about $^1/_8''$ in thickness.

The pellet is placed in position on the spark stand (shown diagramatically in Fig. 6) and held there by the hemispherical end of the spring loaded rod (*E*), inserted in the conical depression, which presses the flat side of the pellet against the eight driving teeth on the spindle (*C*). An $^1/_8''$ diameter boron-free graphite rod having a sharp conical end of 70° included angle is used as counter electrode. This is placed in the spark stand to give a spark gap of 1.5 mm and the gap aligned on the optic axis of the spectrograph 20 cm from the slit with no condensing lens interposed.

Having set the direct reader channels on the appropriate analysis and internal standard lines (see Table 5), appropriately adjusted the sensitivity range of each channel and set the controls to allow a 12 second pre-spark period followed by 100 seconds exposure, the stabilizing air jets and synchronous motor, which should drive the shaft (*D*) at 5 r.p.m., on the spark stand are started. After completion of the automatically controlled sparking run, the voltages on the analysis line and internal standard line condensers are read and noted.

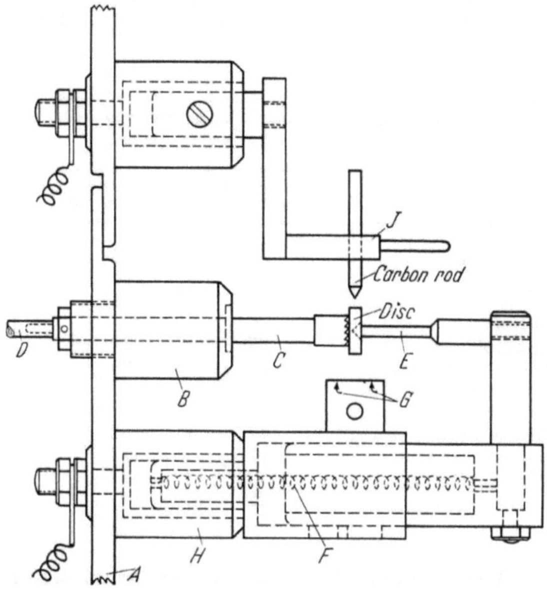

Fig. 6. Diagram of spark stand for use in the rotating disc spark method of spectrochemical analysis. *A*, perspex plate; *B*, nylon bush; *C*, rotating spindle; *D*, flexible polythene shaft; *E*, hemispherical-ended rod; *F*, spring; *G*, air jets; *H*, nylon covered stainless steel bush; *J*, spring-loaded counter-electrode clamp.

Standards are prepared using a base of synthetic "average" plant ash similar to that already described (FARMER, 1950). A series of portions of this base, containing 10—1000 p.p.m. of zinc, boron and copper, and 100—10000 p.p.m. of iron and manganese, at appropriate concentration intervals, is mixed with potassium sulphate, calcium sulphate, graphite powder, cellulose powder and potassium dichromate using the same procedure as that already described for the

Table 5. *The wavelengths of lines used, their spacings and precision attainable at the concentrations given in the determination of a number of elements in plant material by the rotating disc spark method.*

Element	Wavelength Å	Separation from previous line, mm.	Coefficient of variation %	Concentration in dry matter p.p.m.
Zinc	2138.6		1.9	45
Silicon	2435.2	> 20	—	—
Boron	2497.7	7.5	2.8	24
Phosphorus	2553.3	6.2	3.2	2800
Chromium	2677.2[1]	12.5	—	—
Iron	2755.7	7.0	—	—
Magnesium	2802.7	3.9	3.2	2300
Manganese	2949.2	11.0	3.3	111
Aluminium	3082.2	8.6	—	—
Copper	3247.5	9.3	5.4	10.0

[1] Internal standard line.

sample ash. Silicon and aluminium standard mixtures are prepared using chemically analysed plant ash. The standard mixtures are pelleted, the pellets sparked and the line condenser voltages read by the same procedure as that used for sample mixtures. Concentrations of analysis elements in the samples are then calculated from the ratios of the condenser voltages of analysis lines to internal standard lines by means of a calibration curve relating the voltage ratios obtained from the standards to concentration.

Table 5 gives the wave-lengths of lines used, their spacings and the coefficients of variation at the concentrations stated in the analysis of replicates of a pasture herbage sample. The silicon and manganese lines chosen represent a compromise enforced by the fact that two adjacent direct reader slits cannot be placed closer than 4 mm from each other.

Table 6. *The boron concentrations in various plant material samples as determined by the rotating disc spark method and by a colorimetric method.*

Sample	Rotating disc method, p.p.m.	Colorimetric method, p.p.m.
Apple leaf	28	28
Pasture herbage	10	10
Wheat grain	1.4	2
Wheat straw	4.9	4
Broccoli	37	39
Cauliflower	27	29
Lucerne	31	33
Oat grain	4.4	1.5
Tomato leaf	47	52
Pasture herbage	12	12
Kale leaf	27	28
Cabbage leaf	22	21

Coefficients of variation for silicon, iron and aluminium are not given in Table 5 because the suitability of the method for determination of these elements in plant material has not yet been fully examined. If aluminium is to be analysed simultaneously with other elements, it will be necessary to find more suitable source parameters. Contamination introduced by the pelleting die invalidated results for iron.

The accuracy of the method can be assessed from Table 6 in which results for boron in a variety of plant materials are compared with those obtained using a dianthrimide colorimetric procedure.

D. Atomic Absorption Methods.

Since the advantages to be gained from the use of the phenomenon of atomic absorption in chemical analysis have only recently realised (WALSH, 1955; RUSSELL, SHELTON and WALSH, 1957), insufficient investigation has yet been carried out to make possible a complete prediction of its impact on plant analysis. However, the theoretical considerations on which atomic absorption analysis is based and results gained for a few elements suggest that it could completely supplant flame emission analysis and partially supplant arc emission and colorimetric methods for the analysis of plant materials for both major and minor elements.

Investigation into the analysis of plant material by atomic absorption methods has been carried out by ALLAN (1958, 1959) and DAVID (1958, 1959 and unpublished data). The contents of this section are mostly derived from these sources.

Apparatus. Fig. 7 shows, diagramatically, the essentials for an atomic absorption spectrophotometer. Radiation from the source (*A*), which emits the line spectrum of the element to be determined, is passed through the flame (*B*) to the monochromator or light filter (*C*) which isolates the resonance wave-length of the element emitting from the source and passes it to the detector (*D*). The signal from the detector is measured using a galvanometer or amplifier and meter (*E* and *F*).

Measurement of the sample and standard solutions is effected by entraining fine droplets of the solutions in the gases fed to the flame, as in flame photometry, and noting the resulting absorption of the resonance line radiation, as indicated by the meter, on its passage through the flame. Calibration curves are set up using the measurements on the standards and the concentrations of the element to be determined in the sample solutions read off.

Most investigation to date (October, 1960) into atomic absorption analysis has been carried out using apparatus assembled in the laboratory from components now in commercial production, the names of the manufacturers of which are given in the text of this section. It is likely, however, that the range of components, complete instruments and manufacturers of them will increase considerably in the near future.

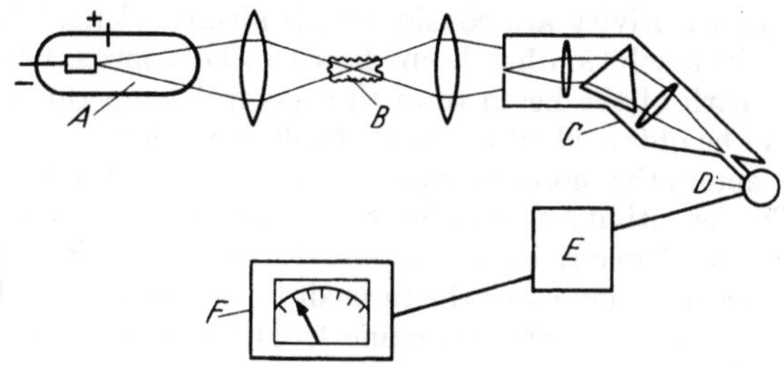

Fig. 7. Diagram of an atomic absorption spectrophotometer.

Two types of atomic absorption apparatus have been employed in analysis, one in which a modulated source is used in conjunction with an a.c. amplifier for measurement and the other in which an unmodulated source and d.c. measuring unit are used. Both types have been found satisfactory in analytical practice, but the a.c. type has the advantage of being unaffected by extraneous light emitted from the flame. This arises from the fact that the d.c. signal from the detector produced by uninterrupted light falling on it is rejected by the a.c. amplifier.

Spectral vapour lamps suitable for use as light sources in the atomic absorption determination of sodium, potassium, rubidium and caesium are manufactured by Phillips Electrical Industries, Holland, and by Osram G.m.b.H., Germany. Transformers or chokes, the circuits of which are available from the manufacturers of the lamps, are necessary to supply these lamps. Hollow cathode discharge tubes are used as light sources emitting the spectra of the less volatile elements. These are manufactured by Ransley Glass Instruments Ltd., Melbourne and Hilger and Watts Ltd., London. Techtron Appliances Ltd., Melbourne manufacture power packs, with 50 or 100 c/s modulated output, for operating hollow cathode discharge tubes in the a.c. type of apparatus and Hilger and Watts Ltd., one for operating them on unmodulated direct current for use in an atomic absorption attachment for the Uvispec spectrophotometer.

The resolution required to isolate the resonance line varies from one element to another, depending upon the complexity of the spectrum of the element. Although light filters have been little used, they would be adequate for the isolation of resonance lines of elements with simple spectra such as the alkalis and alkaline earths. For elements having more complex spectra, such as the transition elements, a monochromator which will resolve about 1 Å at 2500 Å is necessary. The Hilger Uvispec and Beckman DU quartz monochromators have been found adequate in the atomic absorption determination of these elements.

Photoelectric detectors used in flame emission analysis are adequately sensitive for atomic absorption analysis. For production and measurement of a.c. signals from photomultiplier tubes, Techtron Appliances Ltd., Melbourne, Australia, manufacture a combined high tension supply and a.c. amplifier unit. The detector and d.c., null-point measuring system manufactured by Hilger and Watts Ltd.

are integral parts of the Hilger Uvispec Spectrophotometer, the details of which are available from the manufacturer.

The variables under control of the analyst which determine sensitivity in atomic absorption analysis are the current flowing through the vapour lamp or hollow cathode discharge tube, the flame gases used, the spraying efficiency of the atomiser and the length of path through the flame of the beam from the spectral lamp. For maximum sensitivity, the current through the spectral lamp should be reduced to the minimum possible value consistent with adequate signal from the detector. Of the gas mixtures tested so far, the best, when both convenience and sensitivity are considered, is air-acetylene. The magnitude of absorption by a given solution has been shown to be approximately proportional to the length of path of the beam from the spectral lamp through the flame. High sensitivity can be obtained either by multiple reflection of the beam through the flame before it enters the monochromator or by use of a burner having a long, narrow opening. The sensitivity attainable by either method is not unlimited because, in the one case multiple reflection attenuates the beam from the spectral lamp and, at the same time, increases the intensity of extraneous light from the flame while, in the other, an air-acetylene flame length of more than about 12 cm is likely to flash back.

The **procedure** used in making an atomic absorption measurement is, first, to set the meter to read zero (using the zero adjustment on the amplifier) with the slit shutter of the monochromator closed, then to read full scale deflection (by adjustment of the gain of the amplifier or the high tension to the photomultiplier) with the slit shutter open and a mist of pure water entering the flame from the atomiser. Next, the water being fed into the flame by the atomiser is replaced by a sample or standard solution and the meter reading noted. The difference between this reading and full-scale deflection is a measure of the concentration of the analysis element in the solution atomised. Calibration curves are prepared from readings obtained from standard solutions and concentrations read off from them using readings obtained from the sample solutions.

Two types of interference which may affect flame emission analysis cannot affect atomic absorption analysis if a modulated source and a.c. amplifier are used. These are interference due to interchange of excitation energy on collision between excited atoms and electrons or ground-state atoms in the flame and, as already mentioned, interference due to extraneous light reaching the detector.

Two other types of interference which can affect flame emission analysis, those due to the formation of refractory compounds in the flame and to variations in the viscosity and surface tension of the solutions sprayed, also affect atomic absorption analysis. The reason for this is that these properties of the solutions affect the numbers of both ground-state and excited atoms in the flame.

Atomic absorption analytical methods have been applied to the analysis of plant materials for sodium, potassium, calcium, magnesium, zinc, manganese, iron and strontium. The first step in the procedure is, as in flame emission analysis, to get the sample into solution. This can be done in a variety of ways, but the only ones used to date are those of digestion with sulphuric-perchloric-nitric or perchloric-nitric acid mixtures. When filtered and made up to volume, the former digestion produces a solution of inorganic ions in dilute sulphuric acid and the latter one in a dilute solution of perchloric acid.

The atomic absorption determination of sodium, potassium, zinc, manganese and iron has been found to be free from interference by extraneous ions of plant origin up to concentrations of the interfering ions well above those found in plant digest solutions. For these elements, it is only necessary to ensure that the standards each contain the same concentration of residual digestion reagents as do the sample solutions.

Atomic absorption determination of magnesium in plant digest solutions has been found free from interference by extraneous elements of plant origin provided that sufficient calcium is present to suppress potential interferences. The calcium concentration in plants is invariably such that this requirement is met. As is the case in the determination of other elements, it is necessary to have in the standards the same concentration of residual digestion reagents as exists in the sample solutions.

In the atomic absorption determination of calcium in plant digest solutions, it has been found that phosphate, sulphate, aluminium and silicate, if present, can interfere unless preventive steps are taken. The presence of 0.6% magnesium as $MgCl_2$, 2% *v/v* sulphuric acid, 200 p.p.m. sodium as NaCl and 0.15% potassium as KCl in both sample and standard solutions is efficient in suppression of interferences from these ions up to concentrations above those found in plant digest solutions. Magnesium chloride solution prepared by dissolving reagent grade magnesium metal in hydrochloric acid is reliably free from calcium.

It is not possible to recommend definite concentration ranges of analysis elements in the standard solutions because the optimum ranges will vary according to the sensitivity of the equipment used. However, the analyst can easily determine the optimum ranges by trial and error measurements using his equipment.

Table 7 gives the resonance lines used, approximate lower limits of estimation in solution and the precision in the determination of a number of elements by atomic absorption flame spectrophotometric methods. This table deals only with elements which can be determined directly on plant digest solutions and is compiled from results gained in investigation of the analysis of plant material, specifically, by atomic absorption methods. Both precision and sensitivity estimates are tentative pending further research into the design of atomic absorption apparatus which will result in large gains in sensitivity and precision.

Recovery experiments and the comparison of results of analysis of plant samples by atomic absorption and other methods have proved the reliability of atomic absorption results for the elements listed in Table 7.

The data given in Table 7 were all obtained using air-acetylene flames. The flame lengths given in the Table have been found to give adequate sensitivity in the analysis of plant digest solutions for all elements mentioned except, perhaps, zinc. A 10 cm flame gives sufficient sensitivity for analysis of solutions prepared

Table 7. *The wavelengths of resonance lines used, precision and approximate lower limits of estimation of some elements in the atomic absorption analysis of solutions.*

Element	Resonance line Å	Lower limit of estimation, p.p.m. in solution	Flame length, cm.	Coeff. of variation, %, at conc. (p.p.m.) given in brackets
Sodium	5890/6	0.1	2.7	5(0.5); 0.8(5)
Potassium	7665/99	0.1	2.7	7.5(1.0); 1.8(10)
Calcium	4227	0.3	2.0	4(2.5); 2(50)
Magnesium	2852	0.05	2.0	8(0.5); 2.8(5)
Zinc	2139	0.5	2.0	13(1); 2(10)
Manganese	2795	0.1	12	8.8(0.5); 4(25)
Iron	2483	0.05	12	6 5(2.5); 3.3(125)
Strontium	4607	0.05[1]	10	—

[1] 5-fold expanded scale used in measurement.

by the digestion of 2 g of plant dry matter and dilution of the digest to a final volume of 50 ml, whereas a final volume of 10 ml from 2 g of dry matter is necessary in the determination of zinc using a 2 cm flame.

It has recently been suggested (DAVID, 1960) that the addition method should be used in the atomic absorption analysis of plant material for strontium because tests on synthetic solutions have indicated that interference due to the formation of refractory compounds in the flame cannot be sufficiently suppressed to allow the use of conventional standards. The procedure used in applying the addition method is to take two equal aliquots of a sample solution, dilute them with equal volumes of water and strontium chloride solution of known strontium concentration, respectively, then carry out atomic absorption measurements on each. If both of the measurements made fall within the straight-line portion of the calibration curve (0—15% absorption), the strontium concentration in the original solution can be calculated from the two readings, the dilution factor and the magnitude of the addition. In using the addition method, both sample and "standard" are measured under identical conditions of interference resulting from the formation of refractory compounds in the flame.

The following is an *outline of the atomic absorption procedure* used at the C.S.I.R.O. Division of Plant Industry in the determination of sodium, potassium, calcium, magnesium, manganese, iron and zinc in plant material:

Apparatus. A.c. voltage stabiliser; calcium, magnesium, manganese, iron and zinc hollow cathode discharge tubes with power pack for their operation (BOX and WALSH, 1960); sodium and potassium spectral vapour lamps with choke or transformer for their operation; Eel (Evans Electroselenium Ltd.) atomiser and spray chamber; burners having openings for 2.5 cm × 0.5 mm and 10 cm × 0.2 mm, each designed to be interchangeable on the Eel spray chamber; air and acetylene supplies for the burner with associated diaphragm valves and pressure gauges; quartz lenses or adjustable diaphragm for confinement of the beam from the spectral lamp to the flame; quartz monochromator capable of resolving 1 Å at 2500 Å; photomultiplier with H T supply, a.c. amplifier and meter (BOX and WALSH, 1960).

Procedure. Two grams of dry plant material is digested with 1 ml of concentrated sulphuric acid, 4 ml of 60% perchloric acid and about 20 ml of nitric acid until all of the organic matter, perchloric and nitric acids are driven off leaving a residue consisting of the sulphuric acid initially added and the mineral components of the plant matter. This is diluted with about 10 ml of water, filtered and the digestion flask and residue in the filter thoroughly washed with small volumes of water into a 50 ml volumetric flask until full to the mark. This gives a digest solution of the plant material in 2% *v/v* sulphuric acid.

A 5-ml aliquot of each digest solution in a batch, or of a suitable 2% *v/v* sulphuric acid dilution of it, is mixed with 5 ml of a solution containing 12000 p.p.m. magnesium as $MgCl_2$, 2% *v/v* sulphuric acid, 3000 p.p.m. potassium as KCl and 400 p.p.m. sodium as NaCl. These are measured on the atomic absorption apparatus described, using the 2.5 cm burner, against standards containing 0—70 p.p.m. calcium at suitable concentration intervals, 6000 p.p.m. magnesium, 2% *v/v* sulphuric acid, 1500 p.p.m. potassium and 200 p.p.m. sodium. The calcium concentrations in the original plant material samples can then be calculated from the measurements according to the weight of dry material taken and the dilution effected in preparation.

For the determination of sodium, potassium, magnesium, manganese, iron and zinc, **standards,** in 2% *v/v* sulphuric acid solution, are prepared containing 0—10 p.p.m. sodium, 0—20 p.p.m. potassium, 0—10 p.p.m. magnesium, 0 to

20 p.p.m. manganese, 0—20 p.p.m. iron and 0—20 p.p.m. zinc, each at suitable concentration intervals. The digest solutions are diluted, if necessary, with 2% *v/v* sulphuric acid to concentrations of the elements being determined within these ranges and are then measured against the standards using the atomic absorption apparatus described. For sodium, potassium and magnesium, the 2.5 cm burner gives adequate sensitivity, but for manganese, iron and zinc the use of the 10 cm burner is necessary. Generally, little or no dilution of the original digest solution is necessary in the determination of sodium, manganese, iron and zinc. For potassium and magnesium, however, dilution up to about 20-fold will be found necessary.

Atomic absorption measurements are carried out using the appropriate spectral lamp in each case and with the monochromator set on the appropriate resonance line (see Table 7). The pressures of the air and acetylene supplying the burner are set to give a clear blue flame having an inner cone of height about 1.5 mm and are held constant throughout an analytical run. The point of intersection of the beam from the spectral lamp and the flame at which peak sensitivity is attained should be found for each element and measurements carried out at this setting.

A number of elements of interest, other than those listed in Table 7, have either not been examined with a view to their determination in plant material or have been found insufficiently sensitive for determination using atomic absorption equipment at its present stage of development. Notable among these are copper, molybdenum and cobalt. Sensitivity for copper is no doubt adequate using existing laboratory-built equipment incorporating a 10 cm burner and 5-fold expansion of the measuring scale but its determination in plant material using such equipment has not been examined. However, the estimation of molybdenum and cobalt will require gains in sensitivity of about 50-fold. The basis of atomic absorption analysis is such that a future gain in sensitivity of this magnitude is by no means improbable.

References.

AHRENS, L. H.: Spectrochemical Analysis. Cambridge, Mass.: Addison-Wesley Press, Inc. 1950. — ALLAN, J. E.: Analyst **83**, 466 (1958). — ALLAN, J. E.: Spectrochim. Acta 800 (1959). — Assoc. Offic. Agr. Chemists, "Methods of Analysis", 7th ed., 1950.

BERNEKING, A. D., and W. G. SCHRENK: J. Agr. Food Chem. **5**, 742 (1957). — Box, G. F., and A. WALSH: Spectrochim. Acta **16**, 255 (1960).

CHOLAK, J., and D. M. HUBBARD: Ind. Eng. Chem. Anal. **16**, 728 (1944).

DAVID, D. J.: Analyst **83**, 655 (1958). — DAVID, D. J.: Analyst **84**, 536 (1959). — DAVID, D. J.: Nature **187**, 1109 (1960).

FARMER, V. C.: Spectrochim. Acta **4**, 224 (1950).

KNUTSON, K. E.: Analyst **82**, 241 (1957).

LUNDEGARDH, H.: Ark. Kemi. Min. Geol. **10 A**, No. 1 (1928).

MCKENZIE, R. M.: Spectrochim. Acta **8**, 88 (1956). — MCKENZIE, R. M.: Australian J. Appl. Sci. **10**, 488 (1959). — MITCHELL, R. L.: Tech. Comm. No. **44**, Commonwealth Bureau of Soil Science, Harpenden, England (1948). — MITCHELL, R. L., and R. O. SCOTT: J. Soc. Chem. Ind. **66**, 330 (1947). — MITCHELL, R. L., R. O. SCOTT and V. C. FARMER: Macaulay Institute of Soil Research, Aberdeen, Scotland. Ann. Rep. 1943—1944. — MITCHELL, R. L., R. O. SCOTT and V. C. FARMER: Nature **157**, 193 (1946).

OERTEL, A. C.: Australian J. Appl. Sci. **1**, 152 (1950a). — OERTEL, A. C.: Australian J. Appl. Sci. **1**, 259 (1950b). — OERTEL, A. C.: Australian J. Appl. Sci. **6**, 467 (1955). — OERTEL, A. C., and R. M. MCKENZIE: Australian J. Appl. Sci. **5**, 330 (1954). — OERTEL, A. C., and R. M. MCKENZIE: Australian J. Appl. Sci. **6**, 458 (1955). — OERTEL, A. C., and H. C. T. STACE: J. Soc. Chem. Ind. (London) **65**, 350 (1946).

RUSSELL, B. J., J. P. SHELTON and A. WALSH: Spectrochim. Acta **8**, 317 (1957).

SCOTT, R. O.: J. Soc. Chem. Ind. (London) **63**, 25 (1944). — SCOTT, R. O.: J. Soc. Chem. Ind. (London) **65**, 291 (1946). — SCOTT, R. O.: J. Sci. Food and Agriculture **11**, 584 (1960).. — SCOTT, R. O., and R. L. MITCHELL: J. Soc. Chem. Ind. (London) **62**, 4 (1943). — SCOTT, R. O., and A. M. URE: Analyst **83**, 561 (1958).

WALSH, A.: Spectrochim. Acta **7**, 108 (1955). — WILLIAMS, C. H., and J. R. TWINE: C.S.I.R.O. Division of Plant Industry, Divisional Report No. 19 (1960).

Mass Spectrometric Methods.

By

K. Biemann.

With 6 Figures.

Mass spectrometry is commonly known as a method for determination of stable isotopes, detection of impurities in gases, quantitative analysis of hydrocarbons and atomic and molecular weight determinations. Very recently the method has been used for identification and, more important, for the determination of the structure of organic compounds. It is the latter aspect of mass spectrometry this chapter will deal with, particularly since the extreme sensitivity of this tool adds to its value in the investigation of minute amounts of material, an important consideration where compounds have to be isolated laboriously from natural sources like plants. The determination of stable isotopes in gaseous degradation products was discussed in an earlier volume of this series and need not be considered here, with the exception of the measurement of stable isotopes in the intact molecule. The main emphasis will be placed on the qualitative aspects of mass spectrometry, its scope and limitations, and the interpretation of spectra. The principle of quantitative analysis will also be illustrated.

The characteristics of mass spectrometry which make it a useful tool for the investigations of organic compounds are, briefly: The complexity of the spectra combined with high reproducibility allows a very reliable identification of a compound with an authentic sample. The many peaks of the spectrum indicate, in addition to the size of the molecule, the way in which functional groups are connected, the size and structure of side chains, and the presence or absence of certain heteroatoms. The information obtained is quite different in nature from that gained through the commonly used methods like ultraviolet, infrared, and nuclear magnetic resonance spectroscopy. The mass spectrometer is a useful addition to the array of such instruments and is most advantageously used in conjunction with one or more of them. One disadvantage of mass spectrometry is the need to volatilize the compound without decomposition in order to obtain a good spectrum. It is for this reason that the careful selection of the most suitable compound or derivative plays an important role, as will be illustrated in a number of examples throughout the following discussion.

A. Instrumentation.

While space does not permit outlining in detail the working principle and design of mass spectrometers, a subject on which excellent reviews are available (Beynon, 1960; Stewart, 1960), it will be discussed briefly to the extent necessary for the understanding of the applications to be considered.

A mass spectrum is a recording of the abundance of positively charged particles vs. their mass. These particles (positive ions) are produced by the bombardment of the sample vapor with a beam of electrons in the so-called ion source (Fig. 1). Acceleration of these ions by an electrostatic field E and deflection in a magnetic field H forces the particles into well-defined, curved paths, the radii R of which are determined by the mass-to-charge ratio (m/e) of the ions according to the equation

$$\frac{m}{e} = \frac{H^2R^2}{2E}.$$

Continuous scanning of either electric or magnetic field enables ions of increasing m/e to strike successively the collector after travelling a path of fixed radius. Amplification of the current produced between collector and ground recording of this signal vs. H^2/E produces the mass spectrum. The intensity of the ion beams produced is proportional to the intensity of the bombarding electron beam and the concentration (pressure) of the sample in the ion-source. For this reason it is essential to keep these conditions constant during the run to obtain reproducible spectra. The former is easily controlled by electronic means; the latter is achieved by allowing the sample vapor to enter the ion-source through a small leak from a reservoir in which it is kept under somewhat higher pressure (10^{-2} to 10^{-1} mm as compared with 10^{-5} to 10^{-7} mm in the ion-source). Depletion of the sample during the short time (5 to 30 min) required for recording of the spectrum is thus kept small and reproducible with time. The sample is introduced into this reservoir, which with the necessary ports, valves, and vacuum pumps is called the inlet system. Most of the material entering the ion source remains un-ionized and is pumped away by an efficient vacuum system keeping the analyzer tube well evacuated.

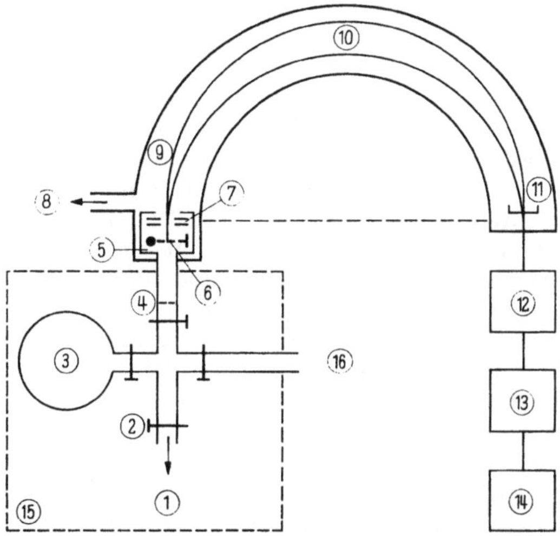

Fig. 1. Diagram of a 180° magnetic deflection mass spectrometer: *1.* Vacuum pump (inlet system), *2.* valves, *3.* reservoir, *4.* leak, *5.* ion source, *6.* electron beam, *7.* electric field (E), *8.* vacuum pump (analyzing system), *9.* deflection (analyzer) tube, *10.* ion beam, *11.* collector, *12.* preamplifier, *13.* amplifier, *14.* recorder, *15.* inlet system (heatable), *16.* sample port. The ion source *(5)*, deflection tube *(9)*, and collector *(11)* are placed in the magnetic field (H).

To be useful, a mass spectrum should exhibit all the peaks characteristic of the compound. Therefore, the *mass range* of the instrument has to encompass all masses beginning with 12 (carbon) to the molecular weight of the compound. The mass spectrometers available for the work on complex organic molecules (Appendix I) record masses up to about 700 and even beyond if certain instrument parameters are changed. This range suffices for most cases because larger molecules are hardly volatile enough to be introduced, with the exception of polyhalogenated compounds.

Even more important is the *resolution* of the instrument because it is imperative for a correct interpretation to be able to determine the masses of all the ions accurate to a single mass unit. This requires a resolution high enough to make the distinction up to the range of the molecular weight. The instrument should, therefore, be able to resolve—at least to an appreciable extent—peaks differing for only one mass unit in the region of the molecular weight of the compound. The instruments referred to above are able to resolve such pairs of peaks almost completely in the region of up to mass 350—400. Mass spectrometers of much higher resolving power and their applications will be discussed in Section C, VIII.

For the applications dealt with in this chapter, the techniques of introducing the sample into the inlet system and, to a certain extent, the conditions in the ion-source are of primary concern.

B. The Sample.

I. Vapor Pressure.

It is evident that one of the most important considerations is the vapor pressure of the material to be investigated, which has to be high enough to give at least 0.01 to 0.02 mm pressure in the inlet system of the mass spectrometer. Obviously the range of applicability may be greatly extended if the inlet system is heated. Temperatures of up to 350° C are being used in the petroleum industry, but one has to keep in mind that non-hydrocarbons are frequently not completely stable at such a temperature, and 200° C might be a safer upper limit. Too low a vapor pressure not only leads to a low sample concentration and thus to a weak spectrum (or none at all) but also makes it difficult to pump off the sample after the run, giving rise to a persistent background which may make the instrument temporarily unuseable.

Since the required sample pressure in the ion source is a few orders of magnitude less than the one in the inlet system, samples of very low volatility may be used if introduced directly into the ion source and permitted to evaporate slowly into the electron beam. This technique, sometimes the only choice, requires a special vacuum lock if the time-consuming venting of this part of the instrument is to be avoided; also the danger of contamination of the source is enhanced.

As a general rule, compounds distilling or subliming fast under 0.02 mm at the temperature of the inlet system (or conditions equivalent to that) are suitable for mass spectrometry.

The amount of material needed for a mass spectrum depends on the volume of the inlet system into which it is expanded, since it is the partial pressure of the sample in the reservoir and not its absolute amount which determines the pressure in the ion source. In commercial instruments the reservoir is about three liters, which requires approximately one micromole of a compound for a pressure of 0.01 mm. A few tenths of a milligram is thus an average. Much smaller amounts (a few micrograms) still give spectra of useful intensity if a smaller reservoir is used as, for example, the volume of only the valve-block after closing off the reservoir. Increasing the intensity of the electron beam is an additional help. It has to be kept in mind that the sample in a small reservoir is depleted much faster than from a large one, a fact which may be neglected for qualitative spectra but not for quantitative analysis.

There is, however, an upper limit to the amount of sample which can be introduced because the ion source functions properly only at relatively low pressure and the path of the ions has to be essentially free of neutral molecules. For practical purposes the pressure in the inlet system should not exceed 0.5 mm. For this reason, a large amount of a dilute solution cannot be introduced to obtain a useful spectrum of the solute, even if the spectrum of the solvent would not interfere. Prior concentration or, better, complete removal of the solvent is then necessary (see Section D, I, 2).

II. Techniques of Introduction

Gases are introduced simply by connecting container (Fig. 2, A) and inlet system with an adapter which can be evacuated. The sample is admitted by opening the valves of container and inlet system. The amount of sample in the reservoir is then measured by a sensitive manometer.

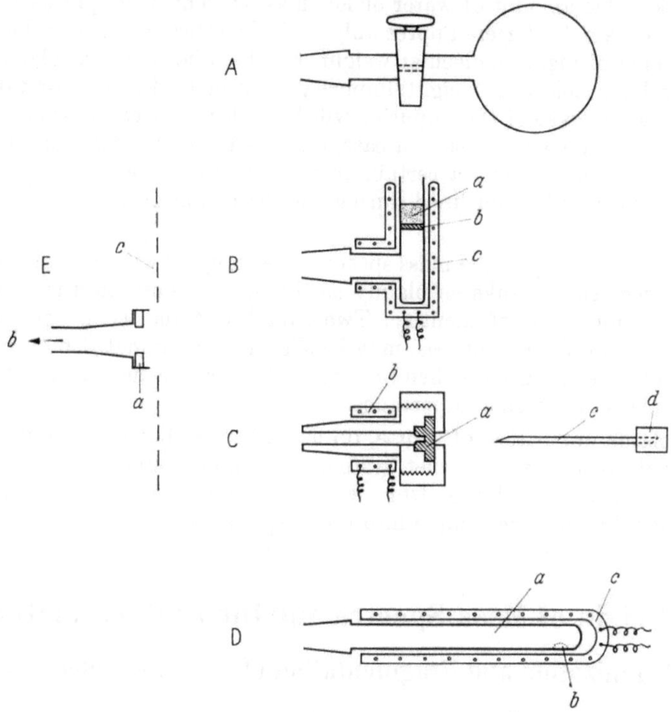

Fig. 2. *A*. Gas sample bottle. *B*. Sintered disc assembly *(a)* molten gallium, *(b)* sintered glass disc, *(c)* heater. *C*. Injection system *(a)* silicon-rubber diaphragm, *(b)* heater, *(c)* piece of hypodermic needle, stoppered with a piece of silicon-rubber *(d)*. *D*. Solids system *(a)* sample tube, *(b)* sample, *(c)* heater. *E*. Port on inlet system into which *A—D* can be inserted. *(a)* Teflon ring, *(b)* to valve-block, *(c)* wall of inlet oven.

Liquids may be admitted from a fine capillary through a disc of sintered glass (Fig. 2, B) covered with mercury, molten gallium, or indium, providing a vacuum tight seal. More preferable is the use of a short piece of a hypodermic needle, pointed on either end, which is filled with the liquid by capillary forces, stoppered on one end with a piece of silicon-rubber. The open end is then pierced through a diaphragm of silicon rubber (Fig. 2, C) in the inlet system, vaporizing the sample into the reservoir. Each of these devices, the sintered disc or the silicon rubber plug, is kept at the temperature of the inlet system.

Solids are placed into a glass tube (Fig. 2, D) which fits through a Teflon O-ring into a port on the inlet system (Fig. 2, E) which can be isolated by a valve and evacuated separately. Part of the glass tube containing the sample extends out of the heated part of the system. After evacuation of the tube, the valve to the reservoir is opened, and the sample is sublimed into it on heating the tube with an external heater. This procedure is convenient for qualitative spectra. For quantitative analysis of a mixture of solids, great care has to be taken to avoid fractionation, either loss of the more volatile components during evacuation of the sample tube or incomplete vaporization of the less volatile components during the introduction.

III. Purity.

The requirements with regard to the purity of the sample differ widely and depend on the particular information sought and the type of impurity.

A considerable amount of water or low-mass solvent, if its presence is clearly recognized, does not interfere appreciably with the interpretation of the spectrum of a compound of higher molecular weight. On the other hand, a relatively small amount of high-molecular weight impurity may make the interpretation very difficult because peaks of the impurity will be scattered over the whole spectrum of the major component. In such a case, one has to try to recognize these interfering peaks, which requires a certain amount of experience. The ways by which this can be done will be outlined during the discussion on mixtures (Section C, VII).

The high sensitivity of the mass spectrometer coupled with a recording system displaying very small peaks as plainly as the most intense ones is sometimes a disadvantage for proof of identity. Two samples of the same compound, but originating from different sources, may be slightly contaminated by different by-products, and the spectra are then no longer identical. In such a case, the purity of the sample is of extreme importance.

Non-volatile impurities, of course, do not interfere, but care should be taken that these do not enter the inlet system; this is accomplished by placing the sample in an external adapter from which the residue is easily removed after distilling or subliming the sample into the inlet system.

C. Origin of Mass Spectra and their Interpretation.

I. Ionization and Fragmentation of Organic Molecules.

If an electron of sufficient energy strikes an atom or a molecule, one of its electrons is removed, leaving an ion with one positive charge (occasionally more than one electron is lost; see Section C, VI). In the mass spectrometer such ions, if of a lifetime of $\geqslant 10^{-6}$ sec, are accelerated and deflected by the electric and magnetic fields and recorded as a peak at the m/e corresponding to their molecular weight, which can thus be determined. The minimum electron energy required for this process is the ionization potential of the compound under consideration, for organic compounds between 7 V and 15 V, in general. The energy of the bombarding electrons used normally is about 50 to 80 V, much above this minimum value.

Use of electrons of an energy higher than the ionization potential forms molecular ions whose excess energy may be sufficient to rupture a bond which, at least in non-cyclic molecules, leads to fragmentation. The positive charge then remains on one of the fragments whereas the other is thought to be a radical or a neutral molecule. These, being uncharged, are not accelerated and thus do not appear in the spectrum. Negative ions formed under certain conditions are also not recorded because they are drawn into the other direction and strike the back wall of the ion source. In a polyatomic molecule a great variety of such fragments can be formed as shown in the following simplified scheme:

$$ABCD + e^- \longrightarrow ABCD^+ + 2e^- \qquad \text{ionization} \qquad (1)$$

$$ABCD^+ \longrightarrow A^+ + BCD\cdot \qquad (2)$$

$$ABCD^+ \longrightarrow AB^+ + CD\cdot \qquad (3)$$

$$(3a)\ AB^+ \longrightarrow A^+ + B\cdot \qquad (3b)\ AB^+ \longrightarrow B^+ + A\cdot \qquad \text{fragmentation}$$

$$ABCD^+ \longrightarrow ABC^+ + D\cdot \qquad (4)$$

$$ABCD^+ \longrightarrow CD^+ + AB\cdot \quad \text{etc.} \qquad (5)$$

$$ABCD^+ \longrightarrow AD^+ + BC\cdot \qquad \text{rearrangement} \qquad (6)$$

$$ABCD^+ + ABCD \longrightarrow [ABCD\ ABCD]^+ \longrightarrow ABCDA^+ + BCD\cdot \qquad \text{ion-molecule collision} \qquad (7)$$

Process (1) depicts the formation of the molecular ion and steps (2) through (5) are examples for such simple fragmentations which may occur successively, as in (3a) and (3b). Their masses, as deduced from the corresponding peaks in the spectrum, indicate the sum of the masses of all atoms in a given group present as such in the molecule. The interpretation of the spectrum in its simplest form would consist in piecing together these fragments to arrive at the structure of the complete molecule. This is somewhat complicated by the two further processes, rearrangements (6) and ion-molecule collisions (7). The former yields fragments consisting of a group of atoms, which is not present as such in the molecule, whereas the latter yields particles larger than the original molecule. Once these rearrangements, which in non-hydrocarbons are rather specific with respect to structure, are well understood, they can be very helpful in the interpretation of the spectrum (see Section C, IV). Ion-molecule collisions, a second-order reaction, are negligible under the sample pressures usually employed. Such peaks may easily be distinguished from normal peaks because the intensities of the former increase with the square of the pressure of the sample. The capture of a hydrogen atom by certain types of molecular ions is the only such process commonly encountered and gives rise to a peak one mass unit above the molecular weight. Examples will be discussed in Sections D, I, 1 and D, II.

A mass spectrum contains, therefore, a peak at almost all possible masses which can be constructed from the atoms present in the molecules. This accounts for the complexity of the spectra, which may contain up to a few hundred, well-defined, sharp peaks. Their intensities, however, vary widely and are, at least on the same instrument, very reproducible. For the interpretation of the spectrum of an unknown compound this would be confusing, were it not for the fact that the intensities of these peaks vary widely over a range of 1:20,000 (the limit of the recording systems commonly employed). Energetically preferred fragmentation processes are much more frequent than those involving less probable breakdown reactions. For this reason the former give rise to intense peaks, whose relation to the structure of the parent molecule may be deduced.

II. The Molecular Weight.

The peak due to the molecular ion is the best starting point for an interpretation because it gives the size of the molecule, and the M^+-ion is the parent of all the fragments (and therefore frequently referred to as the "parent peak"). In general, it is believed to be the peak of highest mass in the spectrum (not counting the isotope peaks discussed below). For a number of reasons this is not always the case, and the identification of the M^+-peak has to be done with care.

Although we can safely say that the majority of compounds exhibit such a peak, there are some whose molecular ion is very unstable and decomposes completely before reaching the collector. The spectrum then consists only of fragment peaks. Groups which can easily accommodate the loss of an electron, like aromatic rings and multiple bonds, stabilize the molecular ion whereas highly branched chains or hydroxyl and amino groups greatly facilitate fragmentation. These influences are additive and a polycyclic aromatic compound like anthracene has a very intense peak corresponding to its molecular weight in contrast to a branched tertiary alcohol, like tert.-butanol, which exhibits no peak at mass 74. Alicyclic molecules have, in general, higher peaks due to the molecular ion, if compared with a similar aliphatic compound, because here even the cleavage of one of the bonds of the ring does not lead to two particles.

Whether or not the peak at highest mass is indeed the molecular weight can be checked by a number of considerations: (1) Unless a molecule contains an odd number of nitrogen atoms (or artificially introduced heavy isotopes), its molecular weight is always an even number, whereas the opposite holds for fragments. Rearrangement ions formed from the molecular species are, however, mostly also of even weight and thus most easily confused with a molecular ion. (2) If the sample is a pure compound and there is a peak less than 15 but more than 3 mass units below the one of highest mass, the latter cannot be the corresponding molecular ion because a methyl group is the smallest entity which can be lost during fragmentation (except one to three hydrogens). For example, a peak 11 mass units below the molecular weight would have to be due to the loss of 11 hydrogen atoms, a very improbable process. It is much more likely that both are fragments of a molecule, the molecular ion of which is very unstable and tends to decompose by loss of H_2O (M-18) and C_2H_5 (M-29), respectively, or of CH_3COOH (M-60) and C_5H_{11} (M-71). The former would indicate a branched alcohol and the latter a branched acetate, loosing H_2O or CH_3COOH in a rearrangement process. (3) Decrease of the energy of the bombarding electrons to a value just above the ionization potential of the compound considerably diminishes the formation of fragments, and the molecular ion becomes relatively more prominent. Its absolute intensity is, however, not increased and this approach is therefore mainly used to distinguish a molecular ion from fragment peaks of an impurity or another component present. (4) Some compounds, particularly amines, esters and ethers form molecular ions which are rather unstable, but have the tendency to abstract a hydrogen from another molecule. One finds in such a case only a small peak at the molecular weight but another at one mass unit higher. Since this "M + 1" peak is due to an ion-molecule collision (Section D, 1), its intensity compared with the other peaks changes with variations of the sample pressure or other conditions in the ion source. Its appearance is an indication of the molecular weight one mass unit lower.

The molecular weight peak permits its determination, which is accurate to the individual mass number in the range in which the instrument gives unit-resolution. Besides this accuracy, which is not reached by other methods, the correct value of all molecular weights of the individual components of a mixture is determined and not an over-all average. This is of greatest importance if one deals with a mixture of homologues or compounds differing only in the number of double bonds present.

Due to the constancy of the abundance of stable, heavy isotopes in nature, certain conclusions as to the elemental composition of the compound may be drawn from the peaks immediately following the molecular weight peak. Up to now we have tacitly assumed the peak under discussion is the one due to a molecule composed only of the most abundant stable isotope, which for the elements present in organic compounds is also the one of lowest atomic weight. Those molecules which happen to contain a heavy isotope will have a correspondingly higher weight and give rise to discrete peaks at higher masses. Their intensity will depend on the statistical probability of the existence of such a combination and is, in turn, related to the abundance of that heavy isotope: C^{13} makes up about 1.1% of natural carbon and methane consists, therefore, of 98.9% molecules with molecular weight 16 and 1.1% with a weight of 17. This is clearly exhibited by the corresponding ratio of the peaks at m/e 16 and 17 in its mass spectrum. The possibility that any one of the four hydrogen atoms is a heavy one can be neglected for most practical purposes because the abundance of deuterium is very small. In ethane each carbon atom has a 1.1% chance to be C^{13} and the peak at m/e 31 is, therefore, 2.2% of the sum of the peaks at 30 and 31; for $C_{17}H_{36}$ the intensity ratios of the peaks 240, 241 and 242 are 100:18.9:1.69 because here the probability that two of the 17 carbon atoms are C^{13} is already considerable. Obviously, the peak one mass unit above the "mono-isotopic" molecular weight can be used to determine the maximum number of carbon atoms possibly present. Similar influence is exerted by the presence of N^{15} and O^{18} (natural abundance 0.36 and 0.20, respectively). The presence of chlorine, bromine and even sulfur in a molecule may be easily spotted because of the abundance of Cl^{37}, Br^{81} and S^{34}. These isotopes give rise to characteristic multiplets of peaks, separated by two mass units, also disclosing the number of such atoms present in the molecule (see Table 1). The above holds in principle also for the elemental composition of fragments, but great care has to be taken not to assume two peaks to be an isotopic multiplet which, in fact, are two different fragments. Decrease of the electron energy can be used also in this case to distinguish between these two possibilities.

Table 1. *Relative intensity of peaks in multiplets due to the presence of chlorine, bromine or sulfur.*

	Relative intensity of masses				
	X	$X+2$	$X+4$	$X+6$	$X+8$
Cl	3	1			
Cl_2	9	6	3		
Cl_3	27	27	9	1	
Cl_4	81	108	54	12	1
Br	1	1			
Br_2	1	2	1		
Br_3	1	3	3	1	
Br_4	1	4	6	4	1
S	1	0.042			
S_2	1	0.084			
S_3	1	0.125	0.005		

III. Simple Fragments.

Previously it had been mentioned that the more intense peaks within a group in the spectrum are the important ones for the interpretation and are due to the energetically preferred fragmentation modes. A discussion of these processes is, therefore, in order.

Since a conventional mass spectrometer records the positively charged particles, the extent of stabilization of such a charge on the ion is of primary importance, and the energy content of the neutral fragment formed in this process is also of considerable influence.

Aliphatic carbon chains fragment mainly at the points of branching, and the positive charge is retained at the particle with the more substituted carbon atom because of the increasing carbonium-ion stability in the series

$$CH_3^+, RCH_2^+, R_2CH^+, R_3C^+ .$$

From fragments of this type, the position and size of branches in aliphatic compounds or side chains on alicyclic molecules can be deduced.

A bond one atom removed from a double bond or an aromatic ring is cleaved with ease because the carbonium ion formed is stabilized by resonance with the unsaturated system. Such fragments are very useful for the interpretation of the spectra of aromatic compounds whereas the specificity of the cleavage of the allylic bond in olefins is not so clear. It appears that double bonds may easily rearrange in the molecular ion and their position in the original molecule is, therefore, not easily deduced from the mass spectrum. The bond directly attached to the aromatic ring is not ruptured to any considerable extent in mono-alkyl substituted benzenes but is cleaved in polysubstituted ones; xylene, for example, has a high peak due to the loss of a methyl group.

The presence of a heteroatom in the molecule introduces a further mode of stabilization of the positive charge by virtue of the availability of their free electron pairs: Alcohols, ethers, amines, thiols, and thioethers exhibit strong peaks due to cleavage of the carbon-carbon bond to which the heteroatom is attached.

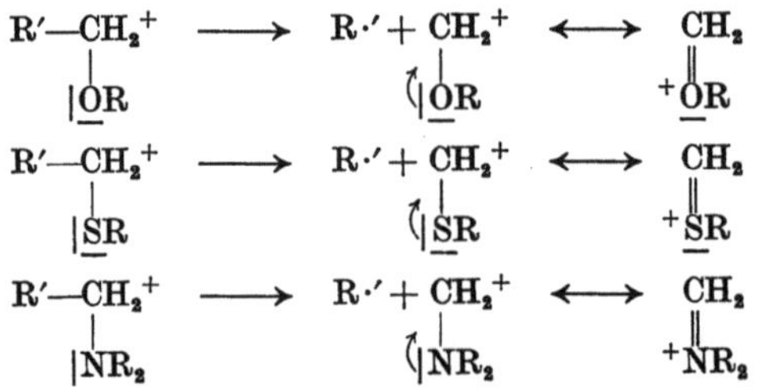

The extent of this stabilization seems to decrease in the order N, S, O and in polyfunctional compounds the N-containing fragment has a better chance to retain the positive charge. It will, therefore, give rise to the more intense peak which is borne out by the spectra of α-amino-alcohols, for example.

The influence of a double bonded heteroatom is a similar one, leading to a triple bond in the resonance form of the ion:

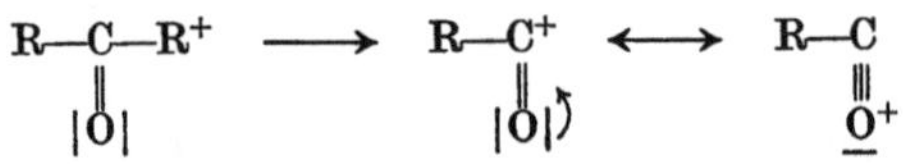

Fragments of this kind are abundant in the spectra of ketones and yield valuable information because they indicate the size of the R-groups.

IV. Rearrangements.

The majority of the processes leading to fragments not present as such in the molecule involve the migration of a hydrogen atom and ejection of a neutral particle. The loss of H_2O from alcohols (a), CH_3COOH from an acetate (b), an alcohol molecule or a ketene from an ester (c) may be pictured simply as a 1,2-elimination

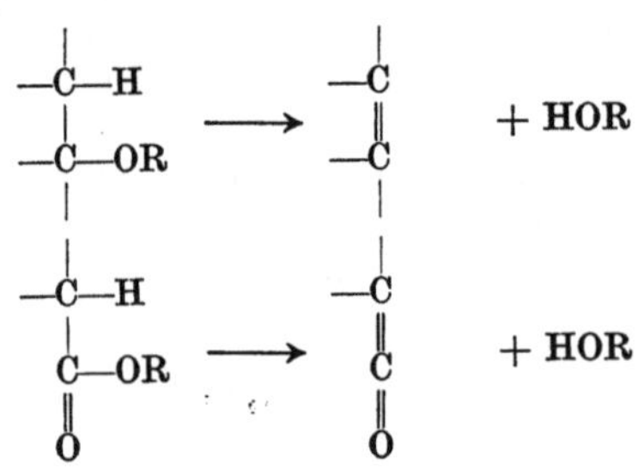

a) R = H
b) R = acyl

c) R = alkyl or aryl

The positive charge remains with the fragment at which it is best stabilized, in general with the fragment containing the newly-formed double bond unless the R-group is aromatic.

Another type of rearrangement is typical for compounds containing a multiple bond connected to a chain of three or more carbon atoms. It was first postulated by McLafferty (1957a) for esters (R = alkoxy) and ketones (R = alkyl or aryl):

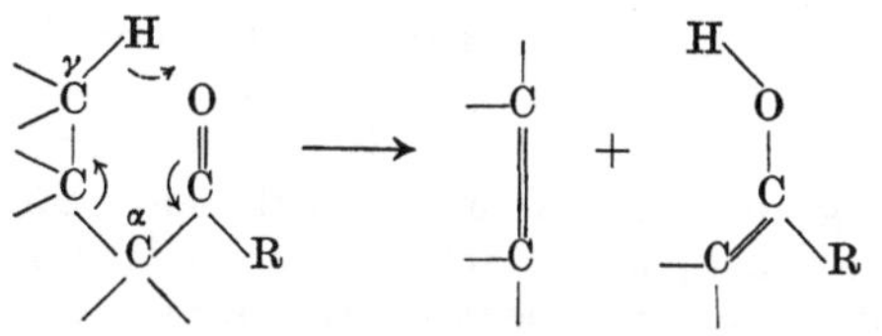

Unless there is an additional stabilizing group present in the olefine fragment, the positive charge remains with the oxygen-containing part. Thus, all aliphatic methyl esters ($R = OCH_3$) which meet the requirement of a hydrogen atom on the γ-carbon exhibit a strong peak at m/e 74, ethyl esters at m/e 88 and so on. Methyl esters with an additional alkyl substituent of n carbon atoms in the α-position exhibit a peak at $74 + 14\,n$. Similarly, methyl ketones of this type ($R = CH_3$), unsubstituted on the α-carbon, show a peak at m/e 58, etc. The above rearrangement is a quite common one for compounds of this general type; since only the multiple bond and the γ-hydrogen is necessary, it is also found in the spectra of amides (Gilpin, 1959), olefins (McLafferty, 1959), and aromatic molecules (McCollum and Meyerson, 1959).

Not only the molecular ion may decompose via a rearrangement process, but also fragment ions, and a probably rather common but rarely recognized process is the following one:

$$R{-}CH_2{-}CH_2^+ \rightarrow R^+ + CH_2{=}CH_2$$

V. Metastable Ions.

The occurrence of a certain fragment in itself does not indicate in which way it was formed from the molecular ion, particularly since the other particle is uncharged and thus not recorded. The so-called "meta-stable" peaks, broad peaks or rather humps of low intensity, may be used to determine whether one intense peak is formed from another intense one in a one-step process. The mass m^* at which this diffuse peak occurs is related to the starting mass, m_i, and the final mass m_f, in the following way:

$$m^* = \frac{m_f^2}{m_i}$$

VI. Multiple-Charged Peaks.

A molecule may on electron bombardment lose more than one electron to form a multiple-charged ion. Since the mass spectrometer resolves the particles according to their mass-to-charge ratio, such an ion will be recorded at its mass divided by the number of positive charges. Under normal ionizing conditions organic compounds, in general, give rise only to single or double charged ions, and the latter appear at half their mass in the spectrum. They are easily recognized only if their mass is an odd number and are, therefore, recorded at half-masses, e.g., the M^{++} of pyridine at m/e 39.5. The removal of two electrons is more probable from molecules of high π-electron density, making double-charged peaks a characteristic of heterocyclic, aromatic or highly unsaturated compounds.

VII. Mixtures.

For the interpretation of a mass spectrum it is of extreme importance to notice whether the sample is a pure substance or a mixture. An attempt to deduce the identity of a compound from a number of fragments which are, in fact, due to different molecules will be fruitless or lead sometimes to a plausible but incorrect assignment. Mixtures are best recognized by the following characteristics: (a) the appearance of peaks below the molecular ion which cannot be formed by the loss of a reasonable fragment (e.g., a mass between 4 and 14); (b) the appearance of a peak 14 mass units below a molecular peak is always due to a lower homologue; (c) the use of low-energy electrons accentuates the molecular species; (d) if the sample is subjected to at least partial fractionation, a mixture will exhibit different spectra before and after such treatment whereas a pure compound remains unchanged. From such changes one often can recognize groups of peaks belonging to the same component.

VIII. High Resolution Spectra.

The single focussing mass spectrometers previously discussed have a resolving power of one in a few hundred. They permit the distinction of ions the weight of which differs by one mass unit. A peak at mass 43 may be due to a fragment C_3H_7 or C_2H_3O, and for peak at mass 250 there are fifty-four possible fragments or molecules of different composition considering only those containing C, H, O and N. The atomic weights are, however, not integral numbers and all these particles have a slightly different mass: $C_3H_7 = 43.0685$, while $C_2H_3O = 43.0321$. In a double-focussing instrument these peaks are not only resolved but their mass can also be determined accurately enough to distinguish between the two. Naturally, this facilitates the interpretation of the spectrum of an organic compound, because fewer possibilities have to be considered (BEYNON, 1959).

Futhermore, the elemental composition of a compound can be arrived at if the exact molecular weight is determined in this way: The molecular weight of tetralin ($C_{10}H_{12}$), indanone (C_8H_9O) and butyl pyrazine ($C_8H_{12}N_2$) are 132.136, 132.010, and 132.110, respectively.

Double-focussing mass spectrometers are just now becoming commercially available (see Appendix I), making the mass spectrometric determination of the empirical formula a distinct possibility in the near future. This approach would be of particular value for the work on very scarce materials which warrant the employment of rather expensive equipment.

The preceding may suffice for an introduction to the interpretation of mass spectra, and some specific examples in the following section may serve as an illustration of the potentialities of this method.

D. Specific Applications.

I. Amino Acids.

1. Qualitative Spectra.

There are a number of procedures presently available to separate and identify or determine quantitatively amino acids on a rather small scale. Mass spectrometry may now be added as an additional method possessing a much higher specificity with regard to their structure. From the discussions in Section C it is to be expected that the spectra of the various amino acids are rather different from each other because they generaly vary in size or at least in the structure of their R-groups in formula I:

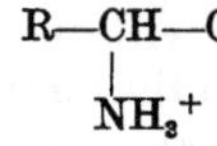

I

The extremely low volatility of such a zwitter-ion seems, however, to eliminate the application of mass spectrometry in this field. A simple chemical reaction, esterification by boiling with ethanolic hydrochloric acid and conversion of the hydrochloride II to the free ester III results in a quite volatile compound which gives excellent mass spectra (BIEMANN et al., 1959).

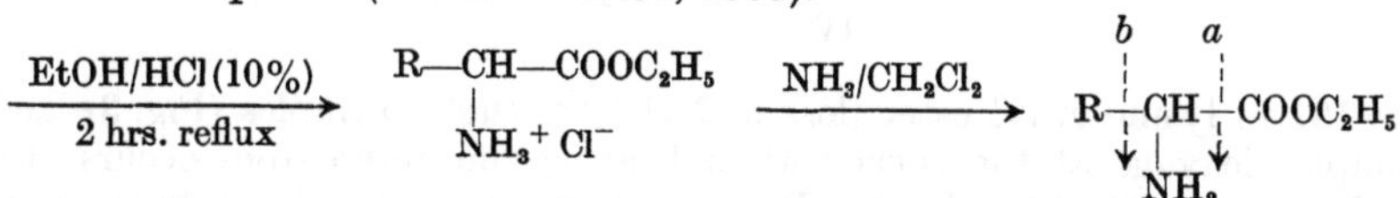

I II III

The major fragmentations observable are due to cleavage (1) at bond *a*, (2) at bond *b*, and (3) at carbon atoms of R which bear either alkyl-branches, hetero-atoms or aromatic rings. The first-mentioned process gives rise to a strong peak at the mass of $\underset{\displaystyle NH_2}{R-CH^+}$, a fragment on which the positive charge is well stabilized (Section C, III). Subtraction of 29 gives the size of the R-group, which in itself would already suffice for the identification of the compound if one knows that the material is one of the known α-amino acids. These peaks fall at *m/e* 30, 44, 58, 72, 86 for the homologous series glycine, alanine, α-amino butyric acid, valine, and leucine, for example.

If fragmentation at bond *b* occurs, the positive charge is again retained at the nitrogen-containing fragment, and a peak at *m/e* 102 is observed in all α-amino acids, unsubstituted at C_α or in the amino group. Proline ethyl ester (Fig. 3b), therfore, does not exhibit such a peak. The intensity of *m/e* 70 is still enhanced because of the lack of competition for the molecular ion by other fragmentation modes.

The fragmentation at b is an interesting one with regard to the previously discussed (Section C, IV) rearrangements taking place in aliphatic esters. This would lead us to expect a peak at *m/e* 103 in the spectra of the ethyl esters of α-amino acids having a hydrogen atom at the γ-carbon. It seems that one of the driving forces leading to the rearranged ion (*m/e* 88 in unsubstituted ethyl esters, see Section C, IV) is the fact that in this way the $-C_\beta-C_\alpha$-bond is broken without formation of an ion $^+CH_2-CO_2C_2H_5$. The latter is a poorly stabilized one because the positive charge is next to a carbonyl-carbon. An amino group on the α-carbon atom sufficiently stabilizes this ion, which is now formed to an appreciable extent. Only in α-amino-β-hydroxy acids, like serine and threonine, is a peak at *m/e* 103 observed, indicating that hydrogens on hydroxyl groups show a greater tendency to rearrange than those attached to carbon. This fragment of *m/e* 102 always decomposes further to give a peak at *m/e* 74 by expulsion of an ethylene molecule, another form of the previously discussed type of rearrangement (Section C, IV):

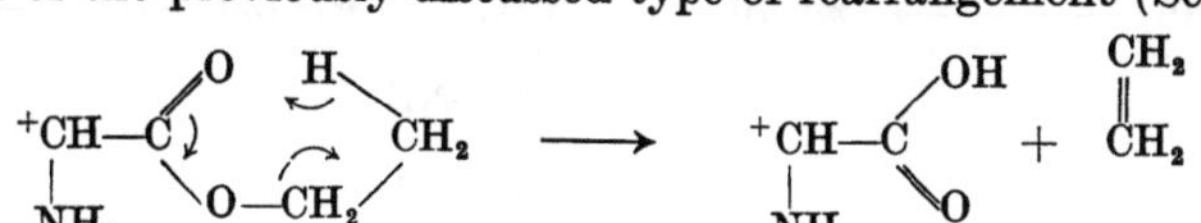

102 74

A number of characteristic peaks are due to the structure of the R-group and to functional groups contained therein. Fragmentation of this part of the molecule may occur in the molecular ion or start with the ion $(R-CH-NH_2)^+$. The former is the case if R contains a heteroatom in the form of hydroxyl, amino, mercapto or thioether group, which can be eliminated as H_2O, NH_3, H_2S or RSH: The peaks at *m/e* 129 in methionine ethyl ester (IV, loss of CH_3SH)

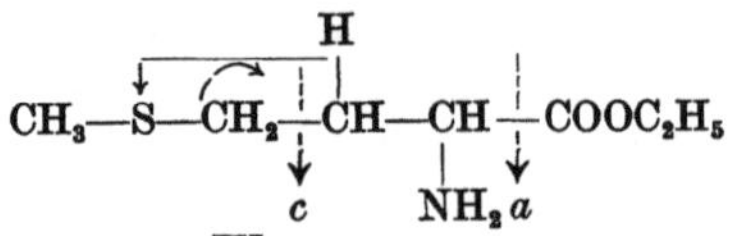

IV

or *m/e* 157 in lysine ethyl ester (loss of NH_3) are such examples (Fig. 3e and f).

Simple cleavage at the carbon atom bearing the heteroatom occurs also as would be expected (see Section C, III). It is only a question of stabilization of the positive charge which fragment will appear in the spectrum: The highest peak in the spectrum of methionine ester is at *m/e* 61, $CH_3-S-CH_2^+$, due to cleavage at *c* in IV. A small peak at *m/e* 132 in the spectrum of threonine ester (V) is formed by the loss of methyl, a process induced by the β-hydroxyl group:

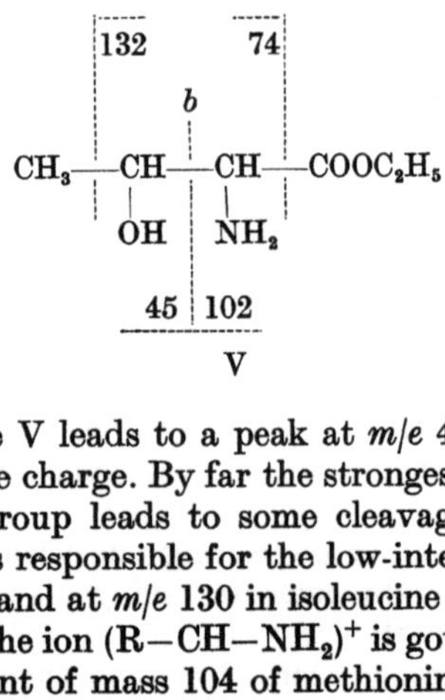

V

Cleavage at *b* in structure V leads to a peak at *m/e* 45 which is, however, competing with *m/e* 102 for the charge. By far the strongest peak is still *m/e* 74.

Branching in the R-group leads to some cleavage at such points, but this effect is rather small. It is responsible for the low-intensity peaks at *m/e* 144 and 116 in leucine ethyl ester and at *m/e* 130 in isoleucine ester (Fig. 3c, d). Similarly further fragmentation of the ion $(R-CH-NH_2)^+$ is governed by functional groups present in R. The fragment of mass 104 of methionine ester (fragmentation at *a* in IV) may eliminate CH_3-S-H, giving rise to the peak at *m/e* 56. The fragment of mass 104 in lysine ester (Fig. 3f) loses NH_3 with great facility to give a peak at mass 84, the highest in the spectrum. Aspartic and glutamic acid ethyl ester (VI and VII, respectively) eliminate the elements of ethanol which gives rise to peaks at mass 70 and 84, respectively, at least the latter being of a cyclic structure.

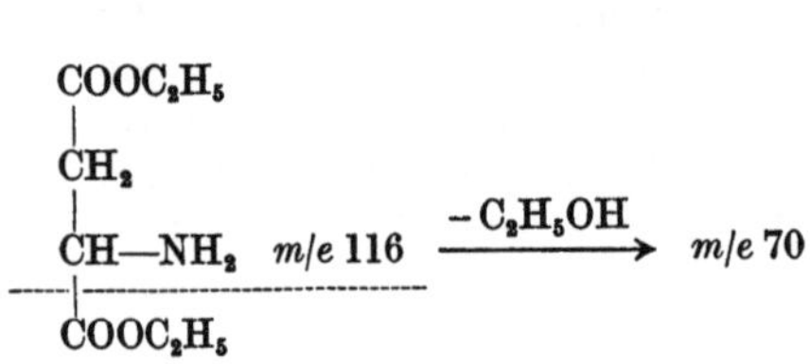

VI

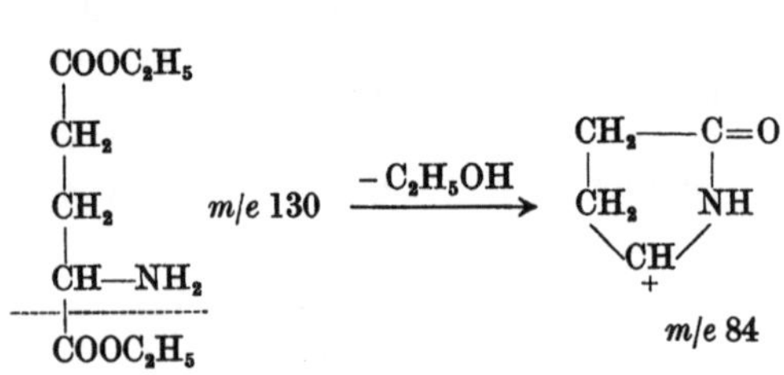

VII

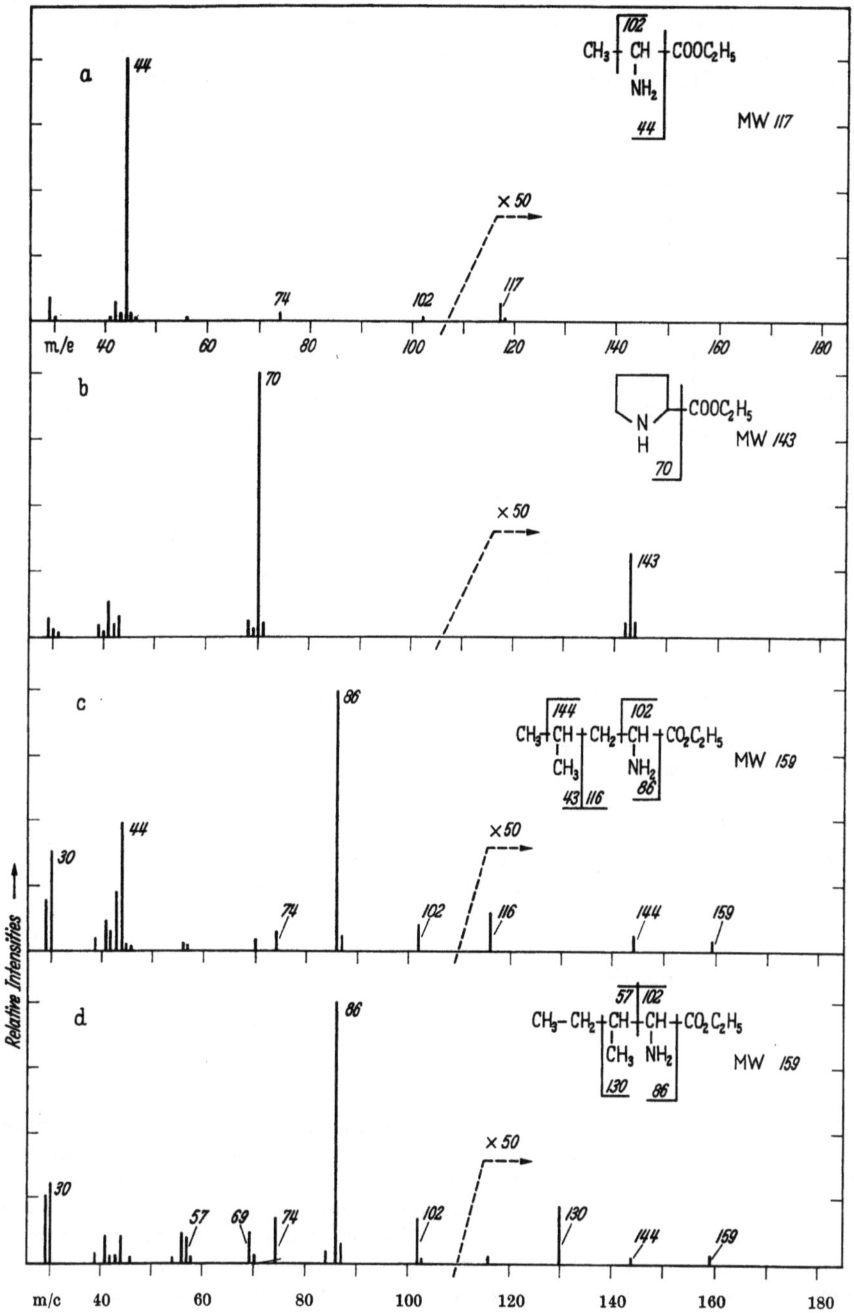

Fig. 3a—d.

Fig. 3a—f. Mass spectra of the ethyl esters of alanine *(a)*, proline *(b)*, leucine *(c)*, isoleucine *(d)*, methionine *(e)* and lysine *(f)* in the region of m/e 29—180.

An aromatic R-group does not fragment easily and, on the other hand, facilitates the stabilization of the positive charge on those particles not containing the nitrogen. Because of the increase in stabilization of the positive charge on the fragment $(Ar-CH_2)^+$ of phenylalanine (VIII), tyrosine (IX), and tryptophan (X),

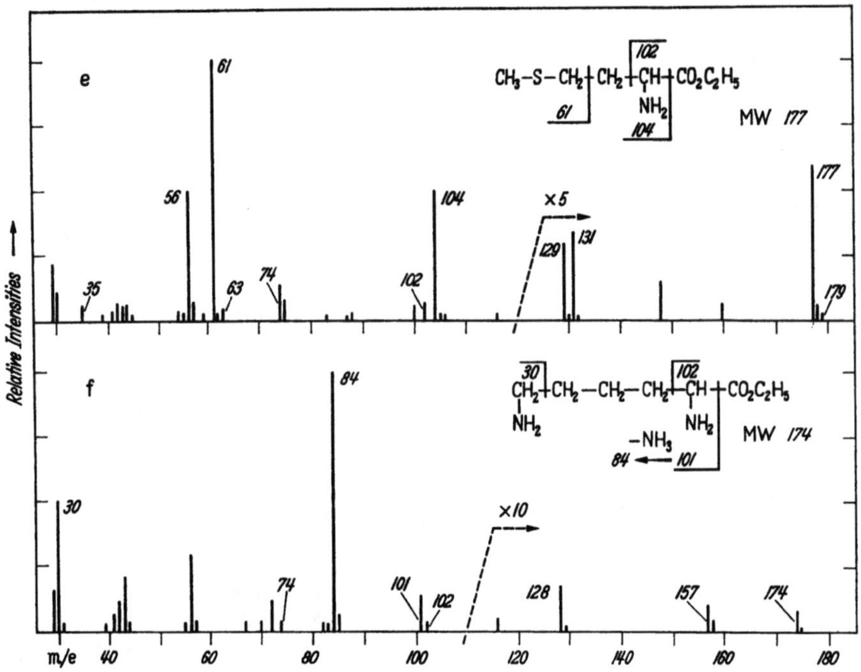

Fig. 3e—f.

the intensity of these peaks (m/e 91, 107 and 130) increases in this series whereas the $Ar-CH_2-CH^+$ peaks (m/e 120, 136 and 159) decrease.
$\quad\quad\quad\quad\quad\quad\quad\quad\quad | $
$\quad\quad\quad\quad\quad\quad\quad\quad NH_2$

$$Ar-CH_2-\underset{NH_2}{\underset{|}{CH}}-COOC_2H_5$$

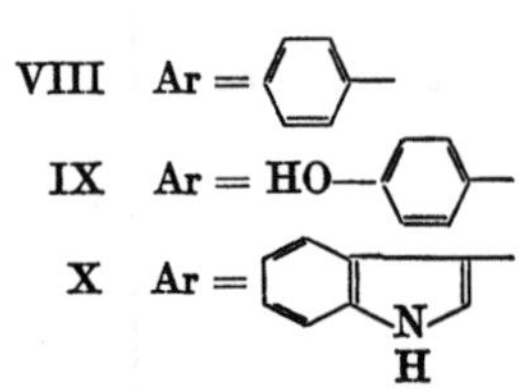

The rigorous identification of an amino acid is, therefore, possible by a brief interpretation of the mass spectrum of its ethyl ester followed by its comparison with the spectrum of an authentic sample. Such a comparison may even reveal the presence of other amino acids as impurities if the sample were prepared and the spectrum obtained in a way which prevents the loss of minor components (for these techniques see Section D,I, 2). Furthermore, the qualitative analysis of a mixture of amino acids can be achieved by an interpretation of the spectrum of the ester mixture making use mainly of the intense peaks. In such a case, the spectrum should also be obtained with low-energy electrons (~ 10 eV). This technique accentuates the peaks due to cleavage at *a* in formula III and thus simplifies the spectrum and its interpretation.

Having shown that mass spectra of esters of amino acids can be explained on the basis of their known structure, the constitution of a new amino acid may be established with the aid of this technique. Recently this was done in the case of an amino acid isolated from the crown gall tissue of plants (BIEMANN et al., 1960). Not enough material was available to establish its structure by conventional means and only the usual analytical characterization was done. The mass spectrum of the esterified material exhibited characteristic peaks at *m/e* 274, 201, 184, 156 and 84, the most intense one. The first was identified as the molecular weight, on the basis of the smaller one at 275, whose intensity ratio compared with the others was influenced by the conditions in the ion-source. This behavior is typical for an "$M+1$" peak, disclosing a molecular weight of one unit less (Section C, II). The even number (274) indicates the presence of 0, 2, 4 . . . nitrogens in the molecule, most probably two. The facile loss of 73 mass units ($-COOC_2H_5$) to give 201 indicates an amino group in the α-position. Mass 201 loses NH_3 leading to the peak at mass 184 as evidenced by the corresponding metastable peak at about mass 169 (calculated 168.4; see Section C, V). An NH_2 grouping, therefore, had to be present which was eliminated in this step. Further fragmentation by loss of 28 mass units (C_2H_4) to mass 156 (metastable peak at *m/e* 133) is reminiscent of the formation of mass 74 from the fragment of mass 102 in α-amino-ethyl esters and indicates the presence of a second carbethoxy-group in the molecule. Finally, the strong peak at mass 84 suggests that also in this compound the two amino groups are linked by a C_5-segment. Structure XI

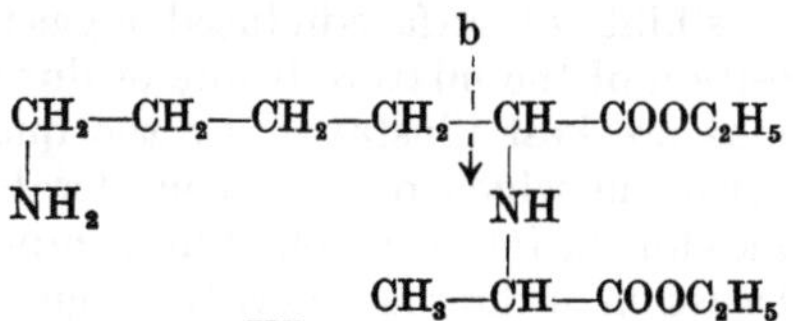

XI

was deduced on the basis of this interpretation, and in consideration of the absence of an appreciable peak at mass 102 (no primary-α-amino acid) plus the presence of a peak at *m*/202 in excess of the calculated C^{13}-isotope peak of mass 201. It should contribute only about 11.7% of the intensity at *m/e* 201 (10×1.1% for the ten carbon atoms plus 2×0.36% for the two nitrogens present in this fragment). The intensity of the peak at mass 202 was, however, 24% of the one at *m/e* 201, thus allowing for a fragment due to cleavage at *b* in XI. The NH_2-group on the terminal carbon atom is borne out by the peak at *m/e* 30. It is higher than the ones at 44, 58 and 72, which would be expected if this group were attached to one of the other carbon atoms. The ester does not spontaneously cyclize under elimination of ethanol, indicating that no amino group is so placed as to give a six-membered lactam with any one of the two carbethoxy groups. Thus, the proposal of a terminal amino group on an unbranched C_5-segment is strengthened further. Ornithine-ester immediately undergoes the cyclization referred to above, and the spectrum obtained is the one due to the lactam. The acid corresponding to ester XI was then synthesized, and one of the two possible diastereomers after optical resolution proved to be identical with the naturally occurring amino acid, which was named lysopine because of its structural relationship to lysine and octopine. It might be pointed out that mass spectra are rather insensitive to stereochemical differences, particularly in aliphatic molecules, and no assignment of this kind could be made on the basis of the mass spectrum of lysopine ester.

Although not necessary in this case, the spectra of additional derivatives such as methyl or propyl esters and acylation or alkylation of the amino groups are often very helpful in such interpretations. The differences between them can be used to confirm or disprove certain assignments. A peak of an ethyl ester supposed to contain the ester grouping must, for example, shift 14 mass numbers in the propyl ester. In comparing methyl esters with higher esters, one has to keep in mind that the former are unable to undergo elimination of olefinic fragments from the alcohol moiety and may, therefore, give spectra of quite different appearance.

2. Quantitative Analysis of Amino Acid Mixtures.

One of the prerequisites for accurate quantitative mass spectrometric analysis is the presence of at least one intense peak in the spectrum of each component to which the other substances do not contribute considerably or at all. Because of the sharpness of the peaks appearing at nominal mass numbers quite widely separated from each other, such contributions are strictly additive and can be easily subtracted. This is quite in contrast to analyses based on ultraviolet or infrared spectra, where the maximum of a band of one component frequently is at a wavelength where another component's absorbancy changes greatly vs. wavelength, making very exact reproduction of the latter necessary. A further advantage of mass spectrometry is the large number of peaks in the spectra which makes it easily possible to find one "analytical" peak for each component even in very complex mixtures. For example, in the petroleum industry mixtures of up to forty components are analyzed routinely. The disadvantage of mass spectrometric analysis is the necessity to vaporize the mixture into the inlet system without any fractionation. Also there is the possibility of surface-induced or gas phase reactions which might change the composition of the mixture before or during the run.

All these considerations are best illustrated by the quantitative analysis of amino acid mixtures, a problem which requires considerable chemical pretreatment and special techniques for the introduction of the sample into the instrument.

On the basis of the discussion of the characteristic fragmentation of amino acid ethyl esters (Section D, I, 1) it may be anticipated that the principal requirement for such an analysis, namely, the presence of a strong peak suffering little interference from other amino esters, is fulfilled quite easily. The strongest peaks in mono-amino monocarboxylic esters are the ones due to the $R-\underset{NH_2}{\underset{|}{CH^+}}$-fragments and are thus always separated by 14 mass units, with the exception of the isomeric leucines. Amino esters with additional functional groups exhibit other strong peaks which are in general also unique for the given compound. Obviously, the quantitative evaluation of the mixture presents no particular problem once the spectrum is obtained and calibration spectra of the pure components are available.

The major difficulties are others: The original sample always is in the form of free amino acids or their hydrochlorides; only about one milligram of the entire mixture, or little more, should be required to make the method applicable to the problems generally encountered; the tendency of amino acid esters to dimerize to diketopiperazines; the wide range in volatility between glycine ester and tryptophan ester.

All these problems are overcome by the following procedure (Biemann and Vetter, 1960a), which will be discussed in more detail. The basic techniques described cannot only be used for the analysis of amino acids. They are generally applicable for the introduction of samples requiring chemical pretreatment or those present in very dilute solution, provided the solvent has a boiling point considerably lower than the solute, and regardless whether quantitative or qualitative information is sought.

The mixture of amino acids (0.5 to 2 micromoles each; i.e., 1 to 2 mg. of a ten-component mixture) is refluxed with 2 ml. of ethanolic hydrochloric acid (10%) for three hours, the solvent evaporated in a dessicator and the remaining ester hydrochlorides dissolved or suspended in 2 ml. of dichloromethane. Dry ammonia is then passed briefly through this mixture liberating the free esters while ammonium chloride precipitates. This is then filtered off and the filtrate collected in the tube shown in Fig. 4 A. The dichloromethane is now distilled off by immersing the bulb in water of about 60°. The tube is then connected with the port on the inlet system of the mass spectrometer and cooled with a mixture of acetone and dry ice kept at —60°. Any remaining solvent is pumped off through the pumping system of the instrument. After closing the pump valves the tube is heated quickly to 160° by replacing the Dewar vessel with an electric heater (Fig. 4B), which quantitatively introduces the esters into the mass spectrometer. The amino esters, when free, are either in a dilute solution at very low temperature or in the vapor phase under low pressure and dimerization is thus prevented. Dichloromethane was chosen as a solvent because it dissolves most of the ester hydrochlorides, has a low boiling point, and its mass spectrum interferes least with the spectra of the esters, important because the last traces of solvent are difficult to remove.

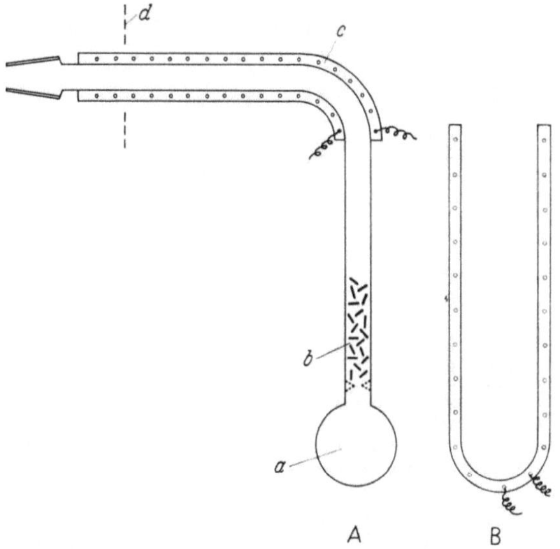

Fig. 4. *(A)* Introduction tube: sample bulb *(a)*, glass rings *(b)* and heater *(c)*; *(d)* front of heated inlet system. *(B)* External heater.

Calibration spectra are obtained of each amino ester by applying the same procedure to samples of the pure amino acids. The intensity of the "analytical peaks" (Table 2, column 2) due to the individual components are obtained by subtraction of the contributions of the others on the basis of the calibration spectra. This is best done by starting with the component whose analytical peak is of highest mass (e.g., phenylalanine in mixture A, Table 2) because it will not have any appreciable contributions from others. In this way one component after another is "peeled off" and there is left at most two or three contributing to each other's analytical peaks. They have to be solved by a set of simultaneous equations or, simpler, by successive approximations.

The molar intensities of all the analytical peaks are, however, different and have to be determined before the results become meaningful. In mass spectrometric analyses this is usually done by measuring exactly the amount of sample when the calibration spectra are run. The intensity of the analytical peak per micron pressure in the inlet system can be used for the comparison of such intensities on a molar basis if the pressures are measured at the same temperature and volume. For this reason mass spectrometric results are always expressed in mole-per cent. In the case of amino esters, this method would require the preparation of a relatively large amount of the pure, distilled ester and introduction as a liquid (Section B, II) because the dichloromethane-method described above is not suited for this purpose: The pressure measurement would not be very accurate due to traces of water, solvent and air, the presence of which cannot be avoided. For this reason, the relative ratios of intensities of the analytical peaks are best determined from several mixtures of accurately known composition.

These mixtures together contain all amino acids to be analyzed and to each mixture is added an internal standard, e.g., aspartic acid.

Table 2. ***Results of quantitative analyses of synthetic mixtures of amino acids*** (Biemann and Vetter, 1960a).

Amino Acid	Mass Number[1]	Mixture A Calcd.	Mixture A Found[8]	Mixture B Calcd.	Mixture B Found[9]	Mixture C Calcd.	Mixture C Found[10]	Mixture D Calcd.	Mixture D Found[10]
gly	30[2]	2.64[7]	2.63	2.16	2.09	—	—	—	—
ala	44[2]	1.73	1.75	1.31	1.28	—	—	—	—
abu	58[2]	1.85	1.84	—	—	—	—	—	—
val	72[2]	1.87	1.83	—	—	—	—	1.60	1.65
leu	44[3]	—	—	—	—	1.41	1.38	—	—
ileu	69[3]	—	—	—	—	1.66	1.68	—	—
pro	70[2]	1.30	1.32	—	—	1.18	1.17	—	—
ser	60[2]	—	—	1.44	1.42	—	—	—	—
thr	74[2]	—	—	0.98	0.98	—	—	—	—
opro	86[2]	1.40	1.41	1.01	1.00	—	—	—	—
met	61[4]	—	—	—	—	1.43	1.44	—	—
phe	120[2]	1.00	1.00	—	—	—	—	1.00	1.03
tyr	107[5]	—	—	—	—	—	—	0.80	0.76
asp	116[2]	1.03	1.05	1.00	1.00	1.00	1.02	1.19	1.21
glu	84[3]	1.40	1.41	—	—	—	—	—	—
orn	69[6]	3.10	3.27	—	—	—	—	—	—

[1] *m/e* of peak used for calculation. — [2] $(R—CH—NH_2)^+$ fragment of $R—CH(NH_2)CO_2C_2H_5$. — [3] Originates through further fragmentation of [2]. — [4] $(CH_3—S—CH_2)^+$. — [5] $(HO—C_7H_6)^+$. — [6] Fragment of orn-lactam. — [7] Values represent molar ratios. — [8] Average of 2 determinations. — [9] Average of 4 determinations. — [10] Average of 3 determinations.

The results of such analyses are summarized in Table 2. The values found agree well with the theoretical ones. The largest deviations (up to 5%) are found with tyrosine and ornithine, the former because of its lower volatility (also the case with tryptophan); the latter because ornithine ester immediately cyclizes to its lactam which, lacking intense peaks, is not very suitable in this analysis. Although methionine can be analyzed without difficulties, the ethyl esters of cysteine, cystine and lanthionine are thermally not stable enough to give reproducible quantitative values. This may be overcome by catalytic desulfurization of the mixture with Raney-nickel prior to esterification. Since the reduction product is alanine, this amino acid has to be absent or must be determined in a separate step. Arginine, the ester of which is still not sufficiently volatile because of the guanidino group, cannot be determined.

II. Amino Acid Sequence in Peptides.

The ester method is not suitable for the determination of the structure of peptides, but they can be made volatile by reduction with lithium aluminum hydride (or deuteride) in tetrahydrofuran (Biemann et al., 1959). The polyamino alcohol (XII) thus obtained from a peptide still has the same structure as the parent peptide as far as type and location of the side chains are concerned except that the $-CO$-groups are replaced by $-CH_2-$:

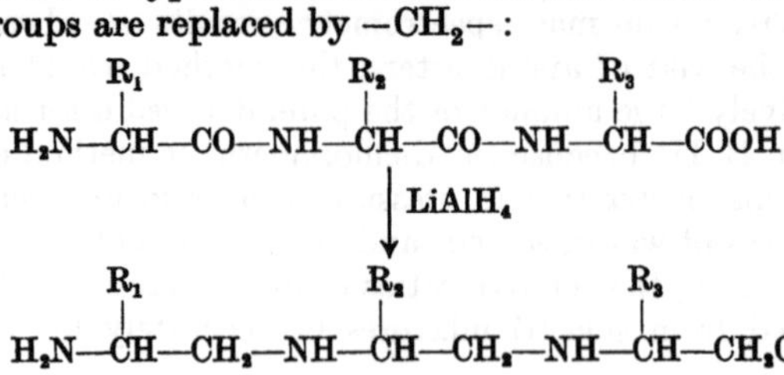

(XII)

In these molecules we now have a succession of –NH–C–C–NH-groups, the carbon-carbon bonds of which are easily cleaved on electron impact (Section C, III). These polyamino alcohols, therefore, give rise to fragments due to cleavage of any one of these bonds and since the positive charge is better stabilized on the more substituted carbon atom, the fragment pertaining to the N-terminal part of the molecule will be somewhat more pronounced. In general, the spectrum of such an amino alcohol will exhibit a very strong peak due to the N-terminal fragment (XIII)

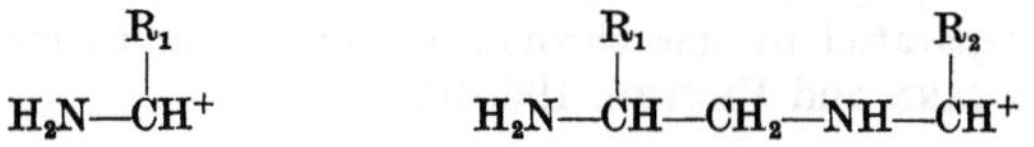

XIII XIV

the mass number of which identifies the N-terminal amino acid (except differentiation of leucine and isoleucine). A somewhat less intense peak will be found $42 + R_2$ mass units higher (fragment XIV) which now establishes the amino acid next to the N-terminal and so forth. The reason for the decrease in intensity towards higher mass numbers is, aside from some instrumental factors, due to the increased number of possibilities for further fragmentation of the larger particles. The sequence deduced in this way is further borne out by the corresponding peaks due to the fragments originating from the C-terminal part of the molecule:

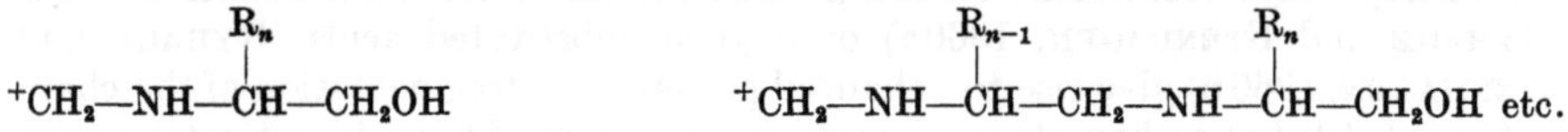

XV XVI

The peak due to fragment XV found at mass $73 + R_n$ reveals the C-terminal amino acid; the one found $42 + R_{n-1}$ mass units higher (fragment XVI), the next amino acid and so on. The molecular weight of the polyaminoalcohol is deduced from the "M + 1 peak" which is small but very characteristic in these compounds and further corroborates the interpretation as do some peaks due to loss of any one of the side chains from the otherwise intact molecule. Such a spectrum is shown in Fig. 5. It should be noted that the spectrum is of the amino alcohol obtained from

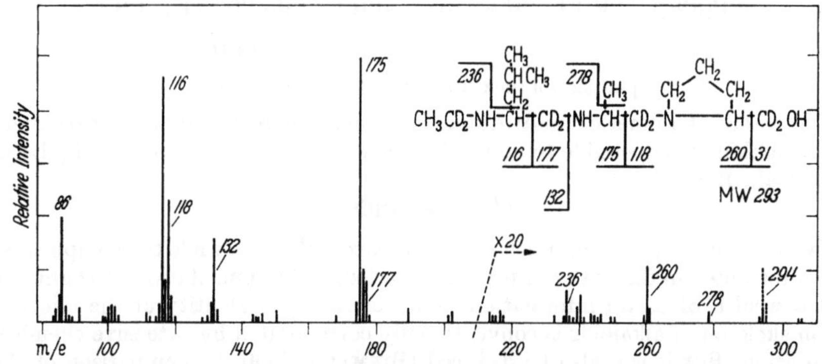

Fig. 5. Mass spectrum of the polyamino alcohol derived from Ac-leu-ala-pro-OEt. The dotted peaks indicate the intensity of the "$M + 1$" peak on switching the ion source from "focused" to "non-focused".

acetyl-L-leucyl-L-alanyl-L-proline by reduction with lithium aluminum deuteride and therefore contains a $-CD_2$-group for each –CO-group of the peptide which thus shifts the mass numbers of the fragments up two mass units per such group contained in it. The acetyl peptides were employed in these studies because they were easier to synthesize as model compounds and also seem to give better results in general. The largest peptide as yet tested successfully with this method was a pentapeptide, Ac-leu-ala-gly-leu-pro.

The reader may predict the corresponding mass spectrum of the compound obtained on reduction of the same peptide with lithium aluminum hydride. It illustrates how advantageous the specific incorporation of heavy isotopes is for the correlation of the peaks in a spectrum with certain parts of the molecule.

Mixtures of such polyamino alcohols obtained by reduction of a mixture of peptides can be separated by gas chromatography prior to mass spectrometric investigation (Biemann and Vetter, 1960b).

III. Fatty Acids and Related Compounds.

While almost all the examples discussed are related to amino acids, considerable work in other areas has also been done. In the fields of research related to materials of the plant kingdom, fatty acids play an important role. The mass spectra of their methyl esters have been thoroughly investigated by Ryhage and Stenhagen (1959) and found to follow the general fragmentation patterns outlined in Section C, III and IV. Particularly important for the spectra of esters of carboxylic acids, is the cyclic rearrangement (Section C, IV), which discloses the size of any alkyl substituent on the α-carbon atom of the acid. Branched acids (Ryhage and Stenhagen, 1960a) or oxygen substituted acids (Ryhage and Stenhagen, 1960b) give rise to enhanced peaks due to fragmentation of the chain at the substituted carbon atoms. Based on the results of these investigations, the structure of mycocerosic acid (Asselineau et al., 1959) and phthiocerol (Ahlquist et al., 1959; Ryhage et al., 1959a, b), compounds occurring in the lipid-fraction of tubercle bacilli have been deduced.

$$CH_3—(CH_2)_{19}—\underset{CH_3}{\underset{|}{CH}}—CH_2—\underset{CH_3}{\underset{|}{CH}}—CH_2—\underset{CH_3}{\underset{|}{CH}}—CH_2—\underset{CH_3}{\underset{|}{CH}}—CO_2H$$

mycocerosic acid

$$CH_3(CH_2)_n—CHOH—CH_2—CHOH—(CH_2)_4—\overset{CH_3}{\overset{|}{CH}}—\underset{OCH_3}{\underset{|}{CH}}—CH_2—CH_3$$

phthiocerol (mixture of $n = 20$ and $n = 22$)

The esters of isomeric, unsaturated fatty acids were found to be extremely similar, making it impossible to determine the location of the double bond (Hallgren et al., 1959).

IV. Alkaloids.

Even the mass spectra of much more complex molecules, while too complicated to be interpreted in full with our present knowledge of fragmentation mechanisms, yield a wealth of useful information. For example, the identity of the alicyclic carbon skeleton of alkaloids is conventionally corroborated by extensive chemical degradation. But it can also be deduced (Biemann, 1960) by comparison of the

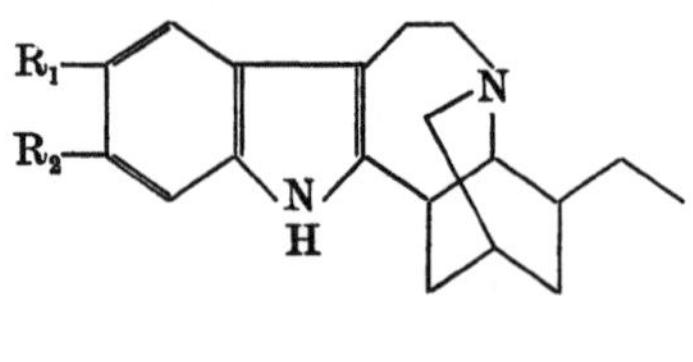

XVII: $R_1 = R_2 = H$
XVIII: $R_1 = OCH_3$, $R_2 = H$
XIX: $R_1 = R_2 = OCH_3$

mass spectra of the compounds in question, as illustrated in Fig. 6, which shows the spectra of ibogamine (XVII), ibogaine (XVIII) and ibogaline (XIX) (BIEMANN and FRIEDMANN-SPITELLER, 1962). The identity of the alicyclic carbon skeleton is borne out by the group of peaks in the region *m/e* 120—150, identical in mass number and very similar in intensity in all three spectra. They are due to the alicyclic ring system, whereas other peaks are shifted 30 and 60 mass numbers, respectively, indicating the presence of one and two methoxyl groups in XVIII and XIX, respectively.

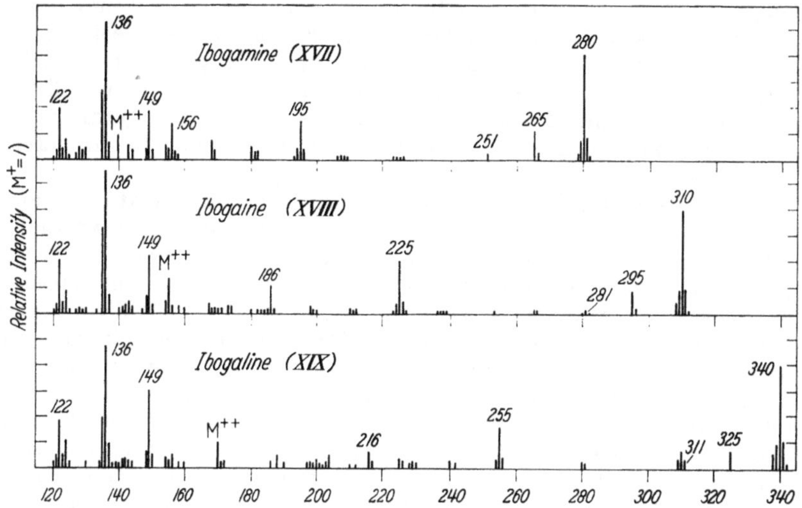

Fig. 6. Mass spectra of ibogamine, ibogaine and ibogaline in the region of m/e 120—350.

V. Miscellaneous Groups.

Besides the types of compounds mentioned, many others have been investigated to a greater or lesser extent: Alcohols (FRIEDEL et al., 1956), ethers (McLAFFERTY, 1957b), acetals (FRIEDEL and SHARKEY, 1956), thiols and thioethers (LEVY and STAHL, 1961), amines (COLLIN, 1952), aldehydes (GILPIN and McLAFFERTY, 1957), ketones (SHARKEY et al., 1956; BEYNON et al., 1960), acids and esters (McLAFFERTY and GOHLKE, 1959; SHARKEY et al., 1959; BEYNON et al., 1961), amides (GILPIN, 1959), steroids (REED, 1958; FRIEDLAND, 1959), terpenes (REED and GILCHRIST, 1960), and indoles (BEYNON and WILLIAMS, 1960).

A group important in the chemistry of plants is sugars. Here the application of mass spectrometry is hampered by the involatility of the compounds, which might be overcome, however, by employing suitable derivatives (ethers, esters). But the structural differences are mostly of a stereochemical nature leading only to slight changes in the spectra which can be interpreted only after an extensive study of model compounds. A limited amount of work in this area has been published (FINAN et al., 1958; FINAN and REED, 1959).

Aliphatic stereoisomers exhibit almost identical mass spectra, while in rigid, cyclic molecules the differences are more pronounced and interpretable. For polycyclic, epimeric alcohols the abundance of the molecular ion was always found to be higher in the spectrum of the less crowded one of the two isomers (BIEMANN and SEIBL, 1959).

VI. Determination of Stable Isotopes in the Intact Molecule.

In volume I of this series the conventional methods for the determination of stable isotopes have been discussed. The principle there is the conversion of the sample to hydrogen, nitrogen or carbon dioxide, gases which can be handled easily in simple mass spectrometers. They also can be obtained in very pure form, making highly accurate determinations possible. These procedures are, however, time-consuming, require a relatively large amount of material and, most important, give a result which is the average isotope ratio of the sample. Mixtures have, therefore, to be separated very carefully into the individual components to obtain correct values for each one of them. On the other hand, a mass spectrum of the complete mixture without degradation will exhibit the peaks of the labeled molecules adjacent to the peaks of the normal species. Their ratios will be an indication of the amount of heavy isotope in this compound, regardless of the presence of other material, as long as the latter does not contribute to the analytical peaks of the former.

This is best exemplified by the determination of N^{15} in a mixture of amino acids (BIEMANN and DEFFNER, 1961). In Section D, I, 1 it was shown that the ethyl esters of amino acids exhibit an intense peak due to the fragment $(R-CH-NH_2)^+$. If the nitrogen atom in this fragment is N^{15}, this peak will appear one mass unit higher: The peak at mass 86 in leucine ethyl ester (Fig. 3, c) will shift to mass 87 and both peaks will be present if only part of the total nitrogen is in the form of the heavy isotope. The ratio of these two peaks corresponds to the ratio of $N^{14}:N^{15}$ in the leucine molecule after correction for the natural abundance of D, C^{13}, and N^{15} as deduced from the spectrum of ordinary leucine. The presence of another amino acid, like proline, cannot only be tolerated, but its N^{15}-content can be determined as well in the same experiment from the ratio of the peaks at m/e 70 and 71. The possible contributions of other amino esters present are eliminated by the use of low-energy electrons (approx. 10 eV), which suppress most of the other fragmentation processes. Furthermore, by stepwise fractionation of the amino esters into the mass spectrometer, even rather complex mixtures may be run without interference. Leucine ester gives a fragment at mass 30 which cannot be suppressed completely by low-energy ionization. If the sample tube used for the quantitative amino acid determination (Section D, I, 2) is warmed first only to 0° C., glycine and alanine ester distil into the inlet system, but not leucine. The spectrum taken at this point will exhibit a peak at m/e 30 due only to glycine. Raising the temperature in three steps permits the remaining amino esters to enter the instrument in groups, eliminating—for example—the interference of the peak at m/e 70 of the aspartic ester with the peak of the more volatile proline ester. A mere arithmetical substraction of these contributions on the basis of calibration spectra, as is done in the quantitative analysis (Section D, I, 2), is not possible here: first, because the peak due to the N^{15}-containing species, while of relatively low intensity, has to be measured as accurately as possible; secondly, spectra obtained under low-energy ionization are not easily reproducible, a fact which does not alter the ratio of two isotopic peaks. Some results obtained by this method on a mixture of eleven amino acids are given in Table 3. One of the two samples was made up of ordinary amino acids and the values indicate the lower limit of detection of excess N^{15} by this technique. The second mixture contained some of the amino acids enriched in N^{15}. Practically all the values for % excess N^{15} fall between $\pm$ 0.3 of the calculated one.

Although this accuracy is far below that for RITTENBERG's (1942) procedure, it is sufficient for all those studies in which only the incorporation of an appreciable amount and not mere traces of N^{15} are significant. Moreover, the method discussed above requires only 3 to 5 micromoles per amino acid instead of about 70, makes the tedious separation into the individual components unnecessary, and permits the determination of all the values in one run, advantages which may play a decisive role in the solution of any problem involving a large number of such analyses. A further extension of this approach to the basic amino acids will probably make it possible to distinguish between the different nitrogens in the molecule.

Table 3. *Determination of N^{15} in mixtures of amino acids.* (BIEMANN and DEFFNER, 1961).

	Mixture A (unlabeled amino acids)	Mixture B (partly labeled) found	given
Glycine	0.22[1]	2.06	2.01
Alanine	0.17	1.86	1.89
Proline	0.19	0.27	0
Valine	0.14	2.83	2.88
Leucine	0.11	2.75	2.90
Serine	0.02	4.58	4.31
Threonine . . .	0.17	0.12	0
Aspartic acid . .	0.08	58.78	58.67
Phenylalanine . .	0.38	3.90	4.03
Glutamic acid . .	0.05	0	0
Tyrosine	—0.11	0.35	0

[1] The values refer to %excess N^{15}.

The method described above is useable not only for N^{15} but also for the determination of D, C^{13} or O^{18}; nor is it restricted to amino esters. The same principle has been used for the determination of deuterium in organic molecules, particularly for locating the isotope in the molecule for the purpose of structure determination (BIEMANN and FRIEDMANN-SPITELLER, 1961).

Appendix I (commercially available mass spectrometers).

The following mass spectrometers, suitable for the type of investigations outlined in this chapter, are presently available commercially (the prices range from $ 25,000 to $ 100,000). Model 21-103 C (Consolidated Electrodynamics Corporation, Pasadena, California). Model MS2H (Associated Electrical Industries, Ltd., Manchester, England). Model CH 4 (Atlas Werke A.G., Bremen, Germany).

For high resolution (Section D, 8): Model 21-110 (C.E.C., Pasadena); Model MS 9 (A.E.I., Manchester).

For work with compounds not exceeding molecular weights of mass 200 to 250: Model 21-130 (C.E.C., Pasadena).

A mass spectrometer of completely different design, based on the "Time of Flight" principle, is produced by Bendix Corporation, Cincinnati, Ohio. The spectra obtained are inferior with respect to accuracy and resolution to the ones obtained with the other instruments. Its advantage is the high speed in which the spectra can be displayed on an oscilloscope screen. It has been used to obtain the spectra of fractions emerging from a gas chromatograph (GOHLKE, 1959).

References.

AHLQUIST, L., R. RYHAGE, E. STENHAGEN and E. VON SYDOW: Arkiv Kemi **14**, 211 (1959). — ASSELINEAU, C., J. ASSELINEAU, R. RYHAGE, S. STÄLLBERG-STENHAGEN and E. STENHAGEN: Acta Chem. Scand. **13**, 822 (1959).

BEYNON, J. H.: in J. D. WALDRON, ed., Advances in Mass Spectrometry, pp. 328—354. London: Pergamon Press 1959. — BEYNON, J. H.: Mass spectrometry and its use in organic chemistry. Amsterdam: Elsevier Publishing Co. 1960. — BEYNON, J. H., R. A. SAUNDERS and A. E. WILLIAMS: Appl. Spectroscopy **14**, 95 (1960); — Anal. Chem. **33**, 221 (1962). — BEYNON, J. H., and A. E. WILLIAMS: Appl. Spectroscopy **13**, 101 (1959). — BIEMANN, K.: Tetrahedron

Letters **15**, 9 (1960). — BIEMANN, K., and G. G. J. DEFFNER: Biochem. Biophys. Res. Comm. **4**, 283 (1961). — BIEMANN, K., and M. FRIEDMANN-SPITELLER: Tetrahedron Letters **2**, 68 (1961). — To be published (1962). — BIEMANN, K., F. GAPP and J. SEIBL: J. Am. Chem. Soc. **81**, 2274 (1959). — BIEMANN, K., C. LIORET, J. ASSELINEAU, E. LEDERER and J. POLONSKY: Biochim. et Biophys. Acta **40**, 369 (1960). — BIEMANN, K., and J. SEIBL: J. Am. Chem. Soc. **81**, 3149 (1959). — BIEMANN, K., J. SEIBL and F. GAPP: Biochem. and Biophys. Res. Comm. **1**, 307 (1959). — BIEMANN, K., and W. VETTER: Biochem. and Biophys. Res. Comm. **2**, 93 (1960a); **3**, 578 (1960b).

COLLIN, J.: Bull. soc. roy. sci. Liége **21**, **446** (1952).

FINAN, P. A., R. I. REED and W. SNEDDEN: Chem. & Ind. (London) **1958**, 1172. — FINAN, P. A., and R. I. REED: Nature (Lond.) **184**, Suppl. 24, 1866 (1959). — FRIEDEL, R. A., J. L. SHULTZ and A. C. SHARKEY, JR.: Anal. Chem. **28**, 926 (1956). — FRIEDEL, R. A., and A. C. SHARKEY, JR.: Anal. Chem. **28**, 940 (1956). — FRIEDLAND, S. S., G. H. LANE, JR., R. T. LONGMAN, K. E. TRAIN and M. J. O'NEAL: Anal. Chem. **31**, 169 (1959).

GILPIN, J. A.: Anal. Chem. **31**, 935 (1959). — GILPIN, J. A., and F. W. MCLAFFERTY: Anal. Chem. **29**, 990 (1957). — GOHLKE, R. S.: Anal. Chem. **31**, 535 (1959).

HALLGREN, B., R. RYHAGE and E. STENHAGEN: Acta Chem. Scand. **13**, 845 (1959).

LEVY, E. J., and W. H. STAHL: Anal. Chem. **33**, 707 (1961).

MCCOLLUM, J. D., and S. MEYERSON: J. Am. Chem. Soc. **81**, 4116 (1959). — MCLAFFERTY, F. W.: Appl. Spectroscopy **11**, 148 (1957a); Anal. Chem. **29**, 1782 (1957b); **31**, 2072 (1959).— MCLAFFERTY, F. W., and R. S. GOHLKE: Anal. Chem. **31**, 2076 (1959).

REED, R. I.: J. Chem. Soc. 3432 (1958). — REED, R. I., and T. GILCHRIST: Experientia (Basel) **16**, 134 (1960). — RITTENBERG, D. in D. W. WILSON and S. P. REIMANN, Preparation and Measurement of Isotopic Tracers. J. W. EDWARDS, Ann Arbor, Mich., 1946, p. 31. — RYHAGE, R., S. STÄLLBERG-STENHAGEN and E. STENHAGEN: Arkiv Kemi **14**, 247, 259 (1959). — RYHAGE, R., and E. STENHAGEN: Arkiv Kemi **13**, 523 (1959); **15**, 291, 333 (1960a); **15**, 545 (1960b).

SHARKEY, JR., A. G., J. L. SHULTZ and R. A. FRIEDEL: Anal. Chem. **28**, 934 (1956); **31**, 87 (1959). — STEWART, D. W.: Mass spectrometry, in A. WEISSBERGER, ed. Physical methods of organic chemistry (Technique of organic chemistry, Vol. I), 3rd. ed., Interscience, Part IV. 3449ff., New York 1960.

Plant Spectra: Absorption and Action.

By

W. L. Butler and K. H. Norris.

With 21 Figures.

This chapter will discuss the information which can be obtained from spectral measurements of plant material. The interaction between light and matter is one of the few non-destructive methods of analysis available to the biological scientist. The spectral measurement of the light transmitted through objects affords a means of looking within those objects. Unfortunately, our prejudices about what is transparent and what is opaque have prevented the full use of these measurements. It is not generally realized that monochromatic light can be transmitted through whole apples, tomatoes, bean seeds, or peanuts and that many biochemical processes which involve light absorbing substances can thus be studied in the intact specimen. *In vivo* spectroscopy is a powerful method for biochemical research which has not been exploited to its fullest extent.

A major barrier to the use of these measurements has been the lack of available instruments. Commercial spectrophotometers are designed for clear solutions and cannot accomodate light-scattering samples. However, it is not difficult to construct a spectrophotometer capable of measurement of very dense materials and the specifications and design of such an instrument will be discussed.

Absorption spectra of light scattering materials must be interpreted with awareness that a number of artifacts may appear. These artifacts can be minimized by proper optical design in the spectrophotometer, but a clear understanding of scattering phenomena is necessary for the interpretation of the absorption spectra. Scatter phenomena and their effect on absorption spectra will be discussed and examples of absorption spectra of different types of plant materials will be presented.

Absorption spectra show the total absorption of light by the pigments present in the plant tissue. Action spectra indicate how the energy may be utilized in those processes which require light. The measurement of action spectra, the interpretation of the results, and examples of what can be learned from these measurements will be included as a part of the discussion of plant spectra.

A. Instrumentation.

Scattering of light by the many cell interfaces present in tissue and culture samples precludes the use of standard spectrophotometers for absorption spectroscopy on these materials. The use of diffusing plates and scatter references has enabled commercial spectrophotometers to provide valid spectra with relatively transparent samples. However, the extremely inefficient energy collection with these techniques prohibits their use with samples of high scatter density. The solid angle intercepted by the receiving element in commercial spectrophotometers is very small, at most no more than a few degrees. This angle can be made to approach 180° by using a large-surface, head-on phototube with the sample in close juxtaposition to the photosurface. This type of sample mounting combined with a high-sensitivity photometer permits spectral measurements on intact leaves, tissue preparations, and dense cultures even though the transmitted energy may be less than 0.001% of the energy incident on the sample.

The best known of the spectrophotometers which employ large-surface phototubes in close juxtaposition to the sample have been developed by CHANCE and coworkers (CHANCE 1951, YANG and LEGALLAIS 1954, YANG 1954). These instruments achieve a high sensitivity and a low noise, coupled with the ability to accomodate light-scattering samples. They are ideally suited for measurement of cultures and tissue preparations having a relatively low scatter loss (optical density of 1 or less). LUNDEGARDH (1956) has described an instrument having high light-collection efficiency with which a point-by-point spectral measurement is made. This instrument, using a scatter blank to compensate for scatter losses, has been used to determine spectra of relatively dense samples such as concentrated yeast suspensions and bundles of wheat roots. A versatile instrument of simple design has been constructed in our laboratory to measure and record the spectra of intact leaves, dense cultures, and tissue slices (BUTLER and NORRIS 1960).

The presence of scatter particles in a sample, in addition to causing a large loss of transmitted energy, causes distortion of the absorption bands. LATIMER (1959) found that a selective scatter on the long-wavelength side of absorption bands caused by anomalous dispersion resulted in a shift of absorption maxima to longer wavelengths. This type of distortion occurs when only a small solid angle of the light in the direction of the incident beam is intercepted by the phototube. PRICE and TETLOW (1948) also showed a shift of the absorption maximum to longer wavelengths and an absorption minimum (CHRISTIANSEN window) on the short wavelength side of the band in spectra of suspensions of light-absorbing particles. Fortunately, the distortions associated with anomalous dispersion are minimized if a large part of the scattered light is measured, as is the case for a head-on phototube close to the sample.

SHIBATA, BENSON and CALVIN (1954) have shown that the distortions from anomalous dispersions can also be minimized by using an opalescent plate placed at the sample to diffuse the light so that the phototube sees a representative portion of the light scattered through the sample in the forward direction. SMITH, SHIBATA and HART (1957) have described the design and construction of an adapter with an opalescent plate for a BECKMAN DK-2 recording spectrophotometer. An opalescent disc is also placed in the reference beam to simulate the scattering characteristics and energy level of the sample beam. The opalescent-plate technique can be satisfactorily used with relatively transparent scattering media. The main drawback is that a large portion of the light is lost at the diffusing plate since it is some distance from the phototube. It should be pointed out that if the sample has sufficient scattering power to diffuse the light an opalescent plate is not needed in the sample beam. However, such samples are generally too opaque for use in commercial spectrophotometers. Visual spectroscopy has been used on scatter samples, but it should be noted that this technique is also subject to distortions from anomalous dispersion if the sample does not produce adequate light diffusion. With relatively transparent suspensions a diffusing plate should probably be used with this technique as well.

An instrument for measuring the spectral absorption of intact leaves and tissues should include the following:

(1) A high-intensity light source with a continuous spectrum throughout the near ultraviolet, visible and near infrared; (2) efficient monochromator with high spectral purity; (3) convenient and versatile sample mounting to take samples of varying size and shape; (4) efficient collection of the light transmitted by the sample; (5) high sensitivity photometer with low noise and wide dynamic range; (6) adequate light shielding of sample and phototube to insure that the radiation received by the photosurface is only the monochromatic radiation transmitted by the sample; (7) provision for eliminating fluorescence, and (8) means for correcting or compensating for the scatter losses.

A double monochromator with prism or grating dispersing elements coupled with a high-intensity, concentrated-filament lamp provides energy of sufficient

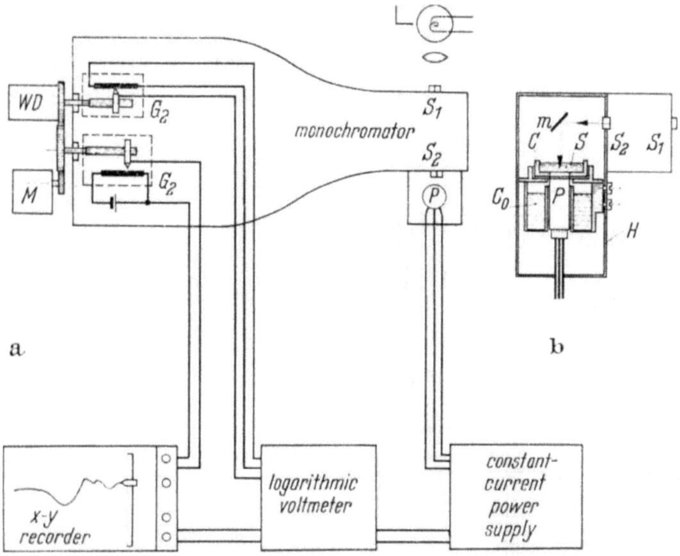

Fig. 1. (a) Block diagram of single-beam spectrophotometer for recording the spectral absorption of dense scattering samples. *WD*, wavelength drum; *M*, motor; G_1, wavelength signal generator; G_2, electrical correction generator; *P*, multiplier-type phototube; *L*, lamp; S_1, entrance slit; S_2, exit slit. (b) End view of monochromator showing optical arrangement: *m*, front-surface mirror; *S*, sample; *C*, cell; *Co*, coolant; *H*, light-tight housing.

intensity and purity for the wavelength region from 350 to 1200 mμ. This type of monochromator used with a sensitive head-on multiplier type phototube can provide a working range of 7 logarithmic decades of light with a 2-mμ bandwidth from 350 to 600 mμ and 4 decades out to 900 mμ with a 10-mμ bandwidth. A block diagram of a spectrophotometer designed specifically for measurements of dense scattering samples is shown in Fig. 1.

A vertical light path, rather than the more commonly used horizontal path, provides a greater versatility and convenience for mounting of samples such as leaves, tissue slices, and tissue cultures. An arrangement with the incident light from the monochromator impinging on the top side of the sample mounted directly above a head-on phototube permits the mounting of samples of a wide range of sizes and shapes. Very simple cells for liquid samples can be constructed by using metal cylinders with a window of plastic, glass, or quartz in the bottom. Glass-walled cells are not suitable for holding dense samples because the light which travels by light-pipe propagation around the sample may be greater than that transmitted through the sample. Open-top cells have the advantage of easy adjustment of path length by changing volume of sample, and they are readily filled and cleaned. The optical requirements on the windows for these cells are not of great consequence unless clear solutions are measured. The windows must, of course, transmit the radiation of interest, but the presence of scratches, striations, etc., are masked by the scatter in the sample. The use of the vertical orientation minimizes the effects from settling of particles since they remain in the light path. Leaves and tissue slices are conveniently mounted between two metal washers with the hole in the center of the washers of the proper size to insure that all the light transmitted to the phototube must pass through the sample. The vertical

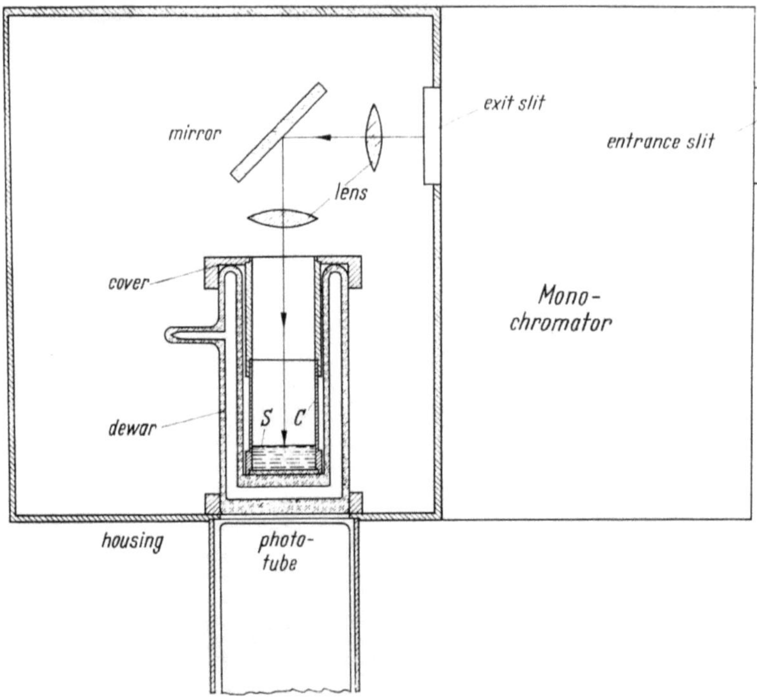

Fig. 2. Arrangement for a spectral measurement on samples maintained at liquid-nitrogen temperature. Sample, S, in cell, C, surrounded by liquid nitrogen in Dewar.

light path with the phototube at the bottom allows the use of a simple open-top Dewar flask with windows at the bottom for holding samples at low temperature (Fig. 2). Samples for measurement are contained in metal-walled cells which are lowered in to the Dewar of liquid nitrogen or other coolant. Mounting of the sample in close juxtaposition to the phototube increases the problem from fluorescence and phosphorescence and in some cases it is necessary to add glass cut-off filters between the sample and the phototube to eliminate fluorescence. On the other hand, it may be desirable to measure the fluorescence. In this case cut-off filters are inserted to eliminate the transmitted energy so that fluorescence excitation spectra can be recorded.

Head-on multiplier type phototubes with different types of photosurfaces are available from several manufacturers. The tri-alkali photosurface (S-20) such as used in the EMI 9558 and RCA 7326 phototubes provides high response throughout the visible region and can be used for wavelengths from 350 to 800 mμ. This photosurface in a tube with a quartz window would be useful down to 250 mμ. Phototubes such as the DuMont 6911 and the RCA 7102 with an S 1 response provide high sensitivity in the 650 to 950 mμ region and may be used out to 1500 mμ.

The photometer circuits used in most commercial spectrophotometers do not provide sufficient range for measurements of dense scatter samples where the transmitted energy may vary more than 6 logarithmic decades in going from a region of low absorption to a region of high absorption and high scatter loss. However, a very simple photometer circuit has been found to have the characteristics desired for these measurements. This photometer utilizes the design of Hariharan and Bhalla (1956) in which the anode current is maintained constant by an electronic control circuit and the dynode accelerating voltage is measured with a logarithmic voltmeter. This type of circuit with an EMI 9558 phototube provides a linear optical-density range of greater than 7 decades with a noise level and sensitivity that makes it possible to measure a change in optical density of 0.005 on samples having an optical density as high as 4. The output voltage from this photometer varies linearly with the logarithm of the light flux on the photocathode as long as this flux generates a current greater than the dark current of the phototube. As the signal level approaches the dark current the photometer output departs from linear and approaches a saturation voltage referred to as the "dark voltage". This deviation from linearity becomes apparent only when the photocurrent is less than ten times the dark current. The maximum density at which measurements can be made is determined by the phototube dark current.

A diffusing disc, filter paper, or some other scattering material can be used as an arbitrary reference in the reference beam of a split-beam spectrophotometer to compensate for scatter losses in the sample. It is also possible to make an arbitrary correction for scatter in a single-beam spectrophotometer by using an electrical correction generated by the wavelength drive. This technique requires a spectrophotometer of good electrical stability, but where this is available the electrical correction offers many advantages. Electrical correction with a multi-turn, tapped potentiometer geared to the wavelength drive provides a readily adjustable correction which can be pre-set and reproduced as desired. It can be used to correct for undesired absorption bands in a sample where it is required to show the presence of a weak absorption near a wide, stronger absorption band of no interest. In addition, it offers the possibility of correcting for the spectral response characteristics of the phototubes, monochromator, and lamp for measurement of samples containing no scatter. Recorder curve followers can also be used for generating an electrical signal for correcting the response characteristics of a single-beam spectrophotometer. The system response curve is prerecorded with conducting ink to provide a chart which determines the curve follower signal of subsequent scans. The curve follower gives a more exact correction for system response since it generates an electrical signal directly proportional to the prerecorded curve. However, it is not as convenient to use and it introduces an additional noise source into the system. The authors have found the use of the tapped potentiometer geared to the wavelength drive to give the most desirable operation.

Derivative spectrophotometry offers another method of studying the spectral absorption changes of light-scattering samples. French and Harper (1957) have described an instrument for automatic recording of derivative spectra. This instrument utilizes the split-beam technique with a reference sample to balance out scatter differences and absorptions which are common to both beams. Derivative spectra can also be obtained with a single-beam spectrophotometer by incorporating an electrical differentiating network into the photometer. The photometer circuit utilizing a multiplier-type phototube operated at constant anode current is very convenient for this application, because the dynode accelerating voltage swings through a large voltage range (300 to 3000 v from full light to dark). This voltage, which is a function of the light energy on the phototube may be connected through a simple resistance-capacitance network to a high-impedance recorder to give a signal proportional to the derivative with respect to time. This does not give a true derivative of absorbancy with respect to wavelength, but the results are adequate for qualitative studies.

B. Light Scatter Phenomena.

There are problems associated with the nonhomogeneity of samples which are not clear solutions even when all of the distortions due to light scatter are eliminated or minimized by collecting most of the scattered light. With a leaf, the transmission of light by nonpigmented areas, e.g., veins and ribs, specifically flattens absorption maxima. Consider a hypothetical leaf which is comprised of 90 percent green areas (A_1) which transmit one percent of the light at the chlorophyll absorption maximum and 10 percent nonpigmented areas (A_2) which transmit 90 percent of the light of all wavelengths. The transmission of the leaf as a whole will be:

$$I(A_1 + A_2) = I_0(0.01\,A_1 + 0.90\,A_2)$$
$$I(1.1\,A_1) = I_0 A_1(0.01 + 0.10)$$
$$\frac{I}{I_0} = \frac{0.1}{1.1} = 0.10\,.$$

Thus, at the absorption maximum the transmission has been increased from one percent to ten percent or the optical density diminished from 2 to 1 by the "light leak" through the nonpigmented areas. At another wavelength which the green areas transmit 20 percent, a similar calculation shows that the light leak decreases the optical density from 0.70 to 0.56. The extent of the flattening depends on the optical density of the absorption band.

The same effect occurs on a microscopic level when one considers pigmented cells or chromatophors suspended in a nonabsorbing medium. A suspension of pigmented spheres is less efficient in absorbing light than a true solution containing the same amount of pigment. This decrease in efficiency which Meistre (1935) has called the "pattern effect" and Rabinowitch (1951) the "sieve effect" is greater at wavelengths where the absorbancy is higher so that a specific flattening of absorption bands occurs. Duysens (1956) has derived the mathematic expression for the flattening as a function of the transmission of the pigmented particle, T_p.

$$\frac{\text{O.D. sus.}}{\text{O.D. sol.}} = \frac{1 - T_p}{\text{Ln}\,\frac{1}{T_p}}$$

Rabinowitch (1956) has derived the same expression from a somewhat different viewpoint by considering a "Beer's law" for a suspension. For a suspension of *Chlorella* cells each of which has a transmission of about 40 percent the optical density of the suspension (O.D. sus.) is 0.65 the optical density of the equivalent solution (O.D. sol.).

These flattening effects are solely the result of macroscopic and microscopic "light leaks" and are not strictly scatter phenomena. A scatter phenomenon which does affect absorption spectra is what Meistre (1935) has called the "detour effect", i.e., the light in traversing a scattering sample takes a path which is longer than the sample thickness. The increased pathlength gives rise to an increased absorption or an intensification of absorption bands. KEILIN and HARTREE (1949), using a microspectroscope, noted the effect experimentally in frozen samples. By carefully freezing pigments in a solution of 50 percent glycerol to produce a maximal turbidity, they were able to intensify the absorption bands approximately 20 fold. BUTLER and NORRIS (1960) intensified the absorption spectrum of clear cytochrome *c* colutions 70 fold by adding $CaCO_3$ scattering particles. The effect is shown in Fig. 3, with a solution of neodymium chloride. This scatter intensification can be a useful analytical tool. The authors have added $CaCO_3$ to mitochondrial preparations and shown the presence of pigments (cytochrome oxidase in particular) which were not detectable in the original suspension.

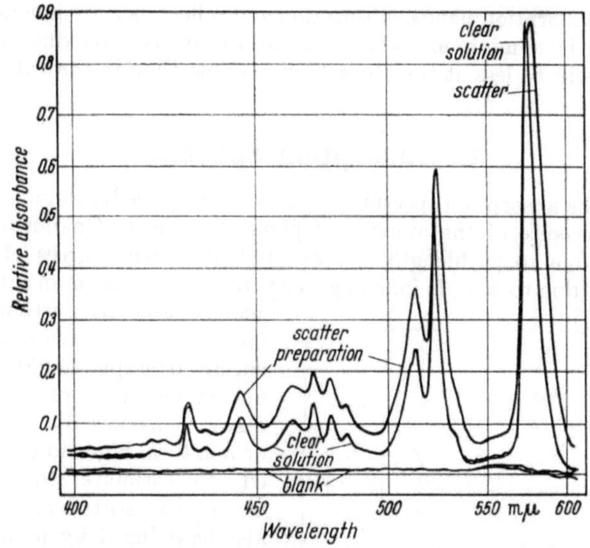

Fig. 3. Scatter intensification with aluminum oxide in a solution of neodymium chloride. Curve for the clear solution obtained on a 1 cm solution containing 1 g of neodymium chloride. Scatter preparation contained 40 mg of neodymium chloride with 2 g of aluminum oxide powder. The clear solution spectrum has been corrected electrically for the system response of the spectrophotometer and the spectrum of the sample in Al_2O_3 has been corrected for the system response and scatter losses.

A mathematical formulation of the optics of turbid materials has been derived by several investigators. The most general of the formulations have been those of DUNTLY (1942) and KUBELKA and MUNK (1931). The latter theory has been used extensively to describe reflectance properties of various surfaces and has found considerable application and experimental verification in the paint and paper industries. The type of approach which has been used starts from a set of differential equations describing the intensity of light traveling away from and toward the illuminated surface of a turbid sample. The analysis is a statistical one in that the formulation is in terms of a scattering coefficient, S, and an absorption coefficient, K, which are averages for the whole sample. The formulas derived are exact if the illumination is diffuse, the sample dull (no specular reflectance), and the scattering particles random and small compared to the sample thickness. The formula for transmittance derived by KUBELKA and MUNK (KUBELKA, 1948),

$$T = \frac{b}{a \sinh bSX - b \cosh b\,SX}$$

(where a and b are constants determined solely by the reflectivity of the sample, S is the scatter coefficient and X is the sample thickness) has been confirmed experimentally by the authors by measuring the transmittance of different thicknesses of dense, light scattering suspensions. With such samples diffuse illumination is not a requirement since the light is completely diffused as soon as it enters the sample. The formula also predicts the intensification of absorption and shows that it is strongly dependent upon the reflectivity of the sample. With a reflectivity of 0.95 a sample with a high scattering power will intensify absorption spectra as much as 40 fold. Reflectivities of 0.5 and 0.98 can lead to intensifications of 3.2 and 100, respectively. A detailed treatment of the theoretical basis of the intensification and the experimental observations will be presented elsewhere. It can be seen in a qualitative manner, however, that the intensification also leads to a flattening of absorption bands. Absorption and reflectance are inversely related: the reflectivity is a minimum where the absorption is a maximum. Thus, the intensification will be less at the absorption maxima than in the valleys.

C. Absorption Spectra.

In general the absorption spectra of light-scattering media cannot be used as a quantitative measure of the amount of pigment present. Unknown degrees of intensification due to pathlength increase and unknown degrees of absorption band flattening due to sample inhomogeneity and the variation of intensification make quantitative estimates hazardous. If quantitative data are needed, an attempt should be made to determine limits of error. In some cases a marker dye can be added or a natural pigment can be extracted to estimate the path-length effect. The scatter of a sample can also be reduced by infiltrating with a medium which closely matches the index of refraction of the scattering particles (BARER 1955).

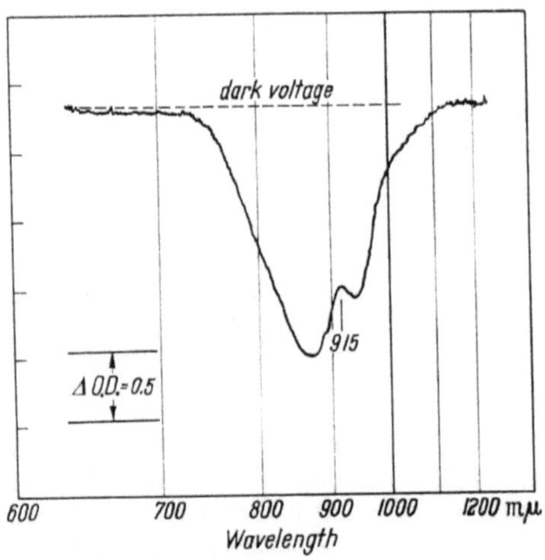

Fig. 4. Spectral absorption of a 4 cm thickness of wood containing a hidden knot. The band at 815 mμ is from an overtone of a C—H stretching frequency of the oil present in the knot. The dark-voltage of the phototube provides a limit to the smallest light level which can be measured. The spectrum has not been corrected for the spectral response of the single-beam spectrophotometer.

It is not generally realized that useful qualitative absorption spectra can be measured with very dense samples, particularly in the red and near-infrared regions of the spectrum. Fig. 4 shows the absorption spectrum of a four-centimeter thickness of wood between 700 and 1000 mμ. This was measured with a DUMONT 6911 phototube cooled to −78° C. This particular sample of wood was more transparent than others because of a knot within the wood (visible in cross-section but not from either surface). The oil in the knot decreases scatter losses. The absorption band at 915 mμ is due to the oil and results from an overtone of a C–H stretching frequency. A stronger absorption band in the region of 1000 mμ makes the sample too dense for measurement.

The absorption spectrum of a 19 mm-thick slice of apple tissue (Fig. 5) demonstrates the scatter intensification of absorption bands which can occur *in vivo*. The absorption band at 675 is due to chlorophyll and the band at 750 mμ due to water results from an overtone of the O—H stretching frequency. If all of the solids were removed from the tissue slice the remaining water would be 1 cm thick. The absorption band of this water in the tissue is slightly stronger than that of 8 cm of clear water. Thus, the scatter in the tissue has intensified the water absorption approximately 10 fold. A similar intensification obtains for the chlorophyll band as well. Extraction of the tissue with methanol showed a chlorophyll absorption band in clear solution that was 1/8 as intense as that in the tissue.

Clearly defined absorption bands through-out the visible region can be shown on tissue homogenates as shown in Fig. 6 for tomato puree. The sample was a 1 cm thickness prepared by homogenizing a red tomato in a blendor. The curve was run on single-beam spectrophotometer with electrical correction to provide a flat base line with no sample. Absorption bands characteristic of carotenoid and lycopene are clearly defined.

The absorption spectra of a suspension such as mitochondrial preparations can be intensified many fold by adding artificial scattering particles. Freezing the sample can also increase the scatter and therefore the optical path length as KEILIN and HARTREE (1949) have shown. Fig. 7 shows the absorption spectrum of a preparation of sweet potato mitochondria at a temperature of $-196°$ C. At such a low temperature the bands are not only intensified due to scatter but they are also sharpened because of less perturbation of the molecules. At higher temperatures the molecular collisions result in a

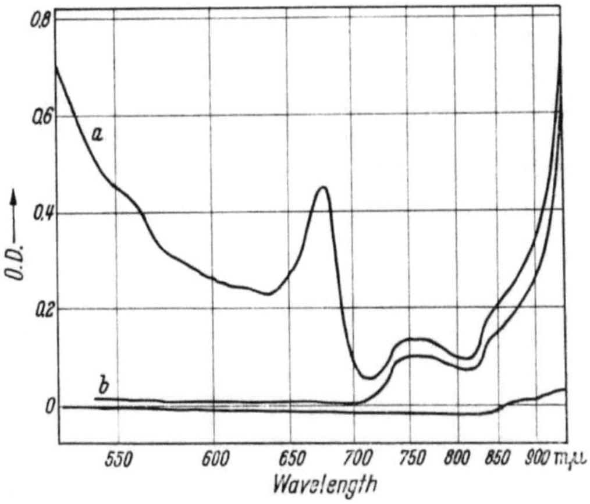

Fig. 5. Absorption spectrum of 19 mm thick slice of apple tissue (a) and 8 cm of clear water (b). The baseline shows the corrected system response.

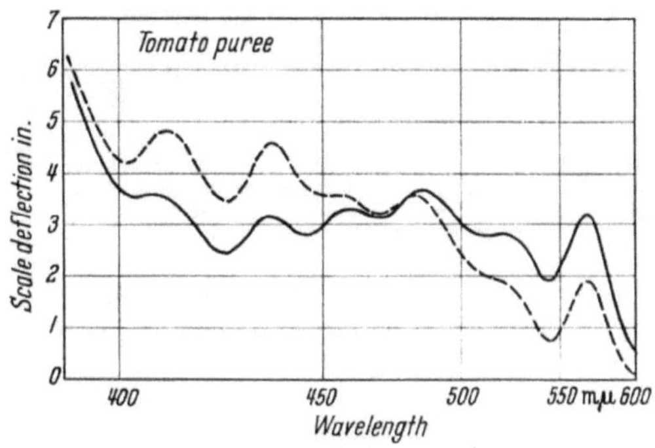

Fig. 6. Spectral absorption (corrected for system response) for homogenates made from tomatoes. Samples 1-cm thick. Sensitivity: 1 inch equals optical density change of 0.2. Solid curve for a full red tomato, dashed curve for a tomato not as fully colored.

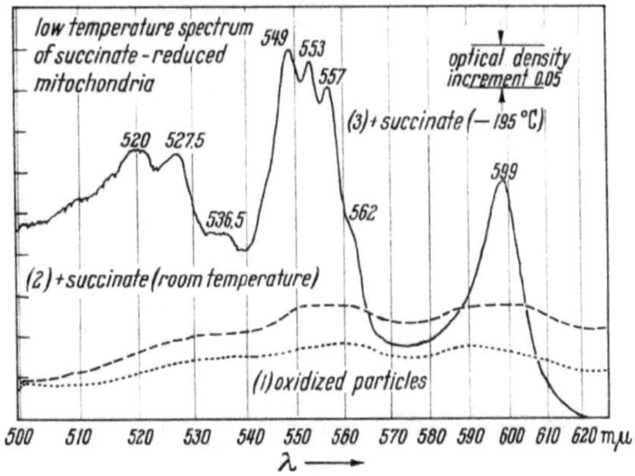

Fig. 7. Spectra of sweet potato mitochondria at room temperature and at $-196°$ C. The spectrum of the oxidized mitochondria (curve 1) was adjusted electrically to give approximately a horizontal line. The reduced spectrum at room temperature shows little change. At $-196°$ C the scatter in the turbid sample intensifies the absorption bands greatly and the low temperature sharpens the bands.

broadening of the bands which is analogous to the pressure broadening of absorption bands in gases. The intensification and sharpening can be achieved with a tissue slice as well as a homogenate by freezing at −196° C (Fig. 8).

Absorption spectra of intact seeds such as beans and peanuts reveal a mixture of pigments which changes during germination. The seeds are mounted in a metal ring with moulding clay so that the light reaching the phototube has passed though both cotyledons. Fig. 9 compares the absorption spectrum of a dry peanut with that of an imbibed peanut just after it has sprouted. The most prominent change is the appearance of a carotenoid pigment with absorption bands at about 480 and 450 mμ. These spectra are very similar to those of lima beans (Butler and Norris, 1960). The synthesis of carotenoid begins just before the radicle appears and is a sign that the seed is about to sprout. As the radicle grows the synthesis and accumulation of carotenoid in the cotyledons continues.

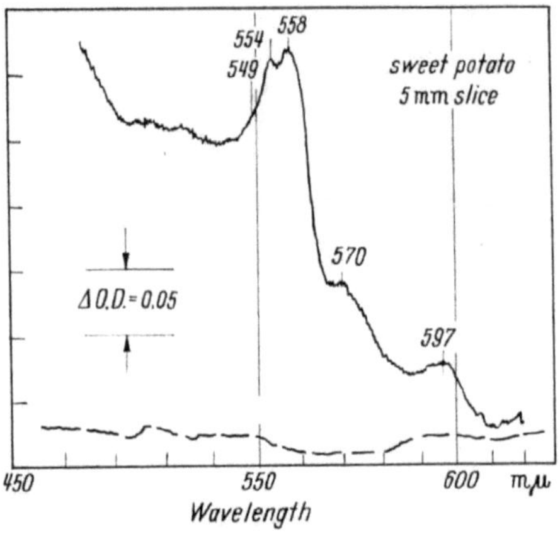

Fig. 8. Spectrum of a 5 mm slice of sweet potato at −196° C. The system was compensated electrically to give approximately a flat base line when a scatter blank of calcium carbonate was used.

There are also smaller absorbancy changes in the region of the spectrum where the cytochromes absorb light. In order to investigate these regions more closely linear electrical compensations were applied from 530 to 575 mμ and from 580 to 700 mμ indicated by the

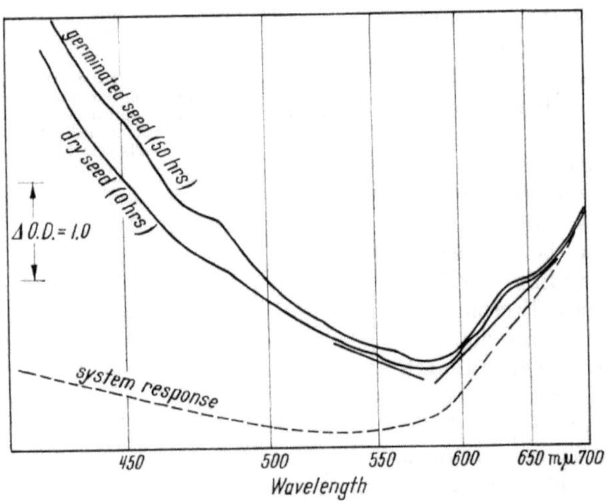

Fig. 9. Absorption spectrum of whole peanuts with skin removed. System response shown in dashed curve. The straight lines below the spectrum of the dry seed are the linear correction which were applied in order to obtain the spectra of Fig. 10 and 11.

straight line just beneath the spectrum of the dry peanut. The spectra of Fig. 10 and 11 represent the absorbancy difference between the spectral curves and the linear-compensation lines of Fig. 9 with a 10-fold increase of sensitivity in the

530 to 575 mμ region and a 5-fold increase in the 580 to 700 mμ region. The spectra of individual peanuts were run at various times after the beginning of imbibition. Imbibition of water, as indicated by weight, was complete in about 12 hours.

The spectral changes of a peanut which germinated (Fig. 10) can be compared with those of a peanut which failed to germinate (Fig. 11.) The two seeds show the same changes up to 30 hours of imbibition. The dry seeds show the presence of reduced cytochrome *c* (λ_{max} = 550 mμ), reduced cytochrome oxidase (λ_{max} = 605 mμ) and an unknown component at 635 mμ. As water is imbibed

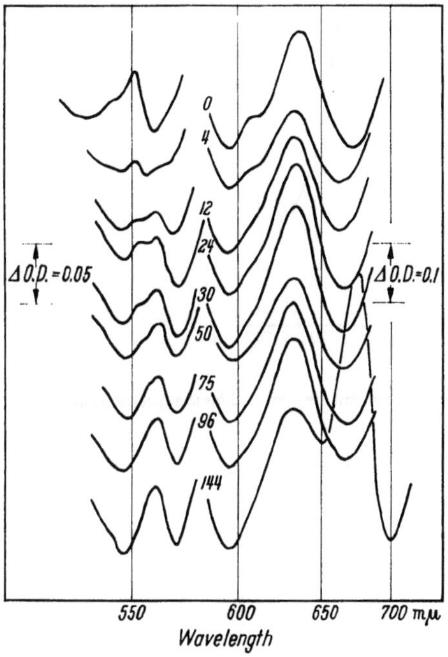

Fig. 10

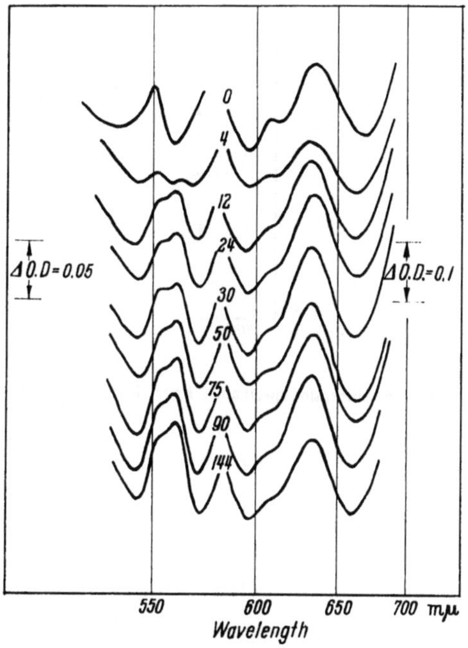

Fig. 11

Fig. 10. Absorption spectra of a germinating peanut at various times (hours) after inhibition with water. The linear corrections shown in Fig. 9 were applied so that the sensitivity of the spectral measurement could be increased.

Fig. 11. Same as Fig. 10 for peanut which failed to germinate.

respiration increases and cytochrome oxidase and cytochrome *c* become more oxidized. A new band at 560 mμ typical of reduced cytochrome *b* also appears. At the completion of imbibition the seeds are in a steady-state, which we may call state 1, which involves a certain level of reduced cytochrome oxidase and cytochrome *c*. This state is maintained until the seed germinates. Sprouting occurred between 30 and 50 hours of imbibition with the peanut of Fig. 10. During this period a new steady-state, state 2, was reached which involved the further oxidation of cytochrome oxidase and cytochrome *c*. This correlates with the increased rate or respiration which is known to occur at the time of germination. The peanut which failed to germinate did not exhibit the transition from state 1 to state 2. The protochlorophyll-chlorophyll system appeared in the cotyledons of the germinated peanut after 96 hours of imbibition.

The red-absorption band of chlorophyll in a leaf can be resolved into several components at $-196°$ C. Fig. 12 compares the absorption spectrum of a leaf at room temperature and at liquid nitrogen temperature. Chlorophyll *b* is clearly resolved from chlorophyll *a*. Of particular interest is the weak band in the 705 mμ region at −196° C. Fluorescence by this component will be discussed below in the section on fluorescence-excitation spectra.

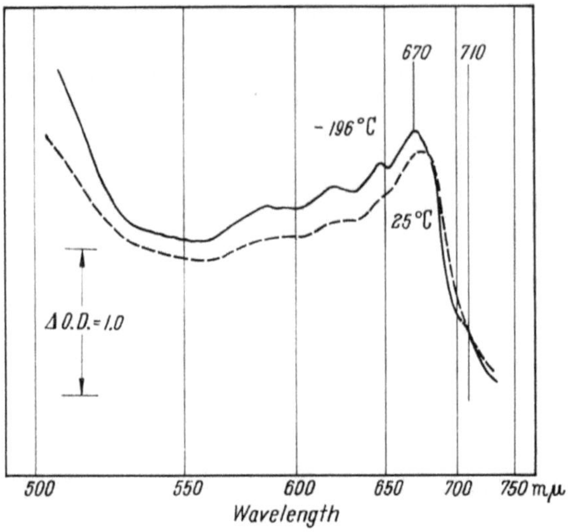

Fig. 12. Absorption spectrum of a bean leaf at 25° C and at —196° C. Spectra are corrected for the spectral response of the instrument.

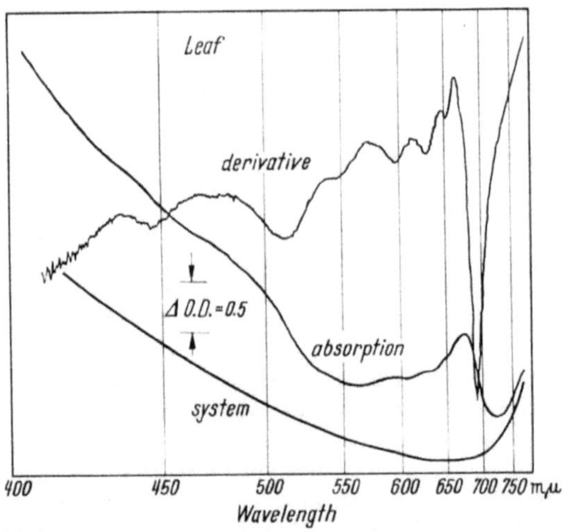

Fig. 13. Absorption and derivative spectra from a green leaf at room temperature.

Giese and French (1955) have shown that derivative spectra may be used to help define overlapping absorption bands. Fig. 13 shows the application of this to a leaf at room temperature. The absorptions spectrum was recorded on the single-beam spectrophotometer without correction for system response. The derivative of this curve was then recorded as indicated. The chlorophyll *b* is indicated as a slight shoulder on the chlorophyll *a* band in absorption while the derivative curve clearly shows the presence of these overlapping bands. The derivative technique can also be used at liquid nitrogen temperatures as shown in Fig. 14. The absorption recorded without correction for system response shows the chlorophyll *b* band separated from the chlorophyll *a*. The derivative improves the resolution of all of the bands, including the absorption shoulder at about 705 mμ.

The first spectrophotometric measurement of phytochrome, the red-far-red phytochromic pigment, was made on intact, dark-grown seedling tissue with the

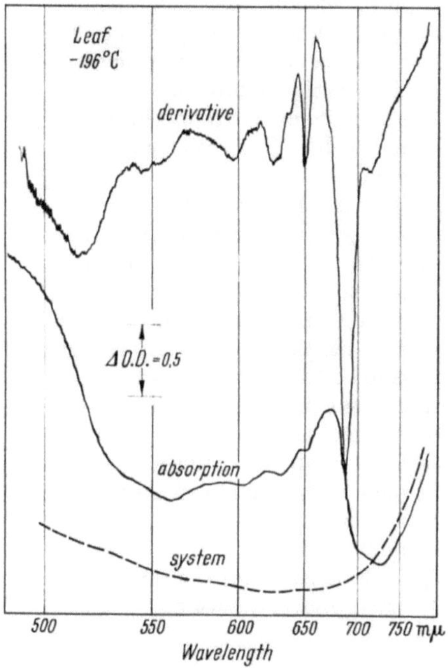

Fig. 14.

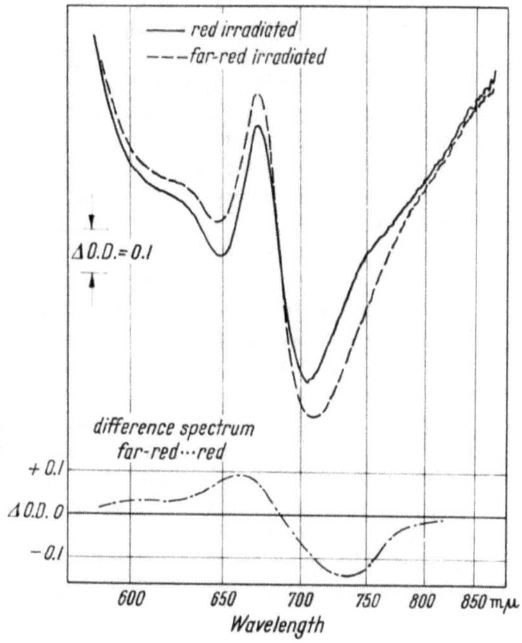

Fig. 15.

Fig. 14. Absorption and derivative spectra from a green leaf at —196° C.

Fig. 15. Absorption spectra of corn coleoptiles 1-cm thick (uncorrected for spectral response of instrument) after irradiation with red light (solid surve) and far-red light (dashed curve). Computed difference spectrum of far-red-irradiated sample minus red-irradiated sample is shown on the lower part of the figure.

single-beam recording spectrophotometer described above (BUTLER, NORRIS, SIEGELMAN, and HENDRICKS, 1959). Action spectra, discussed below, established that the effects of red- and far-red light on plants were mediated through a photoreversible pigment which existed in two forms, one absorbing maximally at 660 mμ, the other at 730 mμ. The absorption of light by either form caused its transformation to the other form. Fig. 15 shows the absorption spectra (uncorrected for system response) of a sample of dark-grown-corn seedlings after red irradiation (solid curve) and again after far-red irradiation (dashed curve). The absorption peak at 672 mμ is due to chlorophyll which has been formed by the initial irradiation with red light. The difference spectrum shows that the far-red irradiation resulted in a decrease in the amount of a substance with an absorption band at 730 mμ and the increase in a 660 mμ absorbing substance. This effect of far-red light was reversed by a subsequent exposure to red light. Thus, the reversible pigment system named phytochrome was demonstrated. The photometric measurement permitted phytochrome to be extracted from the plant. It is a protein which maintains its photochromic behavior after several stages of purification.

D. Action Spectra.

Action spectra afford a means of investigating light-dependent processes in biological systems. The rationale behind action spectra derives from HERSHEL's law which states that light must be absorbed if it is to have a photochemical action. In order for phototropic movements, photoperiodic responses, photosynthesis or photochemical pigment transformation to take place, light must be absorbed by a specific pigment or pigment system. Those wavelengths which are absorbed most readily by the pigment will be most effective in producing the response. An action spectrum is a plot of the relative effectiveness of monochromatic light of different wavelengths to produce a given effect. Ideally, it relates directly to the absorption spectrum of the pigment involved.

In general, the degree of the light-dependent physiological response does not bear a linear relationship to the incident light energy. This often requires that action spectra be plotted as the monochromatic energy required to produce a certain degree of response. This degree should be selected in a region where the response is sensitive to energy rather than in a region of saturation. The response is then measured at several energies at each wavelength so that the energy required for the standard response can be interpolated. In those cases where the response is linear with energy it is sometimes more convenient to plot the action spectra as the degree of response obtained at a constant energy of monochromatic radiation. One should determine, however, whether the response is linear over the entire range tested. If saturation is occurring, the equi-energy action spectrum will show flatter absorption bands than the constant-response presentation.

The energy should always be expressed in quanta/sec. rather than watts or ergs/sec. The photochemical activity of the pigment depends on the number of quanta absorbed, not the number of ergs. In the energy level diagram,

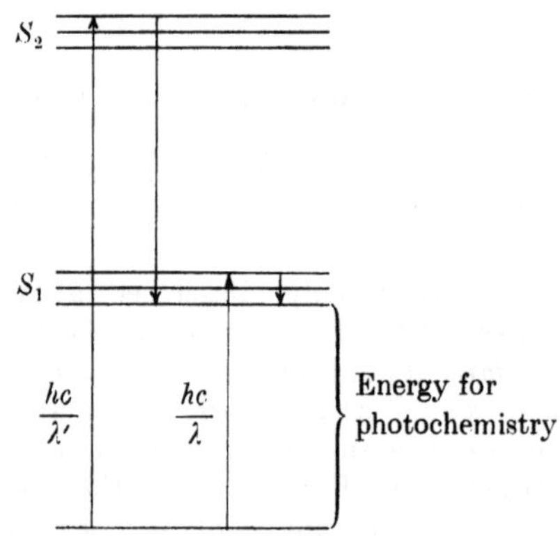

the quantum $\frac{hc}{\lambda'}$ which excites the molecule to second excited singlet level, S_2, is of higher energy than the quantum $\frac{hc}{\lambda}$ which excites the molecule to the first excited singlet state, S_1. However, in both cases the energy in the molecule is rapidly degraded to the lowest vibrational level of the first excited singlet state. Only this energy is available for photochemistry. The excess is lost as heat. The energy of monochromatic radiation can generally be measured with a chemical actinometer which responds according to the quantum flux or with a thermophile which measures the ergs/sec/cm². In the latter case the value obtained should be multiplied by $\frac{\lambda}{hc}$ to convert it to quantum flux.

Monochromators and light sources which provide high intensity monochromatic radiation over relatively large areas of illumination have been described. The large spectrograph at Beltsville (PARKER, HENDRICKS, BORTHWICK, and SCULLY, 1946) uses either one or two large glass prisms to disperse the light from a high intensity carbon arc. A similar spectrograph at the Argonne National Laboratory (MONK and EHRET, 1956) uses a large grating. The display of the entire spectrum has the advantage that a number of wavelength stations can be used simultaneously from a single light source. A possible disadvantage, particularly with large samples, is the variation of wavelength across the sample. Some investigators (WITHROW, 1957, MOHR and SCHOSER, 1959) have used interference filters to obtain monochromatic light of fairly narrow band width. Higher intensities of narrow-band monochromatic radiation can be obtained with a filter than with a spectrograph and the wavelength is constant over the sample but this system has the disadvantage that the action of only one wavelength can be determined in an experiment.

In practice the measurement and interpretation of action spectra present many pitfalls. The presence of non-active pigments is one of the main hazards if one is dealing with a photochemically active pigment which is present in low concentration. If the masking pigment effectively reduces the amount of light available to the active pigment, false minima may appear in the action spectrum at the position of the absorption maxima of the inactive pigment. The absorption spectrum of the sample will generally indicate whether masking effects will be a problem.

The absorption spectrum of the pigment *in vivo* may be different from the known spectrum of the pigment in extracts. Careful measurements of the action spectrum of the phototropic bending with plant tissue which is devoid of masking pigments show three peaks in the blue at 475, 445, and 425 mμ reminiscent of carotenoid pigments (SHROPSHIRE and WITHROW, 1958). There is also a peak in the ultraviolet near 370 mμ, however, which carotenoids in solution do not have. Riboflavin has the ultraviolet peak in about the correct position but only a single broad absorption maximum in the blue. There are examples (CRANE and BEINERT, 1956) of flavins developing three peaks in the blue when dissolved in certain solvents or bound to protein. The possibility also exists that the ultraviolet peak is due to another pigment which transfers its excitation energy to a carotenoid. Examples of this type of energy transfer will be discussed below. Thus, even though the action spectra of phototropism is known to a high degree of accuracy a non-controversial identification of the pigment involved has not been made.

One of the most extensive uses of action spectra has been in the study of the effects of red and far-red light on plant development. Studies by FLINT and MCALISTER (1935) indicated that red light enhanced the germination of lettuce seed while light of somewhat longer wavelength, far-red, inhibited the germination. This phenomenon was studied in detail by BORTHWICK, HENDRICKS, PARKER, TOOLE and TOOLE (1952) at Beltsville. Lettuce seed maintained in the dark showed essentially no germination while those which received light germinated 90 to 100 percent. The action spectrum was measured by placing the seed across the spectrum and determining the energy at each wavelength station which caused 50 percent germination. Red light in the region of 650 to 660 mμ was the most effective (Fig. 16, left side). The stimulatory effect of red-light, however, could be inhibited by a subsequent exposure to light of longer wavelength. The action spectrum for the far-red light was determined by giving the seeds a saturating dose of red light to produce maximal germination before placing them across the spectrum. In this case the spectrum extended out to 850 mμ. A plot of the energy required to inhibit germination back to 50 percent shows an action maximum at about 730 mμ (Fig. 16, right side). The red-far-red system was found to be completely reversible. If imbibed lettuce seed are given a series of alternating exposures to red and far-red light, the light which they received last will determine whether they germinate (Fig. 17).

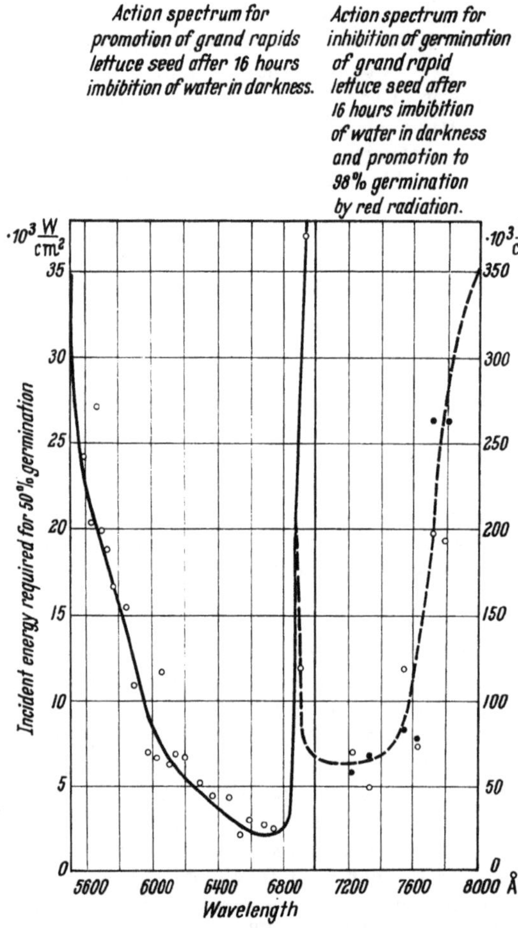

Fig. 16. Action spectra of lettuce seed germination. Right side: Energy of monochromatic radiation required to induce a 50% germination of imbibed seed. Left side: Action spectrum for the inhibition of the stimulatory effect of red light. Energy of monochromatic radiation in the far-red region required to inhibit germination to 50%. The seeds had been previously promoted to 98% germination with red light.

Previous action spectra had shown that a pigment with an absorption maximum at about 660 mμ controlled the flowering of both long day (BORTHWICK, HENDRICKS and PARKER, 1948) and short day (PARKER, HENDRICKS and BORTHWICK, 1946) plants. If plants which were on a photoperiod of short days and long nights were given a brief illumination near the middle of the night, they responded as if they were on the opposite photoperiod of long days and short nights. Short day plants were prevented from flowering while long day plants were induced to flower. Red light near 660 mμ was the most effective night-break light. The same action spectrum had also been obtained for stem elongation and leaf expansion (PARKER, HENDRICKS, and BORTHWICK, 1949) as well as other responses such as the coloration of fruit (PIRINGER and HEINZE, 1954). These action spectra were all very similar to the action spectrum for the effect of light in breaking seed dormancy. The seed germination had provided the additional information that the effect of red light could be reversed by far-red light. If the same pigment system were involved in all of these red-light-sensitive responses, far-red should also be active. It was found that in all of these cases far-red-light with a peak effectiveness at about 730 mμ did counteract the red irradiation.

The reversibility between red and far-red irradiation indicated that the pigment existed in two interconvertible forms with absorption maxima at 660 and 730 mμ (BORTHWICK et al., 1952)

$$P_{660} \underset{730\ \mathrm{m}\mu}{\overset{660\ \mathrm{m}\mu}{\rightleftarrows}} P_{730}\,.$$

Lettuce seed also showed that P_{730} could convert back to P_{660} in the dark. Careful experiments on internode elongation (HENDRICKS, BORTHWICK, and DOWNS, 1956) established that the photoconversion followed first-order kinetics which permitted

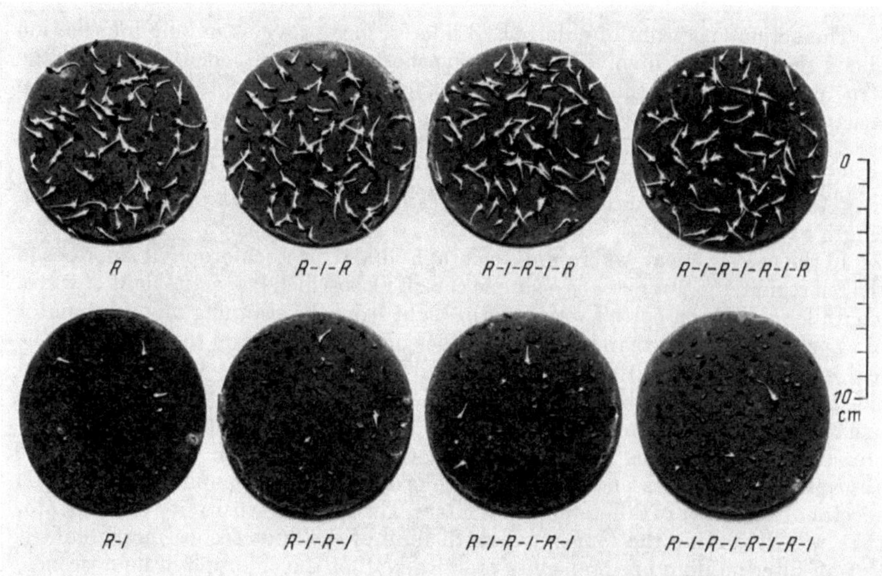

Fig. 17. The imbibed lettuce seed in the petri dishes had been irradiated with red light (*R*) and far-red light (*I*) as indicated. Thee ffects of red and far-red light are repeatedly reversible. If the last irradiation was with the red light, the seeds germinate; if it was with far-red light they do not.

the calculation of the extinction coefficient for both forms of the pigment from action spectra measurements without knowing the concentration of pigment in the tissue (HENDRICKS, et al. ,1956). From the rather high value of the extinction coefficient and the very low absorption of light the concentration of the pigment was shown to be very low. It was postulated that the pigment was an enzyme and as such was a protein. Action spectra established the absorption spectra of the two forms of the pigment, the reversible interconversion between these two forms, the kinetics for both photoreactions as well as the presence of a dark conversion, the extinction coefficients for both forms, and the chemical nature of the pigment. When the pigment was finally measured by spectrophotometric means (Fig. 15) and was isolated from plant tissue as a protein (BUTLER, NORRIS, SIEGELMAN, and HENDRICKS, 1959), the action spectra measurements were confirmed.

The action spectrum for photosynthesis (EMERSON and LEWIS, 1943) has shown the involvement of chlorophyll *a* and accessory pigments such as carotenoids and chlorophyll *b* in the case of green plants and phycobilins in the red and blue-green algae. Careful measurements of the action spectra of photosynthesis with and without background illumination (EMERSON, CHALMERS, and CEDERSTRAND,

1957) has led to greater insights into the action of the pigments. French (1959) has constructed an action spectrophotometer which records the action spectrum of photosynthesis. A platinum electrode which responds to the rate of photosynthetic oxygen production controls the incident light energy from a scanning monochromator so that a constant rate of oxygen evolution is maintained. The light energy, as measured by a photocell which has been corrected for its spectral sensitivity response, is plotted automatically as a function of wavelength.

E. Fluorescence Excitation Spectra.

The action spectrum of chlorophyll fluorescence provides unique information about the pigments involved in photosynthesis and even something about the structural integrity of the photosynthetic apparatus (Butler, 1961 a). The action spectrum of fluorescence, which we shall refer to as the fluorescence-excitation spectrum, can be readily measured with the single-beam recording spectrophotometer described above by placing a cut-off filter between the sample and the phototube. This filter transmits the fluorescent light but not the shorter wavelength exciting light.

In the case of a leaf, where only chlorophyll and protochlorophyll fluoresce in the red region, the filter can be a simple cut-off filter which transmits light of wavelength longer than a cut-off wavelength. Light from the scanning monochromator of wavelength shorter than the cut-off wavelength is incident on the top side of the leaf but is not transmitted to the phototube. The phototube views the bottom side of the leaf through the cut-off filter and measures the intensity of fluorescence excited by the monochromatic light. The flexibility of the measurement is increased greatly by using filters of different cut-off wavelength. Fig. 18 shows the absorption and fluorescence spectra of chlorophyll *a in vivo* and the absorption spectra of a number of different cut-off filters. The spectral sensitivity of the phototube will determine the long wavelength limit of the fluorescence measurement. Cut-off filter number one transmits practicallyall of the chlorophyll fluorescence, including those wavelengths which overlap the chlorophyll absorption spectrum. This filter also transmits a large part of the protochlorophyll fluorescence which has a maximum at about 655 mμ. If one wishes to eliminate those wavelengths which are subject to reabsorption, cut-off filter number two can be used at some sacrifice in the intensity of fluorescence. The cut-off wavelength determines the long wavelength limit of the excitation spectrum as well as the short wavelength limit of the fluorescence measurement. Useful information can be obtained by exploring the excitation spectrum of chlorophyll fluorescence throughout the entire absorption band in the red. Cut-off filter number three transmits sufficient fluorescence to permit measurement and allows the exciting light to scan out to 720 mμ. Each of these cut-off filters has particular advantages for specific applications.

It is often desirable to compare the absorption spectrum with the fluorescence excitation spectrum. The cut-off filter can be removed from between the phototube and the leaf so that the light transmitted by the leaf can be measured. The intensity of light transmitted by the leaf is generally great enough that the fluorescence does not affect the absorption spectrum. The phototube measures $\log \frac{1}{T\lambda}$ (T_λ = transmission of leaf at λ) with the filter removed and $\log \frac{1}{E_{F(\lambda > \lambda_0)}}$ (E_F = intensity of fluorescence, λ_0 = cut-off wavelength) with the cut-off filter in place. Thus, absorption bands appear as maxima and fluorescence excitation bands as minima. The absorption spectra can be corrected electrically for the system spectral response of the single-beam spectrophotometer and the excitation spectra can be corrected for the variation of the intensity of monochromatic exciting light (in quanta/sec/cm^2) with wavelength.

Fig. 19 shows the fluorescence excitation spectra and absorption spectra of two dark grown bean leaves at $-196°$ C. One leaf (dashed curve) had received a a single brief irradiation with red light to convert the protochlorophyll to chlorophyll just prior to freezing. These spectra have not been corrected for energy or system response. There is a good correspondence between excitation peaks and absorption maxima for protochlorophyll ($\lambda_{max} = 650$ mμ) and

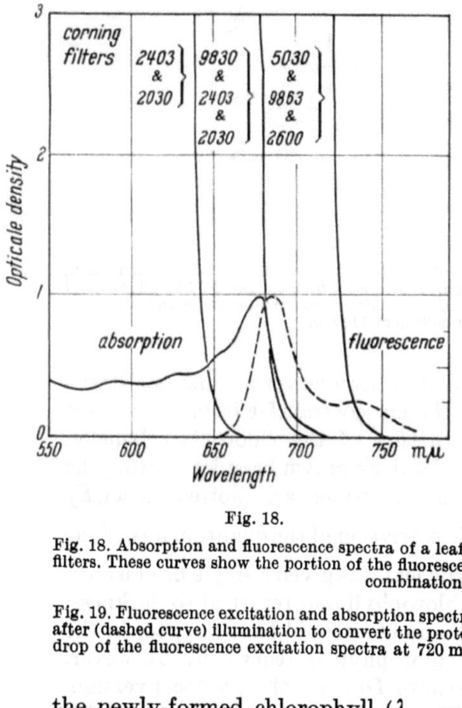

Fig. 18.

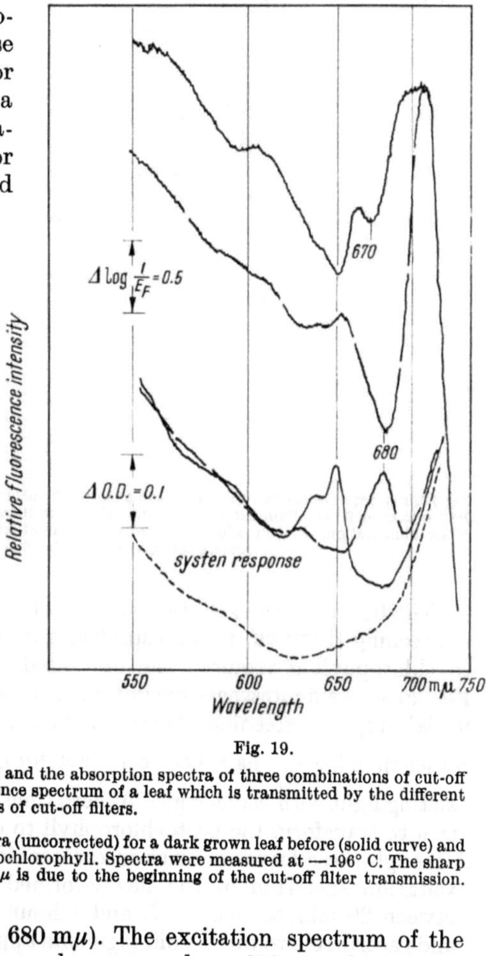

Fig. 19.

Fig. 18. Absorption and fluorescence spectra of a leaf and the absorption spectra of three combinations of cut-off filters. These curves show the portion of the fluorescence spectrum of a leaf which is transmitted by the different combinations of cut-off filters.

Fig. 19. Fluorescence excitation and absorption spectra (uncorrected) for a dark grown leaf before (solid curve) and after (dashed curve) illumination to convert the protochlorophyll. Spectra were measured at —196° C. The sharp drop of the fluorescence excitation spectra at 720 mμ is due to the beginning of the cut-off filter transmission.

the newly-formed chlorophyll ($\lambda_{max} = 680$ mμ). The excitation spectrum of the unilluminated leaf (solid curve), however, shows a peak at 670 mμ which is not apparent in the absorption spectrum. This illustrates the ability of fluorescence-excitation spectroscopy to detect small quantities of fluorescing substances which are in too low a concentration to be detected by absorption spectroscopy. The excitation spectrum is selectively more sensitive to the 670 mμ absorbing component than to protochlorophyll because a larger fraction of the fluorescence emission of the 670 mμ absorbing component extends beyond the 730 mμ cut-off wavelength.

The action spectrum of photosynthesis indicated that light absorbed by carotenoids and chlorophyll *b* as well as that absorbed by chlorophyll *a* was effective. The fluorescence excitation spectra (Duysens, 1952; French and Young, 1951) confirmed that light energy absorbed by these accessory pigments was transferred to chlorophyll a and resulted in chlorophyll a fluorescence as well as photosynthesis. This energy transfer is always toward the pigment with the longer wavelength absorption band and requires some overlap between the fluorescence spectrum of the donor molecule and the absorption spectrum of the acceptor molecule. The greater this overlap and the closer the proximity of the two molecules, the greater the efficiency of transfer.

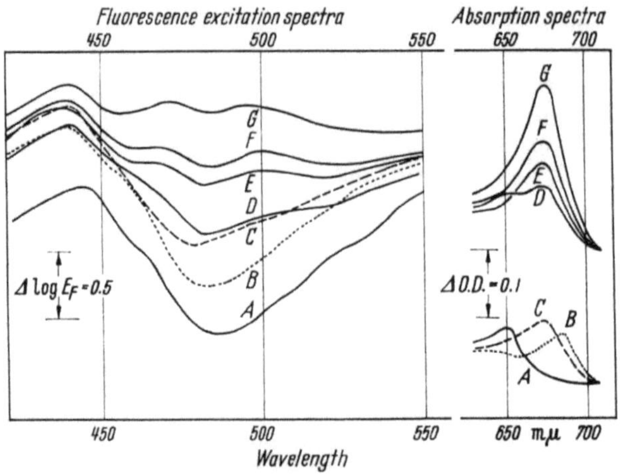

Fig. 20. Fluorescence excitation spectra and absorption spectra of a dark-grown bean leaf. *A:* before illumination, *B:* 1 min after 20 sec illumination, *C:* 10 min after illumination, *D:* 1 hr after illumination, *E:* 1 hr after second 20 sec illumination, *F:* 1 hr after third 20 sec illumination, *G:* 1 hr after 2 hrs continuous illumination. — Absorption spectra have been normalized at 710 mμ.

The fluorescence excitation spectra of etiolated bean leaves at different stages of greening (BUTLER, 1961a) established that the energy transfer from carotenoids to chlorophyll a required the organized structure of the chloroplast lamella. Fig. 20 shows fluorescence excitation spectra of a dark-grown bean leaf during the initial steps of greening. These excitation spectra which are plotted as $\log E_F$ rather than $\log \frac{1}{E_F}$ have been corrected for the variation of the quantum flux of the exciting light with wavelength. At zero time the leaf was given a single brief irradiation to transform the protochlorophyll to chlorophyll and returned to darkness. Although these leaves are rich in carotenoids, no carotenoid bands appear in the excitation spectrum of the newly-formed chlorophyll a (curve B). However, between 20 minutes (curve C) and 1 hour (curve D) after the photoconversion, bands ascribable to carotenoid pigments appear. As more chlorophyll accumulates during subsequent irradiations the carotenoid excitation bands become more prominent, showing an increasing efficiency of energy transfer to chlorophyll a. Electron microscopy shows that at zero time (curve B) the proplastid has no lamellae and that the lamellae develop in response to light as the chlorophyll accumulates. The concomitant development of the capacity for energy transfer from carotenoid to chlorophyll and the formation of lamellae shows that the physical structure which incorporates both pigments is required for energy transfer. These experiments suggest that sensitized fluorescence can be used as an assay for structure in studying the development of greening chloroplasts.

With fully green leaves most of the detail in the fluorescence excitation spectrum is washed out because a large fraction of the incident light is absorbed at all wavelengths. The energy absorbed by the accessory pigments sensitizes the fluorescence of chlorophyll *a* so that all visible wavelengths excite chlorophyll fluorescence. The fluorescence-excitation spectra of a green leaf at room temperature and at $-196°$ C are shown in Fig. 21. Lowering the temperature results in a 25 fold ($\Delta \log \frac{1}{E_F} = 1.4$) increase in the intensity of fluorescence of wavelengths longer than 730 mμ. This is due to the appearance of a fluorescence band at 720 mμ which BRODY demonstrated with algae and concentrated (but not dilute) chlorophyll solutions at $-196°$ C. The 720 mμ fluorescence band is more intense than the 685 mμ emission band of chlorophyll *a* and a greater portion of it extends beyond the filter cut-off wavelength. The absorption spectra of the leaf at $-196°$ C (Fig. 12) showed a small absorption band at about 705 mμ. The strong fluorescence excitation band at 705 mμ at $-196°$ C shows that this small absorption band is responsible for the intense low temperature fluorescence band. Transfer of excitation energy from chlorophyll *a* to the 705 mμ absorbing form of chlorophyll and the quenching of chlorophyll *a* fluorescence by this component have been demonstrated by fluorescence excitation spectroscopy. The significance of this form of chlorophyll to photosynthesis has been discussed elsewhere (BUTLER, 1961 b).

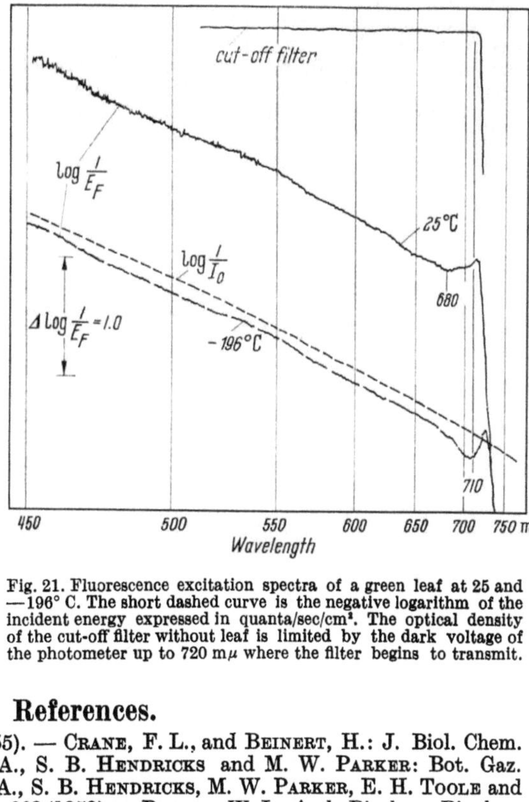

Fig. 21. Fluorescence excitation spectra of a green leaf at 25 and $-196°$ C. The short dashed curve is the negative logarithm of the incident energy expressed in quanta/sec/cm². The optical density of the cut-off filter without leaf is limited by the dark voltage of the photometer up to 720 mμ where the filter begins to transmit.

References.

BARER, R.: Science **121**, 709 (1955). — CRANE, F. L., and BEINERT, H.: J. Biol. Chem. **218**, 717 (1956). — BORTHWICK, H. A., S. B. HENDRICKS and M. W. PARKER: Bot. Gaz. **110**, 103 (1948). — BORTHWICK, H. A., S. B. HENDRICKS, M. W. PARKER, E. H. TOOLE and V. K. TOOLE: Proc. Natl. Acad. Sci. **38**, 662 (1952). — BUTLER, W. L.: Arch. Biochem. Biophys. **92**, 287 (1961 a). — BUTLER, W. L.: Arch. Biochem. Biophys. **93**, 413 (1961 b). — BUTLER, W. L., and K. H. NORRIS: Arch. Biochem. Biophys. **87**, 31 (1960). — BUTLER, W. L., K. H. NORRIS, H. W. SIEGELMAN and S. B. HENDRICKS: Proc. Natl. Acad. Sci. **45**, 1703 (1959).

CHANCE, B.: Rev. Sci. Inst. **22**, 634 (1951).

DUNTLY, S. C.: J. Opt. Soc. Am. **32**, 61 (1942). — DUYSENS, L. M. N.: Thesis Utrecht (1952). — DUYSENS, L. M. N.: Biochem. Biophys. Acta **19**, 1 (1956).

EMERSON, R., R. CHALMERS and C. CEDERSTRAND: Proc. Natl. Acad. Sci. **43**, 133 (1957). — EMERSON, R., and C. M. LEWIS: Am. J. Biol. **310**, 165 (1943).

FLINT, L. H., and E. D. MCALISTER: Smithsonian Misc. Pub. **99**, No. 5 (1935). — FRENCH, C. S.: Carnegie Inst. Wash. Yearbook (1959). — FRENCH, C. S., and G. E. HARPER: Carnegie Inst. Wash. Yearbook **56**, 281 (1957). — FRENCH, C. S., and V. K. YOUNG: J. Gen. Physiol. **35**, 873 (1951).

GIESE, A. T., and C. S. FRENCH: Applied Spectroscopy **9**, 78 (1955).

HARIHARAN, P., and M. S. BHALLA: Rev. Sci. Inst. **27**, 3 (1956). — HENDRICKS, S. B., H.A. BORTHWICK and R. J. DOWNS: Proc. Natl. Acad. Sci. **42**, 19 (1956).

KEILIN, D., and E. F. HARTREE: Nature (Lond.) **164**, 254 (1949). — KUBELKA, P.: J. Opt. Soc. Am. **38**, 448 (1948). — KUBELKA, P., and F. MUNK: Z. techn. Physik **12**, 593 (1931).

Latimer, P.: Plant Physiol. **34**, 193 (1959). — Lundegardh, H.: Biochem. Biophys. Acta **20**, 469 (1956).

Meistre, H.: Coldspring Harbor Symp. Quart. Biol. **3**, 191 (1935). — Mohr, H., and G. Schoser: Planta **53**, 1 (1959). — Monk, G. S., and C. F. Ehret: Radiation Research **5**, 88 (1956).

Parker, M. W., S. B. Hendricks and H. A. Borthwick: Bot. Gaz. **108**, 242 (1946). — Parker, M. W., S. B. Hendricks and H. A. Borthwick: Am. J. Bot. **36**, 194 (1949). — Parker, M. W., S. B. Hendricks, H. A. Borthwick and N. J. Scully: Bot. Gaz. **108**, 1 (1946). — Piringer, A., and P. Heinze: Plant Physiol. **29**, 467 (1954). — Price, W. C., and K. S. Tetlow: J. Chem. Phys. **16**, 1157 (1948).

Rabinowitch, E.: Photosynthesis and related process, Vol. II, pt 1, 672. New York: Interscience (1951. — Rabinowitch, E.: Photosynthesis and related processes, Vol. II, pt 2. New York Interscience 1865 (1956).

Shibata, K., A. Benson and M. Calvin: Biochem. Biophys. Acta **15**, 461 (1954). — Shropshire, W., and R. B. Withrow: Plant Physiol. **33**, 360 (1958). — Smith, J. H. C., K. Shibata and R. W. Hart: Arch. Biochem. Biophys. **72**, 457 (1957).

Withrow, R. B.: Plant Physiol. **32**, 355 (1957).

Yang, C. C.: Rev. Sci. Inst. **25**, 807 (1954). — Yang, C. C., and V. Legallais: Rev. Sci. Inst. **25**, 801 (1954).

Gefriertrocknung[1].

Von

Hans Moor.

Mit 9 Abbildungen.

Die Gefriertrocknung dient der Fixation und Konservierung biologischen Materials auf rein physikalischer Basis: Das Gefrieren ermöglicht die Stillegung des Stoffwechsels, d. h. aller chemischer Prozesse und Diffusionsvorgänge in Bruchteilen von Minuten oder Sekunden. Die anschließende Vakuumsublimation des Eises verleiht dieser Fixation auch bei Temperaturen über dem Nullpunkt Dauerhaftigkeit und umgeht die Trocknung in der flüssigen Phase mit ihren vielfältig störenden Einengungserscheinungen.

Als erster erkannte ALTMANN (1890) die Vorteile der Gefriertrocknung für die Fixation von Zellen und Geweben; später kamen die ersten erfolgreichen Versuche der Konservierung von Mikroorganismen hinzu (HAMMER 1911). Aber erst in den letzten 30 Jahren ist aus dem Laborkuriosum der Jahrhundertwende, auf Grund der Entwicklungen auf dem Gebiete der Vakuum- und Kälte-Technik, eine vielgebrauchte Routinemethode geworden. Sie ist aus der chemischen und pharmazeutischen wie auch aus der Nahrungsmittel-Industrie nicht mehr wegzudenken, ebensowenig aus Medizin und Zoologie. Auf botanischem Gebiet hat die Gefriertrocknung allerdings noch nicht richtig Fuß gefaßt, wohl hauptsächlich darum, weil die an tierischem Material erprobte Methode, unverändert auf pflanzliche Objekte übertragen, in vielen Fällen nicht die gewünschten Resultate zeitigte.

Es hat sich immer wieder gezeigt, daß keine Standardmethode der Gefriertrocknung existiert: Jedes Objekt erfordert, seiner Eigenart entsprechend, eine modifizierte Behandlung, die oft erst noch entwickelt werden muß. Dieser methodische Aufwand im Verein mit der Kostspieligkeit der Apparaturen schränkt die Verwendung der Gefriertrocknung auf diejenigen Gebiete ein, in denen auf die Vorteile der Gefrierfixierung und Trocknung durch Vakuumsublimation nicht verzichtet werden kann.

Die Vorteile der Gefriertrocknung werden am besten als Möglichkeiten formuliert, die sich je nach den Ansprüchen (z. B. Auflösungsvermögen) der Analysenmethode leichter oder schwerer verwirklichen lassen: Die Gefriertrocknung kann

1. die chemische Zusammensetzung eines biologischen Objektes naturgetreu erhalten,
2. darüber hinaus auch noch seine Struktur wahren und letztlich
3. Zellen, Gewebe und ganze Organismen für beinahe unbeschränkte Zeit in lebensfähigem Zustand konservieren.

Die Aussicht auf Erfolg hängt weitgehend davon ab, wie weit wir die physikalischen und chemischen Vorgänge, die sich im Verlaufe des Einfrierens, Trocknens und der Nachbehandlungen abspielen, erkennen und durch präparative Maßnahmen im Sinne einer Begünstigung zu beeinflussen vermögen. Diesem Problemkreis wird in neuester Zeit immer größere Aufmerksamkeit geschenkt, seitdem die technisch-methodische Seite der Gefriertrocknung weitgehend entwickelt vorliegt und auf der ursprünglich rein empirischen Basis keine durchschlagenden Erfolge mehr erzielt wurden.

---

[1] Diese Arbeit wurde durch einen Kredit des schweizerischen Nationalfonds unterstützt.

Die Erörterung der Gesamtheit der sich stellenden Probleme biologischer, chemisch-physikalischer und technischer Art würde weit über den Raum dieser Übersicht hinausführen; im folgenden sollen nur die biologischen Grundlagen und das botanische Anwendungsgebiet der Gefriertrocknung im Vordergrund stehen, während die technisch-physikalische Seite mit dem Hinweis auf einige ausführliche Sammelreferate nur summarisch behandelt wird (GLICK 1949; FLOSDORF 1949, 1950; HARRIS 1951, 1954a; NEUMANN 1952, 1955a, 1958; BELL 1956; REIMER 1959; ROWE 1960; PARKES und SMITH 1960).

A. Die biologischen Probleme der Gefriertrocknung.

Die Konservierung des natürlichen, chemischen Aufbaues biologischer Objekte mit Hilfe der Gefriertrocknung beruht auf der rein physikalischen Art des Eingriffes. Lösungen und Kolloide, z. B. Seren, Homogenate und Zellextrakte irgendwelcher Art, lassen sich ohne weiteres so gut konservieren, daß sie — wieder mit Wasser durchtränkt — selbst nach längerer Zeit noch dieselben Eigenschaften wie die frischen Produkte aufweisen. Auch Viren und Bakterien, deren Lebensfähigkeit zur Hauptsache von der Erhaltung der chemischen Bausteine abhängt, werden ohne große Schwierigkeiten konserviert. Pflanzliche Zellen und Gewebe hingegen können trotz ungestörtem Chemismus durch die physikalischen Vorgänge, hauptsächlich während des Gefrierens, so mannigfaltige Schädigungen an den komplizierten, cytoplasmatischen Strukturen erleiden, daß ein Überleben schwer zu erzielen ist. Selbst eine nur mittelmäßige Strukturerhaltung begegnet bei pflanzlichem Gewebe größeren Schwierigkeiten als bei tierischem: Der viel höhere Wassergehalt, bedingt durch die großen Vacuolen, schafft sehr ungünstige Gefrierbedingungen; die leeren Intercellularräume hindern den raschen Wärmeentzug, und die als Barrieren wirkenden Zellwände verzögern die Trocknung (GOODSPEED und UBER 1935; LEVITT 1960). Wir müssen uns aber auch vergegenwärtigen, daß gerade die Pflanzen in mancher Hinsicht einer Gefriertrocknung entgegenkommen; besitzen sie doch durch die Ausbildung von Dauerorganen oder durch den Erwerb von Frosthärte von Natur aus die Fähigkeit, Kälte und Trockenheit zu überleben! Wenn es auch in vielen Fällen nicht möglich ist, für biologische Untersuchungen resistente Objekte heranzuziehen, so kann doch die Kenntnis der Vorgänge, die mit dem Erfrieren, der Härtung und der Dauerhaftigkeit verbunden sind, Hinweise auf die Ausarbeitung Erfolg versprechender Präparationsmethoden geben (MOOR 1960).

I. Das intracelluläre Gefrieren.

Die letale Wirkung des Einfrierens beruht zur Hauptsache auf den Folgen der Kristallisation des Wassers im Cytoplasma (SMITH 1954; GENEVES 1955; LEVITT 1957b, 1960; ASAHINA 1958; MERYMAN 1960a): Nach einer Unterschreitung des Gefrierpunktes dringen die Eiskeime von außen her durch Wunden oder natürliche Öffnungen in die Zellen ein, oder werden nach einer größeren Unterkühlung im Zellinnern spontan gebildet. Sobald die Kristallisation begonnen hat, schreitet sie über die Plasmodesmen von einer Zelle zur nächsten fort (MODLIBOWSKA und ROGERS 1955). Einerseits üben die mit großer Geschwindigkeit wachsenden, stilettartigen Eiskristalle aufspießende und zerschneidende, die Volumenvergrößerung zersprengende und zerquetschende Wirkungen auf den plasmatischen Feinbau aus (LUYET und GRELL 1936; GENEVES 1955); andererseits bedingt der Entzug reinen Wassers eine Erhöhung der Elektrolytkonzentration in der Restlösung des Zellsaftes, begleitet von p_H-Verschiebungen und Dehydratationen, die sich denaturierend und fällend auf die lebenswichtigen Proteinsysteme auswirken (LOVELOCK 1953a, b; LEVITT 1957; ULLRICH und HEBER 1957). In diesem Zusammenhang scheinen Verletzungen der das Grundplasma begrenzenden Membranen (Plasmalemma, Tonoplast, Kernmembran) besonders schädlich zu sein, während innerplasmatische Desorganisationen bis zu einem gewissen Grade toleriert werden (WEISS and ARMSTRONG 1960).

Viele Untersuchungen haben gezeigt, daß die Struktur und Lebensfähigkeit biologischer Objekte meistens erhalten bleibt, wenn bei einer Temperatursenkung die Eisbildung im Cytoplasma unterbleibt (ILJIN 1934a,b; KESSLER und RUHLAND 1938; LUYET und THOENNES 1938; ULLRICH 1943, 1944). Der Erfolg einer Gefriertrocknung ist deshalb in erster Linie davon abhängig, daß der durch Gefrier- und Rekristallisationspunkt des Zellinhaltes begrenzte, kritische Temperaturbereich gemieden wird.

Die Kristallisation kann auf dem Wege über die Vitrifikation oder durch extracelluläres Gefrieren umgangen werden.

II. Die Vitrifikation.

Eine aus der flüssigen Phase unter ihren Rekristallisationspunkt abgekühlte Substanz erstarrt ohne nennenswerte Kristallisation, wenn die Zeitspanne für die Passage der kritischen Temperaturzone viel kleiner ist als diejenige, in der die vollständige Transformation der Substanz abläuft (STEPHENSON 1960). Auf biologische Objekte übertragen würde diese sog. Vitrifikation oder Verglasung eine Fixierung sämtlicher Moleküle, Atome und Ionen an der natürlichen, dem letzten Augenblick des aktiven Lebens entsprechenden Stelle bewirken und dadurch alle strukturellen und chemischen Denaturierungen verhindern (LUYET 1939a, 1960).

Die *praktische Durchführung der Vitrifikation* pflanzlicher und tierischer Objekte begegenet in vielen Fällen erheblichen Schwierigkeiten: Einerseits kann schon vor der Verfestigung ein Temperaturschock durch Zerreißungen verursachte letale Folgen haben, wenn der Betrag der thermischen Ausdehnung denjenigen der elastischen Dehnbarkeit wichtiger Zellstrukturen überschreitet (SMITH 1954; MERYMAN 1956). Andererseits wird die Vitrifikation reinen Wassers durch seine in dieser Hinsicht ungünstigen Eigenschaften verunmöglicht: Infolge der großen, latenten Kristallisationswärme und der hohen Geschwindigkeit des Kristallwachstums kann das Temperaturintervall von 130° C auch in kleinsten Nebeltröpfchen nicht genügend rasch überbrückt werden (BLACKMAN und LINSGARTEN 1958). — Es hat sich jedoch gezeigt, daß die kritische Zone der Kristallisation in wäßrigen Systemen in der Richtung einer Begünstigung der Vitrifikation eingeschränkt wird: Erstens ist bei höheren Lösungskonzentrationen eine beträchtliche Gefrierpunktserniedrigung zu verzeichnen. Zweitens erfährt die Geschwindigkeit des Kristallwachstums mit zunehmender Wasserbindung eine exponentiell ansteigende Reduktion; Gelatine z. B. vermindert sie in 1%iger Lösung auf die Hälfte und 10%ig auf weniger als 10^{-5} (CALLOW 1925). Drittens wird der Rekristallisationspunkt schon in 1—2 molaren Kohlenhydratlösungen wesentlich erhöht; durch Glycerin z. B. von -130 auf $-70°$, durch Glucose auf $-41°$, durch Dextrin auf $-9°$ und 10%ige Gelatine auf $-10°$ C (LUYET 1939b).

Je mehr die Kristallisationszone eingeengt ist, um so eher kann die Vitrifikation verwirklicht werden. Eine einwandfreie Verglasung ist zwar erst dann möglich, wenn der Rekristallisationspunkt mit der Unterkühlung erreicht werden kann; aber auch schon eine Stufe vorher scheinen die Kristallgröße und das Ausmaß der Kristallisation in biologisch tragbaren Grenzen zu bleiben. Der präparationstechnische Erfolg ist somit in erster Linie von der Erzielung einer hohen Abkühlungsgeschwindigkeit und eines möglichst kleinen Gehaltes der Objekte an freiem Wasser abhängig (LUYET 1951).

III. Das extracelluläre Gefrieren.

Die Natur schlägt zum Schutze der Zellen bei tiefer Temperatur einen anderen Weg ein: Pflanzen, welche Fröste zu überdauern vermögen, zeigen eine, den jahreszeitlichen Temperaturänderungen entsprechende Zu- und Abnahme der Resistenz gegen die intracelluläre Eisbildung (PFEFFER 1904; PISEK 1952, 1958; LARCHER 1954; LARCHER und EGGARTER 1960). Eine extreme Frosthärte ist bei einigen Coniferen zu finden; die Nadeln von *Pinus strobus* z. B. ertragen im Winter eine praktisch unbeschränkte Abkühlung, während sie im Sommer schon bei −7° C absterben (PARKER und SAKAI 1960).

Die Resistenz wird ermöglicht durch die Beschränkung der an und für sich unvermeidlichen Eisbildung auf die Intercellularräume, die als Folge nicht selten eine erhebliche Vergrößerung erfahren (MODLIBOWSKA und ROGERS 1955). Mit der extracellulären Eisbildung ist ein Wasserentzug verbunden, der sich in der Verkleinerung bis Plasmolysierung der Zellen manifestiert. Die im Verlaufe des Herbstes eintretende Härtung der Pflanzenteile äußert sich in einer mit der Reduktion von Stoffwechselaktivität und Wassergehalt Hand in Hand gehenden Konzentrierung des Zellsaftes (ILJIN 1935; LEVITT 1957a; ULLRICH 1957; TERUMOTO 1958; SAKAI 1960b) und Hydratisierung des Plasmas (KESSLER und RUHLAND 1938; LEVITT 1959b): Das frostharte Cytoplasma gestattet durch die erhöhte Permeabilität seiner Membranen den erforderlichen Wasserentzug und schützt sich gleichzeitig durch verstärkte Wasserbindung vor einer zu großen Dehydratisierung (ULLRICH 1943, 1944). An der mit Hilfe der Frosthärte ermöglichten Unterdrückung der Eiskristallbildung im Cytoplasma sind drei Faktoren beteiligt: Die Einengung der Plasmasubstanz durch die mit der extracellulären Eisbildung verbundene Entwässerung bewirkt erstens eine Erhöhung der Zellsaftkonzentration (nicht zu verwechseln mit der durch die Härtung bedingten!), zweitens eine Verdichtung der Capillarsysteme, drittens eine Vergrößerung des Verhältnisses von gebundenem zu freiem Wasser und führt dadurch zur jeweils erforderlichen Gefrierpunktserniedrigung (MOOR 1960). Dem extracellulären Gefrieren wird durch die letalen Folgen einer zu großen Dehydratisierung eine untere Temperaturgrenze gesetzt.

Die verbesserte Vitrifizierbarkeit eines in den Grenzen seiner Frosthärte vorgefrorenen Objektes ist offenkundig (SAKAI 1958b); trotzdem ist dieser Vorteil bis heute noch nicht für die Gefriertrocknung pflanzlicher Objekte ausgewertet worden! Neben der natürlichen Härtung läßt sich auch die bei vielen Objekten künstlich induzierbare Frostresistenz nutzbar machen: Das Adaptionsvermögen biologischer Objekte gestattet einerseits eine Erhöhung der Resistenz durch längere Akklimatisation knapp oberhalb des schädigenden Temperaturbereiches (LARCHER 1954; SAKAI 1958a; LEVITT 1959a) und andererseits bei Mikroorganismen eine selektive Steigerung der Überlebensrate durch abwechselndes Gefrieren und Kultivieren (FIECHTER 1959). Auch eine sich über Tage erstreckende, stufenweise Abkühlung kann in vielen Fällen zu einer sehr schonenden, extracellulären Eisbildung und Dehydratation des Plasmas führen (ILJIN 1933, 1934a; TAYLOR 1960): Die Blätter von *Hedera helix*, *Scolopendrium vulgare* und *Prunus laurocerasus* z. B. überstehen auf diese Weise −70° C (ILJIN 1934a), ohne eigentlich frosthart zu sein. Mit Kohlenhydratlösungen (Glycerin, Glucose, Saccharose) sind ebenfalls Steigerungen der Resistenz pflanzlicher Zellen — sei es durch Plasmolyse oder Permeation (LUYET and HODAPP 1938; LUYET und THOENNES 1938; POLGE 1951; CARVEN 1960) — erreicht worden. Mit einem Welken der Pflanzenteile vor dem Einfrieren werden ähnliche Effekte erzielt: Ein extremes Beispiel sind die Laubblätter von Moosen, welche nach einer Reduktion ihrer

Feuchtigkeit auf mindestens 35% eine beliebig tiefe Abkühlung ertragen (LUYET und GEHENIO 1938). Viele Dauerorgane mit in der Regel noch kleinerem Wassergehalt (Samen, Sporen, Pollenkörner) werden durch die Kälte kaum beeinflußt (BECQUEREL 1925, 1932).

Die Kombination von geeigneten Präparationsmethoden und günstigen Objekten dürfte das Einfrieren sehr vieler pflanzlicher Objekte ohne größere Schädigungen ermöglichen.

IV. Die Trocknung.

Die Trocknung soll dem durch das Gefrieren fixierten Zustand des biologischen Materials Dauerhaftigkeit verleihen, indem das Reaktionsmedium Wasser auf eine möglichst schonende Weise durch Vakuumsublimation entzogen wird. Die Bewahrung des beim Einfrieren erreichten Standes der Fixation ist hauptsächlich von der Trocknungstemperatur abhängig: -20° C genügen in der Regel für die Trocknung chemisch inaktiver Substanzen (z. B. Kultursubstrate, Proteinfraktionen usw.). Erst unter -40° sind Fermentaktivitäten so weit reduziert, daß Stoffwechselvorgänge und autolytische Prozesse nicht mehr ins Gewicht fallen. Die Anwesenheit merklicher Mengen löslicher Substanzen (Elektrolyte, Zucker) verlangt Trocknungstemperaturen unter dem tiefsten eutektischen Punkte bei etwa -60°, um eine im Verlaufe des Wasserentzuges auftretende Verflüssigung und die damit verbundenen Schädigungen zu verhüten. Für die Erhaltung des vitrösen oder feinkristallinen Zustandes sehr schnell eingefrorener Objekte werden mindestens -80° beansprucht, um eine Rekristallisation oder Umkristallisation auszuschalten.

Dem Trocknungsgut wird im Laufe der Vakuumsublimation zuerst das freie, meistens in Form von Eiskristallen vorliegende Wasser entzogen; als zweites folgt das leichter gebundene Quellungswasser aus den Micellarräumen und den Hydratmänteln der hydrophilen Verbindungen, und zuletzt wird das stöchiometrisch gebundene Wasser angegriffen. Wenn auch zwischen diesen drei Stufen alle Übergänge vorkommen, so charakterisieren sie doch im wesentlichen den Trocknungsverlauf: Das freie Wasser wird von der sog. Haupttrocknung erfaßt. Die restlichen 5—10% gebundenen Wassers verursachen keine Rekristallisations- oder Verflüssigungsgefahr mehr und erlauben somit eine Weiterführung der Trocknung bei erhöhter Temperatur. Wird im Verlaufe dieser Nachtrocknung bei zu hoher Erwärmung neben dem Quellungswasser auch am Aufbau des Cytoplasmas beteiligtes Konstitutionswasser entfernt, ist eine Schädigung durch Denaturierungen zu erwarten. Die Erfahrung zeigt, daß für jedes Objekt ein Optimum der Restfeuchtigkeit besteht, in dem sich die konservierende Wirkung der Trockenheit mit dem schädigenden Einfluß eines zu großen Wasserentzuges die Waage halten.

Ob Änderungen der Trocknungsgeschwindigkeit den Erfolg einer Gefriertrocknung zu beeinflussen vermögen, ist noch wenig abgeklärt; wenn wir uns vergegenwärtigen, daß beim Einfrieren mit der Isolierung des Wassers in den Eiskristallen bereits der größte Teil der Trocknung vollzogen wird, so spricht die Erfahrung, daß dem extracellulären Gefrieren der Vorzug zu geben ist, eher für ein möglichst langsames Vorgehen.

V. Fehlerquellen.

Unsachgemäße Behandlungen der Objekte vor und nach dem Einfrieren und Trocknen ziehen durchwegs grobe Verfälschungen nach sich und reduzieren den Wert einer Gefriertrocknung.

Vorbehandlungen. Bei Vorbehandlungen von Mikroorganismen ist zu beachten, daß die Schutzlösungen keine Vergiftungen durch Antikörperreaktionen, Salze, Wasserstoffionenkonzentration und oxydative Prozesse verursachen (FRY 1954; HARRIS 1954b; SMITH 1954); dasselbe gilt für die zur Imitation der Frosthärte verwendeten Imbibitionsmittel. Eine Vitrifikation verlangt kleinste Objektdimensionen und schließt damit die Gefahr in sich, daß die dünnen Gewebeschnitte in der Zeit zwischen der Entnahme vom Objekt und dem Einfrieren vertrocknen.

Einfrieren. Ursprünglich wurde die Vitrifikation durch das Eintauchen der Präparate in flüssige Luft angestrebt; die Objekte bleiben jedoch bis zu sehr tiefen Temperaturen von einem isolierenden Gasmantel umhüllt (LEIDENFROSTsches Phänomen), der den Wärmefluß erheblich verzögert. Mit Hilfe viel höher siedender, durch flüssige Luft gekühlter Gase als Kontaktmedien (Propan, Butagas, Isopentan) konnte eine wesentliche Verbesserung erzielt werden (NEUMANN 1955a); einige Hinweise auf Schädigungen, die auf das Eindringen dieser Substanzen in die Zellen zurückzuführen sind, schränken die Vorteile der erzielten, schnelleren Abkühlung allerdings wieder ein (MERYMAN 1956; REY 1960). — Bei der Übertragung der kleinen, vitrifizierten Objekte vom Einfrierbad in den Trocknungsraum ist der Gefahr einer zu hohen Erwärmung Rechnung zu tragen.

Restfeuchtigkeit. Die dauernde Aufrechterhaltung des bei der Trocknung erreichten Standes der Restfeuchtigkeit ist von größter Bedeutung, da die Lagerfähigkeit gefriergetrockneter Substanzen in erster Linie eine Funktion des Wassergehaltes ist. Um die Wasseraufnahme der sehr hygroskopischen, gefriergetrockneten Objekte aus der Luft zu verhüten, werden sie am besten unter Vakuum in Ampullen eingeschmolzen. Gummiverschlüsse sind weniger geeignet; ihre Wasserhaltigkeit und Wasserdurchlässigkeit bedingt die Vortrocknung der Kappen oder Stöpsel und eine minimale Dimensionierung der Flaschenöffnung. Wenn sauerstoffempfindliche Substanzen (z. B. die mit der Zeit ranzig werdenden Lipoide) nicht unter Vakuum abgeschlossen werden können, ist die Verwendung von Schutzgasen (N_2) angezeigt. Ist eine genügende Reduktion der Restfeuchtigkeit aus irgendeinem Grunde nicht möglich, werden die Konservierungsbedingungen durch eine Lagerung bei möglichst tiefer Temperatur verbessert.

Wiederbefeuchten. Soll ein getrocknetes Objekt wieder mit Wasser durchtränkt werden, ist besonders bei ursprünglich sehr wasserhaltigen Präparaten Vorsicht geboten; eine Überschwemmung kann von tödlichen Zerreißungen gefolgte unregelmäßige Quellungen verursachen. — Unter Umständen sind auch nachträgliche Verquellungen der durch die Eiskristalle vacuolisierten Zellstrukturen für scheinbar gute, bei relativ hohen Trocknungstemperaturen erzielte Gefrierfixierungen verantwortlich zu machen.

VI. Testmethoden.

Konservierung. Die Bestimmung der Qualität einer Gefriertrocknung, was die Konservierung der chemischen Substanzen anbetrifft, stützt sich durchwegs auf den Vergleich mit frischem Material. Proteine zum Beispiel — die bevorzugten Angriffspunkte für Schädigungen — werden auf Veränderungen der Löslichkeit, des isoelektrischen Punktes, des elektrophoretischen Verhaltens (ULLRICH 1941) und der biologischen Aktivität (z. B. bei Fermenten und Antikörpern) untersucht.

Für die Kontrolle der Konservierung von Zellstrukturen kann bis zu einem gewissen Grade der lichtmikroskopische Vergleich mit lebendem Material herangezogen werden; in Spezialfällen sind röntgendiffraktographische Methoden anwendbar. In den meisten Fällen jedoch wird die Qualität der Fixation nach dem Ausmaß der durch die Eiskristalle bewirkten Störungen, d. i. Vacuolisierungen beurteilt.

Lebensfähigkeit. Die Bewahrung der Lebensfähigkeit nach einer Gefrierkonservierung wird durch die Feststellung der Wiederaufnahme von Stoffwechselvorgängen, Wachstum und Fortpflanzung festgestellt: Bei Mikroorganismen werden am ehesten Keimzahlen und Wachstumsgeschwindigkeiten gefriergetrockneter und unbehandelter Kulturen verglichen. Wenn bei höheren Pflanzen auch ein Überleben nach Gefriertrocknung kaum in Frage kommt, so ist die Überprüfung der Strukturerhaltung doch bis zu einem gewissen Grade an Hand der Untersuchung der Lebensfähigkeit gefrorener, noch nicht getrockneter Objekte möglich: Im aufgetauten Objekt zeugt z. B. die Plasmolysierbarkeit von der unveränderten Semipermeabilität der Membranen; als Indicator dienende Vitalfärbungen können die mit einem Absterben verbundenen p_H-Änderungen anzeigen (Luyet 1937); die Eigenfluorescenz der Vacuolen vermittelt Auskünfte über den Zustand der Semipermeabilität des Tonoplasten (Larcher 1953); der Tetrazolium-Test gibt Aufschluß über den Bestand oder das Erlöschen von Fermentaktivitäten (Larcher und Eggarter 1960). Diese Methoden führen allerdings zu Trugschlüssen, wenn der Geschwindigkeit des Auftauens nicht die erforderliche Beachtung geschenkt wird: Vitrifizierte Objekte z. B. verlangen zur Verhütung der Rekristallisation ein sehr rasches Auftauen durch Eintauchen in warmes Wasser, extracellulär gefrorene jedoch eine allmähliche Rehydratisierung durch einen langsamen Übergang zu höheren Temperaturen.

Trocknung. Die für die Beurteilung der Trocknungsqualität maßgebliche Bestimmung des Restwassergehaltes kann 1. durch eine Messung des Wasserdampfdruckes über dem Objekt ausgeführt werden (Beckett 1954[1]). 2. Mit Hilfe einer vollständigen Trocknung, z. B. im Hochvakuum bei 50—60° C bis zur Gewichtskonstanz wird die Restfeuchtigkeit gravimetrisch ermittelt. 3. Der titrimetrischen Methode nach Fischer (1935) und Neumann (1955a) liegt die Reaktion von Jod und Schwefeldioxyd mit Wasser zugrunde. Die Resultate aller Methoden werden jedoch durch das Fehlen einer scharfen Grenze zwischen Quellungs- und Konstitutionswasser und dem Differieren dieser Grenze von Objekt zu Objekt relativiert; mit der Restfeuchtigkeit wird deshalb in der Regel auch die Bestimmungsmethode angegeben.

B. Die Vakuum-Sublimation.

I. Theoretische Grundlagen.

In einem geschlossenen System stellt sich über Eis ein Gleichgewicht zwischen Verdampfung und Kondensation ein, d. h. Wasserdampfsättigung. Wird der Dampfdruck unter die Sättigung reduziert, so beruht die einsetzende Sublimation auf der Herabsetzung der Kondensation bei gleichbleibender Verdampfung. Die Sublimations- bzw. Trocknungsgeschwindigkeit eines gefrorenen Objektes wird deshalb bei einer bestimmten Temperatur einerseits durch den entsprechenden Wasserdampfdruck (Abb. 1) und andererseits durch die Widerstände bestimmt, welche die Entfernung der Dampfphase hindern:

Abgeleitet aus der kinetischen Gastheorie gilt für die Verdampfungsrate Q_n (verdampfte Masse pro Flächen- und Zeiteinheit)

$$Q_n = (p_T - p_a) \cdot (M/2\,\pi\, R \cdot T)^{\frac{1}{2}} \qquad \text{(Malmström 1951)}$$

[1] Ein handelsübliches Gerät (MDA 2, micro-moisture determination apparatus) wird von Edwards & Co., London, hergestellt.

wobei p_T = dem Sättigungsdampfdruck bei der Temperatur T,
p_a = dem effektiven Dampfdruck über dem Eis,
M = dem Molekulargewicht des Wassers,
R = der universellen Gaskonstanten und
T = der absoluten Temperatur
gesetzt sind.

Die maximal erreichbare Verdampfungsrate ist gegeben durch

$$Q_n = 0{,}25 \cdot p_T \cdot T^{\frac{1}{2}} \qquad (p_T \text{ in mm Hg})\,.$$

Sie wird durch die Verkleinerung von p_a unter mindestens $^1/_{10}$ von p_T genügend gut angenähert.

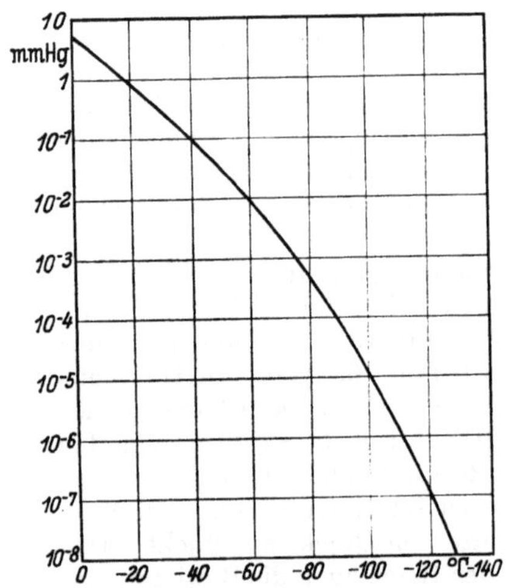

Abb. 1. Sättigungswasserdampfdruck über Eis.

Für die Reduktion des Wasserdampfdruckes über dem Eis bzw. über dem Objekt genügt im Prinzip eine Herabsetzung des Partialdruckes: Die hierauf gründenden Methoden der Vakuumsublimation, welche sich vorgetrockneter, über das Objekt streichender Luft bedienen, haben sich jedoch nicht durchgesetzt (TREFFENBERG 1952; JENSEN 1954; JENSEN und KAVALJIAN 1957; MERYMAN 1959); sie sind zwar viel einfacher als die Hochvakuum-Methoden, ihr Wirkungsgrad ist jedoch in vielen Fällen zu klein.

Der Wasserdampfdruck wird in den meisten Fällen durch Evakuation, d. h. durch Reduktion des Gesamtdruckes gesenkt. Sobald das Vakuum den Sättigungsdampfdruck des Trockengutes unterschreitet, besteht es zur Hauptsache aus Wasserdampf. Die Entfernung dieses Dampfes allein mit Hilfe von Pumpen stößt bei tiefen Temperaturen und demzufolge höheren Vakua ($p_a \leqq p_T/10$) auf die Schwierigkeit, daß sich mit dem Vakuum auch das abzusaugende Dampfvolumen proportional erhöht: 1 ml Eis ergibt schon bei 10^{-3} mm Hg rund 1000 m^3 Wasserdampf! Die Vakuumpumpen können durch die Einführung spezifischer, den Wasserdampf entfernender Medien (hygroskopische Substanzen oder tiefgekühlte Kondensationsflächen) in den Trocknungsraum entlastet werden. Die Saugleistung dieser „Wasserdampfpumpen" wird bei Drucken unter 10^{-2} mm Hg nach der folgenden Gleichung bestimmt:

$$Q_r = (4/3) \cdot (2\,\pi\, M/R \cdot T)^{\frac{1}{2}} \cdot (r^3/L) \cdot (p_a - p_b) \qquad \text{(KNUDSEN, in MALMSTRÖM 1951)}$$

wobei Q_r = der kondensierten bzw. absorbierten Masse pro Flächen- und Zeiteinheit,
r = dem Radius des Vakuumleitungsquerschnittes,
L = der Länge des Weges zwischen Objekt und Wasserdampfpumpe und
p_b = dem Partialdruck über der Wasserdampfpumpe gesetzt sind.

Damit ein Pumpeffekt erzielt werden kann, muß in erster Linie der Wasserdampfdruck (p_b) kleiner sein als das Vakuum (p_a) bzw. als der Dampfdruck des Objektes (p_T). Die Verwendung von Absorptionsmitteln bleibt auf die Aufnahme kleiner Wassermengen bei relativ hoher Temperatur beschränkt, weil der Dampfdruck über ihnen mit der Befeuchtung sehr rasch ansteigt. Die Kondensation des Wasserdampfes an Kühlflächen hat sich als das wirkungsvollste Verfahren erwiesen; schon bei $-100°$ C stellt sich ein Sättigungsdampfdruck von 10^{-5} mm Hg ein (vgl. Abb. 1), der bei Verwendung flüssiger Luft als Kühlmittel ($-190°$) gar eine Reduktion auf 10^{-23} mm Hg erfährt.

Die sich einer Trocknung entgegenstellenden *Widerstände* sind teils apparativ bedingt, teils vom Objekt abhängig (GERSH und STEPHENSON 1954; MATZ 1955):

1. Die KNUDSENsche Gleichung einerseits zeigt den bestimmenden Einfluß des Radius (r) der Vakuumleitungen und des Objektabstandes (L) von der Kondensationsfläche auf die Trocknungsleistung einer Gefriertrocknungsanlage.

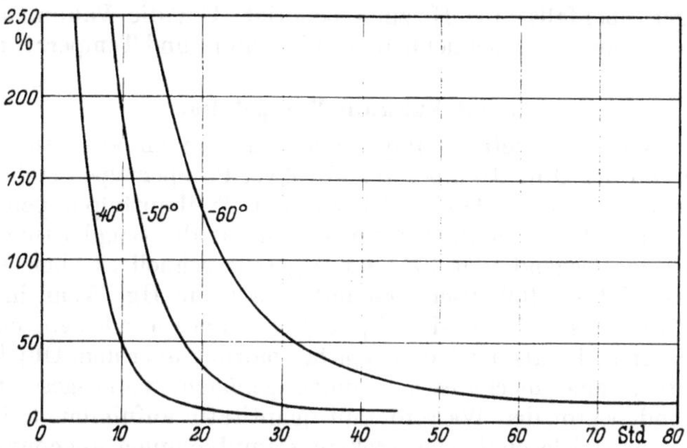

Abb. 2. Entwässerungskurven nach MÜLLER (1957) [der Wassergehalt (in % des Trockengewichtes) als Funktion der Trocknungszeit bei verschiedenen Trocknungstemperaturen].

2. Im Verlaufe der Sublimation wird andererseits die Verdampfungsrate durch den mit der Schichtdicke des bereits getrockneten, äußeren Objektteiles quadratisch anwachsenden Diffusionswiderstand herabgesetzt (GERSH 1951). Nach den Berechnungen von STEPHENSON gewinnt diese Reduktion bei dickeren Objekten ein derartiges Übergewicht, daß der Einfluß des apparativen Widerstandes dagegen bedeutungslos wird. Auch von diesem Gesichtspunkt aus müssen die Objekte einer Gefriertrocknung so klein bzw. so dünn wie möglich sein.

3. Die Wasser bindende Kraft des Objektes setzt der Trocknung bei konstanter Temperatur eine untere Grenze: Die Verkleinerung des Sättigungsdampfdruckes über einem wasserhaltigen System, verglichen mit reinem Wasser, wird durch den Bindungsgrad des Wassers bestimmt. Er reduziert die Sublimation nach der Entfernung des freien Wassers im Maße der Annäherung des Dampfdruckes des Restwassers an das Vakuum der Trocknungsapparatur. Während der Nachtrocknung wird diese Restfeuchtigkeit durch eine den Dampfdruck steigernde Temperaturerhöhung erfaßt und auf das gewünschte Maß reduziert.

Die Entwässerungskurven nach MÜLLER (1957) veranschaulichen den typischen *Trocknungsverlauf* (Abb. 2): Der linke Teil einer Kurve wird, bei gegebener Temperatur, hauptsächlich durch die Verdampfungsrate Q_n (für $Q_n < Q_r$), bzw. die Kondensationsrate Q_r (für $Q_r < Q_n$) bestimmt; der mittlere Teil demonstriert den wachsenden Diffusionswiderstand des getrockneten Objektteiles; der rechte Teil charakterisiert das Wasserbindungsvermögen des Objektes bei der entsprechenden Temperatur.

II. Apparative Ausrüstung.

Die einfachsten Gefriertrocknungsanlagen bestehen aus einem Vakuum-Pumpstand und dem das Objekt enthaltenden Trocknungsraum (BARRETT und BECKETT 1951; GREAVES 1954; MATZ 1955; NEUMANN 1955a; ROWE 1960). Zu diesen Grundeinheiten gesellen sich Einrichtungen, die der Abkühlung bzw. Kühlung von Objekt und Kondensator auf eine konstante, tiefe Temperatur dienen. Bei großen Verdampfungsraten ist zur Kompensation der dem Objekt entzogenen Verdampfungswärme eine Heizeinrichtung notwendig, die auch für eine Temperaturerhöhung über die Zimmertemperatur im Verlaufe der Nachtrocknung herangezogen wird. Für das Schmelzen des im Trocknungstubus erstarrten Einbettungsmittels wird bei der Gefriertrocknung für cytologische Untersuchungen ebenfalls eine Heizung benötigt. Um die Vakuumsublimation überwachen zu können, sind schließlich noch Vakuum- und Temperaturmeßgeräte erforderlich.

1. Der Vakuum-Pumpstand.

Die Wahl eines geeigneten Pumpstandes hängt vom abzusaugenden Gasvolumen und von der Objekttemperatur ab: Zwei Pumpentypen stehen zur Verfügung, mechanische (z. B. Drehschieber-) und Treibmittelpumpen (z. B. Öldiffusionspumpen). Bei Drehschieberpumpen nimmt die Saugleistung (maximal 1—10 l/sec) in der Gegend von 10^{-2} mm Hg sehr schnell ab, bei Öldiffusionspumpen (maximal 100—1000 l/sec) erst unter 10^{-4} mm Hg. Wenn in Rechnung gezogen wird, daß das Vakuum des Trocknungsraumes mindestens eine Zehnerpotenz besser sein sollte als der Sättigungsdampfdruck über dem Objekt, so kann bis etwa $-40°$ C eine mechanische Pumpe genügen, vorausgesetzt, daß ein wirksamer Kondensator den Wasserdampf quantitativ aufnimmt. Sobald tiefere Temperaturen im Spiele sind oder größere Dampfvolumen abgepumpt werden müssen, ist man auf die hundertmal größere Leistung einer Diffusionspumpe angewiesen. Bei Gefriertrocknungsanlagen ohne Wasserdampfkondensator muß darauf geachtet werden, daß keine Schmier- oder Treibmitteldämpfe in den Trocknungsraum gelangen und Schädigungen der Objekte verursachen können (Kühlfallen oder Ölabscheider); im Vorvakuumbereich werden die Pumpen mit Hilfe eines Gasballastventiles vor der Wasserdampf-Kondensation im Rotorraum geschützt.

2. Der Trocknungsraum.

Die große Zahl der beschriebenen, sich hauptsächlich in der Gestaltung und Anordnung des Objekt- und Kondensationsraumes unterscheidenden Gefriertrocknungsanlagen kann in zwei Gruppen unterteilt werden: Die eine umfaßt die Geräte für die Trocknung flüssiger Präparate, aus denen größere Wassermengen bei relativ hoher Temperatur ($> -40°$ C) entfernt werden. Objekt- und Kondensationsraum sind oft voneinander getrennt angeordnet (Abb. 3 und 4), was keine idealen Sublimationsbedingungen ergibt. Das Vakuum im Objektraum wird durch den Wasserdampf so stark beeinflußt, daß der Trocknungsverlauf durch die Anordnung zweier Meßgeräte vor und hinter dem Kondensator (vgl. Abb. 4) registriert werden kann. Kleine Flüssigkeitsmengen und insbesondere empfindliche Präparate werden in der Anordnung der sog. *Rechen-Trocknung* (Abb. 3) gefriergetrocknet, größere Quantitäten durch sog. *Kammertrocknung* (Abb. 4). Die zweite Gruppe umfaßt die Geräte für die Trocknung von festen, wasserhaltigen Objekten, denen nur sehr kleine Wassermengen, jedoch bei möglichst tiefer Temperatur (-60 bis $-80°$ C) extrahiert werden müssen. Zur Erzielung maximaler Sublimationsbedingungen wird die Kondensationsfläche im Trocknungsraum in größtmöglicher Objektnähe (Abb. 5) oder gar das ganze Objekt umgebend

(Abb. 6) angeordnet (MÜLLER 1957; MERYMAN 1960b). Die Wasserdampf-Saugleistung des Kondensators ist gegenüber der Verdampfung derart überdimensioniert, daß der Druck im Trocknungsraum allein von der Leistung des Hochvakuumpumpstandes abhängt.

3. Kühleinrichtungen.

An den Trocknungsraum sind in der Regel Objekt und Kondensator kühlende Einrichtungen angegliedert. Als Kühlmittel werden bei kleineren Apparaturen

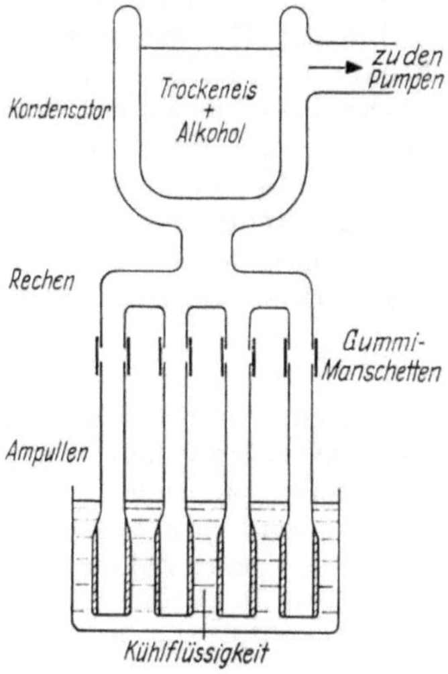

Abb. 3. Rechen-Trocknung.

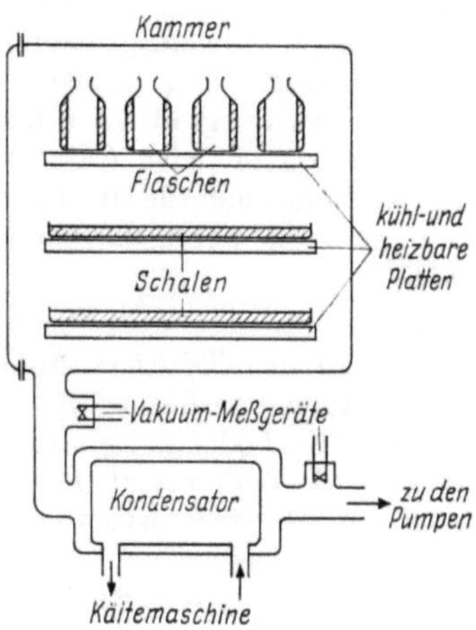

Abb. 4. Kammer-Trocknung.

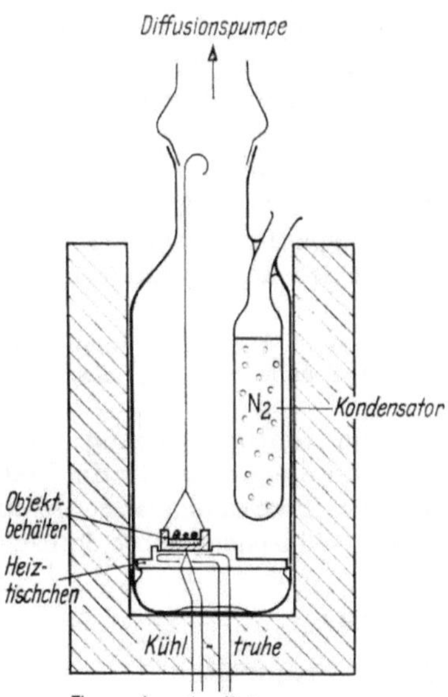

Abb. 5. Trocknung nach MÜLLER.

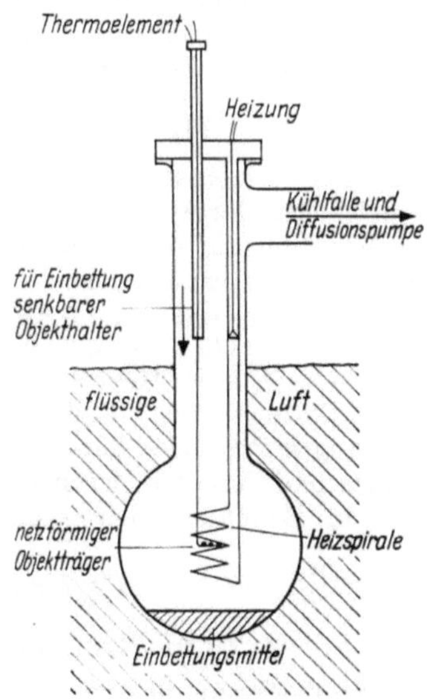

Abb. 6. Trocknung nach MERYMAN.

Gefriergemische, mit Trockeneis gekühlte organische Flüssigkeiten oder flüssige Gase (Luft, Stickstoff) verwendet. Sollen größere Objektmengen und Kondensatorflächen gekühlt (Abb. 4) oder tiefe Temperaturen über längere Zeiträume konstant gehalten werden (Abb. 5), überwiegen die Vorteile der Anwendung von Kältemaschinen den Nachteil der hohen Beschaffungskosten (NEUMANN 1955a).

4. Objektheizung.

Für die Heizung der Objekte stehen drei Methoden zur Verfügung: Die Heizung durch *Wärmeleitung* hat den Vorteil, daß der Objekterwärmung durch die Heizkörpertemperatur eine feste, leicht kontrollierbare obere Grenze gesetzt wird. *Infrarot-* (MERYMAN 1960b) und *Hochfrequenz*-Heizung (MOBERGER et al. 1954; KULENKAMPFF 1956; HOPKINS 1960) ermöglichen durch das Wegfallen der bei der Wärmeleitung notwendigen, großen Kontaktfläche des Objektes mit dem Heizkörper eine allseitige, viel raschere Trocknung. Die im Objekt auftretenden, erheblichen Temperaturgradienten bringen jedoch alle Gefahren einer unkontrollierbaren Erhitzung mit sich (MERYMAN 1960b). Die indirekten Heizmethoden werden in der Regel nur für die Erwärmung des Objektträgers im Trocknungsraum herangezogen, wenn die direkte Wärmeübertragung nicht möglich ist (vgl. Abb. 6).

5. Meßgeräte.

Die Kontrolle der *Objekttemperatur* ist unerläßlich, wenn z. B. eine gewisse Temperatur infolge Verflüssigungs- oder Rekristallisationsgefahr nicht überschritten werden darf oder zu tiefe, die Sublimation verlangsamende Temperaturen unerwünscht sind. Widerstandsthermometer und Thermoelemente haben gegenüber den üblichen Expansionsthermometern den Vorteil kleinerer Maße und Raumbeanspruchung und einer vom Meßort unabhängigen Registrierung der Temperatur. Sie können durch Rückkoppelung mit der Objektheizung direkt als Thermostaten verwendet werden (GREAVES 1954).

Für die *Bestimmung des Vakuums* stehen verschiedene Meßprinzipien zur Verfügung: Für die Messung im Vorvakuumbereich ($1-10^{-3}$ mm Hg) wird die thermoelektrische Methode, für den Hochvakuumbereich ($10^{-3}-10^{-5}$ mm Hg) die Ionisation mit Hilfe einer kalten Kathode (Penning- oder Philips-Ionisationsröhren) bevorzugt, weil beide Bestimmungsarten den Gesamtdruck erfassen.

III. Gefriertrocknungsanlagen.

Für alle üblichen Anwendungsgebiete der Gefriertrocknung sind die wesentlichen Bauelemente der Anlagen oder vollständige Apparaturen käuflich. Folgende Firmen haben sich auf die Produktion der verschiedenen Typen von Gefriertrocknungsanlagen spezialisiert:

Leybold-Hochvakuum-Anlagen GMBH, Köln-Bayental, Deutsche Bundesrepublik; Edwards High Vacuum Ltd., Sussex, England; Secfroid, Lausanne, Schweiz; Canal Industrial Corporation, CANALCO, Bethesda, Md., USA.

Fertigfabrikate werden in der Regel für die Gefriertrocknung (größerer Mengen) flüssiger Präparate herangezogen, während die Geräte für die Fixation von cytologischem Material, hauptsächlich was den Trocknungsraum anbetrifft, meistens vom Benützer nach seinen speziellen Wünschen selbst konstruiert werden.

C. Anwendungen.

I. Gefriertrocknung flüssiger Präparate.

Die Gefriertrocknung flüssiger Präparate hat die Aufgabe, im wäßrigen Milieu nicht haltbare Substanzen wie z. B. Protein- und Fermentfraktionen, Antibiotica oder Vitamine enthaltende Extrakte, Homogenate und Sedimente zu konservieren (FLOSDORF 1951; HETHERINGTON 1954; SINGER 1954; RABINOWITCH 1956). Die ausgeprägte Löslichkeit gefriergetrockneter Präparate, bzw. ihre Durchtränkbarkeit mit einem Lösungsmittel führte zur Bezeichnung „*Lyophilisation*" als Synonym für Gefriertrocknung. Der Wert der Lyophilie vieler reiner, nicht kristallisierbarer Substanzen (Fermente, einige Antibiotica, Proteine usw.), die nur durch Gefriertrocknung erreicht werden kann, liegt auf der Hand. Für die Behandlung von Zellhomogenaten und -fraktionen eröffnet die Lyophilisation neue Wege der fraktionierten Sedimentation und Extraktion, indem sie die Verwendung lipophiler Suspendatoren ermöglicht und dadurch die Ausschwemmung der hydrophilen Substanzen (Salze, Kohlenhydrate, Aminosäuren, Nucleoproteine usw.) ausschaltet (BEHRENS und TAUBERT 1953; PIRIE 1956).

Sowohl der Einfriergeschwindigkeit wie auch der Trocknungstemperatur (−20° C genügt in der Regel) kommt keine große Bedeutung zu. Es darf lediglich im Verlaufe der Trocknung keine Verflüssigung auftreten (ROWE 1960). Um die Trocknung nicht unnötig in die Länge zu ziehen, wird das flüssige Präparat auf einer möglichst großen Fläche, höchstens einige Millimeter dick eingefroren. Verschiedene Einfriermethoden sind vorgeschlagen worden, unter denen sich das mantelförmige Gefrieren (*shell-freezing*, Abb. 7) in Ampullen und Flaschen oder das Gefrieren in flachen Schalen für größere Mengen (*bulk-freezing*, Abb. 4) durchgesetzt haben. Die Nachtrocknung wird je nach der Empfindlichkeit der Substanzen bei 20—50° C durchgeführt; der anzustrebende, kleinstmögliche Restwassergehalt liegt in der Regel bei ungefähr 0,5%. Der Verschluß der Gefäße unter Vakuum ist, wie bereits erwähnt, allen anderen Methoden vorzuziehen; müssen Gummikappen verwendet werden, so empfiehlt sich der wenig durchlässige Butyl-Gummi und ein Fluten[1] der Gefäße mit getrocknetem Stickstoff.

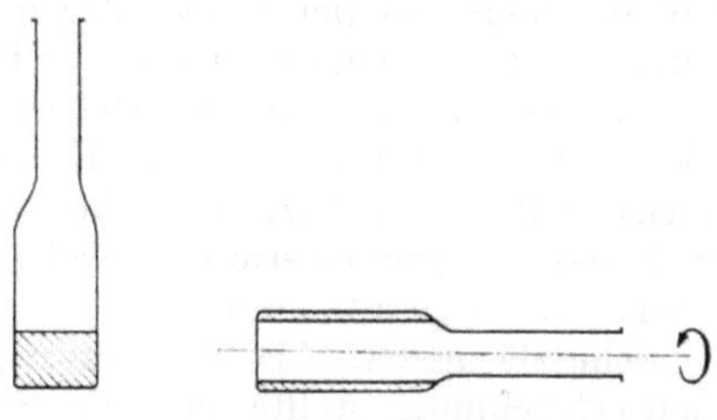

Abb. 7. "shell-freezing". Flaschen oder Ampullen werden liegend in ein Kühlbad getaucht und um ihre Achse rotiert. Die Flüssigkeit friert dabei in dünner Schicht entlang der im Vergleich zur Grundfläche viel größeren Seitenwand ein.

Wie die Erfahrungen mit Seren, Muttermilch, Kulturmedien und anderen eiweißhaltigen Flüssigkeiten lehren, werden Proteine in der Regel bis zu einem hohen Grade naturgetreu konserviert. An Fermenten ließen sich gewisse Aktivitätsänderungen (auch Steigerungen) feststellen, hauptsächlich aber mit der Zeit eine langsame Abnahme (FLOSDORF 1935; NORD 1933; BERSIN 1939; SMITH 1954). Kohlenhydrate und Lipoide ändern sich kaum; einzig Lipoproteine werden zum Teil aufgespalten.

Die Gefriertrocknung flüssiger Präparate kann nur bis zu einer unteren Grenze der Konzentration (einige Prozent) durchgeführt werden; höher verdünnte Substanzen bilden im getrockneten Zustand keine feste Schicht mehr und gehen beim Fluten[1] der Gefäße sehr leicht verloren.

II. Konservierung von Mikroorganismen.

In der Mikrobiologie hat sich die Anwendung der Gefriertrocknung auf vielen Gebieten als die geeignetste und oft praktisch einzig mögliche Konservierungsmethode erwiesen (GREAVES 1944; FLOSDORF 1954; FRY 1954; HARRIS 1954b;

---

[1] Unter Fluten versteht man in diesem Falle das Füllen des evakuierten Raumes mit einem Gas.

FENNELL 1960; MAZUR 1960). Selbst bei hohen Ausfällen (Überlebensrate unter 1%) ist eine Gefriertrocknung wegen der großen Individuenzahl meistens genügend gut, im Gegensatz zu vielzelligen Objekten, in denen ein Verlust von 50% mit dem Tod des Zellverbandes gleichzusetzen ist.

Die Gefriertrocknung weist vor den üblichen Methoden der Subkultivierung mannigfache *Vorteile* auf: Lyophilisierte Sammlungen z. B. benötigen keine Pflege und Ausfälle durch Verschmutzung, Infektion, Variation und Tod sind ausgeschlossen. Gefriergetrocknete Starterkulturen, z. B. für die Milchverwertung (Käse, Buttermilch, Joghurt) oder lebensfähige Viren und Bakterien enthaltende Vaccinepräparate (FLOSDORF 1954) können ohne Schwierigkeit über größte Distanzen verschickt werden. Gefriergetrocknete Präparate garantieren bei Serienversuchen für eine sehr große Konstanz der Anfangsbedingungen.

Der Erfolg einer Konservierung wird bei gleichbleibenden Versuchsbedingungen durch die Frost- resp. Trockenheitsresistenz der Organismen bestimmt: Die höchsten Überlebensraten treten bei den Dauerorganen der Pilze und Bakterien auf, die kleinsten bei Hefen, Algen und Einzelzellen aus Gewebekulturen; bei Viren und Bakterien (in vegetativer Phase) sind alle Übergänge zu finden. Vegetative Phasen werden für die Gefriertrocknung in der Regel im Moment hoher Aktivität aus relativ jungen Kulturen gewonnen, Dauerorgane im Moment höchster Resistenz kurz nach ihrer Reifung. Neben der natürlichen Härte kann auch das Adaptionsvermögen vieler Organismen für die Steigerung der Überlebensrate ausgenützt werden.

Schutzlösungen. Als die wirkungsvollste präparative Maßnahme vor der Gefriertrocknung stellte sich die Suspension der Mikroorganismen in Schutzlösungen heraus, die im allgemeinen zwei Komponenten enthalten: Ein *Kolloid*, meistens ein Protein in 5—20%iger Konzentration; ihm fällt wahrscheinlich die Aufgabe zu, den direkten Kontakt der Organismen mit den Eiskristallen zu verhüten, indem es sich beim Gefrieren als schützender Mantel um die Teilchen legt; ein *Kohlenhydrat* (5—10%ig); es wirkt als Elektrolytpuffer und verhindert eine zu rasche und zu große Dehydratation. Bei sehr empfindlichen Objekten wird für die Stabilisation von Redox-Prozessen im getrockneten Zustand ein Zusatz von Glutaminsäure oder Ascorbinsäure (1%ig) empfohlen. Die Anwesenheit von Salzen, insbesondere NaCl, ist durchwegs äußerst schädlich (FRY 1951, 1954; PROOM 1951; NEUMANN 1955b; GREAVES 1960a, b; SCOTT 1960).

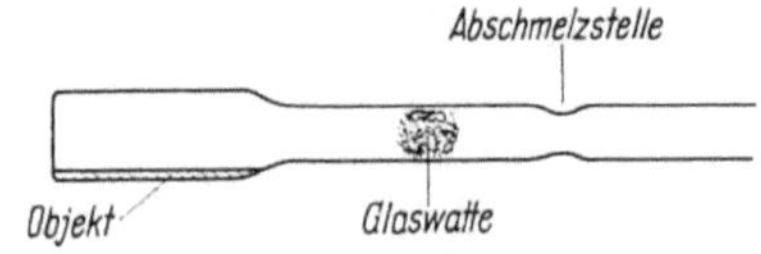

Abb. 8. (Vgl. Text.)

Sehr viele Versuche wurden mit Bakterien durchgeführt: Unter den verschiedensten Schutzkolloiden haben sich inaktiviertes Serum (Mensch, Pferd), Fleischbrühe und Magermilch (Proteingehalt um 7%) oder Pepton, Albumin und Hämoglobin (10—20%ig), als Kohlenhydrat Glucose und Lactose (7,5%ig) am besten bewährt. Mit gutem Erfolg wurden Schutzlösungen dieser Art auch auf andere Mikroorganismen angewendet; es empfehlen sich jedoch kleinere Modifikationen: Virusfraktionen können durch Reinigung inaktiviert werden, so daß die Gefriertrocknung in Gewebesaft oder -homogenat mit einem Kolloidzusatz (1—10%) oft bessere Ergebnisse zeitigt (HARRIS 1954b; HOLLINGS und LELLIOT 1960). Hefen werden am besten in Serum mit 2,5% Glucose (FIECHTER 1959), Sporen und Conidien in Serum mit 10—20% Saccharose lyophilisiert (FENNELL 1960). Für Dauerorgane genügen hin und wieder auch Serum allein oder reines Wasser. Ein längerer Aufenthalt in der Schutzlösung sollte vermieden werden, um den mit einem Auskeimen verbundenen Verlust der Resistenz zu verhüten.

Einfrieren. Für den Gefriertrocknungsprozeß werden Schutzlösung und Mikroorganismen unter sterilen Bedingungen in kleinen, mit Glaswatte verschlossenen Ampullen oder Glasröhrchen deponiert (Abb. 8). Die Konzentration der Teilchen in den in der Regel klein gehaltenen Flüssigkeitsmengen (0,2—0,5 ml) spielt bei der Verwendung einer Schutzlösung keine Rolle. Die Einfriergeschwindigkeit scheint von untergeordneter Bedeutung zu sein; genaue Angaben fehlen. Die Temperatur des Einfrierbades soll für Dauerorgane etwa −20°, für Bakterien und Hefen rund −30° und für Viren wenigstens −40° C betragen. Diese Angaben gelten auch für die Temperatur während der Haupttrocknung; dem Einfluß der Sublimationskälte ist jedoch durch eine Erhöhung der Kühlbadtemperatur um 5—10° C Rechnung zu tragen. Die Lyophilisation wird meistens in Form einer Rechen-Trocknung vorgenommen (Abb. 3). Infolge der Verschiedenartigkeit der bis heute publizierten Präparationsmethoden und Gefriertrocknungsapparaturen können keine allgemein gültigen, zeitlichen Angaben über den Trocknungsverlauf gemacht werden. Nach Abschluß der Haupttrocknung wird in der Nachtrocknung bei 20—30° C die Restfeuchtigkeit von Gewebezellen, Algen und Hefen auf einige Prozent, von Bakterien, Sporen und Conidien auf rund 1% und von Viren auf weniger als 0,5% reduziert. Der ganze Trocknungsprozeß nimmt der Größenordnung nach ungefähr 20 Std. in Anspruch. — Nach beendeter Trocknung werden die unter Vakuum stehenden Glasgefäße an der dafür vorgesehenen Stelle (Abb. 8) abgeschmolzen, um empfindliche Organismen vor den nachteiligen Folgen eines noch so kurzen Vakuumunterbruches zu bewahren.

Eine andere Lyophilisationsmethode bedient sich der Verdampfungskälte für das Einfrieren der Objekte und für die Kühlung im Verlaufe der Haupttrocknung (Greaves 1944): Die Röhrchen werden beim Evakuieren mit mindestens 7facher Erdbeschleunigung zentrifugiert *(centrifugal evaporative freezing)*, um das durch den Ausbruch gelöster Gase verursachte Aufschäumen der Flüssigkeiten zu verhindern (Fry und Greaves 1951). Die Nachteile der teilweisen Trocknung im flüssigen Zustand, der Verschmutzungsgefahr und eines unkontrollierbaren Temperaturverlaufes eröffnen dieser Methode nur ein beschränktes Anwendungsgebiet. — In neuester Zeit hat Meryman (1960c) eine interessante Variante vorgeschlagen: Nylon-Gaze wird mit einer Suspension von Mikroorganismen ohne schützende Beimischung durchtränkt und sofort unter ein gutes Vakuum gebracht. Abgesehen von der Nachtrocknung nimmt der ganze Gefriertrocknungsprozeß infolge der großen Fläche und kleinster Schichtdicke der Flüssigkeit nur einige Minuten in Anspruch. Bei einigen hoch empfindlichen Objekten, die sonst quantitativ absterben, konnten erstaunliche Überlebensraten erzielt werden.

Ausfälle. Die durch den Gefriertrocknungsvorgang bedingten Ausfälle äußern sich bei Viren in der Regel durch eine Senkung des Infektionstiters um eine bis mehrere Zehnerpotenzen. Bakterien, Sporen und Conidien sind meistens dankbare Objekte mit Überlebensraten bis nahezu 100%. Sehr empfindliche Organismen wie z. B. *Neisseria, Vib. cholerae* und Hefepilze überleben mit weniger als 1‰. — Neben dieser primären findet während der Lagerung eine weitere Reduktion statt: Die mit der Zeit logarithmisch gegen Null strebende Überlebensrate (Fry 1954) ist dabei in erster Linie eine Funktion des Restwassergehaltes. Durch die gegensätzliche Abhängigkeit der primären und der sekundären Reduktion vom Restwassergehalt werden die angeführten optimalen Trocknungsgrade bestimmt. Um Ausfälle während der Lagerung zu reduzieren, sind wasserhaltigere Objekte bei möglichst tiefer Temperatur aufzubewahren (Kühlschrank). — Selbst in ungünstigen Fällen bleibt die Virulenz über Jahre und Jahrzehnte unverändert erhalten.

Rekonstitution. Die Rekonstitution gefriergetrockneter Mikroorganismen wird in den meisten Fällen, insbesondere wenn Schutzlösungen verwendet wurden, mit destilliertem Wasser durchgeführt. Für empfindliche Objekte scheint eine vorgängige, langsame Befeuchtung in mit Wasserdampf gesättigter Atmosphäre vorteilhaft zu sein. Für besonders delikate, nach der Nylon-Gaze-Methode lyophilisierte Objekte (ohne Schutzmittel) ist die Rekonstitution mit 40% Dextran enthaltenden, isotonischen Salz- oder Zuckerlösungen unerläßlich.

Sterilität. Während der Gefriertrocknung können Teilchen aus den Gefäßen in den Trocknungsraum geschleudert werden, von wo sie einerseits in andere, fremde Kulturen enthaltende Röhrchen gelangen oder andererseits gar durch den Pumpentrakt nach außen befördert werden. Der Verschluß der Röhrchen mit Glaswatte (Abb. 8) verlängert wohl die Trocknungszeit, er eliminiert jedoch die Infektionsgefahr vollkommen. Die unter Vakuum stehenden Gefäße werden bei Gebrauch auf der Höhe des Wattepfropfes geöffnet, wodurch ein Rutschen der Watte und ein Herausgeblasenwerden von Keimen durch den Lufteinbruch vermieden wird (COWAN 1951; HARRIS 1954b).

III. Fixation für cytochemische Untersuchungen.

Nur in wenigen Fällen ist es möglich, histo- und cytochemische Untersuchungen an lebendem Material durchzuführen. Meistens sind die mit der Schneidetechnik verbundenen, recht komplizierten Eingriffe notwendig, um ein Objekt für eine bestimmte Analysenmethode aufzuschließen: Am Anfang der Präparation steht durchwegs eine Fixation mit der Aufgabe, das metabolisch aktive Objekt in möglichst naturgetreuem Zustand zu stabilisieren und zu konservieren, damit es die weiteren, vorbereitenden Eingriffe ohne Veränderungen übersteht. Die übliche Fixations- und Einbettungstechnik vermag in mancher Hinsicht infolge ihrer denaturierenden, auslaugenden und den Quellungszustand ändernden Einflüsse nicht zu befriedigen (ZEIGER 1938; BAKER 1951; SYLVEN 1951; MÜHLETHALER 1960); die Fixation durch Gefriertrocknung jedoch schaltet gerade diese wesentlichen Artefaktquellen aus (BELL 1956; WHATLEY 1956; NEUMANN 1958):

Das schlagartige Einfrieren verhindert sub- und postmortale Veränderungen chemischer Art beinahe vollständig; die Trocknung durch Vakuumsublimation unterbindet alle Stoffverschiebungen und -verluste, die bei der gewöhnlichen Dehydratation mit Hilfe von Lösungsmitteln unvermeidlich sind. — Ein unveränderter Chemismus ist für viele mikrochemische Nachweisreaktionen (z. B. Fermentnachweise, quantitative Bestimmung chemischer Gruppen durch charakteristische Färbungen, Proteinnachweis mit fluorescierenden Antikörpern) von ausschlaggebender Bedeutung, während der Erfolg mikrophysikalischer Methoden (Mikroradiographie, Autoradiographie und UV-Mikrospektrophotometrie) hauptsächlich von der Verhinderung von Substanzverlusten abhängig ist (DANIELLI 1958; GERSH 1959).

Fixierung. Auch die Fixation durch Gefriertrocknung ist mit Nachteilen behaftet, die im Verlaufe der Präparation berücksichtigt werden müssen: In erster Linie sind die durch die Eisbildung verursachten, strukturellen Schädigungen zu erwähnen, die in vielen Fällen unumgänglich, jedoch stark reduzierbar sind (KULENKAMPFF 1955). In zweiter Linie muß bei den Nachbehandlungen (Einbettung, Nachfixation usw.) dem vollständig erhaltenen Reaktionsvermögen der Objektbestandteile Rechnung getragen werden, damit die Vorteile der Gefriertrocknung nicht wieder rückgängig gemacht werden.

Auf die Abhängigkeit der Vitrifikation, resp. einer maximalen Reduktion der Eiskristallgröße, von der Objektvorbereitung (Adaption durch Härtung, Reduktion des Wassergehaltes u. a.) wurde bereits hingewiesen (siehe S. 76). Durch Eintauchen kleinster Objektstücke in tiefgekühlte Flüssigkeiten besteht die Möglichkeit, hohe Abkühlungsgeschwindigkeiten zu erzielen. In der Praxis werden dünnste Schnitte (0,2—0,1 mm) vom Objekt abgehoben und sofort mit Hilfe von feinen Glaswerkzeugen (Nadeln, Ösen) in die Kühlflüssigkeit eingetaucht. Meistens wird ein Isopentan-Propan-Gemisch (1:4) verwendet (BRETSCHNEIDER und ELBERS 1952), das durch Kondensation des Propans (Sdp. —44,5° C) im tiefgekühlten Isopentan (Sdp. 28° C) hergestellt wird. Auch der Gebrauch reinen Propans ist üblich; er macht jedoch die Verwendung eines speziellen Kondensators notwendig (MOBERGER et al. 1954; MÜLLER 1957) und schließt das LEIDENFROSTsche Phänomen nicht völlig aus. Um die große Explosionsgefahr bei einer allfälligen Mischung der organischen Flüssigkeiten mit flüssiger Luft auszuschalten, dient flüssiger Stickstoff als Kältequelle (NEUMANN 1952). In der Kühlflüssigkeit werden die Objektstückchen in einem flachen Behälter gesammelt, dessen Boden mit einer dicken Schicht des Einbettungsmittels überdeckt ist (Abb. 9). Während der Überführung in den Trocknungsraum (Abb. 5) verhindert die relativ große Masse des Objektbehälters und die das Objekt darin überdeckende Kühlflüssigkeit eine gefährliche Temperaturerhöhung (MÜLLER 1957). Der auf die Betriebstemperatur vorgekühlte Trocknungstubus wird sofort nach der Einsetzung des Objektbehälters in den Heiztisch geschlossen. Eine möglichst sorgfältig eingeleitete Evakuation entfernt zuerst durch langsames Abdampfen die überstehende Kühlflüssigkeit.

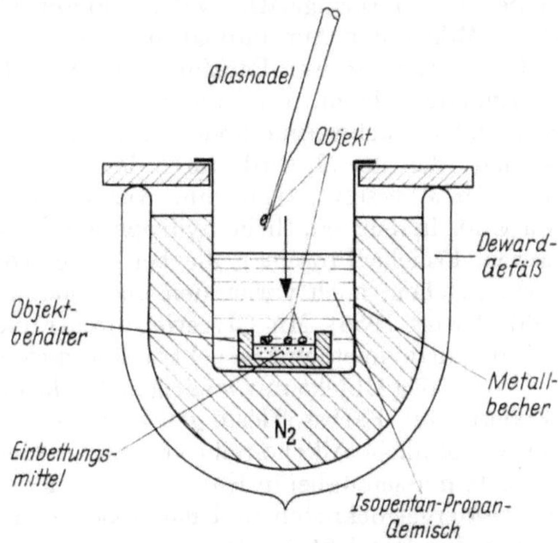

Abb. 9. Anordnung für das sehr schnelle Gefrieren von Gewebestücken: "snap-freezing".

Die zur Verhinderung der Rekristallisation erforderliche, sehr tiefe Temperatur (—60 bis —80° C) zieht den Trocknungsvorgang trotz kleinster Objektdimensionen beträchtlich in die Länge; wenn höchste Anforderungen gestellt werden, umfaßt er in der Regel eine ganze Woche (KULENKAMPFF 1955; MÜLLER 1957). Der Einfluß der Sublimationskälte fällt infolge der sehr kleinen Verdampfungsrate nicht ins Gewicht, so daß die Temperatur des Objektes ungefähr derjenigen des Trocknungsraumes entspricht. Die Nachtrocknung wird bei langsam auf 20° C ansteigender Temperatur durchgeführt. Der genaue Betrag des anzustrebenden Restwassergehaltes ist nebensächlich; er muß lediglich so weit herabgesetzt werden, daß das Einbettungsmittel ungehindert alle Objektteile durchdringen kann.

Einbettung. Für die Einbettung wird der Objektbehälter mit Hilfe des Heiztischchens erwärmt, bis das Einschlußmittel schmilzt und die Objektteilchen darin untersinken. Temperaturen von 60—70° C sollten zur Verhütung von Proteindenaturierungen und Lipoidverflüssigungen nicht überschritten werden. Die Einbettung muß unter Vakuum durchgeführt werden, weil die Anwesenheit von Gasen im Objekt eine vollständige Imbibition verhindert. Es wird darum auch empfohlen, das Einschlußmittel vor seiner Verwendung im Objektbehälter durch kräftiges Erhitzen unter Vakuum zu entgasen. Treten bei einer an sich einwandfreien Einbettung dennoch Blasen auf, so sind sie ein Zeichen für zu große

Restfeuchtigkeit (NEUMANN 1958). Die eingebetteten Objekte werden nach der üblichen Paraffintechnik weiter verarbeitet.

Artefaktbildung. Der jede Nachbehandlung begleitenden Artefaktgefahr muß besondere Aufmerksamkeit geschenkt werden: Schon während der Einbettung können einerseits durch die Oberflächenspannung des eindringenden Einschlußmittels feinste Objektstrukturen zerstört werden und andererseits je nach der Lyophilie der Substanzen Lösungserscheinungen auftreten. Den Schädigungen der ersteren Art scheint keine große Bedeutung zuzukommen. Für die Untersuchung lipophiler Zellkomponenten wird das Paraffin in der Regel durch das hydrophile Carbowax ersetzt (FREDRICSSON 1957). Die Schnittbänder werden vor dem Auflegen auf den Objektträger (Glas, photographische Emulsion) allein mit Hilfe ihres Eigengewichtes in der Wärme gestreckt, um verfälschende Einflüsse der Glyceringelatine oder anderer Gleitmittel zu vermeiden (NEUMANN 1958). Mikro- und Autoradiographie profitieren am meisten von den Vorteilen der Gefriertrocknung, weil Paraffinschnitte direkt verwendet werden können.

Entparaffinieren. Für alle auf die lichtmikroskopische Untersuchung ausgerichteten Nachweismethoden muß das Paraffin aus den Schnitten herausgelöst werden; das Xylol wird durch leichter flüchtige Lösungsmittel (Petroläther, Nonan u. a.) ersetzt, die nachher verdunstet werden. Das Resultat ist ein Schnittpräparat, in dem sämtliche hydrophilen Zellkomponenten (Salze, Zucker, Aminosäuren, lösliche Proteine), die bei der gewöhnlichen Fixation zum größten Teil verlorengehen, noch vorhanden und chemisch unverändert erhalten sind (BELL 1951; SWIFT 1955). Mit Glycerin durchtränkte Präparate eignen sich hauptsächlich für ultraviolett-mikrospektrophotometrische Untersuchungen.

Wenn das Objekt mit wäßrigen Reaktionsmedien in Berührung kommen soll (Fixiermittel, Farbstofflösungen usw.), wird für die Überführung aus dem Xylol eine gewöhnliche Alkoholreihe verwendet. Gewisse Denaturierungen und Substanzverluste müssen dabei in Kauf genommen werden; es hat sich jedoch gezeigt, daß die gefriergetrockneten und nachfixierten Präparate dennoch den gewöhnlichen in mancher Hinsicht überlegen sind. Diese Tatsache hängt hauptsächlich mit dem viel rascheren Eindringen des Fixatives in die gefriergetrockneten, fein geschnittenen Objekte zusammen. Die Verkleinerung des Substanzverlustes ist sehr wahrscheinlich auf eine engere Assoziation der freien Moleküle mit festen Zellstrukturen im Verlaufe der Trocknung zurückzuführen (DANIELLI 1953).

IV. Fixation für elektronenoptische Untersuchungen.

Die mit jeder präparativen Vorbereitung biologischer Objekte verbundenen Artefaktprobleme sind im Prinzip für elektronenmikroskopische wie für cytochemische Untersuchungen dieselben. Wenn die Elektronenmikroskopie auch hauptsächlich der Morphologie verpflichtet ist, so verlangt das bis in die Dimensionen der Makromoleküle erweiterte Auflösungsvermögen doch die Fixation aller Substanzen in unveränderter Lage. Die bereits erwähnten Mängel der chemischen Fixation werden noch durch den Umstand betont, daß bis heute nur zwei sehr ähnlich wirkende Fixative — Osmiumtetroxyd und Kaliumpermanganat — zur Verfügung stehen. Die unzähligen, auf dieser Basis erzielten Resultate sind deshalb mit einer Einseitigkeit behaftet, deren Tragweite nur schwer abgeschätzt werden kann. Die Lebendbeobachtung kommt in der Elektronenmikroskopie infolge präparativer Schwierigkeiten (Hochvakuum, Beschränkung der Objektdicke auf höchstens 1000 Å und die das Objekt zerstörenden Einflüsse des Elektronenstrahles) kaum in Frage (COSSLETT 1951). Die Gefriertrocknung hingegen verspricht, als rein physikalische Präparationsmethode, wertvolle Ergänzungen zu den mit Hilfe chemisch wirkender Fixiermittel erzielten Resultaten zu liefern.

Die Schwierigkeiten, welche sich der Auswertung der Gefriertrocknung für elektronenmikroskopische Zwecke in den Weg stellen, sind eine Folge des gegenüber lichtoptischen Untersuchungsmethoden um das 1000fache gesteigerten Auflösungsvermögens: Die geringsten strukturellen Schädigungen, seien sie durch

Eiskristalle oder das Einbettungsmittel verursacht, werden sichtbar (SJÖSTRAND 1951; BRETSCHNEIDER und ELBERS 1952). Die Entwicklung brauchbarer Präparationsmethoden ist erst in jüngster Zeit in die Wege geleitet worden:

Als erster versuchte MÜLLER (1957) die nach der üblichen Methode (Abb. 5) gefriergetrockneten Objekte in Methacrylat einzubetten, um sie für die Ultramikrotomie zugänglich zu machen. Es zeigte sich, daß das monomere, für die Polymerisation auf 50° C erwärmte Einschlußmittel, wahrscheinlich auf Grund der Auflösung der Lipoide, die Strukturen durch Verquellungen vollständig ruiniert. Eine wesentliche Reduktion dieser Lösungserscheinungen wurde durch die Verwendung von ultraviolettem Licht für die Polymerisation bei einer tieferen Temperatur (0—10° C) erreicht; die Resultate waren jedoch auf Grund des äußerst schwachen Kontrastes der Bilder und umfangreicher Vakuolisierungen der Objektstrukturen immer noch ungenügend.

SJÖSTRAND und BAKER (1958) entwickelten die Methode weiter, indem sie unter Hochvakuum bei möglichst tiefer Temperatur einbetteten und die Schnitte mit Phosphorwolframsäure nachkontrastierten: Einerseits senken Hochvakuum und tiefe Temperatur den Dampfdruck des Einschlußmittels und verhindern damit die Vakuolisierung des Objektes, welche durch die Bildung von Dampfblasen verurscht wird; andererseits wird der Vorteil der geringen Schnittdicke (2—300 Å) für die Färbung der elektronenmikroskopischen Schnittpräparate ohne die sonst unerläßliche Entfernung des Einschlußmittels ausgenützt. Für die Kontrastierung werden die Schnittpräparate mit der Objektseite nach unten 1—2 Std. auf Schwermetallionen enthaltenden Flüssigkeiten flotiert (z. B. Lösungen von Uranylacetat, Phosphorwolframsäure, Phosphormolybdänsäure, Bleiacetat, Bariumhydroxyd, Platintetrabromid: BONADREFF 1957; WATSON 1958a, b; LEVER 1960).

Unabhängig von diesen Arbeiten haben SENO u. Mitarb. (1958) denselben Weg eingeschlagen. — Die in beiden Fällen an tierischem Material erprobte, verbesserte Methode kann der üblichen, chemischen Fixationstechnik vergleichbare Resultate liefern; die Qualität der Bilder läßt jedoch infolge der nur teilweise ausgeschalteten Artefaktquellen oft noch zu wünschen übrig.

Bis heute sind bereits auf drei verschiedenen Gebieten Vorschläge für die Vervollkommnung der Methode veröffentlicht worden: 1. Nach FERNANDEZ-MORAN (1960) kann die Superleitfähigkeit des flüssigen Helium II (1° K), die 10000mal besser als die Wärmeleitfähigkeit von Kupfer ist, für die Verwirklichung einer idealen Vitrifikation herangezogen werden. 2. Der von MERYMAN (1960b) beschriebene Trocknungstubus (Abb. 6) mit maximalem Wirkungsgrad sollte die Gefriertrocknung bei −80° C oder noch tieferen Temperaturen ermöglichen, wodurch die Rekristallisationsgefahr vollständig gebannt wäre. 3. Eine Reihe von hydrophilen Einbettungsmitteln, mit deren Hilfe Lösungserscheinungen umgangen oder doch mindestens geändert werden können, harren ebenfalls der experimentellen Erprobung (GIBBONS 1959; WICHTERLE et al. 1960; STÄUBLI 1960).

Mit der sog. *Gefrier-Ätz-Methode* hat STEERE (1957) einen Präparationsgang vorgeschlagen, der durch die Umgehung der Einbettung eine Reihe von Artefaktmöglichkeiten ausschließt: Das Objekt wird in einem Wassertropfen auf einer gekühlten Metallunterlage eingefroren und mit Hilfe eines feinen Messers durch Spalten und Schaben soweit abgetragen, bis die interessierende Stelle an der Oberfläche liegt. Eine sehr kurzfristige Gefriertrocknung (30 sec) ätzt die Spaltfläche leicht an und enthüllt dadurch die im Eise eingebetteten Objektstrukturen. Das zu Tage tretende Profil wird hierauf durch zwei Aufdampfungen (Schwermetallbeschattung und Kohleverstärkung) ohne eine Unterbrechung des Hochvakuums festgehalten. Der vom Objekt abgelöste und auf ein Trägernetzchen aufgefischte Abdruck stellt ein naturgetreues Bild des gefriergetrockneten Objektteiles dar, das nur noch elektronenmikroskopisch vergrößert werden muß. — Die Gefrierätzung wurde für die vergleichende Untersuchung in vivo und in vitro gezüchteter Viren entwickelt; ein weiteres Anwendungsgebiet blieb ihr in der ursprünglichen Form jedoch auf Grund verschiedener Mängel versagt.

Vom Autor (MOOR et al. 1961) sind auf präparationstechnischem Gebiet einige Verbesserungen erzielt worden, die im wesentlichen auf der Vereinigung eines Ultramikrotomes mit den Bedampfungs- und Gefriertrocknungseinrichtungen im selben Hochvakuumraum beruhen: Mit Hilfe des Ultramikrotomes, das mit einem kühlbaren Messer ausgestattet ist, können trotz der großen Sprödigkeit des Eises glatte Schnittflächen erzeugt werden. Das Schneiden unter Hochvakuum verhindert die Bereifung des Objektes und erlaubt dadurch, im Verein mit genauen Vakuum- und Temperaturmeßgeräten, eine hinreichende Beherrschung kleinster Sublimationsbeträge. Die Beschattung mit einem Platin-Kohle-Gemisch (MOOR 1959) verkürzt die Bedampfungszeit auf etwa 1 sec und reduziert damit die für das gefrorene Objekt gefährliche Wärmestrahlung auf ein Minimum.

Mit der niedrigen, eine Rekristallisation verunmöglichenden Objekttemperatur (−100° C) bleibt als einziges Artefaktproblem die Eiskristallbildung im Verlaufe des Einfrierens übrig. Wie die Untersuchungen von ASAHINA und AOKI (1958) und SAKAI (1960a) zeigen, kann auch hier eine Lösung gefunden werden: Auf dem Umwege über ein vorhergehendes, extracelluläres Gefrieren bei −30° C sind pflanzliche Gewebe, ohne die Lebensfähigkeit zu verlieren, in flüssigem Stickstoff oder Sauerstoff vitrifiziert worden. Die Annahme liegt nahe, daß bei dieser Gefriermethode die Kristallisation im Cytoplasma vollständig unterbunden ist.

Für die *Gefriertrocknung kleinster Teilchen* ist von WILLIAMS (1952) auf Grund der Angaben von WYKOFF (1946) eine sehr einfache Präparationsmethode entwickelt worden: Die Teilchen werden in Lösungen leichtflüchtiger Salze suspendiert und auf einen tiefgekühlten (−120° C), mit einer Collodiumfolie überzogenen Kupferblock aufgesprüht. Auf Grund der kleinen Tröpfchendimension und der großen Wärmeleitfähigkeit von Kupfer können auf diese Art Abkühlungsgeschwindigkeiten von ungefähr 10^5° C pro sec erzielt und eine vollständige Vitrifikation der Objekte erreicht werden (WILLIAMS 1953). Während einer kurzen Gefriertrocknung (10—20 min) bei −60° C wird das freie Wasser wegsublimiert; die Nachtrocknung bei +60° entfernt die Restfeuchtigkeit und die zurückgebliebenen Salze der Suspension. Nach einer Schwermetallbeschattung wird die Collodiumfolie mit den anhaftenden Teilchen vom Kupferblock abgestreift und auf Objektträgernetzchen übertragen.

Gegenüber der gewöhnlichen Eintrocknung aus wäßriger Suspension mit der unumgänglichen Abplattung und Verzerrung der Objekte hat diese Methode den großen Vorteil, daß die natürlichen Formen erhalten bleiben (z. B. bei Viren, Phagen, Bakterien oder Zellfragmenten). Die einzige präparative Schwierigkeit wird durch die Empfindlichkeit der Objekte auf reines Wasser verursacht; die Zahl der osmotisch wirksamen Substanzen, die zugleich metabolisch inaktiv sind und sich unter Hochvakuum bei biologisch tragbaren Temperaturen sublimieren lassen, ist nicht groß: WILLIAMS (1954) empfiehlt vor allem Ammoniumbenzoat in 2%iger Lösung. Der Anwendungsbereich der Methode wird durch das auf einige 1000 Å begrenzte Durchstrahlungsvermögen des Elektronenmikroskopes und die bei größerer Schichtdicke während der Bestrahlung auftretende Objektverbrennung auf Teilchengrößen unter 1 μ eingeschränkt.

D. Verwandte Methoden.

I. Gefrierkonservierung.

Um nicht während der Lagerungszeit an Kältemaschinen gebunden zu sein, ist aus der Gefrierkonservierung mit Hilfe der Vakuumsublimation die Gefriertrocknung entwickelt worden. Kann nun die Gefriertrocknung ihre Zwecke als Konservierungsmethode nicht erfüllen, wird zur Tiefkühlung zurückgegriffen: Wie bereits erwähnt wurde, geht bei einer Lyophilisation von Zellen und Geweben höherer Pflanzen und Tiere die Lebensfähigkeit in der Regel verloren. Die Untersuchungen einer großen Zahl von Autoren (vgl. SMITH 1954) haben gezeigt, daß sehr viele, durch eine Imbibition mit Kohlenhydratlösungen geschützte Objekte ohne wesentliche Schädigung auf -70 bis $-190°$ C abgekühlt werden können. Die Haltbarkeit der Objekte scheint bei diesen Temperaturen nahezu unbeschränkt zu sein. Unter den verwendeten Kohlenhydraten hat sich Glycerin in 15—30%iger Lösung als ein universelles Gefrierschutzmittel erwiesen (PARKES 1951; POLGE 1951; LUYET and GEHENIO 1952; SMITH 1952; LOVELOCK 1953b; ASAHINA 1958; REY 1959; CRAVEN 1960). An die Geschwindigkeit des Einfrierens und Auftauens werden keine besonderen Ansprüche gestellt; die Temperaturänderungen bewegen sich in der Größenordnung von ungefähr 1° C/min. Lagerungstemperaturen unter $-70°$ C sind notwendig, um sowohl den Stoffwechsel wie auch die Giftwirkung des Imbibitionsmittels zu blockieren.

II. Gefriersubstitution.

Die Gefriersubstitution wurde von SIMPSON (1941) als Alternative zur Gefriertrocknung vorgeschlagen: Zur Entfernung des Eises aus den gefrorenen Objekten wird an Stelle der Vakuumsublimation ein Lösungsmittel (Zellosolve, Alkohol, Aceton u. a.) verwendet. Die Einfriertechnik bleibt dieselbe wie in der Gefriertrocknung; die Dehydratisierung beansprucht bei -60 bis $-70°$ C ungefähr eine Woche. Nach der Erwärmung auf Zimmertemperatur werden die Objekte in Paraffin eingebettet und geschnitten. Für die Nachbehandlungen sind dieselben Vorsichtsmaßnahmen wie bei gefriergetrockneten Objekten zu beobachten. Ideale Verhältnisse sind infolge der Kombination der Nachteile der Gefriertrocknung mit denjenigen einer schlechten Fixation durch die Lösungsmittel weder in bezug auf die Strukturerhaltung noch auf die Konservierung des natürlichen Chemismus zu erwarten.

Sehr viel bessere Präparationsbedingungen werden durch die Zusetzung von guten Fixativen (OsO_4, HgCl, Pikrinsäure u. a.) zu den Lösungsmitteln geschaffen (FEDER und SIDMAN 1958): Die auf die Entfernung des Eises unmittelbar folgende Stabilisation der Objektstrukturen in einer wasserarmen Umgebung bewirkt eine ausgezeichnete, chemische Fixation des Objektes. Die Vorteile dieser verbesserten Gefriersubstitution können, wie FERNANDEZ-MORAN (1960) gezeigt hat, auch auf elektronenmikroskopischem Gebiet ausgewertet werden.

III. Gefrierschnitte.

Die Gefrierschnitt-Technik wird hauptsächlich bei Objekten tierischer und menschlicher Herkunft angewendet. Sie hat — insbesondere für den Mediziner — den Vorteil einer gegenüber der gewöhnlichen Schneidetechnik um ein Vielfaches verkürzten Präparationsdauer: 1. fällt die Einbettung durch das Schneiden auf einem Gefriermikrotom weg, 2. wird die Fixationszeit der auf einen Glasobjektträger aufgezogenen, aufgetauten Schnitte durch die Verkleinerung der Schichtdicke erheblich reduziert und 3. kann das Färben und Einschließen der Präparate direkt an die Fixation angeschlossen werden (GATENBY und BEAMS 1950).

Unbehandelte Gefrierschnitte werden auch für cytochemische Untersuchungen herangezogen; die Qualität der Fixation läßt sich jedoch in keiner Weise mit derjenigen gefriergetrockneter Präparate vergleichen, weil die für das Schneiden erforderliche, relativ hohe Temperatur (—10 bis —30° C) eine gute Gefrierfixierung verunmöglicht. Dieser Umstand fällt bei den viel wasserhaltigeren, pflanzlichen Objekten dermaßen ins Gewicht, daß in der Regel auf eine Anwendung der Gefrierschnitt-Technik verzichtet wird.

Literaturverzeichnis.

ALTMANN, R.: Die Elementarorganismen und ihre Beziehung zu den Zellen. Leipzig: Veit 1890. — ASAHINA, E.: Symp. Soc. Cell. Chem. **8**, 1 (1958). — ASAHINA, E., and K. AOKI: Low Temp. Sci. B **16**, 55 (1958).

BAKER, J. R.: Cytological Technique. London: Methuen 1951. — BARRETT, A. S. D., and L. G. BECKETT: Aspects of the design of freeze-drying apparatus, in Freezing and drying. Oxford: Church Army Press 1951. — BECKETT, L. G.: The effects of residual moisture in frozen-dried materials, and its measurement, in Biological applications of freezing and drying. New York: Acad. Press 1954. — BECQUEREL, P.: C. R. Acad. Sci. **181**, 805 (1925); — C. R. Acad. Sci. **194**, 1974 und 2158 (1932). — BEHRENS, M., u. M. TAUBERT: Z. physiol. Chem. **291**, 213 (1953). — BELL, L. G. E.: Some applications of freezing and drying in cytochemistry, in Freezing and drying. Oxford: Church Army Press 1951; — Freeze-drying, in Physical techniques in biological research. New York: Acad. Press 1956. — BERSIN, TH.: Kurzes Lehrbuch der Enzymologie. Leipzig: Akad. Verl. Ges. 1939. — BLACKMAN, M., and N. D. LINSGARTEN: Advances in Physics **7**, 189 (1958). — BONADREFF, W.: Gerontologia **1**, 222 (1957). — BRETSCHNEIDER, L. H., and P. F. ELBERS: Proceedings C **55**, 675 (1952).

CALLOW, E. H.: Proc. Roy. Soc. A **108**, 307 (1925). — COSSLETT, V. E.: Practical electron microscopy. London: Butterworth 1951. — COWAN, S. T.: Infection and drying techniques, in Freezing and drying. Oxford: Church Army Press 1951. — CRAVEN, C.: Exptl. Cell Res. **19**, 164 (1960).

DANIELLI, J. F.: Fixation procedures, in Cytochemistry. New York: Wiley 1953; — General cytochemical methods, Vol. 1. New York: Acad. Press 1958.

FEDER, N., and R. L. SIDMAN: J. Biophys. Biochem. Cytol. **4**, 593 (1958). — FENNELL, D. I.: Bot. Rev. **26**, 79 (1960). — FERNANDEZ-MORAN, H.: Ann. N. Y. Acad. Sci. **85**, 689 (1960). — FIECHTER, A.: Schweiz. Z. allg. Path. **22**, 547 (1959). — FISCHER, K.: Angew. Chemie **48**, 394 (1935). — FLOSDORF, E. W.: Freeze-drying. New York: Reinhold 1949; — Quick-freezing and the freezing-drying process, in Biophysical research methods. New York: Interscience 1950; — Large-scale freeze-drying as currently carried out in the western hemisphere, in Freezing and drying. Oxford: Church Army Press 1951; — The development of freeze-drying, in Biological applications of freezing and drying. New York: Acad. Press 1954. — FLOSDORF, E. W., and S. MUDD: J. Immunol. **29**, 389 (1935). — FREDRICSSON, B.: Verh. anat. Ges. (Jena) **53**, 122 (1957). — FRY, R. M.: The influence of the suspending fluid on the survival of bacteria after drying, in Freezing and drying. Oxford: Church Army Press 1951; — The preservation of bacteria, in Biological applications of freezing and drying. New York: Acad. Press 1954. — FRY, R. M., and R. J. GREAVES: J. Hyg. (Lond.) **49**, 220 (1951).

GATENBY, J. B., and H. W. BEAMS: Bolles Lee's microtomist's vademecum. 11. Aufl. London: Churchill 1950. — GENEVES, L.: Rev. Cyt. Biol. Végétales **16**, 1 (1955). — GERSH, I.: Some recent experiments in fixation by freezing and drying. Oxford: Church Army Press 1951; — Fixation by freezing and drying, in The cell. New York: Acad. Press 1959. — GERSH, I., and J. L. STEPHENSON: Freezing and drying of tissues for morphological and histochemical studies, in Biological applications of freezing and drying. New York: Acad. Press 1954. — GIBBONS, J. R.: Nature (Lond.) **184**, Suppl. 6, 375 (1959). — GLICK, D.: Freezing-drying preparation of tissue, in Techniques of histo- and cytochemistry. New York: Interscience 1949. — GOODSPEED, T. H., and F. M. UBER: Univ. Calif. Publ. Bot. **18**, 23 (1935). — GREAVES, R. I. N.: Nature (Lond.) **153**, 485 (1944); — Theoretical aspects of drying by vacuum sublimation, in Biological applications of freezing and drying. New York: Acad. Press 1954; — Ann. N. Y. Acad. Sci. **85**, 723 (1960) (a); — Some factors which influence the stability of freeze-dried cultures, in Recent research in freezing and drying. Springfield, Ill.: Thomas 1960 (b).

HAMMER, B. W.: J. med. Res. **24**, 527 (1911). — HARRIS, R. J. C.: Freezing and drying. Oxford: Church Army Press 1951; — Biological applications of freezing and drying. New York: Acad. Press 1954 (a); — The preservation of viruses, in Biological applications of freezing and drying. New York: Acad. Press 1954 (b). — HETHERINGTON, D. C.: The preservation of media for the culture of bacteria and tissue, in Biological applications of freezing and drying. New York: Acad. Press 1954. — HOLLINGS, M., and R. A. LELLIOT: Plant Path. 9, 63 (1960). — HOPKINS, A. L.: Ann. N. Y. Acad. Sci. 85, 714 (1960).

ILJIN, W. S.: Protoplasma 20, 105 (1933); — Bull. Assoc. Russe Rech. Sci. (Prague) 1 (IV), 1 (1934) (a); — Bull. Assoc. Russe Rech. Sci. (Prague) 1, (VI), 135 (1934) (b); — Bull. Assoc. Russe Rech. Sci. (Prague) 3, 33 (1935).

JENSEN, W. A.: Exptl. Cell Res. 7, 572 (1954). — JENSEN, W., and L. KAVALJIAN: Stain Techn. 32, 33 (1957).

KESSLER, W., u. W. RUHLAND: Planta 28, 159 (1938). — KULENKAMPFF, H.: Z. wiss. Mikroskopie 62, 427 (1955); — Z. wiss. Mikroskopie 63, 1 (1956).

LARCHER, W.: Mikroskopie 8, 299 (1953); — Planta 44, 607 (1954). — LARCHER, W., u. H. EGGARTER: Protoplasma 51, 595 (1960). — LEVER, J. D.: Nature (Lond.) 186, 810 (1960). — LEVITT, J.: Plant Physiol. 32, 237 (1957) (a); — Science 125, 194 (1957) (b); — Plant Physiol. 34, 401 (1959) (a); — Plant Physiol. 34, 674 (1959) (b); — Ann. N. Y. Acad. Sci. 85, 570 (1960). — LOVELOCK, J. E.: Biochim. Biophys. Acta 10, 414 (1953) (a); — Biochim. Biophys. Acta 11, 28 (1953) (b). — LUYET, B. J.: Science 85, 106 (1937); — Bull. Murithienne 56, 96 (1939) (a); — J. Phys. Chemistry 43, 881 (1939) (b); — Survival of cells, tissues and organisms after ultrarapid freezing, in Freezing and drying. Oxford: Church Army Press 1951; — On the mechanism of growth of ice crystals in aqueous solutions and on the effect of rapid cooling in hindering crystallisation, in Recent research in freezing and drying. Springfield, Ill.: Thomas 1960. — LUYET, B. J., and P. M. GEHENIO: Biodyn. 42, 1 (1938); — Science 116, 526 (1952. — LUYET, B. J., and S. M. GRELL: Biodyn. 23, 1 (1936). — LUYET, B. J., and E. L. HODAPP: Proc. Soc. Biol. Med. 39, 433 (1938). — LUYET, B. J., and G. THOENNES: Science 88, 284 (1938).

MALMSTRÖM, B. G.: Exptl. Cell. Res. 2, 688 (1951). — MATZ, G.: Vak. Techn. 3, 109 (1955) (a); — Vak. Techn. 3, 115 (1955) (b). — MAZUR, P.: Ann. N. Y. Acad. Sci. 85, 610 (1960). — MERYMAN, H. T.: Science 124, 515 (1956); — Science 130, 628 (1959); — Ann. N. Y. Acad. Sci. 85, 503 (1960) (a); — Ann. N. Y. Acad. Sci. 85, 630 (1960) (b); — Ann. N. Y. Acad. Sci. 85, 729 (1960) (c). — MOBERGER, G., B. LINDSTRÖM and L. ANDERSON: Exptl. Cell Res. 6, 228 (1954). — MODLIBOWSKA, J., and W. S. ROGERS: J. exp. Botany 6, 384 (1955). — MOOR, H.: J. Ultrastr. Res. 2, 393 (1959); — Z. Schweiz. Forstv. 30 (Beiheft), 211 (1960). — MOOR, H., K. MÜHLETHALER, H. WALDNER and A. FREY-WYSSLING: J. Biophys. Biochem. Cytol. 10, 1 (1961). — MÜHLETHALER, K.: 4. Intern. Kongr. Elektr. Mikrosk., Berlin 1958. Bd. 2, S. 32. Berlin: Springer 1960. — MÜLLER, H. R.: J. Ultrastr. Res. 1, 109 (1957).

NEUMANN, K. H.: Grundriß der Gefriertrocknung. 1. Auflage (1952), 2. Auflage (1955). Berlin: Musterschmidt 1955 (a); — Vak. Techn. 3, 130 (1955) (b); — Anwendungen der Gefriertrocknung für histochemische Untersuchungen, in Handbuch der Histochemie. Stuttgart: Fischer 1958. — NORD, F. F.: Zum Mechanismus der Enzymwirkung unter besonderer Berücksichtigung der Kryolyse. Stuttgart: Enke 1933.

PARKER, J., and A. SAKAI: Nature (Lond.) 187, 1133 (1960). — PARKES, A. S.: Preservation of spermatozoa, red blood-cells and endocrine tissue at low temperatures, in Freezing and drying. Oxford: Church Army Press 1951. — PARKES, A. S., and A. U. SMITH: Recent research in freezing and drying. Springfield, Ill.: Thomas 1960. — PFEFFER, W.: Handbuch der Pflanzenphysiologie, Bd. 2. Leipzig: Engelmann 1904. — PIRIE, N. W.: Freeze-drying or drying by sublimation, in Moderne Methoden der Pflanzenanalyse, Bd. 1. Berlin: Springer 1956. — PISEK, A.: Naturwiss. 39, 73 (1952); — Gartenwiss. 23, 54 (1958). — POLGE, C.: Nature (Lond.) 167, 949 (1951). — PROOM, H.: Some of the factors affecting the viability of bacterial cultures subjected to the freeze-drying process, in Freezing and drying. Oxford: Church Army Press 1951.

RABINOWITCH, T. H.: Preparation, preservation and activation of chloroplast material, in Photosynthesis, Vol. 2 (Part 2). New York: Interscience 1956. — REIMER, L.: Elektronenmikroskopische Untersuchungs- und Präparationsmethoden. Berlin: Springer 1959. — REY, L.: Conservation de la vie par le froid. Paris: Hermann 1959; — Study of the freezing and drying of tissues at very low temperatures, in Recent research in freezing and drying. Springfield, Ill.: Thomas 1960. — ROWE, T. W. G.: Ann. N. Y. Sci. 85, 641 (1960).

SAKAI, A.: Low Temp. Sci. B 16, 23 (1958) (a); — Low Temp. Sci. B 16, 41 (1958) (b); — Nature (Lond.) 185, 393 (1960) (a); — Nature (Lond.) 185, 698 (1960) (b). — SCOTT, W. J.: A mechanism causing death during storage of dried microorganisms, in Recent research in freezing and drying. Springfield, Ill.: Thomas 1960. — SENO, S., K. YOSHIZAWA, K. UTSUMI and T. MAKI: Symp. Soc. Cell. Chem. 8, 29 (1958). — SIMPSON, W. L.: Anat. Record 80, 173 und 329 (1941). — SINGER, J. H.: The freeze-drying of antibiotics, in Biological applications of freezing and drying. New York: Acad. Press 1954. — SJÖSTRAND, F.: Freeze-drying of tissues for cell analysis by light and electron microscopy, in Freezing and drying. Oxford: Church Army Press 1951. — SJÖSTRAND, F., and R. F. BAKER: J. Ultrastr. Res. 1, 239 (1958). — SMITH, A. U.: Exptl. Cell Res. 3, 574 (1952); — Effects of low temperatures on living cells and tissues, in Biological applications of freezing and drying. New York: Acad. Press 1954. — STÄUBLI, W.: Extrait des comptes rendus des séances de l'Académie des Sciences. Paris, 8. févr. 1960. — STEERE, R. L.: J. Biophys. Biochem. Cytol. **3**, 45 (1957). — STEPHENSON, J. L.: Fundamental physical problems in the freezing and drying of biological materials, in Recent research in freezing and drying. Springfield, Ill.: Thomas 1960. — SWIFT, H.: Cytochemical techniques for nucleic acids, in The nucleic acids. New York: Acad. Press 1955. — SYLVEN, B.: Effects of freezing and drying compared with common fixation procedures, in Freezing and drying. Oxford: Church Army Press 1951.

TAYLOR, A. C.: Ann. Y. N. Acad. Sci. **85**, 595 (1960). — TERUMOTO, J.: Low Temp. Sci. B **16**, 7 (1958). — TREFFENBERG, L.: Ark. Zool. **4**, 295 (1952).

ULLRICH, H.: Kolloid-Z. **96**, 348 (1941); — Protoplasma **38**, 165 (1943); — Klin.Wschr. **23**, 189 (1944). — ULLRICH, H., u. U. HEBER: Planta **48**, 724 (1957).

WATSON, M. L.: J. Biophys. Biochem. Cytol. **4**, 475 (1958) (a); — J. Biophys. Biochem. Cytol. **4**, 727 (1958) (b). — WEISS, L., and J. A. ARMSTRONG: J. Biophys. Biochem. Cytol. **7**, 673 (1960). — WHATLEY, F. R.: Cytochemical methods, in Moderne Methoden der Pflanzenanalyse, Bd. 1. Berlin: Springer 1956. — WICHTERLE, O., P. BARTL and M. ROSENBERG: Nature (Lond.) **186**, 494 (1960). — WILLIAMS, R. C.: Biochim. Biophys. Acta **9**, 237 (1952); — Exptl. Cell Res. **4**, 188 (1953); — The application of freeze-drying to electron microscopy, in Biological application of freezing and drying. New York: Acad. Press 1954.— WYCKOFF, R. W. G.: Science **104**, 36 (1946).

ZEIGER, K.: Physikochemische Grundlagen der histologischen Methodik. Dresden: Steinkopff 1938.

Vapour Phase Chromatography.

By

Stanley P. Burg.

With 7 Figures.

The demand for a highly efficient method for separating and measuring biological substances was an important incentive leading to the development of gas-liquid chromatography by MARTIN and coworkers (MARTIN and SYNGE, 1941; JAMES and MARTIN, 1952a), but subsequent refinement of the art has been largely the responsibility of engineers, physicists and chemists, especially those in the petrochemical industry. Since 1952, caught up in the dazzling emergence of a new and powerful technique, biologists have been slow to appreciate fully the advantages which have become available to them, and reports of application of gas chromatography to biological and biochemical problems have accumulated in a haphazard fashion. This review is intended to bring together the scattered information in this area, and it is particularly directed toward the plant analyst who may wish to apply gas chromatography to his special problem, or who would like to know if it can be applied.

Where it may clarify a discussion recourse is made to theory, the principles which provide the basis for the various types of detection systems are discussed in detail to allow the reader to better evaluate the relative merits of the instruments, and certain other aspects of gas chromatography necessarily common to any review on the subject are also included. It is the author's view that he neither could nor should give these areas the depth of coverage which can be found elsewhere. Rather, these brief discussions are supplemented with appropriate references in the hopes that the interested reader will consult the proper sources. For this purpose the books on gas chromatography by PHILLIPS (1956), PECSOK (1959a), BAYER (1959) and KEULEMANS (1957, 1959) are recommended, and the reports on gas chromatography symposia edited by DESTY (1957, 1958) and COATES, NOEBELS and FAGERSON (1958) will be found valuable.

A. Theoretical Approach.

I. Chromatography in General.

Chromatography always involves two phases in intimate contact with each other: a mobile phase and a stationary phase through which the mobile phase flows or percolates. Since the stationary phase may be either a solid (S) or a liquid (L) and the mobile phase a liquid or a gas (G), four types of chromatography may be envisioned. Denoting the mobile phase first, these are gas-liquid chromatography (GLC), gas-solid chromatography (GSC), liquid-solid chromatography (LSC), and liquid-liquid chromatography(LLC).

1. Nature of the Stationary Phase: Adsorption vs. Partition.

A substance, the solute, will under ideal conditions traverse the chromatographic system only when it is present in the moving phase, and to the extent that the solute has an affinity for the stationary phase the movement of the solute will be impeded and its rate of advance decreased. With the notable exception of paper chromatography, where the stationary phase is either the cellulose fibers of the paper or more generally the water absorbed on the surface or in the microcapillaries of these fibers, in almost all other forms a finely powdered material packed in a long column serves this purpose. The fine particles may themselves constitute the functional surface, or very frequently a thin film of liquid adsorbed on the surface and in the pores of the solid brings about the desired effect. The distinction between these two situations is very important since each provides the basis for a unique type of chromatography. In the first instance the effective phase is a solid towards which certain solutes have an affinity because they are preferentially *adsorbed* on the surface. In the latter case the effective area is a liquid film adsorbed on a solid matrix, and the relative affinity of various solutes for this liquid will depend upon how they *partition* between it and the mobile phase. Although the distinction between *adsorption* and *partition* chromatography is usually definitive, the solid matrix used in the construction of a liquid phase may itself possess some adsorption ability which may supplement the partition process to a greater or lesser extent. This may prove to be advantageous, as in the case of paper chromatography, but generally it will be a distinct disadvantage in gas-liquid chromatography.

2. Mobile Gas Phase: Elution, Displacement and Frontal Analyses.

The factor which makes vapour-phase chromatography unique is not the nature of the stationary phase, for as we have seen this may be either liquid or solid, but rather it is the use of a gaseous mobile phase. Normally the gas is chemically inert, or nearly so, and it has very little affinity for the stationary phase; thus the elution of each solute is accomplished solely by the functioning of the partition process and the passive movement of the mobile gas phase. This is known as *elution* chromatography and today it is favoured to the exclusion of almost all other modifications. If the effluent of the chromatographic column is monitored by some measuring device which is insensitive to solvent molecules, a continuous record of changes in the solute concentration at the exit of the column is obtained (Fig. 1). Initially, with the mobile phase in motion a zero baseline indicates lack of solute in the effluent. Following the introduction at point A of solute mixture, there is a sharp inflection at B denoting the emergence of all unretarded components of the sample. At some later definitive time a symmetrical curve with maximum at point C describes the emergence of a component with a positive affinity for the stationary phase. When this substance has completed its passage through the column the zero base-line is reestablished, and subsequent peaks (not shown) will be traced as solute components with progressively greater affinity for the stationary phase are finally eluted.

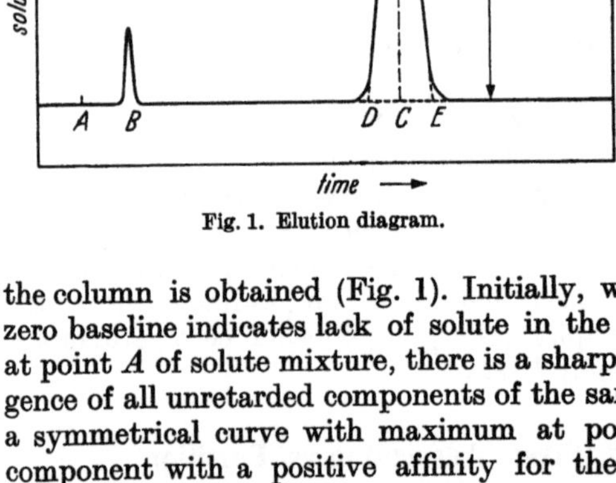

Fig. 1. Elution diagram.

On rare occasions, particularly in the case of adsorption chromatography, it is advantageous after the solute has been added to admix continuously into the gaseous solvent a constant proportion of another vapour which has a higher affinity for the stationary phase than does any member of the solute mixture. The advantage obtained is that the surface of the adsorbent becomes satiated with the tightly bound substance, effectively displacing the various components of solute which emerge as successive bands or steps just ahead of the displacer. This procedure, known as *displacement analysis* is illustrated in Fig. 2. At F the displacer has been introduced into the gas stream and shortly thereafter in the order of increasing adsorbency, fractions a, b, and c emerge, followed by the displacer. In yet another modification the sample may be continuously added to the column mixed into the solvent as a constant contaminant. The least 'sorbed, most rapidly traveling component will emerge first and be enriched at the front of the chromatogram; hence the name *frontal analysis*. In Fig. 3 substances having affinities for the stationary phase in the order $c > b > a$ have been introduced at time A as a continuous stream. The front of the emergent band contains pure "a" mixed with solvent; this is followed by a zone in which the concentration of "a" has decreased approaching its value in the original mixture, while that of "b" is enriched; finally the original mixture breaks through. A fairly accurate, although somewhat cumbersome quantitative evaluation can be made from the heights and lengths of the steps obtained in frontal and displacement analysis, and a good account of the methodology may be found in papers by TISELIUS (1943) and CLAESSON (1946); brief summaries of these are included in the books by KEULEMANS (1959) and PHILLIPS (1956).

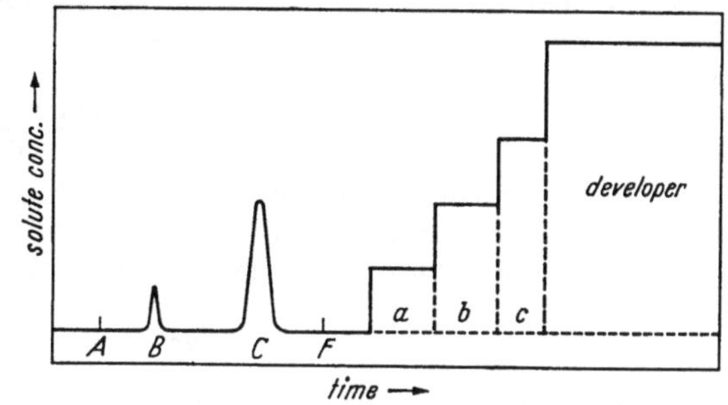

Fig. 2. Displacement analysis.

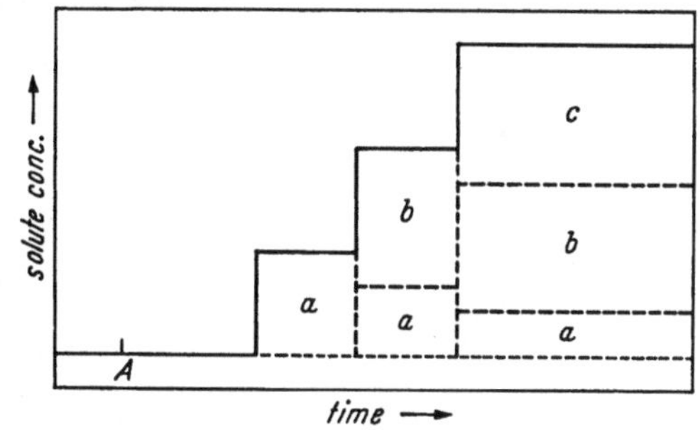

Fig. 3. Frontal analysis.

II. Types of Theories.

Although it is possible to use the technique of gas chromatography with no knowledge of the theory underlying the observed behaviour of gases on a chromatography column, there are distinct limitations to a strictly empirical approach. As is the case with many techniques, theory becomes particularly useful when conventional approaches do not produce the anticipated results, or when unusual problems are attempted. However, once an appropriate system is established, the operation of the assay is completely routine and can be caried out by untrained personnel.

There is no single unifying theory of chromatography because every mathematical approach to the problem requires certain simplifying assumptions which approach validity only in specific, well defined cases. Theories fall into one of four categories depending upon (1) the nature of the distribution isotherm and (2) how closely conditions approximate a hypothetical ideal.

1. Linear vs. Non-Linear Distribution Isotherms.

The relative affinity of a solute for the stationary and mobile phases can be described by a distribution coefficient (K), which defines the situation at complete equilibrium.

$$K = \frac{\text{wt. of solute/ml stationary phase}}{\text{wt. of solute/ml mobile phase}}$$

The distribution coefficient will usually be a function of temperature, decreasing with a rise in temperature, and it also may be dependent upon the solute concentration. A plot of the concentration of the solute in the stationary phase vs the

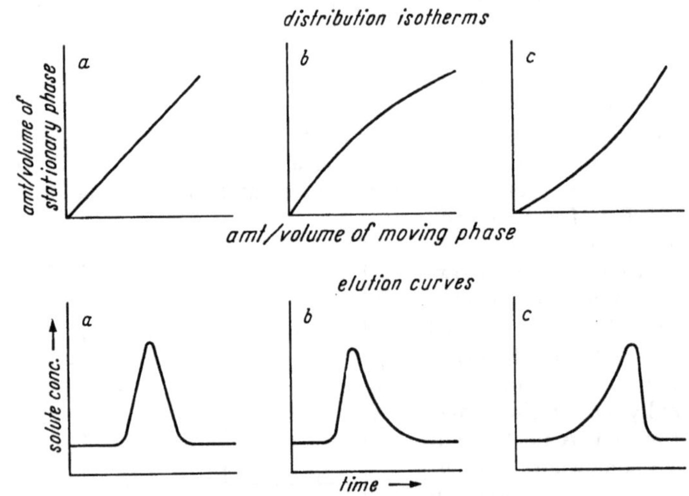

Fig. 4. Distribution isotherms.

concentration or pressure of the solute in the moving phase at a specific temperature, known as a distribution isotherm, will yield a straight line only if the distribution coefficient is independent of solute concentration. The linear isotherm is frequently encountered in gas-liquid chromatography, but it seldom obtains in gas-solid chromatography where it is more commonly observed that as the concentration of solute is increased, a greater proportion of the total solute is found in the gas phase. This type of behavior is shown in Fig. 4b, and may be explained by the fact that as the surface of the adsorbent becomes covered with solute there remain proportionately fewer adsorption sites per unit weight of solid. When this competitive "LANGMUIR" type of isotherm operates, a substance will travel most rapidly in that zone where its concentration is highest. Referring to the elution curve based on a linear isotherm, shown in Fig. 1, the zone of highest concentration is in the vicinity of the peak maximum C and this area will travel faster than either the front or rear of the solute band. The elution curve which results from this behavior is shown in Fig. 4b, where the rapid movement of the central zone has resulted in a sharp front followed by a more slowly moving rear or "tail." Peak asymmetry is undesirable because it leads to overlapping sections on the chromatogram, which is one of the major reasons why GSC is less applicable than is GLC. Another type of non-linear isotherm in which molecules cooperate, leads to an asymmetric peak with diffuse front and sharp rear, as shown in Fig. 4c. In all, five types of isotherms are possible, and each has been observed to operate in GSC, where it produces a predictable type of elution curve (GREGG and STOCK, 1958; PHILLIPS, 1949). The general conclusion to be drawn is that theories describing GLC can assume linear isotherms, whereas those concerned with GSC must postulate non-linear isotherms.

2. Ideal vs. Non-Ideal Chromatography.

In the simplest treatment of chromatography, an instantaneous, reversible equilibrium of solute between the stationary and mobile phases is postulated. Such an approach does not take into account factors which might impede equilibrium, diffusion phenomena occurring in both phases, inconstancy of flow rate, and in general many variables which can complicate the ideal situation. In a few specific types of chromatography some of these assumptions are adequately fulfilled; thus, for instance, in LSC and LLC the flow of solvent may be sufficiently slow to allow for almost total equilibration at all times. However, these various assumptions usually do not pertain to actual experience, and certainly they are not valid in either GSC or GLC, with the result that only the alternative non-ideal approach need be considered. The treatment of GSC is more complicated than that of GLC because the former is based on a non-linear isotherm, and this complication does not clarify an understanding of the basic principles. Therefore the scope of this discussion will be limited to the simpler case postulating linear non-ideal conditions[1], from whence it is possible to set out in one of two directions known alternatively as plate theory and rate theory.

III. Plate Theory.

Although the continual motion of the solvent through the stationary phase precludes the establishment of a complete equilibrium, some finite length of column can be defined in which the equivalent of one equilibration is obtained. The vapour leaving such a section will have that composition which would be in equilibrium with the average concentration of solute in the liquid phase of the section. Using terminology borrowed from distillation and countercurrent distribution, where one complete separation stage is called a "plate," in gas chromatography the length of column which effects a separation equivalent to one stage is defined as the height equivalent to a theoretical plate (HETP). In the plate concept, as first elaborated by MARTIN and SYNGE (1941) the history of a solute sample is described from the moment it is introduced into the first plate until it has passed intermittently through a definitive number of theoretical plates. An alternative approach using the theoretical plate concept is that of GLUECKAUF (1955a) in which he treats a continuous instead of interrupted flow of solvent. The main conclusion which can be derived from both of these formulations is that band broadening is related to the square root of the number of theoretical plates. Thus

$$n = (4\, t_R/w)^2 \qquad \text{(I)}$$

[1] The four possibilities and appropriate sources describing them are: (A) Linear-ideal: WILSON, 1940. (B) Linear non-ideal: *Plate theory*: MARTIN and SYNGE, 1941; MAYER and TOMPKINS, 1947; GLUECKAUF, 1955a. *Rate theory:* LAPIDUS and AMUNDSON, 1952; GLUECKAUF, 1955b; GIDDINGS and EYRING, 1955; DEEMTER, ZUIDERWEG and KLINKENBERG, 1956. (C) Non-linear ideal: WILSON, 1940; DEVAULT, 1943. (D) Non-linear non-ideal: KLINKENBERG and SJENITZER, 1956.

where n is the number of theoretical plates, w the width of the band, (DE in Fig. 1, where $C'D$ and $C'E$ are tangents extended to the base line), and t_R the time between sample insertion and the peak maximum (the interval AC in Fig. 1). This equation predicts that the elution curve for each successive component will tend to be broader and lower than its predecessor. If the column length (L) is known, then the height equivalent to one theoretical plate can be obtained from the expression

$$\mathrm{HETP} = L/n\,. \tag{II}$$

1. Calculation of the Number of Theoretical Plates.

Although for a specific substance t_R and w can be determined with fair precision from the elution curve, this information can not always be applied without reservation to the calculation from Equation I of the number of theoretical plates. One reason for this is that the amount of sample which frequently is used in an analysis will hardly be contained initially in the first plate as the theory assumes. The process may then resemble several overlapping elution analyses originating from adjacent plates near the column inlet, which will result in a widening of the band. This error can be circumvented by plotting HETP vs sample size, in which case a straight line intersecting the axis at zero sample size is often obtained (KEULEMANS, 1959, p. 124). Alternatively, a very small sample will provide a close approximation. A further complication arises because at the moment the sample is introduced the solute concentration at the inlet may be so high that the column initially is operating in a non-linear section of the distribution isotherm. Thus the elution curve may show sufficient asymmetry to displace the peak maximum to a position which does not allow accurate use of the term t_R in Equation I. Normally t_R is "the retention time of the component referred to the peak maximum," but it is also possible to define a final retention time (t_{RF}) equal to the interval AE in Fig. 1, and an initial retention time (t_{RI}) equal to AD in Fig. 1. The length of time required to elute the *entire* sample from the first plate will be constant regardless of the initial concentration in the plate, provided that the isotherm has either remained linear or temporarily converted to the LANGMUIR type (c.f. Fig. 4a and b). Since t_{RF} is a measure of the time taken to just elute the first plate, it can be applied more accurately in this case than t_R, and the number of plates can be calculated from the expression:

$$n = \left[\frac{4\,t_{RF}}{w+2}\right]^2 \tag{III}$$

Conversely, in those few cases where the distribution isotherm has been affected in a manner which causes molecular cooperation (c.f. Fig. 4c), the initial retention time has been found to be the more accurate measure by DIJKSTRA, KEPPLER and SCHOLS (1955).

$$n = \left[\frac{4\,t_{RI}}{w-2}\right]^2 \tag{IV}$$

2. Calculation of the Distribution Coefficient.

By multiplying the retention time (t_R) by the rate of carrier gas flow, the volume of mobile phase expended in eluting a component to its peak maximum can be calculated. This volume is known as the retention volume (V_R) referred to the peak maximum, and plate theory predicts that it is derived in the following way:

$$V_R = (V_G + K V_L)\,n \tag{V}$$

Where V_G and V_L respectively are the volume of the gas and liquid phases in one theoretical plate, and K is the distribution coefficient. The time interval AB in Fig. 1 converted to a gas volume is known as the gas hold-up, and it can be seen that it is equal to nV_G in Equation V. The term $(V_R - nV_G)$, interval BC in Fig. 1, is the apparent retention volume (V'_R) referred to the peak maximum, and it can be read directly from an elution diagram. Since it equals nKV_L if the volume of liquid phase (nV_L) used in the preparation of the column is known[1], it is possible to calculate the distribution coefficient.

3. Evaluation of a Chromatographic Separation.

A mathematical treatment of the degree to which two adjacent peaks contaminate each other has been developed by GLUECKAUF (1955a), and it is applicable when the theoretical number of plates, relative amounts of the two components, and their distribution coefficients are known. The first and last values can be derived by the methods described above, whereas the area under a peak is an accurate measure of quantity even when there is a considerable overlap between two bands (PECSOK, 1959b). GLUECKAUF approached the problem of evaluating the success of a separation by assuming a cut between two overlapping bands made in such a manner that the fractional impurities present in each sector would be equal. On this basis he formulated the relationship between the number of plates (n), separation factor (α), and the degree of fractional band impurity (η) shown in Fig. 5. In this graph, α is the ratio of the distribution coefficients of the two components evaluated so that it is greater than unity, and m_1 and m_2 are the molar amounts or concentrations of the two components in the sample. The use of this graph is best illustrated by an example. Suppose that under a particular set of operating conditions it is necessary to evaluate the success of a separation between two adjacent components of a chromatogram. From the elution diagram it is learned that the substances are present in a 20:1 ratio, there are 100 theoretical plates in the column, and the distribution coefficients are 10 and 16 respectively. The separation factor, $\alpha = K_1/K_2$, is equal to 16/10 or 1.6, and on this line of the graph we find that the point corresponding to $n = 100$ lies on the abscissa at $(m_1^2 + m_2^2/2m_1m_2) = 10^{-2}$. Utilizing the values for the relative amounts of each substance: $m_1 = 20$ and $m_2 = 1$; then $m_1^2 + m_2^2/2m_1m_2 = 10$; hence η equals 10^{-3} (0.1%), which is the extent to which each fraction will contaminate the other. One interesting corollary which emerges from GLUECKAUF's treatment of a chromatographic separation is that under a particular set of operating conditions the largest fractional impurity will be realized when $m_1 = m_2$. This means that it should be possible to decrease the fractional impurity in each of two adjacent bands of a chromatogram by quantitatively balancing the original sample mixture disproportionately in favour of one component.

Theoretically, if the separation factor differs from unity it should always be possible to choose some length of column which will fractionate two components. The reason for this is that the number of theoretical plates is a function of the length of the column, and whereas the absolute distance between two peak maxima increases in direct proportion to the number of plates, band spreading

[1] The volume occupied by the liquid phase at a particular column temperature can be calculated from the expression

$$V_L = W_L/\varrho_L$$

where W_L is the weight of the liquid phase, and ϱ_L the density of the liquid phase at the column temperature.

is related to the square root of this parameter. However, in the case of difficult separations, increasing the length of the column is not always a practical solution because not only may the length of time required for a separation become increasingly long, but eventually factors come into play which cause each additional column increment to contribute a smaller number of effective theoretical plates. The reasons for this will be discussed in the next section, but it should suffice here

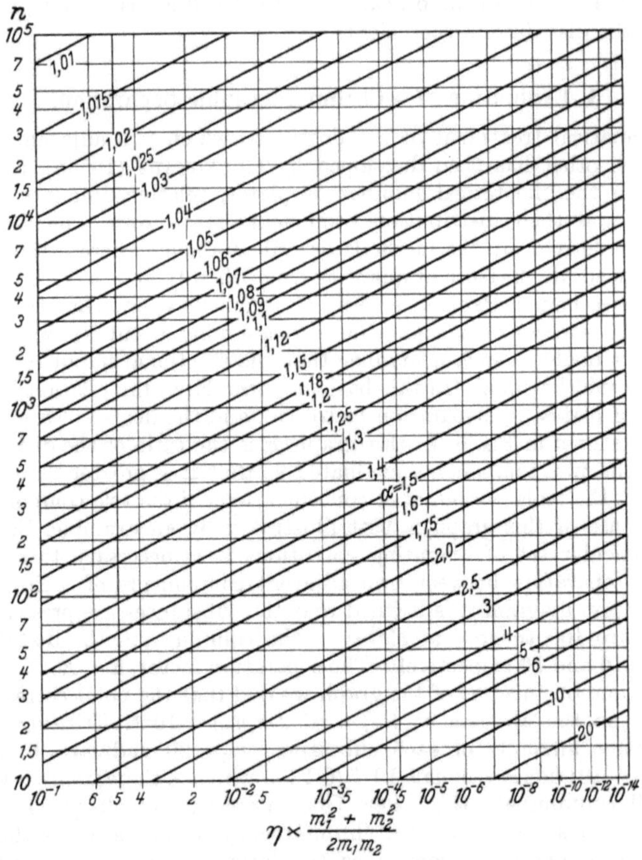

Fig. 5. Relation between fractional band impurity (η), number of plates (n), and separation factor (α) (after GLUECKAUF, 1955 b).

to point out that a packed column offers a considerable resistance to flow with the result that there is a physical limit to the length of column through which carrier gas can be forced at a reasonable inlet pressure. If a satisfactory separation can not be obtained with a practical length of column, two alternatives can be considered; a different liquid phase might yield a higher separation factor, or the column might be made more efficient by increasing the number of theoretical plates per unit length. The first of these possibilities requires a knowledge of the factors which enhance solubility in a liquid, and this is considered in detail in the section devoted to the choice of the proper liquid phase. The second alternative presupposes an understanding of the factors controlling column efficiency and a method for evaluating them. In this context, column efficiency means the ability of the column to produce narrow peaks which are adequately resolved even though their retention times are not very different; i. e. the separation factor is not large. An expression of the "efficacity with which a column design achieves a certain resolution with a given time of waiting" has been introduced by GOLAY (1958a). He defines the performance index (P.I.) as:

$$\text{P.I.} = \frac{\Delta t_x^4 \, t_a \, \Delta p}{t_x^3 \left(t_x - \frac{15}{16} t_a\right)} \tag{VI}$$

where Δt_x is the band width at mid height, t_x the passage time of a band from beginning to end of column, t_a the retention time of an unretarded component of the sample (air), and Δp the pressure drop across the column. If the retention times and band width are expressed in seconds, and the pressure in dynes/cm², the P.I. will have the dimension of a viscosity, poise. Lower values of P.I. indicate more efficient columns, and according to GOLAY the theoretical limit is 0.1 poise. CONDON (1959) has obtained a P.I. of 0.87 poise for *o*-xylene on a 175 foot capillary column coated with poly(propylene glycol), which is as least as low any any other value yet reported. In using the performance index it is essential to note that the expression assumes the column to be operating at its most efficient flow rate. A historically older and more commonly used expression of column efficiency is HETP (see Eq. II), in which case it can be seen that small values indicate highly efficient column performance. Although plate theory provides a method for evaluating HETP quantitatively, it does not tell us very much about the variables which affect column efficiency, and for this information it is necessary to consider rate theory.

IV. Rate Theory.

Rate theory is a kinetic interpretation of the factors which cause the chromatographic process to differ from an ideal situation and which consequently lead to band broadening. Two approaches have been followed: a statistical method by KLINKENBERG and SJENITZER (1956) and a balance sheet evaluating continuous transport phenomena into and out of a thin section of column, by VAN DEEMTER, ZUIDERWEG and KLINKENBERG (1956). Both yield the same result, which is an expression relating band broadening (HETP) to three variables; eddy diffusion, molecular diffusion, and resistance to mass transfer. The complete VAN DEEMTER equation:

$$\text{HETP} = 2\lambda d_p + \frac{2\gamma D_{gas}}{\mu} + \frac{8}{\pi^2}\frac{K'}{(1+K')}\frac{d_f^2}{D_{liq}}\mu \tag{VII}$$

may be simplified to the form:

$$\text{HETP} = \underset{\substack{\text{Eddy}\\\text{Diffusion}}}{A} + \underset{\substack{\text{Molecular}\\\text{Diffusion}}}{B/\mu} + \underset{\substack{\text{Resistance to}\\\text{Mass Transfer}}}{C\mu} \tag{VIII}$$

where λ is a quantity inversely related to the uniformity of particle size and spacing in the column; d_p is the average particle diameter; γ bears an inverse relationship to the tortuosity of the interstitial spaces of the column; D_{gas} and D_{liq} equal the diffusivity of solute in the gas and liquid phases respectively; K' is the ratio of the amount of solute in the liquid phase to the amount of solute in the gas phase; d_f is the average thickness of the liquid film; μ is the linear velocity of the carrier gas; A, B, and C replace the appropriate terms in the complete equation.

1. Eddy Diffusion (A).

Because of the presence of packing in the column, gas molecules will travel by irregular pathways having slightly different average lengths, cross sections, and directional vectors. This is analogous to having a certain amount of turbulence present, and bands can not remain as tight as they would in a more ideal situation such as a flow through a single thin tube. It is important to note that eddy diffusion is considered to be independent of flow rate. However, the eddy diffusion contribution can be minimized by reducing the particle diameter (d_p) or by increasing the uniformity of the particle size and packing (reducing λ). In practice it is observed that as the particle size is reduced beyond some limiting diameter it becomes so difficult to obtain a uniform packing that no improvement of efficiency results (Deemter et al., 1956). Thus Klinkenberg and Sjenitzer (1956) found that a more than ten fold increase in d_p from an average diameter of .055 mm to .6 mm, caused an eight fold decrease in λ, and 300/400 mesh celite was no more efficient as a support than was 50/80 mesh ground firebrick. An always important factor is the uniformity of particle size, as has been amply demonstrated by Boheman and Purnell (1958) who concluded that "highest efficiencies can only be obtained when there is a reasonable degree of uniformity of particle size." It is for this reason that it is important to float off fines and if necessary rescreen the particles to a narrow mesh range before they are used as packing.

2. Molecular Diffusion (B/μ).

Molecular diffusion of solute molecules enhances band broadening since the axial component of diffusion in the direction of carrier gas flow causes a forward spreading of the front, while the axial component in opposition to flow causes a retardation of the solute molecules at the rear. The diffusion process must occur in both the gas and liquid phases, but because diffusivity in a gas is 10^4 to 10^5 times greater than it is in a liquid, the gas phase is of paramount importance. Diffusivity in a gas depends upon the nature of the solute and in particular upon the molecular weight of the gas, and in this respect the high molecular weight gases such as nitrogen, argon and carbon dioxide are preferable to the lighter gases helium and hydrogen. Because the total amount of diffusion which occurs depends among other things upon how long the process operates, an inverse relationship with linear flow rate arises. Another variable which exerts an influence is γ, the tortuosity factor, which approaches a limiting value of unity as the particle size is increased and the interparticular spaces become enlarged.

3. Resistance to Mass Transfer ($C\mu$).

There exists a finite rate at which mass can be transferred between two phases, and this in turn decides how close an approximation to complete equilibrium can be established per unit time interval. When one phase is in motion the time alloted for equilibration varies according to the duration of contact between the effective layers. Thus the resistance to mass transfer contribution is proportional to μ, the gas flow rate. The degree of equilibration will depend to some extent upon how rapidly the molecules toward the centre of the phase can travel to the gas-liquid interface where exchange is occurring. In a typical column the effective thickness of the two phases is roughly equal, but since the rate of diffussion in liquid is ca. 10,000 times slower than in gas only the movement in liquid need be considered, and the thickness of the liquid film (d_f) and the rate of movement of solute in it (D_{liq}) appear as factors affecting mass transfer. An expression of the overall rate of movement also requires a consideration of the concentration "pressure" driving solute between the phases, and this is determined by K':

$$K' = K(F_{liq}/F_{gas}) \tag{IX}$$

where K is the distribution coefficient and F_{liq} and F_{gas} are the volume fractions of liquid and gas in the column. It might be thought that because d_f appears exponentially in the denominator of the VAN DEEMTER equation, the thickness of the liquid coating should exert a very profound influence on column efficiency. KEULEMANS and KWANTES (1957) have experimentally confirmed this prediction; however the effect is much less important than might be expected because K' is dependent upon the ratio of liquid phase/gas phase and hence upon d_f. Any decrease in the volume of the liquid phase (F_{liq}) will be accompanied by a corresponding increase in the volume of the gas phase (F_{gas}), and the combined changes adversely affect K'. One surprising conclusion to be drawn from the VAN DEEMTER equation is that because K' is a property of the solute, column efficiency must vary with the material being analyzed, the column being more efficient for substances having liberal affinity for the liquid phase.

4. Temperature, Flow Rate and Pressure.

Three other factors which have not been discussed, and whose effects are somewhat complex are temperature, flow rate and pressure. (1) **Temperature:** The VAN DEEMTER equation makes no specific mention of temperature, but this factor can affect several of the variables. For instance, D_{liq} and D_{gas} increase with temperature whereas K' decreases, and the net effect is that temperature induced changes in efficiency tend to cancel one another in a somewhat unpredictable manner (c.f. DEWET and PRETORIUS, 1958; DUFFIELD and ROGERS, 1960; HABGOOD and HARRIS, 1959). (2) **Flow:** A cursory examination of the simplified VAN DEEMTER equation (Eq. VIII) reveals that a maximum efficiency will be obtained only when the flow rate is adjusted to yield a minimum contribution from the sum of the mass transfer and diffusional terms. The optimal flow can be determined by plotting HETP vs. μ, in which case, as shown in Fig. 6, a minimum value of HETP will always be obtained. This curve is important in that it provides a graphic method for evaluating the terms of the VAN DEEMTER equation. An extrapolation of the straight part of the curve will intercept the HETP axis at the point corresponding to the eddy diffusion contribution, and the diffusional and mass transfer effects can then be calculated as shown in the Figure. (3) **Pressure:** Although in all forms of chromatography the movement of the mobile phase is contingent upon a pressure gradient, gas chromatography is a special case in that the mobile phase is compressible. The pressure gradient which exists from the column inlet to the column outlet creates a volume gradient in the reverse direction, and for this to be maintained gas molecules which enter the column need to have a lower linear velocity than those leaving it. As there is a single flow rate yielding maximum efficiency, it follows that a column operating with a substantial difference in inlet and outlet pressures can not have an optimal flow throughout

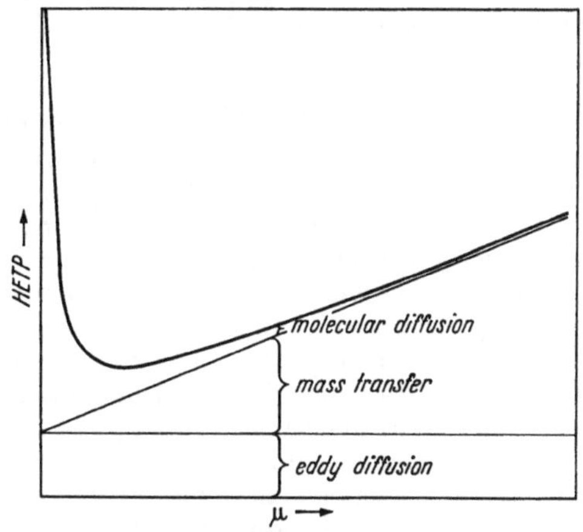

Fig. 6. Plot of HETP vs linear gas velocity.

its length. Fortunately the pressure falls off in a slow, fairly linear fashion in the first part of the column and rather abruptly toward the end, which makes it possible to operate the first 80 to 90 percent of the column at nearly top efficiency by judicious selection of the flow rate. If the outlet/inlet pressure ratio is less than 1.5 or 2 it is hardly necessary to make a correction for pressure in calculating retention volumes, but especially at higher ratios the tables in The Journal of Chromatography (Vol. 2, 1959, pp. D33—D45) or the formula of JAMES and MARTIN (1952b) can be used to convert the observed retention volume (V_R) to a corrected retention volume (V_{R^0}) under standard conditions of pressure:

$$V_R/V_{R^0} = \frac{2}{3}\,\frac{(P_i/P_o)^3 - 1}{(P_i/P_o)^2 - 1} \tag{X}$$

where P_i and P_o respectively are the column inlet and outlet pressures. At a pressure differential greater than approximately 2 between the inlet and outlet, Eq. X reduces to the form:

$$V_R/V_{R^0} = \frac{2\,P_i}{3\,P_o} \tag{XI}$$

It can be seen from this expression that increasing the absolute pressure at both the inlet and outlet should largely eliminate pressure effects. The cause of large inlet/outlet pressure ratios is the resistance to flow afforded by the column packing, which explains why it is often observed that there are less theoretical plates per unit length in longer columns. Columns most commonly are operated with the outlet exposed to the atmosphere, but SCOTT (1958) found that when he increased the average pressure in a long column by placing a restriction across the outlet, not only was the column efficiency improved but also there was created a linear relationship between the number of theoretical plates and the column length up to 50 feet. The belief that low column pressure leads to the specific acceleration of sparingly volatile compounds is in error because "the transport velocity of a solute... at a certain gas rate, is dependent upon the distribution coefficient with respect to the column liquid and is independent of the absolute gas pressure" (KEULEMANS, 1959, p. 151). Any decrease in observed retention time under low pressure conditions should be attributed to a general lowering of column efficiency as a result of an adverse inlet/outlet pressure ratio. Although low pressures increase the concentration of solute in the effluent, which may yield a sensitivity advantage if the detector responds to concentration rather than to absolute amount of solute, they also increase the amount of diffusion taking place in the gas phase, which may cause band broadening. For a very complete discussion of the factors controlling column efficiency KEULEMAN's book (1959) is strongly recommended.

V. Modifications and Additional Theories.

In the preceding sections plate theory has been discussed as a basis for the calculation of distribution coefficients and the number of theoretical plates in a column. After deriving numerical values for these two variables it is a relatively simple matter to evaluate, using terms which can be read directly from an elution diagram, both the success of a separation and the efficiency of the column which produces it. Then, with the insight which rate theory offers, the individual factors which cumulatively determine column efficiency can be examined and altered to produce a predictable change in the column behavior. By proper application of theory the efficiency of columns has been progressively improved until recently this trend has culminated in the development by GOLAY (1958b) of the capillary column which can be constructed with as many as 750,000 theoretical plates (SCOTT, 1959b). The biochemist and biologist can make good use of theory in the design consideration of any chromatographic system which he finds must of necessity diverge from a well defined precedent. Although plate and rate theory are proven, useful concepts which are well substantiated by experimentation, it should not be concluded that they explain all facets of chromatography or that they are without fault. Thus BOHEMAN and PURNELL (1958) and PURNELL (1959) have reported situations in which the eddy diffusion contribution is negative, and they and GIDDINGS (1959) suggest, in accord with observation by GLUECKAUF (1957, pp. 29—30) that eddy diffusion may be sensitive to flow. KIESELBACH (1960) offers the alternative explanation that two or more additional terms may need to be added to the VAN DEEMTER equation; a term expressing diffusion in the moving gas phase to and from the liquid surface ($D\mu$), another describing the relatively stagnant gas on the surface of and within the porous packing ($E\mu$), and perhaps still a sixth term. GOLAY (1958a) has also suggested that the VAN DEEMTER equation lacks a term which he calls dynamic diffusion, and VAN DEEMTER himself has proposed various areas in which further consideration is essential (1958). Extension of already developed theories have been made to cover important new areas, for instance the expression of plate height under conditions of continually changing temperature (STEWART, 1960; HABGOOD and HARRIS, 1960), and finally there is the concept of capillary columns recently developed by GOLAY (1958a, 1958b).

B. Apparatus Requirements.

A basic chromatographic apparatus, as illustrated in Fig. 7, consists of a high pressure source of gas with the appropriate regulators, a sample introduction device, chromatography column, detector, flow meter, and thermostating oven. This is a typical arrangement, but various modifications are possible, and accessories such as fraction collectors and carrier gas purifiers may be added. Since the design requirements of many of the individual component parts are contingent upon the type of detector incorporated into the instrument, it is instructive to begin with a consideration of detection systems.

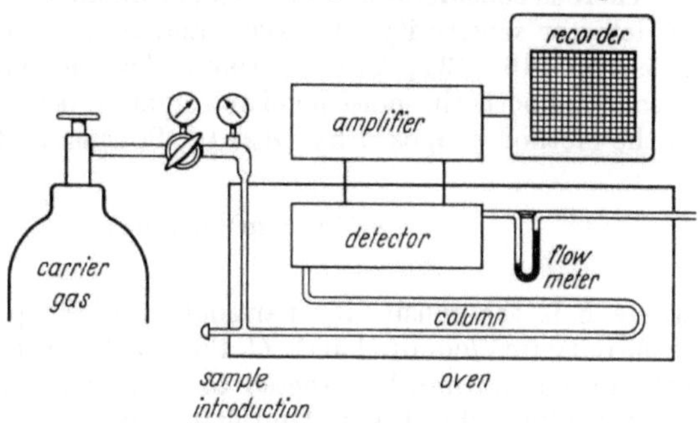

Fig. 7. Gas chromatographic apparatus.

I. Detection Systems.

Any property of gases can be used as a basis for determining the presence of a vapour in a flowing stream of column effluent provided there is some difference between the carrier gas and the sample components with respect to this property. Often the principle of detection is such that the instrument will show a response to the gas used as the mobile phase, but if this background signal is low and reasonably independent of such operational parameters as temperature, pressure and carrier gas flow rate, it is permissible to impose an equal counterbalancing electrical signal on the amplifier or recorder to establish a zero base line. In theory the sensitivity of this type of detector will be limited only by the noise level of the amplifier and recorder units, or by other associated circuitry. On the other hand if the background signal is large and responsive to fluctuations in operational conditions it becomes necessary to make use of a compensating system which usually takes the form of two identical elements, the reference detector located in a stream of pure carrier gas and the sample detector situated at the column exit. The characteristics of the two sensing elements should be reasonably identical, and if operational fluctuations are imposed to the same degree on both then it follows that the instrument will be independent of these variables. Unfortunately, compensation can never be perfect and the residual discrepancy in the response of the two units is sufficient to impart a dependence upon operating factors in the form of an unsteady background signal which invariably establishes the limits of sensitivity. Even in the case where the background signal is low, steady, and independent of environmental changes within the detector cell, compensation still is advantageous since it may partially cancel any electrical noise which happens to be transmitted equally to both sides of the compensating system.

No one detector satisfies all the criteria of a perfect instrument (JAMES, 1957; JOHNS, 1959, p. 18; JOHNSON and STROSS, 1959; HAUSDORFF and BRENNER, 1958; PHILLIPS, 1958, p. 36), nor is any likely to do so, with the result that presently there are available a large array of devices each with its own merits. In recent times emphasis has been directed particularly toward improving sensitivity, not only because such instruments are a great asset in trace analysis but perhaps even more important because they make possible the use of highly efficient capillary columns whose limited capacities impose severe demands upon a detection system. Since different instruments are best applied to different purposes, for experimental work it is always advantageous to have some provision such as a modular design for interchanging detection systems.

There is considerable disagreement about what constitutes a proper expression of detector sensitivity (JOHNSON and STROSS, 1959; MCWILLIAM, 1959; DREW, 1959, pp. 116—134; YOUNG, 1959). An adequate description must contain an account of both the noise level and signal output/weight of sample, as for instance in the method proposed by DIMBAT, PORTER and STROSS (1956):

$$S\,(\text{ml} \times \text{mV per mg}) = \frac{A \times C_1 \times C_2 \times C_3}{W} \qquad \text{(XII)}$$

where S is the sensitivity parameter; A the peak area (cm^2); C_1 the recorder sensitivity (mV/cm of chart); C_2 the chart speed (min/cm); C_3 the flow rate at the column exit (ml/min) corrected to column temperature at atmospheric pressure by multiplying the observed outlet flow by the factor $(273 + T_{column})/(273 + T_{ambient})$; W is the weight of component (mg); and the expression of the noise level (mV) should be included. The difficulty arises because of several factors not taken into account by the expression; for instance, the effect upon sensitivity of the nature of the carrier gas and the substance being analyzed. Moreover it is the ratio of the maximum peak voltage per unit weight of sample vs the noise level which ultimately dictates how small a quantity of material can be detected, and although peak areas may be constant under varied operating conditions, peak heights are affected by flow rate, adequacy of the sample introduction system, sample size, and very importantly by the efficiency of the chromatography column. Admittedly the method of expressing sensitivity as the least number of moles of substance which can be detected, as adopted in this review, suffers in all these respects and it is technically less accurate than certain other definitions, but it has the advantage of giving the reader an uncomplicated approximation of the actual quantities which are needed for analysis.

Detectors are either "*integral*" or "*differential*" in principle depending upon whether they indicate at each moment the integrated value of some property such as the total amount of solute which has previously emerged, or whether they respond to some momentary property of the sample such as its concentration in the carrier gas. In the former additive type of measurement, the elution curve has the form of the schematic representation of a displacement analysis shown in Fig. 2, with the exception that the ascending steps will be somewhat sigmoid in shape. The differential detector will produce the typical elution diagram illustrated in Fig. 1.

1. Integral Methods.

Several integral methods have been applied, but in general they have lower sensitivity than differential techniques, and their accuracy suffers from the unfortunate cumulative contribution of any error in the zero base line stability.

a) Titration (James and Martin, 1952a, 1952b, James, Martin and Smith, 1952).

The method of automatic titration of ionizable substances was first applied by James and Martin (1952a, 1952b) in their pioneering work on gas liquid chromatography. Volatile acids or bases were collected in an aqueous pH indicator, and the changes in color intensity were monitored by a photoelectric cell. The photocurrent was used to drive a recording pen across a revolving drum and to actuate a valve which controlled the rate of back-titration with standard acid or base. Several modifications have been described (McInnes, 1957), and analogous procedures such as coulometric titration (Boer, 1955 and in McInnes, 1957, p. 314) are in use.

b) Electrical Conductivity (Boer, 1955; and in McInnes, 1957, p. 314).

This method is similar in design to the above procedure except that the conductivity of an acidic or basic solution is continuously recorded (Boer, 1955). Aldehydes and ketones can be collected in a hydroxylamine chloro-hydrate absorbent in which changes of conductivity are induced by oxime formation. Other related methods are the use of the polarographic cell (Nederost, 1956) especially for halogens (Janak, Nederost and Bubenikova, 1957) and the measurement of mercaptans by iodine oxidation in a redox half-cell containing iodine-potassium iodide solution in alcohol (Sunner, Karrman and Sunden, 1956).

c) Volume and Pressure Changes (Janak, 1953, 1954, 1957).

When pure carbon dioxide is used as a carrier gas the gaseous sample components can be measured directly in a graduated cylinder by collecting them over base. This system has been successfully employed by Janak (1953, 1954), Leibnitz, Hrapia and Konnecke (1957) and Ray (1954a, 1955); and Janak (1957) and others (Hrapia, 1958; Sevenster, 1958) have adopted elaborate arrangements for automatically recording the volume changes. However pressure changes at a constant volume are easier to record and an ingeneous system to accomplish this has been devised by van der Craats (1956). The pressure and volume systems are particularly applicable to the measurement of hydrocarbons and permanent gases because these classes of compounds have a negligible solubility in aqueous potassium hydroxide. This is particularly true at 60° C which is about the maximum permissible operating temperature. A major difficulty is the requirement for very pure carbon dioxide carrier gas, since commercial tanks may contain as much as 1 percent of non-condensable gas and this would produce an intolerable base line drift. However a method is available for removing non-condensables from cylinders of commercial carbon dioxide by distillation under pressure (Craats, 1959), or else relatively pure carbon dioxide gas can be obtained by the liquification of a well weathered block of solid carbon dioxide in a pressure vessel (Ray, 1955) and by carbonate decomposition.

d) Combustion to Carbon Dioxide.

Several integral methods depend upon the determination of carbon dioxide produced when organic vapours are combusted at the column exit by passage of the effluent through a tube of red hot copper oxide or over a hot platinum wire in oxygen. Titration (BLOM and EDELHAUSEN, 1956), coulometry (LIBERTI, 1957; LIBERTI and CARTONI, 1958) and colorimetry (JURANEK, 1957, 1959) have proven useful as assays.

2. Differential Detectors — which Consume the Sample.

All procedures which consume the sample are subject to the limitation that only combustible organic materials can be detected, and it is impossible to recover components for further analyses. Included in this category are:

a) Combustion to Carbon Dioxide (MARTIN and SMART, 1955).

If organic components are converted to carbon dioxide, any method for the determination of this gas in a flowing stream is a valid detection scheme. The sensitivity is always proportional to the number of carbon atoms in the sample vapour as this determines the amount of carbon dioxide produced. An infra red analyzer was used by MARTIN and SMART (1955) who employed nitrogen as the carrier gas, but carbon dioxide free air might be an even better mobile phase since it would continually regenerate the copper oxide surface, or nitrogen could be used as the carrier gas and a stream of oxygen added at the chromatography column exit. The sensitivity of the infra red analyzer to carbon dioxide is extremely high, showing full scale for as little as .005 percent, but the large volume of the sample cells may cause loss of resolution due to intermixing of adjacent chromatographic bands in the detector. Nevertheless, this detection system has been applied successfully by MARTIN and SMART (1955), LIBERTI, COSTA and PAULUZZI (1955), and HEATON and WENTWORTH (1959). A katharometer has been used to detect the carbon dioxide by MELLOR (1957), NOREM (1958), STUVE (1958), JAMIESON (1959) and NEL and PRETORIUS (1959), in which case it is absolutely essential to remove the water vapour produced during combustion. Alternatively, the water vapour itself may be measured by means of a hygrometer, or GREEN (1957) and FRANC and WURST (1959) have described a method for converting water to hydrogen which can be detected with a katharometer after the carbon dioxide is removed.

b) Hydrogenation to Methane (ZLATKIS and RIDGWAY, 1958; ZLATKIS, ORO and KIMBALL, 1960).

A select group of organic compounds may be converted to methane by using hydrogen as a mobile gas phase and passing the column exit gas over a hot nickel catalyst. Water, which will be produced if oxygen is present, must be removed before the methane can be quantitatively analyzed with a katharometer, but this would not be necessary if certain other types of detectors were employed.

c) Flame Emissivity (GRANT and VAUGHAN, 1957; GRANT, 1958).

In this procedure, based on luminosity measurements, coal gas is mixed with the column effluent and the mixture is combusted at a wide stainless steel jet. The emissivity, which is measured by focussing the light emitted from the flame upon a photocell, increases when a volatile organic compound emerges from the column. Air is supplied to the flame by a blower, and although nitrogen was used as the mobile gas phase in the original apparatus, any carrier gas should be permissable provided that it is non-combustible and can be supplied at a uniform flow rate. The sensitivity of this device varies depending upon the nature of the sample component, but for benzene and for aromatic compounds in general, GRANT (1958) reports that it is at least as good as that of the katharometer ($\sim 10^{-8}$ moles). The advantages of this method are its simplicity, wide operational temperature limits (up to 300° C), linear response over a sufficient range, and small cell volume: disadvantages are the inherently high background signal, the response of the detector to changes in the flow rate of both the carrier gas and coal gas and also to fluctuations in temperature.

d) Hydrogen Flame and Thermocouple (Scott, 1955, 1957).

As originally proposed this method consisted of flowing hydrogen or a mixture of hydrogen and nitrogen through the chromatography column and burning the gas at the column exit in a fixed jet such as a hypodermic needle or fine glass or metal capillary exposed to a stream of air. By means of a thermocouple placed approximately 1.5 cm above the hydrogen flame it is possible to detect the lengthening of the flame (and perhaps a small rise in flame temperature) which accompanies the appearance of an organic sample component in the mixture. The stability of the analyzer depends to an important degree upon the maintenance of a uniform flow rate and low viscosity gases like hydrogen make difficult the satisfying of this requirement. The reason for this is that the column offers very little resistance to hydrogen, but the various high viscosity sample components act as plugs in the column, and as each emerges the resistance to flow decreases and the base line changes. Wirth (1957) substituted pure nitrogen as the mobile phase and added the hydrogen at the column exit, thus eliminating the problem (c.f. also Henderson and Knox, 1956; Primavesi, Oldham and Thompson, 1958). A further advantage of Wirth's modification is that it should permit the use of any non-combustible carrier gas. The flame detector will sense about 10^{-9} moles, although the absolute limit for each substance depends upon its heat of combustion. The response is fairly linear over wide ranges, the instrument has a very small effective volume, a rapid response time, and a high operational temperature limit. The insensitivity of the detector to water, fixed gases and certain chlorinated compounds is fairly complete, and may or may not be advantageous depending upon the intended application.

e) Flame Ionization (McWilliam and Dewar, 1958a, 1958b).

The current flow between two electrodes situated in a partially ionized gas mixture, across which a D.C. potential has been applied, will be determined by the ion flux in the gas occupying the intervening space. If conditions can be so defined that the formation of ions is dependent upon the presence of a sample component admixed with the carrier gas separating the electrodes, then the arrangement becomes a suitable basis for detecting vapours. Ionization may be induced by any of several means, for instance thermionic emission, photoionization, ionizing radiation, or ionization in a flame. The latter, which is in many respects the least understood of the various principles, is utilized in the flame ionization detector. In this instrument the electrical conductivity rather than the temperature and height of flame is measured; otherwise with minor modifications it is similar to Scott's hydrogen flame detector (Scott, 1955). Its operation finds a basis in the well known observation that air in which a flame has burned will conduct electricity to a much greater extent than will fresh air, the reason being that the flame produces a certain number of ionizations. When hydrogen is burned in air or oxygen, very few ions are released, but when an organic compound is present in the flame, many ions are formed. In the flame ionization detector a piece of metal screen or gauze serves as the positive pole, and it is situatedabout 5 to 10 mm above the metal jet which itself constitutes the negative electrode. A potential of about 200 to 300 volts D.C. is applied across the electrodes in series with a load resistor, and the voltage drop across the resistor can be determined with a conventional electrometer-amplifier-recorder combination or with other circuitry (THOMPSON, 1959; HARLEY, NEL and PRETORIUS, 1958). Most organic molecules have ionization potentials in the 9—12 eV range, whereas common permanent gases vary from 13.2 for argon to 24.5 eV for helium; nitrogen and hydrogen ionizing at about 15.5 eV. In a flame, the amount of ionization caused by the addition of an organic molecule is much higher than would be predicted on the basis of the known ionization potentials, and it has been suggested by STERN (1951) that the substances are first burned to carbon aggregates which ionize much more readily, at about 4.6 eV. This explanation finds support in the observations that the sensitivity of the flame ionization detector depends upon the number of carbon atoms present in the sample component and the device is practically insensitive to inorganic substances. The stability of the instrument is upset by the presence of dust particles in the air feeding the flame, but this is alleviated by the expedient of filtering the incoming air stream. A small background signal can be electrically compensated because almost no sensitivity to flow or temperature exists, but for maximum sensitivity McWILLIAM and DEWAR (1958a, 1958b) suggest a dual jet compensating system. The nature of the sample component determines the sensitivity limits, but values approaching 10^{-14} moles have been claimed (McWILLIAM and DEWAR, 1958a) which means that from this standpoint the performance is at least the equal of any other detection system. Although nitrogen is usually used, almost any non-organic carrier gas should be permissible, except that the sensitivity may be adversely affected by too much oxygen (McWILLIAM and DEWAR, 1958a, p. 152) perhaps due to excessive oxidation of the sample. The largest sample which can be successfully passed through the detector is about 100 microliters and this somewhat limited sample size, the loss of sample due to combustion, and the danger inherent in the use of hydrogen gas are the only major objections to this procedure. Both the dependence of sensitivity upon carbon number and the insensitivity of the instrument to permanent gases and water can on occasion be used to advantage.

3. Differential Detectors — which Preserve the Sample.

a) Surface Potential (GRIFFITHS, JAMES and PHILLIPS, 1952; GRIFFITHS and PHILLIPS, 1954).

Use is made of a condensor unit composed of two metal plates, one of which is coated with a surface film of fatty acid or collodion. When either plate is caused to vibrate in close juxtaposition with the other, because of the differences in potential at the two surfaces an alternating e.m.f. is generated in a resistance connecting the plates of the oscillating condensor. The value of this induced voltage is dependent upon the nature of the gas mixture between the two plates, and the background signal observed with the carrier gas flowing can be nullified by adjusting properly the bias of a tube. The main objections to this procedure are that its response is slow and non-linear. About 10^{-8} moles of polar vapours and high molecular weight substances can be detected, but the absolute sensitivity is somewhat lower for other classes of compounds.

b) Dielectric Constant (GRIFFITHS et al., 1952).

The capacity of a condensor depends upon the dielectric coefficient of the insulating material between the plates, and individual gases differ to some extent in this property. If the column effluent is passed through a condensor packed with charcoal, changes in the dielectric coefficient of the gas mixture affect the capacitance, which property can be measured by incorporating the condensor into an oscillating circuit. TURNER (1958) has described a rugged instrument of this type which can detect 10^{-7} moles of sample, and which has a small cell volume and long linear range.

c) Impedance of Gas Flow (GRIFFITHS et al., 1952).

When a restriction is placed in a tube through which a gas flows, a pressure differential is developed across the barrier. In an orifice choke the difference in pressure depends upon the density of the gas, and this is a property determined by the gas mixture. Capillary chokes, in which the pressure differential is a function of viscosity, have proven to be less satisfactory (GRIFFITHS et al., 1952). The method suffers from the necessity for a very accurate control of the flow rate without which background irregularity results, and also it is difficult to calibrate. The sensitivity approaches 10^{-8} moles of substance according to GRIFFITHS et al. (1952).

d) Heat of Vaporization (DUDENBOSTEL and PRIESTLEY, 1956).

Adsorption is an exothermic process, and it results in a momentary heating of the mobile phase whereas vaporization produces the opposite effect. DUDENBOSTEL and PRIESTLEY (1956) used a bed of activated charcoal to readsorb volatile compounds as they emerged from the column, and detected their presence by measuring the differential heat of adsorption and vaporization with a thermocouple. In an even simpler approach, PRIESTLEY (1957) uses the column itself as a detector, and as a component passes, it is sensed by a thermocouple inserted directly in the column end. The signal is proportional to the amount of substance, and by switching the gas stream every few seconds, an oscillating signal is transmitted to the recorder.

e) Interferometer.

These optical instruments have long been used for measuring the refractive indices of gases, and this is a satisfactory basis for a detection system. Although it is relatively insensitive, it has been applied successfully by SCHUFTAN (1931), ROTH, OHME and NICKISCH (1942), TURKEL'TAUB (1955), MEDVEDEVA and TORSUEVA (1955), TOTH and GRAF (1957), GUNESCH and STADTMULLER (1958) and VYAKHIREV and BRUK (1958).

f) Spectroscopy.

Usually ultra violet and infra red spectroscopy are reserved as secondary procedures intended to help in the identification or characterization of samples collected by gas chromatography. However, LIBERTI and CARTONI (1958b) report that absorption at the C—H band frequency of 3330 cm^{-1} can be used to detect hydrocarbons in a flowing gas stream; WHITE, WARD and GALLOWAY (1959) also used a flow through I.R. detector; and U.V. spectroscopy in a stream has been applied occasionally (LUTINSKY, 1957; JOHNSTONE and DOUGLAS, 1959; FRANC and JOKL, 1958, 1959).

Gases fractionated on a chromatography column are of sufficient purity to be assayed further by mass spectroscopy and this instrument frequently has been used in conjunction with but not in lieu of another detector (Benyon, 1954, 1956; Bradford, Harvey and Chalkley, 1955; McNesby, Drew and Gordon, 1955; Drew, McNesby, Smith and Gordon, 1956; Stahl and Levy, 1956; Seligman, 1956; Kebarle and Bryce, 1957; Burr, 1957; Whitham, 1957b; Rippere, 1958; Gordon, Dynken and Doumani, 1958; Turkel'taub and Abramovich, 1958). It is not always necessary to collect samples for a separate analysis because the "Time of Flight" mass spectrometer described by Gohlke (1957, 1959), West (1959) and Galloway and Johns (1957) can be used to monitor directly and continuously the column effluent. This instrument scans the range $m/e = 1$ to 6000 and produces 2000 spectra per second, which can be observed on an oscilloscope screen and if necessary recorded by photography. The T-O-F instrument is less sensitive than a conventional mass spectrometer, and its main advantage is that it provides instantaneous information concerning sample identity, which is a valuable asset in conjunction with quantitative data supplied by some other form of detection device.

g) Radioactivity.

Radioactive components can be detected by means of a Gieger counter (Kokes, Tobin and Emmett, 1955; Evans and Willard, 1956; Roginsky, Ianovsky, Zhabrova, Vinogradova, Kadenatski and Markova, 1958), flow proportional counter (Gordus, Sauer, and Willard, 1957; Wolfgang and Rowland, 1958), scintillation counter (Lowe and Moore, 1958; Popjak, Lowe, Moore, Brown and Smith, 1959; Moussebois and Duyckaerts, 1958; Evans and Willard, 1956; Grandy and Koch, 1960) or ion chamber (Wilzbach and Riesz, 1957; Riesz and Wilzbach, 1958; Mason, Dutton and Bair, 1959; Cacace and Inam-ul-Hag, 1960) placed in series with another detector. Counting systems have been used at temperatures as high as 240° C (Mason et al., 1959) and they can be made sensitive in spite of the fact that the samples reside in the counting chamber for only a limited time interval. In biological work the main value of radiation detectors combined with gas chromatography is that in conjunction with another detection system they constitute a procedure which allows the calculation of specific activities in a single manipulation. Flow-through radiation detectors may also be used qualitatively to scan chromatograms in search of radioactive bands which can then be recovered and assayed more exactly by conventional methods. Wolfgang and Rowland (1958) suggest that at present a flow proportional detector, in which the emerging helium stream is injected with methane and passed through a special counting tube, is about as efficient and applicable as any other type of detection method.

A relatively rapid procedure for the tagging of organic compounds by exchange with tritium gas has been described by Dorfman and Wilzbach (1959), and this method has been applied to fatty acids by Nystrom, Mason, Jones and Dutton (1959) who scanned the radioactivity in the gas stream emerging from a chromatography column. The exchange can be accomplished with a minimum amount of decomposition, which suggests that in certain instances it might prove to be an extremely sensitive assay procedure.

h) Thermal Conductivity — the Katharometer or Diapharometer (Eucken and Knick, 1956: Claesson, 1946: Turner, 1943: Ray, 1954b).

A wire heated by an electric current and suspended in a thin channel will assume an equilibrium temperature when the rate at which heat is supplied to it by the current is equal to the rate of heat loss from it by convection, radiation, and especially conductivity across the gas in the channel to and through the channel walls. Heating is a function of the square of the applied current and the resistance of the wire whereas the heat loss is contingent upon the temperature differential between the hot wire and channel walls, and the thermal conductivity of the gas mixture which the heat must traverse. The ability to conduct heat varies amongst gases from a high value for light molecules such as helium and hydrogen, to conductivities lower by a factor of approximately ten for most organic substances and many of the permanent gases. Tungsten or platinum are usually chosen as filaments because these metals have a high temperature coefficient of resistivity which is almost independent of the absolute temperature over a large range, and there will exist a linear relation between wire temperature and resistance. By incorporating the filament into a WHEATSTONE bridge, it becomes possible to measure its resistance and this property depends ultimately upon the composition (heat conductivity) of the gaseous mixture surrounding the metal. Because the background signal is high and responsive to flow, pressure and temperature changes, it is necessary to compensate for it with an identical wire occupying an adjacent position in the WHEATSTONE bridge, or two sensing and two reference filaments may be arranged as alternating arms of the bridge network. The latter has the advantage of doubling the sensitivity of the detector, and both the four and the two filament systems make the cell idependent to some extent of fluctuations in operating conditions, thermoelectric effects, and the voltage supplying the filament current. Because the rate of heat loss depends not only upon the wire temperature but also upon the temperature of the channel walls, a katharometer's stability can be no better than the temperature constancy of the channels. Therefore it is customary to form the channels as holes drilled through a cylindrical block of some metal possessing high heat capacity (such as aluminum or brass), and if the block is shielded from drafts the temperature of all channel walls will be identical and reasonably constant. To insure proper compensation, all filaments should be equal in resistance, but as exact duplication is never possible, some ancillary system of electrical compensation must be incorporated into the circuit.

The *sensitivity limit* of a thermal conductivity detector ultimately is a characteristic determined by the design of the instrument, and many modifications have been reported (DAYNES, 1920, 1921, 1933; WEAVER, PALMER, FRANTZ, LEDIG and PICKERING, 1920; SLATER, 1926; SHAKESPEAR, 1921; STILES and LEACH, 1931; TURNER, 1943; CLAESSON, 1946; AUFDEMGARTEN, 1939; VEEN, 1949; and recently many others). In order to realize optimal performance certain precautions must be observed, and the conditions under which maximum sensitivity is attained have been thoroughly investigated by DAYNES (1933), WEAVER (1951), and LINDSAY and BROMLEY (1950), and summarized by KEULEMANS (1959). The magnitude of the current supplied to the filaments is an important consideration because the voltage output of the instrument is very nearly a function of the cube of the current and the resistance of the wire at the wire temperature (BURG and STOLWIJK, 1959). The practical limit to which current can be raised is that value at which the noise and/or drift of the apparatus, which are also a function of current, begin to increase more rapidly than the voltage output. In order to achieve this condition it is necessary that the amplifier and recorder circuits have a lower noise level than the katharometer at the filament current which yields maximum sensitivity. The stability of the filament current depends upon the performance of the power supply and although batteries are usually satisfactory, inevitably they result in some drift, whereas electronic D.C. power supplies are only suitable if they are well regulated with low noise and ripple. RAY (1954a) has described a katharometer in which the filaments were heated with an A.C. audio source, and PURCELL and KEELER (1960) report that a stabilized A.C. supply with well designed ancillary equipment for sensing the applied frequency will improve katharometer performance.

In a two component gas mixture of low dilution the change in thermal conductivity of the mixture induced by altering the mole fraction of the lesser component is proportional to the difference in heat conductivity between pure samples of the two components. Since hydrogen and helium possess thermal conductivities greater by an order of magnitude than those of most permanent gases and almost all organic vapours, it follows that the voltage output of a katharometer will be maximized when either of these light gases is used as a mobile phase. RAY (1959a, 1959b) has pointed out that in comparing carrier gases of high and low thermal conductivity, an advantage can only be obtained in favor of the high termal conductivity mobile phase if in both cases the wire temperature is identical. This is not always easy to achieve because much higher currents are required to raise the temperature of a wire bathed in a media which rapidly conducts heat away from it. Another advantage of helium and hydrogen is that the absolute value of thermal conductivity will be almost the same for equivalent concentrations of almost any admixed vapour. DIMBAT et al. (1956) calculate that for most organic compounds a maximum error of 3.5 percent is incurred by assuming that in a katharometer operated with helium carrier gas the quantitative response is independent of the chemical nature of the substance and HAUSDORFF (1955) has estimated the error to be only 0.5 percent. Because of the hazard involved in handling hydrogen, helium is usually preferred if it is available and reasonably priced.

The coefficients of resistivity of platinum and tungsten are approximately equal and both are as high or higher than almost all other metals. The exceptional elasticity and tensile strength of these materials allows them to be drawn into fine wires and these are used as filaments in hot wire katharometers to the exclusion of most other metals. However semi-conductor beads known as thermistors have a temperature coefficient of resistivity about ten times larger than that of platinum and tungsten, although opposite in sign. Unfortunately, thermistors possess relatively low current ratings and their sensitivity reaches a maximum and then declines as the current is further increased (COWAN and STIRLING, 1958; SCHMAUCH and DINERSTEIN, 1960). A further disadvantage is that the temperature coefficient of resistivity decreases by a factor of about 2 for every 30° C increase in temperature, whereas in metals it is fairly independent of temperature. As a result thermistor detectors have a sensitivity advantage over hot wire katharometers up to but not above about 100° C, and in addition their cell volume is smaller and response time shorter; however, thermistors are more difficult to match than are wire filaments.

Katharometers of both the thermistor and hot wire type can sense about 10^{-8} mole of substance, although a few instruments of special design have been described with improved detection limits (BURG and STOLWIJK, 1959; BENNETT, NOGARE, SAFRANSKI and LEWIS 1958; PURCELL and KEELER, 1960). Thermal conductivity cells have a wide dynamic temperature range and instruments have been operated satisfactorily at 500° C (FELTON, 1957, 1958). The major disadvantages of katharometers are that they are susceptible to fluctuations in temperature, carrier gas flow, and chamber pressure; their sensitivity is limited and declines at high temperatures, in part due to the lowered temperature differential between the filaments and block and the tendency for thermal conductivities to converge (DAYNES, 1933; SCHMAUCH and DINERSTEIN, 1960), and at very high temperatures sample pyrolysis may occur with the result that carbonaceous materials will be deposited on the filaments and reduce the stability of the instrument.

i) Gas Density Balance (JAMES and MARTIN, 1954).

The variation in density of a gas mixture as a function of its composition was first used as a detection system by CLAESSON (1946) who employed a liquid manometer for this purpose. A much more elaborate instrument, capable of high sensitivity, was developed by JAMES and MARTIN and its characteristics were thoroughly investigated by MUNDAY and PRIMAVESI (1957). The gas density balance consists of a network of channels cut into a copper block in a form mechanically equivalent to a Wheatstone bridge. The corners of the bridge inside the block are connected by an anemometer, a channel through which is strung a copper wire having a small central segment of constantan. A hairpin heater, mounted in the center of the anemometer and rotated, serves to warm the two copper-constantan thermojunctions to exactly the same temperature. Pure nitrogen carrier gas enters through an inlet on one side of the block after which the channel divides to form two adjacent arms of the bridge which pass around the anemometer continuing as the opposite arms of the bridge and eventually reforming at a combined outlet. The sample effluent is carried into the block in a channel which bifurcates and each segment feeds into one of the outlet arms of the bridge, with one connecting much closer to the outlet than the other. This means that column effluent never reaches the anemometer, but if its density is higher than that of the reference gas it will unbalance the pressure at the ends of the anemometer channel and cause a slight flow of reference gas across the thermojunctions. This results in a shift of heat toward one thermocouple at the expense of the other, and since only pure carrier gas passes across the anemometer, the response is independent of the thermal conductivity of the sample effluent and linear over an adequate range. Provision is made for mechanically balancing the gas density meter by the adjustment of rods which are fitted into the four arms of the bridge network in such a manner that they control the flow resistance of the channels. The thermoelectric output can be amplified and recorded.

The major advantages of this meter are that it responds to a simple property of gases, the density, which is a function of molecular weight, and also the instrument is relatively sensitive (10^{-8} moles). At first it was thought that the construction was so intricate that the expense would be prohibitive, but this problem has been circumvented to a considerable extent (c.f. MUNDAY and PRIMAVESI, 1957) and several commercial firms now offer the instrument. A major disadvantage is that while it can be operated at temperatures as high as 300° C, (HAWKS, 1957) its sensitivity decreases as the temperature is raised.

j) High Voltage Ionization (HARLEY and PRETORIUS, 1956).

HARLEY and PRETORIUS evolved a detector in which a portion of the column effluent is passed at a few mm pressure between two electrodes maintained at an applied potential of about 900 volts D.C. Changes in electrical conductance between the electrodes is caused by the ionization of the sample molecules and this in turn is dependent upon the ioization potential of the components. PITKETHLY (1958) has modified this method slightly by using neon glow lamps as electrodes arranged in a bridge circuit. The sensitivity limits of both the original method and modification is about 10^{-11} moles of sample, but the absolute value depends upon the nature of both the carrier gas and sample. The procedure requires a controlled vacuum, it is restricted to small sample sizes in the order of 10 to 20 micrograms, and the instrument's characteristics are not sufficiently stable to allow it to be used without extreme caution.

High voltages at atmospheric pressure can also be used to induce ionization in a gas, for which purpose a four inch long copper tube of one-quarter inch O.D. may serve as cathode, and a tungsten wire, somewhat less than 1 mil in diameter, suspended axially and well insulated from the copper, is the anode (STOLWIJK, personal communication). The interior of the tube should be polished, and the gas inflow and outflow are outside the area containing the exposed wire. The field in the vicinity of a wire is very concentrated and exponentially related to the "thinness" of the wire, and if an extremely fine filament is used there will exist some applied D.C. voltage at which the field gradient will be steep enough to exceed the dielectric strength of the gas surrounding it (for air 36 kilovolt/cm) and the gas will be ionized. With helium surrounding the filament, an ionization current of 20 μA will be realized at about 690 V, and some recombination will occur resulting in photon emission. Although the mechanism by which the detector operates is uncertain, it may depend upon the fact that helium photons are energetic enough to ionize most organic compounds. A quantitative measure of the presence of vapours admixed with the helium can be made by recording changes in the voltage output of a current stabilized power supply, adjusted to deliver about 20 μA. Helium is most suitable as a carrier gas because with air, 2420 V are required to produce an ionization current of 20 μA, whereas with carbon dioxide the voltage required to maintain a stable current is higher than the spark-over voltage between wire and cylinder. The absolute sensitivity of the instrument has not been determined, but it is high.

k) Thermionic Emission (RYCE and BRYCE, 1957).

Electrons emitted by thermionic emission from a hot filament serve as the source of ionizing energy in a detection system designed by RYCE and BRYCE. An ionization pressure gauge was modified for this purpose, and with helium as a carrier gas a small amount of column effluent is leaked into the tube at a low pressure with 18 volts D.C. applied potential. This voltage is too low to ionize helium but more than sufficient to ionize nearly any other substance so that the electrical output of the gauge is almost exclusively a result of the presence of substances other than helium. This detection system has many favorable qualities. It is sensitive to all classes of compounds including permanent gases (for instance the response to hydrogen in a helium carrier gas is as high as the response to nitrogen), changes in flow and temperature have no effect, response is linear up to 10 ml of gas (HINKLE, TUCKER, WALL and COMBS, 1959), the detector is unhindered by sample overloads, the effective cell volume is small, and the vacuum does not have to be controlled. The limit of detection is about 10^{-11} moles, depending upon the substance analyzed, and for permanent gases this is about as sensitive as any other method. The major difficulty seems to be the susceptibility of the filament to poisoning by products formed at its surface, with the result that the characteristics of the instrument may constantly change.

l) β-Ray Ionization (DEAL, OTVOS, SMITH and ZUCCO, 1956; BOER, 1957; LOVELOCK, 1958a).

Conductivity between two electrodes maintained at a differential D.C. potential in a gas bombarded by ionizing radiation will depend among other things upon the ionization potential and ionization cross section of the vapour in the intervening space. In theory any of the various types of ionizing radiation might be used, but β- radiation has frequently been singled out as the most applicable. Differential detectors based on this principle, using Sr^{90} as a source of β-radiation and nitrogen or hydrogen as a gas phase have been described by DEAL et al. (1956) and BOER (1957). The sensitivity obtained is slightly better than that of the katharometer, but in addition the detector is insensitive to flow variations and is practically unresponsive to water vapour. Although fluctuations in temperature and pressure are sensed, this disadvantage is negated to a considerable degree by the differential system. The sensitivity to various substances differs and to some extent it is dependent upon molecular weight since the β-ray ionization cross section of a molecule is a function of this physical property. This detection system might have realized greater popularity in its original form except for a remarkable modification introduced by LOVELOCK (1958a).

The basis of this modification is the observation that when gases absorb ionizing radiation, in addition to the ions which are created, there also are formed short lived metastable atoms which are excited but not ionized. However, the metastable atoms of the rare gases such as helium and argon are long lived, and when these gases are exposed to radiation, the concentration of excited atoms approaches that of the ions. By using one of the rare gases in the β-ray detector, the advantage gained is that while the metastable atoms will not contribute to the background current, they possess sufficient energy to ionize by collision any component in the same effluent which has an ionization potential less than that of the metastable state of the carrier gas. Helium (I.P. 24.5 eV, metastable at 19.7 eV) was found to be unsatisfactory, presumably because commercial cylinders contained sufficient impurity to discharge the metastable atoms as fast as they were formed (LOVELOCK, 1958a). However, commercial argon (I.P. 15.7 eV, metastable at 11.6 eV) was completely satisfactory. Low applied voltages (< 300 volts D.C.) yield an equilibrium concentration of metastable argon which can be completely discharged by a relatively small organic vapour concentration, with the result that operating under this condition the detector has a linear response of limited range. At about 400 volts D.C. the linearity is optimal and the sensitivity improved, while at higher voltages (< 1200 volts D.C.) the sensitivity is maximum and the linearity disappears in favour of excess signals for high sample concentrations, presumably due to numerous ionizations caused by the secondary electrons as they are accelerated through the large voltage. However by proper selection of a resistance in series with the ionization chamber, it is possible to obtain linearity at relatively high voltages (LOVELOCK, JAMES and PIPER, 1959).

All compounds having molecular weights in excess of about 100 have ionization potentials sufficiently below that of metastable argon to be ionized to equal extents by collision. Therefore, with the exception of light vapours, the β-ray detector displays an almost equivalent sensitivity to all compounds, which is a distinct advantage (LOVELOCK, 1958a). The instrument is almost unaffected by fluctuations in temperature, gas flow rate, and pressure, and it is only slightly responsive to water. An earlier objection, that the large chamber volume caused some lack of resolution and made the detector inapplicable to capillary chromatography, has now been overcome by the development of devices having effective volumes a as low as few microliters (LOVELOCK, 1958b; LIPSKY, LANDOWNE and LOVELOCK, 1956; CONDON, 1959). One of the few disadvantages which the instrument possesses is that it responds negatively or not at all to permanent gases, and it is only slightly sensitive to light molecular weight organic compounds. One way to circumvent this problem might be to use very pure helium instead of argon, but research grade helium is extremely expensive. However, both the impurities present in commercial helium and ways of removing them have been investigated and the two most promising methods for obtaining sufficient purity in the laboratory are the trapping of contaminants in brominated charcoal maintained at liquid nitrogen temperature (DREW, 1959, pp. 97—115) and the use of 5 A molecular sieves at room temperature as a purifying column (GUILD, BINGHAM and AUL, 1958). In an entirely different approach, WILLIS (1959) found that the sensitivity of a β-ray detector to hydrogen, oxygen and methane was increased by bleeding an organic vapour (ethylene) continuously into the argon.

The ultimate *sensitivity* of the β-ray detector is limited by random fluctuations in background current caused by the unsteady nature of the β-emission from the source, and these fluctuations depend upon both the input time constant and the intensity and nature of the radioactive source. Alpha emitters have been used, such as Ra^{226}, but they give somewhat lower sensitivities than do β-sources. Amongst the latter, T^3 provides the highest sensitivity, is free of γ-rays that need shielding, and any radioactive hydrogen desorbed from the film escapes harmlessly into the atmosphere. However its use imposes a 225° C temperature maximum, whereas Sr^{90} allows operation of the instrument up to 275° C, which is the flow point of the teflon insulators. Kr^{85} has been substituted for Sr^{90} in some applications because, being a gas, if it is accidentally liberated it is simply diluted in the atmosphere. LOVELOCK et al. (1959) state that they have been able to detect 2×10^{-13} M with a β-ray instrument, whereas they believe the actual sensitivity limit may be as much as 1000 times greater.

m) Radio Frequency Detector (KARMEN and BOWMAN, 1959a, 1959b).

KARMEN and BOWMAN have described a detector based on the principle that a radio-frequency incited glow discharge in a stream of helium at atmospheric pressure produces a current, the magnitude of which is inversely proportional to the solute concentration. A R.F. detector cell rectifies the discharge and without additional amplification the difference between its D.C. output and a fixed bias current can be recorded. This detector is relatively insensitive to pressure and flow changes, and only slightly affected by temperature fluctuations. A sensitivity of 10^{-12} M was stated by KARMEN and BOWMAN (1959b), and a commercial instrument which now is available claims a detection limit, under certain conditions, of 10^{-15} M. The absolute sensitivity varies depending upon the particular vapour, but it is high for fixed gases and water as well as organic compounds, with the limitation that the dynamic range is so restricted that only small samples may be analyzed.

4. Summary of Detector Characteristics.

At present, for routine analysis in which moderate sensitivity will suffice, the katharometer and gas density balance will usually be found satisfactory and relatively trouble free. When high sensitivity and small cell volume are required, the β-ray and flame ionization detectors are preferred. The former is less suitable for light molecular weight compounds than is the latter, and has a slightly lower temperature range, but possesses the advantage that the sample is not destroyed. Neither of these instruments is applicable to the measurement of permanent gases with the sensitivity shown by the low pressure D.C. discharge instrument. The radio frequency detector looms as an unknown quantity, perhaps capable of equalling or surpassing other detectors in certain respects.

II. Gas Phase.

The choice of the proper carrier gas is made on the basis of a number of considerations, foremost among which is the type of detection system employed. The requirements of the different instruments have been discussed in the preceding section, and it will be noted that in many instances there is a certain amount of leeway, in which case the following considerations become important. *1.* In heavy gases such as argon, nitrogen and carbon dioxide, diffusivity is less and column-efficiency greater than in lighter gases. *2.* The more cryogenic gases, such as are commonly used in gas chromatography, are insoluble in the stationary liquid phase and hence have no direct bearing on the retention time of solutes. However in gas-solid chromatography, a considerable quantity of the mobile phase may be adsorbed on the surface of the stationary adsorbent which diminishes the column's effective length and ability to retain solutes. In this respect helium has been found to be preferable to most other gases (GREENE and ROY, 1957) because it is adsorbed to the least extent. *3.* Packed columns offer a considerable resistance to flow, which may create a pressure differential between inlet and outlet of sufficient magnitude to cause an unfavorable flow rate through a significant length of the column. A reduced inlet/outlet pressure ratio can be obtained by using light molecular weight gases toward which the column packing shows the greatest permeability.

The *flow rate* of the mobile phase is normally adjusted by altering the column inlet pressure, for which purpose commercial pressure regulators of sufficient accuracy are available. Quantitative measurements of the flow rate can be made by a number of methods, including rotameters, orifice meters, soapfilm flow meters and displacement of water. The former two methods are the most convenient but the least accurate; moreover they create a back pressure and are temperature dependent whereas although the moving soap bubble is cumbersome to employ and unusable for continuous readings, it is preferred when the highest accuracy is required. For $^1/_4$ inch columns an HETP minimum usually occurs with a volume flow between 25 and 200 ml/minute; larger columns require an increased flow in direct proportion to the square of their diameter, and capillary columns with I.D. of about 0.01 inch function most efficiently with flow rates between 0.5 and 1 ml/minute.

It is often advantageous to gradually increase the temperature of the column from a low to a high value during the course of an analysis, but this technique, known as temperature programming, causes flow changes which may induce base line drift in detection systems which are flow sensitive. Several methods for preventing the occurrence of this undesirable effect have been developed including the use of a flow controlling device in place of a pressure regulator (CLOUGH, 1959; KNOX, 1959; GUILD et al., 1958) and the operation of the column at high pressure through its length (DREW, 1959b, pp. 97—115) although SULLIVAN, WALSH and MERRITT (1959) state that flow changes during temperature programming do not seriously affect the separations obtained. LIPSKY et al. (1959) have described the use of pressure programming in which a gradual alteration in flow rate is intentionnally induced; this procedure was used in the separation of lipids, and base line stability was maintained by the use of the flow insensitive β-ray ionization detector.

III. Sample Introduction.

The method of sample introduction influences markedly the efficiency of a column by affecting both the retention time and the peak width. The plate theory assumes that the entire sample arrives instantaneously and is contained in the first plate of the column. This ideal is usually only approximated, and the actual situation will prevail between one of two extremes. *1.* "*plug*" *flow*, in which the sample reaches the head of the column as a plug of gas, diluted only by the volume of carrier gas into which it was introduced initially, and *2. exponential flow*, in which the sample mixes freely with the carrier gas, with the result that the concentration arriving at the column head falls exponentially towards zero as a function of time. Both cases have been examined mathematically in terms of plate theory by PORTER et al. (1956), who show that plug flow gives rise to symmetrical peaks, while exponential flow inevitably results in asymmetry and tailing. The nearest approximation to plug flow will be realized when the sample is introduced as close to the head of the column as possible, preferably in a small narrow space which promotes movement of the contained gas as a "slug". In the case of solutes which are gases at the time of introduction, provided the sample size is not excessively large, this condition can be approximated closely; liquids and solids however create special problems since they must be vaporized prior to introduction into the column, at which time their volume increases many hundred fold. The largest sample which can be introduced in a short time interval, admixed as a constant concentration in the mobile phase, without adversely affecting the efficiency of the column by causing band broadening, was calculated by VAN DEEMTER et al. (1956) to be less than $0.5\sqrt{n}\cdot(V_g + KV_L)$ for columns with a large number of theoretical plates. However, KEULEMANS (1959, p. 194) points out that this calculation is based on a technique which was a combination of frontal analysis and elution development, whereas for a straight forward elution development in which the sample is instantaneously introduced, the factor 0.5 should be replaced by 0.02 or a figure in that range of magnitude. Both VAN DEEMTER et al. (1956) and PORTER et al. (1956) found that theory predicts an improvement in column efficiency as a result of an increased concentration of a fixed quantity of sample, or a reduction in sample size while holding the concentration constant, and this interpretation was confirmed experimentally by PORTER et al.

1. Gaseous Samples.

The introduction of gaseous samples presents few technical problems. Two techniques are in common use; *1.* injection through a self-sealing silicone rubber septum by means of a hypodermic syringe and *2.* trapping of a known amount of gas in a tube between two stopcocks, and introduction of this gas by diverting the carrier stream through the tube so as to sweep the sample into the column. By either of these means, samples as large as 20 ml of gas at S.T.P. may be analyzed by using $^1/_4$ inch columns (HARDY and POLLARD, 1959). Sample injection must be performed at the column inlet where a high pressure may exist and although conventional syringes are usually adequate, very high pressures may make necessary the use of special gas-tight syringes. Care must be taken in trace analysis to use a greaseless system or, if the syringe plunger must be greased, to select a lubricant in which the various solutes are insoluble. The accuracy of delivery suffers from needle end effects, but although the total quantity may not be precise to closer than $\pm$ 5 percent, it is usually possible to obtain reproducibility within much finer tolerances. For high pressure work it also may be convenient to replace the serum cap with a pressure "tight" device, such as a piece of rubber compressed by a butt-joint (LITTLEWOOD, PHILLIPS and PRICE, 1955).

Numerous types of gas sampling valves and by-pass systems are in use, and most commercial instruments provide a modification of a four, six or eight port gas sampling valve. All these methods have in common provision for flushing a calibrated gas or metal tube with the sample, turning a valve(s) or stopcock(s) to simultaneously enclose the sample while allowing the flow of carrier gas to by-pass the calibrated sample "loop" until sample introduction is effected by diverting the gas stream through the sampling chamber to flush its contents into the column. In the simplest by-pass system, two or three stopcocks are needed (PERCIVAL, 1957; HARVEY and CHALKLEY, 1955), but more elaborate procedures employing a single stopcock have been devised (GLEW and YOUNG, 1958; VAN DER CRAATS, 1956), in which the sample loop may be filled without interrupting the flow of carrier gas. PERCIVAL (1957) has described a system in which a calibrated sample loop is evacuated and gas is allowed to enter from a cylinder until some desired pressure is established which can be measured by an attached manometer. This procedure can be made applicable to many problems by substituting in place of the sample cylinder a gas generating system of the type commonly used in vacuum manifolds. The various gas sampling valves are easily adapted for the periodic analysis of gas streams for which purpose they have been automated (AGEIKIN, ITSKOVICH and MIKHARLOV, 1957; GRUNE, CARTER and KEENAN, 1957; HOOIMEIJER, KWANTES and CRAAFTS, 1958); this makes them ideally suited for process stream analysis and control. Greaseless valves have been described (RAATS, HARLEY and PRETORIUS, 1957) and help to prevent cross-contamination. When substances are present in trace amounts, the sample loop can be filled with an adsorbent, for instance silica gel, and immersed in a liquid nitrogen bath, in which case it acts as a condensor if the trace substance is condensed and the main component of the stream passed at the low temperature (BRENNER and ETTRE, 1959). When the sample loop is warmed again and the contents flushed into the chromatography column, the concentration of the trace impurity can be determined from a knowledge of the rate and duration of flow through the condensor unit. Separate low temperature columns which can be transferred to the chromatographic instrument have been used for air pollution studies (WEST, SEN and GIBSON, 1958; EGGERTSEN and NELSEN, 1959) and for collection of volatiles produced by plant material (MEIGH, 1959; RHOADES, 1958).

2. Solid and Liquid Samples.

Solid and liquid samples will be considered as a single case because solids are usually dissolved in a volatile liquid prior to analysis. The introduction of a liquid or dissolved solid may be accomplished with a syringe, and OGILVIE, SIMMONS and HINDS (1958) suggest that viscous samples be heated with an infra red lamp until they are fluid, or else they may be diluted with a small amount of lower molecular weight solvent. Although samples between 0.5 and 100 μl in size have been analyzed with a $^1/_4$ inch analytic column, a size of about 10 μl is nearer to the average which these columns can handle efficiently, and this small volume creates special problems. The dead volume in the needle bore and hub, creeping of liquid and expansion of the needle give rise to errors of sufficient magnitude in comparison to the total volume of liquid involved to make impossible the estimation of the size of the sample which has been introduced, even though successive introductions may be reproducible. The CHANEY adapter will improve the accuracy of a conventional syringe, but for best results microliter syringes accurate to 0.1 percent are commercially available (LANGER and PANTAGES, 1958; CARLE, 1957; SWEETING, 1959) and these may be used for gases as well as liquids. A liquid sample may be injected directly onto the surface of the column where it will dissolve in the stationary phase, and if the sample is very small and introduced rapidly, an ideal situation can be closely approximated. It often is sufficient to allow the heated carrier gas to volatilize the solutes, but in an alternative procedure the liquid is introduced into a vaporization chamber consisting of a heated block or flash vaporizer through which carrier gas flows. The latter has the disadvantage that the situation may resemble an exponential rather than a plug sample introduction, but the vapourization can be made extremely rapid.

A unique micro-dipper system for liquids has been developed by TENNEY and HARRIS (1957), and this device finds application in many commercial instruments. A special pipette (1 μl to 25 μl) is filled by capillary action and inserted into a tube which continues to be flushed by carrier gas leaked intentionally through a slightly open large bore cock, until the handle of the pipette forms a seal with an O-ring. The cock is now opened wide to allow the passage of the pipette body through its bore. When the pipette has been inserted to its full length the tip becomes seated in the channel through which carrier gas flows to the column, in such a manner that the carrier gas passes through a hole in the side of the pipette and flushes its contents into a heated vaporizer block and ultimately to the column. This same system can accommodate various gas pipettes, and its precision with both liquids and gases is more than adequate for most purposes.

In another method the weighed sample is placed in a thin walled glass ampoule which is introduced onto a heated by-pass chamber, and the latter is either flushed with carrier gas or evacuated before the bulb is exposed to a bellows crusher (SORENSEN and SOLTOFT, 1958; DIMBAT et al., 1956). SCOTT (1958) has described a device for introducing a liquid sample under conditions of extremely high inlet pressure.

Capillary columns have a very limited capacity, operating efficiently when the total sample is not larger the ca. 10^{-6} or 10^{-7} moles, and each individual component less than about 10^{-9} moles. The most exacting devices can not deliver these quantities of liquid in a reproducible manner, but it is possible to modify conventional procedures of sample introduction by diverting a known percentage of the sample before it reaches the column. This is accomplished by splitting the stream just before the column entrance and balancing the capillary flow resistance against a variable or fixed resistance in the alternate stream leading to the atmosphere. By this procedure a very small but well defined fraction of the sample amounting to as little as 0.02 to 0.1 percent can be introduced (LIPSKY et al., 1959; CONDON, 1959; KAISER and HOLZHAUSER, 1959).

IV. Stationary Phase.

1. Column Construction.

Columns have been packed in tubes of glass, stainless steel, copper, aluminum, monel, nickel and nylon. Except in very special circumstances, for instance when a corrosive material is a sample component, there is very little difference in the performance of these various materials. Aluminum and copper may be coiled after they are packed, which is a convenience when very long columns are needed, and it has been shown by VAN DER CRAATS (1957, p. 331) that coiling hardly effects column efficiency if the coil diameter is at least ten times greater than the tube diameter. Glass has the advantage that the column packing can be inspected, and if the column is straight or U-shaped it can be repacked whenever cracks appear. However, metal columns tend to have better temperature constancy than glass, and their temperature may be changed more rapidly which makes them better suited for programmed temperature chromatography. Monel and copper occasionally cause difficulties, for instance with phenols (LANGENAU and ROGERS, 1960), and at high temperatures copper and aluminum may be reactive and catalytic, whereas glass and stainless steel are not likely to be so. GOLAY capillary columns were originally fabricated from glass, copper or stainless steel, but more recently nylon has proven satisfactory (SCOTT, 1959a) and there is no reason why larger bore nylon tubes might not be used for packed columns.

2. Column Dimensions.

Within reason the column diameter seems not to have a very profound effect on column efficiency. Thus CARLE and JOHNS (1958) found that for a constant sample quantity/cross sectional area ratio, little change in efficiency was noted with columns as large as $^5/_8$ inch, and EVANS, MASSINGHAM, STACEY and TATLOW (1958) had a similare experience with column diameters as large as 3inches. However theory predicts and SCOTT (1958) has demonstrated an increase in the diffusion contribution as a function of column diameter. If time for separation is not a consideration, increased column lengths produce more theoretical plates and very little change in HETP, provided that the inlet/outlet pressure ratio is not excessively elevated (SCOTT, 1958). Moreover, the sample capacity of a column is enhanced in proportion to the square root of its length (KEULEMANS, 1959, p. 26), so that by a combination of increased diameter and length it is possible to increase the load which can be successfully chromatographed to a degree which makes possible the use of this technique on a preparatory scale.

3. Capillary Columns (GOLAY, 1958a, 1958b).

The improved efficiencies obtained with capillary columns are not so much a result of the greatly reduced diameter as of the removal of packing from the column. In the absence of packing irregularities, the eddy diffusion contribution of the VAN DEEMTER equation disappears, and the tortuosity factor of the molecular diffusion term is reduced. GOLAY analyzed the theoretical performance of capillary chromatography columns and predicted that the excess gas resistance of a packed column should make it a far less efficient instrument for chromatographic separations than the capillary tube. Early attempts with coated capillaries were disappointing because of the difficulty of loading the minute samples which are required, and because of the lack of analyzers of sufficient sensitivity and low enough effective volume to detect and resolve the emergent bands. The advent of the flame ionization and low volume β-ray ionization instruments overcame these difficulties and in most cases capillaries have been found to give about five times greater resolution and a shorter time of analysis. Columns are prepared by filling them with a dilute solution of 1 to 20 percent stationary phase dissolved in a volatile solvent, and dispelling the solvent by allowing it to evaporate, or by placing the column in an oven to leave a film approximately 3500 to 5000 Å thick of the non-volatile liquid coated on the walls. Diameters vary from 0.005 to 0.02 inches, the most common being 0.01 inches, and lengths from 60 to 1000 feet have proven useful (SCOTT, 1959b). Longer narrower columns have improved efficiencies, whereas shorter wider capillaries produce quicker separations with some loss of resolution. GOLAY (1958a) has predicted that even greater efficiencies may be possible if the inner walls of the capillary are replaced by a roughened surface or porous layer, and if tubular columns with rectangular cross sections can be constructed.

4. Solid Support.

Present indications are that the site occupied by the liquid phase is to a large extent the fine pores in the surface of the solid support (BAKER, LEE and WALL, 1959), with the additional contribution of a thin film over most of the remaining surface. Non porous supports do not provide sufficient surface area to prevent clogging of the interparticular spaces with liquid, but if the depth of the pores is too great the time required for diffusion between the liquid and the gas phase is sufficient to cause band broadening. The necessity for small particles of uniform size was discussed in connection with the VAN DEEMTER rate theory, where it was pointed out that the main difficulty is the problem of packing a very fine support uniformly. An additional advantage of slightly coarser particles is that they have a higher permeability to the mobile phase, with the result that a lower pressure of gas is required to obtain a specific rate of flow. Indeed JOHNS (1958) has shown that at a constant flow rate the pressure drop across a column is inversely proportional to the particle diameter. However, as would be expected, large particles have a lower specific area, and JOHNS found that at a constant flow rate the retention time of a component was inversely related to particle diameter. According to BAKER et al. (1959) the optimum support is a hard material, with a pore size of one micron or less, and a uniform diameter of between 275 and 300 microns. Other desirable qualities are chemical inertness and the lack of adsorptive properties.

Celite 545, a brand of diatomaceous earth, was selected as a support by JAMES and MARTIN (1952a). This and the closely related Celite 535 have a range of particle size from about 10 to 40 microns, and they usually must be screened to a closer mesh and/or subjected to elutriation for removal of fines. Apparently after coating with the liquid phase, some uncovered areas may remain (SZEKELY, KORMANY, RACZ and TRAPLY, 1958), which gives a residual adsorptive character to the support especially toward acidic, basic and other polar compounds. To a considerable degree this condition can be alleviated by chemical treatment. For instance, JAMES and MARTIN (1952a) washed celite with dilute acid, rinsed it with water and dried it for use in the separation of fatty acids, but they found that an intermediate treatment with 5 percent methanolic sodium hydroxide was necessary before use with amines (1952b). Iron salts which are present in celite, firebrick and related materials are excellent isomerization catalysts and the slightly acidic nature of these substances can cause dehydrations, for instance citral to *p*-cymene or alpha terpineol to dipentene (LAGENAU and ROGERS, 1960). In the case of essential oil analysis, LAGENAU and ROGERS report that these difficulties can be overcome by washing the support with concentrated HCl, rinsing with water, and neutralizing with 1 percent alcoholic KOH or Na_2CO_3 which will leave the material slightly alkaline when dry. In a separation of isomeric xylenes, ZLATKIS, LING and KAUFMAN (1959) obtained improved results with aqua regia treated celite. A coarser grade of diatomaceous earth, the screen fraction of ground firebrick between 30 and 80 mesh, was suggested as a support by KEULEMANS and KWANTES (1955) and subsequent investigations established that JOHNS-MANVILLE C-22 insulating brick and STERMACHOL No. 22 firebrick are superior in performance to Celite 545 (DIMBAT et al., 1956; KEULEMANS and KWANTES, 1957) although DESTY, GODFREY and HABOURN (1958) found little difference between narrow mesh range fractions of the two supports. In a comparison of various diatomaceous earths and their derivatives, JOHNS (1958) reported that 42 to 50 mesh C-22 firebrick produced superior resolution and peak shape than did Dicalcite, Lompoc Celite and Celite 410 of the same particle size. The chemical composition of ground firebrick differs slightly from that of Celite because the former contains small quantities of mineral bonding agents. The firebrick tends to be more polar and both it and some forms of Celite can catalyze the dehydration of certain oxygenated compounds, for instance n-butyl alcohol and cyclohexanol (ANON, 1958). According to OMEROD and SCOTT (1959), gold or silver plating C-22 firebrick will reduce the peak asymmetry associated with polar compounds.

5. "Active" Solid Adsorbents.

Gas-solid chromatography has been reviewed by JANAK (1959) and the uses of solid adsorbents by PATTON (1956); KEULEMANS (1959) and PHILLIPS (1956) give the overall subject excellent coverage in their books. Adsorbent columns have been found to possess efficiencies closely similar to those of the average partition column, and the same general considerations, with minor differences, apply to the behavior of both. One of the more important differences is the affinity of the adsorbent for many of the gases used commonly as mobile phases. Although this is problematical in a few instances, more generally it is an important reason for the continued use of the adsorption technique, because the permanent gases N_2, O_2, Kr, A, H_2, He, and Ne can be fractionated only by adsorption techniques, whereas other fixed gases such as N_2O, CO_2, CO, and NO and low molecular weight olefins and paraffins can be separated on solids at least as well as on partition columns. The major limitations of G.S.C. are the non linear adsorption isotherm and the restriction of the technique to low boiling compounds because of the very high energy of the adsorption bond formed with sparingly volatile substances.

a) Charcoal, Alumina, Silica Gel.

Carbon, alumina and silica gel will fractionate roughly the same spectrum of compounds, which encompasses most of the permanent gases and light molecular weight hydrocarbons, the range being greater for paraffins than for olefins. Alumina and silica gel are polar adsorbents and exhibit a selectivity on the basis of solute polarity as well as molecular weight. They are both useful with hydrocarbons containing as many as four or five carbons (HRAPIA and KONNECKE, 1956; SOKOL, 1957; SCOTT, 1959b) although alumina has been found to be preferable for these compounds (GREENE and PUST, 1957), whereas charcoal, which discriminates between substances entirely on the basis of molecular weight, is restricted to a limit of about two or three carbons (JANAK and RUSEK, 1953; EUCKEN and KNICK, 1936). Activated carbon is particularly suited for the analysis of permanent gases, separating nitrogen from oxygen at room temperature (MADISON, 1958; WENCKE, 1957), a fractionation which can be accomplished only at dry ice/acetone temperature with silica gel (SZULCZEWSKI and HIGUCHI, 1957). Alumina irreversibly binds carbon dioxide and its adsorbency and that of silica gel are affected by their water content, although in different ways (SCOTT, 1959b; CREMER, 1959; TURKEL'TAUB, PORSHNEVA and KANCHEEVA, 1956).

b) Molecular Sieves.

Molecular sieves, which probably offer the most successful approach to the separation of many permanent gases, belong to a family of minerals known as zeolites and consist of hydrated silicates of aluminum with alkalis or alkaline earths or both. The water which they contain can be readily expelled by outgassing at high temperatures, with a resultant intumescence; the name zeolite means literally "to boil a stone" from the Greek. By the action of various reagents, crystalline substitution products are obtained in which the alkalis or alkaline earths are replaced by cations. Unlike most hydrated materials when the water of hydration is expelled from crystalline zeolites they do not collapse or rearrange, but rather the physical structure remains intact and a network of internal channels is produced comprising about one half the volume of the material. Not only can many crystalline zeolites of natural occurrence be synthesized in the laboratory, but also many species not known to exist in nature have been produced (BARRER, 1948; BARRER and BAYNHAM, 1956; BARRER, BAYNHAM and McCALLUM, 1954), and in particular a species known as zeolite type A (BRECK, EVERSOLE, MILTON, REED and THOMAS, 1956) is widely used in gas chromatography. The intercrytalline voids of zeolites have dimensions about the size of molecules, and depending upon the exact size of these pores, various molecular species may be readily or slowly occluded or completely excluded (BARRER, 1949). Not only the size but also the shape of a molecule is important in determining whether it can enter a zeolite channel, in which case it may be strongly adsorbed. Derivatives of the synthetic type A zeolite have a cubic A crystal structure with a pore diameter of 4.2 angstroms, but because of the position sodium occupies in the voids, type 4A will admit only molecules smaller than 4 angstroms, whereas type 5A with sodium exchanged by calcium which is situated differently, has an effective diameter of 5 angstroms (BRECK et al., 1956); 10 Å (type 10 A) and 13 Å (type 13 X) sizes are also available. All of the cavities and pores of a molecular sieve are exactly identical because of the unique crystal structure of the synthetic zeolites, which means that the sieve will produce a sharply delineated dimensional effect not possible when there is a distribution of pore sizes. To illustrate the "sieving effect," BARRER and IBBITSON (1944a, 1944b) and others (MAIR and SHAMAIENGAR, 1958; BRENNER and COATES, 1958; ZIEGENHAIN, 1957) have shown that n-paraffins boiling in the gasoline range have molecular diameters less than 5 angstroms and hence will enter and be adsorbed in the pores of a 5A sieve, whereas the branched paraffins, cycloparaffins and aromatics are too large and pass through unimpeded. Generally speaking, the larger molecular weight hydrocarbons are so strongly adsorbed if they enter the pores of the sieve that they can only be desorbed by drastic methods, for instance with super heated steam or warm liquid ethanol (MAIR and SHAMAIENGAR, 1958), and with these substances zeolites are used primarily as precolumns to effect a crude separation by class. However if all or most of the molecules in the gas mixture can pass freely through the pores, selective separation can be made on the basis of adsorptability. Although their capacities may differ, adsorption characteristics of all the molecular sieves are the same, and they show a preference for carbon unsaturation and molecular polarity (ANON 1957a) as well as molecular weight (KNIGHT, 1959, p. 23). A separation of hydrogen, oxygen, nitrogen, methane and carbon monoxide was reported by KRYACOS and BOORD (1957) and other gases for which zeolites subsequently have been found useful are He, A, Ne, hydrogen isotopes, Kr, NO_2, D_2O, NO, NO_2, methane, ethane and acetylene. Water is so strongly adsorbed that it will replace all other adsorbents, and the water content of a zeolite can affect markedly its retention characteristics (JANAK, KREJCI and DUBSKY, 1959; BARRER and ROBINS, 1953) as can the addition of other polar sorbates such as ammonia and methylamine (BARRER and REES, 1954a, 1954b). Working at low temperatures, BARRER and coworkers (BARRER and BROOK, 1953; BARRER, BULTITUDE and SUTHERLAND, 1957) found that gases which had relatively slow and closely similar rates of 'sorption' at higher temperatures now showed marked differences in the kinetics of their adsorption processes, which should be a useful consideration in gas chromatographic separations.

c) Tailing Reducers.

The surface of an adsorbent may be modified by the addition of a small amount of non-volatile liquid which results in an enhancement of both the range of applicability with respect to carbon number, and the symmetry of the resultant peaks. Coating Pelletex with 1.5 percent squalene (*w/w*) improves separations of C_5 and C_6 saturates and olefins (KNIGHT, 1958; EGGERTSEN, KNIGHT and GROENNINGS, 1956), and SCOTT (1959b) and MORROW and BUCKLEY (1957) modified alumina with 2 percent silicone oil (*w/w*) as a "tailing" reducer and reported good resolution for C_1 to C_5 saturates. WHITE and COWAN (1958) examined from a thermodynamic viewpoint the possibility of obtaining symmetrical elution curves and upon this basis they investigated and reported successful the use of montmorillonite with organic molecules base exchanged on its surface.

d) Chromathermography.

Another approach to improving the separations obtained with G.S.C. has been selective desorption induced by the imposition of a temperature gradient along the column, a technique known as chromathermography which was first introduced by EUKEN and KNICK (1936) and subsequently used by many others (ZHUKHOVITSKII, TURKEL'TAUB and SOKOLOV, 1953; HAMMAR, 1951; TURKEL'-TAUB, 1954; BUCUR, 1956; ZHUKHOVITSKII and TURKEL'TAUB, 1954, 1956; GREENE, MOBERG and WILSON, 1956). This method is based on the lowered adsorptivity which solids exhibit at elevated temperatures, and it may be implemented either by moving a heater along the stationary phase or maintaining a temperature gradient across the length of the column. GREENE et al. (1956) report that chromathermography reduces the time required for analyses and improves peak sharpness and symmetry. A slightly different technique known as programmed temperature gas chromatography (PTGC), in which the temperature of the entire column is gradually raised is often used in G.L.C.

e) Multiple Columns.

Although one particular adsorbent may not separate a complex mixture of components, a satisfactory combination of materials can often be found. Either fixed proportions of adsorbents may be mixed together (KREJCI and TESARIK, 1959), or else two-stage or multiple columns may be used (BRENNER and CIEPLINSKI, 1959; SEVENSTER, 1958b). There are no reports of capillary adsorption chromatography, but in principle this should be possible either by coating the surface of a capillary with an adsorbent, for instance Pd has been coated on quartz (THOMAS and SMITH, 1959), or by using a small bore aluminum capillary in which case the surface aluminum oxide might act as the adsorbent.

6. Liquid Phase for GLC.

a) Column Preparation.

The inert support is coated by dissolving the liquid phase in a volatile solvent such as methylene chloride, acetone, methanol or pentane, wetting the solid support with this solution and then allowing it to evaporate. Although air drying is frequently used, it is permissible to hasten the process by stirring the particles over a steam bath, or by spreading them on a tray which may be heated with an I.R. lamp or in a vacuum oven (if the liquid phase has a low enough vapour pressure). The optimal liquid phase/solid phase ratio will vary depending upon the nature of the non-volatile liquid and the amount of sample. A thin coating reduces the distance through which the solute must diffuse and thus minimizes the mass transfer contribution, but the amount of sample which can be handled without overloading the column is diminished and the likelihood of exposing "bare" surface with its possible catalytic or adsorptive properties is enhanced. Coatings of between 15 and 50 percent liquid phase (*w/w*) have often been used, but 20 to 25 percent is usually satisfactory, and approximately 75 ml of liquid will be needed to wet 100 ml of solid support (JOHNS, 1959). The column is prepared by plugging one end with glass wool or a metal filler, pouring in the packing while constantly vibrating the column by tapping it or holding it against an electric vibrator, and plugging the remaining end when the packing has filled to within about one-half inch of the top. After the column has been packed it is customary to condition it with a high temperature treatment, especially if it is intended for use at elevated temperatures. Baking the column at some temperature in excess of that which it will experience during operation serves to drive off residual solvents and other volatile contaminants, helps to distribute the stationary phase over the support, and hastens the completion of incipient chemical reactions. Heating for one hour with a normal or supranormal flow of carrier gas is often sufficient, after which the column may be connected to the analyzer and the base line stability checked. As an example of successful column conditioning, FELTON (1958) baked a column containing 20 percent D.C. high vacuum silicone grease at 400° C for 2 to 3 hours with a flow of 25 ml/minute of helium, and then operated it successfully at 320 to 370° C for extended periods of time.

b) Selection of the Stationary Phase.

The stationary liquid is selected primarily on the basis of requirements imposed by the sample composition, but certain other factors are of fundamental importance, and in general they are related to the temperature of operation. The vapour pressure of the liquid should not exceed a certain maximum value, estimated at 0.01 to 0.1 mm mercury at the column temperature (ANON., 1957b), which condition will be satisfied if the column is operated 70° below the stationary phase's "high vacuum" boiling point (i.e. boiling point at 0.1—0.5 mm mercury) (ANON., 1958) or about 250° C below its boiling point at atmospheric pressure (JOHNS, 1959). It is important that the liquid phase not be excessively viscous at the temperature of column operation, as this condition will amplify the mass transfer contribution by making less rapid the movement of the sample components in and out of the liquid phase. Also the stationary liquid must be stable at all temperatures to which it is exposed. If these conditions are adhered to closely, the lifetime of a column may exceed several thousand analyses.

In order to obtain a proper separation of two components, it is necessary that they have different distribution coefficients in the chromatographic system and sufficient solubility in the liquid phase to make practicable a fractionation based upon whatever distinction in distribution coefficients exists. The likelihood that a separation will occur depends also upon the vapour pressures of the sample components. These relationships can be summarized in a equation relating α, the relative volatility, to the vapour pressures of the pure components (P_1° and P_2°) and the tendencies of the molecules to escape from the solution in a manner different than would be expected from their concentrations and the vapour pressure of the pure solutes (γ_1 and γ_2):

$$\alpha_{1,2} = \frac{\gamma_1 P_1^0}{\gamma_2 P_2^0} = \frac{K_2}{K_1} \qquad \text{(XIII)}$$

where K_1 and K_2 are the distribution coefficients. The more the relative volatility differs from unity, the greater is the likelihood that a successful separation will occur. In this context, γ is the RAOULT's law deviation factor, and the γ ratio is indicative of the selectivity which the liquid phase exhibits for the two components. Various cohesive forces which may contribute to keeping a molecule in solution by solute-solvent interaction are: (1) non-polar dispersion forces (LONDON), sometimes called VAN DER WAALS forces, (2) induction forces (DEBYE) between permanent and inducible dipoles and (3) orientation forces (KEESOM) between permanent dipoles, the hydrogen bond being a special case. In particular it can be seen that polarity is of extreme importance, and this fact gives rise to a general "rule of similarity" which states that "for an efficient, normal separation the components of the mixture (the solutes) and the stationary liquid (the solvent) should preferably show some resemblance" (KEULEMANS, 1959, p. 20). If a homologous series of non-polar compounds is partitioned against a non-polar stationary liquid, the separation which occurs will be based entirely upon boiling points (i.e. in general, molecular weight), whereas the retention times of polar compounds, fractionated on a polar stationary phase, will be related to both the boiling point and polarity. As a corrolary it is important to bear in mind that molecules with inducible dipoles will be detained to a greater extent by a polar phase than completely non-polar compounds. Quantitatively the energy of the hydrogen bond is intermediate between that of most cohesive forces and the chemical bond, with the result that hydrogen bonding, when operative, can be an overriding consideration. For instance on lubrol MO, a liquid phase containing a polyethylene-oxide sequence, primary and secondary amines are retarded whereas tertiary amines which can not form a hydrogen bond are not (JAMES and MARTIN, 1952a). Occasionally the postulate that the liquid phase be chemically inert has been deliberately broken, and use made of specific interactions such as adduct formation. For instance the weak complex formed between unsaturated hydrocarbons and the silver ion contributes to the separation of olefins on columns containing silver nitrate dissolved in a suitable organic solvent (BRADFORD, HARVEY and CHALKLEY, 1955; BEDNAS and RUSSELL, 1958).

In order to make maximum use of the "*rule of similarity*" it is advantageous to classify compounds, albeit crudely, into categories of similar cohesive energy. The following system has been suggested (KEULEMANS, 1959, p. 169), in which the classes are arranged in order of decreasing cohesive force.

I. Water, polyalcohols, aminoalcohols, oxy-acids, polyphenols, di- and tri-carboxylic acids.

II. Alcohols, fatty acids, phenols, primary and secondary amines, oximes, nitro compounds and nitriles with α-H atoms.

III. Ethers, ketones, aldehydes, esters, tertiary amines, nitro compounds and nitriles without α-H atoms.

IV. Chlorinated hydrocarbons, aromatic and olefinic hydrocarbons.

V. Saturated hydrocarbons, carbon disulfide, tetrachloromethane.

If the solute and solvent are members of the same or adjacent classes there will occur a selective retention, whereas if they differ widely in class the stationary phase will have little attraction for the sample component. Although this classification takes into account forces between permanent dipoles, for instance the extreme case of hydrogen bonding, and also non-polar forces between solute and solvent, it does not treat orientation effects or the special cases of chemical bonding and adduct formation. However if it can be predicted that the interaction of the solvent and solute will lead to dipole induction, the classification will indicate what effect mutual polarizability is likely to have on the retention time of the sample relative to other components of the mixture.

The number of stationary phases available is large enough to render a thorough discussion of their properties beyond the scope of this review. An especially complete treatment of this topic may be found in a book published by MAY and BAKER Ltd. called "Materials for Gas Chromatography" (*Anon.*, 1958) which lists more than 50 examples with appropriate references, and this information can also be culled from the indices of the Gas Chromatography Abstracts (KNAPMAN, 1958, 1959) and tables of chromatographic data in supplements to each of the volumes of The Journal of Chromatography. Guides to column selection are provided by many instrument manufacturers and several reviews and books give noble coverage to this topic (c.f. JOHNS, 1959; HARDY and POLLARD, 1959; KEULE-

MANS, 1959 — in appendix). The multiplication of the number of substances used as stationary phases makes difficult the indexing of retention volumes. Because this tendency toward diversification is rendered unnecessary to some extent by the improved efficiencies of both packed and capillary columns, the "Committee on Nomenclature of the London Symposium" on vapour phase chromatography recommended that whenever possible the stationary phase be chosen from six substances listed in Table 1 (DESTY, 1957, p. xii) to which has been added dimethylsulfolane which in 1958 the committee suggested for use in gas analyses as shown in Table 2 (KEULEMANS, 1959, p. 220).

It will be noted in Table 1 that the recommendations encompass a selection of compounds ranging from negligible to high polarity, and this choice is also offered for both applications in Table 2. For specific problems it will often be necessary to extend beyond the limited scope of the tables, and in extreme cases use has been made of two stage or multiple columns (FITZGERALD, 1959; HURN, CHASE and HUGHES, 1959; SKIRVAN and HOELSCHER, 1959; AUSLOOS and MURAD, 1958, FREDERICKS and BROOKS, 1956; NUNEZ, ARMSTRONG and COGSWELL, 1957; TAYLOR and DUNLOP, 1958; McFADDEN, 1958).

Table 1.

Solvent	Temperature range (° C)	Polarity
n-hexadecane	20 to 50	none
squalene	20 to 150	none
benzyldiphenyl	80 to 150	slight
dinonyl phthalate	20 to 100	moderate
dimethylformamide	—20 to 20	high
diglycerol	20 to 100	high
(dimethylsulfolane)	0 to 25	high

Table 2.

Application	Substance	Temperature (°C)
gas analyses	squalene	0 to 25
gas analyses	dimethylsulfolane	0 to 25
other analyses	squalene	50 to 150
other analyses	dinonyl phthalate	50 to 130
other analyses	diglycerol	50 to 150

Temperatures higher than 150° C are not included in the recommendations because the committee did not consider that this range had been adequately investigated. Several difficulties are incurred at higher temperatures, most important of which are volatization or decomposition of the stationary phase, convergence of vapour pressure and γ ratios, and catalytic decomposition of the sample. Thermal agitation is the main force driving molecules out of solution, and although a limited rise in temperature may affect the volatility of two compounds only slightly, the absolute values of the partition coefficients can decline to such an extent that the samples may not be retained sufficiently long to permit a separation. Thus even fractionations which can be accomplished successfully at low temperatures may become more difficult as the temperature is raised. For temperatures approaching 250° C, silicone oils, apiezon greases, arylalkylsulfonates ("Tide"), bitumen (KEULEMANS, 1959, appendix), salts of fatty acids (BARBER, PHILLIPS, TUSA and VERDIN, 1959; PHILLIPS, 1958), asphaltene (KNIGHT, 1959, p. 52), polyethylene and stearone (TAYLOR, 1958) are suitable. However for temperatures above 250° C, Apiezon L and high vacuum silicone greases (ADLARD and WHITHAM, 1958; NOGARE and SAFRANSKI, 1958; FELTON, 1958), polyphenol tars (BAXTER and KEEN, 1959), inorganic eutectics (PHILLIPS, 1958; WACHI, 1959; HANNEMAN, SPENCER and JOHNSON, 1960) and silicone gum rubber (MARTIN, BENNETT and MARTINEZ, 1960) have been successfully employed. A very serious problem often encountered in high temperature chromatography is migration of the liquid phase through the column packing by a process known as "weeping." Explanations such as an evaporation-condensation cycle and decreased viscosity coupled with sample friction, capillary attraction and liquid diffusional effects have been advanced, but the cause is still uncertain. The best solution is to reduce the amount of liquid phase to between a few and 10 percent (DREW, 1959, p. 100), which should give adequate surface coverage even though it entails a considerable loss in capacity. It is possible to work at temperatures 100° C above or below the boiling point of a sample component, which considerably extends the useful range of gas chromatography and makes possible the separation of fairly complex mixtures at a single column temperature. However, towards the limits of this temperature allowance, retention times are so long or short as to make it highly desirable to employ programmed temperatures for mixtures of compounds having widely differing boiling points.

c) Improvement of Peak Symmetry.

Peak distortion in GLC is almost invariably a result of adsorption by the support, and several methods have been described previously by which the support may be treated to reduce its polarity. Generally the effect is negligible with non-polar compounds, but it can be a serious problem with polar substances (KWANTES and RIJNDER, 1958), especially when using ground firebrick or non treated celite and a stationary liquid having a lower polarity than the sample component. The effect can be diminished by increasing the polarity of the liquid phase, but other considerations may make this an untenable alternative. Another approach based on the same principle is that of JOHNS (1958) who found that the addition of 2 percent of several highly polar materials to 30 percent silicone 550 partition liquid reduced tailing on the column. A similar result was obtained by BAYER (1958) although he used a higher concentration of the tailing reducer. In a unique approach to this problem, KNIGHT (1958) added a volatile tailing reducer continuously to the mobile phase by saturating the gas at room temperature with a substance similar in type to the sample. Good results were obtained for alcohols when the gas stream was saturated with water, and the improvement was even more marked at higher temperatures. Temperature programming has a similar effect because it increases the volatility of the "tail" during the analysis, and NOGARE and BENNETT (1958) cite the use of this technique in obtaining symmetrical peaks for alcohols. Tailing can come about for other reasons, such as exponential sample introduction, in which case the cure is an entirely different matter.

V. Temperature Control.

The retention time of a sample component will decrease about 5 percent per 1° C increase in temperature, or more exactly stated the log of the retention time is inversely proportional to the absolute temperature. MARTIN (referred to in JOHNS, 1959, p. 21) estimates that the vapour pressure of the sample may range from 10 to 1000 mm mercury, and KEULEMANS (1959, p. 22) has proposed that as a rule of thumb it is best to aim at an elution period of between 10 and 30 minutes, which most often can be obtained by setting the column at or slightly above the average boiling point of the sample. Added to this is the consideration that certain detection systems are sensitive to temperature changes, which again implies the necessity for accurate control of temperature. In order that condensation of the material in the detector chamber may be prevented, it is necessary that it be maintained at a temperature at least as high as that of the column, which is accomplished most frequently by housing both, along with the sample injection system, in close proximity inside a thermostatically controlled chamber. Occasionally if there is danger of condensation from high boiling materials it is necessary to supplement the detector with an auxiliary heater, or as in the case when column temperature has to be changed during a run, the column and detector may be housed in separately controlled environments. The first columns to be used in gas chromatography were constructed from glass, and since it is not feasible to coil these either before or after packing, rather long narrow thermostated ovens were required. These have the disadvantage of a geometry which is not particularly suited to the regulation of heat input and distribution (DREW, 1959, p. 103). However good control has been achieved by several methods, and there is a tendency even in the era of coiled columns to maintain the straight configuration for the convenience it offers in column storage and quick interchange.

1. Types of Heating Units.

a) An oil bath (KEULEMANS and KWANTES, 1957) provides very accurate temperature control (ca. ± 0.05° C), but its heat capacity is sufficiently high to make difficult a rapid change in temperature, as required in programmed temperature applications. KEULEMANS (1959, p. 63) suggests that a bath oil be selected which has the lowest possible viscosity and still acceptable volatility characteristics at the intended bath temperature.

b) A vapour jacket surrounding the column becomes a suitable thermostat when a constantly boiling liquid of good thermal stability is refluxed into the closed system at a controlled pressure (LITTLEWOOD et al., 1955). The apparatus may be operated below, at, or above atmospheric pressure and a few of the different liquids which can be used to cover select temperature ranges are acetone (30 to 50° C), benzene (50 to 80° C), toluene (80 to 110° C), p-xylene (110 to 140° C), 2-octanol (150 to 180° C), propylene glycol (160 to 190° C) and ethylene glycol (170 to 200° C). Like the previous method this procedure is not applicable to programmed temperature chromatography, and indeed the extremely long time which is consumed in bringing the vapour bath to temperature equilibrium is a distinct disadvantage. Also it is difficult to find stable liquids which can be used at high temperatures, and when available and used their hot vapour constitutes a distinct hazard.

c) An air bath provided with a rapid circulation (MARTIN et al., 1960; ASHBURY, DAVIES and DRINKWATER, 1957; WHITHAM, 1957, p. 195) provides slightly less rigid temperature control (ca. ± 1° C) than does an oil bath, but it permits rapid adjustment of the temperature. A standard laboratory drying oven may be modified for this purpose, or ovens have been designed to permit heating as rapidly as 35 to 48° C per minute and cooling at nearly twice that rate with temperatures up to 500° C (MARTIN et al., 1960).

d) A block heater can be used conveniently with coiled columns. There are various constructional modifications, but generally a heating assembly is imbedded in a mass of metal which is selected for its high heat conductivity and so shaped that the coiled column may wrap around or slip inside and in close proximity to the block (DREW and MCNESBY, 1957). Temperature regulation using a metal heat sink may be accurate to m 0.5° C, but the rate of heating and cooling is slowed by the heat capacity of the large mass of metal.

2. Programmed Temperature Chromatography (PTGC).

Programmed temperature gas chromatography allows complex mixtures whose components differ widely in boiling point to be fractionated with excellent peak symmetry and sharpness, and it reduces the time required for analyses. These advantages were first demonstrated for GLC by GRIFFITH et al. (1952), although chromathermography had been described for GSC at a much earlier date. Although

PTGC was not fully appreciated for many years, now a programming feature is available in many commercial instruments. The subject apparently has not been reviewed comprehensively although good accounts can be found in DREW (1959, pp. 97—115), a brief coverage is available in MARTIN et al. (1960) and DREW and McNESBY (1957), while the theory of PTGC has been discussed by HABGOOD and HARRIS (1960), NOGARE and LANGLOIS (1960), and STEWART (1960).

Temperatures may be programmed from the freezing point to just below the boiling point of the stationary liquid. At the freezing point, the solubility of low boiling substances in the liquid phase is so reduced that they pass through the column without any retention, whereas if the boiling point is approached a sudden break in base line stability is indicative of liquid phase volitization. At a sufficiently low starting temperature all components of the sample may be condensed in the first plate of the column, which insures a "plug" type of sample introduction irrespective of the method used. A very considerable thermal separation is realized as the column is warmed and each component brought separately to a temperature at which it can begin to migrate. The length of the column should be so chosen that at about the time any component has been warmed to a temperature which causes its rate of movement to approach that of the carrier gas, it arrives at the exit. Fortunately each successive component will reach its limiting velocity after traversing about the same length of column, so that band spreading due to samples passing unhindered through an ineffective length of column can be avoided. As the temperature increases, distribution coefficients of each sample fraction approach very low values at which the distribution isotherms almost invariably are linear, with the result that "tails" literally catch up to produce symmetrical bands. The quantitative interpretation of PTGC is made easy by the circumstance that although the peak heights will vary with temperature, the area under a peak is not related to either temperature or flow rate and hence will be equally valid for all components regardless of the temperature region in which they were eluted. If a compensating differential detector is used, base line stability can be improved by temperature programming both a sample column and an identical control column (KNOX, 1959), as this assures that the response to stationary phase vaporization will be equalized. The merits of several forms of heating systems were studied by MARTIN et al. (1960). Methods investigated were the air bath, a nichrome wire heating tape wrapped directly around the column, and self heating by passing high current at low voltage directly through the resistance of the metal column itself. The conclusion was reached that while the air bath gave the lowest temperature limit and slowest rate of warming, it was still more than satisfactory in both these respects and surpassed the other methods in almost every criterion. DREW and McNESBY (1957) and DREW (1959, pp. 97—115) have described apparatus using either a block heater or a combination air bath and self heating column, in which provision is made for the immersion of the column in a liquid nitrogen bath at the start of the analysis. When the bath is removed, the heater is turned on and the temperature can be programmed starting from a very low value. Linear programming of the temperature has been accomplished by driving up scale the set point of a temperature controller which was actuated by a thermocouple or resistance thermometer on the column (HARRISON, KNIGHT, KELLY and HEATH, 1958; NOGARE, 1958; NOGARE and BENNETT, 1958; MURRAY, 1958). A linear rate of temperature increase has the advantage that it produces more reproducible separations than uncontrolled heating, and probably a slightly better separation since the relationship between the temperature and carbon number is nearly linear (NOGARE and LANGLOIS, 1960) in marked contrast to the correlation between log retention time and carbon number in isothermal chromatography.

C. Techniques.

I. Sample Preparation.

1. Removal of Water.

Whenever possible a sample should be completely rid of all water before it is analyzed by gas chromatography, in spite of the fact that certain detectors show verly little response to the vapour. A major difficulty is that water is often adsorbed by the stationary support in which case it may tail so badly that it interferes with other substances. Moreover, many partition liquids are completely incompatible with it; for instance small quantities of water will render such phases as high boiling hydrocarbons or esters completely unsuitable. Frequently in biological samples water unavoidably must remain a major component, and aqueous solutions have been resolved in a few cases (BENNETT et al., 1958; HUNTER, ORTEGREN and PENCE, 1960; BODNAR and MAYEUX, 1958; MURRAY and WILLIAMS, 1956; BROWNING and WATTS, 1959), for which purpose glycerol and polyglycol columns frequently are adequate. MADISON (1958) and SIMMONS and SNYDER (1958) have used preparatory columns to affect crude separations of the bulk material, after which the fraction containing the trace component was trapped and reintroduced. Such a method should work well for water, especially if it is coupled with a "cutting" device which makes unnecessary the intermediate steps. Several commercial instruments are provided with plumbing arrangements which allow a preliminary fractionation to be carried out on a pre-cut column, and after the sample fraction has passed from this into the main column, by proper arrangement of valves it is possible to divert the carrier stream around the pre-cut column and continue the chromatographic development. The undesired bulk fraction can then be backflushed off the pre-cut columns by reversing the direction of carrier gas flow through it. CHUNDELA and JANAK (1959) determined alcohols in blood using as a stationary phase the hydrophilic substances diglycerol or Carbowax, and to minimize the analysis time the water, which chromatographed very slowly, was backflushed off the column. Small amounts of water vapour may be removed by a precolumn in which they are irreversibly adsorbed; for instance both water and carbon dioxide were removed by ascarite in a method developed by EGGERTSEN and NELSON (1958) and it was shown by FARRINGTON et al. (1959) that anhydrous potassium carbonate removes water vapour without adsorbing even traces of other substances.

2. Esterification.

If the sample is esterified, the column temperature may be reduced, and also substances which otherwise would decompose before volatizing occasionally can be analyzed. Three methods are in frequent use. (a) **Methanolic HCl** (c.f. JAMES, 1960, p. 19). This procedure, based on the classic method of E. FISCHER, is time consuming and not applicable to very small samples. (b) **Diazomethane** (c.f. JAMES, 1960, p. 19). The reagent can be prepared from Diazold (N-methyl-N-nitroso-p-toluene sulfonamide, known as MNSA) by a well known procedure (for instance, see KIRKLAND, 1960). The major drawbacks are the toxicity and danger of explosion of the reagent, the necessity for using freshly prepared diazomethane, and the possibility of loss of polyunsaturated acids by reaction with diazomethane. The reaction is fairly rapid, but SCHLENK and GELLERMAN (1960) have noted that it can be made instantaneous if gaseous diazomethane is introduced into a solution containing acids in 10 percent methanolic ether, and they describe a micro-esterification procedure. The preparation of M-C^{14}-NSA from $BaC^{14}O_3$ has been reported by STOLL, RUSCHMANN, WARTBURG and RENZ (1956), and this precursor could be used to form labeled methyl esters. (c) **Methanolic BF_3** (MITCHELL, SMITH and BRYANT, 1940). This modification of a FISHER esterification has the advantage that it can be accomplished in about 10 minutes, utilizes a reagent of more than 4 months shelf-life, and avoids some of the dangers inherent in the preparation and use of diazomethane.

II. Fraction Collection.

With minor exceptions, all fraction collectors are composed of a series of interchangeable U-tube freezing traps immersed in liquid nitrogen. However, it has proven surprisingly difficult to obtain efficient recoveries by this method. CRAATS (1957, p. 96) has proposed that upon rapid cooling there often occurs a formation of fog or minute crystals which are flushed out of the trap by the non-condensable carrier gas. According to CRAATS this can be circumvented by heating the gas stream between several successive traps, and he cites recoveries of 99 percent for C_2 hydrocarbons using a glass tube bent into 3 hairpin turns. GLUECKAUF (1957, p. 29) found that upon filling the empty space in a U-trap with cotton wool, complete condensation up to the vapour pressure of the sample was assured, presumably because fog droplets collect on the filling material. ATKINSON and TUEY (1958) and KRATZ, JACOBS and MITZNER (1959) solved the problem of fog formation by exposing the exit arm of the U-tube to an electrostatic precipitator consisting of two electrodes maintained at a very high potential. When the vapour pressure was too high at liquid nitrogen temperatures, GLUECKAUF (1957, p. 97) made use of a small plug of charcoal, and found it to be a very efficient trap, whereas activated alumina is recommended by AMBROSE (1957, p. 210) in lieu of charcoal which he claims has too strong a retention for organic compounds. ATKINSON and TUEY (1958) experienced difficulty in obtaining sufficient cooling of the hot exit gases, but were able to design a trap with small clearances which gave acceptable recoveries when operated with a turbulent flow. It has been claimed that the maintenance of a low outlet pressure increases the efficiency of recovery (CVETANOVIC and KUTSCHE, 1957), but this inconvenience probably is unnecessary if the precautions described above are observed.

III. Sample Identification.

Gas chromatography is primarily a procedure for separating a mixture into purified fractions. Since it is carried out with minute quantities of material, the identification of the purified components demands procedures that can cope with equally limited amounts of compounds. However, additional information concerning sample identity can be derived from the chromatograms themselves because the structural properties of molecules determine their chromatographic behavior and certain types of detection systems exhibit a specificity toward molecular classes or physical properties.

1. Use of Standards.

The retention times of unknown compounds may be compared to known standards on columns having different liquid phases and at several temperatures. Cochromatography of a mixture containing the unknown and a known quantity of a pure compound thought to be identical with it, will either show an additive enhancement of the peaks or a resolution into two semi-distinct bands. Tables of retention volumes are a regular feature in The Journal of Chromatography (and see HARDY and POLLARD, 1959; SCHOLLY and BRENNER, 1959; RAUPP, 1958; KELKER,1959; and the indices of the Gas Chromatography Abstracts edited by

KNAPMAN, 1958, 1959), and these greatly simplify the tracing down of a compound although, whenever possible, it is best to verify such conclusions by other methods of sample identification.

2. Homologous Series Plots.

For a homologous series of compounds, log retention volume is related linearly to the number of carbon atoms, excluding the lowest members of the series. This was first demonstrated by JAMES and MARTIN (1952b) for normal and iso-fatty acids, and later by RAY (1954a), LEWIS, PATTON and KAYE (1956), and others for a variety of classes of compounds. These plots can usually be established by comparison of a few well selected members of a series, and the results obtained may be extended to unknowns. Compounds belonging to different homologous series may display nearly identical retention volumes, and the technique of homologous series plots is ambiguous unless the results obtained with at least two different stationary phases are compared. JAMES, MARTIN and SMITH (1952) plotted retention times for one stationary phase vs. those for another and obtained a series of straight lines each passing through the origin with a slope characterictic of the class of compound. The data can also be plotted as log retention $time_1$ vs log retention $time_2$ (PIEROTTI, DEAL, DERR and PORTER, 1956; LEWIS et al., 1956), in which case the intercept of the compounds behavior on each of the two columns will be characteristic of the class. For comparing results obtained at different temperatures, it is permissible to presuppose a linear relationship between log retention time and the reciprocal of the temperature (LITTLEWOOD et al., 1955).

3. Detectors with Different Responses.

If two detectors have responses which are dependent upon different physical properties, then it is possible to determine empirically a relationship between individual compounds or classes of compounds and the ratio of response in the two measuring systems relative to some standard. This was successfully demonstrated by GRANT (1958) for aromatic, paraffinic and naphthenic hydrocarbons using a flame emissivity detector in series with a katharometer. It should be possible to calculate the carbon number and molecular weight of an unknown by combining a detector which measures very nearly the absolute molar quantity (for instance a β-ray instrument or a katharometer using helium as carrier gas) with an instrument whose response is dependent upon carbon number (a combustion technique or flame ionization detector) or molecular weight (gas density balance). LIBERTI, CONTI and CRESCENZI (1956) have developed equations from which molecular weights can be calculated by comparing the peak areas obtained with a gas density meter when a known and unknown vapour are eluted using mobile phases of very different molecular weight.

4. Auxiliary Instruments.

Although the mass spectrometer, infra red and ultraviolet spectrometers, and radiation detectors have been successfully modified for stream analysis, they are capable of much higher sensitivity and resolution when used in a static system. The qualitative analysis of multicomponent mixtures by these methods, especially the mass spectrometer, usually results in information which is too complex to be readily interpreted, but the fractions collected after gas chromatography are of sufficient purity to be analyzed by any of these means, although for U.V. and I.R. analysis the amounts may not always be sufficient. For infra red spectroscopy, micro cells of 0.002 ml liquid volume are available, small gas cells of 7.5 ml volume have been described by RUSSELL (1958), ANDERSON (1959) has discussed the advantages of converting liquids to vapours prior to spectroscopic analysis, and GALLAWAY, JOHNS, TIPOTSCH and APLIN (1959) have developed microtechniques.

5. Electron Affinity Spectroscopy.

When a β-ray ionization chamber is operated at low voltages, electrons released by argon are collected on the anode before they have time to recombine with positive ions. An organic vapour added to the argon can capture the electrons to form negative ions which recombine with positive ions, thus preventing electrons from reaching the anode. Electron capture is recognized by a decreased base current. LOVELOCK and LIPSKY (1960) noted that the ability of a molecule to capture electrons is related to its predominant functional group and not to its size or hydrocarbon configuration. By varying the operating voltage of the detector, it is possible to identify a voltage at which electron capture predominates for a particular species with a resulting negative peak.

6. Functional Group Classification.

Color tests with classification reagents can detect as little as 1 μg of substance (DUBOIS and MONKMAN, 1959) depending upon the particular assay. WALSH and MERRIT (1960) bifurcated the gas line emerging from a detector with a three way stopcock and used stream splitting devices to divide both segments into five equal streams terminating in syringe needles. Each needle dipped into a vial containing an appropriate reagent, and by means of the stopcock a fraction was diverted into one set of reagents while the alternate channel was closed to permit replacement of the used reagent vials. The sensitivity of this procedure would not be suitable for capillary chromatography.

The presence of a particular component in a gaseous mixture can sometimes be substantiated by specific absorption or reaction of the suspected vapour before the sample is analyzed. For instance, aqueous base may be drawn into a syringe to remove carbon dioxide, and either mercuric perchlorate reagent or bromine water will remove ethylene, so that when the sample is chromatographed the presence of the particular gase is confirmed by the disappearance of the appropriate peak. A useful procedure for identifying compounds with a carbonyl group has been suggested and tested by BASSETTE and WHITNAH (1960). By adding aqueous sodium bisulfite to liquid samples before chromatography, they were able to convert the carbonyl compounds to less volatile products.

IV. Quantitative Analysis.

The quantitizing of results obtained with integral detectors presents no technical problems because by definition these devices indicate at each moment the total amount of sample that has previously emerged. It is a simple a matter to convert the units in which the results have been recorded to weights of sample. Differential detectors do not per se indicate weights of sample, but this can be inferred from the information they provide.

1. Peak Heights.

The equating of peak heights with sample weight is a rapid and fairly accurate approach provided that several precautions are observed. The sample size must be reasonably small, perhaps less than about 2 μL of liquid, because a linear relationship between peak height and quantity exists only over a limited range; the temperature must be accurately controlled as it and peak height are related as a non linear function (JOHNS, 1959, p. 42), and a separate calibration must be made for each component due to the dependence of the absolute height of a peak upon retention volume. Peak height has been found to be fairly independent of carrier gas flow rate (DIMBAT et al., 1956), and it should be noted that highly efficient columns such as capillaries produce very narrow bands which make easy the measurement of heights, but render difficult that of areas.

2. Peak Areas.

The major advantages of area as opposed to height measurements are linearity with sample size over much greater ranges, lack of dependence upon operating conditions except flow (there is a linear inverse relationship), and exact proportionality with respect to sample quantity regardless of retention time. Methods which are useful in integrating the area under a curve are: (a) *geometric approximation*, by multiplying the peak height by the width at halfheight (x times y in Fig. 1), (b) *counting the number of squares* contained under a curve which has been retraced onto mm graph paper, (c) *weighing the peaks* after cutting them out, (d) *a planimeter*, and (e) *automatic integrating devices* based on mechanical, electromechanical or electronic principles. Popular designs are the disk integrator, watthour meter, integrating amplifier, integrating motor (NOGARE, BENNETT and HARDEN, 1958; KEULEMANS, 1957, p. 209), and electronic pulse counters (DREW, 1959, p. 134). Usually the integrated result is expressed with a pipping pen mounted in the margin of the recorder, or as a read-out on a digital counter or dial, although the design and performance of a printing integrator for gas chromatography has been described (CLAUDY, WATSON, COATES, KAYE and DAVIS, 1959).

3. Overlapping Peaks.

Although CREMER and HAUPT (1958) found it permissible to use heights as a measure of quantity when the overlap between two peaks was not severe, frequently it is better to consider peak areas. Using as a dividing line the perpendicular extended to the base line from the minimum between two overlapping peaks, a surprisingly accurate correlation has been found to exist between the amount of each component and the integrated area so obtained (BRACE, referred to in PECSOK, 1959b). An equally valid method which may prove slightly more accurate for asymmetric peaks is to use as a dividing line the perpendicular dropped from the point of intersection of the tangents to the inside slopes of the overlapping curves. In this latter case, extensions of the inner tangents from their point of intersection to the base line form a triangle with the base line as one side, and the perpendicular subdivides this triangle into the respective contributions of the "overlap area" which can be attributed to each of the components. In the method of "approximating triangles" these areas are used as correction factors which are substrated from the total areas under each peak as determined by multiplying one half the base times the height. If the separation is sufficiently incomplete, one curve may appear as a shoulder superimposed upon the other, with no obvious minimum and no way to determine the slope of the inner tangent of the lesser component. Under these conditions it is fairly accurate to construct the outer tangent and assume an isosceles triangle. ZHUKHOVITSKII and TURKEL'TAUB (1958) consider the problem of overlapping peaks in detail, and GLUECKAUF's treatment has been discussed previously (see Fig. 5).

4. Sloping Base Line and Secondary Peaks.

When operated near their sensitivity limit, certain detectors exhibit a continuous slight drift. Ideally this drift will be constant or nearly so, and a simple correction factor can be applied. A related problem is the estimation of the peak area or height of a secondary curve superimposed on the trailing edge of a major component. Usually the base line can be reconstructed artificially by disregarding the minor constituent and continuing the steep downward slope of the major component's trailing edge with a smooth curve connecting it to the slow after-drift (JOHNS, 1959, p. 47; BURG and STOLWIJK, 1959).

5. Instrument Correction Factors.

When the response of a detector differs from one molecular species or class to another, in order to equate the peak area to an absolute quantity it will be necessary to apply a suitable correction factor for each compound. If the instrument responds to a simple molecular parameter, for instance molecular weight, carbon number or heat of combustion, a fairly accurate correction can be made without empirical calibration if the sample identity is known with at least sufficient accuracy to estimate the value of the pertinent physical property. For instance, MESSNER, ROSIE and ARGABRIGHT (1959) found that the relative response of a thermal conductivity detector can be predicted from a consideration of the molecular weight and class of compound (c.f. also LITTLEWOOD, 1959 and MASON and SAXENA, 1958), and ROSIE and GROB (1957) caution that considerable error can result even when helium is used unless these corrections are applied. HINKLE and JOHNSON (1958) using a thermistor detector substantiated their derivation of a relationship between change in mole concentration, change in peak area, and component vapour density raised to the two-thirds power, and if the vapour density of the pure compound is known, its peak area can be quantitized by comparison with a suitable standard. Various other procedures by which correction factors can be derived, are discussed by KEULEMANS, KWANTES and RIJNDERS (1957) and DIMBAT et al. (1956).

6. Internal Standard and Internal Normalization.

By adding to the sample a known amount of substance for which the instrument has been calibrated, it is possible to calculate from the elution curve exactly how much of the sample mixture was introduced. Either the peak height or area of the internal standard will suffice, provided that the position of the substance on the chromatogram does not overlap that of any other component. An additional advantage gained with the internal standard is that it acts as a check upon the constancy of operating conditions and, within the limits of accuracy of the sample introduction, upon the sensitivity constancy of the detector. If the integrated peak areas of all components are summated, provided that the instrument responds equally to all substances, the total will be a measure of the amount of sample which was introduced. This procedure, known as internal normalization, can also be used when all the components and correction factors are known or can be estimated.

D. Applications.

This discussion will be limited to analytic and preparative applications of gas liquid chromatography. A third area which is particularly interesting to physical chemists is the interpretation of chromatographic data by well developed theories which make possible its use in the calculation of such physical properties as activity coefficients, heats and entropies of solution and adsorption, distribution isotherms and surface area. Accounts of the procedures involved are included in the books by KEULEMANS (1959, pp. 182—190) and PHILLIPS (1956, pp. 90—93), reviews by HARDY and POLLARD, (1959) GREENE (1959), as well as in papers by ANDERSON (1956), PORTER et al. (1956) and WHITE and COWAN (1958).

I. Analytic Applications.

A section in a recent review of the applications of partition chromatography to plant sciences has been dedicated to a discussion of gas chromatography (THOMPSON, HONDA, HUNT, KRUPKA, MORRIS, POWELL, SILBERSTEIN, TOWERS and ZACHARIUS, 1959) and LIPSKY and LANDOWNE (1960) reviewed the biochemical applications.

1. Carbon Dioxide and Oxygen: Respiration and Photosynthesis.

Because of the fortuitous circumstance that nitrogen and oxygen have thermal conductivities which are nearly identical, a katharometer may be used to determine carbon dioxide concentration changes in an air stream passing over a photosynthesizing or respiring tissue. This, in fact, was one of the first applications of the instrument (NOYONS, 1922; SLATER, 1927) and it is based on the observation that changes in oxygen content of the air stream will affect very little the thermal conductivity of a mixture which is predominantly oxygen and nitrogen, whereas changes in the content of carbon dioxide, which has only slightly more than half the thermal conductivity of oxygen and nitrogen, will cause a marked alteration. The amounts of carbon dioxide produced or consumed by a plant tissue quantitatively are far in excess of all other volatile exchanges with the exception of water vapour, so that if the air stream is either dried or saturated, the lack of specificity can be disregarded. If an equivalence of formation and consumption of carbon dioxide and oxygen is assumed, a 10 percent correction for the contribution of oxygen changes to the observed reading must be applied (WALLER, 1926). The sensitivity of this method is roughly comparable to that of the infra-red gas analyzer, but the measuring chamber of the katharometer is vastly smaller so that changes can be recorded with a very brief lag period. For this reason the instrument is particularly suited to the study of rapid induction phenomena in photosynthesis, to which it has been successfully applied by AUFDEMGARTEN (1939), VEEN (1949) and SEMENENKO (1958). The measurement of rates of respiration and photosynthesis of leaves (WALLER, 1926) and respiration in fruits (GANE, 1936; BURG and THIMANN, 1960) and other plant tissues (STILES and LEACH, 1931) has been accomplished with a katharometer. In all of these procedures a by-pass around the sample chamber is used to check occasionally the validity of the base line. A method for simultaneously determining oxygen and carbon dioxide with a katharometer, involving the use of a helium/oxygen (80/20) carrier gas and a short silica gel column has been described by BURG and STOLWIJK (1959), and a review by BEHMANN and KERCKOFF (1957) discusses the use of the katharometer in the analysis of biological gas mixtures.

A very straight forward approach to determining carbon dioxide production is to enclose a known weight of tissue in a chamber of calibrated volume, provided with a side arm fitted with a serum cap. Small gas samples can be removed at intervals with a syringe and gas chromatographed on silica gel, charcoal, propylene carbonate or dimethyl sulfoxide. The amount of sample introduced need not be controlled accurately because if the total concentration of carbon dioxide is small, the air present in the sample provides an internal standard. A more elegant method would be to construct a closed circuit gas train in which air is continuously circulated with a push-pull pump. The sample loop of a six-way gas valve could be made an integral part of the system, and periodic measurements might be made with an automatized valve.

2. Hydrogen, Hydrogen Sulfide, Methane, and other Fermentation Gases.

These gases which arise along with carbon dioxide during the fermentative processes of microorganisms may be separated on molecular sieves, charcoal and a benzyl cellosolve partition column (EDWARDS, NICHOLSON and REIDLAND, 1959). In a study on the reduction of carbon dioxide to methane during anaerobic fermentation, CHMIELOWSKI and ISAAC (1959) determined hydrogen, nitrogen, methane and carbon dioxide on an activated charcoal column in a single analytic procedure, and methane production during the digestion of sewerage sludge was investigated over a period of many days (CHMIELOWSKI, SIMPSON and ISAAC, 1959).

3. Olefins and Saturated Hydrocarbons.

An aluminum oxide column was used by BURG and THIMANN (1959, 1960) to separate ethylene from various volatile products of the apple, other fruits, and the fungus *P. digitatum*, and their results were in agreement with those of previous workers, using other techniques, who found ethylene to be the only volatile saturated or olefinic hydrocarbon produced by these tissues. HEULIN and KENNET (1959) used a mercuric perchlorate absorbent to collect the olefins produced by apples, released the gas from the reagent, and chromatographed it on a tricresyl phosphate column. It was concluded that if olefins other than ethylene were present, quantitatively they could not exceed $^1/_{1000}$ of its concentration. However, MEIGH (1959) used a flame ionization detector and liquid paraffin phase to detect acetylene, ethane, propylene and propane at less than $^1/_{1000}$ the concentration of ethylene in the volatiles collected in a freezing trap from a stream of air passed over apples. Suitable precautions were taken to purify the incoming air stream, and it is mentioned that a similar range of compounds were also found amongst the emanations from dormant potatoes.

A procedure involving absorption on mercuric perchlorate, by which ethylene can be concentrated from as much as a liter of air, has been used to extract olefins from the air surrounding apple tissue to which a carbon labeled isotope had been administered (BURG, 1959, 1960). A radioactivity scan of the chromatographic effluent confirmed that all activity in the purified sample traveled in the ethylene band, and the relative contribution and rate of incorporation of the various carbon atoms of glucose into carbon dioxide and ethylene was determined. Changes in the concentration of ethylene in the internal atmosphere of individual apples were studied for several consecutive days by gas chromatographing air samples extracted with a syringe. It was found that fluctuations in the concentration of ethylene in the internal atmosphere of the fruit were correlated exactly with changes in the rate of production of the gas, and there was evidence for a 24 hour cycle of production in at least one apple variety (BURG, 1958). The investigation of various aspects of the biochemistry and physiology of ethylene production in apples and tomatoes, such as the effects of temperature, various gas mixtures, inhibitors, and tissue damage, have been facilitated by gas chromatography (BURG and THIMANN, 1958, 1959; MEIGH, 1959; MEIGH, NORRIS, CRAFT and LIEBERMAN, 1960; CRAFT, 1960). Although it could not be demonstrated in mitochondria, BURG and THIMANN presented evidence that this probably is the cite of ethylene production, and MEIGH et al. (1960) showed that actively respiring mitochondria did not produce ethylene in contradistinction to claims from other sources (SPENCER, 1959). Recently LIEBERMAN (personal communication), using a gas chromatographic method, has claimed to obtain very slight production of ethylene from the mitochondria of several fruits.

A few other hydrocarbons, for instance n-heptane which is a major constituent in the turpentine of certain pine species, are known in plants and admirably suited for analysis by gas chromatography.

4. Nitrogen, Nitrous Oxide, Nitrogen Dioxide and Nitric Oxide.

With the exception of nitrogen dioxide, the various nitrogen containing gases which arise during the denitrification process carried out by many bacteria can be fractionated on a molecular sieve column (GRAVEN, 1959) and on other adsorbent columns (JANAK and RUSEK, 1954). Nitrogen dioxide tails badly on partition columns and does not emerge from adsorbent columns, but GREENE and PUST (1958) used a thoroughly wetted Molecular Sieve 5A column to convert the nitrogen dioxide to nitric oxide, the latter emerging as a distinct peak. Using both silica gel and alumina, the nitrous oxide production of *E. freundii* and *E. coli* was studied at various temperatures and as a function of the composition of the growth medium (BURG, unpublished data) and the identity of the gas was confirmed independently by infra red spectroscopy.

5. Ammonia, Organic Amines and Amino Acids.

Fractionation procedures and retention volume data for ammonia (BURKS, BAKER, CLARK, ESSLINGER and LACY, 1959; JAMES, MARTIN and SMITH, 1952; RICHARZ, LUTZ and GUYER, 1959) and ethanolamine (JAMES, 1952) have been reported, and a few applications to biological problems are pertinent although they are not related to plant analysis per se (GRONINGER, 1958; BURKS et al., 1959). Polar stationary phases are particularly suitable because of the ability of primary and secondary amines to hydrogen bond, and these solvents may prove valuable in the analysis of dimethyl-ethanolamine and perhaps choline, but probably not betaine and sarcosine without further modification because they decompose before boiling.

Several low boiling derivitives of amino acids have been prepared and their suitability for analysis by GLC investigated. Methyl esters may be formed from the free acids (BAYER, REUTHER and BORN, 1958) or directly from proteins (BAYER, 1958), and BAYER and coworkers have separated the esters of aliphatic, acidic, aromatic, sulfur and heterocyclic amino acids including several structural isomers using a silicone high vacuum grease partition phase with 10 percent sodium caproate as a tailing reducer. The N-trifluoroacetyl derivitives produced still sharper separations and have the advantage that they are less prone to polymerize than are the simple methyl esters (BAYER, 1958). YOUNGS (1959) described the preparation and fractionation of the N-acetyl butyl esters of several amino acids, including some derived from a gelatin hydrolyzate, using hydrogenated vegetable oil as a stationary phase. The reaction of ninhydrin with α-amino acids yields aldehydes, usually volatile and smaller by one carbon atom than the parent compound. The separation of the ninhydrin oxidation products of leucine, isoleucine and valine was accomplished by HUNTER, DIMICK and CORSE (1956) and various other amino acids have been separated in this way (BIER and TEITELBAUM, 1958, 1959; ZLATKIS and ORO, 1958; ZLATKIS, ORO and KIMBALL, 1960). For effecting the oxidation, ZLATKIS and coworkers developed a microreactor consisting of a heated column of diatomaceous earth impregnated with 30 percent ninhydrin, and the volatile aldehydes emitted upon the introduction of a liquid mixture of ninhydrin and amino acids were flushed in a stream of hydrogen directly into and through the chromatography column. Alanine, α-amino-n-butyric acid, valine, norvaline, leucine, isoleucine and norleucine have been quantitatively analyzed by this procedure, but of the other amino acids which produce volatile aldehydes, glycine is unsatisfactory because the product formaldehyde readily polymerizes and the aldehydes of phenyl alanine and methionine have low volatilities and impose a requirement for higher column operating temperatures. In the reaction of amino acids with alkaline hypochlorite a coupled decarboxylation and deamination occur to yield aldehydes similar to those resulting from ninhydrin oxidations. BAYER (1958) used a dinonyl phthalate column to separate the degradation products of amino acids treated in this manner, and found that although neutral aliphatic amino acids were converted quantitatively to their aldehydes, acidic and sulfur containing amino acids gave complex mixtures of volatile substances. The VAN SLYKE nitrogen determination, in which the reaction between an α-amino group and nitrous acid yields nitrogen gas and the corresponding hydroxy acid, has been adopted by LIBERTI (1958, pp. 341—342) as a method for preparing volatile derivitives of α-amino acids. A quantitative deamination was obtained, the resultant hydroxy acids were esterified with diazomethane, and by programming the temperature of a silicone D.C. 550 column between 80 and 140° C an excellent separation of the hydroxy acid methyl esters derived from glycine, valine, alanine, leucine, serine, aspartic and glutamic acids was produced.

6. Alkaloids, Indoles, Purines and Related Compounds.

The three principle tobacco alkaloids, nicotine, nornicotine and anabasine, as well as lesser constituents such as anatabine, myomine, metanicotine, nicotyrine, cotinine, nornicotyrine, N-methyl nicotinamide and several 3-pyridyl compounds were separated at 180 to 190° C on three different stationary phases by QUIN and HOBBS (1958) and QUIN (1959). Retention data is listed for polyethylene glycol, polypropylene glycol and polybutylene glycol columns (see also the review by JOHNSTONE and PLIMMER, 1959). The related compounds pyrrolidine and N-methyl pyrrolidine which are also contained in the tobacco plant were not identified, but N-methyl pyrrolidine has been assayed by gas chromatography (FRANZEN and KUNTZE, 1959; DECORA and DINNEEN, 1959). The technique should be equally useful for many other alkaloids, for instance atropine (sub. 118° C).

Indole has been separated from coal tar fractions by JANAK and HRIVNAC (1959) who used as a stationary phase either Reoplex 400, Apiezon L or silicone elastometer E 301 at 200° C. STOWE (personal communication) has chromatographed indole derivitives of biological importance by employing as a stationary phase substances containing aromatic nuclei, for instance polyethyleneglycolisophthalates. He also achieved reasonably good separations using silicone derivitives, although in all cases rather high temperatures are required. Since nearly all indoles are steam volatile and molecular distillation or sublimation has often been a means of purifying indole derivitives, it is likely that GLC will eventually encompass a broad spectrum of indole compounds.

A few methylated purines sublime at temperatures which suggest that GLC might prove useful in their analysis, for instance theobromine (sub. 290—295° C) and caffeine (sub. 180° C).

7. Carbohydrates.

The methyl ethers of methyl glycosides have relatively low boiling points, and McINNIS, BALL, COOPER and BISHOP (1958) successfully fractionated fully methylated derivitives of β-methyl-D-xylopyranoside, α and β-methyl-L-arabopyranoside, α-methyl-D-mannopyranoside, α-methyl-D-galactopyranoside, and α-methyl-D-glucopyranoside. It was ascertained that no structural changes occurred in the sugar molecules during their separation at 170° C on an Apiezon M stationary phase. Using similar procedures, BISHOP and COOPER (1960) separated partially methylated glucoses and sugar alcohols whose free hydroxy groups were acetylated. This work was extended by KIRCHER (1960; c.f. also HAMILTON and KIRCHER, 1958) who tested various columns of methylated carbohydrates at temperatures between 190 and 225° C and concluded that methylated hydroxyethylcellulose was very satisfactory. A complete separation of the α and β anomers of fully methylated methyl gluco- and mannopyranosides was demonstrated, as well as synthetic and biologically derived di- and trimethyl hexoses. The significance of this procedure as it relates to the end group assay for determining carbohydrate linkages was indicated by the fractionation and estimation of several methyl glycosides which were synthesized from the hydrolyzates of methylated cellulose, starch, guar gum and a dextran.

8. Lipids, Fatty Acids.

An especially thorough account of the application of GLC to lipid biochemistry and the analysis of fatty acids is that of JAMES (1960), and LIPSKY and LANDOWNE (1960) discuss many biological applications. To obtain fatty acid esters from a lipid it is not necessary to pass through a saponification step, for by ester interchange with or alcoholysis of the original lipid the esters can be derived directly (JAMES, 1960, pp. 18—19). Short chain acids from C_1—C_6 are separated conveniently on a stationary phase of 10 percent stearic acid in D.C. 550 silicone at 137° C (JAMES and MARTIN, 1952b) but there is some tendency for dimerization (JAMES and MARTIN, 1956). The methyl esters are preferable in this respect, and both straight and branch chain esters fractionate well at 78° C on dinonyl or dioctyl phthalate. Fatty acid esters of chain length C_6—C_{22} are conveniently separated on Apiezons and polyesters at about 200° C, whereas for longer chain acids up to C_{30} the Apiezons are preferred because of their temperature stability (BEERTHIUS, DIJKSTRA, KEPPLER and RECOURT, 1959), and esters up to C_{35} have been analyzed at still higher temperatures (ADLARD and WHITHAM, 1958). The use of polar stationary phases in the separation of long chain fatty acids was introduced by ORR and CALLEN (1958) who investigated the applicability of a commercial polymer Reoplex 400, and by LIPSKY and LANDOWNE (1958) who worked with an adipate polyester of diethylene glycol. The advantage gained is that there is a striking selective retardation due to unsaturation, which allows rapid analyses to be performed with excellent peak sharpness (LIPSKY, LANDOWNE and GODET, 1959).

Methods of identifying fatty acids recovered after gas chromatography are discussed by JAMES (1960) and LIPSKY and LANDOWNE (1960), and it might be noted that in several cases gas chromatography is itself a useful tool for the identification of the products of certain qualitative tests.

The even numbered C_{16}—C_{24} unsaturated acids of vegetable oils (CRAIG and MURTY, 1959) and various components of olive kernel oil (JART, 1959) have been assayed by GLC, which is indicative of the value of this technique in food technology. Similarly, gross analyses of the fatty acids in several plant parts such as tamarak heartwood (NAIR and RUDLOFF, 1959), sorghum leaf and stem (BURNET and LOHMAR, 1959), and the oil extracted from seeds of *Ulmus* (SORENSON and SALTOFT, 1958) have been reported, and the rate of formation of fatty acids in growing linseed was investigated (*Anon.*, 1959). In the tubercle bacillus, various fatty acids including stearic, palmitic, oleic and tuberculostearic were studied by CASON and TAVS (1959) and myoceric acid by ASSELINEAU, ASSELINEAU, RYHAGE, STALLBERG-STENHAGEN and STENHAGEN (1959).

9. Mono and Dicarboxylic Acids of Low Molecular Weight, and their Derivatives.

Irrespective of the success which has been achieved in separating organic acids by paper chromatography and on ion exchange resin columns, the rapidity with which GLC can fractionate and quantitatively evaluate these compounds makes it likely that there will be a shift in emphasis toward the latter technique. A few analyses of these compounds by GLC have been reported, for instance the identification of acetic, propionic, pyruvic and succinic acids amongst the fermentation products of *Leishmania donovani* (CROWTHER, FULTON and JOYNER, 1954), but in general there is a surprising scarcity of reports. This, in spite of the fact that the methyl esters of glycolic, oxalic, glyceric, fumaric, succinic, lactic, oxaloacetic, malonic, malic, tartaric, citric and undoubtedly other pertinent acids are stable and volatile at temperatures commonly used in GLC. Indeed, LIBERTI (1958, p. 341) has described a very clean separation of several of these acids using a 30 percent silicone D.C. 550 column and a temperature programmed from 100 to 160° C.

10. Alcohols, Aldehydes, Ketones and Miscellaneous Esters.

The separation of alcohols from methyl to nonyl on a silicone column with programmed temperature through the range 48 to 285° C was described by NOGARE and BENNET (1958) and alcohols up to C_{20} have been run on an Apiezon column at 196° C (BROUGHTON and WHEATLEY, 1959). RAY's results (1954a) indicate that a phthalate liquid phase is adequate for alcohols, aldehydes and ketones. Members of these classes of compounds are included in most plant fragrances, and especially the aroma of fruits, so that GLC is very applicable in this area. The volatile flavours of strawberries were studied by DIMICK and CORSE (CORSE and DIMICK, 1959; DIMICK and CORSE, 1956, 1958; DIMICK and MAKOWER, 1956) who prepared a concentrated oil of the volatile components and chromatographed it on Carbowax and silicone columns. Amongst the components identified were isoamyl alcohol, trans-2-hexene-1-ol, n-hexanol and trans-2-hexen-1-yl-acetate, and retention times are listed for several ethyl esters, iso-acids and aldehydes of biological interest. After a preliminary chemical fractionation, CALVARANO (1958a) used GLC to identify propionic, iso-valeric and nonyl-aldehydes, as well as several terpenoid and phenyl propanoid compounds in the distilled oil of *Rosa centifolia*. Other reported applications are the analysis of the methyl esters of fatty alcohols present in linseed and soy oils (LINK, HICKMAN and MORRISSETTE, 1959), isoamyl and D-amyl alcohol, ethanol, ethyl acetate and acetone produced during the fermentation of beer (VAN DER KLOOT, TENNEY and BAVISOTTO, 1958), ethanol, propanol, isopropanol, sec-butanol, 3-methyl-2-butanol, isoamyl alcohol and hexanol in various fermentation fusel oils (BARAUD and GENEVOIS, 1958), and isoamyl and isobutyl alcohols in beer and wine (FLEISCHMANN and DAGHETTA, 1958).

11. Phenyl Propanoid Compounds, Aromatic Acids, Phenols and Related Substances.

This diverse group of oxygenated aromatic compounds represents a biochemical homogeneity in the sense that its various members are very likely derived from common precursors or in some other way they are metabolically closely related (NEISH, 1960; BOGORAD, 1958).

The benzyl propanoid compounds have been implicated in the synthesis and structure of lignin, and a commonly used analytic tool in the study of lignin is the measurement of vanillin and syringaldehyde liberated during nitrobenzene oxidation of extractive-free plant material, or from sulfite lignin during refluxing with hot alkali. The boiling point of vanillin (ca. 295° C) suggests that it, and perhaps syringaldehyde can be assayed by gas chromatography. Esters of ferulic acid have been detected by GLC in acetone extracts of tamarac heartwood (NAIR and RUDLOFF, 1959); eugenol, isoeugenol, and their methyl ethers and acetates can be separated on Apiezon and silicone stationary phases (STAHL and TRENNHEUSER, 1960; CALVARANO, 1958; NAVES and ODERMATT, 1958; NAVES, 1958a); and ethyl cinnamate (DIMICK and MAKOWER, 1956), anetholes (NAVES, 1958b), phenylacetaldehyde and phenyl ethanol (CALVARANO, 1958; NAVES, 1958a) have been identified in various essential oils. It is likely that gas chromatography will come to play an increasingly important role in the study of lignin structure and biosynthesis, as well as in the analysis of various related compounds such as the coumarins.

Amongst the phenolic substances, thymol and carvacrol (NAVES, 1959), guaiacol (DOMANGE and LONGUEVALLE, 1958) and salicylic aldehyde (CALVARANO, 1958a) have been identified in essential oils. Retention data for other phenols on eight different stationary phases is listed by JANAK and KOMERS (1958), and for a column containing n-decyl phthalate by BERMANN and JENTZSCH (1958).

A few other compounds belonging in this general group which are of interest in plant analysis are the methyl ester of 4-chloro-2-methyl phenoxy acetic acid (HADDOCK, HILL and JONES, 1959); methyl ester of salicylic acid (DOMANGE and LONGUEVALLE, 1958), benzoic acid (KOMERS and BAZANT, 1959), and tocopherols (*Anon.*, 1960a, p. 3). The low boiling point of vitamin K_1 (140—145° C) makes likely its analysis by gas chromatography.

12. Terpenes.

Probably more attention has been directed toward the analysis of terpenes with gas chromatography than toward any other group of substances found in plants. In part this is a result of the economic importance of these compounds in relation to the odor of essential oils and perfumes, and the flavour and odor of foods. The number of types of terpenes is extremely large, and the different representatives are sufficiently common to make the analysis of any essential oil a very complex problem. An excellent review of the inherent difficulties is that of GUENTHER, KULKA and ROGERS (1959), who also discuss the role of gas chromatography in this area, as do LIBERTI and CARTONI (1958). Several instances have been noted in which isomerizations of terpenes occur during their passage through a column due to interaction with the stationary phase (VILKAS and ABRAHAM, 1959; NAVES, 1958c; LANGENAU and ROGERS, 1960) and although NAVES suggested the use of sea sand or sodium chloride as a support, apparently it is sufficient to use specially treated celite (LANGENAU and ROGERS, 1960). The separation and identification by gas chromatography of well over a dozen terpenes in a variety of essential oils has been reported by NAVES (1958a, 1958b), retention data and the identification of monoterpenes in several essential oils is discussed by STAHL and TRENNHEUSER (1960) and LIBERTI and CARTONI (1958); retention volumes of various terpenoid compounds are listed by BAYER, KUPFER and REUTHER (1958), substances found in pine oil were investigated by CURKAL and JANAK (1959), and other useful sources are the papers of CALVARANO (1957, 1958a, 1958b), ZUBYK and CONNER (1960) and LIBERTI and CARTONI (1958b). Gas chromatography has proven useful as an assay for α-pinene in connection with a study of terpene formation in pine from mevalonic acid (STANLEY, 1958). Commonly used partitioning agents for terpenes are silicones (NAVES, 1958a) polyesters (NAVES and ODERMATT, 1958), glycol polymers (CALVARANO, 1958b) and sucrose acetate isobutyrate (SMITH, BARTLET and LEVI, 1960) at temperatures in the range between 130 and 170° C.

It is interesting to note that perhydrolycopene, a carotinoid derivitive, has been analyzed by GLC (*Anon.*, 1960a, p. 3).

13. Sulfur Compounds.

Essential oils frequently contain several non-terpenoid constituents amongst which various thiocyanates, organic sulfides and mercaptans are encountered. Mercaptans and other sulfur compounds also contribute to the odors of foods and fungi, but until the advent of gas chromatography there existed no practical assay for any of these substances. In onion oils, which are especially rich in sulfur compounds, various sulfides, disulfides and mercaptans have been identified by GLC (JACOBS, 1957; CARSON and WONG, 1957; NEIGISCH and STAHL, 1956), and propyl disulfide was found, rather than allyl propyl disulfide which was long considered to be a constituent. Allylisothiocyanate and other thioisocyanates are important constituents of certain oils, for instance Crucifer oil, and KJAER and JART's results (1957) show that alkyl isothiocyanates can be analyzed by GLC. Columns which have proven successful in the separation of sulfur compounds are tricresyl phosphate (AMBERG, 1958; RYCE and BRYCE, 1957), glycols and glycol polymers (CARSON and WONG, 1957, 1959; JACOBS, 1957), dioctyl or dinonyl phthalate (SUNNER et al., 1956; SPENCER, BAUMANN and JOHNSON, 1958), Triton S (ADAMS and KOPPE, 1959) and non-polar oils (KARCHMER, 1959; LIBERTI and CARTONI, 1957; SULLIVAN et al., 1959).

14. Steroids.

The first direct analysis of steroids by gas chromatography was reported by EGLINGTON, HAMILTON, HODGES and RAPHAEL (1959) using an Apiezon L column at temperatures above 200° C. Since this rather crude beginning, in which long retention times and poor selectivity were encountered, rapid strides have ensued and a wide variety of steroids, including various sterols and hormones, are now encompassed by the technique (SWEELEY and HORNING, 1960; HORNING, SWEELEY and VANDER-HEUVAL, 1960; *Anon.*, 1960a, p. 3; *Anon.*, 1960b, pp. 1—4). Temperatures in the vincinity of 300° C are required and low vapour pressure stable silicone compounds make successful liquid phases.

Gas chromatography also offers great promise for simplifying the determination of steroid structures. Pyrolysis of these compounds gives rise to a mixture of hydrocarbons, C_8 in the case of cholesteryl chloride, and COX, HIGH and JONES (1958) investigated the possibility of coupling gas chromatography with a degradative technique. The side chain was split off with an acid catalyst, and the volatile products were identified by chromatographing them on an Apiezon L column at 100° C. Results with a wide variety of steroids demonstrated that the carbon number of the derived products agreed with that of the side chain of the parent compound. Included in the steroids examined were several C_{28}—C_{29} substances characteristic of the types commonly found in plants.

The gibberellins bear a structural resemblance to the steroids, and it is not unlikely that methyl esters of these compounds, or fully methylated derivitives, might be assayed by gas chromatography.

15. Miscellaneous Compounds.

A streptomycin antagonist derived from *Pseudomonas pyocyanea* was oxidized by CORNFORTH and JAMES (1956), and the derivitives were compared chromatographically with various synthetic substances, by which means it was established that the parent compound was a mixture of alkyl-hydroxyquinoline-N-oxides. The red pigment of *Penicillium herquei* known as herqueinone, when degraded with alkali yields low molecular weight carbonyl compounds which were identified by gas chromatography of their oximes (CASON and HARRIS, 1959). Derivitives of acids extracted from olive leaves have been investigated by MEAKINS and SWINDELLS (1959). Classical methods of ellucidating the structure of porphyrins depend upon the oxidative or reductive degradation of the compound and the isolation and characterization of the products, and it is in the latter respect that MORELY, COOPER and HOLT (1959) found gas chromatography to be particularly helpful.

II. Preparative Gas Chromatography.

There are two basically different philosophies concerning the best way to maximize the amount of pure substance which can be prepared and recovered by GLC, although these are not mutually exclusive and a compromise might be satisfactory. Either larger and larger columns can be constructed, which is the approach taken by EVANS and TATLOW (1955, 1957) who used columns measuring 16 ft. by 3 in. (EVANS et al., 1958), or an automatic recycling instrument can be used to continuously fractionate and collect smaller samples which are accumulated (ATKINSON and TUEY, 1958; AMBROSE and COLLERSON, 1956; KOVATS, SIMON and HEILBRONNER, 1958). For laboratory preparations, the single cycle method using a large column undoubtedly will be preferred because it is the simpler procedure and as much as 50 gm (or even 250 gm!) of material can be passed through a 3 in. column in a single cycle (EVANS, 1958, p. 286). However, it may be that the recycling system will prove to be more satisfactory for large scale ventures where price per pound is an important consideration (ATKINSON and TUEY, 1958). As indicative of the use of preparative procedures, quantities of several pure fractions have been prepared from essential oils by TEISSEIRE (1959) and submitted to further analysis, and the purification of C^{14} p-cymene (GUERMONT and PICHAT, 1959) and tritiated compounds (WILZBACH, 1957) has been reported.

E. Conclusion.

In an address at a recent symposium, A. J. P. MARTIN (1958, pp. 241—242) discussed the past, present and future of gas chromatography and with wry humor remarked "No one is ever content with what a good method will do. There are always attempts to push it to what it won't do, so, we are all going to be asked to run non-volatile substances. The only way we can attempt to do it, as far as I can see at the moment, is to run the columns hotter and hotter. Eventually, thermal decomposition will overcome us and the components will break up before they get to the other end of the column. Not only do we have to contend with the substances themselves breaking up, but with the stationary phase decomposing as well." The last problem has been partially solved even since these remarks were made, but in general this brief paragraph is a very apt summary of the scope and limitations of gas chromatography. It is likely that any substance which at some temperature reaches a suitably high vapour pressure without decomposing will ultimately be analyzed by GLC, in amounts varying from fractions of a μg to hundreds of grams. If a substance does not possess this property, then it is certain that attempts will be made to convert it to a more volatile derivitive or to quantitatively decompose it to a volatile product. For the plant scientist, where the subject compliantly provides aromas, flavours, oils and essences of quantities and complexities which defy resolution, gas chromatography offers new vistas of opportunity where previously even the most arduous and laborious efforts often came to naught.

References.

ADAMS, D. F., and R. K. KOPPE: Tappi **42**, 601—605 (1959). — ADLARD, E. R., and B. T. WHITHAM: in DESTY pp. 351—368 (1958). — AGEIKEN, D. I., E. L. ITSKOVICH and S. P. MIKHAILOV: Zavodskaya Lab. **23**, 492—496 (1957). — AMBERG, C. H.: Can. J. Chem. **36**, 590—592 (1958). — AMBROSE, D.: in DESTY (1957). — AMBROSE, D., and R. R. COLLERSON: Nature **177**, 84 (1956). — ANDERSON, D. M. W.: Analyst **84**, 50—55 (1959). — ANDERSON, J. R. A.: J. Am. Chem. Soc. **78**, 5692—5693 (1956). — *Anonymous:* Petrol. Refiner, July 2—6 (1957a); — Burrell Corp. Bulletin No. 835, Pittsburgh (1957b); — Materials for gas chromatography. Dagenham, England: May and Baker Ltd. 1958; — Rev. Current Lit. Paint, Colour, Varnish and Allied Ind. **206**, 552 (1959); — Aerograph Research Notes, Summer (1960a); — Aerograph Research Notes, Winter (1960b). — ASHBURY, G. K., A. J. DAVIES and J. W. DRINKWATER: Anal. Chem. **29**, 918—925 (1957). — ASSELINEAU, C., J. ASSELINEAU, R. RYHAGE, S. STALLBERG-STENHAGEN and E. STENHAGEN: Acta Chem. Scand. **13**, 822—824 (1959). — ATKINSON, E. P., and G. A. P. TUEY: in DESTY pp. 270—287 (1958). — AUFDEMGARTEN, H.: Planta **29**, 643—678 (1939). — AUSLOOS, P., and E. MURAD: J. Am. Chem. Soc. **80**, 5929—5932 (1958).

BAKER, W. J., E. H. LEE and R. F. WALL: ISA Symposium on Gas Chromatography, Michigan (1959). — BARAUD, J., and L. GENEVOIS: Compt. Rend. **247**, 2479—2481 (1958). — BARBER, D. W., C. S. G. PHILLIPS, G. F. TUSA and A. VERDIN: J. Chem. Soc. **1959**, 18—24. — BARRER, R. M.: J. Chem. Soc. **1948**, 127—132; — Discussions Faraday Soc. **7**, 135—141 (1949). — BARRER, R. M., and J. W. BAYNHAM: J. Chem. Soc. **1956**, 2882—2891. — BARRER, R. M., J. W. BAYNHAM and N. MCCALLUM: J. Chem. Soc. **1953**, 4035—4041. — BARRER, R. M., and D. W. BROOK: Trans. Faraday Soc. **49**, 1049—1059 (1953). — BARRER, R. M., F. W. BULTITUDE and J. W. SUTHERLAND: Trans. Faraday Soc., **53**, 1111—1123 (1957). — BARRER, R. M., and D. A. IBBITSON: Trans. Faraday Soc. **40**, 195—206 (1944a); **40**, 206—216 (1944b). — BARRER, R. M., and L. V. REES: Trans. Faraday Soc. **50**, 852—863 (1954a); **50**, 989—999 (1954b). — BARRER, R. M., and A. B. ROBINS: Trans. Faraday Soc. **49**, 807—815 (1953). — BASSETTE, R., and C. H. WHITNAH: Anal. Chem. **32**, 1098—1100 (1960). — BAXTER, R. A., and R. T. KEEN: Anal. Chem. **31**, 475—476 (1959). — BAYER, E.: in DESTY pp. 333—342 (1958); — Anleitungen für die chemische Laboratoriumspraxis, Vol. X, Berlin, Göttingen, Heidelberg: Springer-Verlag 1959. — BAYER, E., G. KUPFER and K. H. REUTHER: Z. Anal. Chem. **164**, 1—9 (1958). — BAYER, E., K. H. REUTHER and F. BORN: Angew. Chem. **69**, 640 (1957). — BEDNAS, M. E., and D. S. RUSSELL: Can. J. Chem. **36**, 1272—1276 (1958). — BEERTHIUS, R. K., G. DIJKSTRA, J. G. KEPPLER and J. H. RECOURT: Ann. N.Y. Acad. Sci. **72**, 616—632 (1959). — BEHMANN, F. W., and W. G. KERCKOFF: Arch. tech. Messen, Lfg. **263**, 271—274 (1957). — BENNETT, C. E., S. D. NOGARE, L. W. SAFRANSKI and C. D. LEWIS: Anal. Chem. **30**, 898—902 (1958). — BENYON, J. H.: Nature **174**, 735 (1954); — Mikrochim. Acta **1956**, 437—453. — BERGMANN, G., and D. JENTZSCH: Z. Anal. Chem. **164**, 10—29 (1958). — BIER, M., and P. TEITELBAUM: Fed. Proc. **17**, 191 (1958); — Ann. N.Y. Acad. Sci. **72**, 641—648 (1959). — BISHOP, C. T., and F. P. COOPER: Can. J. Chem. **38**, 388—395 (1960). — BLOM, L., and L. EDELHAUSEN: Anal. Chim. Acta **15**, 559—566 (1956). — BODNAR, S. J., and S. J. MAYEUX: Anal. Chem. **30**, 1384—1387 (1958). — BOER, H.: Proc. 4th World Petrol. Congr., Section V/A, Paper 1, Rome (1955). — BOER, H.: in DESTY pp. 169—184 (1957). — BOGORAD, L.: Ann. Rev. Plant Physiol. **9**, 417—448 (1958). — BOHEMAN, J., and J. H. PURNELL: in DESTY pp. 6—22 (1958). — BRADFORD, B. W., D. HARVEY and D. E. CHALKLEY: J. Inst. Petrol. **41**, 80—91 (1955). — BRECK, D. W., W. G. EVERSOLE, R. M. MILTON, T. B. REED and T. L. THOMAS: J. Am. Chem. Soc. **78**, 5963—5971 (1956). — BRENNER, N., and E. CIEPLINSKI: Ann. N.Y. Acad. Sci. **72**, 705—713 (1959). — BRENNER, N., and V. J. COATES: Nature **181**, 1401—1402 (1958). — BRENNER, N., and L. S. ETTRE: Anal Chem. **31**, 1815—1818 (1959). — BROUGHTON, B., and V. R. WHEATLEY: Biochem. J. **73**, 144—149 (1959). — BROWNING, L., and J. O. WATTA: Petrol. Refiner **38**, 302 (1959). — BUCUR, R.: Rev. Chim. (Bucharest) **7**, 163—165 (1956). — BURG, S. P.: Ph. D. Dissertation, Harvard University (1958); — Arch. Biochem. Biophys. **84**, 543—546 (1959); — Plant Physiol. Proc. **35**, vii—viii (1960). — BURG, S. P., and J. A. J. STOLWIJK: J. Biochem. Microbiol. Techn. and Eng. **1**, 245—259 (1959). — BURG, S. P., and K. V. THIMANN: Proc. Natl. Acad. Sci. **45**, 335—344 (1959); — Plant Physiol. **35**, 24—35 (1960). — BURKS, R. E., E. B. BAKER, P. CLARKE, J. ESSLINGER and J. C. LACY: J. Agr. Food Chem. **7**, 778—782 (1959). — BURNET, M. C., and R. L. LOHMAR: J. Agr. Food Chem. **7**, 436—437 (1959). — BURR, J. G.: J. Phys. Chem. **61**, 1477—1480 (1957).

CACACE, F., and INAM-UL-HAG: Science **131**, 732—733 (1960). — CALVARANO, M.: Essenze deriv. agrumari **27**, 208—220 (1957); **28**, 157—163 (1958a); **28**, 107—118 (1958b). — CARLE, D. W.: in COATES et al.: pp. 67—72 (1958). — CARLE, D. W., and T. JOHNS: ISA Proc. Natl. Symp. on Instrumental Methods of Analysis, Houston, Texas (1958). — CARSON, J. F., and F. F. WONG: Am. Chem. Soc. D-115 (1957). — CASON, J., and E. R. HARRIS: J. Org. Chem.

24, 676—679 (1959). — CASON, J., and P. TAVS: J. Biol. Chem. 234, 1401—1405 (1959). — CHMIELOWSKI, J., and C. G. ISAAC: Nature 183, 1120—1121 (1959). — CHMIELOWSKI, J., J. R. SIMPSON and C. G. ISAAC: J. Appl. Chem. 9, Abstr. ii—531 (1959). — CHUNDELA, B., and J. JANAK: Section on Gas Chromatog., 3rd Conference on Analytical Chem., Prague (1959). — CLAESSON, S.: Arkiv. Kemi, Mineral. Geol. 23A, 133pp. (1946). — CLAUDY, J. H., E. S. WATSON, V. J. COATES, M. KAYE and J. J. DAVIS: Pittsburgh Conf. on Anal. Chem. and Appl. Spectroscopy (1959). — CLOUGH, K. H.: Southeastern Regional ACS Meeting, Gainesville, Florida (1959). — COATES, V. J., H. J. NOEBELS and I. S. FAGERSON: Gas Chromatography, New York: Academic Press 1958. — CONDON, R. D.: Anal. Chem. 31, 1717—1722 (1959). — CORNFORTH, J. W., and A. T. JAMES: Biochem. J. 63, 124—130 (1956). — CORSE, J., and K. P. DIMICK: Flavor Research and Food Acceptance, 302—314 (1958). — COWAN, C. B., and P. H. STERLING: in COATES et al. pp. 165—190 (1958). — COX, J. S. G., L. B. HIGH and E. R. H. JONES: Proc. Chem. Soc. 1958, 234—235. — CRAATS, F. VAN DE: Anal. Chim. Acta 14, 136—149 (1956); — in DESTY (1957); — in KEULEMANS pp. 227—228 (1959). — CRAFT, C. C.: Plant Physiol. Proc. 35, vii (1960). — CRAIG, B. M., and N. L. MURTY: J. Am. Oil Chemists' Soc. 36, 549—552 (1959). — CREMER, E.: Arch. Biochem. Biophys. 83, 345—349 (1959). — CROWTHER, S., J. D. FULTON and L. P. JOYNER: Biochem. J. 56, 182—185 (1954). — CURKAL, H., and J. JANAK: Parfumerie und Kosmetik 40, 518 (1959). — CVENTANOVIC, R. J., and K. O. KUTSCHE: in DESTY pp. 87—97 (1957).

DAYNES, H. A.: Proc. Roy. Soc. A. 97, 273—286 (1920); — Proc. Phys. Soc. 33, 165—169 (1921); — Gas analysis by measurement of thermal conductivity, London: Cambridge University Press 1933. — DEAL, C. H., J. W. OTVOS, V. N. SMITH and P. S. ZUCCO: Anal. Chem. 28, 1958—1964 (1956). — DECORA, A. W., and G. U. DINNEEN: ISA Proc. Anal. Instr. Div. 2nd Int. Gas Chromatog. Symposium, Preprints 2, 12—15 (1959). — DEEMTER, J. J. VAN: in DESTY pp. 3—5 (1958). — DEEMTER, J. J. VAN, F. J. ZUIDERWEG and A. KLINKENBERG: Chem. Eng. Sci. 5, 271—289 (1956). — DESTY, D. H.: Vapour phase chromatography. London: Academic Press 1957; — Gas chromatography. London: Butterworths 1958. — DESTY, D. H., F. M. GODFREY and C. L. A. HARBOURN: in DESTY pp. 200—215 (1958). — DE VAULT, D.: J. Am. Chem. Soc. 65, 532—540 (1943). — DE WET, W. J., and V. PRETORIUS: Anal. Chem. 30, 325—329 (1958). — DIJKSTRA, G., J. G. KEPPLER and J. A. SCHOLS: Rec. Trav. Chim. 74, 805—812 (1955). — DIMBAT, M., P. E. PORTER and F. H. STROSS: Anal. Chem. 28, 290 to 297 (1956). — DIMICK, K. P., and J. CORSE: Food Technol. 10, 360—364 (1956); — Am. Perfumer Aromat. 71, 45, 48, 53 (1958). — DIMICK, K. P., and B. B. MAKOWER: Food Technol. 10, 73—75 (1956). — DOMANGE, L., and S. LONGUEVALLE: Compt. Rend. 247, 209—211 (1958). — DORFMAN, L. M., and K. E. WILZBACH: J. Phys. Chem. 63, 799—801 (1959). — DREW, C. M.: in PECSOK (1959a). — DREW, C. M., and J. R. MCNESBY: in DESTY 213—221 (1957). — DREW, C. M., J. R. MCNESBY, S. R. SMITH and A. S. GORDON: Anal. Chem. 28, 979—983 (1956). — DUBOIS, L., and J. L. MONKMAN: ISA Proc. Anal. Instr. Div. 2nd Int. Gas Chromatog. Symposium. Preprints 2, 96—104 (1959). — DUDENBOSTEL, B. F., and W. PRIESTLY: Ind. Eng. Chem. 48, 49A (1956). — DUFFIELD, J. J., and L. B. ROGERS: Anal. Chem. 32, 340—343 (1960).

EDWARDS, W. R., D. E. NICHOLSON and G. A. REIDLAND: U.S. 2, 875, 849 (1959); — Chem. Abstr. 53, 9906 (1959). — EGGERTSEN, F. T., H. S. KNIGHT and S. GROENNINGS: Anal. Chem. 28, 303—306 (1956). — EGGERTSEN, F. T., and F. M. NELSEN: Anal. Chem. 30, 1040 (1958). — EGLINGTON, G., R. J. HAMILTON, R. HODGES and R. A. RAPHAEL: Chem. and Ind. (London) 1959, 955—957. — EUCKEN, A., and H. KNICK: Brennstoff-Chem. 17, 241—244 (1936). — EVANS, D. E. M.: in DESTY (1958). — EVANS, D. E. M., W. E. MASSINGHAM, M. STACEY and J. C. TATLOW: Nature 182, 591—592 (1958). — EVANS, D. E. M., and J. C. TATLOW: J. Chem. Soc. 1955, 1184—1188; — in DESTY pp. 256—265 (1957). — EVANS, J. B., and J. E. WILLARD: J. Am. Chem. Soc. 78, 2908—2909 (1956).

FARRINGTON, P. S., R. L. PECSOK, R. L. MEEKER and T. J. OLSON: Anal. Chem. 31, 1512—1516 (1959). — FELTON, H. R.: Chem. Age (London) 78, 563 (1957); — in COATES et al. pp. 131—143 (1958). — FELTON, H. R., and A. A. BUEHLER: Anal. Chem. 30, 1163 (1958). — FITZGERALD, J. S.: Australian J. Appl. Sci. 10, 169—189 (1959). — FLEISCHMANN, L., and A. DAGHETTA: Ricerca sci. 28, 2286—2290 (1958). — FRANC, J., and J. JOKL: Chem. Listy 52, 276—282 (1958); — Collection Czechoslov. Chem. Commun. 24, 144—151 (1959). — FRANC, J., and M. WURST: Section on Gas Chromatog. 3rd Conference on Anal. Chem., Prague (1959). — FRANZEN, V., and H. KUNTZE: Ann. 627, 15—22 (1959). — FREDERICKS, E. M., and F. R. BROOKS: Anal. Chem. 28, 297—303 (1956).

GALLAWAY, W. S., and T. JOHNS: Appl. Spectroscopy 11, 102 (1957). — GALLAWAY, W. S., T. JOHNS, D. G. TIPOTSCH and R. J. APLIN: Alberta Gas Chromatog. Symposium, Edmonton (1959). — GANE, R.: New Phyt. 35, 383—402 (1936). — GIDDINGS, J. C.: Nature 184, 357 to 358 (1959). — GIDDINGS, J. C., and H. EYRING: J. Phys. Chem. 59, 416—421 (1955). — GLEW, D. N., and D. M. YOUNG: Anal. Chem. 30, 1890 (1958). — GLICK, D.: Methods of Biochemical Analysis, Volume 8, Interscience, New York 1960. — GLUECKAUF, E.: Trans.

Faraday Soc. **51**, 34—44 (1955a); — Soc. Chem. Ind. (London) **1955b**, 34—36; — in Desty (1957). — Gohlke, R. S.: Div. Petrol. Chem. ACS. Vol. 2, No. 4, D-77, New York (1957); — Anal. Chem. **31**, 535—541 (1959). — Golay, M. J. E.: in Desty pp. 36—55 (1958a); — in Coates et al. pp. 1—13 (1958b). — Gordon, S., A. R. van Dyken and T. F. Doumani: J. Phys. Chem. **62**, 20—24 (1958). — Gordus, A. A., M. C. Sauer and J. E. Willard: J. Am. Chem. Soc. **79**, 3284—3285 (1957). — Grandy, G. L., and R. C. Koch: Rev. Sci. Instr. **31**, 786 (1960). — Grant, D. W.: in Desty pp. 153—164 (1958). — Grant, D. W., and G. A. Vaughan: in Desty pp. 413—421 (1957). — Graven, W. M.: Anal. Chem. **31**, 1197 to 1199 (1959). — Green, G. E.: Nature **180**, 295—296 (1957). — Greene, S. A.: in Pecsok pp. 28—47 (1959a). — Greene, S. A., M. L. Moberg and E. M. Wilson: Anal. Chem. **28**, 1369—1370 (1956). — Greene, S. A., and H. Pust: Anal. Chem. **29**, 1055 (1957); **30**, 1039 (1958). — Greene, S. A., and H. E. Roy: Anal. Chem. **28**, 1369—1370 (1957). — Gregg, S. J., and R. Stock: in Desty pp. 90—124 (1958). — Griffiths, J. H., D. J. James and C. S. G. Phillips: Analyst **77**, 897—904 (1952). — Griffiths, J. H., and C. S. G. Phillips: J. Chem. Soc. **1954**, 3446—3453. — Groninger, H. S.: Comm. Fisheries Rev. **20**, 23—26 (1958). — Grune, W. N., J. V. Carter and J. P. Keenan: Sewage and Ind. Wastes **28**, 1433—1442 (1956). — Guenther, E., K. Kulka and J. A. Rogers: Anal. Chem. **31**, 679—687 (1959). — Guermont, J. P., and L. Pichat: Bull. soc. chim. France **1959**, 580—582. — Guild, L. V., S. Bingham and F. Aul: in Desty pp. 226—247 (1958). — Gunesch, H., and R. Stadtmuller: Rev. Chim. Bucharest **9**, 35—38 (1958).

Habgood, H. W., and W. E. Harris: 2nd Alberta Gas Chromatog. Symposium, Edmonton (1959); — Anal. Chem. **32**, 1206 (1960). — Haddock, L. A., R. Hill and A. G. Jones: Mfg. Chemist **30**, 57—60 (1959). — Hamilton, J. K., and H. W. Kircher: J. Am. Chem. Soc. **80**, 4703—4709 (1958). — Hammar, C. G. B.: Svensk Kem. Tid. **63**, 125—135 (1951). — Hanneman, W. W., C. F. Spencer and J. F. Johnson: Anal. Chem. **32**, 1386—1388 (1960). — Hardy, C. J., and F. H. Pollard: J. Chromatog. **2**, 1—43 (1959). — Harley, J., W. Nel and V. Pretorius: Nature **181**, 177—178 (1958). — Harley, J., and V. Pretorius: Nature **178**, 1244 (1956). — Harrison, G. F., P. Knight, R. P. Kelly and M. T. Heath: in Desty pp. 216—225 (1958). — Harvey, D., and D. E. Chalkley: Fuel (London) **34**, 191—200 (1955). — Hausdorff, H. H., and N. Brenner: Oil and Gas J. **56**, 122—124 (1958). — Hawkes, J. C.: in Desty pp. 266—276 (1957). — Heaton, W. B., and J. T. Wentworth: Anal. Chem. **31**, 349—357 (1959). — Henderson, J. I., and J. H. Knox: J. Chem. Soc. 2299—2302 (1956). — Heulin, F. E., and B. H. Kennett: Nature **184**, 996 (1959). — Hinkle, E. A., and S. E. J. Johnsen: in Coates et al. pp. 25—30 (1958). — Hinkle, E. A., H. C. Tucker, R. F. Wall and J. F. Combs: ISA Proc. Anal. Instr. Div. 2nd Int. Gas Chromatog. Symposium, Preprints **2**, 27—32 (1959). — Hooimeijer, J., A. Kwantes and F. van de Craafts: in Desty pp. 297—299 (1958). — Horning, E. C., C. C. Sweeley and W. J. A. Vander-Heuval: J. Am. Chem. Soc. **82**, 3481 (1960). — Hrapia, H.: Presented at the Symposium on Gas Chromatog. Leipzig 1958. — Hrapia, H., and H. G. Konnecke: J. Prakt. Chem. **3**, 106—112 (1956). — Hunter, I. R., K. P. Dimick and J. W. Corse: Chem. and Ind. (London) **1956**, 294—295. — Hunter, I. R., V. H. Ortegren and J. W. Pence: Anal. Chem. **32**, 682—684 (1960). — Hurn, R. W., J. O. Chase and K. J. Hughes: Ann. N.Y. Acad. Sci. **72**, 675—684 (1959).

Jacobs, M. B.: Am. Perfumer Aromat. **70**, 53—56 (1957). — James, A. T.: Biochem. J. **52**, 242—247 (1952); — in Desty pp. 127—130 (1957); — in Glick pp. 1—59 (1960). — James, A. T., and A. J. P. Martin: Analyst **77**, 915—932 (1952a); — Biochem. J. **50**, 679 to 690 (1952b); — Brit. Med. Bull. **10**, 170—176 (1954); — Biochem. J. **63**, 144—152 (1956). — James, A. T., A. J. P. Martin and G. H. Smith: Biochem. J. **52**, 238—241 (1952). Jamieson, G. R.: Analyst **84**, 74 (1959). — Janak, J.: Chem. Listy. **47**, 464—467 (1953); — Collection Czechoslov. Chem. Commun. **19**, 684—699 (1954); — in Desty pp. 247—255 (1957); — Ann. N.Y. Acad. Sci. **72**, 606—612 (1959). — Janak, J., and M. Hrivnac: Section on Gas Chromatog. 3rd Conference on Anal. Chem. Prague (1959). — Janak, J., and R. Komers: in Desty pp. 343—350 (1958). — Janak, J., M. Krejci and H. E. Dubsky: Chem. Listy. **52**, 1099—1107 (1958). — Janak, J., M. Nederost and V. Bubenikova: Chem. Listy. **51**, 890—894 (1957). — Janak, J., and M. Rusek: Chem. Listy. **47**, 1190—1196 (1953); **48**, 397—400 (1954). — Jart, A.: Acta Chem. Scand. **13**, 1723—1724 (1959). — Johns, T.: in Coates et al. pp. 31—39 (1958); — Gas chromatography applications manual. Fullerton: Beckman Instruments Inc. 1959. — Johnson, H. W., and F. H. Stross: Anal. Chem. **31**, 1206—1211 (1959). — Johnstone, R. A., and A. G. Douglas: Chem. and Ind. (London) **1959**, 154. — Johnstone, R. A., and J. R. Plimmer: Chem. Reviews **59**, 885—936 (1959). — Juranek, J.: Chem. Listy. **51**, 2280—2286 (1957); — Collection Czechoslov. Chem. Commun. **24**, 135—143 (1959).

Kaise, R., and H. Holzhauser: 3rd Conference on Anal. Chem., Prague 1959. — Karchmer, J. H.: Anal. Chem. **31**, 1377—1379 (1959). — Karmen, A., and R. L. Bowman: Ann. N.Y. Acad. Sci. **72**, 714—719 (1959a); — ISA Symposium on Gas Chromatog., Michigan 1959b. — Kebarle, P., and W. A. Bryce: Can. J. Chem. **35**, 576—579 (1957). — Kelker, H.:

Angew. Chem. 71, 218—220 (1959). — KEULEMANS, A. I. M.: Gas chromatography. New York: Reinhold 1957; — Gas chromatography. New York: 2nd ed. Reinhold 1959. — KEULEMANS, A. I. M., and A. KWANTES: World Petroleum Congress Proc. Rome (1955); — in DESTY pp. 15—34 (1957). — KEULEMANS, A. I. M., A. KWANTES and G. W. A. RIJNDERS: Anal. Chim. Acta 16, 29—39 (1957). — KIELSELBACH, R.: Anal. Chem. 32, 880—881 (1960). — KIRCHER, H. W.: Anal. Chem. 32, 1103—1106 (1960). — KIRKLAND, J. J.: Anal. Chem. 32, 1388—1393 (1960). — KJAER, A., and A. JART: Acta Chem. Scand. 11, 1423 (1957). — KLINKENBERG, A., and F. SJENITZER: Chem. Eng. Sci. 5, 258—270 (1956). — KNAPMAN, C. E. H.: Gas chromatography abstracts, London: Butterworths 1958; — Gas chromatography abstracts. London: Butterworths 1959. — KNIGHT, H. S.: Anal. Chem. 30, 2030—2032 (1958); — in PECSOK (1959a). — KNOX, J. H.: Chem. and Ind. (London) 1959, 1085. — KOKES, R. J., H. TOBIN and P. H. EMMETT: J. Am. Chem. Soc. 77, 5860—5862 (1955). — KOMERS, R., and V. BAZANT: Doklady Akad. Nauk. S.S.S.R. 126, 1268—1269 (1959). — KOVATS, E., W. SIMON and E. HEILBRONNER: Helv. Chim. Acta 41, 275—288 (1958). — KRATZ, P., M. JACOBS and B. M. MITZNER: Analyst 84, 671—672 (1959). — KREJCI, M., and K. TESARIK: Section on Gas Chromatog. 3rd Conference on Anal. Chem., Prague (1959). — KRYACOS, G., and C. E. BOORD: Anal. Chem. 29, 787—788 (1957). — KWANTES, A., and G. W. A. RIJNDERS: in DESTY pp. 125—136 (1958).

LANGENAU, E. E., and A. R. ROGERS: Am. Perfumer Aromat. March (1960). — LANGER, S. H., and P. PANTAGES: Anal. Chem. 30, 1889—1890 (1958). — LAPIDUS, L., and N. R. AMUNDSON: J. Phys. Chem. 56, 984—988 (1952). — LEIBNETZ, E., H. HRAPIA and H. G. KONNECKE: Brennstoff-Chem. 38, 14—16 (1957). — LEWIS, J. S., H. W. PATTON and W. I. KAYE: Anal. Chem. 28, 1370—1373 (1956). — LIBERTI, A.: Anal. Chim. Acta 17, 247—253 (1957); — in DESTY (1958). — LIBERTI, A., and G. P. CARTONI: Chim. e ind. (Milan) 39, 821—824 (1957); — in DESTY pp. 321—329 (1958); — Ricerca Sci. 28, 1192—1198 (1958b). — LIBERTI, A., L. CONTI and V. CRESCENZI: Nature 178, 1067—1069 (1956). — LIBERTI, A., G. COSTA and E. PAULUZZI: Chim. e ind. (Milan) 38, 674—677 (1955). — LINDSAY, A. L., and L. A. BROMLEY: Ind. Eng. Chem. 42, 1508 (1950). — LINK, W. E., H. M. HICKMAN and R. A. MORRISSETTE: J. Am. Oil Chemists' Soc. 36, 300—303 (1959). — LIPSKY, S. R., and R. A. LANDOWNE: Biochim. Biophys. Acta 27, 666—667 (1958); — Ann. Rev. Biochem. 29, 649 to 668 (1960). — LIPSKY, S. R., R. A. LANDOWNE and M. R. GODET, Biochim. Biophys. Acta 31, 336—347 (1959). — LIPSKY, S. R., R. A. LANDOWNE and J. E. LOVELOCK: Anal. Chem. 31, 852—856 (1959). — LITTLEWOOD, A. B.: Nature 184, 1631—1632 (1959). — LITTLEWOOD, A. B., C. S. G. PHILLIPS and D. T. PRICE: J. Chem. Soc. 1955, 1480—1489. — LOVELOCK, J. E.: J. Chromatog. 1, 35—46 (1958a); — Nature 182, 1663—1664 (1958b). — LOVELOCK, J. E., A. T. JAMES and E. A. PIPER: Ann. N. Y. Acad. Sci. 72, 720—729 (1959). — LOVELOCK, J. E., and S. R. LIPSKY: J. Am. Chem. Soc. 82, 431—433 (1960). — LOWE, A. E., and D. MOORE: Nature 182, 133 (1958). — LUTINSKI, C.: Abstr. Appl. Spectroscopy 11, 100 (1957).

MCFADDEN, W. H.: Anal. Chem. 30, 479—481 (1958). — MCINNES, A. G.: in DESTY pp. 304—315 (1957). — MCINNIS, A. G., D. H. BALL, F. P. COOPER and C. T. BISHOP: J. Chromatog. 1, 556—557 (1958). — MCNESBY, J. R., C. M. DREW and A. S. GORDON: J. Phys. Chem. 59, 988—989 (1955). — MCWILLIAM, I. G.: J. Appl. Chem. (London) 9, 379—388 (1959). — MCWILLIAM, I. G., and R. A. DEWAR: in DESTY pp. 142—152 (1958); — Nature 181, 760 (1958b). — MADISON, J. J.: Anal. Chem. 30, 1859—1862 (1958). — MAIR, B. J., and M. SHAMAIENGAR: Anal. Chem. 30, 276—279 (1958). — MARTIN, A. E., and J. SMART: Nature 175, 422—423 (1955). — MARTIN, A. J. P.: in COATES et al. pp. 237—247 (1958). — MARTIN, A. G. P., C. E. BENNET and F. W. MARTINEZ: 3rd Symposium Gas Chromatog. Edingburgh 1960. — MARTIN, A. J. P., and R. L. M. SYNGE: Biochem. J. 35, 1358—1368 (1941). — MASON, E. A., and C. S. SAXENA: Phys. Fluids 1, 361—369 (1958). — MASON, L. H., H. J. DUTTON and L. R. BAIR: J. Chromatog. 2, 322—323 (1959). — MAYER, S. W., and E. R. TOMPKINS: J. Am. Chem. Soc. 69, 2866—2874 (1947). — MEAKINS, G. D., and R. SWINDELLS: J. Chem. Soc. 1959, 1044—1047. — MEDVEDEVA, N. I., and E. S. TORSUEVA: Trudy Komissii Anal. Khim., Akad. Nauk. S.S.S.R., Inst. Geokhim. i. Anal. Khim. 6, 88—96 (1955). — MEIGH, D. F.: Nature 184, 1072 (1959). — MEIGH, D. F., K. H. NORRIS, C. C. CRAFT and M. LIEBERMAN: Nature 186, 902—903 (1960). — MELLOR, N.: in DESTY pp. 63—73 (1957). — MESSNER, A. E., D. M. ROSIE and P. A. ARGABRIGHT: Anal. Chem. 31, 230—233 (1959). — MITCHELL, J., jr., D.M. SMITH and W. M. D. BRYANT: J. Am. Chem. Soc. 62, 4 (1940). — MORLEY, H. R., F. D. COOPER and A. S. HOTT: Chem. and Ind. (London) 1959, 1018. — MORROW, H. N., and K. B. BUCKLEY: Petrol. Refiner 36, 157—161 (1957). — MOUSSEBOIS, C., and G. DUYCKAERTS: J. Chromatog. 1, 200—201 (1958). — MUNDAY, C. W., and G. R. PRIMAVESI: in DESTY pp. 146—153 (1957). — MURRAY, K. E.: in DESTY pp. 245—246 (1958). — MURRAY, W. J., and A. F. WILLIAMS: Chem. and Ind. (London) 1956, 1020—1021.

NAIR, G. V., and E. VON RUDLOFF: Can. J. Chem. 37, 1608—1613 (1959). — NAVES, Y. R.: Perfumery Essent. Oil Record 49, 290—296 (1958a); — Compt. Rend. 246, 1734—1736 (1958b); — J. Soc. Cosmetic Chemists 9, 101—103 (1958c); — France et ses parfums 2,

23—27 (1959). — NAVES, Y. R., and A. ODERMATT: France et ses parfums 1, 10 (1958). — NEDEROST, M.: Chem. Listy. 50, 317—318 (1956). — NEIGISCH, W. D., and W. H. STAHL: Food Res. 21, 657—665 (1956). — NEISH, A. C.: Ann. Rev. Plant Physiol. 11, 55—80 (1960). — NEL, W. M., and V. PRETORIUS: S. African Ind. Chemist 13, 68—70 (1959). — NOGARE, S. D.: in DESTY pp. 241—244 (1958). — NOGARE, S. D., and C. E. BENNETT: Anal. Chem. 30, 1157—1158 (1958). — NOGARE, S. D., C. E. BENNETT and J. C. HARDIN: in COATES et al. pp. 117—129 (1958). — NOGARE, S. D., and W. E. LANGLOIS: Anal. Chem. 32, 767—770 (1960). — NOGARE, S. D., and L. W. SAFRANSKI: Anal. Chem. 30, 894—898 (1958). — NOREM, S. D.: in COATES et al. pp. 191—194 (1958). — NOYONS, A. K.: Arch. néerl. physiol. 7, 488 to 495 (1922). — NUNEZ, L. J., W. H. ARMSTRONG and H. W. COGSWELL: Anal. Chem. 29, 1164 to 1165 (1957). — NYSTROM, R. F., L. H. MASON, E. P. JONES and H. J. DUTTON: J. Am. Oil Chemists' Soc. 36, 212—214 (1959).

OGILVIE, J. L., M. C. SIMMONS and G. P. HINDS: Anal. Chem. 30, 25—27 (1958). — OMEROD, E. C., and R. P. W. SCOTT: J. Chromatog. 2, 65—68 (1959). — ORR, C. H., and J.E. CALLEN: J. Am. Chem. Soc. 80, 249 (1958).

PATTON, H. W.: ACS Meeting, Texas, Abstracts p. 14B, No. 37 (1956). — PECSOK, R. L.: Principles and practice of gas chromatography. New York: Wiley 1959a; — (1959b), in PECSOK pp. 135—150 (1959a). — PERCIVAL, W. C.: Anal. Chem. 29, 20—24 (1957). — PHILLIPS, C. S. G.: Disc. Faraday Soc. 7, 241—248 (1949); — Gas chromatography. London: Butterworths 1956; — in COATES et al. pp. 51—65 (1958). — PHILLIPS, G.: J. Sci. Instr. 28, 342—347 (1952). — PIEROTTI, G. J., C. H. DEAL, E. L. DERR and P. E. PORTER: J. Am. Chem. Soc. 78, 2989—2998 (1956). — PITKETHLY, R. C.: Anal. Chem. 30, 1309—1314 (1958). — POPJAK, G., A. E. LOWE, D. MOORE, L. BROWN and F. A. SMITH: Proc. Biochem. Soc. in Biochem. J. 73, 33 pp. (1959). — PORTER, R. S., C. H. DEAL and F. H. STROSS: J. Am. Chem. Soc. 78, 2999—3006 (1956). — PRIESTLEY, W.: in DESTY p. 165 (1957). — PRIMAVESI, G. R., G. F. OLDHAM and R. J. THOMPSON: in DESTY pp. 165—177 (1958). — PURCELL, J. R., and R. N. KEELER: Rev. Sci. Instr. 31, 304—306 (1960). — PURNELL, J. H.: Ann. N.Y. Acad. Sci. 72, 592—605 (1959).

QUIN, L. D.: J. Org. Chem. 24, 911—916 (1959). — QUIN, L. D., and M. E. HOBBS: Anal. Chem. 30, 1400—1405 (1958).

RAATS, E., J. HARLEY and V. PRETORIUS: J. Sci. Instr. 34, 510 (1957). — RAUPP, G.: Z. Anal. Chem. 164, 135—146 (1958). — RAY, N. H., J. Appl. Chem. (London) 4, 21—25 (1954a); 4, 82—85 (1954b); — Analyst 80, 853—860 (1955); — Nature 184, 54 (1959a); — ISA Proc. Anal. Instr. Div. 2nd Int. Gas Chromatog. Symposium. Preprints 2, 66—71 (1959b). — REISZ, P., and K. E. WILZBACH: J. Phys. Chem. 62, 6—9 (1958). — RHOADES, J. W.: Food Research 23, 254—261 (1958). — RICHARZ, W., M. LUTZ and A. GUYER: Helv. Chim. Acta 42, 2212—2218 (1959). — RIPPERE, R. E.: in COATES et al. pp. 223—224 (1958). — ROGINSKY, S. Z., M. I. IANOVSKY, G. M. ZHABROVA, O. M. VINOGRADOVA, B. M. KADENATSKI and Z. A. MARKOVA: Doklady Akad. Nauk. S.S.S.R. 121, 674—677 (1958). — ROSIE, D. M., and R. L. GROB: Anal. Chem. 29, 1263—1264 (1957). — ROTH, F., W. OHME and A. NICKISCH: Oel u. Kohle 38, 1133—1143 (1942). — RUSSELL, D. S.: Can. J. Chem. 36, 1745—1746 (1958). — RYCE, S. A., and W. A. BRYCE: Nature 179, 541 (1957).

SCHLENK, H., and J. L. GELLERMAN: Anal. Chem. 32, 1412—1414 (1960). — SCHMAUCH, L. J., and R. A. DINERSTEIN: Anal. Chem. 32, 343—352 (1960). — SCHOLLY, P. R., and N. BRENNER: 2nd International Gas Chromatog. Symposium, Michigan (1959). — SCHUFTAN, P.: Gasanalyse in der Technik. Leipzig 1931. — SCOTT, C. G.: J. Inst. Petrol 45, 118—122 (1959a). — SCOTT, R. P. W.: Nature 176, 793 (1955); — in DESTY pp. 131—145 (1957); pp. 189—199 (1958); — Nature 183, 1753—1754 (1959b). — SELIGMAN, R. B.: Abstract in Anal. Chem. 28, 1058 (1956). — SEMENENKO, V. E.: Fiziol. Rastenii, Akad. Nauk. S.S.S.R. 5, 561—568 (1958). — SEVENSTER, P. G.: S. African Ind. Chemist 12, 75—76 (1958a); 12, 151—158 (1958b). — SHAKESPEAR, G. A.: Proc. Phys. Soc. 33, 163—164 (1921). — SIMMONS, M. C., and L. R. SNYDER: Anal. Chem. 30, 32—35 (1958). — SKRIVAN, J. F., and H. E. HOELSCHER: A. I. Ch. E. Journal 5, 348—353 (1959). — SLATER, W. K.: J. Sci. Instit. 3, 177—181 (1926); — Biochem. J. 21, 198—203 (1927). — SMITH, D. M., J. C. BARTLET and L. LEVI: Anal. Chem. 32, 568 (1960). — SOKOL, L.: Chem. Prumysl. 7, 189—190 (1957). — SORENSEN, I., and P. SOLTOFT: Acta Chem. Scand. 12, 814—822 (1958). — SPENCER, C. F., F. BAUMANN and J. F. JOHNSON: Anal. Chem. 30, 1473—1474 (1958). — SPENCER, M. S.: Nature 184, 1231—1232 (1959). — STAHL, W. H., and E. J. LEVY: Abstract in Anal. Chem. 28, 1058 (1956). — STAHL, E., and L. TRENNHEUSER: Arch. Pharmaz. 9, 826—837 (1960). — STANLEY, R. G.: Nature 182, 738—739 (1958). — STERN, O.: quoted by LEWIS, B., and G. VON ELBE in: Combustion flames and explosions of gases, p. 206. New York 1951. — STEWART, G. H.: Anal. Chem. 32, 1205 (1960). — STILES, W., and W. LEACH: Annals Bot. 45, 461—488 (1931). — STOLL, A., J. RUSCHMANN, A. VON WARTBURG and J. RENZ: Helv. Chim. Acta 39, 993—999 (1956). — STUVE, W.: in DESTY pp. 178—188 (1958). — SULLIVAN, J. H., J. T. WALSHE and C. MERRITT: Anal. Chem. 31, 1826—1828 (1959). — SUNNER, S., K. J. KARRMAN and V. SUNDEN: Mikrochim. Acta 7—8, 1144—1151 (1956). — SWEELEY, C. C., and E. C. HORNING: Nature 187, 144—145 (1960). — SWEETING, J. W.: Chem. and Ind. (London) 1959, 1150. — SZEKELY, G., T. KORMANY, G. RACZ and G. TRAPLY: Periodica Polytech. 2, 269—274 (1958). — SZULCZEWSKI, D. H., and T. HIGUCHI: Anal. Chem. 29, 1541—1543 (1957).

TAYLOR, B. W.: in COATES et al. pp. 155—164 (1958). — TAYLOR, G. W., and A. S. DUNLOP: in COATES et al. pp. 73—85 (1958). — TEISSEIRE, P.: Recherches 9, 10—21 (1959). — TENNEY, H. M., and R. J. HARRISS: Anal. Chem. 29, 317—318 (1957). — THOMAS, C. O., and H. A. SMITH: J. Phys. Chem. 63, 427—432 (1959). — THOMPSON, A. E.: J. Chromatog. 2, 148—154 (1959). — THOMPSON, J. F., S. I. HONDA, G. E. HUNT, R. M. KRUPKA, C. J. MORRIS, L. E. POWELL, O. O. SILBERSTEIN, G. H. N. TOWERS and R. M. ZACHARIUS: Botan. Review 25, 1—263 (1959). — TISELIUS, A.: Arkiv. Kemi, Mineral. Geol. 16A, 11 pp. (1943). — TOTH, J., and L. GRAF: Magyar Kem. Folyoirat 63, 216—221 (1957). — TURKEL'TAUB, N. M.: Neftyanoe Khoz. 32, 72—77 (1954); — Trudy Komissii Anal. Khim., Akad. Nauk. S.S.S.R. Inst. Geokhim i Anal. Khim. 6, 146—161 (1955). — TURKEL'TAUB, N. M., and L. Y. ABRAMOVICH: Zavodskaya Lab. 22, 1032—1039 (1956); — Zhur. Anal. Khim. 13, 43—47 (1958). — TURKEL'TAUB, N. M., N. V. PORSHNEVA and O. A. KANCHEEVA: Zavodskaya Lab. 22, 735 to 738 (1956). — TURNER, D. W.: Nature 181, 1265—1266 (1958). — TURNER, N. C.: Petrol. Refiner 22, 140—144 (1943).

VAN DER KLOOT, A. P., R. I. TENNEY and V. BAVISOTTO: Proc. Am. Soc. Brewing Chemists 1958, 96—103. — VEEN, R. VAN DER: Physiol. Plantarum 2, 217—234 (1949). — VILKAS, M., and N. A. ABRAHAM: Bull. soc. chim. France 1959, 1651—1652. — VYAKHIREV, D. A., and A. I. BRUK: Uch. Zap. Gor'kovsk. Inst. 32, 43—53 (1958).

WACHI, F. M.: Dissert. Abstr. 20, 53 (1959). — WALLER, E. R.: New Phyt. 25, 109—118 (1926). — WALSH, J. T., and C. MERRIT: Anal. Chem. 32, 1378—1381 (1960). — WEAVER, E. R.: Physical methods in chemical analysis, Vol. 2 pp. 387 et seq. New York: Academic Press 1951. — WEAVER, E. R., P. E. PALMER, H. W. FRANTZ, P. C. LEDIG and S. F. PICKERING: Ind. Eng. Chem. 12, 359—366 (1920). — WENCKE, K.: Chem. Tech. (Berlin) 9, 404—406 (1957). — WEST, P. W., B. SEN and N. A. GIBSON: Anal. Chem. 30, 1390—1397 (1958). — WEST, T. S.: Chem. Age (London) 81, 733 (1959). — WHITE, D., and C. T. COWAN: in DESTY pp. 116—124 (1958). — WHITE, J. U., W. M. WARD and W. S. GALLAWAY: Anal. Chem. 31, 1267—1270 (1959). — WHITHAM, B. T.: in DESTY pp. 194—212 (1957); pp. 395—412 (1957). — WILLIS, V.: Nature 184, 895 (1959). — WILSON, J. N.: J. Am. Chem. Soc. 62, 1583—1591 (1940). — WILZBACH, K. E.: Proc. Symposium Tritium Tracer Appl., N.Y.C. (1957). — WILZBACH, K. E., and P. RIESZ: Science 126, 748 (1957). — WIRTH, M. M.: in DESTY pp. 154—168 (1957). — WOLFGANG, R., and F. S. ROWLAND: Anal. Chem. 30, 903 to 906 (1958).

YOUNG, I. G.: ISA Proc. Anal. Instr. Div. 2nd Int. Gas Chromatog. Symposium. Preprints 2, 40—47 (1959). — YOUNGS, C. G.: Anal. Chem. 31, 1019—1021 (1959).

ZDERIC, J. A., W. A. BONNER and T. W. GREENLEE: J. Am. Chem. Soc. 79, 1696—1698 (1957). — ZHUKHOVITSKII, A. A., and N. M. TURKEL'TAUB: Doklady Akad. Nauk. S.S.S.R. 94, 77—80 (1954); — Uspekhi Khim. 25, 859—871 (1956); — Zavodskaya Lab. 24, 796—798 (1958). — ZHUKHOVITSKII, A. A., N. M. TURKEL'TAUB and V. A. SOKOLOV: Doklady Akad. Nauk. S.S.S.R. 88, 859—862 (1953). — ZIEGENHAIN, W. C.: Petrol. Eng. 29, C6—C12 (1957). — ZLATKIS, A., and H. R. KAUFMAN: Nature 184, 2010 (1959). — ZLATKIS, A., S. LING and H. R. KAUFMAN: Anal. Chem. 31, 945—947 (1959). — ZLATKIS, A., and J. F. ORO: Anal. Chem. 30, 1156 (1958). — ZLATKIS, A., J. F. ORO and A. P. KIMBALL: Anal. Chem. 32, 162—164 (1960). — ZLATKIS, A., and J. A. RIDGWAY: Nature 182, 130—131 (1958). — ZUBYK, W. J., and A. Z. CONNER: Anal. Chem. 32, 912—917 (1960).

Ion-Exchange Chromatography.

By

N. K. Boardman.

With 15 Figures.

Chromatographic methods using either the synthetic ion-exchange resins or other ion-exchange materials are widely used for the separation of many classes of ionized or ionizable substances. Substances of both high and low molecular weight have been successfully chromatographed on ion-exchange materials. Ion-exchange chromatography is useful both for preparative and analytical work, but it is particularly suitable for analytical work with complex mixtures of ionizable substances, where separation of the individual components is frequently a prerequisite to their quantitative determination by standard procedures. Ion-exchange resins may also be used to separate the non-ionized or neutral substances in an extract from the ionized components.

The main problems facing the user of ion-exchange materials are the choice of a suitable resin from the many which are now available commercially and the selection of suitable conditions for carrying out the chromatogram. For a few classes of substances, e.g. the amino acids, reliable methods are described in detail in the literature; in other cases, details of particular separations are available but the user may need to modify the described procedure to suit a particular batch of resin. However, in many instances, the researcher will be required to devise a method to suit his particular problem and this article has been written with this thought well in mind.

It is intended to outline briefly the principles governing ion-exchange chromatography; for a fuller discussion on chromatographic principles the reader is referred to the section in Vol. 1 of this handbook (p. 95). There is some repetition in the present article but this seemed desirable as one of the principle aims of the present author was to emphasize those aspects of ion-exchange chromatography, both theoretical and practical, which would assist the research worker to fully exploit the use of ion-exchange resins for both preparative and analytical work with the minimum number of exploratory experiments. A general survey of the chemical forces involved in the interactions between the ion-exchange resin and a variety of solutes is given. An appreciation of these interactions is fundamental to an understanding of the factors, e.g. pH of eluting buffer, salt concentration, temperature etc., which influence separations on columns of ion-exchange resins and it should prove invaluable to the plant analyst or plant chemist in choosing the most suitable resin and in selecting suitable conditions for a particular separation.

Apparatus and experimental procedures will be considered in some detail again with emphasis on these points which will assist the research worker to select the optimum size of column, the correct flow rate, etc. By way of illustration, some specific applications of ion-exchange resins will be described and this is followed by a summary of the classes of substances which have been separated by ion-exchange chromatography.

Definition. Ion-exchange chromatography may be considered as a special form of adsorption chromatography where the adsorbent or stationary phase is an ion-exchange material and the mobile phase is an ionizing solvent. In ion-exchange chromatography, solute ions are exchanged for counter ions present on the ion-exchange material. It should not be inferred from this definition that the forces involved in ion-exchange chromatography are purely coulombic; there are important contributions from secondary forces such as VAN DER WAAL's and hydrogen bonding. In fact, there are numerous examples of separations which would not be expected to occur if the adsorption were purely electrostatic in nature. It is difficult in some instances to distinguish between ion-exchange chromatography and adsorption chromatography. In this article it is proposed to classify as ion-exchange chromatography those separations which are achieved on ion-exchange materials which are at least partly ionized under the conditions of the experiment.

A. Ion-Exchange Materials.

Ion-exchange materials of importance generally include the clay minerals, the synthetic alumino-silicates, the synthetic organic ion-exchange resins and the modified celluloses. Although differing greatly in detailed behaviour these materials have certain characteristics in common. They contain ionized or ionizable groups which serve as exchange sites and their structure is such that diffusion of ions through the bulk material may occur.

Although there are numerous reports in the literature of the use of the inorganic ion-exchangers for chromatography, the most important group by far for the practical chromatographer is the synthetic ion-exchange resins and this group will be treated in some detail. Another group, the modified celluloses have assumed considerable importance in recent years as ion-exchangers suitable for the chromatography of proteins and nucleic acids and this group also will receive detailed treatment.

Ion-exchange resins. An ion-exchange resin consists essentially of a cross-linked polymeric structure to which is attached ionized or ionizable groups. These groups are usually attached as substituents to the backbone structure, but they may be integral with it as units in the chain. The polymer thus carries an electrostatic charge, but this is counterbalanced exactly by counter ions and it is these counter ions which will exchange reversibly with other ions in the medium without any chemical changes occurring in the charged polymer. Cation-exchange resins contain non-diffusible acidic groups which may be strong ($-SO_3H$) or weak [$-OH$, $-COOH$, $-PO(OH)_2$] and diffusible cations, and similarly anion-exchange resins contain non-diffusible basic groups, either strong or weak, and diffusible anions.

If a cation-exchanger containing counter ions A^+ is added to a solution containing cations B^+, the latter diffuse into the resin matrix and counter ions A^+ originally present diffuse out until equilibrium is reached. At equilibrium, the ions usually are not distributed uniformly between the resin and the ambient solution. The resin exhibits a preference for some ions rather than others, an effect called the selectivity of the resin. This principle of ionic selectivity is the important factor underlying ion-exchange chromatography. Thus the order of separation on columns of ion-exchange resins may be predicted from data of "relative affinities" obtained from equilibrium studies.

The simplest theoretical treatment of ion-exchange equilibria is to apply the law of mass action to the following type of equation.

$$R^-A^+ + B^+X^- \rightleftharpoons R^-B^+ + A^+X^-$$

where R^- represents the fixed charge on the resin, A^+, B^+ are the exchangeable cations and X^- the corresponding non-exchangeable anion in solution.

Thus

$$\frac{[B_R^+]\,[A_S^+]}{[A_R^+]\,[B_S^+]} = K_C$$

where [R] represent concentrations in the resin.
[S] represent concentrations in the solution.

A full discussion of the applicability of this type of equation where the mass action terms are interpreted as concentrations is beyond the scope of this article. Many workers have found K_C values to be approximately constant for a given exchange and their values give a useful practical measure of the relative affinities of many ions. A theoretical improvement over the simple mass action equation is obtained by inserting activities instead of concentrations in the mass action equation, but there are practical difficulties of determining activity coefficients for different ions in the resin. BAUMAN and EICHHORN (1947) arrived at a similar equation by considering the exchange between resin and ambient solution as a DONNAN equilibrium. These workers proposed that a highly cross-linked resin should be regarded as a concentrated electrolyte solution (of the order of 5 molar) having activity coefficients similar to those obtained in ordinary solution.

GREGOR (1951) proposed an alternative theory of selection. In his theory, GREGOR introduced the concept of strain energy or swelling pressure to account for the fact that the resin will prefer the smaller of a pair of ions. It is now clear that allowance should be made for both an activity coefficient effect and a swelling pressure effect in any theory of ion selectivity.

I. Fundamental Properties of Ion-Exchange Resins.

1. Chemical Structure.

The original synthetic ion-exchange resins were formed by condensing polyhydric phenols or mixtures of polyhydric phenols and phenol sulphonic acids with formaldehyde (ADAMS and HOLMES 1935). The structure of a typical resin of this type is shown in Fig. 1. The particular resin shown is polyfunctional as it contains

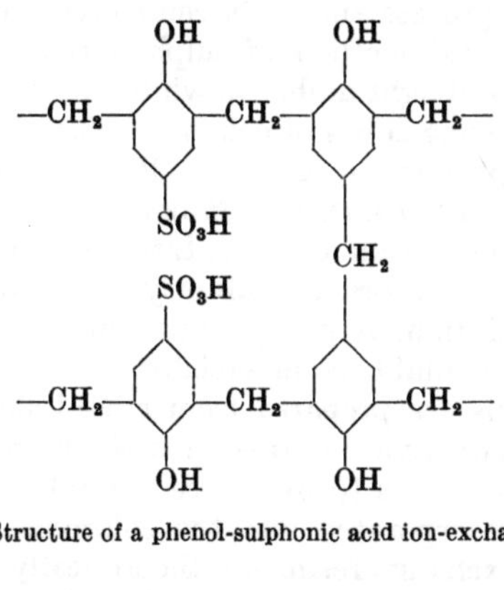

Fig. 1. Structure of a phenol-sulphonic acid ion-exchange resin.

both the strongly acidic sulphonic acid group and the very weakly acidic phenolic hydroxyl group. Polyfunctional anion-exchange resins may be obtained in a similar way by condensing m. phenylene diamine with formaldehyde.

These polyfunctional condensation resins have limited uses for chromatographic purposes, since they behave as mixtures of ionizable groups. They also suffer from the practical disadvantage of limited chemical stability and many of them tend to leak coloured breakdown products into solution. Except in a few special instances, research workers will have little use for this type of resin.

The most useful resins for chromatographic purposes are the polystyrene resins. An inert, three dimensional, cross-linked hydrocarbon network is formed by copolymerising styrene with divinyl benzene (DVB) or some other compound having two unsaturated bonds. Ionic groups are then attached to confer the property of ion-exchange. The usual amount of DVB is 5—10% on the weight of styrene, although resins with DVB contents as low as 1% or as high as 16% are commercially available. The copolymerisation of styrene and DVB is usually carried out as a suspension process in which the monomers are vigorously agitated in an aqueous medium containing a stabilizing agent. This serves to prevent the spherical droplets formed from adhering to one another. The degree of agitation determines the size of the droplets and hence the size of the polymer beads. Resins prepared in this way are called bead resins.

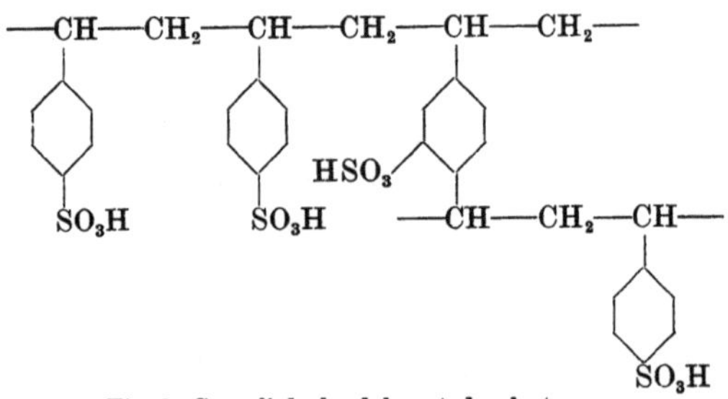

Fig. 2. Cross-linked sulphonated polystyrene.

The polystyrene polymer may be treated in a number of ways to produce a variety of ion-exchange resins. The benzene rings may be sulphonated to produce a strongly acidic ion-exchange resin of high capacity (Fig. 2). Measurement of the exchange capacity or sulphur content shows that there is one sulphonic acid group per benzene ring. The polystyrene polymer tends to darken considerably during the sulphonation process and early commercial resins were dark brown in colour. Due to an improved method of sulphonation, sulphonated polystyrene resins may now be prepared light amber to white in colour. These resins are particularly valuable for following zones of coloured substances.

A resin containing phosphonic groups $-PO(OH)_2$ has been prepared from cross-linked polystyrene (BREGMAN and MURATA, 1952) and at least one phosphonic acid resin is available commercially. Little is known about their properties although one would expect an order of affinity for cations similar to that shown by phosphate ions in solution. A phosphorous cation-exchange resin containing $-PO.OH$ groups is also available commercially.

Anion exchange resins are prepared from cross-linked polystyrene by a two step process. First, the copolymer is treated with chloromethyl ether in the presence of a swelling agent, so as to introduce methylene chloride groups (Fig. 3) and then an amine is incorporated into the chloromethylated product. A very large number of anion-exchange resins are theoretically possible since any amine can be incorporated, although in practice only a limited number of resins are manufactured. Introduction of a tertiary amine will produce a quaternary ammonium group, a base comparable in strength to sodium hydroxide, whereas treatment with a primary or secondary amine will produce a weakly basic resin.

Anion-exchange resins containing the trimethylbenzylammonium chloride grouping (called Type I) resins are so strongly basic that they are difficult to convert to the free base form, even with sodium hydroxide. A reduction in basicity takes place when one of the methyl groups (Fig. 3) is replaced by a hydroxyethyl ($-C_2H_4OH$) group. Such a resin (Type II resin) still behaves as a strongly basic resin, but it is easier to regenerate.

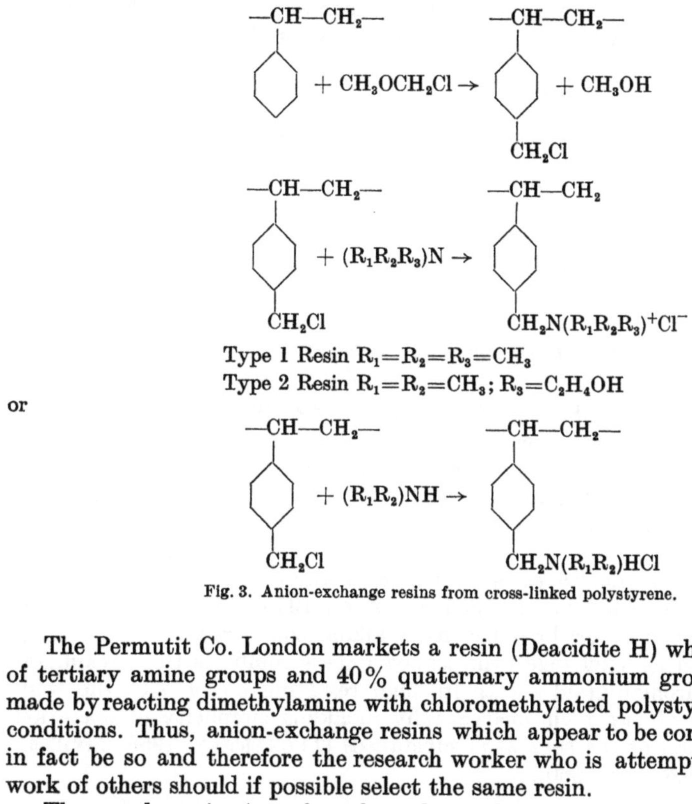

Fig. 3. Anion-exchange resins from cross-linked polystyrene.

The Permutit Co. London markets a resin (Deacidite H) which contains 60% of tertiary amine groups and 40% quaternary ammonium groups. This resin is made by reacting dimethylamine with chloromethylated polystyrene under special conditions. Thus, anion-exchange resins which appear to be comparable may not in fact be so and therefore the research worker who is attempting to repeat the work of others should if possible select the same resin.

The copolymerization of methacrylic acid and DVB or other suitable cross-linking agent yields a cross-linked polymethacrylic acid and this acts as a weakly acidic cation-exchange resin (Fig. 4).

Chemical stability. The condensation resins have limited chemical stability, showing a strong tendency to decompose at temperatures above 50° C for the hydrogen or free base forms and above 90° C for the salt forms. The polystyrene resins and the polymethacrylic one are very stable, and although the free base forms of the anion exchange resins should not be used above 60° C, the salt forms, as well as the salt and hydrogen forms of the cation-exchange resins are stable up to 120° C. The condensation resins are sensitive to oxidising agents, but the polystyrene resins and the polymethacrylic acid resin resist all but the strongest oxidising agents.

Chelating resins. Resins containing $—CH_2N(CH_2COOH)_2$ groups have been prepared by PEPPER and HALE (1955) from chloromethylated cross-linked polystyrene and by GREGOR, TAIFER, CITAREL and BECKER (1952). Such a resin is now available commercially from Bio-Rad Laboratories. The order of affinity for metal ions is similar to the stabilities of the EDTA-metal complexes in solution. Divalent ions are held selectively and can be removed from high concentrations of monovalent ions. Polymers containing hydroxamic acid groups selectively exchange ferric iron (CORNAZ and DEUEL, 1954).

Some special resins. A resin suitable for the resolution of optical isomers was prepared by GRUBHOFER and SCHLEITH (1954) by combining quinine with the carboxyl groups of an ion-exchange resin. In a similar way, ISLIKER (1953) chemically combined antigens with a carboxylic acid resin to produce a resin which was

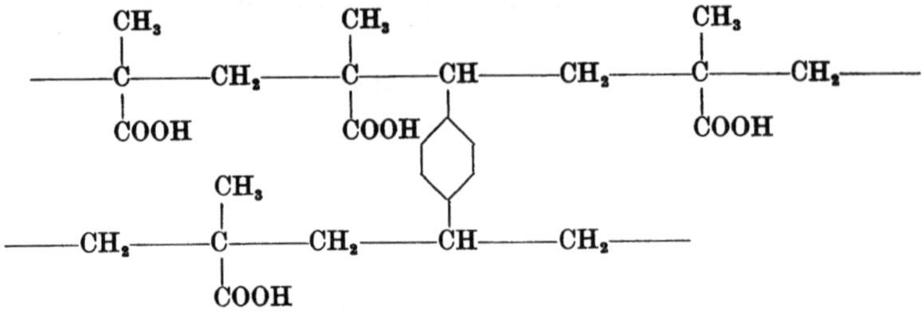

Fig. 4. Cross-linked polymethacrylic acid.

used for purification of antibodies. KENT and SLADE (1959) reacted antibody proteins with diazotized polystyrene to produce a resin for separating antigens. Special resins for the separation of proteins and large peptides respectively were prepared by BOARDMAN (1955, 1959) and FEITELSON and PARTRIDGE (1956) by coating the surface of a diatomaceous earth with cross-linked sulphonated polystyrene.

Ion-exchange resin – loaded papers. The preparation of ion-exchange resin – loaded papers were described first in 1955 (LEDERER, 1955; HALE, 1955), but it is only recently that such papers have become available commercially (see Table 1). Chromatography on ion-exchange resin-loaded papers combines the advantages of ion-exchange chromatography with its high resolving power with the well-known advantages of paper chromatography including simplicity of operation and ease of detection of the separated substances (TUCKERMAN, 1958; MYHERE and SMITH, 1958; KNIGHT, 1960).

2. Physical Properties.

a) Cross-Linking.

The fixed ionizable groups on the polymeric matrix are hydrophilic groups and thus they tend to draw the whole polymer into solution. This is prevented by inserting the cross-links into the polymer so as to form a 3 dimensional network which is insoluble in water. However, the ion-exchange resins behave as hygroscopic gels, swelling and shrinking with adsorption and desorption of water. The degree of swelling is dependent on the amount of cross-bonding, the water filling the pores formed by the criss-crossing network. Thus, as the degree of cross-linking is lowered, the resin swells more in water and the pores become larger. For satisfactory chromatography, the rate of penetration of the solute into the resin should be rapid; this will not be so if the degree of cross-linking is too high. On the other hand if the degree of cross-linking is too low, the resin will swell too much, making column operation difficult. The useful range of cross-linking varies between 20% and 2%, the pore size varying from about 5 Å to 35 Å or larger.

Thus large ions cannot pass down the pores of the more tightly cross-linked resins and in fact, these resins may be used to separate ions of different sizes. This process termed "molecular sieving" is treated in a separate article (p. 205).

The degree of cross-linking of the monofunctional cation-exchange resins is controlled by the concentration of DVB in the original mixture of monomers; side reactions leading to additional cross-links are insignificant. The percentage of DVB is a good measure of the amount of cross-linking. However, in the case of the monofunctional anion-exchange resins further cross-links in addition to those formed by DVB are introduced during the chloromethylation procedure. These additional cross-links can form a substantial part of the total cross-linking and therefore the DVB content cannot be regarded as a measure of the cross-linking. With these resins there is no satisfactory way of defining total cross-linking, but in practice it is found that determination of the weight swelling of a resin (PEPPER, REICHENBERG and HALE, 1952) gives a satisfactory measure of the degree of cross-linking. The weight swelling is defined as the weight of water associated with 1.0 gm dry hydrogen resin (cation-exchanger) or chloride form (anion exchanger) and it does not include interstital water which is first removed by centrifugation. The weight swelling is directly related to the pore size of the resin. When comparing different resins a more significant parameter is the weight of water associated with 1 mg equivalent of dry resin (specific water regain).

b) Exchange Capacity.

The total exchange capacity of an ion-exchange resin for simple ions is independent of the particle size of the resin and it agrees closely with the theoretical value. It thus appears that all the ionizable groups are accessible as exchange sites and the capacity of the resin is determined by the concentration of ionizable groups. However, with large molecules such as the proteins, exchange appears to take place on the surface of the resin since the protein ions are too large to permit their entry into the resin matrix. The capacity of an ion-exchange resin for a large ion will depend to a large extent on the nature of the surface of the resin. Ions of intermediate size between those of proteins and simple ions may not be able to penetrate the pores of a highly cross-linked resin, but they may be able to penetrate the pores of a resin of low cross-linking. The exchange capacity of a resin for these ions will be dependent on the degree of cross-linking.

The exchange capacity of an ion-exchange resin under a particular set of conditions is dependent not only on the concentration of ionizable groups in the resin but also on the nature of the ionizable groups. Thus, if a resin contains only the strongly acidic sulphonic acid group, the capacity of the resin for simple ions is constant and independent of pH above pH 3, (Fig. 5a). In contrast a resin containing only carboxyl groups has a lower capacity in neutral and acid solutions than in alkaline ones. In acid solution, the carboxyl groups on the resin are mainly in the undissociated form and therefore not able to accept other cations (Fig. 5b).

Similarly, the capacity of a resin containing basic quaternary nitrogen groups for simple anions is constant over a wide range of pH, (Fig. 5c), but resins containing the less basic amino groups exhibit an increasing capacity as the acidity of the solution is increased (Fig. 5d).

In other words, the capacity of an ion-exchange resin is determined by the concentration of its ionizable groups which are actually ionized under the particular conditions of the experiment. The capacity may be obtained by carrying out a titration curve on the resin. However, it should be remembered that the exchange capacity of a weakly acidic or weakly basic ion-exchange resin is dependent not only on the pH but also on the cation or anion concentration respectively in the ambient solution. This point is more clearly seen by reference to Fig. 5b, which shows the titration curve of Amberlite IRC-50, a carboxylic acid ion-exchange resin, at two different concentrations of sodium chloride. In 0.1 M NaCl the resin has a pK (pH of half ionization) of about 6.8 whereas in 1.0 M NaCl, the pK is about 5.7. The titration curve of IRC-50 also shows that not all of the carboxyl groups have the same pK; the titration curve is much flatter than for a monocarboxylic acid such as acetic acid.

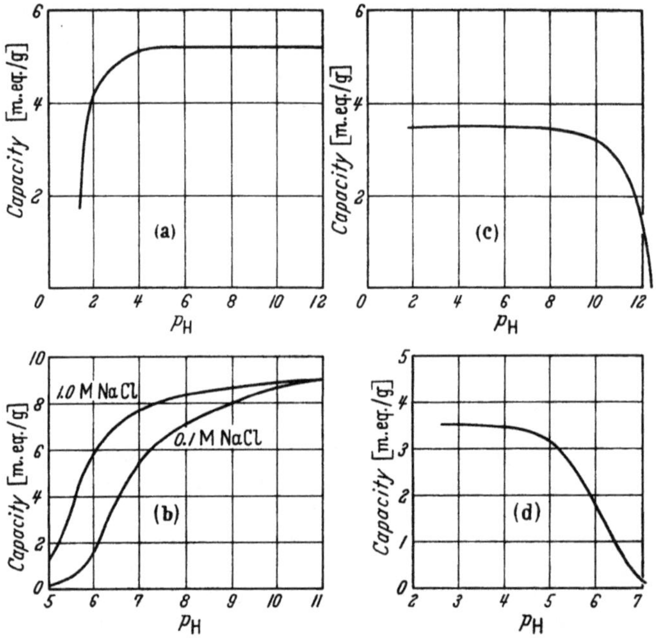

Fig. 5. Titration curves of a number of ion-exchange resins. (*a*) Sulphonated polystyrene. (*b*) Polymethacrylic acid. (*c*) Strongly basic resin. (*d*) Weakly basic resin.

There are two ways of expressing the capacity of a resin. It may be expressed as milligram equivalents/gm dry resin or as milligram equivalents/ml of wet resin. The latter figure is very often more useful to the chromatographer as the capacity of a packed column may be calculated from the dimensions of the resin bed.

c) Particle Size.

Ion-exchange resins are usually manufactured in a variety of particle sizes. The commercial resins designed primarily for industrial uses are supplied with particle sizes in the range 10—50 mesh/inch, while resins prepared for chromatographic use in the laboratory have particle sizes down to 200—400 mesh/inch. Unfortunately, two systems are in use for designating the particle size of the resin and the research worker should take this into account when ordering resins from different manufacturers. Thus, the range of resins supplied by the Dow Chemical Co. are sieved in the dry state and the manufacturer's mesh designation approximates the size of the particle of dry resin. On the other hand, the Permutit Co. (London) supply chromatographic grade resins which have been graded in the wet state. The chromatographic grade resins supplied by Rohm & Haas Co. are prepared by grinding bead material and sieving dry to the size indicated.

For analytical work, the resin should be as fine as possible, but 200—400 mesh (wet) probably represents the practical limit for column work. Finer particles than this give columns with very slow flow rates. There are also manufacturing difficulties in producing bead resins with smaller particle sizes. As mentioned previously, the particle size of the polystyrene beads is governed by the type and degree of agitation during the polymerisation process. As the ionizable groups are incorporated the beads swell with a result that the final bead size may be 2 or 3 times that of the original polystyrene beads. With resins of very low cross-linking (1—2%) where the hydrocarbon beads swell considerably during the attachment of the ionic groups it is difficult to produce fine bead resins.

The sharpness of the separation which may be achieved with a particular ion-exchange resin depends not only on the average particle size of the resin, but also on the range of particle sizes. The rate for elution of an ion from the resin is normally controlled by particle diffusion and the width of the elution band is determined by the largest particles present rather than the average.

A series of analytical grade resins are now available from the Bio-Rad Laboratories (A.G. series). These resins are processed from the Dowex ion-exchange resins. Each resin is sized to exact wet mesh specification and purified by cycling with appropriate acids, bases and organic solvents for 7—10 days. Resins are then converted to their final form and washed thoroughly with de-ionized water. Each lot is analysed for capacity, moisture content, cross linkage (swelling) and metal content. Bio-Rad Laboratories also market resins which have been prepared specially for amino acid analysis.

II. Ion-Exchange Celluloses.

The ion-exchange celluloses are widely used for the separation and purification of high molecular weight substances, including proteins, nucleic acids and acidic polysaccharides. Acidic or basic groups are incorporated into celluloses by reacting alkali swollen wood or cotton cellulose powder with the appropriate chloro-compound. The reaction must be carefully controlled, since substitution of the hydroxyl groups in cellulose with ionic groups causes swelling of the cellulose by interfering with the interchain hydrogen bonding. Products with capacities much greater than 1 milli-equivalent (m.eq.)/gm have a tendency to be gelatinous with a high resistance to the flow of water. In spite of this limitation on the number of ionic groups which may be incorporated, the ion-exchange celluloses have a high capacity for proteins and nucleic acids.

For the detailed preparation of a number of cellulose ion-exchangers, the reader is referred to the original paper by Peterson and Sober (1956). A brief outline of the methods of preparation are given in the following list of ion-exchange celluloses, many of which are available commercially (see Table 1).

Cation-exchange celluloses.

Carboxyl methyl cellulose (CM cellulose), active group $-OCH_2COOH$, prepared by reacting cellulose with chloroacetic acid.

Phosphate cellulose (P cellulose), active group $-OPO_3H_2$ prepared from cellulose and $POCl_3$.

Sulphoethyl cellulose (SE cellulose) $-OC_2H_5SO_3H$ or

Sulphomethyl cellulose (SM cellulose) $-OCH_2SO_3H$ prepared from cellulose and chloromethane sulphonate (Porath 1957).

Anion-exchange celluloses.

Diethylamino-ethyl cellulose (DEAE cellulose) $-OC_2H_4N(C_2H_5)_2$ prepared from cellulose and 2 chloroethylamine.

Triethylamino-ethyl cellulose. (TEAE cellulose) $-OC_2H_4\ N^+\ (C_2H_5)_3\ Br^-$ formed from DEAE cellulose by refluxing with ethyl bromide (PORATH 1957).

ECTEOLA cellulose. Exact structure not known, because of many reactions including polymerisation which are possible. ECTEOLA cellulose is a reaction product of cellulose, epichlorhydrin and triethanolamine.

Para-aminobenzyl cellulose. (P.A.B. cellulose) $-OCH_2 \cdot C_6H_4NH_2$, formed by reacting cellulose with p-nitrobenzylchloride, followed by reduction of the nitro group (CAMPBELL, LUESCHER and LERMAN, 1951).

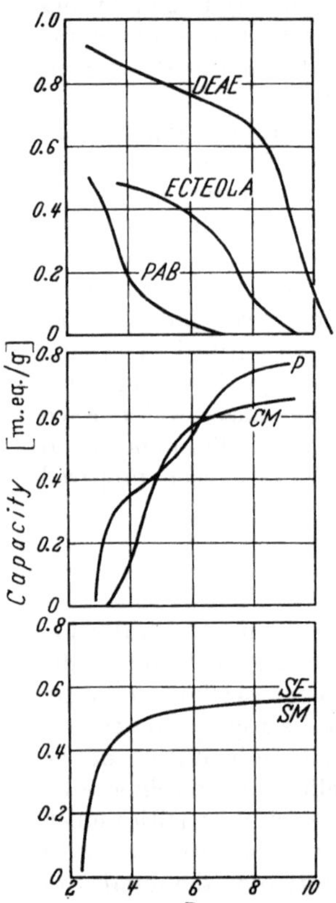

Fig. 6. Titration curves of a number of cellulose ion-exchangers.

The ion-exchange celluloses usually have capacities in the range 0.3—1.0 meq per gm. Some typical titration curves are shown in Fig. 6. As is the case with the weakly acidic or weakly basic ion-exchange resins, the proportion of ionized groups depends on the salt concentration in the ambient solution. The pK of ECTEOLA varies from about 5.5—7 and DEAE 8—9.5. TEAE is ionized over a wide range of pH since the ionic groups are strong bases. CM cellulose has a pK which varies between 3.5—4.5 and thus it has stronger acidic groups than the comparable ion-exchange resin, IRC-50. P cellulose is a dibasic cation-exchanger with a first pK between 2—3 and a second at 6—7. SE and SM cellulose contain the strongly acidic sulphonic acid group and they are completely ionized over a wide range of pH down to a value of 4.

The capacities of the ion-exchange celluloses may be determined by nitrogen, phosphorus or sulphur analyses as well as by titration. The capacity of CM cellulose is determined usually by titration.

Stability. PETERSON and SOBER (1956) found that the cellulose ion-exchangers, with the probable exception of P cellulose, could be used repeatedly and suffered no apparent deterioration upon months of storage in a cold room in contact with aqueous solutions ranging from 0.05 M Na_2HPO_4 to 1 N NaOH. Prolonged exposure to solutions more acid than pH 4 was avoided, although brief contact with 1 N HCl was not harmful. Long exposure of TEAE cellulose to sodium hydroxide solutions should be avoided.

Ion-exchange papers. A range of Whatman ion-exchange papers comparable in chemical structure to the cellulose ion-exchange powders has become available recently (KNIGHT 1959). (See Table 1 of page 204).

B. Theory of Chromatographic Procedures.

The procedures developed by TISELIUS (1943) and CLAESSON (1946) for column development with molecular adsorbents such as charcoal or alumina are applicable also to the field of ion-exchange chromatography. These workers distinguished 3 main ways of developing a chromatogram which they called frontal analysis, elution analysis and displacement development. It is beyond the scope of this article to discuss the theoretical basis of these procedures in detail, but it is important that the user of ion-exchange resins for chromatography should have some appreciation of the processes which govern not only the separations obtained but also the sharpness of such separations.

I. Elution Analysis.

Since the method of elution analysis has wider application in ion-exchange chromatography than either the frontal analysis or displacement methods and is of far greater importance to the analyst, more space will be devoted to a discussion of this technique. In the elution method, a small quantity of the solution to be analysed is passed into the column and the solutes are adsorbed as a narrow band at the top of the column. The chromatogram is then developed by passing pure solvent through the column or perhaps a new solvent which tends to alter the distribution of the solutes in favour of the liquid phase. In this way the components are made to move down the column at relative rates which will depend on their distribution coefficients with respect to the fixed phase and the solvent used to develop the chromatogram. If the distribution coefficients are sufficiently different, the bands due to each component will separate and an analysis of the effluent will show results similar to that shown in Fig. 7.

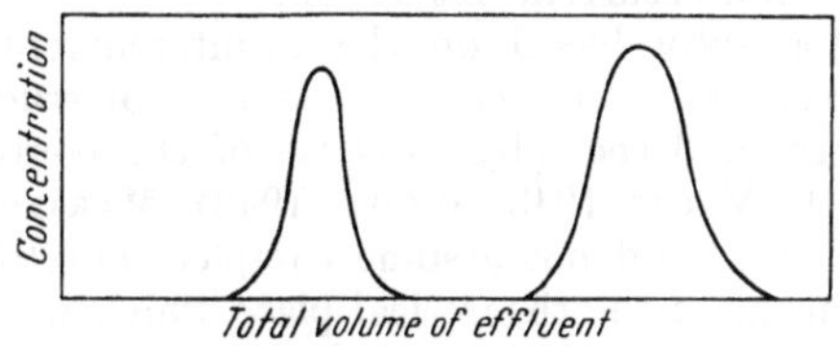

Fig. 7. Elution analysis.

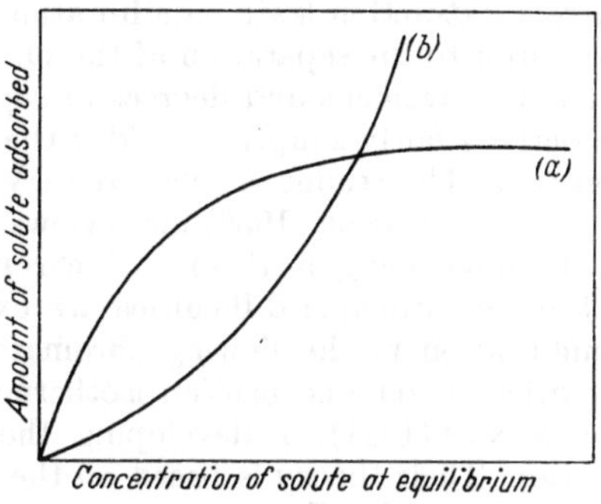

Fig. 8. Adsorption isotherms.

In ion-exchange chromatography, the solutes to be separated are either cations or anions and the solvent is usually water or a salt solution containing an ion or ions which will exchange with the ions in the mixture to be separated.

Symmetrical elution peaks are obtained only when the distribution coefficient of the solute is independent of the concentration of that solute and is also independent of the concentration of other solutes in the system. This means that the adsorption isotherm of the solute (the curve relating the amount of the solute adsorbed by a constant amount of resin to the concentration of solute in the ambient phase at equilibrium) is a straight line passing through the origin. If the isotherms are of the Freundlich or Langmuir type (Fig. 8, curve *a*) then the elution peaks will show sharp fronts and diffuse rear boundaries. An adsorption of the type shown in Fig. 8, curve *b*, will give rise to bands with diffuse fronts and sharp rear boundaries. This can be readily understood by considering the behaviour of a small quantity of solute assumed to have diffused to a position ahead of the solute forming the peak. Thus, for good elution chromatograms linear isotherms are necessary and the adsorption must be reversible.

However, in the case of solutes having Freundlich isotherms (curve *a*), tailing can be greatly suppressed by eluting with a buffer of gradually increasing eluting power. This method introduced by Alm, Williams and Tiselius (1952) is known as gradient elution analysis. In this way, each component will gradually find itself surrounded by an eluent of the critical concentration or pH which will make it move in an almost ideal fashion. Tailing will be greatly suppressed as the rear of the band is always in a stronger eluting medium than the front. The gradient method has some advantages over the older method of applying a succession of different solutions, since a sudden change of solvent introduces the danger of producing a false peak. Material which is not eluted by one solvent is picked up by the second.

1. Theory of Elution Analysis.

The essential principle of elution analysis is that of multistage countercurrent distribution, a principle well illustrated by the countercurrent extraction train i.e. a long train of vessels each containing a system of two phases. In the countercurrent extraction train, equilibration is theoretically achieved at each agitation step prior to the separation of the phases and the essential theory which predicts rates of movement and degrees of separation from the partition coefficients can be derived fairly simply, provided the partition coefficient of each solute remains constant. The reader is referred to Vol. 1 of this handbook (p. 66) and to Von Tavel and Signer (1956) for reviews on countercurrent extraction.

In developing a theory of chromatography based on the countercurrent extraction principle difficulties are experienced, first, because of a lack of true equilibration in the flowing chromatogram and secondly, because of the complexities of some adsorption isotherms (De Vault 1943; Weiss, 1943). Martin and Synge (1941), in developing their theory did not assume complete equilibration. Their theory is based on the concept of the theoretical plate which was well known in distillation theory. Martin and Synge (1941) considered the column as a large number of thin layers each of which is equivalent to one theoretical plate and the height of such a layer is called the HETP or "height equivalent to one theoretical plate". The HETP was defined as the thickness of a layer such that the solution issuing from it is in equilibrium with the mean concentration of solute in the non-mobile phase throughout the layer. In order to make the equations manageable, Martin and Synge (1941) made certain simplifying assumptions viz. that the diffusion of solute from plate to plate is negligible, and that at equilibrium the distribution coefficient of a solute between the two phases (stationary and mobile) is independent both of the absolute value of its concentration and the presence of other solutes. The separations obtained in practice are rather less than the theory predicts mainly because of non-uniformity of flow in the column. Nevertheless their theory is useful for predicting suitable conditions for separations from the predetermined distribution coefficients.

Martin and Synge's theory was developed for partition chromatography and some of their notations are not suited to ion-exchange chromatography. Mayer and Thompkins (1947) applied the "theoretical plate" concept to ion-exchange chromatography replacing many of the notations of Martin and Synge (1941) with more suitable ones. Dixon and Stack-Dunne (1955) also derived equations similar to those of Mayer and Thompkins. The reader is referred to the original papers for the derivation of the equations; only a brief discussion of their significance is given below. The equations are useful for analysing the results of an ion-exchange chromatogram and for testing the homogeneity of any particular peak.

The following is a list of definitions which are used later in the experimental section.

Total volume of effluent (V') is equal to the total volume of liquid which flows from the column from the time of application of the solute until the maximum concentration of solute emerges from the column.

Hold-up volume (V_0) of column is defined as the volume of solvent which is held between the resin particles.

Elution volume ($V = V' - V_0$) is equal to the total volume of effluent minus the hold-up volume.

$$R_F \textit{ value} = \frac{\text{Hold up volume}}{\text{Total volume of effluent}} = \frac{V_0}{V'} .$$

It is identical with the definition by MARTIN and SYNGE (1941) for use in partition chromatography.

Distribution coefficient (C). In ion-exchange chromatography, it is more convenient to define the distribution coefficient as follows:

$$C = \frac{\text{fraction of solute in a certain mass of resin}}{\text{fraction of solute in the liquid phase in contact with that resin}}$$

This avoids the necessity to determine the volume of the resin phase. (In terms of the usual distribution coefficient K, which equal the mass of solute/ml resin divided by the mass of solute/ml of liquid phase.

$$C = \frac{b\,K}{V_0}$$

where b is the volume of the resin phase in the column).

The following equations are derived from the theoretical plate theory.

1. $$\frac{V}{V_0} = C \qquad \text{or} \quad \frac{1}{R_F} - 1 = C$$

in other words, if we know the distribution coefficient C of a particular solute, we can determine the rate of movement of its peak, compared with the total volume of the effluent.

2. The concentration profile of a band in the direction of movement is symmetrical and has the slope of the normal curve of error.

3. The number of theoretical plates (N) in a column (a measure of the resolving power of the column) may be determined by applying the following equation to the normal curve of error of a solute

$$N = \frac{V'(V' - V_0)}{S^2}$$

where S is the standard deviation of the distribution curve.

2. Conditions for Successful Elution Analysis on Ion-Exchange Resins.

As pointed out in the above theoretical discussion, symmetrical curves are obtained only if the distribution coefficients of the solutes are independent of their concentrations. However, several workers have shown that the exchange data obtained with ion-exchange resins may be expressed in terms of a LANGMUIR isotherm, (curve a, Fig. 8) giving rise to unsymmetrical peaks on elution chromatography. BOYD, SCHUBERT and ADAMSON (1947) derived the following equation.

$$\left(\frac{\varkappa}{m}\right)_{A^+} = \frac{k\,b_1\,{}^cA^+}{b_1\,{}^cA_+ + b_2\,{}^cB_+} = \frac{\dfrac{k\,b_1}{b_2}\left(\dfrac{{}^cA_+}{{}^cB_+}\right)}{1 + \dfrac{b_1}{b_2}\left(\dfrac{{}^cA_+}{{}^cB^+}\right)}$$

where $\left(\frac{\varkappa}{m}\right)_{A^+}$ is the quantity of an ion A^+ adsorbed per unit weight of ion-exchange resin, ${}^CA^+$ and ${}^CB^+$ are the equilibrium concentrations or activities of the ions A^+ and B^+ is solution and b_1, b_2 and k are constants. Their treatment depends on the assumption that only one type of exchange site is present in the resin, as in a monofunctional resin, and it also implies that there is no interaction between the adsorbed ions.

However, if the concentration in solution of an adsorbed ion A^+ is small compared with the concentration of ion B^+ (the ion originally present on the resin) the adsorption of A^+ is directly proportional to its concentration and conditions are ideal for elution chromatography. For example, in amino acid analysis on sulphonated polystyrene, the resin is originally in the sodium form and adsorption takes place from a large excess of a sodium buffer.

On the other hand, if the concentration in solution of an adsorbed ion A^+ is very much greater than the concentration of the ion originally present on the resin, the adsorption of A^+ is nearly constant. The ion B^+ is displaced completely from the exchangeable sites on the resin by the ion A^+ and the resin becomes saturated with A^+. These conditions are ideal for displacement chromatography.

II. Displacement Development.

1. Completely Ionized Components.

In the method of displacement development, a solution containing a mixture of ions is first run onto a column of an ion-exchange resin until the top half or two-thirds of it is saturated. The column is then developed by passing through the column a solution of an ion which is strongly adsorbed compared with the ions in the mixture. The developing ion forms a band at the top of the column and it displaces before it all the ions in the mixture to be separated. The component ions displace each other also, so that they are arranged in order of increasing affinity, the one with the strongest affinity for the resin moving as a band immediately in front of the band of the developing ion. With continued passage of the developing ion, a situation is reached in which separation is complete and a steady state results in which each component is displaced by a more strongly adsorbed ion directly above it. Eventually, the component ions will emerge from the column and an analysis of the effluent will give results similar to those shown in Fig. 9. This is the displacement diagram obtained with completely ionized components e.g. the alkali metal ions; the behaviour of incompletely ionized components will be considered shortly.

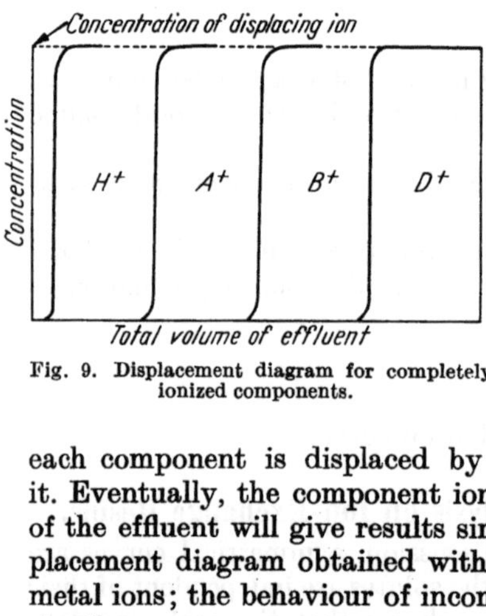

Fig. 9. Displacement diagram for completely ionized components.

In displacement development the bands are contiguous and in practice it is not possible to isolate, in a pure condition, the whole of the material in a band due to overlapping with adjacent bands.

2. Incompletely Ionized Components.

The ease with which an ion is displaced by another ion depends not only on the relative affinities of the two ions for the resin, but also on their relative concentrations in solution. Thus displacement will occur if the concentration in solution of an adsorbed ion, A^+ is very much greater than the concentration of the ion, B^+ originally present on the resin. The concentration of B^+ may be reduced to a low value compared with A^+, if B^+ complexes specifically with some substance in solution or, of more general importance, if B^+ is only partly ionized in the presence of A^+. This leads us to a brief discussion of displacement development with incompletely ionized components. For a fuller treatment, the reader is referred to the papers of PARTRIDGE and coworkers (see particularly PARTRIDGE and WESTALL (1949), PARTRIDGE and BRIMLEY (1952)] on the displacement chromatography of amino acids on ion-exchange resins.

Consider a column of sulphonated polystyrene ion-exchange resin in the H^+ form and suppose that a mixture containing a number of bases of different strengths is passed into the column until the column is half saturated. The ionization of the bases may be represented in order of increasing basicity as follows:

$$AOH \rightleftharpoons A^+ + OH^-$$
$$BOH \rightleftharpoons B^+ + OH^-$$
$$COH \rightleftharpoons C^+ + OH^-$$

H^+ will be displaced from the resin by the cations A^+, B^+ and C^+, the displaced H^+ combining with the OH^- from the bases and only water will appear in the effluent. The column may then be developed by passing through a solution of a strong base $D^+\,OH^-$, which is fully ionized over a wide range of pH. The ions, D^+ will be adsorbed as a band at the top of the column and they will displace the ions, C^+ of the weaker base COH from the exchange sites on the resin, since the ionization of C^+OH^- is depressed by the high concentration of OH^- associated with the strong base, D^+OH^-. The C^+ will travel down the column in the form of the uncharged base COH until it reaches sites on the resin occupied by ions, B^+ of the even weaker base, BOH. B^+ will then be displaced by C^+, since the ionization of B^+OH^- is depressed in the presence of C^+OH^-. Thus the components in the mixture are arranged in order of basic dissociation constant, the one with the largest constant moving as a band in front of the strong base used to develop the column. However, the total concentration of solute in each band may not be constant as is the case with completely ionized solutes and the effluent diagram may show a series of steps (Fig. 10).

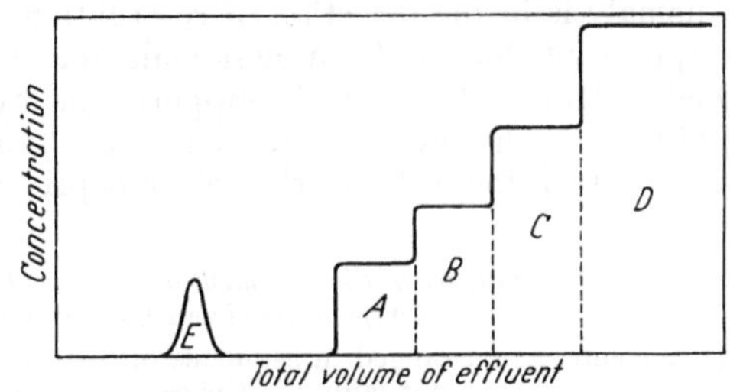

Fig. 10. Displacement diagram for incompletely ionized components.

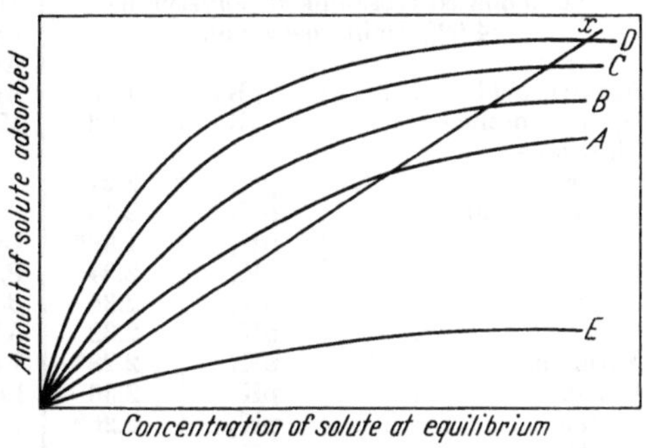

Fig. 11. Displacement isotherms.

The concentration at which a particular solute is eluted from the column depends on the nature of its adsorption isotherm and on the concentration of the developer. For good displacement chromatograms, the adsorption isotherms of the solutes should be convex curves as shown in Fig. 11. A, B, C and E are adsorption isotherms of components A, B, C and E and D is the isotherm of the developing solute. If X is the concentration of the developer, then the concentrations at which the various components appear in the effluent are obtained from the intersections of the respective isotherms with a line drawn from X to the origin O. If the line XO does not intersect the adsorption isotherm (as is the case with solute E), the solute concerned is not displaced from the column, but instead it appears as a separate elution peak in front of the displacement bands.

A strong base has a straight line parallel to the concentration axis as its adsorption isotherm, the amount of cation adsorbed being independent of the concentration of the base except at very low values. Adsorption of a very weak base will be lower at low concentrations of the base due to hydrolysis of the salt formed between the strongly acidic resin and the weak base. In this case the isotherm will be a highly convex curve. Thus weak bases of different basicity are particularly suited to chromatography on a strongly acidic cation-exchange resin by the displacement method. Similarly, weak acids may be separated by displacement chromatography on a strongly basic anion-exchange resin, the dissociation of the weaker acids being depressed by the stronger acids.

Zwitterions, such as the amino acids' may be separated by displacement on either type of resin as they may be regarded either as weak bases or weak acids. The order of displacement of the amino acids is roughly in the order of their relevant pK values (Partridge and Brimley, 1952) as may be seen from Table 2. It will be noticed that a few amino acids fall out of line, but in all cases the anomaly is in the direction of retardation, suggesting that the effect is due to a larger contribution from non-coulombic forces. Experiments carried out at a higher temperature, 60° C support this hypothesis (Partridge and Brimley, 1951) since the order of displacement at the higher temperature conforms more closely with the order of the relevant pK values.

Table 2. *The sequence of displacement of a range of solutes from two ion-exchange resins at room temperature* (from Partridge and Brimley, 1952).

(The solutes are arranged in sequence of displacement from the resins. Relevant pK values not falling in the sequence are marked with asterisks.)

Sulphonated cross-linked polystyrene (4.5% divinylbenzene).		
Aspartic acid	pK_1	1.88
Hydroxyproline	pK_1	1.92
Threonine	—	—
Serine	pK_1	2.21
Glutamic acid	pK_1	2.19
Proline	pK_1	1.99*
Glycine	pK_1	2.34
Alanine	pK_1	2.34
Valine	pK_1	2.32
Methionine	pK_1	2.28*
Leucine	pK_1	2.36
Cystine	pK_2	2.26*
Creatine	pK_1	approx. 3.0
Phenylalanine	pK_1	1.83*
β-Alanine	pK_1	3.60
Trimethylamine oxide	pK	approx. 4.5
Creatinine	pK	approx. 4.8
Histidine	pK_2	6.0
Methylhistidine	pK_2	6.48
Carnosine	pK_2	6.83
Anserine	pK_2	7.04
Hydroxylysine	—	—
Lysine	pK_2	8.95
Ammonia	pK	9.27
Arginine	pK_2	9.04
Methylamine	pK	10.64
Dowex 2		
Lysine	pK_3	10.53
Proline	pK_2	10.60*
β-Alanine	pK_2	10.19
Alanine	pK_2	9.69
Valine	pK_2	9.62
Leucine	pK_2	9.60
Glycine	pK_2	9.60
Carnosine	pK_3	9.51
Threonine	—	—
Serine	pK_2	9.15
Histidine	pK_3	9.17*
Methionine	pK_2	9.21*
Methylhistidine	pK_3	8.85
Cystine	pK_3	7.85
Tyrosine	pK_2	9.11*
Acetic acid	pK	4.75
Glutamic acid	pK_2	4.25
Aspartic acid	pK_2	3.65
Hydrochloric acid	—	—

Comparison of Elution and Displacement Methods.

The elution method has advantages for analytical work. It is the more powerful method of separation and provided the distribution coefficients are sufficiently different, each component can be completely separated from the others. Secondly, only a small amount of material is required. Elution development, however, has disadvantages as a preparative method since the components are obtained in relatively low concentrations and they are mixed with a large volume of eluant. Another disadvantage is that only a small fraction of the total exchange capacity of the column is used.

Displacement development is well suited to preparative work, as pure materials are obtained in high concentrations. High column loadings of as much as 50% or more of the total capacity are possible. The displacement method is not suited to quantitative work since the bands are contiguous and not all the material in a band is obtained in a pure condition. Displacement development is unsatisfactory when the solutes are adsorbed independently of one another.

III. Frontal Analysis.

The third method of column development is called frontal analysis. In this method, the solution containing the mixed components is passed continuously through the column of the resin. At first, only pure solvent flows from the column, then a solution containing only one component, then one containing 2 components and so on until the solution flowing from the column has the same composition as the solution flowing onto it. Thus only the first component emerges from the column in a pure state. Frontal analysis is useful for qualitative analysis, particularly for the preliminary examination of an unknown mixture. It may be used for the qualitative analysis of substances that show some degree of irreversible adsorption.

C. Apparatus.

The essential part of the apparatus for column chromatography is the filtration or chromatographic tube. The ion-exchange resin is usually supported on a plug of glass wool or a sintered glass disc sealed into the glass column. The simplest form of column (Fig. 12a) consists of a piece of glass tubing which is restricted at its lower end to take a plug of glass wool. Care should be taken to see that the upper surface of the plug is reasonably flat. A length of glass rod with the cut edge left sharp can be used to cut off loose strands of glass wool by rolling the glass rod. However, for many separations, it is more satisfactory to support the column of ion-exchange resin on a perfectly flat sintered glass disc, as shown in Fig. 12b. Although the ratio of the height of the resin column to its diameter is not critical, a ratio of about 10:1 or 20:1 is usually convenient. However, for analytical work involving the separation of a large number of closely substances on the one column, a ratio of up to 150:1 may be employed. There should be ample space in the tube above the resin column to allow application of the sample to the column and for general manipulations such as the application or change of the eluant.

The eluting solution is passed into the column from a tap funnel or other suitable container which may be connected directly to the column or by a length of plastic tubing. The second method is preferable for analytical work, since the rate of flow of solution through the column can be controlled simply by varying the height of the reservoir. In order to ensure that air is not sucked into the column if left unattended for long periods the lowest part of the plastic tubing should be below the level of the outlet from the column. An alternate method is to have from the column a capillary outlet which rises above the level of the top of the resin column (Fig. 12d).

For analytical work, it is not advisable to have a tap in the outlet line from the column, as this usually produces a small mixing compartment between the tap and the sintered disc which leads to the spreading of the bands after they leave the column. Hence for analytical work or for fine separations an apparatus similar to that shown in Fig. 12b is recommended. A ground glass joint may be used to connect the reservoir to the column. Ground glass connections are recommended in those cases where it is necessary to operate the column under pressure. The reservoir may be connected to a supply of compressed air through a reducing

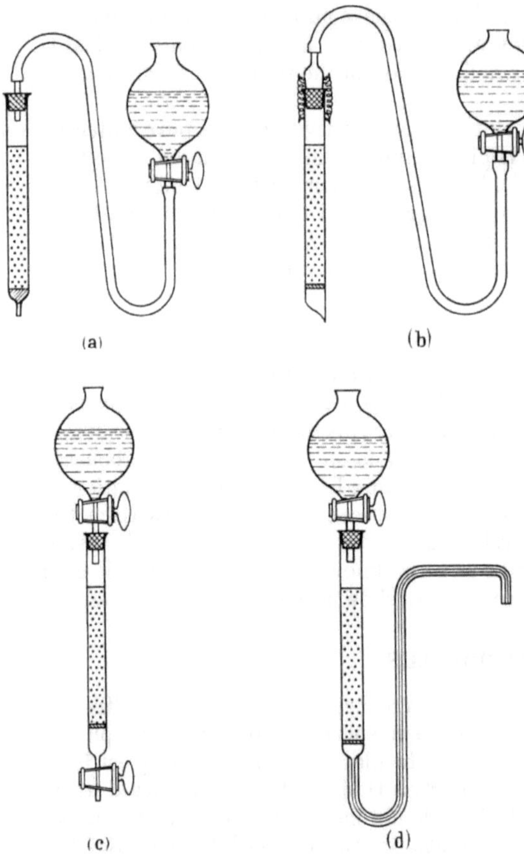

Fig. 12. Chromatographic tubes.

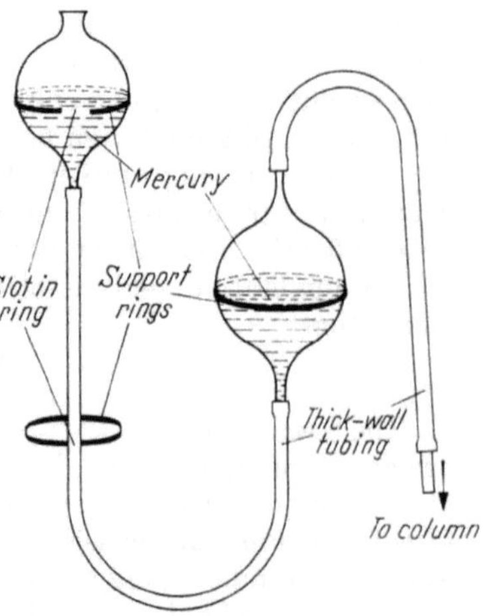

Fig. 13. Device for applying pressure to a small column.

valve. The author has found a simple form of mercury reservoir very convenient for packing small columns under constant pressure (Fig. 13). Several types of chromatographic tubes are available commercially and most are fitted with ground glass joints and sintered glass discs. For micro-analytical work, a glass pipette may be used as a column and the ion-exchange resin may be contained between two glass wool plugs, the upper plug preventing air from entering the column (DJURFELDT and SAMUELSON, 1950). For experiments on a large scale, columns constructed from industrial glass pipeline are suitable (Fig. 14b). It is sometimes desirable to operate a column at constant temperature and in these cases, the column is fitted with a glass jacket through which water is circulated from a constant temperature bath.

In displacement development, where the fronts of the bands are sharp it is desirable to use a series of columns of different size. Because of some non-uniformity of flow, the sharp front of a boundary passing down the first column becomes distorted. However, once separation of the components has taken place on the first column, the effluent may be passed to a column of smaller diameter, which again starts with sharp fronts. A third column of still smaller diameter gives a further improvement in the sharpness of the separation. Coupled columns are of no advantage in elution analysis where self-sharpening boundaries are not formed.

Coupled columns may be arranged side by side or one above the other, depending on convenience. A suitable arrangement for multiple columns is shown in Fig. 14a. Chromatographic tubes made by sealing glass taps onto glass tubing of suitable diameter are suitable and the resin may be supported on plugs of glasswool carefully packed into the bases of the tubes; stopcocks and inlets are made of capillary tubing to minimize mixing. For larger scale work, columns of industrial glass pipeline with tapered ends have been found suitable (PARTRIDGE and BRIMLEY, 1952). The type of construction adopted by PARTRIDGE and BRIMLEY is shown in Fig. 14b. Standard metal flanges (*K.K*)

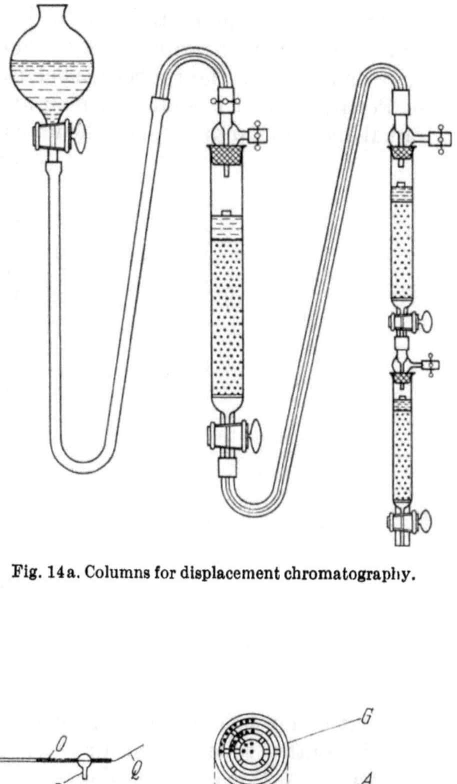

Fig. 14a. Columns for displacement chromatography.

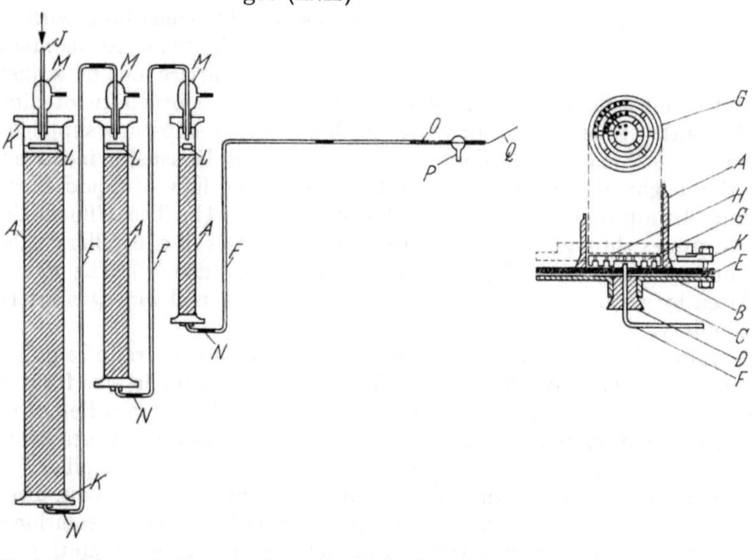

Fig. 14b. Columns for displacement chromatography (from PARTRIDGE and BRIMLEY, 1952).

provided by the manufacturer of industrial pipeline are employed for bolting on the end plates in the construction of chromatographic tubes. Fig. 14b shows the endplate *B* which is constructed in metal and provided with a collar *C* to hold the cork *D*. The end of the chromatographic tube is closed by a thick rubber gasket E, and the outlet tube F is forced through a small hole in the gasket, thus forming a tight seal. The outlet tube is constructed from capillary tubing and is supported by a cork D Closure of the top of the chromatographic tube is carried out in a similar way.

In order to support the resin and to secure uniformity of flow, the column is provided with a filtration plate G which fits loosely inside the chromatographic tube and rests on the rubber gasket E. The filtration plate is turned from a disc of Perspex and it is shown in plan and section in Fig. 14b. It is provided with radial and annular grooves on the underside and is perforated with a large number of evenly spaced holes which pass through the thin sections of the plate left by the grooves. The filtration disc is covered by a piece of woven glass fabric to prevent small particles of resin from clogging the small holes in the disc; alternatively, filter paper may be used.

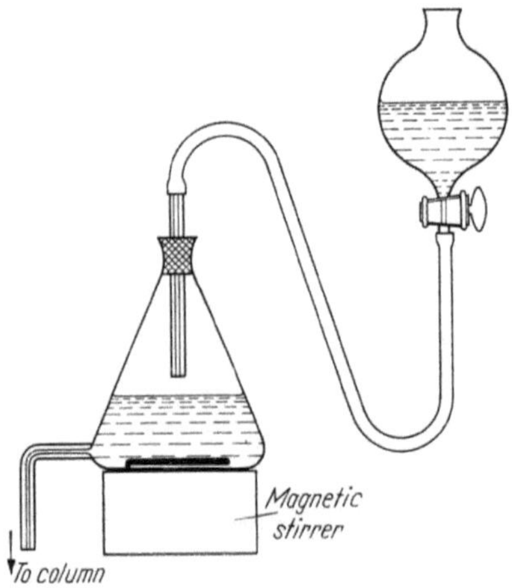

Fig. 15. A simple apparatus for gradient elution work.

The arrangement of the apparatus is shown in Fig. 14b. The chromatographic tubes are arranged side by side and connected by capillary tubing ($F.F.$). The solution passes into the apparatus at J from a reservoir. Each column is provided with a float ($L.L$) made from a disc of polyethylene which prevents the disturbance of the top of the resin by the dropping liquid. The height of liquid above the resin is kept as small as possible consistant with the proper action of the float. Adjustment of the liquid level is carried out by adjusting the amount of air held at the top of each filtration tube by means of clips placed on the outlet tube to the air traps ($M.M$). Further clips are supplied at the rubber connections ($N.N$). For large scale work, it is usually inconvenient to vary the height of the reservoir in order to vary the flow of liquid through the column. An alternative method is shown in Fig. 14b. The effluent from the column passes through a flow regulator which consists of a capillary tube blown to a bulb at one end and provided with an outlet tube (P). The flow may be regulated by varying the depth to which the stainless steel wire Q is inserted into the capillary.

Column units consisting of an outer shell of stainless steel around a polyethylene column are available commercially. The units, which vary in size from 0.5 ml—16 ml may be screwed together to form a series of columns of different size.

Gradient elution analysis. The simplest type of apparatus for gradient elution work is shown in Fig. 15. It consists of a conical flask with a capillary side arm attachment which leads to the column, and a tap funnel as reservoir. The conical flask, termed the mixing chamber, is partly filled with the eluting solution initially used to develop the column; the tap funnel contains a second eluting solution, usually a solution of higher concentration or one of different pH from that placed in the mixing chamber. A magnetic stirring device agitates the solution in the mixing chamber and as liquid flows from this chamber to the column, it is replaced dropwise by an equal volume of solution from the tap funnel.

With this simple type of apparatus containing initially two different concentrations of a particular solution in the two chambers, it can be shown (CHERKIN,

MARTINEZ and DUNN, 1953; BOARDMAN, 1953) that the concentration of eluting solute entering the column increases linearly with the total volume of eluant at first and later it approaches asymptotically the concentration of the solute in the tap funnel.

$$C = C_1 - (C_1 - C_0)\, e^{-\frac{V}{Y}}$$

where C is the concentration of eluting solute leaving the mixing chamber when a total volume V has passed through the column, C_1 is the concentration of the eluting solute in the tap funnel and C_0 its initial concentration in the mixing chamber. Y is the volume of liquid in the mixing chamber. Provided the total volume of eluant is small compared with the volume of solution in the mixing chamber, the concentration of the eluting solute will increase in a linear fashion with the total volume of eluant.

$$C = C_0 + \frac{(C_1 - C_0)\, V}{Y}\, .$$

In many instances it is found that the resolution of a particular separation may be increased by using a gradient which does not vary linearly with volume of eluant. Many mixing devices have been described in the literature for the production of non-linear gradients (BOCK and LIND, 1954; LAKSHMANAN and LIEBERMAN, 1953; DRAKE, 1955; BOMAN, 1955) but the most versatile is one devised by PETERSON and SOBER (1959) and it is now available commercially. It consists of a series of 9 identical mixing chambers which contain eluants of varying composition in hydrostatic equilibrium. The shape of the resulting gradient is determined by the initial concentration of eluant in each chamber and its position in the series. The reader should consult the original paper for full details of the construction of the apparatus.

D. Experimental Procedures.

I. Purification of Ion-Exchange Resins.

Specially purified resins for analytical work are available commercially (Table 1) and for most purposes these may be used without further treatment. Other batches of resins obtained from the manufacturers should be purified before use.

Grinding and Grading. It is preferable to purchase material in the appropriate mesh size. However, if the industrial resin (10—50 mesh) is the only one immediately available, this may be ground and graded into the desired particle size and then washed to remove impurities of low molecular weight and heavy metal ions. Small quantities of resin may be ground with a pestle and mortar either in the dry or wet state; for larger quantities a small laboratory swing out hammer mill has been found suitable, although a rather large proportion of fine material is produced.

The ground material may be graded by sieving in the dry state. Particles which jam in the screen openings may be dislodged by rubbing a cork or rubber bung over the screen. The resin may be graded further, if desired, by sedimentation. The resin is stirred with water and then allowed to settle; after a certain time which may vary from a few minutes to half an hour depending on the particle size required, the supernatant which contains the finest particles is removed. The settling process is repeated several times until the supernatant is clear. It is recommended that the sedimentation method be used to remove the fine particles from all batches of commercial resins except the analytical grade resins.

Purification. Cation-exchange resins have a high affinity for iron, but this can be removed by allowing the resin to stand in 6.N HCl. If iron is present, the supernatant will appear yellow and treatment with 6.N HCl should be repeated until the supernatant is colourless. An alternative method for removing small amounts of heavy metal is to wash the resin with 2 M NaCl containing ethylene diamine tetra-acetic acid (EDTA).

After washing with distilled water, the resin is treated alternately with 2 N NaOH and 2 N HCl, the resin being washed with distilled water between each treatment. 2 or 3 cycles with acid and alkali should be sufficient to remove the impurities of low molecular weight. If it is desired to have the resin in the H^+ form it is treated finally with 2 N HCl, if in the Na form with 2 N NaOH. Finally the resin is washed thoroughly with distilled water or with the buffer initially used to develop the chromatogram.

Anion exchange materials contain fewer ionic impurities than cation exchangers. They should be cycled several times between N HCl and N NaOH with water washes between each treatment. If the resin is required in the free base form, it is finally treated with NaOH and then washed with water, if it is required in the chloride form it is treated finally with HCl, if in the formate form with formic acid and so on. It should be mentioned that anion-exchange resins have a greater tendency to break down than cation-exchangers and it is difficult to remove the last traces of low molecular weight organic matter. If a strongly basic anion-exchange resin is to be used in displacement chromatography the resin should be treated with CO_2-free NaOH during the purification steps, since the presence of bicarbonate ions on the resin may cause gas formation during the displacement process.

II. Preparation of the Column.

Columns of ion-exchange resins are prepared by suspending the resin in water or the buffer used to develop the chromatogram, allowing it to swell freely and pouring the resultant slurry into the chromatographic tube. The sintered disc is covered with a small circle of filter paper as fine particles of resin may enter the sinter and restrict the flow of solution through the column. For analytical work using the elution method, the following procedure has been found satisfactory for producing regular and well-packed columns. The resin is suspended in about 2 volumes of water or buffer and the resulting slurry poured into the chromatographic tube in small portions at a time with sufficient in each portion to form a column of resin about 5—10 cm high. Each portion is poured in through a funnel, the tip of which is bent to direct the stream against the side of the tube. The liquid above the resin is allowed to fall to within 5—10 cm of the resin surface before the next portion is added. Ion-exchange resisns usually pack sufficiently well without application of outside pressure, but for long columns of fine grade resins it may be necessary to apply a pressure of up to 15 cm mercury. If the surface of the resin is irregular after settling of the last portion, it is stirred with a glass rod or preferably by rotating near the surface a perforated metal disc attached to a metal rod. The surface of the resin should not be dry for more than a few minutes, as air may enter the column.

The column is then equilibrated with the buffer to be used in applying the mixture to the column. Weakly acidic or weakly basic ion-exchange resins are themselves good buffers and large volumes of buffer are required to ensure equilibration of these resins.

Packing of cellulose ion-exchange columns. It is rather difficult to prepare well-packed columns of the cellulose ion-exchangers. Care should be taken to ensure that air bubbles are not trapped in the column during its preparation. In general, the material obtained commercially may be used without further treatment. A slurry of the ion-exchanger is prepared by evacuating the dry powder and mixing it under reduced pressure with distilled water or the buffer used as the eluting solution. If necessary, very fine particles should be removed by sedimentation. After pouring the slurry into the column, a pressure of up to 10 lb/sq inch is applied and the procedure as outlined for the ion-exchange resins is followed.

III. Operation of the Column.

Since the procedures for operating a column are somewhat different for the elution and displacement methods, it will be more convenient to cover the two methods separately.

1. Elution Analysis.

a) Column Loading.

The first question to be decided is the amount of material to be applied to the column and this in turn depends on the size of the column. For analytical work it is usual to use a column of about 1 cm in diameter and of a length sufficient to separate the components in the mixture. For many separations a column of 15—30 cm length is usually sufficient, although in some analyses columns of up to 150 cm have been used. The resolution obtained is dependent also on the total amount of material applied to the column. Conversely, the amount of material which may be applied to a particular column is related to the difficulty of the separation which in turn is related to the relative affinities of the various components in the mixture for the ion-exchange resin.

As a rough guide, the amount of material which may be applied to a column of 1 cm diameter (corresponding to a cross sectional area of 0.78 cm^2) is of the order of 0.05—0.15 milliequivalents, although in some instances the loading may be much higher. The maximum permissible loading for a particular separation or separations may be found by running a number of chromatograms with different loadings and examining the resolution obtained. For columns of larger diameter, the loading will be proportional to the cross-sectional area, although it should be remembered that with columns above 3 cm in diameter, the resolution is usually lower because of a greater non-uniformity of flow.

The substances to be separated should be adsorbed initially as a narrow band at the top of the column. If the material is in a fairly large volume, it should be applied to the column in a solvent of very low eluting power (usually a buffer of low ionic strength) so that the solutes are adsorbed strongly to the resin, the resin column having been equilibrated previously with the same buffer. However, it is preferable to apply the substances to be separated in a small volume of buffer; 0.3—1.0 ml for columns of 1 cm diameter and proportionately larger samples for columns of larger cross-sectional area. The band of adsorbed material is more likely to be uniform when the sample is applied from a small volume.

The sample is added to the column by a pipette with a bent tip to direct the stream against the side of the tube. This ensures that the surface is not disturbed during addition of the sample. The time taken for the sample to drain into the column should correspond to a flow rate which is equal or less than the flow rate to be used in the development of the chromatogram. Rapid application of the sample to the column can lead to a spreading of the initial band and poorer resolution on development. A simple method for reducing the flow rate, if necessary, during the application of the sample is to immerse the lower end of the chromatographic tube in a beaker or cylinder of water.

After the sample has drained into the column, the wall of the tube above the resin is washed with a small amount of solvent and this is allowed to drain into the resin. Again the surface of the resin should not be allowed to remain dry. The chromatographic tube above the level of the resin is then completely filled with the first eluting solvent and the tube connected to the tap reservoir. During this operation, it is important also to ensure that the flow rate does not increase markedly; lowering the tube further into the cylinder of water is a convenient method of achieving this.

The flow rate of solvent through the column is controlled by adjustment of the height of the tap reservoir. The flow rate used will depend on a number of factors such as particle size of resin, degree of cross-linking, molecular weight and nature of the solutes. The optimum flow rate is the fastest rate which gives the desired resolution. For low molecular weight solutes, flow rates for analytical work will usually be in the range 10—100 ml/h for a column of 1 cm diameter, 40—400 ml/h for a column of 2 cm diameter and so on.

When changing from one eluting solvent to another, the same precautions should be observed as in the application of the first solvent. The lower end of the chromatographic tube may again be immersed in a cylinder of water. Most of the first solvent remaining in the tube is sucked off with the aid of a pipette with a bent tip and the small amount remaining is allowed to drain into the resin. The procedure is then identical to that used for the application of the first solvent.

b) Choice of Solvents.

For completely ionized solutes, it is usual to start with a salt or buffer solution of low ionic strength and then to increase the concentration of the salt or buffer successively at each change of eluant. For separations on cation-exchange resins, the cation concentration of the eluting buffer is the important parameter and for separations on anion-exchange resins, the anion concentration. Divalent ions are more effective than monovalent ions. Thus, it is usual to start with a monovalent eluting ion and if too high concentrations are necessary to elute all the components, a change is made to a divalent ion. For incompletely ionized solutes or for separations on the weakly acidic or weakly basic ion-exchangers, changes in pH in conjunction with changes in ion concentrations are usually used to elute the components. Alternations in pH which depress the ionization either of the solute or the groups on the resin will usually assist the elution. However, with the weakly acidic or weakly basic ion-exchange resins (but not the ion-exchange celluloses), the range of buffers which may be applied to the column is limited somewhat by the swelling of the resin and the resultant increase in the resistance to flow.

The choice of suitable eluants is usually a matter of trial and error. For an initial survey, time will be saved if a relatively short column of ion-exchanger is used (5×1 or 10×1 cm). An alternative method of selecting eluants is to carry out a series of distribution experiments in test tubes using a fixed volume of resin and a series of eluants, containing a fixed amount of solute. The rate of movement of a solute on a column may be calculated from the distribution coefficient. This method is particularly valuable for solutes which may show some irreversible adsorption on the ion-exchanger e.g. the proteins. Distribution coefficients between 2 and 5 are the most favourable for elution chromatography. The author has found the following procedure satisfactory; a very similar procedure is described by Hirs, Moore and Stein (1953).

A graduated centrifuge tube is filled with sufficient of a suspension of ion-exchanger to give a volume of sediment after centrifugation of 2 ml. The ion-exchanger is then equilibrated with the trial buffer by the procedure of shaking it with the buffer, centrifuging and carefully sucking off the supernatant. This procedure should be repeated until the pH of the supernatant is the same as the pH of the buffer. A 2 ml aliquot of a solution of the solute (containing 2—4 mg/ml of solute) in the same buffer is then added, the centrifuge tube is shaken and allowed to stand for about 5 minutes with occasional shaking. The ion-exchanger is sedimented by centrifugation and a 1 ml aliquot of the supernatant is withdrawn for analysis. Knowing the volume of liquid between the particles of ion-exchanger, a distribution coefficient of the solute may be determined.

However, for successful column experiments where many equilibrations of the solute between the ion-exchanger and the ambient phase take place, it is necessary for the solute to be reversibly adsorbed and desorbed at each equilibration. Hence, in test-tube experiments it is important that a reequilibration experiment is carried out. After removing the aliquot for analysis at the first equilibration, 1 ml of the buffer is added back, the tube is shaken and allowed to stand as described for the first equilibration. After centrifugation, a further 1 ml aliquot is withdrawn for analysis of the solute.

The distribution coefficient (C) may then be calculated in the following way.

Let the volume of liquid between the particles of ion-exchanger per ml of sediment $= V_e$ [For a fine grade form of the ion-exchange resin, Amberlite IRC-50, V_e is approximately 0.56 ml (Boardman, 1953) for DEAE-Cellulose it is about 0.7 ml (Thompson and O'Donnell, 1960). These values were determined by the centrifuge method of Pepper, Reichenberg and Hale, 1952].

Then the total volume of ambient phase in the test-tube $= 2\,V_e + 2.0$.

Let $a_0 =$ concentration of solute in the solution added to the resin.

$a_1 =$ concentration of solute in ambient phase after first equilibration.

$a_2 =$ concentration of solute in ambient phase after second equilibration.

At the first equilibration:

Amount of solute added to tube $= 2.0\,a_0$.

Amount of solute in ambient phase after equilibration $= (2\,V_e + 2.0)\,a_1$.

Amount of solute adsorbed by ion-exchanger $= 2.0\,a_0 - (2\,V_e + 2.0)\,a_1$.

Amount of solute in ambient phase between particles of ion-exchanger $= 2\,V_e a_1$.

$$\therefore \text{ the distribution coefficient } (C_1) = \frac{2.0 a_0 - (2\,V_e + 2.0)\,a_1}{2\,V_e a_1} = \frac{2.0\,(a_0 - a_1)}{2\,V_e a_1} - 1\,.$$

At the second equilibration:

Amount of solute in ambient phase after equilibration $= (2\,V_e + 2.0)\,a_2$.

Amount of solute withdrawn for analysis at first equilibration $= 1.0\,a_1$.

Amount of solute adsorbed by ion-exchanger $= 2.0\,a_0 - a_1 - (2\,V_e + 2.0)\,a_2$.

Amount of solute between particles of ion exchanger $= 2\,V_e a_2$.

$$\therefore \text{ distribution coefficient } (C_2) = \frac{2.0 a_0 - a_1 - (2\,V_e + 2.0)\,a_2}{2\,V_e a_2} = \frac{2.0\,(a_0 - a_2) - a_1}{2\,V_e a_2} - 1\,.$$

If the adsorption of the solute on the ion-exchanger is reversible, then $C_1 = C_2$.

The rate of movement of the solute on a column is given by

$$R_F = \frac{1}{C+1}.$$

In gradient elution analysis the degree of separation obtained and the sharpness of the bands is a function both of the nature of the gradient and the steepness of the gradient. Selection of a suitable gradient is usually based on trial experiments. Experiments with stepwise changes in buffer will indicate the range of salt concentration or pH over which the gradient should extend. When a mixture is being chromatographed for the first time, it is convenient to use a linear gradient with a limit concentration well beyond the anticipated maximum elution requirements (Peterson and Sober, 1959). If the chromatogram resulting from the use of a linear gradient is crowded in some parts and spread out in others the gradient can be decreased or increased at the appropriate places in the next experiment. In cases where the early parts are crowded, whereas the later peaks need sharpening, it is convenient to retain the linear gradient initially, but with a lower slope, and increase to the original limit in the later parts (Peterson and Sober, 1959).

Volatile buffers. For preparative work, using the elution method it is advantageous to use volatile eluants which can subsequently be removed by evaporation or sublimation. Some examples of volatile eluants are listed in the Table 3.

Table 3. *Some volatile eluants.*

	pK_A	Range of pH
Hydrochloric acid		< 1—3
Ammonium formate-formic acid	3.8	2.7—4.8
Ammonium acetate-acetic acid	4.7	3.7—5.8
Pyridinium acetate-acetic acid	4.7	3.7—5.8
Pyridine-pyridinium acetate	5.2	4.2—6.2
Collidinium acetate-acetic acid	4.7	3.7—5.8
2:4:6 Collidine-collidinium acetate	7.3	6.3—8.3
Ammonium acetate-ammonium hydroxide	9.3	8.3—10.3
Triethylamine carbonate-triethylamine	10.7	9.7—12.7

The composition of buffers can be calculated from the formula:

$$\mathrm{pH} = \mathrm{pK} + \log \frac{\text{conc. of salt}}{\text{conc. of acid}}$$

c) Selection of Temperature.

The selection of the most suitable temperature or temperatures is usually a matter of trial and error. The upper limit is governed by the stability of the resin (Table 1, p. 198). For most separations, temperatures between room temperature and 60° C have proved satisfactory. Protein chromatograms are usually conducted at temperatures near 0° C. A discussion on the effect of temperature in ion-exchange chromatography is included in the section dealing with mechanisms of adsorption towards the end of this article.

d) Analysis of Effluent.

The effluent from a column is usually analysed by collecting it in discrete fractions and estimating the amount of material in each fraction by some standard procedure. A suitable size of fraction would be 1—2 ml for a column of 1 cm diameter, 4—8 ml for a column of 2 cm diameter and so on. Frequently it is not necessary to analyse each fraction from the column; a preliminary survey of every second or third fraction to locate the positions of the solutes may save time.

In the majority of experiments it is usual to use an automatic device called a fraction collector or fraction cutter to divide the effluent into equal fraction. Most fraction collectors consist of a circular rack containing holes at regular intervals near the circumference to take test tubes. The rack is rotated about a vertical axis through its centre and the effluent is passed into each tube in turn as it moves beneath the column outlet. The movement of the rack is usually in discrete steps regulated by an escapement mechanism. Alternatively the rack may be held stationary and the effluent is led to the tube by means of a rotating distribution arm.

It is not proposed to describe in detail any particular fraction collector, but only to state the different principles on which fraction collectors usually operate and to evaluate the advantages and disadvantages of each type. Fraction collectors may work on one of the four principles.

(1) Equal volumes of liquid,
(2) equal weight of liquid,
(3) equal time intervals,
(4) by counting drops.

(1) **"Equal volume" fraction collectors** may be divided into two types, those in which the effluent is collected in a self-emptying syphon and those in which a light source and photocell are used to detect a pre-set volume. In the latter type, the effluent from the column falls either into a calibrated glass vessel fitted with glass valves and placed between the column outlet and the tubes, or directly into the tubes. When the liquid in the glass vessel or tube reaches a certain pre-set height, determined by setting the position of the light source and photocell, the photocell is triggered and this either actuates the glass valves in the intermediate vessel machine or it causes the fraction collector to pass the next tube under the column in the type where the fractions are collected directly into the tubes. Dropping the effluent directly into the tubes has the advantage that a fraction is not contaminated by a small amount of the preceding fraction as occurs in the intermediate vessel machine.

Syphons are well suited for the collections of reasonably large fractions, 5—100 ml, but they do not appear satisfactory for quantitative work requiring the collection of accurate fractions of smaller size. The performance of a siphon is dependent on the surface tension of the liquid. Thus a small siphon that works well on water may not work at all with an organic liquid. It is a matter of the diameter of the side arm; 3—4 mm for water, 2 mm for alcohol.

(2) One type of **"equal weight collector"** consists essentially of a reasonably sensitive balance and each tube in turn is brought onto the scale of the balance and filled to the weight required. The main disadvantage of this type of collector is that it requires a set of tubes of identical weight. It depends on the performance of a sensitive balance and it only collects fractions of equal volume if the density of the effluent remains constant. Another type of collector in this group employs a magnetic balance (Simmonds and Wood, 1954) and in the apparatus described the liquid is collected in a small polythene cup until its weight counterbalances the attraction of the magnet. The contents of the cup are then emptied into a tube. The main disadvantage of this type of collector is incomplete transfer of liquid from the polythene cup to the tube.

(3) A large number of **time-based fraction collectors** have been described. They have the advantage of greater flexibility in the range of fraction sizes which can be collected. However, the fractions tend to vary somewhat in size due to changes

in flow rate and therefore time-based fraction collectors are not suitable usually for quantitative work.

(4) In the usual type of **drop-counting fraction collector** each drop as it leaves the column interupts a beam of light falling on a photocell which operates a counter which in turn actuates the movement of the turntable. The uniformity of the fraction depends on a constant surface tension of the effluent.

The choice of a suitable fraction collector depends partly on the separations to be carried out. For large scale work, a syphon type or time-based collector may be used. For quantitative work in which the surface tension of the effluent remains constant, a photoelectric drop-counting collector where the effluent drops directly into the tubes is the most accurate. If there are likely to be changes in surface tension then a volume measuring collector employing a photocell will prove the most satisfactory.

Continuous analysis of effluent. Quantitative analysis of the effluent may be carried out continuously and automatically, the complexity of the apparatus depending on the exact analytical procedure. Apparatus for continuous analysis is now available commercially. The effluent from the chromatographic column is mixed with the appropriate reagents in the correct proportions with the aid of accurate pumps. Colour development is carried out automatically and the solution then passes continuously through a flow-through cell of a colorimeter. The results are plotted on the moving chart of a recorder. The effluent from a column may also be monitored for radioactive substances, the levels of radioactivity being plotted on a moving chart.

e) Regeneration of Ion-Exchangers.

Cation-exchange resins are usually regenerated for further use by passing a solution of 2 N NaOH or other strong alkali through the column and then washing well with water; anion-exchange resins are usually regenerated with NHCl followed by water. Weakly acidic or weakly basic ion-exchange resins usually cannot be regenerated in the column, as they swell greatly in passing from the uncharged state to the ionic form. Resins of low degrees of cross-linking also cannot be regenerated in the column because of swelling changes. Microbiological growth on columns may be controlled by autoclaving the resin at intervals (Dickman, Morrill and Trupin, 1960).

After use, the ion-exchange celluloses may be treated with 1 N NaOH, but it is recommended that a mixture of 0.5 M NaCl—0.5 N NaOH be used to regenerate CM cellulose in order to control the swelling effect of strong alkali on this adsorbent (Peterson and Sober, 1956).

2. Displacement Development.

The columns are prepared simply by pouring a slurry of the resin into the tube and allowing it to settle. As the particles of resin fall to the bottom of the tube, water is allowed to flow from the outlet tube in order to avoid the disturbance caused by the upward displacement of water by the falling resin. Some space should be left above the resin for possible volume changes in the resin during development of the chromatogram. Cation-exchange resins will be in the H^+ form and anion-exchange resins in the free base forms and they should be washed free of acid and alkali respectively. With the strongly basic anion exchange resins it is necessary to use distilled water which has been boiled to remove CO_2.

a) Selection of Column Size.

As mentioned previously, the resolving power in displacement chromatography is greater if coupled columns are used. A suitable range of columns is shown in

Table 4. *Data for a range of laboratory columns* (from PARTRIDGE and BRIMLEY, 1952).

Filtration tube		Sulphonated polystyrene (capacity, 5.25 mg equiv./dry g.)					Dowex 2 (capacity, 3.7 mg equiv./dry g.)				
(1) No.	(2) Diameter (cm)	(3) Depth of bed (cm)	(4) Dry wt. of resin (g)	(5) Particle size (mesh/in.)	(6) Volume 0.075 N-NaOH (ml)	(7) Flow rate 0.075 N-NaOH (ml/min)	(8) Depth of bed (cm)	(9) Dry wt. of resin (g)	(10) Particle size (mesh/in.)	(11) Volume 0.1 N-HCl required (ml)	(12) Flow rate 0.1 N-HCl (ml/min)
I	7.6	46	830	60—100	58000	—	—	—	—	—	—
II	5.1	31	255	60—100	17500	50	—	—	—	—	—
III	3.8	23	106	60—100	7300	25	23	100	100/200	3700	7.5
IV	2.5	15	30	60—100	2100	12	15	28	100/200	1000	3.0
V	1.7	10	9	60—100	620	6	10	8.3	250/500	310	1.5
VI	1.2	7.2	3.2	80—100	220	2	7.2	3.0	250/500	111	0.75
VII	0.8	4.8	1.0	100—120	70	0.6	4.8	0.9	250/500	33	0.33

Table 4, which is taken from the paper of PARTRIDGE and BRIMLEY (1952) on the displacement chromatography of amino acids. The scheme is flexible since any set of 3 consecutive tubes may be chosen from the range. Column 5 in the table shows the particle size of resin recommended for each chromatographic column, 60—80 mesh/inch resin is suitable for columns of 2″ (5.1 cm) and 3″ (7.6 cm) diameter, 80—100 mesh for from $^1/_2$″ to $1^1/_2$″ and a smaller mesh size for below $^1/_2$″.

The quantity of material applied to the largest column of the three should be such that it is sufficient to saturate $^1/_2$ to $^2/_3$ of the column. The size of the largest column required may be estimated from a preliminary experiment in which a small aliquot of the solution to be chromatographed is passed through a small column of resin and the acid or alkali which appears in the effluent determined by titration. The rate of application of the sample may be higher by a factor of 2 or 3 than the rate of application recommended for the developer solution (columns 7 and 12 in Table 4). After application of the sample, the column is washed with a little water before application of the developer. The rate of flow of the displacement developer must be that listed for the smallest column of the set of three.

b) Concentration of Developer.

The higher the concentration of developer, the higher the concentration of displaced substance in the effluent and if the adsorption isotherms of the solutes are highly convex, the concentration of displaced solute will approach that of the developer. Therefore, each solute in the mixture should be completely soluble at a concentration equal to that of the developer, or material may come out of solution during the displacement process. Under these circumstances, the working of the column may be impaired or the flow may be stopped completely. Similar considerations apply to the bicarbonate ion on an anion-exchange column. If the concentration of displacer is too high, the solubility of CO_2 will be exceeded and the column may be broken by gas bubbles. A slight amount of gas formation can be tolerated, but more than this will diminish the resolving power of the column by upsetting the uniform flow of liquid or it may stop the flow altogether.

c) Size of Fraction.

Knowing the total quantity of solute to be separated and the concentration of displacing liquid, the total volume of the effluent can be calculated, assuming that the solutes are displaced at concentrations equal to the concentration of the displacer. This total volume may be divided conveniently into a number of fractions depending on the number of solutes to be separated. Overlap of two solutes should be confined to 2 or 3 fractions.

d) Regeneration of Resins.

Cation-exchange resins are regenerated for further displacement experiments with 2 N HCl. The amount of regenerating liquid depends on the resin used and the rate of flow. Columns of sulphonated polystyrene seem to require about ten times the amount of acid required to titrate the sodium on the column when it is saturated. Anion-exchange resins are regenerated with N NaOH; the amount of soda required for a strongly basic resin is about twenty timis the amount necessary to titrate the hydrochloric acid from the resin when saturated. After regeneration the resin is washed with distilled water to remove regenerating liquid. With basic resins the distilled water should be boiled and cooled before use.

E. Some Applications of Ion-Exchange Chromatography.

I. De-Ionization and Preliminary Group Separation of Extracts from Plant Tissues.

Paper chromatography is the most versatile method for the qualitative examination of the solutes obtained by extraction of plant tissues (Linskens, 1959), but paper chromatography can rarely be used on a crude extract as one class of compound may interfere with the separation of another class. Thus, even small concentrations of inorganic salts may interfere with the chromatography of such substances as the amino acids or organic acids. Therefore it is usual to carry out a preliminary separation of the extract into groups and then each group is examined by paper chromatography. Displacement chromatography from ion-exchange resins offers a simple and rapid technique for fractionating a plant extract into groups of acidic, basic and neutral substances (Partridge, 1952).

Two methods are in general use for the extraction of the non-lipid solutes from plant materials; these are extraction into water (cold or hot) or extraction with alcohol (cold or hot) to a final concentration of 80%. Cold water extracts require de-proteinisation and the simplest laboratory procedure is dialysis through cellophane. Another procedure involves the use of a highly cross-linked ion-exchange resin (see separate article on molecular sieving, p. 205). Extracts are centrifuged and the supernatant liquid decanted off. Aqueous extracts may be added directly to a column of an ion-exchange resin; alcoholic extracts are reduced to a small volume on a rotary film evaporator, diluted with water and freed from colloidal matter by filtration with the aid of a diatomaceous earth such as celite.

Separation into Four Groups.

a) Aromatic Substances.

Polyphenolic and aromatic substances may be removed from the extract by adsorption on acid-washed charcoal. [The carbon is pretreated by shaking with $5\% \left(\frac{V}{V}\right)$ acetic acid for 1 hour and washing thoroughly with water.] In this process, the amino acids tyrosine, phenylalanine and tryptophan are removed, but other amino acids are not adsorbed (Partridge, 1949). Phenolics are desorbed from the carbon with hot 95% ethanol or with aqueous phenol-acetic acid mixtures. If polyphenolic substances are not required as a separate fraction, the adsorption on carbon may be omitted, provided that the concentration of polyphenols is not too high. Polyphenolic substances behave irregularly on ion-exchange columns owing to molecular adsorption and if they are present in large amounts it is preferable to remove them.

b) Cationic or Basic Groups.

The filtrate is passed through a column of a cation-exchange resin in its hydrogen form, usually a column of cross-linked sulphonated polystyrene. The resin removes the cations from the solution (both inorganic and organic and including the zwitterionic amino acids) and the effluent from the column contains the anions as free acids, and the neutral substances such as sugars.

In order to determine the size of column required for the removal of all the cations in the extract, it is usual to take a small aliquot of the extract and pass it onto a small column of the cation-exchange resin in the H^+ form. Sufficient water is then passed through the column to remove the anions and the total amount of acid in the effluent is determined by titration against sodium hydroxide. From the titre, one can calculate the total amount of cations in the extract and hence the amount of ion-exchange resin required. A column which can handle the whole extract may now be constructed, allowing at least a 3 fold excess of resin over the calculated value. The concentration of cations in the solution passed onto the column should not exceed 0.2 N and should it do so, the solution should be diluted.

The cationic substances may be recovered from the column by displacement. If bases stronger than ammonia are not required, it is convenient to displace with 0.1 N ammonium hydroxide and the ammonia can be removed easily afterwards. Ammonia will only partly displace lysine, and strong bases such as arginine are not displaced. Sufficient ammonia should be passed onto the column to completely saturate it (see Table 4) and in fact it is usual to pass through an excess of ammonia. If the strongly basic components in the extract are required, the column is displaced with 0.1 N sodium hydroxide instead of ammonia. In this case a sharp fractionation is required and if necessary a column of narrow diameter may be coupled to the outlet of the main column and the effluent is collected in small fractions. If ammonia is used to displace the cations, then the commercial grade, large bead form of the resin is quite satisfactory. However, if sodium hydroxide is used, a resin of fine particle size is used. Usually it is found that only one fraction containing organic material is contaminated with sodium hydroxide. A small amount of sodium hydroxide can usually be tolerated in the subsequent paper chromatography and the last fraction containing organic material may be mixed with the other fractions or it may be retained for separate treatment.

c) Acidic Substances.

The effluent from the cation-exchange column prior to displacement is passed through a column of an anion-exchange resin and the column is washed with water. Since the solution contains no cations, a weakly basic or strongly basic resin may be used, but since strongly basic resins may cause epimerisation of sugars it is recommended that a weakly basic resin be used. The amount of anion exchange resin should be roughly equivalent to the amount of cation-exchange resin in the first column. The anionic substances can be recovered from the column in a salt free condition by displacement with 0.1 N hydrochloric acid and again it is usual to use a small excess over that required to saturate the column.

d) Neutral Substances.

The neutral substances are contained in the effluent from the anion-exchange resin prior to displacement. If only the neutral fraction is required, the extract may be passed through a mixed bed resin consisting of a mixture of a sulphonated polystyrene cation-exchange resin and a weakly basic anion-exchange resin.

II. Separation of a Group of Closely Related Solutes by the Elution Method.

The high resolving power which can be attained with the elution method is very well illustrated by the separation of mixtures of amino acids into their individual components (MOORE and STEIN 1951, see Vol. I. of this handbook p. 122). In their more recent method (MOORE, SPACKMAN and STEIN, 1958) improved resolution and speed of operation has been obtained by the use of a finer resin of a standard particle size distribution. The acidic and neutral amino acids are separated on a column of 8% cross-linked sulphonated polystyrene, 150×0.9 cm, using a single operating temperature of 50° C and 2 buffers as eluants, and the basic amino acids are separated on a column of the same resin, 15×0.9 cm, using a single buffer. A complete amino acid analysis may be accomplished in 48 h with a fraction collector or in 24 h with automatic recording equipment.

III. Separation of Compounds of High Molecular Weight.

(Peptides, Proteins and Nucleic Acids).

Solutes which are not able to diffuse readily into the pores of the ion-exchange resins of normal cross-linkage (8—12% DVB) offer special problems in ion-exchange chromatography. Substances with molecular weights in excess of about 300 usually show broader peaks and poorer resolution compared with substances of lower molecular weight. However, if a resin of low cross-linkage is used it is found that satisfactory chromatograms can be obtained from solutes with molecular weights up to about 3,000.

Thus, DOWMONT and FRUTON (1952) showed that Dowex 50×4 was suitable for the separation of a series of di, tri and tetrapeptides (Molecular Wts. 200—500) and even the hexapeptide, glutathione in the oxidised form (M. Wt. 612) gave a sharp peak. HIRS, MOORE and STEIN (1956) using a resin with 2% DVB (Dowex 50×2) obtained satisfactory chromatograms from peptides ranging in size from 2 to 22 residues (M. Wt. 3120).

Substances of still larger molecular weight such as the proteins and nucleic acids are not able to penetrate even the pores of an ion-exchange resin of low cross-linkage. Interaction between protein and resin takes place mainly at the surface of the ion-exchange resin with a result that only those ion-exchangers with large surface areas per gm are able to adsorb appreciable quantities of protein. Thus ion-exchange resins prepared from cross-linked polystyrene have proved, in general, unsuitable for the chromatography of proteins because of their low capacity for protein. However, if sulphonated polystyrene is deposited as a film on the surface of an inert material possessing a large surface area (e.g. Celite, a diatomaceous earth) then an ion-exchanger resin with a high capacity for protein may be obtained (BOARDMAN, 1959). Of the synthetic ion-exchange resins, the most successful for protein chromatography has been the methacrylic acid resin, Amberlite IRC-50. The particles of IRC-50 are found to have a sintered structure under the microscope in contrast to the smooth beads of the polystyrene resins. Several basic proteins and a few proteins with isoelectric points in the region of pH 7 have been chromatographed on IRC-50 (see review by MOORE and STEIN, 1956), but IRC-50 does not appear suitable for separating acidic proteins. Since the majority of proteins have isoelectric points below 7, IRC-50 has limited use in protein chromatography.

The cellulose ion-exchangers possess a high capacity for the adsorption of proteins and the anion-exchange celluloses in particular have been widely used for the purification of proteins and nucleic acids. The reader is referred to the section on enzyme purification by H. BOMAN in Vol. 5 of this handbook for details on the use of ion-exchange materials for protein purification.

Another factor which limits the choice of suitable ion-exchangers for protein chromatography is the instability of most proteins. Proteins are readily denatured when strongly adsorbed on many surfaces; to reduce this hazard, protein chromatography is usually carried out at temperature near 0° C.

A third complication in protein chromatography arises from the polyvalent nature of proteins. It is almost certain that a protein molecule is bound to an ion-exchanger by several linkages, which must be broken simultaneously for the protein to move down the column. The rate of movement of such a polyion will be very dependent on the concentration of the small ions used to elute it (BOARDMAN, 1953; TISELIUS, 1954; BOARDMAN and PARTRIDGE, 1955). A protein will tend to stay in a zone at the top of the column ($R_F = 0$) or to move directly through the column ($R_F = 1$) and it will tend to change from one position to the other with small changes in the concentration of the eluting ion. A similar effect will be observed with small changes in pH of the eluting buffer over certain critical ranges since one of the main effects of pH is to change the charges on the protein molecule.

The fact that the R_F of a protein can vary from zero to one over a narrow range of salt concentration and pH and that this range varies from protein to protein makes the use of the method of elution with a buffer of constant composition unsuitable for the separation of all the proteins in the mixture. A buffer which elutes one of the proteins in the mixture is unlikely to effect a satisfactory separation of the other proteins in the mixture. The elution of all the proteins, individually, will require a series of buffers of increasing eluting power or a buffer of gradually increasing eluting power.

Here we should draw an important distinction between the ion-exchange chromatography of solutes of low molecular weight, compared with the chromatography of proteins. With low molecular weight solutes, separations are effected by means of the countercurrent principle and the degrees of separation obtained are dependent on differences in distribution coefficients of the solutes and the number of theoretical plates in the column. Resolution can be increased by increasing the length of the column. In the majority of experiments reported in the literature, the separation of proteins on ion-exchange materials is governed largely by a principle of "selective desorption" rather than countercurrent distribution. The mixture of proteins is adsorbed in a band at the top of the column and each protein is desorbed in turn by passing a series of buffers or a buffer of gradually increasing power. Once a protein is desorbed from the top of the column it may pass down the column with an $R_F = 1$ and hence the remainder of the column is not used for the separation. Increased resolution is obtained not by increasing the length of the column but by using a more gradual change in the eluting power of the buffers.

It should be emphasized that a protein which shows marked tailing on the column due to some irreversible adsorption may appear in several peaks, as each change of the eluting power may desorb some of the protein from the column. This appearance of a single protein in several peaks may occur even when gradient elution is used. If the results are to be interpreted in terms of number of components or homogeneity of a particular component, it is very important that each peak should be rechromatographed under the same conditions. Unfortunately, this is not always possible with the very labile proteins.

F. Ion-Exclusion.

A new technique, using ion-exchange resins was devised by Wheaton and Bauman (1953). Known as ion-exclusion, it is used for the separation of electrolytes and non-electrolytes. The process does not involve ion-exchange; it is a type of partition chromatography using ion-exchange resins.

If an ion-exchange resin is placed in a solution containing an electrolyte possessing an ion in common with the ion-exchanger, it is found that the concentration of electrolyte inside the beads of ion-exchange resin is low (due to the Donnan effect) compared with the concentration of electrolyte in the outside solution; in some cases, the ratio may be 1:10 (Wheaton and Bauman, 1953). Thus, the ions are excluded from the ion-exchanger. On the other hand, the concentration of non-electrolyte in the resin either will be equal to its concentration in the outside solution (a partition effect) or it will be higher due to molecular adsorption of the non-electrolyte. If a mixture containing an electrolyte and a non-electrolyte is added to a column of an ion-exchange resin containing a common ion with the electrolyte, a separation will take place with the electrolyte emerging from the column ahead of the non-electrolyte. The terms electrolyte and non-electrolyte are relative terms. Weak acids may be regarded as non-electrolytes and, in fact, acetic acid may be separated from hydrochloric acid by the ion-exclusion process. Good separations are obtainable if the electrolyte has a dissociation constant greater than 2×10^{-1} and the non-electrolyte a dissociation constant of less than 5×10^{-2} (Reichenberg, 1957). The separation of an electrolyte such as sodium chloride and a non-electrolyte such as ethyl alcohol can be carried out, of course, by conventional ion-exchange as well as by ion-exclusion. Where it is desired to separate a small amount of an ionized substance from a large amount of an unionized one, the ion-exchange method is preferable. In the reverse situation, the ion-exclusion method would be better. The chief use of the ion-exclusion method is for the separation by the elution method of weak and strong electrolytes. In the ion-exclusion technique, the resin does not require regeneration.

G. Mechanism of Adsorption of Solutes on Ion-Exchange Resins.

The adsorption affinity of a solute for an ion-exchange resin may be regarded as made up of an electrostatic term and a van der Waal's term (Davies, 1949). The electrostatic term which results from the Coulomb attraction of the electric charges on the solute ion and the fixed ions on the resin makes a very important contribution to the total adsorption affinity. But whereas ions of different valency show large variations in adsorption affinity, ions of the same valency and of similar size are unlikely to show a sufficiently large variation for this to be the basis of a separation.

The van der Waal's term results from the interaction between the hydrocarbon part of the solute and the hydrocarbon matrix of the ion-exchanger. Variations in the hydrocarbon part of the solute are likely to result in considerable changes in the adsorption affinity. In an homologous series the van der Waal's term will increase with molecular weight, while the presence of an aromatic group in a solute will considerably enhance its van der Waal's term and hence its adsorption to the ion-exchange resin.

With weakly acidic or weakly basic ion-exchangers, interaction between the uncharged polar groups on the exchanger and polar groups on the solute can form an important contribution to the adsorption affinity. For example, the uncharged groups on a carboxylic acid ion-exchanger can form strong hydrogen bonds with similar groups on the solute or with $—NH_2$ groups on the solute.

Effect of temperature on separations. Here we are concerned with the effect of temperature on the relative affinities of the solutes for the ion-exchanger. Temperature changes would not be expected to influence the electrostatic terms of the ions relative to one another, nor would they effect greatly the relative dissociation constants of weak electrolytes. But temperature could have a considerable relative effect on the hydrogen bond term and particularly the VAN DER WAAL's term. Thus solutes which are not separated at one temperature may be easily separated at another. For example, MOORE and STEIN (1951) report that the aromatic amino acids, tyrosine and phenylalanine are completely resolved on a column of sulphonated polystyrene at 60° C whereas they run together at 25° C and 37° C. Conversely temperature may be used as a test for homogeneity. A substance which appears homogeneous at one temperature may be split into two bands at another temperature.

There is another way in which temperature may effect a chromatogram. Solutes which are bound to an ion-exchanger by a strong VAN DER WAAL's contribution exhibit slow rates of exchange. An increase in temperature will diminish the VAN DER WAAL's contribution and increase the rate of exchange of those solutes. This will result in a considerable sharpening of the bands.

H. Applicability of Ion-Exchange Chromatography.

This final section is intended as a general guide to the user or potential user of ion-exchange chromatography. It is written to assist the research worker to evaluate the possibilities of the method and to facilitate the selection of suitable ion-exchangers and experimental conditions for separating the substances in which he is interested.

Ion-exchange resins should be useful for separating a variety of substances provided they are water soluble and ionic in character. With crude mixtures containing several classes of compounds, it is usually advisable to carry out a preliminary group separation before attempting a separation of closely related substances. Aromatic compounds and other substances which can be strongly adsorbed to the ion-exchanger by molecular forces (VAN DER WAAL and polar forces) are likely to interfere with the separation of ionic constituents and they should be removed prior to ion-exchange chromatography.

I. Choice of Ion-Exchanger.

Cross-linked sulphonated polystyrene has proved to be the most versatile of the cation-exchangers, because of the wide range of pH over which it is ionized. The degree of cross-linkage become important for substances with molecular weights in excess of 300—400. When chromatographing substances which may be labile over certain ranges of pH, it should be remembered that the pH on the resin will be 1—2 units below the pH in the ambient solution by virtue of the DONNAN equilibrium. Ionic substances which are also aromatic in character may be strongly adsorbed on the polystyrene type resins, but may show a lower adsorption on an ionized polymethacrylic acid resin or a cellulose ion-exchanger. Higher operating temperatures will usually give better chromatograms with this type of substance.

Table 5. *A selection of compounds which have been successfully subjected to chromatography on ion-exchange materials.*

Class of substance	Ion-exchange material	References
Amino acids	Dowex 50	Moore and Stein (1951)
	Amberlite IR–120	Moore, Spackman and Stein (1958)
	Zeo-Karb 225	Partridge and Brimley (1952)
	Dowex 2	Partridge and Brimley (1952)
Organic acids	Amberlite IRA–400	Bryant and Overall (1953)
	Dowex 1	Busch, Hurlbert and Potter (1952)
	Deacidite E (a polystyrene, weakly basic resin would be preferable now)	Hulme (1951)
Organic bases	Zeo-Karb 225	Shewan et al. (1952)
Purines and pyrimidines	Dowex 1	Cohn (1951)
	Dowex 50	Wall (1953)
Nucleosides	Dowex 1	Cohn (1955)
	Amberlite IRA–400	
(as borate complexes)	Dowex 1	Jaenicke and von Dahl (1952)
Nucleotides	Dowex 1	Bergkvist and Deutsch (1954)
		Hurlbert et al. (1954)
Oligonucleotides	Dowex 1 (2% DVB)	Volkin and Cohn (1953)
Nucleic acids	ECTEOLA cellulose	Bendich et al. (1953)
Peptides	Dowex 50×8	Schroeder et al. (1954)
		Moore and Stein (1956) (Review)
	Dowex 50×4	Thompson (1955)
	Dowex 50×2	Hirs, Moore and Stein (1956)
Peptides (large)	Amberlite IRC–50	Bailey, Moore and Stein (1956)
Non-steroid hormones (peptides)	Amberlite IRC–50	Dixon and Stack-Dunne (1955)
	Amberlite IRC–50	Taylor (1954)
Proteins and enzymes	Amberlite IRC–50	Moore and Stein (1956) } Reviews
	CM-cellulose	Sober and Peterson (1957) } Reviews
	DEAE cellulose	Sober and Peterson (1958) } Reviews
	TEAE cellulose	Boman (This handbook) Vol. VI } Reviews
Viruses	Amberlite IRC–50	Muller and Rose (1952)
		Muller (1957) (a review)
Tobacco mosaic virus and potato X virus	DEAE cellulose	Levin (1958)
Influenza virus	Amberlite IRA–400	Matheka and Armbruster (1958)
Carbohydrates (as borate complexes)	Dowex 1	Khym and Zill (1952)
Phosphate esters	Dowex 1	Khym and Cohn (1953)
Uronic acids	Dowex 1	Khym and Doherty (1952)
Amino sugars	Dowex 50	Gardell (1953)
Aldehydes and ketones (bisulphite forms)	Strong base and weak base resins	Gabrielson and Samuelson (1952)
Streptomycin	Amberlite IRC–50	Howe and Putter (1951)
Inorganic cations		
Rare earth metals	Amberlite IR–100	Spedding (1949)
Alkali metals	Dowex 50	Cohn and Kohn (1948)
	Amberlite IR–100	Kayas (1950)
Alkaline earth metals	Dowex 50	Lerner and Rieman (1954)
Transition elements	Dowex 1	Kraus and Moore (1953)
Inorganic acids	Dowex 2	Atteberry and Boyd (1950)

It should be emphasized that at pH values below about 5.5 (the exact pH will vary with the ionic strength) a reasonable proportion of the carboxyl groups on a methacrylic acid resin will be uncharged and these uncharged groups may bind substances by polar bonding. In fact, there are many examples in the literature of the chromatography of substances on methacrylic acid resins at pH values where the resin is largely uncharged; in these cases we are dealing with adsorption chromatography on an ion-exchange resin. Some substances may be bound to the ionized carboxyl groups by electrostatic forces and to the uncharged groups by polar forces and their successful chromatography may depend on achieving the correct balance between these forces. Separations may be very dependent on the pH and cation concentration of the eluting buffer.

Substances which are polyvalent or polyionic usually are strongly adsorbed on ion-exchange resins and high concentrations of a monovalent ion may be required to elute them. Polyionic substances which cannot be desorbed from sulphonated polystyrene even at high concentrations of the eluting cation may chromatograph satisfactorily on a polymethacrylic acid resin at pH values where the resin is only partly ionized. The fixed charges on the partly ionized resin would be further apart and the polyion may be held to the resin by fewer links. The cellulose ion-exchangers with their much lower charge density than the synthetic resins may well prove to be eminently suited for the chromatography of polyionic substances. The cellulose ion-exchangers should be suitable also for the chromatography of substances of intermediate molecular weight which diffuse into the pores of ion-exchange resins only with difficulty.

For the separation of anionic substances, the most suitable resins generally have been the strongly basic resins made from cross-linked polystyrene. Similar considerations apply to chromatography on the weakly basic resins as were discussed for polymethacrylic acid, except that the weakly basic resins prepared from cross-linked polystyrene are more or less equivalent to the strongly basic resins over those ranges of pH where both types of resin are fully ionized. If the weakly basic resin is only partly ionized under the conditions of the experiment, then the solutes may be adsorbed to the uncharged hydroxyl groups on the resin by polar forces, as well as by normal coulombic forces to the charged groups. With fully ionized anion-exchange resins, the pH on the resin will be 1—2 units higher than the pH in the ambient phase. Substances which are labile at high pH values should not be placed in contact with a strongly basic resin in its free base form.

A selection of compounds which have been successfully subjected to chromatography on ion-exchange materials is included in Table 5. Neither the list of substances nor the references cited should be regarded as comprehensive. Table 5 is included merely to demonstrate the classes of substances to which ion-exchange chromatography can be applied.

II. Conditions for Chromatography.

The most suitable conditions for the separation of the components in a mixture are usually determined by trial and error. However, an appreciation of the possible variables in the eluting conditions will be helpful in carrying out preliminary experiments.

The following is a list of parameters which may be varied.

(1) Concentration and nature of eluting cation or anion;
(2) valency of eluting cation or anion;
(3) pH of the solution;
(4) temperature;

(5) use of complexing agents;
(6) addition of organic solvent to eluting solution.

Several of these variables have been discussed in some detail earlier in this article. Mass action theory tells us that the affinity of a solute ion for the resin should be decreased by increasing the concentration or valency of the eluting ion. However, it should be remembered that the number of ionizable groups on the resin may be affected by the concentration of the eluting ion.

A variation of pH may be used either to suppress the ionization of the groups on the resin or the groups on the solute. If the adsorption is governed mainly by ion-exchange, depression of either the charge on the resin or on the solute should lead to a desorption of the solute. The optimum pH's for the separation of a number of weak electrolytes may be predicted from the relevant pK values. For example, consider the adsorption of two weak acids, HA_1, HA_2 on an anion-exchange resin under acidic conditions. It may be shown (Davies, 1949) that the ionized forms of these acids will show their maximum difference at a pH equal to $\frac{1}{2}(pK_1 + pK_2)$ and the best separation of the two anions will be achieved when the pH of the solution is 1—2 units lower than the value $\frac{1}{2}(pK_1 + pK_2)$. Another method of altering the relative concentrations of two ions is to add an agent which will specifically form a complex with one of the ions.

Temperature is important in separations which are governed by differences in van der Waal's adsorption. Another method of facilitating the elution of ions which are bound by strong van der Waal's forces is to add a miscible organic solvent to the eluting solution. For example, Moore and Stein (1951) found that the presence of 1% benzyl alcohol increases the elution rate of the aromatic amino acids relative to the other amino acids and Hale, Hawdon, Jones and Packham (1952) used 75% aqueous alcohol to elute β and γ resorcyclic acids from an anion-exchange resin.

III. Rechromatography.

As mentioned in the section on protein chromatography, solutes which show some degree of irreversible adsorption on the resin may appear in several peaks in the effluent curve. Hence, before any conclusion can be reached regarding the homogeneity of the material in a particular peak, the material from that peak should be rechromatographed under the same conditions as was used in the first chromatogram. With complex solutes, e.g. peptides, a single symmetrical peak is not sufficient evidence alone for a homogeneous substance. Rechromatography under different conditions or at a different temperature will help in establishing the homogeneity of the material in a single symmetrical peak.

References.

Adams, B. A., and E. L. Holmes: J. Soc. Chem. Ind. **54**, 1T (1935). — Alm, R. S., R. J. P. Williams and A. Tiselius: Acta Chem. Scand. **6**, 826 (1952). — Atteberry, R. W., and G. E. Boyd: J. Am. Chem. Soc. **72**, 4805 (1950).

Bailey, J. L., S. Moore and W. H. Stein: J. Biol. Chem. **221**, 143 (1956). — Bauman, W. C., and J. Eichhorn: J. Am. Chem. Soc. **73**, 642 (1951). — Bendich, A., J. R. Fresco, H. S. Rosenkranz and S. M. Beiser: J. Am. Chem. Soc. **77**, 3671 (1955). — Bergkvist, R., and A. Deutsch: Acta Chem. Scand. **8**, 1877 (1954). — Boardman, N. K.: Ph. D. Dissertation, University of Cambridge (1953); — Biochim Biophys. Acta **18**, 290 (1955); — J. Chromatography **2**, 388 (1959). — Boardman, N. K., and S. M. Partridge: Biochem. J. **59**, 543 (1955). — Bock, R. M., and N. S. Lind: Anal. Chem. **26**, 1543 (1954). — Boman, H. G.: Biochim. Biophys. Acta **16**, 245 (1955). — Boyd, G. E., J. Schubert and A. W. Adamson: J. Am. Chem. Soc. **69**, 2818 (1947). — Bregman, J. E., and Y. Murata: J. Am. Chem. Soc. **74**, 1867 (1952). — Bryant, F., and B. T. Overell: Biochim. Biophys. Acta **10**, 471 (1953). — Busch, H., R. B. Hurlbert and V. R. Potter: J. Biol. Chem. **196**, 717 (1952).

CAMPBELL, D. H., E. LUESCHER and L. S. LERMAN: Proc. Natl. Acad. Sci. U.S. **37**, 575 (1951). — CHERKIN, A., F. E. MARTINEZ and M. S. DUNN: J. Am. Chem. Soc. **75**, 1244 (1953). — CLAESSON, S.: Ark. Kemi Min Geol. 23 A No. 1 (1946). — COHN, W. E.: J. Cell. Comp. Physiol. **38** Supp. 1, 21 (1951); — The nucleic acids. Ed. E. CHARGAFF and J. W. DAVIDSON, p. 211. New York: Academic Press 1955. — COHN, W. E., and H. W. KOHN: J. Am. Chem. Soc. **70**, 1986 (1948). — CORNAZ, J. P., and H. DEUEL: Experientia (Basel) **10**, 137 (1954).

DAVIES, C. W.: Biochem. J. **45**, 38 (1949). — DE VAULT, D.: J. Am. Chem. Soc. **65**, 532 (1943). — DICKMAN, S. R., G. A. MORRILL and K. M. TRUPIN: J. Biol. Chem. **235**, 169 (1960). — DIXON, H. B. F., and M. P. STACK-DUNNE: Biochem. J. **61**, 483 (1955). — DJURFELDT, R., and O. SAMUELSON: Acta Chem. Scand. **4**, 165 (1950). — DOWMONT, Y. P., and J. S. FRUTON: J. Biol. Chem. **197**, 271 (1952). — DRAKE, B.: Arkiv Kemi 8, 1 (1955).

FEITELSON, J., and S. M. PARTRIDGE: Biochem. J. **64**, 607 (1956).

GABRIELSON, G., and O. SAMUELSON: Acta Chem. Scand. **6**, 729 (1952). — GARDELL, S.: Acta Chem. Scand. **7**, 207 (1953). — GREGOR, H. P.: J. Am. Chem. Soc. **73**, 642 (1951). — GREGOR, H. P., M. TAIFER, L. CITAREL and E. I. BECKER: Ind. Eng. Chem. **44**, 2834 (1952). — GRUBHOFER, N., and L. SCHLEITH: Z. Physiol. Chem. **296**, 262 (1954).

HALE, D. K.: Chem. & Ind. (London) **1955**, 1147. — HALE, D. K., A. R. HAWDON, J. I, JONES and D. I. PACKHAM: J. Chem. Soc. **1952**, 3503. — HIRS, C. H. W., S. MOORE and W. H. STEIN: J. Biol. Chem. **200**, 493 (1953); **219**, 623 (1956). — HOWE, E. E., and I. PUTTER: U.S. Patent 2, 541, 320 (1951) See C. A. **45**, 4893f.. — HULME, A. C.: J. Exp. Bot. **2**, 298 (1951). — HURLBERT, R. B., A. F. SCHMITZ, A. F. BRUMM and V. R. POTTER: J. Biol. Chem. **209**, 23 (1954).

ISLIKER, H. C.: Ann. N. Y. Acad. Sci. **57**, 225 (1953).

JAENICKE, L., and K. VON DAHL: Naturwiss. **39**, 87 (1952).

KAYAS, G.: J. Chem. Phys. **47**, 408 (1950). — KENT, L. H., and J. H. R. SLADE: Nature **183**, 325 (1959). — KHYM, J. X., and W. E. COHN: J. Am. Chem. Soc. **75**, 1153 (1953). — KHYM, J. X., and D. G. DOHERTY: J. Am. Chem. Soc. **74**, 3199 (1952). — KHYM, J. X., and L. P. ZILL: J. Am. Chem. Soc. **74**, 2090 (1952). — KNIGHT, C. S.: Nature **183**, 165 (1959); **189**, 739 (1960). — KRAUS, K. A., and C. E. MOORE: J. Am. Chem. Soc. **73**, 1460 (1953).

LAKSHMANAM, T. K., and S. LIEBERMAN: Arch. Biochem. Biophys. **45**, 235 (1953). — LEDERER, M.: Anal. Chim. Acta **12**, 142 (1955). — LERNER, M., and W. RIEMAN: Anal. Chem. **26**, 610 (1954). — LEVIN, O.: Arch. Biochem. Biophys. **78**, 33 (1958). — LINSKENS, H. F.: Papierchromatographie in der Botanik. ed. H. F. LINSKENS. 2nd ed. Springer Verlag 1959.

MARTIN, A. J. P., and R. L. M. SYNGE: Biochem. J. **35**, 1358 (1941). — MATHEKA, H. D.. and O. ARMBRUSTER: Virology **6**, 584 (1958). — MAYER, S. W., and E. R. THOMPKINS: J. Amer. Chem. Soc. **69**, 2866 (1947). — MOORE, S., and W. H. STEIN: J. Biol. Chem. **192**, 663 (1951); — Adv. Protein. Chem. **11**, 205 (1956). — MOORE, S., D. H. SPACKMAN and W. H. STEIN: Anal. Chem. **30**, 1185 (1958). — MULLER, R. H.: In Ion-exchangers in organic and biochemistry p. 248, Ed. CALMON, C., and T. R. E. KRESSMAN, Interscience, New York 1957. — MULLER, R. H., and H. M. ROSE: Proc. Soc. Exptl. Biol. Med. **80**, 27 (1952). — MYHRE, D. V., and F. SMITH: J. Org. Chem. **23**, 1229 (1958).

PARTRIDGE, S. M.: Biochem. J. **44**, 521 (1949); — Analyst **26**, 955 (1952). — PARTRIDGE, S. M., and R. C. BRIMLEY: Biochem. J. **48**, 313 (1951); **49**, 153 (1951); **51**, 628 (1952). — PARTRIDGE, S. M., and R. G. WESTALL: Biochem. J. **44**, 418 (1949). — PEPPER, K. W.: J. Appl. Chem. (London) **1**, 124 (1951). — PEPPER, K. W., and D. K. HALE: In Ion-exchange and its applications. Soc. Chem. Ind. Symposium. London 1955. — PEPPER, K. W., D. REICHENBERG and D. K. HALE: J. Chem. Soc. **1952**, 3129. — PETERSON, E. A., and H. A. SOBER: J. Am. Chem. Soc. **78**, 751 (1956); — Anal. Chem. **31**, 857 (1959). — PORATH, J.: Arkiv Kemi **11**, 97 (1957).

REICHENBERG, D.: In Ion-exchangers in organic and biochemistry. p. 184. Ed. C. CALMON and T. R. E. KRESSMAN. Interscience. New York 1957.

SCHROEDER, W. A., L. M. KAY, J. LE GETTE, L. HONNEN and F. C. GREEN: J. Am. Chem. Soc. **76**, 3556 (1954). — SHEWAN, J. M., L. I. FLETCHER, S. M. PARTRIDGE and R. C. BRIMLEY: J. Sci. Food Agr. **1952**, 394. — SIMMONDS, D. H., and K. I. WOOD: Anal. Chem. **26**, 1860 (1954). — SPEDDING, F. H.: Discus. Faraday Soc. **7**, 214 (1949). — SOBER, H. A., and E. A. PETERSON: In Ion-exchangers in organic and biochemistry, p. 318. Ed. C. CALMON and T. R. E. KRESSMAN, Interscience. New York 1957; — Fed. Proc. **17**, 1116 (1958).

TAVEL, P. VON, and R. SIGNER: Adv. Protein Chem. **11**, 237 (1956). — TAYLOR, S. P.: Proc. Soc. Exptl. Biol. Med. **85**, 226 (1954). — THOMPSON, A. R.: Biochem. J. **61**, 253 (1955). — THOMPSON, E. O. P., and I. J. O'DONNELL: Aust. J. Biol. Sci. **13**, 393 (1960). — TISELIUS, A.: Arkiv Kemi Min. Geol. **16** A, No. 18 (1943); — Arkiv Kemi **7**, 443 (1954). — TUCKERMAN, M. M.: Anal. Chem. **30**, 231 (1958).

VOLKIN, E., and W. E. COHN: J. Biol. Chem. **205**, 767 (1953).

WALL, J. S.: Anal. Chem. **25**, 950 (1953). — WEISS, J.: J. Chem. Soc. **1943**, 297. — WHEATON, R. M., and W. C. BAUMAN: Ind. Eng. Chem. **45**, 228 (1953).

Table 1. Chromatographic grade resins and celluloses.

(For a complete list of industrial grade resins refer to The Merck Index, 7th Edition, 1960).

A. Cation-Exchange resins.

(Some cation-exchange resins of German origin are described in Vol. I, p. 119—120).

1. Dowex series supplied by the Dow Chemical Company, Midland, Mich. U.S.A. and A.G. series (analytical grade resins processed from Dow resins) supplied by Bio-Rad Laboratories, 32nd and Griffen Avenue, California, U.S.A. Resins normally supplied in H^+ forms.

Dowex resin and cross-linkage[1]	A.G. resin	Cat. No. (A.G. resin)	Chemical type	Dry mesh designation	Actual U.S. standard mesh range	Maximum temperature	Capacity m.eq./g. (dry)	Capacity m.eq./ml (wet)
Dowex 50–X1	AG 50W–X1	B512	a	50—100	20—50	up to 150°	5.0	0.4
Dowex 50–X2	AG 50W–X2	B522	a	50—100	20—50	up to 150°	5.2	0.7
Dowex 50–X2	AG 50W–X2	B523	a	100—200	50—100	up to 150°	5.2	0.7
Dowex 50–X2	AG 50W–X2	B524	a	200—400	80—200	up to 150°	5.2	0.7
Dowex 50–X4	AG 50W–X4	B541	a	20—50	20—50	up to 150°	5.2	1.2
Dowex 50–X4	AG 50W–X4	B542	a	50—100	40—80	up to 150°	5.2	1.2
Dowex 50–X4	AG 50W–X4	B543	a	100—200	60—140	up to 150°	5.2	1.2
Dowex 50–X4	AG 50W–X4	B544	a	200—400	100—200	up to 150°	5.2	1.2
Dowex 50–X4	AG 50W–X4	B545	a	Minus 400	200—400	up to 150°	5.2	1.2
Dowex 50–X5	AG 50W–X5	B555	a	Minus 400	200—400	up to 150°	5.2	1.3
Dowex 50–X8	AG 50W–X8	B581	a	20—50	20—50	up to 150°	5.1	1.7
Dowex 50–X8	AG 50W–X8	B582	a	50—100	40—80	up to 150°	5.1	1.7
Dowex 50–X8	AG 50W–X8	B583	a	100—200	60—140	up to 150°	5.1	1.7
Dowex 50–X8	AG 50W–X8	B584	a	200—400	100—230	up to 150°	5.1	1.7
Dowex 50–X8	AG 50W–X8	B585	a	Minus 400	230—400	up to 150°	5.1	1.7
Dowex 50–X12	AG 50W–X12	B5122	a	50—100	40—80	up to 150°	5.0	2.3
Dowex 50–X12	AG 50W–X12	B5123	a	100—200	60—140	up to 150°	5.0	2.3
Dowex 50–X12	AG 50W–X12	B5124	a	200—400	140—230	up to 150°	5.0	2.3
Dowex 50–X12	AG 50W–X12	B5125	a	Minus 400	230—400	up to 150°	5.0	2.3
Dowex 50–X16	AG 50W–X16	B5161	a	20—50	20—50	up to 150°	4.9	2.6
	AG 50W–X16[2]	B5162	a	50—100	40—80	up to 150°	4.9	2.6
	AG 50W–X16[2]	B5163	a	100—200	60—140	up to 150°	4.9	2.6
	AG 50W–X16[2]	B5164	a	200—400	140—230	up to 150°	4.9	2.6

[1] Cross-linkage denoted by number after the X e.g. –X4 means that 4% divinyl benzene is incorporated into polymer beads.

[2] Prepared by grinding 20—50 mesh material.

a Polystyrene sulphonic acid.

2. Other cation-exchange resins supplied by Bio-Rad Laboratories.

Resin	Catalogue No.	Parent material	Chemical type	Wet mesh designation	Cross-linkage	Maximum temperature	Capacity m.eq./gm.	Capacity m.eq./ml.
Bio-Rex 70	B701	Duolite CS–101	Acrylic carboxylic acid	20—50	10	100	10.2	3.5
Bio-Rex 70	B702	Duolite CS–101	,,	50—100	10	100	10.2	3.5
Bio-Rex 70	B703	Duolite CS–101	,,	100—200	10	100	10.2	3.5
Bio-Rex 70	B704	Duolite CS–101	,,	200—400	10	100	10.2	3.5
Bio-Rex 70	B705	Duolite CS–101	,,	Minus 400	10	100	10.2	3.5
Bio-Rex 62	B621	Duolite C–62	Polystyrene phosphorus acid	20—50	5.5	100	6.0	2.8
Bio-Rex 62	B622	Duolite C–62	,,	50—100	5.5	100	6.0	2.8
Bio-Rex 62	B623	Duolite C–62	,,	100—200	5.5	100	6.0	2.8
Bio-Rex 62	B624	Duolite C–62	,,	200—400	5.5	100	6.0	2.8
Bio-Rex 63	B631	Duolite C–63	Polystyrene phosphoric acid	20—50	5.5	100	6.0	3.1
Bio-Rex 63	B632	Duolite C–63	,,	50—100	5.5	100	6.0	3.1
Bio-Rex 63	B633	Duolite C–63	,,	100—200	5.5	100	6.0	3.1
Bio-Rex 63	B634	Duolite C–63	,,	200—400	5.5	100	6.0	3.1
Chelex 100	B1002	Dowex A1	Polystyrene imino diacetate[1]	50—100		75	Divalent ions held selectively	
Chelex 100	B1003	Dowex A1	,,	100—200		75	,,	
Chelex 100	B1004	Dowex A1	,,	200—400		75	,,	
Bio-Rex 40	B401	Duolite C3	Phenol-sulphonic acid	20—50		40	2.9	1.2
Bio-Rex 40	B402	Duolite C3	,,	50—100		40	2.9	1.2
Bio-Rex 40	B403	Duolite C3	,,	100—200		40	2.9	1.2
Bio-Rex 40	B404	Duolite C3	,,	200—400		40	2.9	1.2

[1] 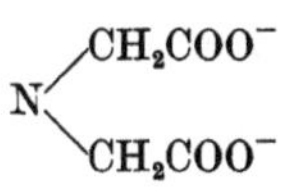

3. Resins supplied by The Permutit Company Ltd., Chiswick, London, England.

Zeo Karb 225. (Catalogue numbers SRC 1—24). Polystyrene sulphonic acid; sodium form. Capacity: 4.5—5.0 m.eq./g.

Mesh BSS wet grading	% DVB 1%	2%	4.5%	8%	12%	20%
Water regain[1]	6—12	3—5	1.5—2	0.9—1.1	0.6—0.9	0.4—0.6
14— 52	SRC 1	SRC 5	SRC 9	SRC 13	SRC 17	SRC 21
52—100	SRC 2	SRC 6	SRC 10	SRC 14	SRC 18	SRC 22
100—200	SRC 3	SRC 7	SRC 11	SRC 15	SRC 19	SRC 23
< 200	SRC 4	SRC 8	SRC 12	SRC 16	SRC 20	SRC 24

[1] Water regain, defined as: The weight of water associated with 1.0 gm of dry hydrogen resin measured by drying the resin after centrifuging to remove interstitial water.

Zeo Karb 226. (SRC 41—48). Cross-linked polymethacrylic acid; hydrogen form. Capacity: 9—10 m.eq./g.

Mesh BSS wet grading	% DVB	
	$2^1/_2$	$4^1/_2$
	Water regain	
	0.8—1.0	0.6—0.8
14— 52	SRC 41	SRC 45
52—100	SRC 42	SRC 46
100—200	SRC 43	SRC 47
< 200	SRC 44	SRC 48

4. Amberlite ion-exchange resins manufactured by Rohm & Haas Company, Philadelphia U.S.A. Laboratory grades supplied by Mallinckrodt Chemical Works 5 Pa. New York, U.S.A

Resin	Code	Parent material	Chemical type	Approximate total cross-linkage %	Dry mesh designation	Maximum temperature	Minimum total exchange m.eq./g.	Capacity m.eq./ml
IR–120 A.R.	3323		a	8	20—50	100	5.0	1.9
IR–120 C.P.	3347		a	1	20—50	120	4.2	0.3
IR–120 C.P.	3348		a	5	20—50	120	4.3	1.2
IR–120 C.P.	3349		a	8	20—50	120	4.4	1.9
IR–120 C.P.	3350		a	10	20—50	120	4.3	2.1
IR–120 C.P.	3351		a	12	20—50	120	4.0	2.1
CG–120 A.R.[1]	3337	IR–120 A.R.	a	8	100—200	120	4.5	
CG–120 A.R.[1]	3338	IR–120 A.R.	a	8	200—400	120	4.5	
CG–120 A.R.[1]	3339	IR–120 A.R.	a	8	400—600	120	4.5	
IRC–50 A.R.	3324		b	5	20—50	120	10.0	3.5
IRC–50 C.P.	3398		b	5	20—50	120	10.0	3.5
CG–50 A.R.[1]	3340	IRC–50 A.R.	b	5	100—200	120	10.0	
CG–50 A.R.[1]	3341	IRC–50 A.R.	b	5	200—400	120	10.0	
CG–50 A.R.[1]	3342	IRC–50 A.R.	b	5	400—600	120	10.0	
XE–89 C.P.			c	—	20—50	120	11.0	

[1] Chromatographic grade resins milled from A.R. beads.
a Polystyrene sulphonic acid.
b Polymethacrylic acid (pH range 5—14).
c Polyacrylic acid (pH range 4—14).

Laboratory grade resins formerly supplied by Rohm & Haas in XE series.

Name of resin	Parent material	Mesh designation
XE–64	IRC–50	100—325
XE–97	IRC–50	100—325 (Pharmaceutical grade)
XE–69	IR–120	100—325
XE–112	Carboxylic acid resin stronger than XE–64	100—325

B. Anion-exchange resins.

(Some anion-exchange resins of German origin are described in Vol. I, p. 119—120).

1. *Dowex series and A.G. series* (normally supplied in chloride form).

Dowex resin and cross-linkage	A.G. resin	Catalogue No. (A.G. resin)	Chemical type	Dry mesh designation	Actual U.S. standard wet mesh range	Maximum temperature (°C)	Capacity m.eq./g.	Capacity m.eq./ml (wet)
Dowex 1–X1	AG1–X1	B112	d	50—100	20—60	50 (OH) 150 (Cl^-)	3.2	0.4
Dowex 1–X2	AG1–X2	B122	d	50—100	20—60	50 (OH) 150 (Cl^-)	3.5	0.8
Dowex 1–X2	AG1–X2	B123	d	100—200	60—140	50 (OH) 150 (Cl^-)	3.5	0.8
Dowex 1–X2	AG1–X2	B124	d	200—400	80—200	50 (OH) 150 (Cl^-)	3.5	0.8
Dowex 1–X4	AG1–X4	B141	d	20—50	20—50	50 (OH) 150 (Cl^-)	3.5	1.2
Dowex 1–X4	AG1–X4	B142	d	50—100	40—80	50 (OH) 150 (Cl^-)	3.5	1.2
Dowex 1–X4	AG1–X4	B143	d	100—200	60—140	50 (OH) 150 (Cl^-)	3.5	1.2
Dowex 1–X4	AG1–X4	B144	d	200—400	100—230	50 (OH) 150 (Cl^-)	3.5	1.2
Dowex 1–X8	AG1–X8	B181	d	20—50	20—50	50 (OH) 150 (Cl^-)	3.2	1.4
Dowex 1–X8	AG1–X8	B182	d	50—100	40—80	50 (OH) 150 (Cl^-)	3.2	1.4
Dowex 1–X8	AG1–X8	B183	d	100—200	80—140	50 (OH) 150 (Cl^-)	3.2	1.4
Dowex 1–X8	AG1–X8	B184	d	200—400	140—325	50 (OH) 150 (Cl^-)	3.2	1.4
Dowex 1–X8	AG1–X8	B185	d	Minus 400	—230	50 (OH) 150 (Cl^-)	3.2	1.4
Dowex 1–X10	AG1–X10	B1102	d	50—100	40—80	50 (OH) 150 (Cl^-)	3.0	1.5
Dowex 1–X10	AG1–X10	B1103	d	100—200	60—140	50 (OH) 150 (Cl^-)	3.0	1.5
Dowex 1–X10	AG1–X10	B1104	d	200—400	140—325	50 (OH) 150 (Cl^-)	3.0	1.5
Dowex 1–X10	AG1–X10	B1105	d	Minus 400	—230	50 (OH) 150 (Cl^-)	3.0	1.5
Dowex 2–X4	AG2–X4	B241	e	20—50	20—50	30 (OH) 150 (Cl^-)	3.2	1.2
Dowex 2–X8	AG2–X8	B281	e	20—50	20—50	30 (OH) 150 (Cl^-)	3.2	1.4
Dowex 2–X8	AG2–X8	B282	e	50—100	40—80	30 (OH) 150 (Cl^-)	3.2	1.4
Dowex 2–X8	AG2–X8	B283	e	100—200	60—140	30 (OH) 150 (Cl^-)	3.2	1.4
Dowex 2–X8	AG2–X8	B284	e	200—400	140—325	30 (OH) 150 (Cl^-)	3.2	1.4
Dowex 2–X10	AG2–X10	B2102	e	50—100	40—80	30 (OH) 150 (Cl–)	3.0	1.5
Dowex 2–X10	AG2–X10	B2103	e	100—200	60—140	30 (OH) 150 (Cl^-)	3.0	1.5
Dowex 2–X10	AG2–X10	B2104	e	200—400	140—325	30 (OH) 150 (Cl^-)	3.0	1.5
Dowex 3–X4	AG3–X4	B341	f	20—50	16—40	65	5.5	2.5
Dowex 3–X4	AG3–X4[1]	B342	f	100—200	60—140	65	5.5	2.5
Dowex 3–X4	AG3–X4[1]	B343	f	200—400	140—325	65	5.5	2.5
Dowex 21K[3]	AG21K[3]		d		16—20	50 (OH) 150 (Cl^-)		1.2
Dowex 21K[3]	AG21K[3]		d		20—50	50 (OH) 150 (Cl^-)		1.3

Dowex resin and cross-linkage	A.G. resin	Catalogue No. (A.G. resin)	Chemical type	Dry mesh designation	Actual U.S. standard wet mesh range	Maximum temperature (°C)	Capacity m.eq./g.	Capacity m.eq./ml. (wet)
Dowex 21K[3]	AG21K[3]		d		50—100	50 (OH) 150 (Cl^-)		1.3
Duolite A–30	Bio-Rex 5	B51	g		20—50	60	8.7[2]	
Duolite A–30	Bio-Rex 5	B52	g		50—100	60	8.7[2]	
Duolite A–30	Bio-Rex 5	B53	g		100—200	60	8.7[2]	
Duolite A–30	Bio-Rex 5	B54	g		200—400	60	8.7[2]	

[1] These resins are prepared by grinding 20—50 mesh material.
[2] Salt splitting capacity 2.0 m.eq./g.
[3] Because of its novel cross-linking, the resin has many of the desirable properties of a lower cross-linked resin (X4) including reduced preference for chloride ion over hydroxide ion, greater permeability and faster equilibrium rate. At the same time its capacity and volume changes are comparable to an 8 per cent cross-linked resin.
d Type 1 quaternary ammonium (Polystyrene-trimethylbenzyl).
e Type 2 quaternary ammonium (Polystyrene-dimethylhydroxyethylbenzyl).
f Polystyrene polyamine (weakly basic).
g Aliphatic amine, tertiary and quaternary.

2. Anion-exchange resins supplied by the Permutit Company, London.

De-Acidite FF. (Catalogue numbers SRA 61—72). Polystyrene-trimethylbenzyl (Type I) ammonium groups (strong base). Capacity 4.0 m.eq./g. Chloride form.

Mesh BSS wet grading	Approx % cross-linking 2—3	3—5	7—9
	Water regain[2] 1.5—2.0	1.0—1.5	0.6—1.0
14— 52	SRA 61	SRA 65[1]	SRA 69[1]
52—100	SRA 62	SRA 66	SRA 70
100—200	SRA 63	SRA 67	SRA 71
< 200	SRA 64	SRA 68	SRA 72

De-Acidite G (SRA 91—102). Polystyrene-diethyl-benzylamine groups (weak base). Capacity 4.0 m.eq. per g. Chloride form.

Mesh BSS wet grading	Approx % cross-linking 2—3	3—5	7—9
	Water regain[2] 1.5—2.0	1.0—1.5	0.6—1.0
14— 52	SRA 91	SRA 95	SRA 99[1]
52—100	SRA 92	SRA 96	SRA 100
100—200	SRA 93	SRA 97	SRA 101
< 200	SRA 94	SRA 98	SRA 102

[1] These resins are from the Permutit Industrial range of Ion Exchange Resins.
[2] This is defined as the weight of water associated with 1.0 gm dry resin chloride and is measured by drying the resin after centrifuging to remove interstitial water.

De-Acidite H (SRA 121—132). Polystyrene — mixed base (tertiary and quaternary see page 163). Capacity 3.8 m.eq./g. Chloride form.

Mesh BSS wet grading	Approx % cross-linking 2—3	3—5	7—9
	Water regain[2] 1.5—2.0	1.0—1.5	0.6—1.0
14— 52	SRA 121	SRA 125	SRA 129[1]
52—100	SRA 122	SRA 126	SRA 130
100—200	SRA 123	SRA 127	SRA 131
< 200	SRA 124	SRA 128	SRA 132

[1] This resin is from the Permutit Industrial Range of Ion Exchange Materials.
[2] This is defined as the weight of water associated with 1.0 gm dry resin chloride and is measured by drying the resin after centrifuging to remove interstitial water.

3. Amberlite anion-exchange resins, supplied by Mallinckrodt Chemical Works.

Resin	Code	Parent material	Chemical type	Approximate cross-linkage	Dry mesh designation	Maximum temperature °C	Minimum total exchange capacity	
							m.eq./gm.	m.eq./ml.
IRA-400 AR	3335		d	8	20—50	75	3.3	1.2
IRA-400 CP	3403		d	8	20—50	75	3.3	1.2
CG-400 AR[1]	3345	IRA-400	d	8	100—200	75	3.3	
CG-400 AR[1]	3346	IRA-400	d	8	200—400	75	3.3	
IRA-401 AR	3336		d	4	20—50	75	3.4	0.8
IRA-402 CP	3402		d	6	20—50	75	4.3	1.2
IRA-410 CP	3404		e	8	20—50	75	3.3	1.3
IR-45 AR	3328		h	—	20—50	75	5.0	2.0
IR-45 CP	3400		h	—	20—50	75	5.0	2.0

[1] Chromatographic grade resins milled from AR beads.
d Type 1 quaternary ammonium (polystyrene-trimethylbenzyl).
e Type 2 quaternary ammonium (polystyrene-dimethyl hydroxyethylbenzyl).
h Polystyrene-polyamine (weakly basic).

Anion-exchange resins formerly supplied in XE series.

Name of resin	Parent material	Mesh designation
XE-67	IRA-400	100—325
XE-75	IRA-401	100—325
XE-98	IRA-411 (lightly cross-linked IRA-410) 4—6% DVB	100—325
XE-76	IR-45	100—325

C. Ion-exchange resin-loaded papers.

Supplier	Ion exchange paper	Amberlite resin incorporated	Functional exchange group	Ionic[1] form (as supplied)	Approximate ion exchange capacity	
					m equiv/g	m equiv/cm²
H. Reeve Angel & Co., New York	Amberlite SA-2	IR-120	Sulphonic acid	Na^+	1.9	0.03
	Amberlite WA-2	IRC-50	Carboxylic acid	H^+	4.6	0.06
British Drug Houses Poole, Dorset, England	Amberlite SB-2	IRA-400	Quaternary ammonium	Cl^-	1.5	0.02
	Amberlite WB-2	IR-4B	Primary and secondary	OH^- (free base)	3.0	0.04

[1] The ion exchange resin in the papers is readily converted from one ionic form to another by soaking the paper in a 4% solution of the appropriate acid, base, salt, or suitable buffer solution, followed by rinsing with water to remove excess salts.

D. Ion-exchange celluloses.

Supplier	Ion-exchanger	Form	Code number	Capacity (m.eq./g.) (a)[1]	(b)[1]
Brown Company 500 Fifth Avenue, New York 36, N.Y. (U.S.A.)	CM cellulose (types 20 and 40)				
	DEAE cellulose (types 20 and 40)				
	ECTEOLA cellulose (types 20 and 40)				
Eastman Kodak Rochester 3, New York (U.S.A.)	DEAE cellulose				
	ECTEOLA cellulose				
Serva-Entwicklungslabor, Heidelberg (Germany)	CM cellulose				
	P cellulose				
	SE cellulose				
	PAB cellulose				
	DEAE cellulose				
	TEAE cellulose				
	ECTEOLA cellulose				
Bio-Rad Laboratories, 32nd and Griffin Avenue, Richmond, California (U.S.A.)	CM cellulose (Cellex CM)				
	P cellulose (Cellex P)				
	DEAE cellulose (Cellex D)				
	TEAE cellulose (Cellex T)				
	ECTEOLA cellulose (Cellex E)				
Whatman[2]	P cellulose	3 flocs, 3 powders	P10	1.0	0.5
			P40	4.2	2.1
			P70	7.4	3.7
H. Reeve Angel & Co., 9 Bridewell Place, London E.C.4. (England)		1 paper	P20	2.1	1.0
and	CM cellulose	2 flocs, 2 powders	CM30	0.3	
			CM70	0.7	
52 Duane Street, New York, 7. (U.S.A.)		1 paper	CM50	0.5	
	Citrate cellulose (CT)	1 floc, 1 powder	CT60	2.4	
		1 paper	CT30	1.2	
	Amino ethyl (AE)	1 floc, 1 powder	AE50	1.0	
		1 paper	AE30	0.6	
	DEAE cellulose	1 floc, 1 powder	DE50	1.0	
		1 paper	DE20	0.4	
	ECTEOLA cellulose	1 floc, 1 powder	ET30	0.5	
		1 paper	ET20	0.3	

[1] P cellulose is bifunctional (a) maximum theoretical capacity (b) capacity of strongly acidic groups.

[2] Available in 3 forms: floc, powder and paper. A number of chemical forms are available depending on the derivative: A code letter and number have been assigned to each material. The principal code numbers are in steps of ten from 10 to 90 and give an indication of the nominal ion-exchange capacity, 10 being the lowest and 90 the highest.

Molecular Sieving other than Dialysis.

By

N. K. Boardman.

With 3 Figures.

The concept of molecular sieving is best illustrated by the well known process of dialysis, although a consideration of dialysis itself is excluded from this article. The dialysis membrane consists essentially of a series of pores and molecules which are not able to enter the pores because of their size may be separated from the smaller molecules which are able to pass through the membrane.

In the processes to be considered here, molecules of different sizes are separated by virtue of their relative abilities to penetrate to the inside of an adsorbent or gel. The adsorbent or gel may be considered as an impermeable solid phase possessing a network of interconnected pores of molecular size. It is desirable to distinguish between two types of molecular sieving, which may be termed "complete molecular sieving" and "partial molecular sieving." In "complete molecular sieving", the flow of at least one of the solutes into the adsorbent or gel is completely prevented while the flow of the other solutes occurs at a finite rate. On the other-hand, in "partial molecular sieving", all solutes enter the adsorbent or gel and the separation of molecules of different sizes is due either to differences in their rate of movement (diffusion) down the pores of the adsorbent or to differences in their depth of penetration into the pores. In this article, molecular sieving from aqueous solutions only will be considered, although it should be mentioned that gases have been effectively separated by molecular sieving (Barrer, 1949).

A. Ion-Exchange Materials as Ionic Sieves.

In the preceding article on ion-exchange chromatography, (p. 160) it was pointed out that an ion-exchange resin consists essentially of a cross-linked hydrocarbon matrix to which are attached ionizable groups in a rather uniform manner. When the resin is immersed in an aqueous solution, water enters the pores formed by the criss-crossing polymer network. This causes the resin to swell until an equilibrium is reached between, on the one hand, the tendency of the resin to completely dissolve in the water and, on the other hand, the restricting influence of the cross-bonds. Thus the degree of swelling of the resin and hence the size of the pores are dependent on the amount of cross-bonding. For the synthetic ion-exchange resins, the pores are of molecular size varying from about 5 Å for a tightly cross-linked resin (10% divinyl benzene) to 35 Å or larger for a lightly cross-linked one (1% divinyl benzene). By selecting a resin with suitable cross-linking it is possible to separate ions of different size by "complete molecular sieving".

Cernescu (1933) was probably the first to demonstrate that it was possible to separate ions of different sizes by molecular sieving on an adsorbent. He showed that the natural zeolite, chabazite will adsorb ammonium ions, but that only negligible amounts of the tetramethyl ammonium ions were adsorbed. KUNIN (1949) and KRESSMAN (1952) obtained similar effects with the synthetic ion-exchange resins. Fig. 1, taken from KUNIN's paper shows how the exchange capacity of the resins, Amberlite IR-120 and Amberlite IRC-50 decreases as the ionic diameter is increased. KRESSMAN's experiments were performed with the Permutit resins, Zeo Karb 215 (phenol sulphonic) and Zeo Karb. 225 (sulphonated polystyrene, approximately 10% divinyl benzene). The smallest pores in Zeo Karb 215 are larger than the effective size of the phenylbenzyldimethyl-ammonium ion (major diameter 11.2 Å) although they are smaller than the tetrabenzyl-ammonium ion. The smallest pores of Zeo Karb 225, on the otherhand, are smaller that the tetramethyl-ammonium ion (4.6 Å).

Similar studies on the adsorption of large anions (dyes) on the anion-exchangers De Acidite B and De Acidite E respectively were carried out by RICHARDSON (1949) and KRESSMAN (1952). RICHARDSON's results suggest a pore size for De Acidite B of between 10—20 Å.

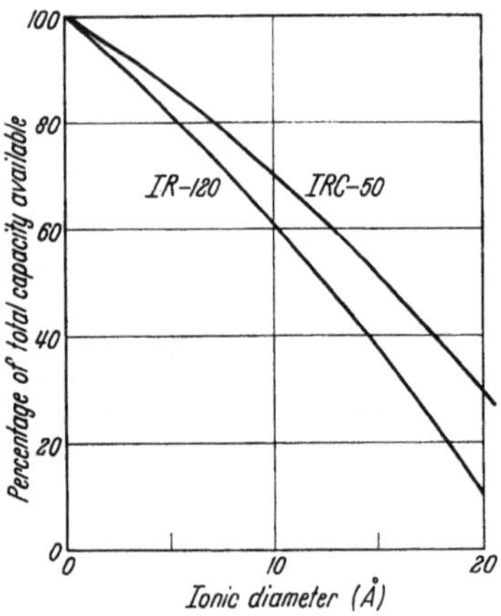

Fig. 1. Effect of ionic diameter on the total available exchange capacity. (From KUNIN, 1949).

The relative sizes of the pores of the resin and of the diffusing ion will determine not only the equilibrium position of the exchange process but also the rate at which equilibrium is attained. For example, KRESSMAN (1952) showed that the times of half exchange of various ions under comparable conditions with the ammonium form of Zeo Karb 225 were: Na^+, 1.25 min; $N(CH_3)_4^+$, 1.75 min; $N(C_2H_5)_4^+$, 3.0 min and $N(CH_3)_2(C_6H_5)_2^+$ ca. 1 week. This means that certain ions are able to enter the pores of the ion-exchange resin but they do so only with difficulty. It should also be remembered that the pores of an ion-exchange resin will vary in size and an ion of critical size may be able to enter the large pores readily but will only diffuse into the smaller pores with difficulty.

Separations by Ionic Sieving.

The use of ion-exchange resins to separate quantitatively large ions from smaller ones was reported first by DUEUL, SOLMS and ANYAS-WEISZ (1950) and by RICHARDSON (1951). The former group of workers used columns of Amberlite IR-4B in its OH form and solutions containing a polymeric anion and its related monomeric anion were passed into the column. The polymeric anion was not adsorbed to the resin to any significant extent and its passed directly through the column, but the monomeric anion was adsorbed. In this way polygalacturonic acid (pectic acid) was separated from galacturonic acid, polymannuronic acid (alginic acid) from mannuronic acid, and polyphosphoric acid from phosphoric acid and pyrophosphoric acid. Similarly, columns of the cation-exchanger, Amberlite IR-120 in its H^+ form were used to separate the protein, clupeine, from the amino acid, arginine.

RICHARDSON (1951) made use of columns of the cation-exchanger, Zeo Karb 215 (H form) and the anion-exchangers, De Acidite B and De Acidite F (OH forms) for the purification of a number of acid dyes, which themselves were not able to enter the pores of the resins. The purification consisted of a deionization of the crude salted out dyes; sodium ions were removed first on a column of the cation-exchange resin and then the inorganic acids on a column of one of the anion-exchanger resins.

The monofunctional, polystyrene type, ion-exchange resins offer many advantages over the earlier resins for ion-exchange chromatography (p. 162). They have further advantages when used as molecular sieves. The polystyrene resins are usually supplied in bead form which means that their surface areas and hence their tendency to adsorb large ions on their surface are kept as low as possible. Furthermore, the polystyrene resins are now available in a wide range of cross-linking and hence pore size.

The use of cross-linked sulphonated polystyrene as an ionic sieve was reported by THOMPSON (1952) and PARTRIDGE (1952). The experiment of THOMPSON (1952) will be described in detail as the method is generally applicable either for the removal of a protein impurity from a mixture of ionic constituents (deproteinization) or for the removal of ionic impurities from a protein preparation (de-salting). In THOMPSON's experiment the protein, lysozyme, was treated with another protein carboxypeptidase, which has the ability to liberate the terminal amino acid (in a stepwise manner) from the carboxyl end of the protein chain. The next step was to separate the liberated amino acids from the large peptide chains still remaining.

An aliquot of the reaction mixture was added to 50 mg. of sulphonated polystyrene (divinyl benzene content, 10—12%) in its H^+ form; this represented a five-fold excess in terms of the capacity of the resin to adsorb the free amino acids liberated by the carboxypeptidase. The resin was in the form of large beads, 15—30 mesh/inch, to reduce the surface adsorption of protein. The resin was shaken mechanically for one hour with the solution, the supernatant was then decanted off and the resin beads washed once or twice with distilled water. The adsorbed amino acids were displaced from the resin with 5 N ammonium hydroxide. It should be mentioned that this type of experiment also can be carried out by a column procedure.

For the complete desalting of a protein it is necessary to use both a cation-exchange resin and an anion-exchange resin and in order to avoid large changes in pH which might have a deleterious effect on the protein, a mixed bed of the two types of resins is used. Thus, JANUS, KERCHINGTON and WARD (1951) produced isoionic gelatin by passage of a 1—10% solution of gelatin through a column of mixed resins (2.5 parts by volume of Amberlite IRA-400 and 1 part of Amberlite-IR-120) at 40° C. However, some proteins may not be amenable to desalting by ion-exchange resins; for instance, TALLAN and STEIN (1953) found that lysozyme was extensively altered on passage through a column of the mixed resins, Amberlite IR-120 and Amberlite IRA-400. Other proteins may be precipitated at their isoionic point. The use of large beads of resin will minimize any configurational changes in a protein as such changes would be expected to occur only when the protein is adsorbed to the surface of the resin.

Complete molecular sieving is of course an alternative process to dialysis, but it appears to have some advantages over dialysis. Separations can be carried out more rapidly and, on a very small scale, more easily. Small ions such as amino acids may be displaced from the resin in high concentration whereas in dialysis they are obtained as dilute solutions. By a suitable choice of a resin, it is possible to make the method selective for anions or cations. Finally, a small ion (e. g. cofactor) which is strongly bound to a macromolecule such as a protein may not be removed by dialysis, but it may be removed by binding to an ion-exchange resin.

In "complete molecular sieving", the size of column is not critical, but the amount of resin in the column should represent at least a 3-fold excess over that required theoretically to adsorb all the ions of low molecular weight.

The maximum rate of flow of the solution through the column will depend to some extent on the relative sizes of the ions to be separated and the degree of cross-linking of the resin. Normally, high flow rates many times those recommended for ion-exchange chromatography may be used. However, if one of the ions being adsorbed by the resin shows a slow rate of exchange, because its size is comparable with the pore size of the resin, then a lower flow rate will be needed for sharp separations.

Separation of ions of intermediate size. If an ion of intermediate size (Mol. Wt. 2000—3000) is present in a mixture containing also ions of high and ions of low molecular weight it is possible by a judicious selection of ion-exchange resins to separate the ion of intermediate size from the other ions by "complete molecular sieving". The mixture is passed first through a column of tightly cross-linked resin, which adsorbs only the small ions and then through a column of lightly cross-linked resin which retains the ion of intermediate size.

Partial molecular sieving. It is difficult to decide from the experimental evidence whether "partial molecular sieving" plays any role in the separations obtainable on columns of lightly cross-linked ion-exchange resins. A slow rate of diffusion of ions into the pores of a resin would mean non-attainment of equilibrium resulting in a broadening of the peaks and poor separations. The fact that peptides containing 20—30 residues may be chromatographed satisfactorily on columns of Dowex 50 (2% D.V.B.) without evidence of undue spreading suggests that these peptides are able to diffuse readily in and out of the pores of the ion-exchange resin (Moore and Stein 1956).

B. Molecular Sieving on Starch.

Lathe and Ruthven (1956) demonstrated that solutes vary greatly in their ability to penetrate the pores of hydrated starch grains and they concluded that the extent of penetration is primarily a function of the molecular size of the solute. Columns of starch in water thus act as molecular sieves and Table 1, compiled from data in the paper by Lathe and Ruthven shows that the order of elution of neutral substances from a starch column is mainly in a decreasing order of molecular weight. With acidic or basic substances, factors other than molecular size also affect their rate of movement on starch columns. Basic substances are retarded and the extent of retardation is dependent on the pH, suggesting that starch can act as a cation-exchange resin.

Table 1. *Elution of neutral molecules from columns of 75 g of potato starch in water*

Solute	Mol. Wt.	Volume of Effluent (ml.)
Amylase	160,000	28.5
Soluble laminarin	3,600	29
Inulin	5,700	30
Cyanocobalamin	1,300	32
Cyclomaltoheptaose	1,134	33
Cyclomaltohexaose	972	34
Raffinose	504	37
Sucrose	342	42
Lactose	342	42
Maltose	342	42
Galactose	180	49
Glucose	180	49
Fructose	180	51

Lathe and Ruthven proposed that as far as neutral substances are concerned columns of starch in water may be regarded as a special type of partition system, based on the molecular sieve effect. The fact that sugars of intermediate size could be separated from the larger amylose and the smaller glucose led LATHE and RUTHVEN to the conclusion that molecules vary in the depth to which they penetrate the starch grains. A possible explanation (LATHE and RUTHVEN, 1956) for varying depths of penetration of molecules of different size is that the amylopectin chains in starch radiate from the centre of the starch granule and the distance between neighbouring chains is very small at the centre and increases towards the periphery. Solute molecules may penetrate only to the depth at which their diameters are equal to the distance between amylopectin chains. As a result the volume of stationary phase in which solute becomes distributed is determined by the molecular size of the solute. Thus with starch, we appear to have true examples of "partial molecular sieving".

Applicability of method. Columns of untreated starch are useful for separating substances in the molecular weight range 100–1000 (LATHE and RUTHVEN 1956; LINDQVIST and STORGARDS 1955). The behaviour of charged molecules is complicadet by ion-exchange effects, but these may be minimized by the selection of an appropriate pH and by the addition of salt. Proteins do not penetrate the grains of untreated starch and they appear with the leading edge of the solvent. However, if maize starch is heated in water, the grains swell and a considerable increase in pore size is obtained (LATHE and RUTHVEN, 1956). Partial separation of proteins with molecular weights below 100,000 is now possible. Thus, LATHE and RUTHVEN (1956) found that with columns of threefold swollen maize starch, cytochrome c (Mol. Wt 13,000) is eluted in 18.8 ml, lactoglobin (38,000) in 16,7 ml, haemoglobin (67,000) in 15.0 ml and globulin (150,000) in 14.2 ml. In order to minimize adsorption effects, elution was carried out in the presence of 0.25 M sodium chloride and usually at a pH 8.5.

Starch gel electrophoresis, immuno-diffusion and immuno-electrophoresis.

Molecular size plays an important role in determining the movement of proteins in starch gel electrophoresis and in immuno-diffusion and immuno-electrophoretic techniques. For example, SMITHIES (1959) found that the relative rates of migration of β-lipoprotein (Mol. Wt 1,300,000) transferin (90,000) and haemoglobin (67,000) in starch gel at pH 8.5 were approximately 0.14, 0.88 and 1.0 whereas in filter paper electrophoresis they migrate at the same rate. Other examples which support the idea of a molecular sieve effect in starch gel electrophoresis are given by SMITHIES (1959).

C. Dextran Gels as Molecular Sieves.

Disadvantages of starch for molecular sieving are its poorly defined composition and its high resistance to flow, and these led to the development of synthetic gels made by cross-linking the polysaccharide, dextran, in such a way that a three dimensional network of polysaccharide chains is obtained. (PORATH and FLODIN, 1959). The network has a non-ionic character, the polar properties being due mainly to the high content of hydroxyl groups. However, a few carboxyl groups (10 μ equivalents/g dry wt) are present. As with the ion-exchange resins, the degree of swelling and hence the average pore size can be varied by increasing or decreasing the amount of cross-linking. Determination of the "water regain" or "weight swelling", defined as the weight of water per gram of dry gel gives a measure of the cross-linking (p. 165).

Three grades of dextran gels (designated Sephedex G-25, G-50 and G-75) are now available commercially from Pharmacia, Uppsala, Sweden. Actual analytical values for water regain and dry sieve analysis are provided with each batch of gel.

G-25 will completely exclude neutral polysaccharides with molecular weights greater than 3,500—4,500, G-50 will exclude those with molecular weights above 8,000—10,000 and G-75 those with molecular weights above 40,000—50,000.

The process of separating molecules of different size by passage through a column of a dextran gel is termed gel filtration (PORATH and FLODIN, 1959).

1. Theory of Gel Filtration.

A theory of gel filtration has been presented by GELOTTE (1960). A column of dextran gel contains two kinds of aqueous phase, one within the gel grains (inner phase) and one surrounding the grains (outer phase). If the volume of the inner phase is denoted by Vi, the volume of the outer phase by Vo and the partition coefficient for a substance between these two phases as K_D, then the total volume of liquid which flows from the column from the time of application of the substance until the maximum concentration of the substance appears in the effluent (Ve) is given by:

$$Ve = Vo + K_D Vi$$

A substance submitted to gel filtration may be characterized by its K_D value:

$$K_D = \frac{Ve - Vo}{Vi}$$

In terms of the R_F concept (p. 171):

$$K_D = \frac{Vo}{Vi}\left(\frac{1}{R_F} - 1\right)$$

For large molecules which are completely excluded from the inner phase $Ve = Vo$ and $K_D = 0$. For substances which can diffuse freely into the inner phase, K_D should have a value of 1, but in practice it is found that a K_D value of about 0.8 indicates non-restricted diffusion (GELOTTE, 1960). This discrepancy between theoretical and observed values can be explained by assuming that part of the water of hydration is firmly bound to the polysaccharide framework and is inaccessible to the solute molecules (GELOTTE, 1960).

Vo is determined experimentally by passing through the column a solution of a coloured substance which is excluded from the inner phase. Haemoglobin may be used in salt or water solutions and colloidal india ink in water solutions. Vo may also be calculated from the equation.

$$Vo = Vt - \frac{a(1 + W_R)}{d}$$

where Vt is the total bed volume,
W_R is the water regain, d the wet density and
a the dry weight of the gel.

Vi is calculated from the water regain, since $Vi = a W_R$.

2. Preparation and Operating of Columns.

For good results it is important to prepare well-packed columns. The dry gel is allowed to swell in .05 M sodium chloride for half an hour and the fine material is removed by repeated sedimentations and decantations. Well-packed columns may be prepared by the procedures described by FLODIN and KUPKE (1956) and PORATH (1956) for packing cellulose columns for zone electrophoresis. A glass tube (Fig. 2) approximately 1 meter long with a large funnel at its upper end is connected to the chromatographic tube either by a ground glass joint or through a rubber stopper; for good packing both the chromatographic tube (a) and the sedimentation tube (b) should be vertical. The tubes are partly filled with water, which has been de-aerated by evacuation under reduced pressure and then the suspension of the dextran gel is poured into the funnel and stirred mechanically. When a sediment layer of 1—2 cm has formed, the stopcock at the bottom of the chromatographic tube is opened carefully and water is allowed to flow at a slow rate. The rising boundary of packed gel should be distinctly visible and horizontal, otherwise the column will not be packed uniformly. When all the gel has settled, a circular filter paper is placed on the surface of the gel column to prevent disturbances. The column is washed with distilled water or an electrolyte solution, care being taken to ensure that the water does not drain below the top surface of the column.

The procedure for operating the column are similar to those described for ion-exchange chromatography.

Size of columns. The size of column will depend on the particular problem. The diameter to height ratio is not critical, but values of 1—10 to 4—10 are preferable. The capacity of a column in gel filtration is large compared with the capacities of the usual types of partition columns. On columns, 3 × 60 to 70 cm, PORATH (1960) applied quantities of material of the order of 50—100 mg, but in many separations much higher loadings may be possible. Thus in one experiment PORATH applied over a gram of material to a column of gel, 3 × 65 cm. The sample is usually added to a column of this size in a volume of 5—10 ml, although sometimes up to 100 ml may be applied. The filtration rate will depend on the particular separation; for a column of 3 cm diameter, rates of the order of 25—100 ml/hr are usual.

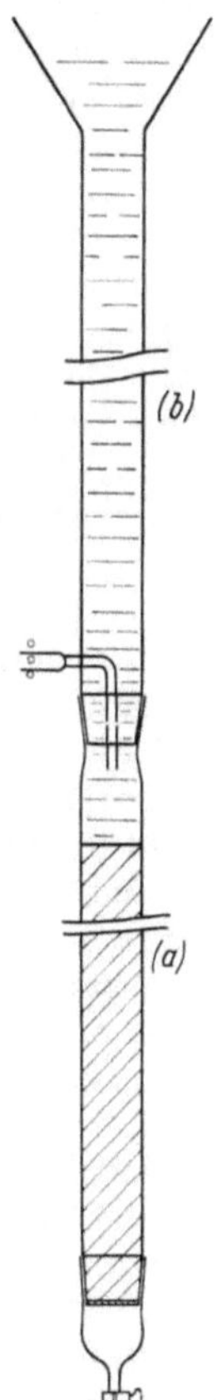

Fig. 2. Apparatus for packing dextran gel columns (After FLODIN and KUPKE, 1956; PORATH, 1956).

3. Some Applications of Dextran Gels as Molecular Sieves.

The studies of PORATH and FLODIN (1959), PORATH (1959) and PORATH (1960) clearly show that dextran gels may be used as molecular sieves to separate molecules of different size.

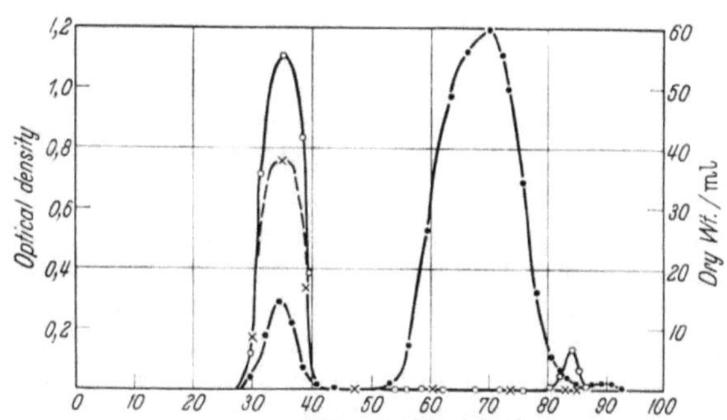

Fig. 3. Separation of ammonium sulphate from serum proteins by gel filtration through a column of dextran gel, 2.5 × 17.5 cm. Fractions were analysed by the method of FOLIN-LOWRY (---x---), absorption at 280 mμ (—o—) and weight distribution (—·—). (From PORATH and FLODIN, 1959).

PORATH and FLODIN (1959) separated glucose and two soluble dextran fractions of mean molecular weight 1,000 and 20,000 respectively on a gel with a water regain of 2.5 g water/g.

Proteins may be desalted by passage through a column of dextran gel. Fig. 3, which is taken from the paper by PORATH and FLODIN (1959) shows the separation of ammonium sulphate from serum proteins. Elution was carried out with water at a flow rate of 25 ml/hr, but the flow rate could have been increased many-fold without significant spreading of the peaks. The main advantage of gel filtration over dialysis for desalting is the speed with which it can be performed.

Gel filtration may be used to change a protein from one buffer to another. The column is first equilibrated with the new buffer; the protein solution is applied to the column in the old buffer and the protein then eluted with the new buffer. When a volume, Vo has passed into the column protein will begin to appear in the effluent, whereas the old buffer will not appear until a volume, $Vo + Vi$, has passed.

Protein solutions may be concentrated by treatment with dry dextran gel (GELOTTE and KRANTZ, 1959; FLODIN, GELOTTE and PORATH, 1960). The gel takes up water corresponding to its water regain, leaving the protein in the external solution. At the same time, salts and solutes of low molecular weight are removed to an extent that the concentration and pH remains approximately constant. For protein concentration it is recommended that a gel in the size range 50—100 mesh/inch be used, so that the gel can be removed easily by centrifugation; with large volumes of liquid, a basket centrifuge is convenient and for small volumes, filter adaptors in centrifuge tubes may be used. Recoveries of protein were of the order of 90% or more for 10—20 fold concentration, but in the case of cytochrome *c* and a phosphatase from *E. coli*, a large amount of the activity was removed by the gel and it could be recovered only after extensive washing (FLODIN, GELOTTE and PORATH, 1960). Dextran gel may be regenerated for further use by washing with water followed by 2 suspensions in ethanol and drying at 80° C.

A good example of the effect of a variation in cross-linking of the gel is provided by PORATH (1960). Whereas cytochrome *c* (Mol Wt 13,000) and phycocyanin (270,000) move at the same rate on a tightly cross-linked gel (water regain 1.9 g/g), they are easily separated on a loosely cross-linked gel (water regain 5.5).

Gel filtration is a convenient method for removing a proteolytic enzyme from the products of its action and at the same time a separation of some of the peptides in the mixture may be obtained. For example, PORATH (1960) passed a peptic digest of cytochrome *c* through a column of a loosely cross-linked gel and obtained a separation of the coloured ferriporphyrin peptide both from the pepsin and from the bulk of the uncoloured peptides. Other examples of the use of dextran gels for purifying peptides are provided by LINDNER, ELMQVIST and PORATH (1959) and by HILL and KONIGSBERG (1960).

KISLIUK (1960) demonstrated that dextran gels may be useful for removing bound cofactors from enzyme preparations. On a lightly cross-linked gel, KISLIUK (1960) was able to remove completely the folic acid cofactor from an enzyme preparation from *E. Coli*, whereas a similar dialysed preparation retained 54% of the activity. The cofactor was eluted from the gel in a relatively small volume.

Scope of gel filtration. Dextran gel filtration should have wide application for separating water-soluble molecules with molecular weights up to 50,000. Solutes larger than this are not retained on any of the gels available commercially at present and they appear with the leading edge of the solvent.

Molecules which enter the gel grains may interact with the gel material and this leads to chromatographic separations. A discussion of the chromatographic aspect of gel filtration is beyond the scope of this article, although it should be mentioned that both adsorption and ion-exchange play a role. The chromatographic peaks are usually symmetrical, indicating reversible interaction of solute and gel (PORATH, 1960). If necessary, these interactions may be minimized by the proper choice of pH of the solution and by equilibrating the column with an electrolyte instead of water (PORATH, 1960; GELOTTE, 1960).

References.

BARRER, R. L.: Quart. Rev. Chem. Soc. **3**, 293 (1949).

CERNESCU, N.: Dissertation, 661 Eidg. Technische Hochschule, Zürich (1933).

DUEUL, H., J. SOLMS and L. ANYAS-WEISZ: Helv. Chim. Acta. **33**, 2171 (1950).

FLODIN, P., B. GELOTTE and J. PORATH: Nature (Lond.) **188**, 493 (1960). — FLODIN, P., and D. W. KUPKE: Biochim. Biophys. Acta **21**, 368 (1956).

GELOTTE, B.: J. Chromatography **3**, 330 (1960). — GELOTTE, B., and A. B. KRANTZ: Acta Chem. Scand. **13**, 2127 (1959).

HILL, R. J., and W. KONIGSBERG: J. biol. Chem. **235**, PC 21 (1960).

JANUS, J. W., A. W. KERCHINGTON and A. G. WARD: Research (Lond.) **4**, 247 (1951).

KISLIUK, R. L.: Biochim. Biophys. Acta **40**, 531 (1960). — KRESSMAN, T. R. E.: J. Phys. Chem. **56**, 118 (1952). — KUNIN, R.: Anal. Chem. **21**, 87 (1949).

LATHE, G. H., and C. R. J. RUTHVEN: Biochem. J. **62**, 665 (1956). — LINDNER, E. B., A. ELMQVIST and J. PORATH: Nature (Lond.) **184**, 1565 (1959). — LINDQVIST, B., and T. STORGÅRDS: Nature (Lond.) **175**, 511 (1955).

MOORE, S., and W. H. STEIN: Adv. Protein Chem. **11**, 192 (1956).

PARTRIDGE, S. M.: Nature (Lond.) **169**, 497 (1952). — PORATH, J.: Biochim. Biophys. Acta **22**, 151 (1956). — PORATH, J.: Clin. Chim. Acta **4**, 776 (1959). — PORATH, J.: Biochim. Biophys. Acta **39**, 193 (1960). — PORATH, J., and P. FLODIN: Nature (Lond.) **183**, 1657 (1959).

RICHARDSON, R. W.: Nature (Lond.) **164**, 916 (1949). — RICHARDSON, R. W.: J. chem. Soc. 901 (1951).

SMITHIES, O.: Adv. Protein Chem. **14**, 65 (1959).

TALLAN, H. H., and W. H. STEIN: J. biol. Chem. **200**, 507 (1953). — THOMPSON, A. R.: Nature (Lond.) **169**, 496 (1952).

Dünnschicht-Chromatographie.

Von

Egon Stahl.

Mit 16 Abbildungen.

Zur Auftrennung und Charakterisierung von Naturstoffgemischen sind Methoden, die eine schnelle Identifizierung kleinster Mengen zulassen, von besonderem Wert. Einen Fortschritt brachte auf diesem Gebiet die Dünnschicht-Chromatographie, die im Prinzip durch den Übergang von der geschlossenen zur „offenen" Säule gekennzeichnet ist. Bereits 1938 wurde dieser Weg von Ismailow und Schraiber beschritten, die auf Objektträgern ein Adsorbens in dünner Schicht auftrugen und hierauf eine Reihe von Arzneipflanzenauszügen in Art

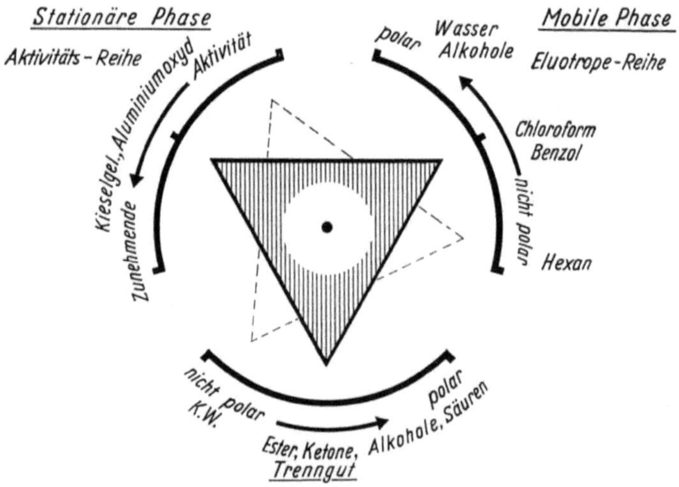

Abb. 1. Zusammenhang (grob-schematisch) zwischen den 3 variablen Hauptgrößen bei der Adsorptionschromatographie. Das schraffierte Dreieck denke man sich zunächst einmal im Schwerpunkt drehbar. Ist eine Größe gegeben und somit festgelegt, so stelle man eine Dreieckspitze hierauf ein und verlege dann den Drehpunkt in diese Spitze. Die beiden anderen ergeben die anwendbaren Möglichkeiten. Sind zwei Größen vorgegeben und die zwei Dreieckspitzen hierauf eingestellt, so zeigt die noch freie in den zu wählenden Bereich.

der zehn Jahre später durch Rutter[1] beschriebenen Rundfiltertechnik trennten. Wenn man von der durch Kirchner u. Mitarb. induzierten, gelegentlichen Anwendung auf dem Gebiet der ätherischen Öle absieht, geriet diese Methode wieder in Vergessenheit. Erst 20 Jahre später, nachdem die verfahrenstechnischen Schwierigkeiten gelöst, die Bedingungen standardisiert und die Anwendungsbreite gezeigt wurde, gewann die Methode unter der Bezeichnung Dünnschicht-Chromatographie (*Thin-layer Chromatography* = TLC) schnell an Bedeutung (Stahl 1958—1961).

[1] Siehe Bd. I dieses Handbuchs, S. 95—148.

Für den Einsatz des Verfahrens sprechen folgende Gründe: Die Trennzeit ist kurz; sie liegt je nach dem Elutionsmittel zwischen 20 und 100 min. Im Vergleich zu den TSWETTschen-Säulen[1] ist die Trennschärfe wesentlich besser, und man hat die Möglichkeit, mit den aggressivsten Sprühreagentien den Trennerfolg sofort festzustellen. Auf den fluorescenzfreien, rein anorganischen Sorptionsschichten liegt einerseits die untere Nachweisgrenze unter der Papierchromatographie, und andererseits lassen sich oft noch Milligramm-Mengen trennen. Der Einsatzschwerpunkt der Dünnschicht-Chromatographie lag zunächst im Bereich der lipophilen Stoffgemische. Der Anwendungsbereich läßt sich jedoch bei Verwendung polarer Elutionsgemische bis in den stark hydrophilen Stoffbereich ausdehnen. Ob eine Trennung der Gemische auf dem Wege der Adsorption und/oder der Verteilung zustande kommt, hängt von dem verwendeten Sorptions- und Elutionsmittel ab. Demzufolge gelten die bereits behandelten theoretischen Grundlagen[1]. Die Zusammenhänge und gleitenden Übergänge sind in der Abb. 1 zusammengefaßt.

A. Methode und Geräte.

I. Die Herstellung dünner Sorptionsschichten.

Die erforderlichen dünnen Sorptionsschichten lassen sich auf indifferenten Trägerscheiben nach verschiedenen Verfahren aufbringen. Da eine Reihe von Faktoren, u. a. auch die Schichtdicke, einen Einfluß auf die Trennung hat, ist

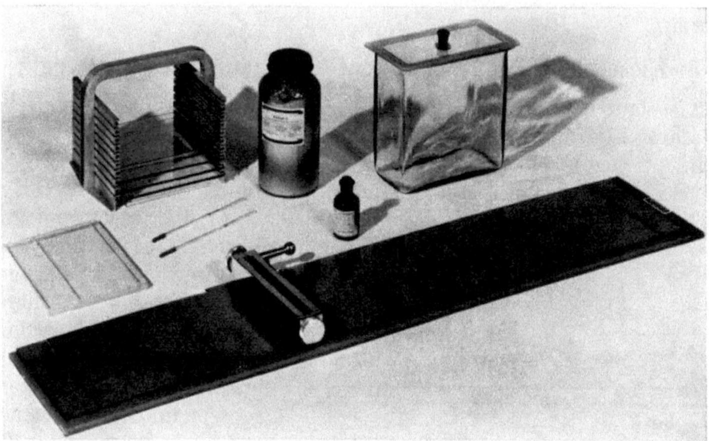

Abb. 2. Grundausrüstung zur Dünnschicht-Chromatographie (DESAGA, Nr. 600). Im Vordergrund die Arbeitsschablone mit aufgelegten Scheiben und das Streichgerät. Links dahinter die durchsichtige Beschriftungsschablone, Pipetten und Trockengestell. In der Mitte: Testgemisch und Kieselgel G; rechts daneben die Trennkammer (Aufnahme DESAGA).

es zweckmäßig, unter standardisierten Bedingungen zu arbeiten. Eine rationelle Herstellung der etwa 250 μ dünnen Schichten ist nach folgendem Verfahren möglich:

Auf einer 110 cm langen Schablone werden planparallele Glasscheiben (200 × 200 mm) oder Streifen (50—100 × 200 mm) genau gleicher Stärke zu einer Bahn ausgelegt und mit dem Dünnschicht-Streicher die dünnflüssige Sorptionsmasse in einem Zug aufgetragen (Abb. 2,3). Zum Trocknen werden die beschichteten Scheiben in einem Leichtmetall-Trockengestell gestapelt und in horizontaler Lage

[1] Siehe Bd. I dieses Handbuchs, S. 95—148.

etwa 30 min bei 110—140° getrocknet. Danach sind sie gebrauchsfertig und können im gleichen Gestell in einem Exsiccator (30 cm ⌀) über Blaugel aufbewahrt werden. Die zu einer „Grundausrüstung" (Abb. 2) zusammengefaßten Standardgeräte werden von der Firma DESAGA, Heidelberg, Hauptstr. 60, hergestellt.

II. Sorptionsmittel und Bereitung der Streichmasse.

Die zur Herstellung dünner Schichten verwendeten Sorptionsmittel müssen in ihrer Korngrößenzusammensetzung, Oberflächenstruktur und Haftfestigkeit der Methode angepaßt sein. Diesen Erfordernissen entsprechen zumeist nur die speziell für diesen Zweck hergestellten und standardisierten Sorbentien wie z. B.:

1. Kieselgel G für Dünnschicht-Chromatographie „Merck".

Die Kieselgel G-Schichten haben eine BROCKMANNsche Aktivitätsstufe, die nahe bei II liegt. Auf diesen Schichten lassen sich die meisten adsorptionschromatographischen und auch die bisher auf formamidimprägnierten Papieren durchgeführten verteilungschromatographischen Trennprobleme lösen.

2. Aluminiumoxid G für Dünnschicht-Chromatographie „Merck".

Die schwach alkalischen Aluminiumoxid G-Schichten eignen sich vor allem zur Auftrennung basischer Stoffgemische (Alkaloide, Amine usw.). Außerdem lassen sich alle bisher für Aluminiumoxidsäulen beschriebene Trennungen hierauf durchführen.

3. Kieselgur G für Dünnschicht-Chromatographie „Merck".

Auf Kieselgur G-Schichten lassen sich verteilungschromatographische Trennungen stark hydrophiler (z. B. Zucker) und amphoterer Substanzgemische erreichen.

Zum Beschichten von fünf 20×20 cm-Scheiben (Fassungsvermögen der Arbeitsschablone) werden 30 g Sorbens (1, 2 oder 3) in einer trockenen Porzellanreibschale (etwa 10 cm ⌀) mit der doppelten Gewichtsmenge destillierten Wassers zu einer gleichmäßigen Masse verrührt. Die dünnflüssige Suspension wird dann sofort in das Streichgerät gegossen und ausgestrichen. Man kann auch das Sorbens direkt in den Dünnschicht-Streicher einfüllen, das Wasser zugeben, die Hülse (Abb. 3a) um 90° drehen und 30 sec kräftig durchschütteln.

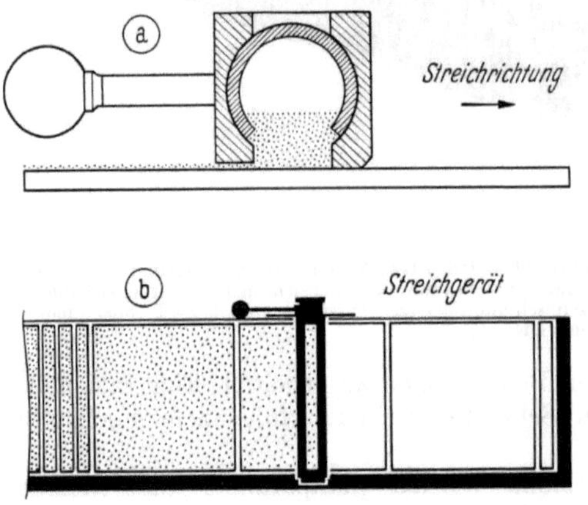

Abb. 3. a) Arbeitsweise des Dünnschichtstreichgerätes. b) Arbeitsschablone mit aufgelegten Glasplatten. Die links vom Streichgerät liegenden Glasplatten sind bereits beschichtet.

Verwendet man zur Herstellung der Streichmasse statt Wasser verdünnte Säuren (z. B. 0,5 n-Oxalsäure), Laugen, Puffer usw., so erhält man Schichten, auf denen sich spezielle Trennprobleme lösen lassen (STAHL 1959a). Eine „*Phasenumkehr*" der Schicht erreicht man durch Eintauchen der beschichteten Platte z. B. in 5—10%ige Silicon- oder Paraffinlösungen und anschließendes

Entfernen des leichtsiedenden Lösungsmittels. Neben den beschriebenen Sorptionsmitteln (1—3) kann man Schichten, z. B. aus Calcium- und/oder Magnesiumphosphaten, -oxyden oder -silicaten, Cellulosepulver[1], Stärke, Zucker und Perlon-(Polyamid)-pulver, Harnstoff usw. herstellen.

III. Auftragen der Substanzen und Auswahl des Elutionsmittels.

Um die Sorptionsschicht beim Auftragen und Beschriften nicht zu beschädigen, verwendet man eine sog. Beschriftungsschablone (Abb. 2). Zunächst markiert man sich 15 mm vom unteren Rand durch Einstiche mit einem spitzen, harten Bleistift oder einer Nadel die Startpunkte im Abstand von 15—20 mm, außerdem die Laufstrecke (zumeist 100 mm). Auf die so markierten Startpunkte trägt man mit einer 10 mm³-Blutzuckerpipette die gelösten Substanzen wie bei der Papierchromatographie auf. Es ist vorteilhaft, sich Lösungen mit einem Gehalt von 5, 1 und 0,1 % des zu untersuchenden Gemischs herzustellen und hiervon zwischen 1 und 5 mm³ aufzutragen. Der Durchmesser der Startpunkte sollte etwa gleich groß sein und 5 mm nicht überschreiten. Zur genauen Festlegung der Bedingungen sollte man immer ein sog. Testgemisch, das aus mindestens 2 verschiedenen, leicht zugänglichen Farbstoffen besteht, mitchromatographieren. Bei der Auswahl des Elutionsmittels richtet man sich zunächst unter Berücksichtigung des in Abb. 1 gegebenen Schemas nach der eluotropen, bzw. mixotropen Reihe (s. Bd. I, S. 88 u. 109). Bei Substanzgemischen unbekannter Zusammensetzung kann man zur Orientierung mit Chloroform oder Benzol als Elutionsmittel beginnen und dann, je nach der Lage der Zonen, ein stärker oder schwächer eluierendes Solvens verwenden. Erfahrungsgemäß läßt sich zuweilen eine Trennung verbessern, wenn man Gemische von 2 bis höchstens 3 Lösungsmitteln unterschiedlicher Polarität verwendet. Es ist anzunehmen, daß es hierdurch zur Ausbildung eines mehr oder weniger kontinuierlichen Polaritätsgefälles in der Sorptionsschicht kommt.

IV. Trennkammer und deren Sättigungszustand.

Nach dem Auftragen der Substanzen wird die Platte in eine Trennkammer gestellt, deren Boden 5—8 mm hoch mit dem Elutionsmittel beschickt ist. Man

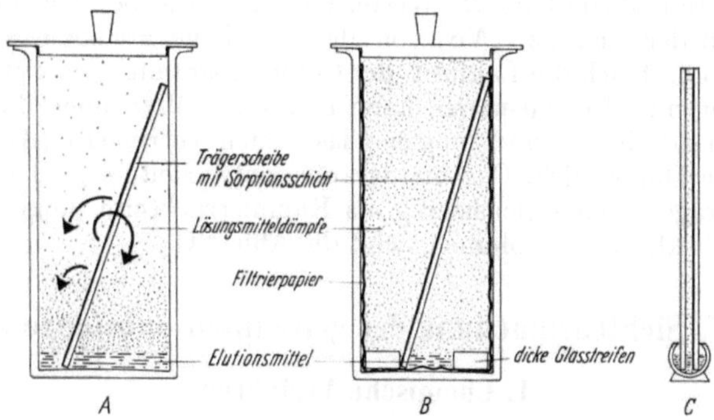

Abb. 4. Trennkammern und Sättigung. a) Trennkammer mit normaler Sättigung (= NS). Die von der Sorptionsschicht weggehenden Pfeile sollen das Abdampfen des Lösungsmittels symbolisieren, die Punktierung die Dampfdichte. b) Trennkammer durch Filtrierpapiereinlage mit dem Lösungsmittel gesättigt (= KS). c) S-Trennkammer und Spezial-Trog nach STAHL für 20—100 cm breite Scheiben.

[1] Die Fa. Macherey, Nagel & Co. Düren stellt 5 verschiedenartige Cellulosepulver, z. B. acetylierte-, Carboxymethyl-, Diäthylaminoäthyl-ECTEOLA-Cellulose mit und ohne Gipszusatz speziell für die Dünnschicht-Chromatographie her.

beobachtet, daß in den ersten Zentimetern die Lösungsmittelfront schnell steigt und im oberen Teil langsamer wandert. Die Wanderungsgeschwindigkeit ist einmal von der Polarität der Elutionsmittel abhängig und zum anderen vom Kammervolumen. Auf den üblichen Schichten wandern die nicht oder wenig polaren Lösungsmittel schnell, stark polare — wie z. B. Alkohole und Wasser — langsam. Der Einfluß des Volumens der Trennkammer auf die Wanderungsgeschwindigkeit erklärt sich aus dem Befund, daß ein nicht unerheblicher Anteil des bereits in der Sorptionsschicht aufgestiegenen Lösungsmittels wieder verdampft. Dieser Abdampfeffekt ist im Bereich der Lösungsmittelfront am stärksten ausgeprägt und nimmt mit der Steighöhe zu (Abb. 4 A). Je größer, bzw. undichter die Kammer ist, desto mehr muß abdampfen, bis es zu einer völligen Sättigung der Kammer kommt. In der Regel ist dieser Vorgang unerwünscht; die Ergebnisse sind schlecht reproduzierbar, und es tritt bei Elutionsgemischen ein unerwünschter Randeffekt auf. Man chromatographiert daher vorteilhaft in einer völlig mit Lösungsmitteldämpfen gesättigten Kammeratmosphäre (STAHL 1959a). Das läßt sich leicht bewerkstelligen, indem man an den beiden Innenwandbreitseiten der Kammer einen etwa 15 cm breiten, mit dem Elutionsgemisch getränkten Filterpapierstreifen adhäsiv so fixiert, daß er noch in das Solvens eintaucht (Abbildung 4 B). Bei einer derartigen *Kammersättigung* (früher zur Abgrenzung gegen das Klimatisieren als „Kammerübersättigung" bezeichnet) erhält man gut reproduzierbare R_F-Werte, die nicht wie bei dem normalen Sättigungszustand der Kammer (*NS*) von der Steighöhe abhängig sind (Abb. 5). Außerdem verkürzt sich die Laufzeit der Chromatogramme um etwa ein Drittel.

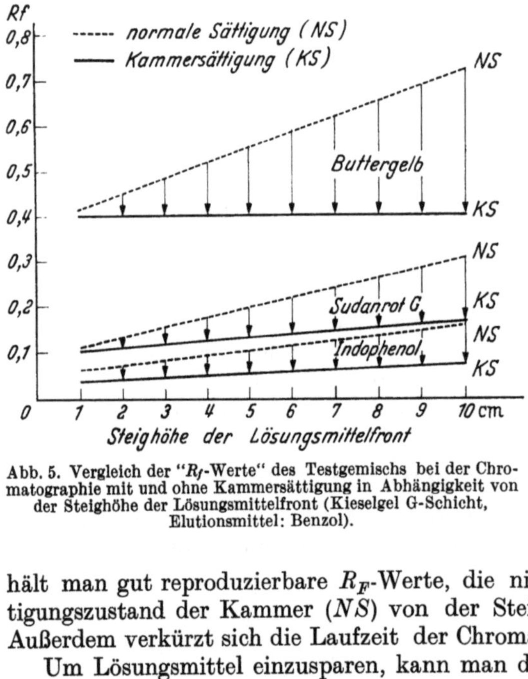

Abb. 5. Vergleich der "R_f-Werte" des Testgemischs bei der Chromatographie mit und ohne Kammersättigung in Abhängigkeit von der Steighöhe der Lösungsmittelfront (Kieselgel G-Schicht, Elutionsmittel: Benzol).

Um Lösungsmittel einzusparen, kann man das Füllvolumen durch Einlagen von entsprechend dicken und langen Glasstreifen verringern (Abb. 4 B). Eine speziell für die Dünnschicht-Chromatographie entwickelte sog. S-Trennkammer, bei der die Trägerscheibe gleichzeitig als Kammerrückseite fungiert, und einen Spezialtrog mit kleinstem Volumen, zeigt die Abb. 4 C.

V. Sichtbarmachung der getrennten Substanzen.

1. Chemische Verfahren.

Die rein anorganischen, fluorescenzfreien Sorptionsschichten lassen zur Sichtbarmachung farbloser Substanzen die Verwendung aggressiver Sprühreagentien und ein anschließendes Erhitzen bis zur Verkohlung zu. Man wird bei Naturstoffgemischen unbekannter Zusammensetzung zunächst mit Universalreagenzien beginnen, um möglichst alle darin enthaltenen Verbindungen zu erkennen. Hierzu eignen sich z. B. konzentrierte Mineralsäuren, denen man zur Farbverstärkung

0,5—1% eines Aldehyds (z. B. Formaldehyd, Anisaldehyd, Vanillin, Furfural oder p-Dimethylaminobenzaldehyd) zusetzen kann. Bewährt haben sich ferner etwa 20%ige Antimon-III-(bzw. V-)chloridlösungen in wasserfreiem Chloroform (bzw. Tetrachlorkohlenstoff) und 0,5%ige Jod-Chloroformlösung. Weitere Möglichkeiten ergeben die von LEROSEN u. Mitarb. zur Kennzeichnung von Adsorptionszonen zusammengestellten Reagentien (Mod. Meth. Pfl.-Analyse Bd. I, S. 104). Außerdem ergeben die in der Papierchromatographie erprobten Sprühreagentien nur in

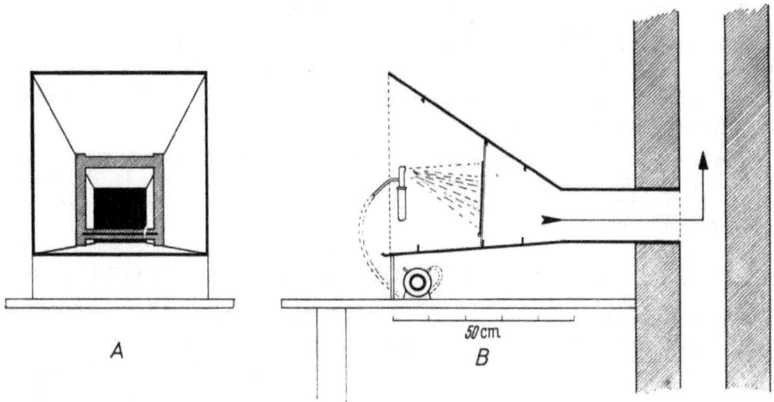

Abb. 6. Trovidur-Sprühkabine an den Abzugskamin angeschlossen. a) Vorderansicht (schraffiert: Halterahmen zum Einstellen der Chromatogramme); b) Seitenansicht (Schnitt), zusätzlich ist die Druckluftsprühvorrichtung eingezeichnet.

Ausnahmefällen auf anorganischen Schichten schlechtere Ergebnisse. Es zeigte sich, daß dann zumeist die Cellulose (bzw. deren Abbauprodukte) an der Farbreaktion mitbeteiligt ist. In diesen Fällen setzt man dem Reagens Cellulosehydrolysat, bzw. Zucker zu.

Das Aufsprühen erfolgt am besten mit Druckluft und sehr fein vernebelnden Sprühern. Ein gut ziehender Abzug ist bei den aggressiven Sprühreagentien unbedingt erforderlich. Bewährt haben sich kleine, ganz aus Trovidur geschweißte, trichterförmige *Sprühkabinen*, die direkt mit dem Abzugskamin oder dem Ventilator verbunden sind (Abb. 6).

2. Physikalische Verfahren.

a) Aktivitätsmessung.

Eine Auswertung von Dünnschicht-Chromatogrammen, auf denen radioaktive Verbindungen getrennt wurden, ist — wie MANGOLD (1959) und GRISEBACH zeigten — möglich. Nach BILLEK, KRATZL und PUSCHMANN bereitet auf den relativ festhaftenden Sorptionsschichten weder die Autoradiographie noch die Aktivitätsmessung unter einem Endfensterzählrohr Schwierigkeiten. Im letzteren Fall ist eine Anpassung an die Fensterform durch bandförmiges Auftragen des zu untersuchenden Substanzgemisches zweckmäßig.

b) Fluorescenzverfahren.

Bevor man chemische oder biologische Nachweisverfahren einsetzt, sollte man das Chromatogramm sowohl im lang- als auch im kurzwelligen ultravioletten Licht betrachten und die Zonen mit einer Eigenfluorescenz markieren. Neben dem Photoprintverfahren ist analog dem von BROCKMANN u. Mitarb. für Aluminiumoxid-Säulen vorgeschlagenen Verfahren (B. I, S. 103) ein Zusatz von Leuchtfarbstoffen zum Sorptionsmittel möglich. GÄNSHIRT u. Mitarb. haben diese Methode für die Dünnschicht-Chromatographie ausgearbeitet und schlagen einen 2%igen Zusatz des Leuchtstoffes ZS-Super „*Riedel de Haën*" zum Kieselgel G vor. Im kurzwelligen ultravioletten Licht (Schwerpunkt 254 mμ) zeigen diese Schichten eine starke Grün-Fluorescenz. Die getrennten Substanzen treten hierauf als dunkle Zonen hervor.

3. Biologische Verfahren.

Zur Erkennung der Wuchsstoffwirksamkeit von einfachen Indolkörpern, sowie von Pflanzen- und Tierextrakten wurden nach der Chromatographie die abgeschabten Zonen direkt auf Agarwürfelchen gestreut und im *Avena*-Krümmungstest die Aktivität bestimmt (KALDEWEY und STAHL).

Zum Nachweis der insecticiden Wirkung von Chromatogrammzonen wurde der entsprechende Bereich abgeschabt (Abb. 12) und in kleine Mikroküvetten (z. B. Hohlschliffobjektträger u. ä.) gestreut, in denen die gegen Insecticide sehr empfindlichen *Aedes*-Larven schwammen (STAHL 1960a).

VI. Untere Erfassungsgrenze im Vergleich zur Papierchromatographie.

Die untere Erfassungsgrenze der getrennten Substanzen ist einerseits von der Empfindlichkeit der verwendeten Nachweismethode und andererseits von der Konzentration (Verteilung) der Substanz pro Flächeneinheit (richtiger: Volumeneinheit) abhängig. Vergleichsversuche ergaben, daß bei gleicher Auftragemenge auf den Kieselgel G-Schichten nach der Chromatographie die Konzentration der Substanzen pro Flächeneinheit wesentlich höher ist als auf Cellulosepapier (Abb. 7). Das ist eine zwanglose Erklärung für den Befund, daß die Nachweisempfindlichkeit auf den Dünnschicht-Chromatogrammen um den Faktor 10—100 höher liegt als auf dem Papier. Hinzu kommt, daß auf anorganischen Schichten ohne Störungen kräftigere Reaktionsbedingungen angewendet werden können.

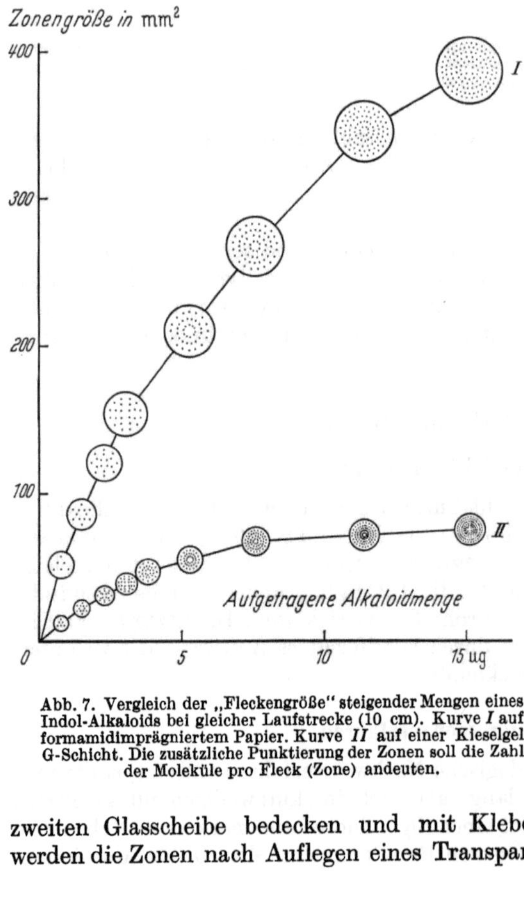

Abb. 7. Vergleich der „Fleckengröße" steigender Mengen eines Indol-Alkaloids bei gleicher Laufstrecke (10 cm). Kurve *I* auf formamidimprägniertem Papier. Kurve *II* auf einer Kieselgel G-Schicht. Die zusätzliche Punktierung der Zonen soll die Zahl der Moleküle pro Fleck (Zone) andeuten.

VII. Dokumentation und allgemeine Auswertung.

Das fertige, auf der Trägerscheibe befindliche Chromatogramm wird man nur selten zur Aufbewahrung mit einer zweiten Glasscheibe bedecken und mit Klebeband einrahmen. In der Regel werden die Zonen nach Auflegen eines Transparentpapiers im Durchlicht nachgezeichnet und hierzu gleich die den Reaktionsfarben entsprechenden Farbstifte verwendet. In Spezialfällen kann man auf die Schicht auch eine

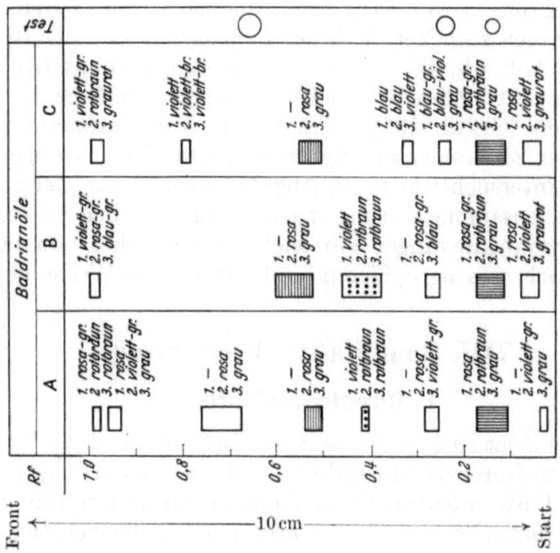

Abb. 8b. Darstellung des Chromatogramms 8a mit Angabe der Reaktionsfarben. *1* = einige Minuten nach dem Besprühen (ohne Erhitzen); *2* = nach dem Erhitzen (10 min 100°); *3* = 24 Std. später.

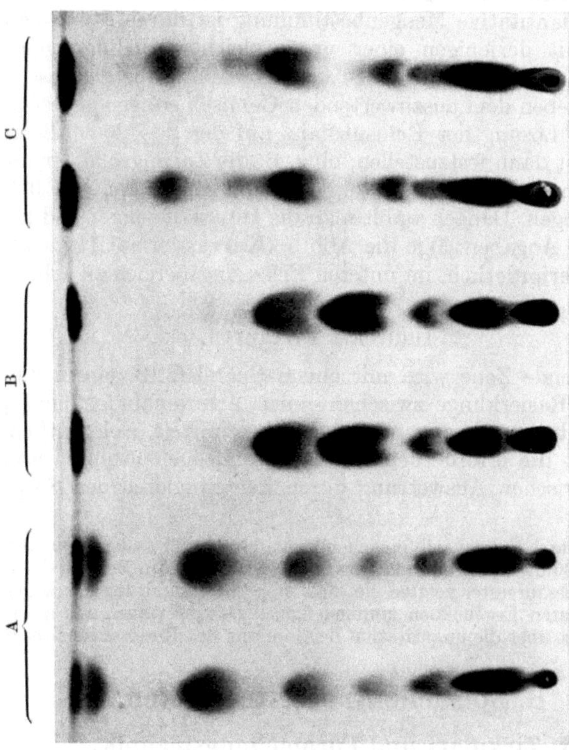

Abb. 8a. Fotografie eines Dünnschicht-Chromatogramms verschiedener ätherischer Baldrianöle (A, B, C) auf einer Kieselgel G-Schicht, Elutionsmittel Benzol (NS). Von den ätherischen Ölen wurden je 400 µg (10%ig gelöst) aufgetragen. Die Sichtbarmachung erfolgte durch Aufsprühen von Antimon-III-chloridlösung und Erhitzen auf 100°.

durchsichtige Klebefolie auflegen, diese gleichmäßig und fest mit einem planen Stück Holz andrücken und dann mit der Folie die daran haftende Schicht abziehen. Außer dem direkten Weg der Herstellung von „Kontaktabzügen" wird zur Dokumentation häufig die Fotografie mit Schwarzweiß- oder Farbfilm eingesetzt. Man verwendet hierzu eine Durchlicht-, bzw. eine Kombination mit der Auflichtbeleuchtung (Abb. 8a). Eine übersichtliche Darstellungsmethode von Chromatogrammen zum Vergleich der Zonengröße und der Reaktionsfarben ergibt sich aus der Abb. 8b. Die Zonen des Chromatogramms (Abb. 8a) wurden hierzu ausplanimetriert und mit gleicher Grundkantenlänge in ein R_F-Wert-Schema eingetragen (STAHL 1958). Trägt man zur Chromatographie genaue Mengen eines in der Zusammensetzung unbekannten Naturstoffgemischs auf, so kann ein späterer Untersucher hiernach Angaben über die Art der Substanzen und die mengenmäßige Zusammensetzung machen. Erforderlich ist es, zur genauen Festlegung der Trennbedingungen ein leicht zugängliches, am besten farbiges Testgemisch mitzuchromatographieren und dessen R_F-Werte ebenfalls einzutragen (Abb. 8b).

VIII. Quantitative Auswertung.

1. Direkte Verfahren.

a) Nach MANGOLD ist eine *direkte photometrische Auswertung* der Dünnschicht-Chromatogramme im Durchstrahlverfahren, z. B. mit dem „Photovolt-Electronic-Densitometer" Mod. 525 möglich. Es sind nur kleine Änderungen an der Haltevorrichtung des Gerätes nötig; ein Transparentmachen der Sorptionsschicht, etwa mit einer Paraffin-Äther-Lösung, ist nicht erforderlich.

b) Eine halbquantitative Mengenbestimmung ist durch *Vergleich der Zonen-(Flecken-)größe* mit derjenigen einer unter gleichen Bedingungen chromatographierten bekannten Substanzmenge möglich. Man gabelt gewissermaßen ein und trägt hierzu neben dem auszuwertenden Gemisch steigende Mengen einer im Gehalt bekannten Lösung der Reinsubstanz auf der 20×20 cm-Schicht nebeneinander auf. Es ist dann festzustellen, ob z. B. die Zonengröße der unbekannten Substanz zwischen denjenigen der Vergleichssubstanzflecken von 0,5 und 1 μg oder 1 und 5 μg liegen. Danach wählt man die Intervalle enger und kommt so zu recht brauchbaren Angaben. Wie die Abb. 7 (Kurvenverlauf II) zeigt, ist es bei diesem Verfahren erforderlich, im unteren Erfassungsbereich zu arbeiten.

2. Indirekte Verfahren.

Die auszuwertende Zone wird mit einem Spatel (z. B. einem 2 mm breiten Bruchstück einer Rasierklinge zwischen einem Arterienabklemmer) quantitativ abgeschabt, die Substanz eluiert und der Gehalt ermittelt. Gelegentlich kann man zur Elution direkt die erforderliche Menge der Reagenzlösung verwenden und vor der colorimetrischen Auswertung durch Zentrifugieren das Sorptionsmittel entfernen.

GÄNSEHIRT u. Mitarb. haben sich eingehend mit diesen und anderen quantitativen Auswertemethoden der Dünnschicht-Chromatographie beschäftigt. Am Beispiel der Gallensäuren und p-Hydroxybenzoesäureester zeigten sie, daß man bei Einhaltung der angegebenen Bedingungen zu sehr guten Ergebnissen kommen kann. Das geht auch aus einer Arbeit von SCHLEMMER und LINK über die quantitative Bestimmung des Reserpingehaltes hervor.

B. Spezielle Arbeitstechniken.

I. Zirkular- und Formgebungstechnik.

Außer der zumeist bevorzugten aufsteigenden Trenntechnik läßt sich auch die in der Papierchromatographie häufiger verwendete Zirkular-(„Rundfilter"-)

Methode und die Formgebungs-(Keilstreifen-)Technik durchführen. Für die horizontal arbeitende Zirkularmethode sind zwei Geräteanordnungen in der Abb. 9 wiedergegeben (STAHL 1958b).

Auch das Prinzip der Keilstreifenmethode läßt sich auf den Schichten leicht realisieren. Mit einem 0,5 mm breiten Spatel oder einer entsprechenden Schreibfeder und einer Plexiglasschablone, in der Fünfecke in entsprechender Größe ausgespart sind, zieht man sich die Trennungslinien und hebt dabei automatisch die

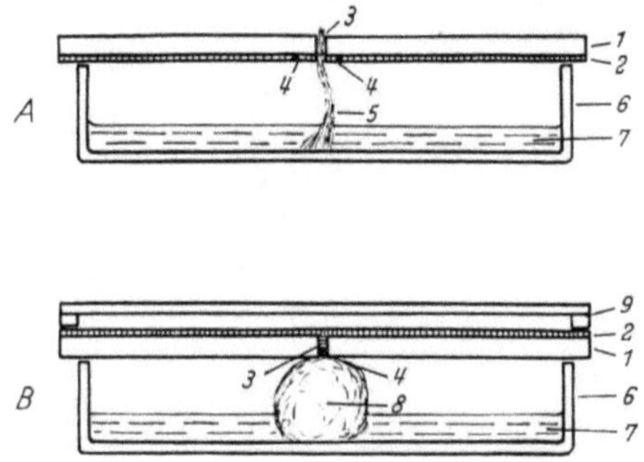

Abb. 9. Schematische Darstellung von zwei Geräteanordnungen zur Zirkulartechnik auf Sorptionsschichten, *A* zur Trennung mehrerer Substanzgemische (bis 8). *B* mit kleiner Vortrennsäule für ein Substanzgemisch. *1* = Trägerscheibe 20×20 cm, *2* = Sorptionsschicht, *3* = Bohrung 2 mm ⌀, *4* = Auftragestelle der Substanz. *5* = Wattedocht, *6* = Petrischalendeckel, *7* = Elutionsmittel, *8* = Wattebausch.

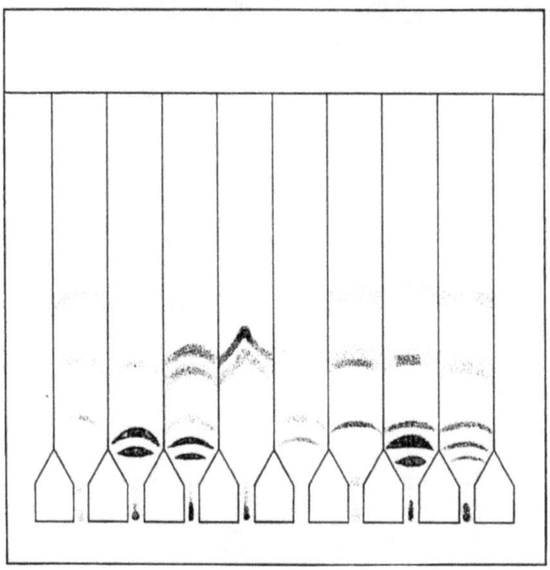

Abb. 10. Sorptionsschicht mit Keilstreifeneinteilung. Die Trennung erfolgt bandförmig, nur bei zu großen Mengen (Mitte) tritt eine ∧-Form auf.

Schicht ab. Aus dem Chromatogramm (Abb. 10) ist eine zweckmäßige Form der Keilstreifen sowie der Trennerfolg zu erkennen.

II. Stufentechnik.

Bei einer Reihe von Naturstoffgemischen beobachtet man, daß sich in einem Elutionsgemisch nur ein Teil der Substanzen gut auftrennt, während in einem anderen Gemisch sich nur die restlichen Bestandteile trennen lassen. Um die Vorteile beider Lösungsmittelgemische zu vereinigen, chromatographiert man mit beiden *nacheinander* bei verschieden langer Trennstrecke (STAHL 1959a).

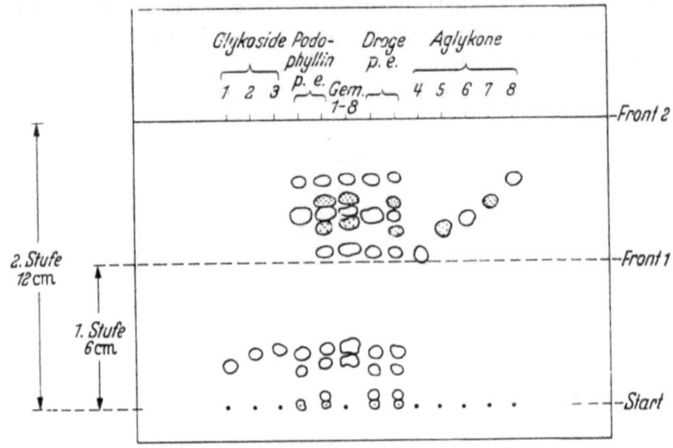

Abb. 11. Dünnschicht-Chromatogramm von Podophyllum-Inhaltsstoffen in der Stufentechnik durchgeführt. Schicht: Kieselgel G. 1. Stufe: Chloroform + 10% Methanol, 2. Stufe: Chloroform + 35% Azeton. *1* α-Peltatin-β-glucosid, *2* Podophyllotoxin-glucosid, *3* β-Peltatin-β-D-glucosid, *4* 4'-Demethylpodophyllotoxin, *5* α-Peltatin, *6* Podophyllotoxin, *7* β-Peltatin, *8* 1-Deshydroxy-podophyllotoxin. Von den Reinsubstanzen wurden je 1,0 μg, von den Podophyllinen und der Droge wurde 0,5 mm³ der 10%igen alkoholischen Auszüge aufgetragen (KS). Sichtbarmachung: Schwefel-Essigsäureanhydrid (1 + 3), 15 min, 100°. Die punktiert gezeichneten Zonen reagieren zusätzlich mit Diazoniumsalzen unter Farbbildung (*e.* = emodi, *p.* = peltatum).

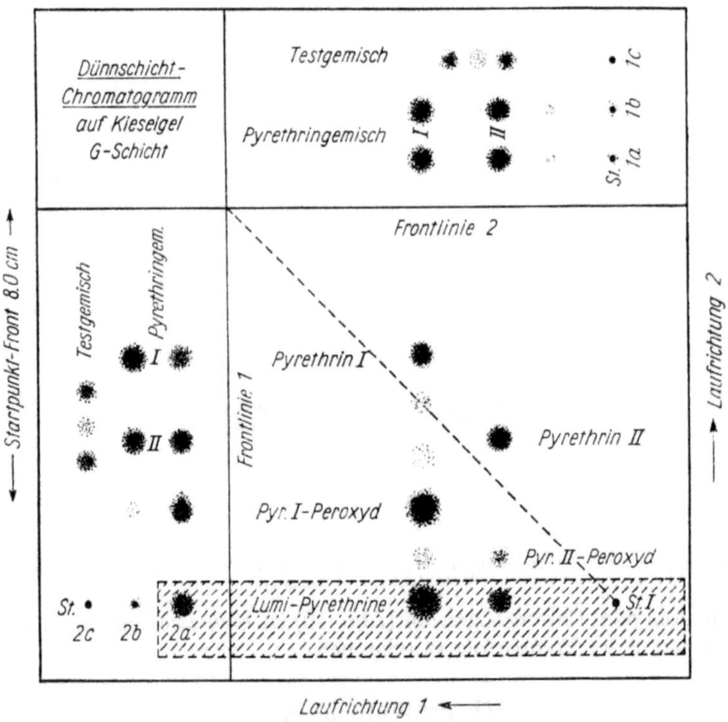

Abb. 12. Dünnschicht-Chromatogramm (TRT-Technik) eines Pyrethrumkonzentrates. Die Bestrahlung (gestricheltes Feld) erfolgte nach der Trennung in Laufrichtung 1. Der Reaktionsausfall der Zonen mit Antimon-III-chlorid, 2,4-Dinitrophenylhydrazin und Kaliumjodid-Essigsäure/Stärke wurde aufeinander projiziert.

Diese Stufentechnik wurde bei Carotinoid- und Alkaloidtrennungen mit Erfolg angewendet und soll am Beispiel der Dünnschicht-Chromatographie von *Podophyllum*-Inhaltsstoffen veranschaulicht werden. Mit dem in der 1. Stufe verwendeten Gemisch werden die Glykoside getrennt (Abb. 11, *1—3*). In der 2. Stufe die Aglykone. Vergleichsweise sind neben den Reinsubstanzen auch die Alkoholextrakte aus der Droge und dem zugehörigen Harz mitchromatographiert. Am Rande sei erwähnt, daß eine Verbesserung des Trenneffektes gelegentlich auch durch ein mehrmaliges Aufsteigenlassen des gleichen Elutionsmittels zu erreichen ist.

III. Zweidimensionale Arbeitsweise ohne und mit Zwischenreaktion.

In der Papierchromatographie wählt man bei der zweidimensionalen Arbeitsweise (Bd. I, S. 129) für die beiden Laufrichtungen möglichst verschiedenartig trennende Lösungsmittelsysteme. Dieses Verfahren läßt sich auf die 20×20 cm-Sorptionsschichten übertragen, wie am Beispiel der Trennung von Antioxydantien (SEHER), Aminosäuren (Abb. 14) und Indolkörpern (Abb. 15) gezeigt wurde.

Verwendet man jedoch für beide Laufrichtungen das gleiche Elutionsmittel, so müßten die Substanzen in einer Reihe etwa in der Diagonalen (Abb. 12; gestrichelte Linie) liegen. Schaltet man nun nach dem Durchlauf in Richtung 1 eine Zwischenreaktion (z. B. Bestrahlung mit γ-, Röntgen- oder UV-Strahlen) ein, so kann man nach der Chromatographie in Laufrichtung 2 erkennen, welche Substanzen des Gemischs sich verändert haben. Am Beispiel einer Studie über die Inaktivierung der verschieden stark insecticiden Pyrethrine bei Lichteinwirkung wurde diese Arbeitsweise von STAHL (1960) eingesetzt (Abb. 12).

Aus dem Auftreten von Peroxydzonen und an der Größe der Zonen ist zu erkennen, daß das stärker insecticide Pyrethrin I schneller über Peroxyde zu Lumipyrethrinen inaktiviert wird als das schwächer wirkende Pyrethrin II.

C. Anwendungsbereich und Einsatzmöglichkeiten.

Die zahlreichen experimentellen Variationsmöglichkeiten, die eine Chromatographie auf dünnen Sorptionsschichten bietet, erschließen einen großen Anwendungsbereich. Auf dem Gebiet der Pflanzenanalyse wurde die Dünnschicht-Chromatographie wie folgt eingesetzt:

1. **Als Vergleichsverfahren** zur schnellen Erkennung von Veränderungen der sekundären Pflanzeninhaltsstoffe. Zur Kennzeichnung „chemischer Rassen", z. B. bei *Acorus calamus* (WULFF u. STAHL) und zur Identifizierung morphologisch-anatomisch schwer unterscheidbarer Drogen (Abb. 11, e., p. und Abb. 13, R, C), ferner zum Nachweis von Drogenverfälschungen (s. auch HÖRHAMMER u. Mitarb.).

2. **In Kombination mit der Gasphasenchromatographie und der TSWETT'schen Säulentrennung.** Bei der Gasphasenchromatographie (s. S. 97) von Naturstoffgemischen ergeben sich gelegentlich erhebliche Schwierigkeiten bei der Identifizierung der mehr oder weniger zahlreichen "*peaks*". Fängt man die fraglichen Fraktionen auf, löst und chromatographiert sie z. B. auf Kieselgel G-Schichten mit Benzol oder Chloroform-Äthylentrichlorid (1 + 1), so läßt sich schnell entscheiden, in welche Gruppe (z. B. Terpen-Sesquiterpen-, Ester-, Phenol- oder Alkohol-Gruppe usw.) die Substanz gehört (STAHL und TRENNHEUSER). Außerdem erkennt man hiermit nicht selten, daß unter einem "*peak*" sich weitere Substanzen verbergen und daß bei labilen Inhaltsstoffen (ätherische Öle) Zersetzungen eintreten.

3. **Bei der Isolierung von Naturstoffen** zur Klärung, ob und unter welchen Bedingungen ein Trennproblem chromatographisch zu lösen ist, im weiteren Verlauf zur Verfolgung der Anreicherung (STAHL 1959a u. 1959b) und mikropräparativen Trennung. Hierzu wird das Substanzgemisch als Startband (bis 10 und mehr Milligramm pro 20×20 cm-Schicht) aufgetragen, nach der Chromatographie die gewünschte Zone abgeschabt und eluiert.

4. **Zur Identitäts- und Reinheitsprüfung.** Je zahlreicher und je verschiedenartiger die Identitätsprüfungen sind, um so größer ist bei einer Übereinstimmung

der Ergebnisse mit authentischen Substanzen die Sicherheit. Die hohe Nachweisempfindlichkeit der Methode läßt zumeist Verunreinigungen von unter 0,1 % gut erkennen. Man trägt zur Reinheitsprüfung steigende Mengen auf und ermittelt so zunächst die maximal mögliche Auftragemenge. Liegt sie z. B. bei 100 μg und ist die Nachweisempfindlichkeit für die unerwünschten Begleitsubstanzen hoch (z. B. bei 0,005 μg), so lassen sich Angaben über die Größenordnung der Verunreinigung machen.

5. Zur Untersuchung der Inhaltsstoffe kleiner Zellkomplexe. Auf 20—50 μ dünnen Kieselgel-Schichten konnte erstmals der Inhalt einzelner Exkretbehälter aufgetrennt und mit den auf dem Wege der Wasserdampfdestillation erhaltenen Produkten verglichen werden (STAHL 1956).

D. Anwendungsbeispiele.

1. Alkaloide.

Die Auftrennung von Alkaloidgemischen gelingt zumeist auf Kieselgel G- und/oder Aluminiumoxid G-Schichten. In der Regel beginnt man mit Chloroform als Elutionsmittel, dem dann je nach Alkaloidgruppe 5—20% Methanol zugesetzt

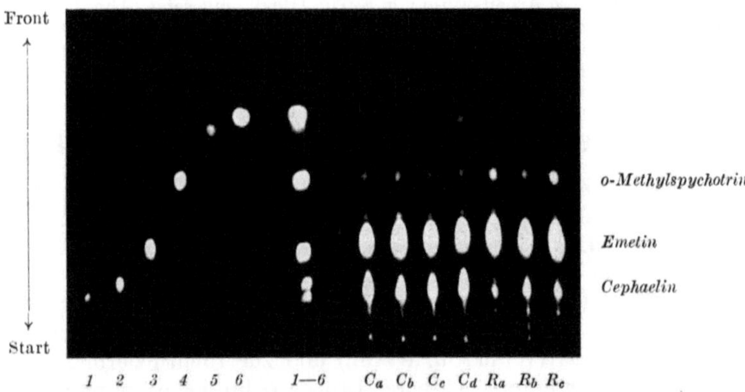

Abb. 13. Ipecacuanha-Alkaloide (~ 0,1 μg) und Drogenauszüge getrennt auf einer Kieselgel G-Schicht. Nach der Jodreaktion im langwelligen UV-Licht fotografiert. *1* Cephaelin (blaue Fluorescenz), *2* Psychotrin (gelb-blaue Fluorescenz), *3* Emetin (gelb-blaue Fluorescenz), *4* o-Methylpsychotrin (gelbe Fluorescenz), *5* Protoemetin (blaue Fluorescenz), *6* Emetamin (gelbe Fluorescenz), *R* = Rio-Ipecacuanha-Auszüge verschiedener Drogen (*a*—*c*), *C* = Carthagena-Ipecacuanha-Auszüge verschiedener Drogen (*a*—*d*).

werden. Eine geringfügige Beigabe einer kleinen Menge (0,05—1 %) Diäthylamin ist vorteilhaft, um die Alkaloide in die Basenform zu drängen.

Die Abb. 13 zeigt ein Dünnschicht-Chromatogramm von Ipecacuanha-Alkaloiden und Drogenauszügen der Wurzeln von *Uragoga ipecacuanha* BAILL. (= Rio-Ipecacuanha) und der nicht offizinellen *Uragoga acuminata* KARSTEN (= Carthagena-Ipecacuanha). Als Fließmittel diente Chloroform + 15% Methanol. Zur Sichtbarmachung wurde mit einer 0,5%igen Jod-Chloroform-Lösung besprüht und 15 min auf 60° erhitzt. Im ultravioletten Licht treten dann die Ipecacuanha-Alkaloide als stark gelb, bzw. blau fluorescierende Zonen hervor, und man kann noch bis 0,001 μg erkennen.

Einen Analysengang zur Identifizierung von Alkaloiden haben WALDI u. Mitarb. ausgearbeitet. Sie bevorzugen ebenfalls Kieselgel G-Schichten und verwenden zur Sichtbarmachung vor allem Jodplateatreagens.

2. Aminosäuren.

Nachdem 1959 sowohl NÜRNBERG als auch MUTSCHLER und ROCHELMEYER über die Trennung von Aminosäuren auf Kieselgel G-Schichten berichtet haben, stellten BRENNER und NIEDERWIESER eingehende Vergleiche mit den gebräuchlichen papierchromatographischen Trennverfahren an (Abb. 14). BRENNER und NIEDERWIESER kamen zu der Schlußfolgerung, daß eine Chromatographie der Aminosäuren auf Kieselgel G-Schichten, neben einem erheblichen Zeitgewinnl

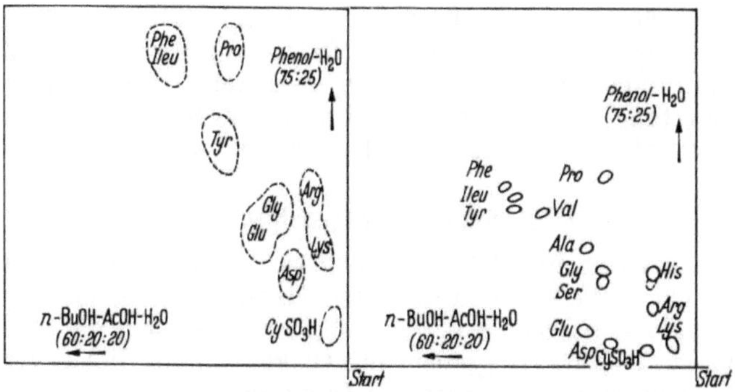

Abb. 14. Vergleich zwischen einem Papier- (links) und einem Dünnschicht-Chromatogramm (rechts) von 10 Aminosäuren. Aufgetragene Menge 1 μg pro Substanz, Sichtbarmachung mit einem modifizierten Ninhydrin-Reagens- (BRENNER und NIEDERWIESER 1960).

eine Erhöhung der bisherigen Nachweisempfindlichkeit um den Faktor 10 bringt, und daß die Trennungen den papierchromatographischen ebenbürtig oder sogar überlegen sind.

3. Indol-Auxine.

Auf Kieselgel G-Schichten läßt sich eine Spurenanalyse der „einfachen" Indolderivate schnell durchführen. STAHL und KALDEWEY trennten mit einem

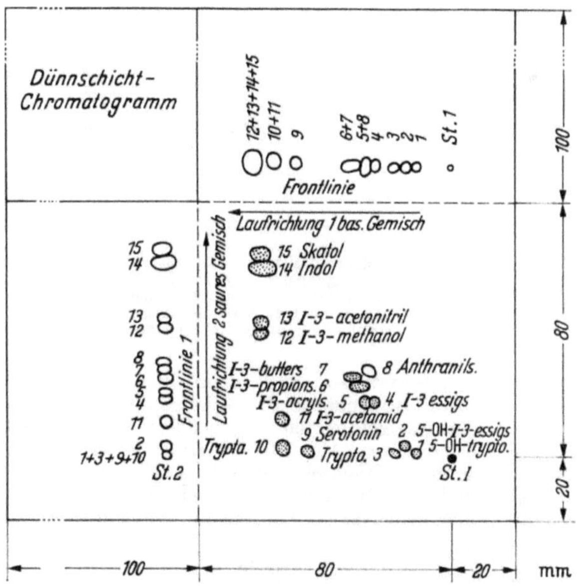

Abb. 15. Zweidimensionales Dünnschicht-Chromatogramm 14 „einfacher" Indolderivate + Anthranilsäure. Das aufgetragene Gemisch (St. I, 1 und 2) enthielt 0,5 μg von jeder Substanz. Aus dem Vergleichschromatogramm über Startpunkt 1 (St. 1) ist der Trennerfolg des basischen Elutionsmittels *A* und über St. 2 derjenige des sauren Elutionsmittels *S* zu erkennen. *1* 5-Hydroxytryptophan, *2* 5-Hydroxyindol-3-essigsäure, *3* Tryptophan, *4* Indol-3-essigsäure, *5* Indol-3-acrylsäure, *6* Indol-3-propionsäure, *7* Indol-3-buttersäure, *8* Anthranilsäure, *9* Serotonin, *10* Tryptamin, *11* Indol-3-acetamid, *12* Indol-3-methanol, *13* Indol-3-acetonitril, *14* Indol, *15* Skatol.

Gemisch (*S*) aus Chloroform + 5% Essigsäure (96%ig) unter anderem die Indol-3-carbonsäuren in 15—40 min. Als zweites Gemisch (*A*) schlagen sie Methylacetat + Isopropanol + Ammoniaklösung, 25%ig, (45 + 35 + 20 Vol.-Teile) vor. Die Vorteile beider Systeme vereinigt das zweidimensionale Chromatogramm (Abb. 15).

Bei einer modifizierten VAN URK-Reaktion liegt die untere Nachweisgrenze (Trennstrecke 6 cm) für das Heteroauxin (Indol-3-essigsäure) um 0,02 μg und bei der PROCHAZKAschen Fluorescenzreaktion im Bereich von 0,001 μg. Eine biologische Auswertung durch direktes Übertragen der abgeschabten Kieselgel-Zone auf Agar gelingt ohne Schwierigkeiten (KALDEWEY und STAHL).

4. Steroide und Steroidglykoside.

Die Arbeiten von STAHL (1959b), BARBIER u. Mitarb., VAN DAM u. Mitarb. sowie WALDI und Mitarb. lassen die guten Trennmöglichkeiten auf dem Gebiet der Steroide und nahestehender Verbindungen erkennen. Für die Pflanzenanalyse ist die Dünnschicht-Chromatographie von Glykosiden aus dieser Klasse im Hinblick auf die herzwirksamen Drogen von Interesse.

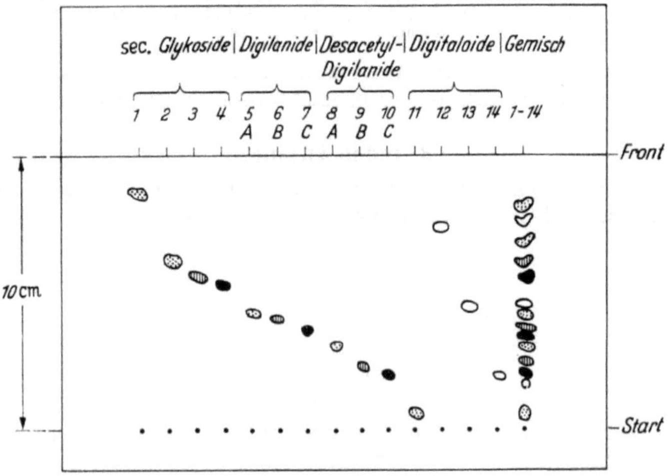

Abb. 16. Dünnschicht-Chromatogramm von Herzglykosiden (je 1 μg) auf einer Kieselgel G-Schicht (KS). *1* Acetyldigitoxin, *2* Digitoxin, *3* Gitoxin, *4* Digoxin, *5* Digilanid A, *6* Digilanid B, *7* Digilanid C, *8* Desacetyldigilanid A, *9* Desacetyldigilanid B, *10* Desacetyldigilanid C, *11* k-Strophantosid, *12* Cymarin, *13* Proscillaridin A, *14* Scillaren A. Gemisch aus gleichen Teilen der Substanzen *1—14*. Fluorescenz im langwelligen UV-Licht nach der Trichloressigsäure-Chloramin-Reaktion: hellgelb = □, braungelb = ▒, hellblau = ▥, violettblau = ■.

STAHL und KALTENBACH beschrieben die Trennung eines Gemisches von 14 Herzgiftglykosiden. Als Elutionsmittel verwendeten sie Methylenchlorid + Methanol + Formamid (80 + 19 + 1 Vol.-Teile) und zur Sichtbarmachung das Trichloressigsäure-Chloraminreagens oder Antimon-III-Chlorid-Chloroform (Abb. 16). Die Laufzeit der Chromatogramme betrug eine halbe Stunde, und die untere Nachweisgrenze der Glykoside lag bei etwa 0,01 μg. Im Hinblick auf den Zuckeranteil der Glykoside ist zu erwähnen, daß von den gleichen Autoren auch eine Möglichkeit zur Spurenanalyse von Zuckern auf Kieselgur G-Schichten ausgearbeitet wurde.

5. Lipide.

Auf dem Gebiet der Fettchromatographie wurde die Methode zunächst von MANGOLD und seinem Arbeitskreis eingesetzt, den besonderen Erfordernissen angepaßt und mit bestem Erfolg zur Trennung von höheren Fettsäuren und deren Alkoholen und Estern verwendet.

KAUFMANN und MAKUS bestätigten diese Ergebnisse und erweiterten sie; JATZKEWITZ und MEHL erarbeiteten die Trennbedingungen für spezielle physiologisch-chemische Probleme. Über die Dünnschicht-Chromatographie anorganischer Verbindungen berichten H. und M. SEILER. Die bis Mitte 1961 bekannt gewordenen weiteren Anwendungsgebiete der Dünnschicht-Chromatographie sind mit den entsprechenden Literaturangaben in zwei Übersichtsberichten zusammengestellt (STAHL 1961 a).

Literatur.

BARBIER, M., H. JÄGER, H. TOBIAS u. E. WYSS: Helv. Chim. Acta **42**, 2440 (1959). — BILLEK, V.: (Privatmitteilung). — BRENNER, M., u. A. NIEDERWIESER: Experientia (Basel) **16**, 378 (1960) s. a. Experientia (Basel) **17**, 145 (1961).

DAM, VAN M. J. D., G. J. DE KLEUVER u. J. G. DE HEUS: J. Chromatogr. **4**, 26 (1960).

GÄNSHIRT, H., u. Mitarbeiter: Arch. Pharmaz. *293*, 65, 925, 1065 (1960). — GRISEBACH, H.: Z. Naturforsch. **14b**, 802 (1959); — Chem Ber. **93**, 2326 (1960).

HÖRHAMMER, L., H. WAGNER u. B. LAY: Pharmazie **15**, 645 (1960).

ISMAILOW, N. A., u. M. S. SCHRAIBER: Farmazia (russ.) No. **3**, 1—7 (1938).

JATZKEWITZ, H., u. E. MEHL: Hoppe Seyler's Z. physiol. Chem. **320**, 251 (1960).

KALDEWEY, H., u. E. STAHL: Planta (im Druck) 1962). — KAUFMANN, H. P., u. Z. MAKUS: Fette, Seifen, Anstrichmittel **62**, 1014 (1960). — KIRCHNER, J. G., JOHN M. MILLER u. G. J. KELLER: Analyt. Chem. **23**, 420 (1951). — KRATZL, K., u. G. PUSCHMANN: Holzforschung **14**, Nr. 1 (1960).

MANGOLD, H. K.: Fette, Seifen, Anstrichmittel **61**, 877 (1959) u. Privatmitteilungen. — MANGOLD, H. K., u. DONALD C. MALINS: J. Am. Oil Chem. Soc. **37**, 383, 576 (1960). — MUTSCHLER, E., u. H. ROCHELMEYER: Arch. Pharmaz. **292**, 449 (1959).

NÜRNBERG, E.: Arch. Pharmaz. **292**, 610 (1959).

SCHLEMMER, F., u. L. LINK: Pharmaz.-Ztg. **104**, 1349 (1959). — SEILER, H. u. M.: Helv. chim. Acta **43**, 1939 (1960). — SEHER, A.: Fette, Seifen, Anstrichmittel **62**, 345 (1959). — STAHL, EGON: (I. Mitt.) Pharmazie **11**, 633 (1956); — (II. Mitt.) Chem. Ztg. **82**, 323 (1958 a); — (III. Mitt.) Parfüm. u. Kosm. **39**, 564 (1958 b); — (IV. Mitt.) Arch. Pharmaz. **292**, 411 (1959 a); — (V. Mitt.) Arch. Pharmaz. **293**, 531 (1960); — Pharmaz. Rundsch. **1**, No. 2, 1 (1959 b); — Z. analyt. Chem. (Übersichtsbericht) **181**, 303 (1961 a) u. Angew. Chem. **73**, 646 (1961 a). — STAHL, EGON, u. H. KALDEWEY: (VII. Mitt.) Z. physiol. Chem. **323**, 182 (1961). — STAHL, EGON, u. U. KALTENBACH: (VI. u. XI. Mitt.) J. Chromatogr. **5**, 351 u. 458 (1961). — STAHL, EGON, u. L. TRENNHEUSER: Arch. Pharmaz. **293**, 826 (1960).

WALDI, D., K. SCHNACKERZ u. F. MUNTER: J. Chromatogr. **6**, 81 (1961). — WULFF, H. D., u. EGON STAHL: Naturwissenschaften **47**, 114 (1960).

Paper Chromatography on a Preparative Scale.

By

F. A. Hommes and H. F. Linskens.

With 23 Figures.

Among the techniques which recently have been developed in biochemical analysis, paper chromatography takes a first place. It is as common as is classical elementary analysis for a laboratory of organic chemistry. The reason why this technique has assumed such enormous proportions is twofold:

Firstly, because of its high resolving power, which allows handling of impure samples.

Secondly, because the technique is easy to perform and requires relatively inexpensive equipement.

Although originally developed as an analytical tool paper chromatography has been extended for preparative purposes.

Many reviews of paper chromatography have appeared in recent years. For a description of the principles of paper chromatography the reader is referred to the article of HELLMANN (1955) in volume I of this series, and to the monographies of CRAMER (1953), LEDERER (1957), HAIS and MACEK (1958) and LINSKENS (1959). In this article a review will be given on paper chromatographic techniques for preparative purposes. This technique, which in most cases is an extension of the analytical method, may be applied to any group of substances, which can be separated in the analytical method in microgram quantities only. In addition, special techniques which have been developed for larger quantities will be reviewed.

A. General.

The same precautions which are recommended for analytical paper chromatography, must also be kept in mind for the preparative technique.

I. Impurities.

In most cases the sample to be chromatographed must be protein, lipid and salt free.

Proteins can be removed best by addition of ethanol to a final concentration of 80% (DE VENDIER and ÅGREN 1948), which denatures the proteins. The insoluble proteins can be centrifuged off and the supernatant evaporated to dryness

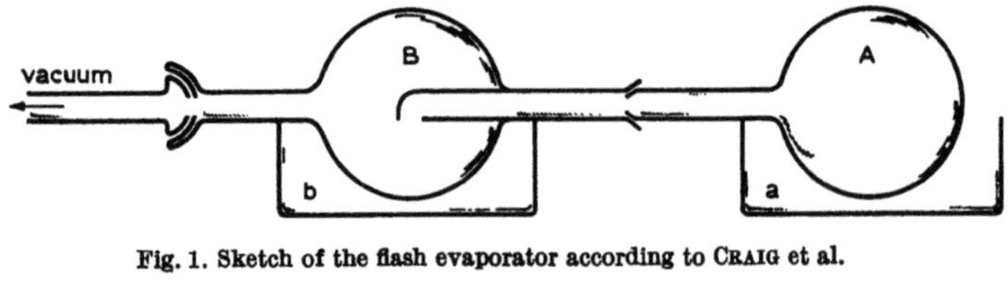

Fig. 1. Sketch of the flash evaporator according to CRAIG et al.

in vacuo and redissolved in a small amount of a suitable solvent. This evaporation can be carried out best in the so-called flash evaporator, developed by CRAIG (1950). The apparatus consists of two rotating flasks (Fig. 1), *A* and *B*. The sample to be evaporated is put into flask *A*, which is placed in a thermostatically controlled water bath *a*. The system is evaporated (in most cases an ordinary water aspirator is sufficient), while flasks *A* and *B* are slowly rotated along their horizontal axes by an electromotor. Flask *B* rotates in an ice-cooled water bath *b*. By the rotation, a thin film of solvent is produced on the entire surface of flask *A*, which allows a rapid distillation. The solvent is condensed in flask *B*. The apparatus, which permits rapid distillation at relatively low temperatures, is now commercially available, as well as several modifications[1]. This method is recommended for stable compounds. For unstable substances, or in the case of materials of unknown stability, lyophilisation is recommended (MOOR 1961).

Lipids can be removed by extraction of the dry residue with a suitable lipid solvent, such as ethanol-chloroform, ether etc. Care must be taken that no other compounds go into solution together with the lipids. The choice of the extraction solvent depends thus on the compounds which are chromatographed.

Inorganic salts greatly interfere with paper chromatography. Irregular chromatograms, tailing, non-reproducible R_F-values are often the results of the presence of inorganic salts. They must therefore be removed, for instance by ion exchange chromatography (BRAUNITZER 1955) or electrodialysis.

The chromatography chamber should have a constant temperature (within 1° C) in order to get reproducible results.

II. Choice of Solvent System.

In most cases the solvent system which proves to give good results in the analytical method, may be also applicable for preparative purposes. Each problem will require its own solvent system, which in most instances is known from the literature.

III. Quantities.

The quantities which can be separated depend upon the paper to be used. For preparative work it is best to choose a thick paper, because more material can be applied to one paper.

Table 1 gives a summary of types of papers of different factories.

Table 1.

Factory	Number	Thickness mm	Property of paper
Whatman	3	0.36	rapid
Whatman	3	0.31	rapid
Whatman	31	0.50	rapid
Machery, Nagel & Co.	214	0.30	medium
Schleicher & Schüll. .	2315	0.30	rapid
Schleicher & Schüll. .	2316	0.30	rapid
Schleicher & Schüll. .	2071	0.65	cardboard
Schleicher & Schüll. .	2230	0.90	cardboard
Schleicher & Schüll. .	2181	4.00	cardboard
Binzer	225	—	rapid

Approximately 100 mg quantities may be separated on papers of 0.3—0.9 mm thickness. For larger amounts (approximately up to 1 g quantities) a cardboard has to be used.

[1] Instrumentation Associates Inc., New York, U.S.A., U. W. Towers & Co. Ltd., Widnes, England, W. Buchi, Flawil, Switzerland.

IV. Application of the Sample.

This is in principle a problem of evaporation. In the analytical method, one drop of the solution to be analysed is sufficient in most cases. In preparative paper chromatography a larger volume of solution must be applied to one spot, which means that the solution has to be applied to the paper in small portions with frequent evaporation of the solvent. This may be performed by a hair drier or an infrared lamp. The solution can be applied manually; however it is a time consuming job.

A special apparatus has been constructed for this purpose, for instance the apparatus described by DU RUISSEAU (1958). It consists of a hypodermic syringe with an electrically driven plunger. The syringe is filled with the solution, the needle of the syringe is placed on the paper and the plunger is forced down at an adjustable speed. This speed is given by the rate of evaporation of the solvent by the hair drier. Approximately 1 ml of an aqueous solution can be applied to a chromatogram in 2 hours. If only a one-dimensional technique is employed, the sample may be applied across the entire width of the paper, which increases the capacity of the system. The sample is delivered as a streak of uniform width by moving a pipette equipped with a rubber bulb along the starting line.

V. Localisation of Bands.

As the irrigated chromatogram has to undergo further treatment (elution of the separated compounds) the chromatogram cannot be coloured as such.

The paper is placed on a glass plate, covered with a sheet of Whatman No. 1 paper (or an other equivalent paper) and the whole is covered with another glass plate. Under slight pressure small amounts of the resolved bands are absorbed by the Whatman No. 1 paper, which can now be coloured by spraying with, or dipping in a suitable colour producing reagent. The outlines of the spots are traced on the chromatogram by superimposing the print. This method permits a good isolation of the compounds even if the spots are badly distorted by irregularities in the paper. In the one-dimensional technique a small slip of the paper may be cut off and developed in the usual way.

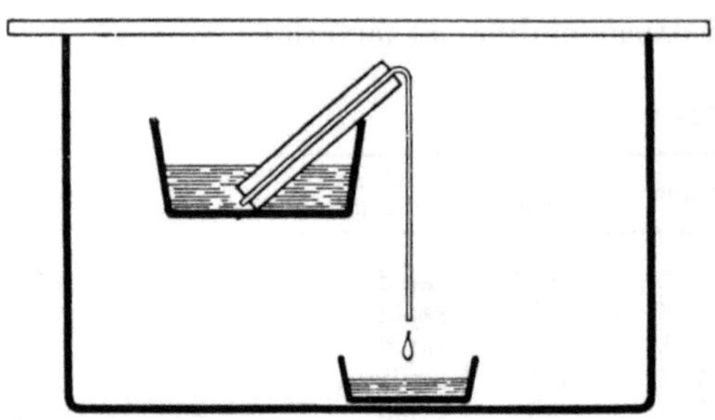

Fig. 2. Elution of a compound from a paper strip. The paper strips containing the spots or bands are placed between two pieces of glass resting in a container, v.a. Petri dish. The wash liquor rises between the plates by capillary attraction and drops off the end of the paper into an other vessel.

VI. Elution of the Resolved Spots.

The traced spots are cut out, snipped in small pieces and put in a centrifuge tube. A suitable solvent is added — dependent on the particular compound, however in most cases water will dissolve most biological substances — the mixture is stirred occasionally for one hour and centrifuged. The supernatant is removed and the extraction is repeated twice. The combined supernatants are evaporated to dryness, as indicated above, and tested for homogeneity.

Although this method is easy to perform it gives the compounds in rather diluted solution. The following procedure avoids this disadvantage. The areas containing the traced bands are cut from the chromatogram as narrow rectangular strips. At one of the narrow ends of the strip a tip is cut, the other end is placed between two glass plates, which are put into a container with a suitable elution solvent (see Fig. 2). The compounds are eluted in a closed chamber, for instance in a chromatography jar, to avoid evaporation, by a descending flow. It has been shown that the compounds are eluted in the first volumes of eluate which comes off from the strips. This method thus gives a more concentrated solution. In using cardboards a small strip of Whatman No. 3 MM has to be attached to the cardboard for fixation between the two glass plates.

This method of elution may also be employed for a one-dimensional technique in which the sample is applied in bands. After the bands have been allocated in the first chromatogram, they are cut out as shown in Fig. 3. The strip is fixed between two glass plates and the band is eluted in the same direction as the

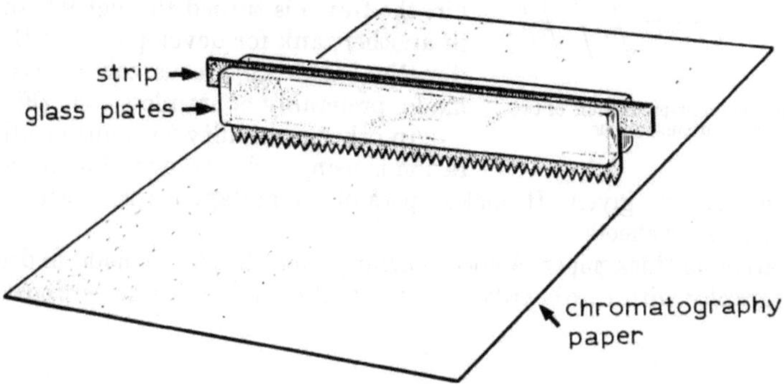

Fig. 3. Elution of a band from a paper strip and simultaneous application for one-dimensional paper chromatography.

chromatography was performed. The eluate is collected on a line of a second chromatography paper, which after drying, is developed in another system (Honegger 1955).

This technique may also be employed in a combination of paper electrophoresis and paper chromatography.

B. Multisheet- and Cardboard-Chromatography.

Multiplicative separation. The capacity of the chromatographic system can be enlarged by running simultaneously several papers. This has been done by Fischer (1954) in a method which is known as *multiplicative chromatography.* In this method the same paper as in the analytical method is used. It has the advantage of getting well resolved spots. With thick papers and cardboards the resolution of the individual compounds is in general not so good as with thinner papers.

The solution is applied as a band, approximately 8 cm from the margin of the paper. Many papers, up to 100, are spotted in this way. The sheets are piled up, between two consecutive papers, two strips of plastic, 2 mm thick, 2 cm broad and a few cm larger as the papers, one on top and one on the lower side of the paper. On both sides of the paper pile the projecting parts of the plastic strips are pressed together by rubber rings. Hanging on the plastic strips the pile is placed in a chromatography jar for ascending chromatography. One of the papers may be used for the localisation of the bands.

A *multisheet chromatography* tank, developed by Datta, Dent and Harris (1950), is commercially available[1]. It consists of a glass tank, an aluminium frame for holding 12 filter paper squares, a solvent disk and a glass lid. Whatman No. 1 or No. 4 filter papers, with punched holes in each corner, suited for the aluminium frame, are also commercially available. The mixture to be analysed is applied in one corner of the papers, which are then threaded over the 4 rods, so that the corners with the mixtures are all at the same corner of the frame. Tubes between two consecutive papers prevent the papers from touching each other. The loaded frame is placed in the dish for ascending chromatography. After development the papers are dried in a current of warm air, the frame is turned through 90° and placed in another tank for development in the second direction. This apparatus can be easily home made, preferably of stainless steel (Fig. 4).

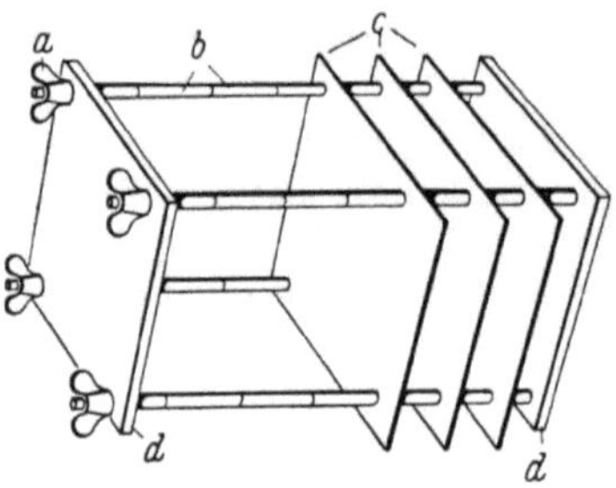

Fig. 4. Frame for mass-production of two-way paper chromatograms.

An other possibility to multiplicative separation is using a slit trough (Fig. 5), for which the dimensions are given. It makes possible simultaneous separation of two-dimensional paper sheets.

Separation on thick paper. A special arrangement has to be made in descending chromatography with cardboards. It is very difficult to fold the cardboard in the

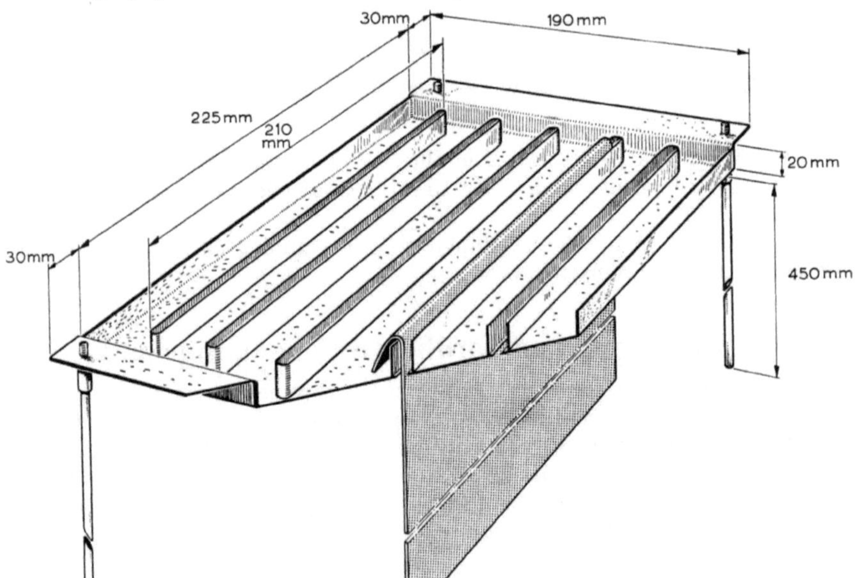

Fig. 5. Slit trough for simultaneous separation of two-dimensional paper sheets.

elution trough. This problem is solved best as proposed by Brownell, Hamilton and Casselman (1957), who sewed with a sewing machine and heavy cotton

[1] Shandon Scientific Company Ltd., Landow, England.

thread, a Whatman No. 3 MM paper strip to one of the edges of the cardboard, lengthwise. The paper strips is then looped over the end of the cardboard and sewn along the back to form a stirrup. If necessary, a second strip is sewn, face to face, along the same edge of the cardboard to permit a constant flow of solvent. In using cardboards it is advisable to prewash the papers chromatographically with 0.01 N hydrochloric acid for 3 days, with distilled water for 4 days and finally with the developing solvent.

C. Separation on Paper-Packs.

I. Circular Chromatopack Procedure.

The multi-sheet chromatography can be modified into the *chromatopack*-procedure which was first applied for circular chromatography by BROCKMANN and PATT (1953) (Fig. 6).

A pack of 15 sheets (e.g. Schleicher & Schüll 2043 b) is folded twice, keeping the machine direction uniform throughout the pack. A hole, 1 cm in diameter, is drilled in the centre and a special screw-assembly is inserted which compresses the centre of the paper-pack and serves also as an inlet for the mobile phase. For the latter purpose, the central part of the screw has 4 lateral indentations (Fig. 6c). The chromatopack is supported by a flat dish or a porcelain ring and is operated in a desiccator. Buffer solution is introduced via the lid of the desiccator and into a funnel, until the pack of filterpaper is entirely moistened. Then the mixture to be analysed is slowly applied in the same way. Before this mixture has dried up, the addition of the separating solvent from an automatic pipette is started.

A certain modification is represented by the procedure of GIRI and PARIHAR (1955): 3—5 filterpaper disks, from thick Whatman No. 3 paper, placed between glass sheets, 30 cm in diameter, are cut out along the edges. Pointed tips are shaped which are bent into the mouth of a wide funnel. Solvent is introduced in the centre of the filterdisks, the separated fractions trickle down from the pointed tips and can be collected in a row of reagent tubes. This procedure has not been applied extensively.

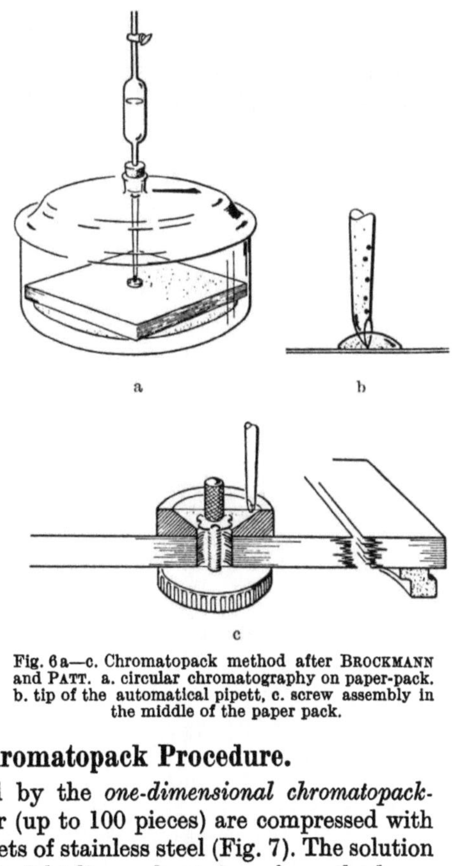

Fig. 6a—c. Chromatopack method after BROCKMANN and PATT. a. circular chromatography on paper-pack. b. tip of the automatical pipett, c. screw assembly in the middle of the paper pack.

II. One Dimensional Chromatopack Procedure.

A very good separation is achieved by the *one-dimensional chromatopack*-procedure. Strips or sheets of filter paper (up to 100 pieces) are compressed with moderate, uniform force between two sheets of stainless steel (Fig. 7). The solution which must be separated is applied in straight lines, about 4 cm from the lower edges. The pack is put into a dish containing the solvent, and separation is performed by the ascending technique in the saturated atmosphere of a suitable container (PORTER 1951).

The *chromatopack* can also be modified for *descending* separation. Therefore the upper edges of the paper sheets are cut out in a crescent-shape. In this way, a container is shaped between the compressing sheets of steel, the bottom of which consists of the upper edges of the paper sheets (Fig. 8). The mixture is again applied in straight lines, ca. 3 cm from the upper edges, and the solvent is applied from above (ARX-NEHER 1955). After the separation, the outermost sheets are used in order to localize the zones. The remaining zones are then cut out, eluted and the separated substances can be recovered for further analysis.

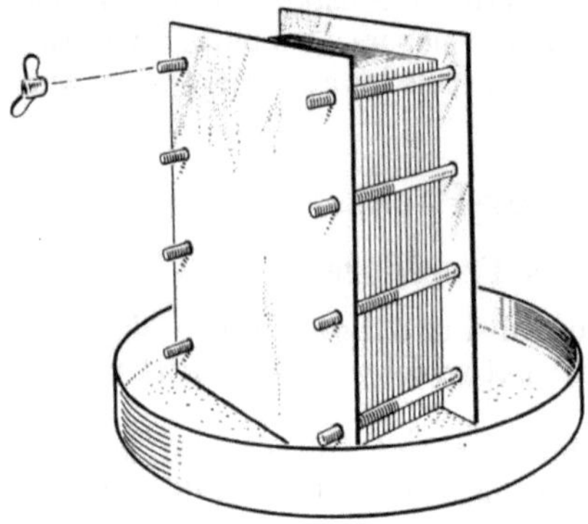

Fig. 7. One-dimensional chromatopack-procedure, ascending.

III. Chromatopile Procedure.

When the chromatopile-procedure is used, a large number (200—500) of filter paper disks, of equal diameter and quality are piled up into a column and com-

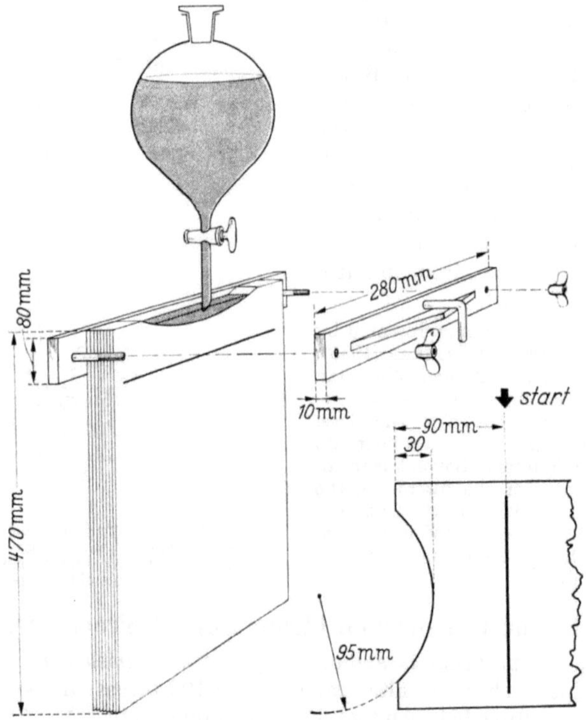

Fig. 8. One-dimensional chromatopack-procedure, descending.

pressed between two plates (Fig. 9). The mixture to be analysed is absorbed in 10 similar disks, dried and put on top of the column, which is finally covered with

20 more disks and put in a chamber with a saturated atmosphere. After *descending* development, the column can be easily taken apart. Batches of 5, 10 or 20 disks can be analysed separately, depending on the sharpness of separation that is desired (MITSCHELL-HASKINS 1949, MITCHELL, GORDON and HASKINS 1949).

An *ascending* modification of the chromatopile-procedure has been used by VAN DER WANT-PETERS-INKLAAR (1952). This has the advantage that the development comes to a stop automatically, after all disks have become wetted; elution is thus avoided. The modification proposed by ZECHMEISTER (1951) consists of using a glass cylinder, which is filled with small paper disks. However, this method proves to be rather uncertain. The glass tubes must be of very constant diameter in order to avoid channelling of the solvent and edge effects.

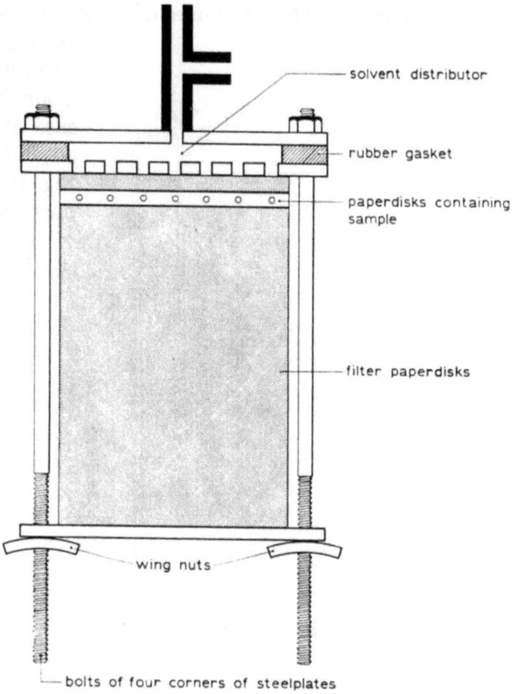

Fig. 9. Diagram of a filter paper column chromatopile.

D. Column Chromatography.

Separation of a mixture of compounds may also be accomplished with columns of cellulose powder. Examples of this technique are given by DREW, MARSHALL and FRIEDBERG (1952) for the separation of pyruvic, oxalo-acetic and α-ketoglutaric acids, DENANDER (1952) for the separation of glucofructosones, BLACKBURN (1951) for the fractionation of α-amino acids and DEKKER and TODD (1950) for the separation of thymidine and uracil deoxyribosides.

I. Columns of Cellulose Powder.

The chromatogram tubes are usually made of a glass tube, drawn out at one end or equipped with a porous plate at the bottom. A drawn out chromatogram tube may be closed at the bottom by a small plug of glass wool. The length of the column is approximately 10× its diameter, but is mostly determined by the separation of the mixture.

CROWELL and KÖNIG (1944) have developed tubes of glass or plastic with a thin sector cut out along their length. Removal of this sector facilitates the localisation of the bands of colourless substances after development of the chromatogram by streaking techniques (see below).

1. Filling of the Column.

The column is filled with cellulose powder, for instance Whatman cellulose powder of Schleicher & Schüll cellulose powder 123 or 123a (the latter has been washed with hydrochloric acid and distilled water) in small portions. After each addition the column is shaken in order to get a uniform distribution of particles. According to SYNGE (1951) the cellulose powder may be suspended in the solvent to be used for development of the chromatogram. The suspension is poured into the column to the desired height and washed with the solvent. This method is recommended because it will give a more uniform column.

The solution to be analysed is applied on top of the column by careful addition of the solution to the surface of the cellulose. The solution should be as concentrated as possible. This will yield the compounds in a narrow starting zone and therefore result in better separation of the individual compounds. It is to be remembered that the surface of the cellulose column should remain untouched, because otherwise pits will be formed and an irregular starting zone is obtained. If this happens the cellulose may be stirred carefully up to the lowest limit of the pit, if this does not exceed 0.5 cm.

2. Elution.

After complete soaking in of the solution the compounds are eluted with the solvent. The first volumes of the solvent are applied to the column in the same way as described for the mixture. In most cases a pressure device is needed in order to get a sufficiently rapid flow of the solvent, especially in case of volatile solvents where suction is not possible. This may be performed as shown in Fig. 10a or Fig. 10b. The latter gives a constant pres sure, independent of the level of the solvent.

The most elegant solution of this problem is the use of the LKB *Miniflow* precision micropump[1]. This pump gives a constant flow of solvent, without pulsations (up to 60 ml per hour) by the simultaneous action of two syringes: if one is pumping, the other is being filled, etc (Fig. 11).

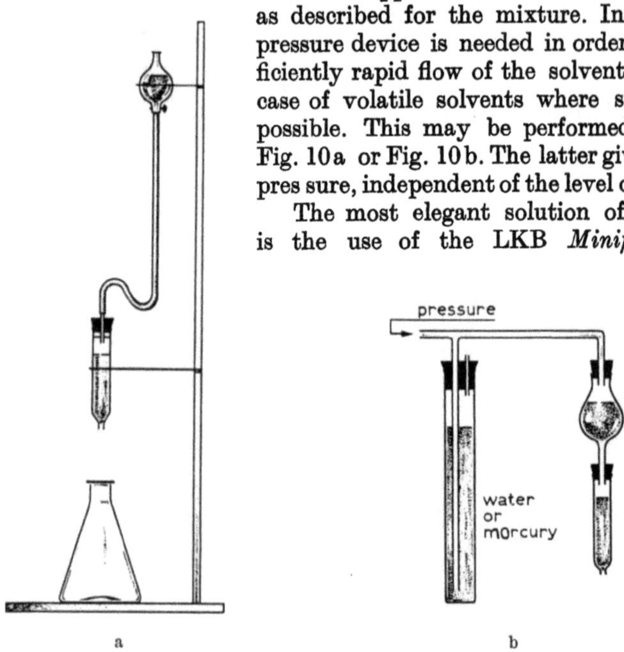

Fig. 10 a and b. Pressure devices for column chromatography.

3. Fraction collector.

The eluate is collected in fractions, preferably with the aid of an automatic fraction collector, the first of which was constructed by MOORE and STEIN (1948).

A fraction collector may operate on a basis of *time* — which yields a given number of fractions per hour — of *volume* — a constant volume per fraction — of *drop counting*, which gives a constant number of drops per fraction. In as much as the flow of solvent is constant, fraction collectors on a time basis will also yield a constant volume per fraction. If a constant flow rate of the solvent can be achieved, these fraction collectors are recommended, because they are mechanically more reliable (Fig. 12). Examples of fraction collectors are described by JAMES, MARTIN and RANDALL (1951) and EDMAN (1948). A relatively simple and

[1] LKB, Stockholm, Sweden, Fig. 9.

inexpensive apparatus, operating on time basis and suited for the at the same time running of two columns, has been described by HUISMAN and KRAUS (1952).

If the bands are only separated on the column, but not eluted from the column, colourless compounds have to be localised on the column by streaking techniques. This technique consists in streaking a suitable reagent with a brush over the length of the column, after it has been removed from the tube. It is, however, very difficult to obtain the column in an intact form outside the tube. For this

Fig. 11. Precision micropump (type: MiniFlow LKB). The precision pump consists of two main parts: the driving mechanism, which is located in a cast metal housing, and the pump system proper in its holder attached to the above-mentioned housing. The pumps, driven by a 2-watt synchronous motor, are two all-glass injection syringes, each permanently sealed to a valve centerpiece with optically flat valve plates and seats. Separate suction and discharge ends are attached to each centerpiece by rigid metal clamps. The syringe plungers are mounted in self-aligning holders which are in turn attached to the vertically moving couplings of the drive unit.

reason the above mentioned tubes of CROWELL and KÖNIG (1944) may be useful. After the bands have been located, that part of the column is isolated and eluted.

In case of complete elution of the bands *each fraction has to be tested* for the presence of a compound. An aliquot part of each fraction may be removed and treated with a suitable colour producing reagent. The optical density of each fraction is measured and plotted versus the tube number, which yields a diagram from which the individual peaks can be determined. This method yields quantitative results concerning the amounts of each compound. In the case of ultraviolet absorbing compounds each fraction may be measured in a spectrophotometer. This can be done automatically with the so-called *Uvicord* (LKB, Stockholm, Sweden), see Fig. 13, 14. This apparatus consists of an optical unit, which is

Fig. 12. Automatic fraction collector, collecting effluents by timed flow, volumetric or drop-count method.

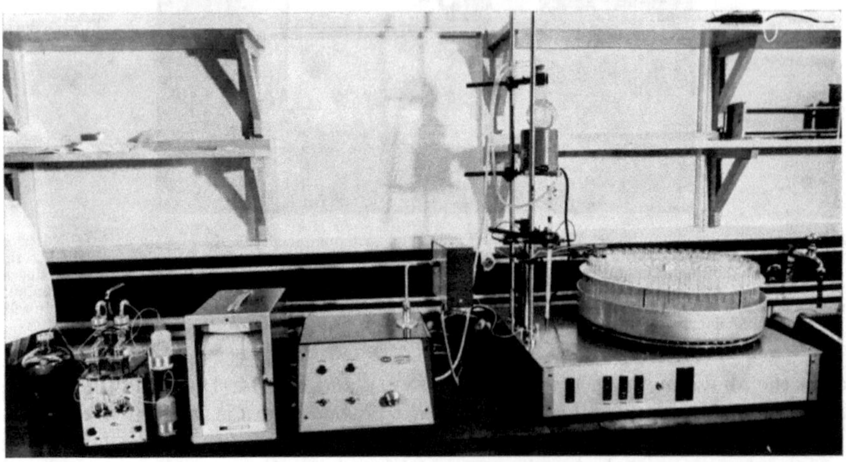

Fig. 13. Instrumental set-up for column chromatography. From left to right: Micropump, recorder, power-supply-amplifier-unit, optical unit UV-absorptiometer, drop counter, column with container for eluant (above), fraction collector.

connected with the outflow of the chromatography column. The eluate flows through a cuvette and the percentage transmission is continuously measured.

The output of the photocell is amplified by the control unit and printed with a recorder. From the known paper velocity of this recorder and the number of fractions per hour, the peaks can be localised (Fig. 13, 14).

An elegant qualitative method has been described by VAN DER SCHAAF (1955), who brings a drop of each fraction on a small strip of Whatman No. 1 paper which is then sprayed with a suitable reagent. The combined fractions of one peak are concentrated as described above and tested for homogeneity.

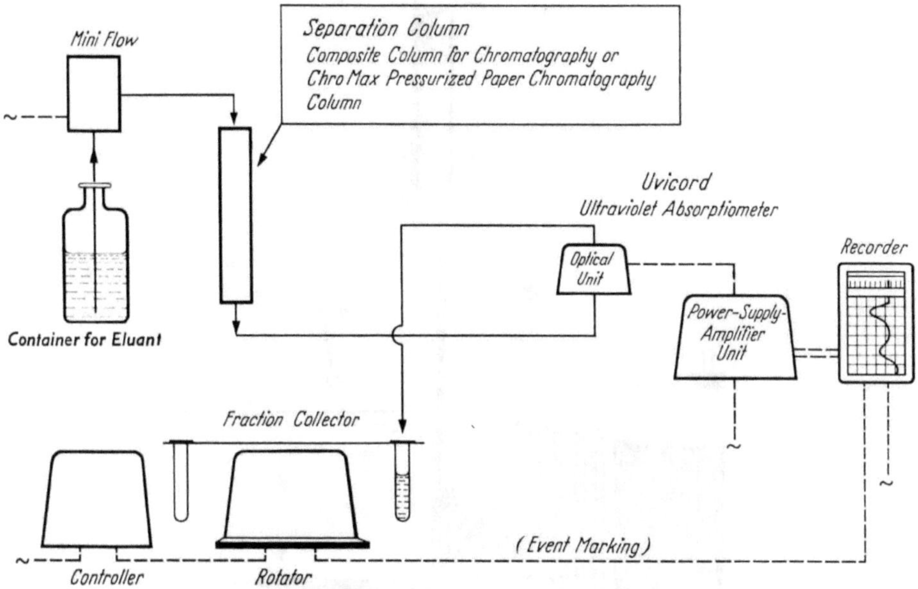

Fig. 14. A typical instrument setup for chromatographic separation on a preparative scale, using compact column or pressurized paper chromatography column.

II. Paper Roll Column Chromatography.

A preparative separation similar to analytical chromatography on a single sheet of paper has been obtained on paper rolls. Paper roll columns have been originally designed by HAGDAHL and DANIELSON (1954, 1955), who used a cylindrical core of polythene in which the swelling of the paper roll resulting from the absorption of the solvent produced an internal pressure, which evened out irregularities in the porosity. The most important advantage of the ready-made columns is that the difficult and time-consuming packing procedure can be bypassed by their use. The analytical separation methods, developed for a single sheet, may be applied directly on a preparative scale. Somewhat later the same method was further developed to the *ChroMax-Method*[1] by the introduction of a pressure mantle (HAGDAHL and LERNER 1958) (Fig. 15).

For the ChroMax columns paper rolls are used, which consist of a long sheet of paper tightly wound on a polythene rod and enveloped in a seamless polythene skin. This roll is put into a pressure mantle consisting of a steel tube with a rubber tube inside (Fig. 16a). Air can be pumped through a valve into the space formed between both tubes by means of a simple hand pump or by a pressure ballast flask. In this manner isoporosity is obtained by means of a flexible mantle which

[1] LKB-Produkter, P.O. Box 12220, Stockholm 12, Sweden.

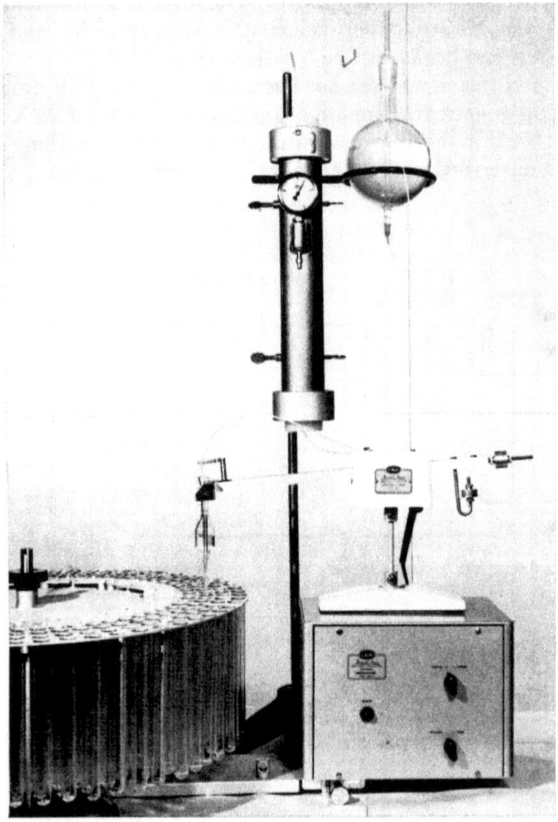

Fig. 15. Preparative paper chromatography by ChroMax column in combination with a solvent reservoir of the Mariotte flask type. Below: volume-controlled fraction collector.

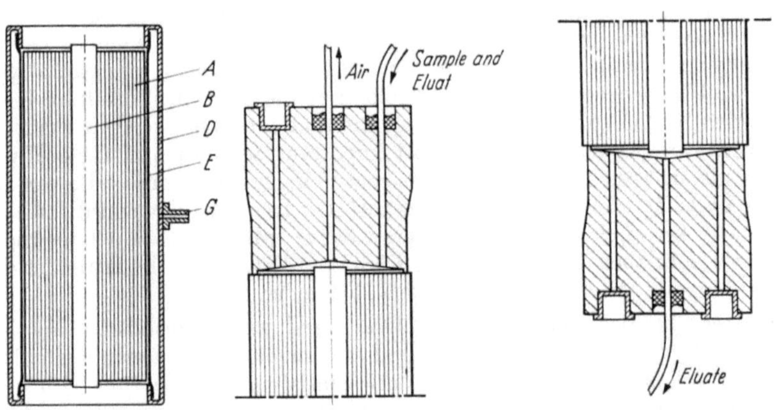

a b c

Fig. 16a—c. a Longitudinal cross sections through a ChroMax column (schematic). *A* paper, *B* polythene rod, *D* steel tube, *E* rubber tube, *G* air inlet valve. b Polythene inlet plug of the ChroMax column, connected by a length of narrow bore polythene tubing to the solvent reservoir. An air vent is left clear. c Polythene outlet plug, connected to the fraction collecting device.

surrounds the paper roll and exerts a constant pressure over its entire curved surface. This pressure can be adjusted to the exact value demanded and controlled by the pressure gauge.

Experimental procedure. The column should be mounted in a vertical position, for example on an ordinary retort stand. The paper roll inside is saturated with the solvent to be used in the experiment. Then the valve rubber is replaced on the air inlet valve (Fig. 16b) and the mantle is inflated to an excess pressure of about 0,5 kg/cm². The pressure is controlled by the *dye test:* A small quantity of the insolvent in question containing a suitable dye is passed through the column. The appearence of the lower surface of the paper rolls at the moment when the coloured solution first reached this surface, is a good indication for the correct pressure. This is achieved, when the whole surface is instantaneously coloured. If the colour first appears at the circumference, the pressure is too low, and if it first appears towards the centre, the pressure is too high. Once the balancing pressure is found for the actual column-system, this value will hold for several weeks. It is necessary only to keep the pressure in the mantle constantly at this value.

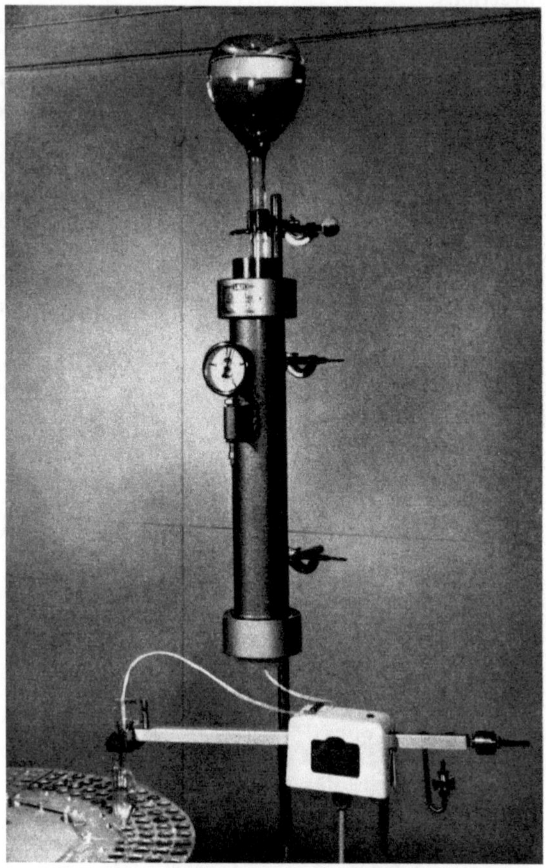

Fig. 17. Paper roll column mounted in a stand. Eluting liquid is supplied from the inverted bottle whose neck is situated some distance (about 2 cm) above the upper end of the paper roll.

After the roll has been saturated with solvent, a solution of the mixture to be separated is poured into the compartment formed by the acid-proof cylinder, mounted on the top of the paper roll. When the sample solution has soaked into the paper the compartment at the top is rinsed with a small quantity of the solvent, which is allowed to penetrate into the paper. Lastly, from an inverted bottle a steady downward flow of solvent through the column is started. The liquid emerging from the lower end of the roll is collected in a funnel and carried to a fraction collector for further analysis (see Fig. 16a, 17).

This method is at present *not* suitable for separations with non-polar solvents (benzene, chloroform, collidine, ethers, pyridine, toluene etc.). Such solvents diffuse through the thin polythene skin of the paper rolls and damage the rubber tube of the mantle.

E. Continuous Paper Chromatography.

For the continuous chromatographic separation of a mixture of substances the simultaneous separation of the components in two directions is required. This can be accomplished in two different ways:

a) By the combination of two different separating principles in different directions,

b) by the movement of the solid phase during the process of separation.

A combination of two principles is found in *paper electrochromatography:* An electric current is applied horizontally to a paperchromatogram, on which a mixture of substances is being chromatographed by descending technique; the action of the electrical field on the descending compounds results in a separation of their paths depending on the mobilities in the field. The mixture to be analysed can be continuously applied, separated and collected into fractions.

The continuous separation by movement of the solid phase of a column, in a direction other than the direction of flow, is as yet only possible with columns of ion exchange resins; this procedure, called counter current chromatography, requires homogeneity of the solid phase, under the conditions of movement.

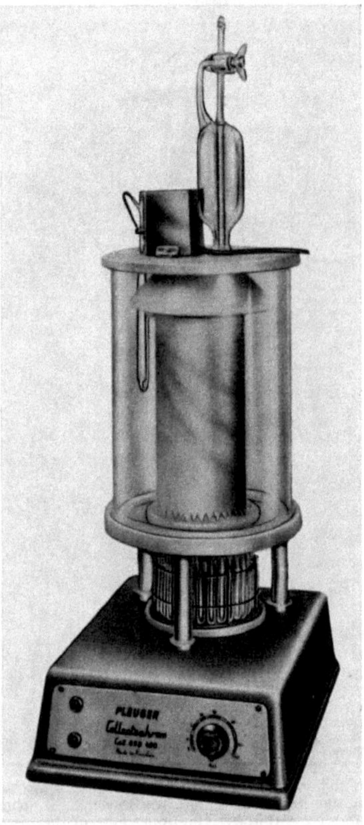

Fig. 18

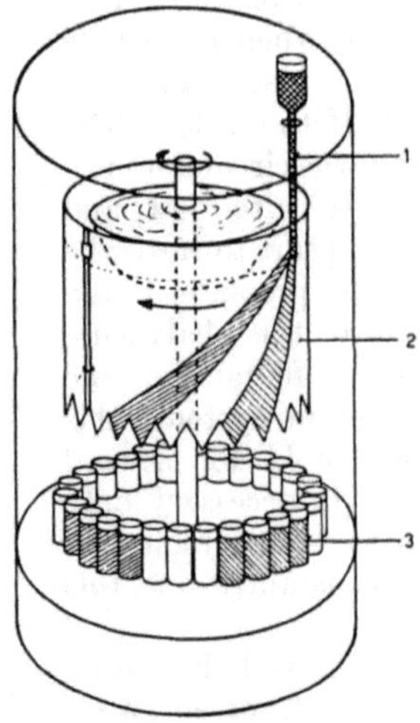

Fig. 19

Fig. 18. Collectochrome apparatus. The chromatographic chamber is shut off by a glass-cylinder, covered by a glass plate. At its lower end, it is lapped flat so as to fit on the flange. The latter is supplied with a groove, that should be filled with solvent, so as to facilitate the saturation inside the chamber.

Fig. 19. Collectochrom (schematic). The central pivot carries the solvent trough, into which the top strips of the paper are hung. The sampler (*1*) consists of a capillary system, consisting of a trough containing the solution to be separated, from which it is sucked up by a filter paper strip into a capillary tube, which deposits it on the paper cylinder (*2*). Underneath of the lower rim of the paper, there is a circular holder for some 30 tubes (*3*). These are supplied with broadened and overlapping rims, thus obviating loss of drops between the tubes along the holder.

Collectochrome procedure[1] (Fig. 18). When the solid phase is moved perpendicularly to the flow direction of the mobile phase, a continuous separation of many com-

[1] Ets. G. Pleuger, S.A. Wijnegem-Antwerpen, Belgium.

pounds can be achieved. This is a case of *elution chromatography with moving solid phase.* The instrumentation of this continuous paper chromatography was first

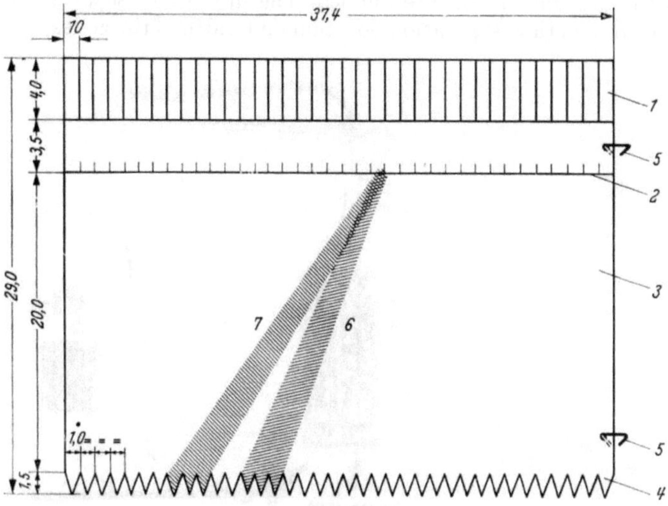

Fig. 20. Dimension of the paper sheet for Collectochrom technique.

elaborated by SOLMS (1954, 1955): The solid phase (a sheet of chromatographic paper) is made into a hollow cylinder and put around a vertical shaft. At distances of 1 cm, 4 cm deep notches are made in the upper edge of the sheet and the resulting strips are bent inwards into an annular trough (Fig. 19). Then the sheet is percolated downward with a suitable solvent. The mixture to be analysed is applied continuously at a fixed point, the solid phase being rotated with constant speed (1,3 cm/h) around the axis (Fig. 19).

The components of the applied mixture move vertically downward with the eluting solvent, at speeds determined by their R_F values. Horizontally they participate in the constant movement of the carrier sheet. The developed paper then will show one straight path for every component of the mixture. The inclination "α" being a function of the speed of rotation of the cylinder "a", (Fig. 19/2) of the speed of the solvent "b", and of the R_F value:

$$\operatorname{tg}\alpha = a \cdot \frac{R_F}{b} .$$

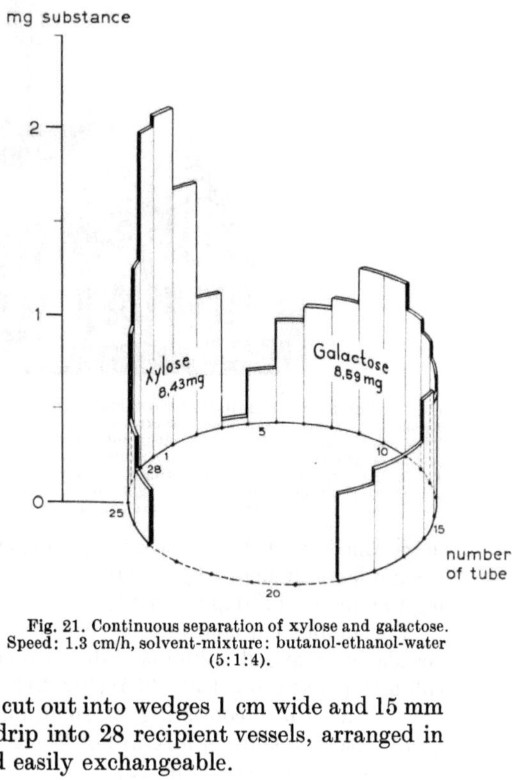

Fig. 21. Continuous separation of xylose and galactose. Speed: 1.3 cm/h, solvent-mixture: butanol-ethanol-water (5:1:4).

The bottom edge of the rotating paper is cut out into wedges 1 cm wide and 15 mm long (Fig. 20). The separated fractions drip into 28 recipient vessels, arranged in a circle. These vessels are stationary and easily exchangeable.

The quality of the separation depends largely upon the paper used, the constancy of temperature, the saturation of the atmosphere around the cylinder, and upon the amount of mixture applied. The number of separable compounds and the sharpness of their separation (location and width of the zones) can be largely

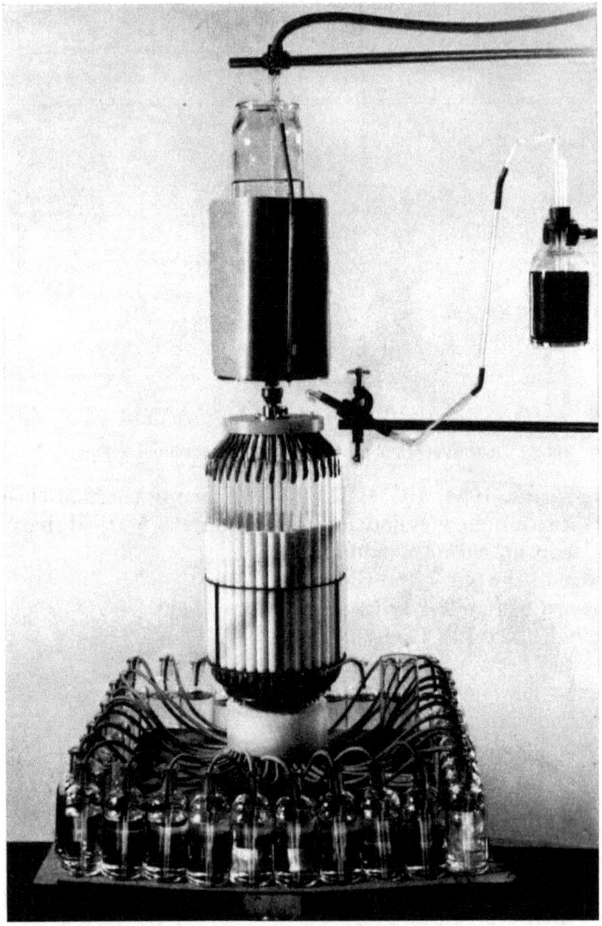

Abb. 22. Experimental apparatus for continuous chromatographic separation. On a wooden frame, beneath which are situated 2 synchronous motors, is arranged the fraction collector, consisting of a flow distributor, and of rubber tubes connecting each hole of the distributor with each flask. The collector arrangement does not rotate. The column is composed of 36 component columns arranged in a large cylinder and kept in position by a brass ring. The feeding arrangement for the eluting agent consists of a flask on the top with a siphon with capillary tube. For the sample solution, there is on the right side a small Mariotte flask and a long capillary tube. The composite column is slowly rotating, the siphon for the eluting liquid is faster rotating.

determined by variation of the kind of paper used which determines the vertical speed of sample components; by varying the speed of rotation of the solid phase (determining the horizontal deplacement) and by variation in the rate of application.

Continuous chromatography can also be achieved with the aid of a battery of circularly arranged columns. Filter paper rolls of Munktell-paper are used as columns (Svensson et al. 1955) (Fig. 21). Continuous chromatography is suitable only for mixtures consisting of small numbers of components (2—3) but gives quite reliable results. A mixture of 17 mg xylose and galactose could be completely separated in 21 hours (Fig. 22).

F. Accelerated Chromatography.

The separation of larger amounts of substances can also be achieved by speeding up the process of separation. Two possibilities are discussed:

a) increasing temperature during separation,

b) increasing the flow rate of the solvent by centrifugal forces.

I. High Temperature Paper Chromatography.

Separation at temperatures exceeding room temperature requires special attention for constancy of temperature, and for the saturation of the atmosphere.

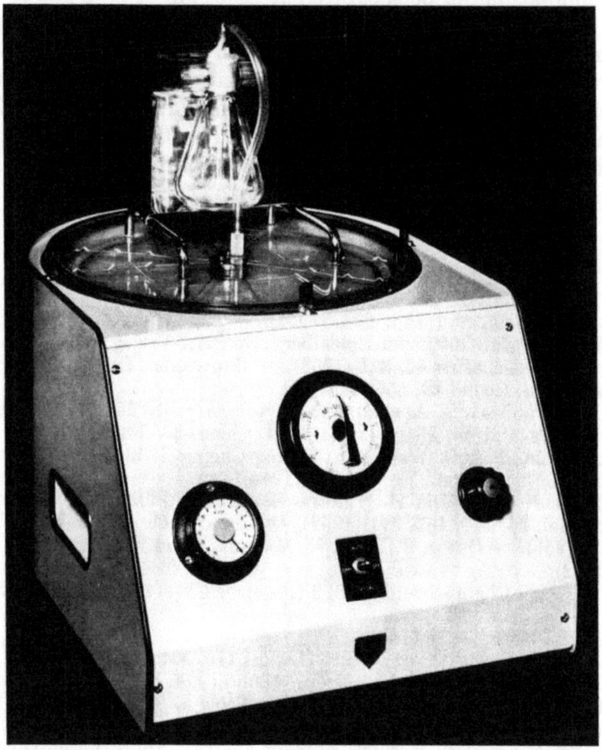

Fig. 23. Centrifugal chromatograph, with head and cover plate for paper discs. Application of the solvent in the centre.

Chambers of sealed stainless steel[1] and horizontal filter papers are generally used. In order to compensate for the increased vapour pressure, the cover lid is fixed with special locking devices. The apparatus is operating in a reliable thermostat. At 60° C a 30 cm displacement of the solvent requires only 35 minutes. Quantitative separations of amino acid mixtures were found to be reproducible within ± 5% (Roberts 1957, Roberts-Bucek 1957).

It is also possible to bring up the whole chamber to a high temperature. Because fluctuations of a few degrees in ambient temperature within the cabinet during the chromatographic run interfere with separation of the spots and the

[1] "Speedi-cell", Labline J. D. Marshall Intern. Inc. Chicago 2, Ill. U.S.A.

reproducibility of R_F values, it may be necessary to use double-walled chromatographic cabinets, designed to keep such temperature changes at a minimum (Ensslin 1956).

II. Centrifugal chromatography.

A considerable saving of time can be achieved when the solvent does not run in the normal gravity field, but in an artificially increased gravity field. Horizontally placed filterpaper disks (32 cm, or radially arranged strips up to 5 cm wide) are rotated in an atmosphere saturated with solvent vapour[1]. The motor speed, variable between 250 and 1500 rpm is controlled via a time switch (Fig. 23). Mixtures of amino-acids can be separated at 1200 rpm in 4—5 minutes (Whatman No. 1 paper) or in 8—10 minutes (Whatman No. 3 paper). The R_F-values obtained on the rotating paper disk were in all cases found to be slightly higher than those obtained by the conventional descending technique (McDonald, Ribeiro and Banaszak 1959). Time-saving centrifugal chromatography is especially indicated for the separation of substances labelled with radioactive tracers of short half-life (McDonald-Bermes-Shepherd 1957, McDonald-McKendell 1957, McDonald-McKendell-Bermes 1958, Carriker u. Applegate 1960).

References.

Arx, E. von, u. R. Neher: Helv. Chim. Acta **39**, 1664 (1956).

Blackburn, S.: Chem. and Ind. **294**, (1951). — Braunitzer, G.: Modern methods of plant analysis, Vol. I. Berlin-Göttingen-Heidelberg: Springer-Verlag 1960. — Brockmann, H., u. P. Patt: Naturwissenschaften **40**, 221 (1953). — Brownell, H. H., J. G. Hamilton and A. A. Casselman: Anal. Chem. **29**, 550 (1957).

Carriker, R. C., and H. G. Applegate: Phyton (Argent.) **15**, 67—78 (1960). — Craig, L. C., J. D. Gregory and W. Hausmann: Anal. Chem. **22**, 1462 (1950). — Cramer, F.: Papierchromatographie, 2. Aufl. Weinheim: Verlag Chemie 1953. — Cramell, W. R., and O. König: Ind. Eng. Chem. Anal. Ed. **16**, 347 (1944).

Datta, S. P., C. E. Dent and H. Harris: Science **112**, 621 (1950). — Dedander, R.: Bull. Soc. Chim. Biol. **34**, 144, 157, 171 (1952). — Dekker, C. A., and A. R. Todd: Nature (Lond.) **166**, 557 (1950). — Drew, F. D., L. M. Marshall and F. Friedberg: J. Am. Chem. Soc. **74**, 1852 (1952).

Edman, P. L.: Acta Chem. Scand. **2**, 592 (1948).— Ensslin, W.: Z. analyt. Chem. **153**, 1—4 (1956).

Fischer, A. L.: Planta **4**, 288 (1954).

Giri, K. V., and D. B. Parihar: Nature (Lond.) **175**, 304 (1955).

Hagdahl, L., and C. E. Danielson: Nature (Lond.) **174**, 1062 (1954); — Sci. Tools **2**, 13 (1955). — Hagdahl, L., and K. D. Lerner: Sci. Tools **5**, 23 (1958). — Hais, J. M., und K. Macek: Handbuch der Papierchromatographie, Bd. I. Jena: Gustav Fischer Verlag 1958. — Hellmann, H.: Modern methods of plant analysis, Vol. I. Berlin-Göttingen-Heidelberg: Springer-Verlag 1955. — Honegger, C. G.: Helv. Chim. Acta **39**, 1671 (1956). — Huisman, T. H. J., and S. Kraus: Chem. Weekbl. **48**, 1007 (1952).

James, A. T., A. J. P. Martin and S. S. Randall: Biochem. J. **49**, 293 (1951).

Lederer, E., and M. Lederer: Chromatography, 2nd ed. Amsterdam: Elseviers Publ. Comp. 1957. — Linskens, H. F.: Papierchromatographie in der Botanik, 2nd ed. Berlin-Göttingen-Heidelberg: Springer-Verlag 1959.

Moor, H.: Modern methods of plant analysis, Vol. V. p. 73 Berlin-Göttingen-Heidelberg: Springer-Verlag. — Maare, S., and W. H. Stein: Ann. N.Y. Acad. Sci. **49**, 265 (1948). — McDonald, H. J., E. W. Bernes and H. G. Shepherd: Naturwissenschaften **44**, 9 (1957); — Chromatogr. Methods **2**, 1 (1957). — McDonald, H. J., and L. V. McKendell: Naturwissenschaften **44**, 616 (1957). — McDonald, H.J., L. V. McKendell and E. W. Bermes: J. Chromatogr. **1**, 259 (1958). — McDonald, H. J., L. P. Ribeiro and L. J. Banaszak: Anal. Chem. **31**, 825—829 (1959). — Mitchell, K. H., M. Gordon and F. A. Haskins: J. biol. Chem. **180**, 1071 (1949). — Mitchell, H. K., and F. A. Haskins: Science **110**, 278 (1949).

[1] B.T.L. Centrifugal Chromatograph, Baird & Tatlock, Ltd, Chadwell Heath, Essex, England.

NEHER, R.: J. Chromatogr. **1**, 122 (1958).

OSHIMA, Y., and T. NAKARAYASHI: J. agricult. chem. Soc. Japan **27**, 759 (1953).

PORTER, W. L.: Anal. Chem. **23**, 412 (1951).

ROBERTS, H. R.: Anal. Chem. **29**, 1443 (1957). — DU RUISSEAU, J. P.: Anal. Chem. **30**, 455 (1958).

SCHAAF, P. C. VAN DER: Thesis Univ. Groningen 1955. — SOLMS, J.: Helv. Chim. Acta **38**, 1127—1133 (1955); — Kalium Symposium 291—314. Bern 1954. — SVENSSON, H., C. AGRELL, S. DEHLEN and L. HAGDAHL: Sci. Tools **2**, 17 (1955). — SYNGE, R. L. M.: Biochem. J. **48**, 429 (1951).

TATA, J. R., and A. W. HEMMINGS: Discovery **20**, 278 (1959); — J. Chromatography **3**, 225—229 (1960).

DE VERDIER, C. H., and G. ÅGREN: Acta Chem. Scand. **2**, 783 (1948).

WANT, G. H. VAN DER, H. PETERS and P. INKLAAR: Rec. trav. chim. Pays-Bas **71**, 1221—1227 (1952).

ZECHMEISTER, L.: Science **113**, 35 (1951).

Determination of Size, Shape and Homogeneity of Macromolecules in Solution.

By

I. J. O'Donnell and E. F. Woods.

With 20 Figures.

A. Average Molecular Weights.

If solute particles are heterogeneous with regard to size, shape or configuration then physical measurements give an average value of the molecular weight and size. Consider a heterogeneous system of N molecules where N_i is the number of molecules of each component, c_i the weight concentration in g/ml, and M_i the molecular weight of each component. The number fraction n_i of molecules having molecular weight M_i will be N_i/N and $N = \Sigma N_i$, $c = \Sigma c_i$.

In colligative methods such as osmotic pressure all solute molecules contribute equally despite differences in weight; thus the property measured depends only on the number of molecules present. The average molecular weight in such cases is called a number average molecular weight and can be described by several alternative expressions —

$$M_n = \frac{\Sigma N_i M_i}{\Sigma N_i} = \frac{\Sigma N_i M_i}{N} = \sum n_i M_i = \frac{\Sigma c_i}{\Sigma c_i/M_i}\,.$$

If we take as the weighting factor the weight fraction w_i of all molecules having molecular weight M_i, $w_i = \frac{N_i M_i}{\Sigma N_i M_i}$, then

$$M_w = \Sigma\, w_i M_i = \frac{\Sigma N_i M_i^2}{\Sigma N_i M_i} = \frac{\Sigma c_i M_i}{\Sigma c_i}\,.$$

In a similar way higher averages can be defined

$$M_z = \frac{\Sigma N_i M_i^3}{\Sigma N_i M_i^2} = \frac{\Sigma c_i M_i^2}{\Sigma c_i M_i}$$

$$M_{z+1} \text{ or } M_4 = \frac{\Sigma N_i M_i^4}{\Sigma N_i M_i^3} = \frac{\Sigma c_i M_i^3}{\Sigma c_i M_i^2}\,.$$

For a polydisperse system M_w will always be greater than M_n, M_z will be greater than M_w etc.; thus in the higher averages more emphasis is placed on the heavier molecules of the system. If we have a system with equal numbers of molecules of molecular weights $M_1 = 10{,}000$ and $M_2 = 50{,}000$ then $M_n = 30{,}000$; $M_w = 43{,}300$; $M_z = 48{,}500$, $M_{z+1} = 49{,}700$. If the same two species were present in equal parts by weight then $M_n = 16{,}600$; $M_w = 30{,}000$; $M_z = 43{,}300$; $M_{z+1} = 48{,}500$. It is obvious that we may have similar averages of any other property of the system such as length, sedimentation coefficient, intrinsic viscosity, diffusion coefficient etc.

B. Osmotic Pressure.

The main purpose of osmotic pressure measurements is to determine the molecular weight of a macromolecular solute. It also gives information on thermodynamic interactions by measurement of the virial coefficients and these are related to those found in light scattering and sedimentation equilibrium measurements.

Osmotic pressure measurements are most useful for determining molecular weights in the range about 5,000—500,000. For solutes of molecular weight less than 8—10,000 it is difficult to get membranes completely impermeable to the solute molecules (particularly for aqueous solutions) and above 500,000 the precision of measurement of osmotic pressure is small.

If the macromolecular solution is heterogeneous the molecular weight obtained from osmotic pressure measurements is a number-average, M_n (see section A.).

Osmotic pressure measurements of a macromolecular solution are capable of a degree of precision about 10^3 times greater than boiling-point elevation or freezing-point depression which also depend on the colligative properties of a solution.

The main emphasis in this section will be placed on aqueous colloid solutions but the method has also been used extensively for synthetic polymers in organic solvents.

Some of the good reviews and general articles on osmotic pressure are those by ALEXANDER and JOHNSON (1949), EDSALL (1953), FLORY (1953), OVERBEEK (1956), ADAIR (1961) and WAGNER and MOORE (1959), the last-named giving a comprehensive experimental coverage.

I. General Theoretical Considerations.

1. Definition of Osmotic Pressure.

If a solvent and a solution of macromolecules are separated by a solid membrane impermeable to macromolecules there will be a flow of solvent into the solution in an attempt to equalize the chemical potential of the solvent on both sides of the membrane. The flow can be prevented by the application of a pressure to the solution and this pressure is called the osmotic pressure of the solution[1].

The osmotic pressure of a macromolecular solution depends on temperature, composition of solution, activities of the various components present and their charge.

2. VAN'T HOFF's Law and Derivation of the Osmotic Pressure Equation.

The analogy of osmotic pressure in solutions to pressure in gases can be seen in VAN'T HOFF's law which is the exact counterpart of the ideal gas law. It can be derived thermodynamically (see for example ADAIR, 1961) and can be written

$$p = p^i - p^0 = RTC_2/1000 = RTc_2/M_2 \tag{1}$$

where p = observed osmotic pressure, p^i = pressure on the solution, p° = pressure on the solvent, T = absolute temperature, C_2 = molar concentration of macro-

[1] This is also sometimes (but not commonly) called the "colloid osmotic pressure" or "oncotic pressure" to distinguish it from "total osmotic pressure" which is the osmotic pressure of a solution with a gaseous membrane; this membrane would be impermeable to everything except volatile solvent. Such "total osmotic pressure" would be obtained from vapour pressure measurements.

molecules and c_2 = concentration of macromolecules in g/ml of solution[1]. If p is in atmospheres then $R = 82.07$ c.c.-atmospheres/degree/mole.

Equation (1) in the first place only holds for an *ideal solution of an uncharged macromolecule*; as will be seen later it also holds for ideal charged macromolecules under certain conditions. It is used to determine the molecular weight of the macromolecule being considered.

a) Ideal Charged Macromolecule — DONNAN Equilibrium.

For an *ideal solution of a charged macromolecule* there is also a contribution to the observed osmotic pressure arising from osmotic activity of counter-ions on the charged macromolecule and, if salt is present, from an unequal distribution of diffusible ions across the membrane. These phenomena are described by the DONNAN equilibrium and the correct osmotic pressure equation can again be obtained thermodynamically (see for example OVERBEEK, 1956). Two extreme cases may be considered (it will be assumed that the salt used is uni-univalent). If the concentration of macro-ion, zC_2, (where z is the valency of the charged macro-ion) is much less than the concentration C_1 of the uni-univalent neutral salt, i.e. $zC_2 \ll C_1$ then the observed osmotic pressure is that which would be caused by the macro-ions alone and $p = RTC_2/1000$. On the other hand if the concentration of the salt is much less than that of the charged macromolecule, i.e. $C_1 \ll zC_2$ the osmotic activity of the counter-ions is effective and $p = RT(z+1)C_2/1000$. This is dealt with fully by OVERBEEK (1956). At intermediate salt concentrations the contribution of diffusible ions to the observed osmotic pressure can be calculated (see below). In practical measurement of osmotic pressure for the determination of molecular weight of a charged macromolecular solute, salt is added and p/C_2 or p/c_2 is extrapolated to zero colloid concentration to overcome the DONNAN effect.

b) Non-Ideal Charged or Uncharged Macromolecule, i. e. the General Case.

Solutions of macromolecules are in fact not ideal and an equation for the expression of osmotic pressure results should take this into account. There are two essentially similar approaches. For any macromolecule (charged or uncharged) the experimentally observed osmotic pressure can be expressed in the form

$$p = g' R T c_2/M_2 = (\nu + 1)\, \Phi\, R T c_2/M_2 \,.$$

Here Φ is the practical or molal osmotic coefficient and $(\nu + 1)$ is the number of ions into which the macromolecule dissociates. g' is an osmotic coefficient which includes the effect of non-ideality and the effect due to counter-ions and unequal distribution of diffusible ions if the colloid is charged. If the macromolecule is uncharged $g' = \Phi$. ALEXANDROWICZ (1959), among others, has used a similar approach to this. Or one may write an equation of the type

$$p = R T c_2 (1/M_2 + A_2 c_2 + A_3 c_2^2 + \cdots)\,. \qquad (2)$$

Here, $A_2, A_3, \ldots$ are called second, third, ... virial coefficients and include the activity and charge effects. SCATCHARD (1946) and EDSALL (1953, p. 584) have used this equation for proteins and FLORY (see FLORY 1953, p. 280, 513) has used a similar equation for polymers.

[1] The use of a subscript 2 for the macromolecular component is in accordance with SCATCHARD (1946) and EDSALL (1953) who used even numbers for the non-diffusible components and odd numbers for the diffusible ones.

For synthetic polymers in the range $p/c < 3(p/c)_0$ FLORY states that in general the second and third virial coefficients should be enough to describe the behaviour, $(p/c)_0$ being the value of p/c at infinite dilution. For globular (i.e. not too asymmetric) proteins which behave more ideally than synthetic polymers (this implies that some neutral salt is present if the protein is charged) the A_3 term may be neglected giving

$$p/RTc_2 = 1/M_2 + A_2 c_2 \,. \tag{3}$$

This equation is sufficient for quite considerable concentration ranges. For bovine plasma albumin SCATCHARD et al. (1946) showed it described the experimental data in the range 1—6 per cent protein concentration in 0.15 molar sodium chloride concentration and pH range 5.37—7.0.

Since these two approaches are essentially the same the "virial coefficient" one will be followed here and from now on consideration will be restricted to the concentration region where the third and higher virial coefficients can be neglected, (i.e. equation 3). The discussion will be mainly on proteins as the macromolecules but the considerations apply equally well to other molecules provided their activity coefficients in the solution do not differ greatly from unity.

A_2 is in general positive; a negative value indicates dissociation of the macromolecule on dilution. The osmotic coefficients and virial coefficients are related to those which appear in the corresponding light-scattering and sedimentation equilibrium equations. For instance for a *homogeneous* macromolecular solute the A_2 in equations (2) and (3) is identical with the A_2 in equations for light scattering (equn. 16) and for sedimentation equilibrium (equn. 67).

SCATCHARD (1946) has derived general equations for the osmotic equilibria of proteins, and for the case of a three-component system of a protein in a solution of a uni-univalent salt the osmotic pressure may be expressed by the following equation (SCATCHARD, 1946, p. 2318; EDSALL, 1953, p. 592):

$$\frac{p}{RTc_2} = \frac{1}{M_2} + \frac{1000}{2M_2^2}\left(\frac{Z_2^2}{2m_3} + \beta_{22} - \frac{\beta_{23}^2 m_3}{2+\beta_{33} m_3}\right) c_2 \tag{4}$$

where $c_2 = g$ of protein per cc of solution, M_2 = molecular weight of protein and Z_2 its charge, m_3 = molarity of the uni-univalent salt, p = osmotic pressure. β_{22}, β_{23} and β_{33} are activity terms representing protein-protein, protein-salt and salt-salt interactions respectively. We will use this particular three-component system for illustrative purposes here.

Comparison of equations (3) and (4) shows that A_2, i.e. the coefficient of c_2, has been split into three terms — a DONNAN term and two activity terms. For the system being considered the DONNAN term, $\frac{1000}{2M_2^2}\cdot\frac{Z_2^2}{2m_3}c_2$, caused by the charge of the protein molecule is zero if the net charge on the protein is zero, i.e. for a protein at its isoelectric point; it is diminished by increasing the salt concentration m_3 and is very large for a charged solute at low salt concentration. To get an idea of the magnitude of this term it is found that an approximately 0.86 per cent solution of bovine plasma albumin (M.W. 69,000) in 0.15 M sodium chloride at pH 7.4 has a net charge of approximately -15; the isoelectric point at this sodium chloride concentration is at pH 5.3 approximately (SCATCHARD et al., 1946). These figures show that the DONNAN term is about 5 per cent of $1/M_2$. If the salt concentration is 0.05 M it would be approximately 15 per cent and if it were 0.3 M it would be approximately 2 per cent of $1/M_2$. This term will vanish at infinite dilution of the protein provided sufficient salt is present (see OVERBEEK, 1956). This cannot be seen from equation (4) because this equation is an approximation.

The second term $\beta_{22} \cdot \frac{1000}{2 M_2^2} c_2$ is a function of the variation of the activity coefficient of the protein with its own concentration and must vanish at infinite dilution of protein. It is large for very asymmetrical particles but small for globular proteins. The third term is an expression of the interaction between protein-salt and salt-salt molecules. It is generally small but will not vanish at infinite dilution of the protein.

On the whole when considering globular proteins in aqueous solutions the Donnan term is the only one that is important when measuring osmotic pressure with protein concentrations of 2 g/100 ml solution or less.

3. Extrapolation to Zero Concentration.

To determine molecular weight the value of p/c_2 at $c_2 = 0$ is required and then M_2 is calculated from equation (1). To extrapolate p/c_2 it is expedient to find conditions where A_2 is small so as to increase the certainty of extrapolation. In the case of synthetic polymers A_2 is small in poor solvents or in theta solvents near the "Flory temperature" (see E. III); for proteins it is seen from the previous paragraphs that A_2 can be made small by having salt present and/or zero charge on the protein. For rod-like macromolecules in good solvents A_2 perhaps cannot be made small and all that can be done is to extrapolate to zero concentration from the lowest experimentally-feasible concentrations of macromolecules.

II. Experimental Method.

1. Types of Osmometers.

There are many types of osmometers and a number of them are described in the article by Wagner and Moore (1959) and Adair (1961). The main thing desired in a good osmometer is the rapid attainment of equilibrium. This aim is assisted if the volume of the macromolecular solution is small and the membrane area large, the membrane is porous, the capillary diameter in which the pressure is developed is small, and the pressure is initially set near the final expected pressure.

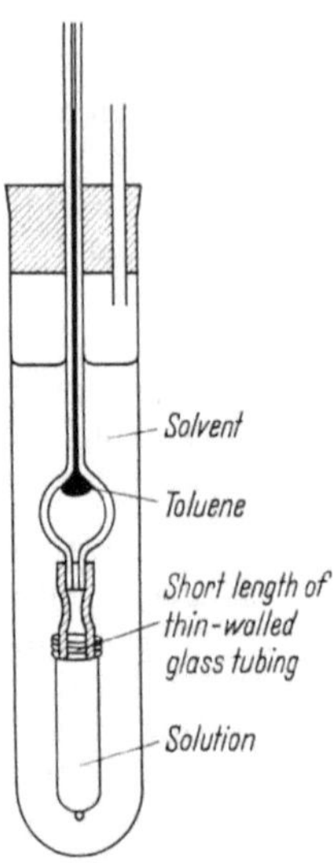

Fig. 1. Simple osmometer.

Mention will be made here of only four types of osmometer. The *simple osmometer* which has been used extensively by Adair (see Adair 1961) is made from a thin bag of collodion or cellulose tubing. The open top of the bag is slipped over a rubber sleeve and fixed with a rubber band or piece of cotton. The bag is then filled with colloid solution (thoroughly dialyzed previously in the case of charged molecules such as proteins) and a capillary of 0.5—1 mm bore is inserted into the rubber tubing, care being taken to avoid air bubbles. For aqueous solutions toluene is often put on top of the buffer in the capillary so that the glass is wet easily (see Fig. 1). This is then immersed in a large test tube (4.5 cm diameter) of solvent and held in place by pushing the capillary through a stopper, the solvent being higher than the solution in the bag. There should be a hole in the stopper so that the solvent and solution are at the same external pressure. The whole osmometer may be enclosed in an air-tight vessel if desired. The whole is then put in a water bath at constant temperature

($\pm 0.01°$) and the experiment is allowed to proceed until the difference in levels between the liquid in the capillary and the outer solvent is constant. If in the case of charged colloids the dialysis has been sufficient, the membrane is thin, and the initial pressure be close to the equilibrium pressure, then readings close to the final pressure are obtained in a few hours but small changes due to capillarity may continue for several days. Equilibrium can be attained more quickly if the whole apparatus is gently rocked while in the bath. [BURK and GREENBERG (1930) did this when measuring osmotic pressure of proteins in concentrated urea solutions.] The capillary rise of solution itself in the capillary must be determined and subtracted to give the correct osmotic pressure (see ADAIR, 1961).

The second type of osmometer is exemplified by the ZIMM MYERSON (1946) osmometer. This instrument (as modified by RIESEL and BERGER, 1959, who incorporated also some modifications by STABIN and IMMERGUT, 1954) would appear to be a satisfactory and convenient instrument. It uses two flat membranes which allows fairly rapid equilibrium attainment. It is shown in Fig. 2 and is almost self-explanatory. RIESEL and BERGER found the C-clamp to be a big improvement over the original design. Its use eliminated accidental leakages at the places where the membrane forms a seal with the glass inner compartment. The solution-containing compartment holds about 2 ml and is separated on both ends from the outer solvent by two membranes about 1.5 cm in diameter, the distance between the membranes being about 1 cm. The osmometer is assembled with its membranes (which have been soaked in the outer solvent). Following RIESEL and BERGER (1959) the inner compartment is nearly filled with solution through the 2 mm diameter filling capillary using a syringe fitted with a polyethylene capillary tube. If aqueous solutions are being used it is then topped in both capillaries with an organic solvent such as decalin or toluene. The rod F to position the meniscus is put in place and sealed with a drop of mercury. The osmometer can then be immersed in solvent and this may be covered with 2—3 mm of decalin or toluene. The whole is immersed in a constant temperature bath ($\pm 0.01°$ or better) and equilibrium is allowed to be attained. The height of the liquid in the capillary above that of the outer liquid is then read. It is reproducible to within ± 0.2 to ± 0.5 mm. The capillary rise with outside solvent on both sides of the membrane must be determined and subtracted from the height obtained with the macromolecular solution. If aqueous solutions are being used and these contain salt, as is usual when measuring the osmotic pressure of protein solutions, it is advisable to dialyze the colloid solution against the salt solution before putting it in the osmometer.

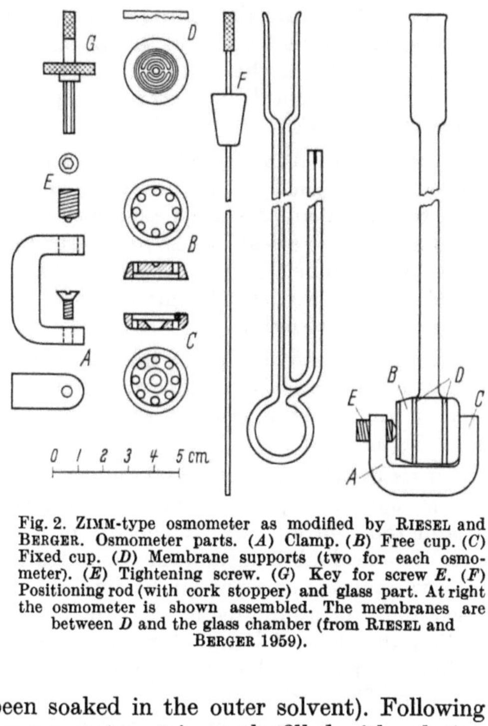

Fig. 2. ZIMM-type osmometer as modified by RIESEL and BERGER. Osmometer parts. (A) Clamp. (B) Free cup. (C) Fixed cup. (D) Membrane supports (two for each osmometer). (E) Tightening screw. (G) Key for screw E. (F) Positioning rod (with cork stopper) and glass part. At right the osmometer is shown assembled. The membranes are between D and the glass chamber (from RIESEL and BERGER 1959).

In aqueous solutions the stainless steel C-clamp used by Riesel and Berger corroded in the presence of chloride ions. Sodium acetate and acetic acid were inert as also was a mixture of nitric acid and sodium nitrate at pH 3. A different stainless steel containing more molybdenum would probably be satisfactory in the presence of chloride ions or tantalum, platinum-rhodium, or possibly monel metal could be used.

A concentration osmometer has been developed by Alexandrowicz (1959) and its principle is similar to that involved in the isopiestic method except that equilibrium is maintained in the liquid phase through a membrane rather than via the vapour phase. In the concentration osmometer a solution of known activity is separated from a solution of unknown activity by a semipermeable membrane. These are allowed to come to equilibrium under the same hydrostatic pressure. The concentrations of the solutions are then determined and the apparent molecular weight of the unknown solute can be calculated from the known activity of the standard. By extrapolation to infinite dilution the true molecular weight is obtained.

According to Alexandrowicz the advantages of the osmotic concentration method are simplicity of apparatus (concentration measurements are in general easier to make than pressure measurements), lack of need for temperature control, and the fact that a leak in the apparatus involving loss of solution is much less serious (and can be taken into account) than when measuring pressures. This last-named point is particularly important when studying concentrated solutions of ionized macromolecules.

The main disadvantage of this method is that the instrument must be calibrated using a suitable standard. Alexandrowicz used Carbowax 6000 polyethylene glycol and determined its activity at various concentrations by the light-scattering method.

An *electronic osmometer* has been designed by Rowe and Abrams (1957) and is now manufactured commercially by Nash and Thompson, Oakcroft Road, Chessington, Surrey U.K. It consists of a U-tube with a tap between the two limbs. In one limb a membrane separates the colloid solution from the solvent and a variable air pressure can be applied to the top colloid solution. If the tap is closed and a pressure greater than the osmotic pressure is applied to the colloid solution the flow of liquid down through the membrane displaces a piece of platinum foil inserted in the lower wall of the tube between the tap and the membrane. This deflection is measured by a galvanometer via a mechano-electronic transducer. At the equilibrium pressure there is no deflection and this gives the osmotic pressure of the solution.

The advantages of this system of measuring osmotic pressure are that it does not require thermostatic control, 0.5 ml of colloid solution is sufficient for a determination, and osmotic pressures can be determined in 15—30 minutes.

a) Dynamic and Static Methods.

Osmotic pressure may be determined by either static or dynamic methods. In the static method the instrument is allowed to stand at constant temperature until equilibrium is reached. In the dynamic method (which can be applied with the Zimm-Myerson osmometer) the pressure is varied and the rate of flow of liquid at various pressures is measured by following the movement of liquid in the capillary. The figures can be interpolated to find the pressure (osmotic pressure + + capillary rise) at which no movement occurs.

b) Membranes.

The biggest difficulty in obtaining exact and reliable osmotic pressure measurements is getting a satisfactory membrane which will allow salts and solvent to diffuse through at a reasonable rate but which is completely impermeable to the macromolecules being considered. If a membrane allows appreciable amounts of macromolecules to pass through in the time it takes for equilibrium to be attained in the osmometer it is fairly useless. The variety of results obtainable on one non-homogeneous polymer with different membranes is well illustrated by the work of DONNET, ROTH and MEYERHOFF (1958).

Both flat and bag-shaped membranes of varying average pore sizes can be made from collodion (see WAGNER and MOORE 1959 and ADAIR 1961) but the commercial availability of cellulose tubing, sheet cellophane and graded cellulose membranes is a big attraction. Visking cellulose tubing (Visking Corporation, Chicago, Ill., U.S.A.) is satisfactory for bag-type osmometers and work by CRAIG and co-workers (see CRAIG, KONIGSBERG, STRACHER and KING 1958) has given figures for their permeability to various proteins. Visking 18/32 tubing was the one CRAIG found to have the finest pore size, but 8/32 is satisfactory for many purposes. Of the flat membranes the *non-waterproofed* cellophane sheet Nos. 300 and 600 (obtainable from E. I. Du Pont de Nemours and Co., Wilmington, Delaware, U.S.A., or British Cellophane Co., England), are satisfactory. They probably have an average pore size of 2—3 mμ. This cellophane sheet can also be obtained in the undried state and is called gel cellophane and apparently has a lower porosity than the dried material.

Graded cellulose membranes are commercially available from Membranfilter-Gesellschaft, Sartorious-Werke A.G. Göttingen, Germany, and from Schleicher and Schull Co., Keene, New Hampshire, U.S.A. There are a variety of membrane sizes having average pore diameters down to less than 5 mμ. The membrane filters for aqueous solutions are the "Ultrafine" series. With the cellophane sheet and the "Ultrafine" filters the lower limit of permeability is a molecular weight of around 10,000. The pore sizes of the cellophane sheet membranes can be changed by swelling (see WAGNER and MOORE, 1959).

Usually the membranes can be stored in water containing 1 per cent formaldehyde. Membranes should be soaked for several hours in changes of the solvent in which they are to be used.

2. Difficulties and Precautions.

(i) Membranes. The membrane must be impermeable to the large solute molecules and this can best be determined by examining the solvent at the end of an osmotic pressure experiment to see if any macromolecules have diffused through.

(ii) Ballooning is the term used for slow stretching of the membrane and will vary with the size of the membrane. It is particularly troublesome in dynamic methods but in static methods it is overcome if the equilibration is allowed to go on long enough. Collodion bag-type membranes are probably less troublesome than cellulose tubing ones. Its elimination is effected by clamping a flat membrane firmly between surfaces which are relieved to allow free movement of the solvent and macromolecular solution.

(iii) It is important that the **temperature** of the osmometer be kept constant to within about 0.01°. This is not so much because the osmotic pressure is directly proportional to the absolute temperature but because of the thermometer effect which causes the liquid in the capillary to oscillate with changes in temperature.

The temperature control must be good enough so that this does not happen. The osmometer is usually immersed in a liquid bath.

(iv) The **solute concentration** in the osmotic pressure equation must be that existing at equilibrium. It can either be determined on the macromolecular solution after equilibrium has been attained or it can be calculated from the initial concentration by correcting it for the amount of solvent that has entered; the extent of dilution can be determined from the movement of liquid in the capillary during the experiment. If there is adsorption of macromolecular solute on the membrane the only safe procedure is to determine the concentration at equilibrium.

(v) It must be certain that the apparatus is not **leaking**; if it is, a stable height of meniscus over a long period will not be maintained. The testing of this can be done by applying a pressure 5—10 times the one to be observed and looking for leaks — use of a solution of a dye can be helpful in this regard.

(vi) If an osmometer is used which requires more than a day to allow equilibrium to be attained precautions should be taken against **bacterial growth** particularly for proteins in aqueous solutions. This can usually be achieved by working at low temperatures (2—4°) or by adding a preservative.

(vii) No **gas bubbles** should be present in the osmometer.

(viii) With considerable care and trouble the height of the column of liquid in the equilibrated osmometer can be measured to ± 0.1 mm. Ordinarily, the precision is ± 0.2 to ± 0.5 mm. Increased precision can be obtained with the osmotic balance developed by JULLANDER (1945) and modified by ENOKSSON (1948) but this has not yet achieved a wide use.

(ix) ADAIR (1925) states that three criteria must be satisfied to be sure a **true osmotic equilibrium** has been attained:

(i) The pressures must be permanent.

(ii) The pressure must be reversible, i.e. the same final pressure should be approached from either side.

(iii) The pressures must be reproducible, i.e. different membranes in different experiments should give the same answer.

C. Light Scattering.

The scattering of light by the molecules of a gas and by dispersed particles such as smokes or colloidal particles suspended in a liquid medium (RAYLEIGH and TYNDALL scattering) is a well-known phenomenon. Although first used by PUTZEYS and BROSTEAUX in 1935 to determine molecular weights of proteins on a relative scale, the main impetus to the application of the method stemmed from the work of DEBYE (1944, 1947) who applied the method to determine the molecular weights of high polymers in solution. Light scattering has become in recent years one of the most used methods for the study of natural and synthetic macromolecules. In the most favourable cases the measurement of the intensity of the scattered light by dilute solutions of macromolecules can furnish the weight of the molecule, a measure of its maximum dimension, and also information on thermodynamic interactions. An added advantage is that the solution remains undisturbed during the measurement so that it is possible to follow changes with time.

Some knowledge of the nature of scattering is essential for the interpretation of light scattering data; however for a more comprehensive account of the theory of light scattering the reader is referred to the following reviews — OSTER (1948), EDSALL and DANDLIKER (1951), DOTY and EDSALL (1951), STACEY (1956), GEIDUSCHEK and HOLTZER (1958), and PETERLIN (1959).

I. Fundamental Concepts of the Theory of Light Scattering.

In terms of the electromagnetic theory, a light beam represents an electromagnetic field and is characterised by an electric and magnetic vector oscillating in a direction always perpendicular to the direction of the propagation of the wave.

For the phenomena considered here the effects of the magnetic component are negligible compared to those of the electric field and will not be considered. A perfectly homogeneous medium e.g. a vacuum, does not scatter light. However when light falls on matter a disturbance of the wave motion results. The oscillating electric field of the light wave causes a displacement of the electrons of the substance from their unperturbed positions, i.e. causes polarization, the particle now possessing an induced dipole moment. The induced dipoles oscillate with the same frequency as the electric field of the light wave, and these act as secondary sources of electromagnetic radiation and light of the same frequency as the incident radiation is scattered in all directions. The intensity of the scattered radiation increases with the size of the molecules, the number of scattering centres per unit volume, and their electric polarizability since the more easily the electrons are displaced the greater will be the scattering. The interaction of radiation with matter is also responsible for many other phenomena such as refraction of light and this enables the polarizability to be determined experimentally from the magnitude of the refractive index. Due to the scattering of light there is an attenuation of the light beam as it passes through an inhomogeneous medium. The scattered light may thus be either measured directly or by the decrease in intensity of the incident beam after passing through the medium, provided there is no absorption of light.

The calculation of the scattering of a system can be approached in two ways (a) by a detailed calculation of the electromagnetic field surrounding a particle in order to derive the loss of light energy due to scattering or (b) by treating the scattering as the result of fluctuations in density or concentration as a result of thermal motions of the molecules. We will consider first some fundamental relations which begin with the work of Lord RAYLEIGH in 1871 who studied the scattering from a dilute gas composed of small isotropic molecules (RAYLEIGH scattering). The electric moment induced in an isotropic molecule is always parallel to the electric field vector of the applied field. For isotropic particles which are smaller than one twentieth of the wave length of light, the intensity of light, i_θ, per unit volume of the scattering medium scattered at an angle θ from the incident beam, relative to the intensity of the incident light I_0, is given by the expression —

$$\frac{i_\theta}{I_0} = \frac{8\,\pi^4 \nu \alpha^2}{\lambda^4 r^2}(1 + \cos^2\theta)\,, \tag{5}$$

when the incident light is unpolarised. α is the polarisability ($p = \alpha F$ where p is the induced electric moment and F the electric vector of the incident light), r is the distance from the centre of scattering to the observer, ν is the number of molecules per cc, and λ is the wave length of light in the medium; $\lambda = \lambda_0/n$ where λ_0 is the wave length in vacuo and n is the refractive index of the medium. It is convenient to define the reduced intensity of scattering, R_θ, called RAYLEIGH's ratio, as follows:

$$R_\theta = i_\theta r^2/I_0 = R_0(1 + \cos^2\theta)/2 = R_{90}(1 + \cos^2\theta)\,. \tag{6}$$

For unpolarised incident light if the light source, scattering object and observer are in a horizontal plane, the scattered light consists of two components — a component with the electric vector vertical (called vertically polarised) whose intensity is independent of angle and a horizontally polarised component, the intensity of which is proportional to $\cos^2\theta$. The total intensity of scattering is thus proportional to $(1 + \cos^2\theta)$ as shown by equation (5). The radiation diagram characterising the secondary wave scattered from an incident unpolarised beam is depicted in Fig. 3, the scattering envelope approximating the shape of a dumb-bell.

The characteristics of RAYLEIGH scattering for unpolarised incident light can be summarised as follows:

(1) The scattering diagram is symmetrical i.e. just as much light is scattered in the backward direction as in the forward direction i.e. the scattering at $(90 + \theta)$ equals the scattering at $(90 - \theta)$.

(2) The scattered light is unpolarised in the forward and backward directions and partially polarised at all intermediate angles between 0° and 90°, and 90° and 180°.

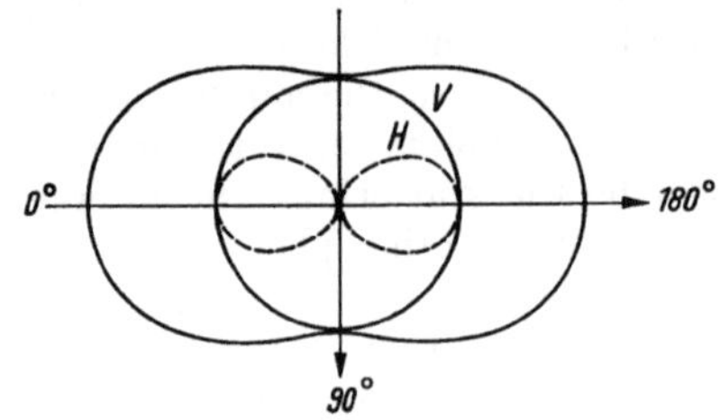

Fig. 3. Scattering envelope for small isotropic molecule. The molecule is taken as located at the origin and the forward direction of the incident beam is indicated by the arrow. The length of the line drawn from the origin to a point on the curve is proportional to the intensity of scattered light. The continuous line represents the vertically polarised component (V) whose intensity is independent of angle (electric vector perpendicular to plane of paper), and the dotted line is the horizontally polarised component (H) which varies as $\cos^2\theta$. The scattering observed (R) at any angle is the sum of the two and is given by the dumb-bell shaped curve.

(3) At $\theta = 90°$ to the incident beam the light is totally polarised and its electric vector is perpendicular to the plane containing the direction of the incident beam and that of the scattered light.

(4) The intensity is greatest in a direction parallel to that of the incident beam and weakest in a direction perpendicular to this: the intensity at 90° is half that at 0°.

The scattering may be expressed in terms of the decrease in intensity of the incident beam as it passes through the medium. If I_0 is the intensity of the incident beam, I that of the transmitted beam, l the path length of the scattering system then

$$I = I_0 \exp(-\tau l)$$

or in terms of logarithms

$$\tau = \frac{1}{l} \ln \frac{I_0}{I} \tag{7}$$

where τ is the turbidity of the medium. If the total radiation intensity is integrated over all angles of scattering it can be readily shown that

$$\tau = \frac{8\pi}{3} R_0 = \frac{16\pi}{3} R_{90}\,. \tag{8}$$

The turbidity, τ, and the reduced intensity have dimensions of reciprocal length and are expressed in cm^{-1}. It has become widespread in light scattering measurements to use the quantity τ following the work of DEBYE. However R_θ, or some function of it is generally the quantity which is measured in light scattering investigations. The turbidity, τ, as defined only expresses the diminution in the incident intensity at 0° and is difficult to relate to other angles so it is preferable to use R_θ which refers to the actual quantity measured. The distinction becomes even more important when the angular intensity is unsymmetrical as in the case of macromolecules whose dimensions are not small compared with the wave length of light.

For the practical application of equation (5) the polarisability α is replaced by the well-known relations between polarisability and dielectric constant which is equal to the square of the refractive index if there is no absorption of light. Thus we may write

$$\frac{n^2 - n_0^2}{n^2 + 2n_0^2} = \frac{4}{3}\pi \nu \alpha \tag{9}$$

where n is the refractive index of the particle, and n_0 that of the reference medium which for a dilute gas is a vacuum for which $n_0 = 1$, and therefore

$$\frac{n^2 - 1}{n^2 + 2} = \frac{4}{3}\pi \nu \alpha$$

and since n differs little from unity for gases $n^2 + 2 \approx 3$ and $n + 1 \approx 2$, thus $2\pi \nu \alpha \approx n - 1$, and equation (5) now becomes

$$R_\theta = \frac{2\pi^2 (n-1)^2}{\lambda^4 \nu} (1 + \cos^2\theta). \tag{10}$$

II. Scattering by Dilute Solutions.

The preceding treatment can be extended to dilute solutions if the solvent is considered homogeneous and we are interested only in the additional scattering due to a small amount of solute in a much larger amount of solvent. In equation (9) n is now the refractive index of the solution and n_0 that of the solvent. Making the approximations $n + n_0 = 2n_0$, $n^2 + 2n_0^2 = 3n_0^2$ then

$$\frac{n - n_0}{n_0} = 2\pi \nu \alpha .$$

If c is the concentration in g/ml, then $\nu = cN/M$ where M is the molecular weight of the particle, N is Avogadro's number, and substituting dn/dc for $(n - n_0)/c$, $\lambda = \lambda_0/n_0$, equation (5) takes the form

$$R_\theta = \frac{2\pi^2 n_0^2 (dn/dc)^2}{N\lambda_0^4} \cdot c \cdot M(1 + \cos^2\theta) \tag{11}$$

and for scattering at 90°

$$R_{90} = \frac{2\pi^2 n_0^2 (dn/dc)^2}{N\lambda_0^4} \cdot c \cdot M = KcM \tag{12}$$

where

$$K = \frac{2\pi^2 n_0^2 (dn/dc)^2}{N\lambda_0^4} .$$

Alternatively, combining with equation (8)

$$\tau = \frac{32\pi^3 n_0^2 (dn/dc)^2}{3N\lambda_0^4} \cdot c \cdot \mathrm{M} = HcM \tag{13}$$

where

$$H = \frac{32\pi^3 n_0^2 (dn/dc)^2}{3N\lambda_0^4} .$$

The constants K and H are characteristic for a given solvent-solute system and can be readily evaluated. Thus the molecular weight can be calculated provided the measurements are made at sufficiently low concentrations to justify the assumption that the particles scatter independently of each other. It is sufficient to plot $\frac{Kc}{R_{90}}$ or $\frac{Hc}{\tau}$ against c and extrapolate to zero concentration. The intercept at $c = 0$ is equal to $1/M$. The extrapolation should be linear if the measurements are made at low enough concentration.

III. Fluctuation Theory of Scattering.

In the preceding treatment it has been assumed that the scattering particles are sufficiently far apart to scatter independently of each other. These conditions do not hold for liquids and solids where there is destructive interference from light scattered from different molecules. In a completely ordered state, for example a perfect crystal at the absolute zero, no light should be scattered. In liquids the intensity of the scattered light is only about one fiftieth of that expected for independent particles. The direct calculation of the scattered light for liquids from interference theory is too difficult but the problem was solved by SMOLUCHOWSKI and EINSTEIN who considered the observed scattering to be due to non-uniformities arising from minute fluctuations in density. In a solution we have in addition inhomogeneities due to fluctuations in concentration. The concentration of solute molecules in any small region of the solution fluctuates constantly, although the average concentration taken over a long period of time is the same as the solution as a whole. The fluctuations in solvent density are much smaller than the concentration fluctuations and need not be considered here since their contribution is eliminated by deducting the scattering intensity observed for the pure liquid from that of the solution. The chance of a fluctuation in concentration is proportional to the change of GIBB's free energy that would be involved in producing the same concentration change and this is in turn related to the change in osmotic pressure that would be produced by a change in concentration of solute. The concentration fluctuations can also be related to corresponding fluctuations in the optical dielectric constant. These considerations lead to the equation

$$R_\theta = \frac{2\pi^2 n_0^2 (dn/dc)^2}{N\lambda_0^4 (\partial P/\partial c)} \cdot R \cdot T \cdot c(1+\cos^2\theta) \tag{14}$$

or for the scattering at 90°

$$R_{90} = \frac{K \cdot R \cdot T \cdot c}{(\partial P/\partial c)} \tag{15}$$

where P is the osmotic pressure of the solution. A fuller account of the derivation of these equations is given in the review by OSTER (1948). The osmotic pressure may be expressed in powers of concentration by means of the equation

$$\frac{P}{cRT} = \frac{1}{M} + A_2 c + A_3 c^2 + \cdots \tag{16}$$

Inserting equation (16) into equation (15)

$$\frac{Kc}{R_{90}} = \frac{1}{M} + 2A_2 c + 3A_3 c^2 + \cdots \tag{17}$$

or

$$\frac{Hc}{\tau} = \frac{1}{M} + 2A_2 c + 3A_3 c^2 + \cdots.$$

At $c = 0$ equations (17) are the same as equations (12) and (13). In practice Kc/R_{90} is usually measured in dilute solutions so that only the first two terms on the right-hand side of equations (17) need be considered and Kc/R_{90} (or Hc/τ) plotted against concentration and the data extrapolated linearly to zero concentration. As in section C II the intercept is equal to the reciprocal of the molecular weight. The quantity, A_2, is called a second virial coefficient and is given by half the slope of the light scattering plot. It is the same as that obtained from an osmotic pressure-concentration plot and is a measure of the interactions of the particles, and their shape. In general large values of A_2 result from the steric consequences of an expanded or asymmetric shape, or from the existence of a net repulsion between the particles. For compact molecules like globular proteins in aqueous solutions in the absence of high net charge (for example in the presence of sufficient neutral salt) the value of A_2 is low. For rod-like molecules such as tobacco mosaic virus the interaction constant, A_2, is large and its effect is apparent at much lower concentrations. For long coiled polymer molecules the value of A_2 depends mainly on the solvent. In good solvents the molecules are highly expanded even in the absence of electrostatic repulsions and large values of A_2 are observed whereas in poor solvents such as solvent-precipitant systems low values of A_2 are obtained. Greater intermolecular attraction leads to negative values of A_2 which indicates that solute-solute contacts are sufficiently preferred over solute-solvent contacts and that phase separation is close.

IV. Systems of Isotropic Particles Comparable in Size to the Wave Length: Internal Interference.

In the preceding sections equations were derived on the assumption that all the light scattered from a given particle is in phase, the scattering particles being considered as point scatterers which could be represented by single dipoles. For

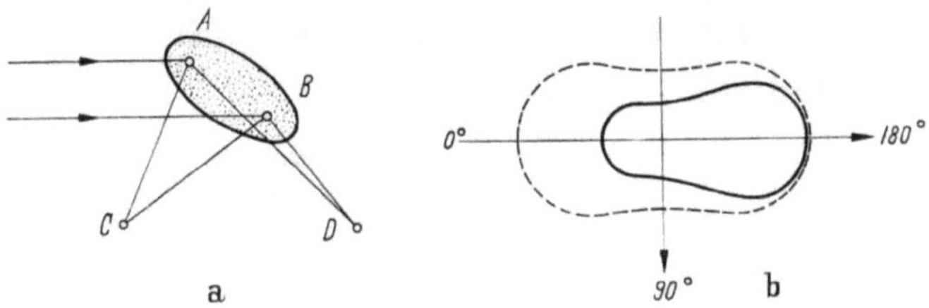

Fig. 4. (a) Scattering from a large particle. Destructive interference occurs between light scattered from different volume elements A and B. The path length difference for light scattered in the backward direction ($AB + BC — AC$) is greater than for light scattered in the forward direction ($AB + BD — AD$) where C and D are at angles symmetrical about the perpendicular to the incident beam. This leads to a dissymmetry in the angular distribution of intensity as illustrated in Fig. (b) which compares the scattering envelopes for small particles (dotted line) and large particles (continuous line).

particles whose largest dimension exceeds approximately one twentieth of the wave length of the light these assumptions are not valid. For such particles light rays arriving at the observer from different points in the one molecule have travelled different path lengths; consequently they are out of phase and interference results. There is an all round decrease in scattered intensity with a greater reduction in the backward direction because of the greater path difference. The result is a dissymmetry in the angular distribution of intensity as illustrated in Fig. 4. For $\theta = 0$ there is no interference and as the angle of scattering increases the interference increases also because of the increase in the path length difference. The intensity of the 90° scattering is less than that of corresponding particles of the same molecular weight but smaller size; hence a correction for this loss in intensity due to interference must therefore be applied before the molecular weight can be calculated from equation (17). The effect of the internal interference for dilute solutions can be described by inserting a particle scattering factor $P(\theta)$ into the light scattering equations. Equations (6) and (11) now become

$$R_\theta = R_0 \frac{1 + \cos^2\theta}{2} \cdot P(\theta) \tag{18}$$

$$(R_\theta)_{c=0} = KcM(1 + \cos^2\theta)\, P(\theta)\,. \tag{19}$$

For small particles $P(\theta) = 1$ and for particles greater than $1/20\lambda$, $P(0) = 1$ and $P(\theta)$ decreases with increase of θ. The scattering envelope for such a particle can be evaluated by summing up the waves from the scattering points taking into account their phase differences. This summation can be performed if a number of assumptions are made. The solution should be sufficiently dilute so that there is no interference of light from different molecules, i.e. no external interference, but only interference from different points on the same molecule (internal interference). The molecules are assumed to be isotropic, randomly oriented in space, not to absorb light, and the effect of variations of refractive index inside the particles can be neglected. Finally it is assumed that the electric field of the incident beam is not significantly distorted by the particle having a different refractive index to the solvent. The problem is exactly analogous to that encountered in X-ray scattering and yields the same general solution

$$P(\theta) = \sum_i \sum_j \frac{\sin h r_{ij}}{h r_{ij}} \tag{20}$$

where the double summation is performed over all pairs of scattering elements i and j, $h = \frac{4\pi}{\lambda} \sin\theta/2$, and r_{ij} is the distance between the ith and j^{th} scattering element. $P(\theta)$ can be readily calculated for various geometrical shapes and the expressions for three of the most common shapes are as follows.

Rods (NEUGEBAUER 1943) of length L

$$P(\theta) = \frac{1}{x} \int_0^{2x} \frac{\sin w}{w} \cdot dw - \left(\frac{\sin x}{x}\right)^2 \tag{21}$$

where

$$x = hL/2 = \frac{2\pi L \sin\theta/2}{\lambda}$$

and w is an integration variable. ($w = hr$, where r is the distance between any pair of scattering centres and need not be evaluated since the integral is a tabulated function).

Spheres (RAYLEIGH 1914) of diameter D

$$P(\theta) = \left[\frac{3}{x^3}\ (\sin x - x\cos x)\right]^2 \tag{22}$$

where $x = hD/2$.

Random coils (DEBYE 1947) of root mean square end-to-end length R

$$P(\theta) = \frac{2}{x^2}\,[e^{-x} - (1 - x)] \tag{23}$$

where $x = h^2R^2/6$.

The particle scattering factors for other shapes are tabulated by GEIDUSCHEK and HOLTZER (1958) together with the original references to their derivation. Graphs and tables of $P(\theta)$ as a function of x can be readily constructed for the various shapes (see Fig. 5) and these are given by DOTY and STEINER (1950) and STACEY (1956) for the three most commonly encountered shapes.

DEBYE (1947) has shown that the particle scattering factor $P(\theta)$ given by equation (20) and its reciprocal $P^{-1}(\theta)$ can be expanded in a power series to give the entirely general result, independent of the shape of the particle,

$$\begin{aligned} P^{-1}(\theta) &= 1 + h^2\varrho^2/3 - \cdots \\ &= 1 + \left(\frac{4\pi}{\lambda}\sin\theta/2\right)^2 \frac{\varrho^2}{3} - \cdots \end{aligned} \tag{24}$$

where ϱ is the radius of gyration of the particle, i.e. the square root of the mean value of the square of the distance between the centre of mass and any mass element of the particle. Thus for a given wave length the initial slope of $P^{-1}(\theta)$ plotted against $\sin^2(\theta/2)$ is directly proportional to the square of the radius of gyration of the particle. It is also evident from equation (24) that if $P^{-1}(\theta)$ is plotted against $h^2\varrho^2$ the curves for all particle shapes have the same initial slope of 1/3 but at higher values of $h^2\varrho^2$ the form of the curve is determined by the particle shape. Equations (19) and (24) combined form the basis for determining the molecular dimensions of macromolecules in solution. One can determine the radius of gyration directly without any assumption as to the shape of the particle. If the shape is known or assumed then a characteristic dimension of the particle can be calculated. The relation between radius of gyration and dimensions of the three most common models are — rod $\varrho^2 = L^2/12$, sphere $\varrho^2 = 3D^2/20$, and random coil $\varrho^2 = R^2/6$. There are two methods in general use for utilising the measurements of angular distribution of light scattering — the dissymmetry method and the ZIMM method.

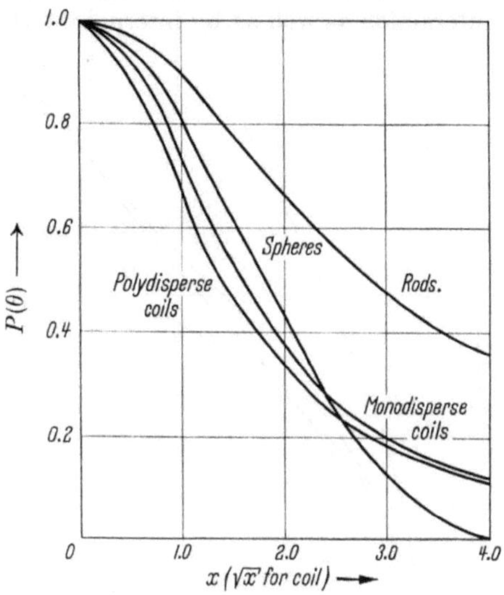

Fig. 5. Variation of particle scattering factor $P(\theta)$ with x for rods and spheres and with $\sqrt{x}$ for random coils.

1. Dissymmetry Method.

Instead of measuring the whole angular dependence of the scattering and extrapolating to $\theta = 0$ the intensity may be measured at 90° and at two angles symmetrical to 90°. In this case the $(1 + \cos^2\theta)$ term cancels out in equation (18) and the ratio $i_\theta/i_{180-\theta}$ is known as the dissymmetry Z. For very small particles Z is unity and increases with the size of the particle. It is evident from the expressions for the various models (equations 21—23) that $P(\theta)$ is a function of both θ and D/λ where D is a characteristic dimension i.e. length of rod, diameter of sphere or root mean square end-to-end length of random coil. Thus equation (18) can be written

$$i_\theta = i_0 \frac{1 + \cos^2\theta}{2} P(D/\lambda, \theta)$$

so that

$$\frac{i_{45}}{i_{135}} = \frac{P(D/\lambda, 45)}{P(D/\lambda, 135)} = Z\,. \tag{25}$$

This equation shows that from a measurement of Z and a knowledge of the functional form of P, D/λ can be calculated and hence D. Graphs of D/λ as a function of Z are readily plotted for various models (Fig. 6) and also $P(90)$ [or better still $1/P(90)$] as a function of Z (Fig. 7). These functions have been tabulated by DOTY and STEINER (1950), STACEY (1956) and by BEATTIE and BOOTH (1960). Thus having measured Z the correction factor $1/P(90)$ to be used in equation (19) can be readily found, thus,

$$(Kc/R_{90})_{c=0} = 1/M\,P(90)\,. \tag{26}$$

The measured values of the dissymmetry must be extrapolated to zero concentration to remove interaction between solute particles and the extrapolated value is called the intrinsic dissymmetry $[Z]$ and is used to determine $1/P(90)$ in equation (26). Since Z is not in general a linear function of c, a better extrapolation is obtained by plotting $1/(Z-1)$ against c.

The intensity of scattered light at finite concentrations is also altered by solute interaction as well as by internal interference. Solute interaction is characterised

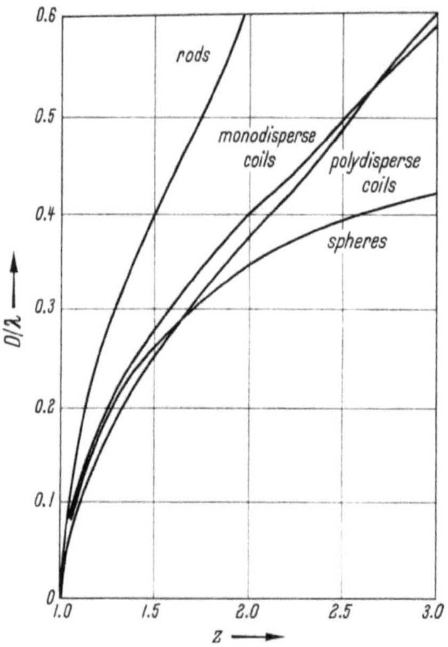

Fig. 6. Particle dimensions D/λ as a function of Z for various models.

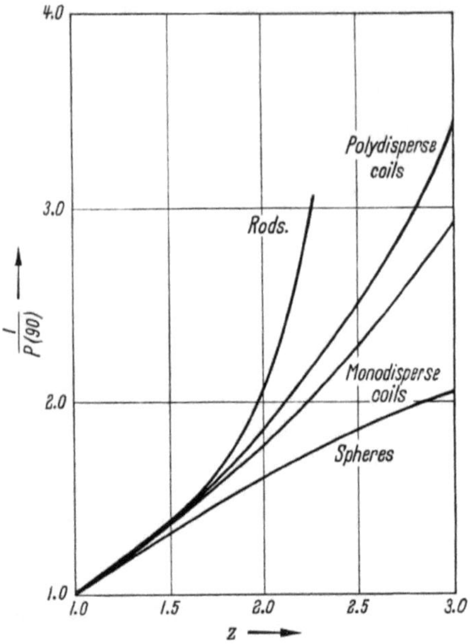

Fig. 7. Correction factor $\frac{1}{P(90)}$ as a function of Z for various models.

by the second virial coefficient A_2 (equation 17) and ZIMM (1948) has shown that in such cases the following equation approximately holds

$$\frac{Kc}{R_{90}} P(90) = \frac{1}{M} + 2A_2 P(90) \cdot c\,. \tag{27}$$

Thus $P(90)$ is determined from the intrinsic dissymmetry $[Z]$, this correction factor then used to multiply the observed values of Kc/R_{90} and these corrected values plotted as a function of concentration. The intercept and slope of the line are then given by equation (27).

The advantages of the dissymmetry method are that the scattering needs to be measured at only three angles (45°, 90°, 135°) and that polarisation factors and most other corrections cancel out or are substantially reduced in the ratio i_{45}/i_{135}. The main disadvantage is that a model of known $P(\theta)$ must be assumed. If the shape deviates from that of the model or the material is polydisperse the exact form of $P(\theta)$ is uncertain and the method will no longer give the correct size or molecular weight, the deviations from the true values being greater the larger the particles. Where the molecular size is small so that $P(\theta)$ is approximately equal for all shapes at all angles (see Fig. 5—7) then the dissymmetry method may be used — this corresponds to dissymmetries less than approximately 1.4. In other cases it is best to measure the complete angular range and use the more general ZIMM (1948) method.

2. ZIMM Method.

In this method no assumptions are necessary about the shape of the particle since the method is based on the fact that at $\theta = 0$, the particle scattering factors are unity. The scattering at zero angle is estimated by extrapolation of the scattering at observable angles. In addition the effects of intermolecular interference need to be removed by extrapolation of the angular scattering to zero concentration. Equations (17) and (24) show that Kc/R_{90} may be expressed as a power series in c, and $P^{-1}(\theta)$ can be expressed as a power series in $\sin^2(\theta/2)$. For small values of θ and c, $1/R_\theta$ is linear both in $\sin^2(\theta/2)$ and in c. The data may be conveniently represented by the equation

$$\frac{Kc}{R_\theta}(1+\cos^2\theta) = \frac{P^{-1}(\theta)}{M} + 2A_2c + \cdots. \tag{28}$$

To perform the extrapolation $Kc(1+\cos^2\theta)/R_\theta$ is plotted as a function of $(\sin^2(\theta/2) + kc)$, where k is an arbitrary constant chosen to spread out the concentration data and facilitate the extrapolation to zero angle and zero concentration. Smooth lines are drawn through the points at constant angle and constant concentration forming a grid of parallel lines. Lines of constant angle are extrapolated to zero concentration, and lines of constant concentration are extrapolated to zero angle. These extrapolated points are then joined by two lines which should meet at the same intercept of $Kc(1+\cos^2\theta)/R_0$ at zero concentration.

At $\theta = 0$, $c = 0$,

$$Kc(1+\cos^2\theta)/R_\theta = \frac{1}{M}, \tag{29}$$

i.e. the molecular weight is equal to the reciprocal of the intercept.

For $c = 0$

$$P^{-1}(\theta) = \frac{Kc}{R} M\,(1+\cos^2\theta). \tag{30}$$

For $\theta = 0$ and small c

$$2A_2 = \frac{d(Kc(1+\cos^2\theta)/R_\theta)}{dc} \tag{31}$$

i.e. A_2 is obtained from the slope of the zero angle line. The radius of gyration can be calculated from the ratio of the initial slope of the zero concentration line to the intercept as follows.

From equations (24) and (30)

$$\begin{aligned} P^{-1}(\theta) &= 1 + \frac{16\pi^2}{3\lambda^2}\sin^2(\theta/2)\,\varrho^2 \ldots \\ &= \frac{Kc}{R_{90}} \cdot M(1+\cos^2\theta) \end{aligned} \tag{32}$$

$$\frac{d[(Kc/R_\theta)(1+\cos^2\theta)]}{d\sin^2(\theta/2)} = \frac{16\pi^2}{3\lambda^2}\,\frac{\varrho^2}{M}$$

and
$$\varrho = \frac{\lambda}{4\pi}\sqrt{3M\left.\frac{d\,Kc(1+\cos^2\theta)/R_\theta}{d\sin^2(\theta/2)}\right|_{c=0}} = \frac{\lambda}{4\pi}\sqrt{\frac{3\text{ limiting slope}}{\text{intercept}}}. \tag{33}$$

The dimensions of the scattering particle may thus be determined from the radius of gyration if the shape is known. A plot of $P^{-1}(\theta)$ as a function of $h^2\varrho^2$ (equation 24) has an initial slope of 1/3 but the form of the curve at higher values may allow identification of the particle as a sphere, random coil, or rod. If the particle does not fit these models then other more complex models may be investigated or the deviation may be due to polydispersity of the sample.

The advantage of the ZIMM method over the dissymmetry method is that it gives a more complete picture of the scattering. The method is more difficult because it depends for extrapolation on data obtained at low angles. A rapid increase of scattering at low angles indicates the presence of material of large size. This may be an indicator of accidental contamination such as dust or a true heterogeneity due to a heavy component whose presence may be overlooked by the dissymmetry method. For particles whose dissymmetry is less than 1.4 the $P^{-1}(\theta)$ curves are almost coincident (see Fig. 7) and for these cases the dissymmetry method is satisfactory.

V. Polydispersity.

It is readily shown that the molecular weight obtained by light scattering is the weight average molecular weight M_w. For each component at $c = 0$,

$$\frac{K c_i}{(R_{90})_i} = \frac{1}{M_i}$$

so that

$$(R_{90})_i = M_i K c_i$$

$$R_{90} = \Sigma (R_{90})_i = \Sigma K c_i M_i, \quad \text{also} \quad R_{90} = K c M = K M \Sigma c_i$$

so that

$$M = \frac{\Sigma c_i M_i}{\Sigma c_i} = M_w \,.$$

The above derivation assumes that all the solute species have the same refractive index increment and that therefore K is independent of molecular weight and shape. This is generally true for most polymer systems but should different molecular species have different refractive index increments, as may be possible in some protein mixtures, appropriate corrections need to be applied in calculating M_w. These corrections may be important since dn/dc occurs as the square in the constant K.

Next we need to consider the effect of polydispersity on the limiting slope of $P^{-1}(\theta)$ plotted against $\sin^2(\theta/2)$. The nature of this average has been discussed by ZIMM (1948) for the case of randomly coiled molecules and more generally by EHRLICH and DOTY (1954). The dimension found in light scattering is the square of the radius of gyration and it can be shown that it is the z-average of this quantity which is determined for a polydisperse system. The dependence of the mean square radius of gyration, $\overline{\varrho^2}$, on the molecular weight depends on the shape of the particle and may be rather complicated since it involves the ratio of ϱ to M. The simplest result is for randomly coiled molecules discussed by ZIMM (1948) where we have ϱ^2 proportional to M and $(\overline{\varrho^2})^{\frac{1}{2}}$ corresponds to the radius of gyration of the molecule having the z-average molecular weight. For spheres the $(\overline{\varrho^2})^{\frac{1}{2}}$ corresponds to a molecule less than the z-average molecular weight; for rods $(\overline{\varrho^2})^{\frac{1}{2}}$ corresponds to a molecule greater than the z-average, in this case the $\sqrt{z(z+1)}$ average.

Further information about the molecular weight distribution in polydisperse systems may be obtained from the shape of the curve relating the particle scattering factor $P^{-1}(\theta)$ with $h^2\varrho^2$. $P^{-1}(\theta)$ is obtained from the zero concentration points on the ZIMM plot by means of equation (30). The $P^{-1}(\theta)$ vs. $h^2\varrho^2$ curve for all polydisperse systems will always be below the $P^{-1}(\theta)$ curve for the monodisperse case (GOLDSTEIN 1953) — the slope of the $P^{-1}(\theta)$ curve decreases at higher angles and this downward curvature increases with the degree of polydispersity. At low angles the larger molecules dominate the scattering and $P^{-1}(\theta)$ behaves like the scattering function of a large molecule whereas at higher angles the smaller molecules are contributing proportionately more and $P^{-1}(\theta)$ will rise less steeply, since the slope of the $P^{-1}(\theta)$ curve is less for small molecules. Unfortunately this downward curvature cannot be interpreted unambiguously since it can be due to polydispersity of molecular weight (BENOIT 1953), polydispersity of molecular shape or size (RICE 1955), or to molecular stiffness (PETERLIN 1953). Polydispersity of molecular weight has been treated for the cases of random coils (BENOIT 1953; BENOIT, HOLTZER and DOTY 1954) and rods (HOLTZER 1955) by the use of the asymptotic forms of the $P(\theta)$ curves for large values of x. In the most favourable circumstances one may obtain M_n, M_w, and the number-average and z-average of ϱ^2.

A particularly common form of distribution encountered in random coil polymers is the exponential distribution $f(M) = 4M/M_w^2\, e^{-2M/M_w}$ and this has been found to be a good approximation to more complicated distributions. In this expression $f(M)$ is the fraction of molecules with molecular weights between M and $M + dM$, and $\int_0^\infty f M \cdot dM = 1$. In this special case $M_z{:}M_w{:}M_n{::}3{:}2{:}1$ and $\left[\frac{Kc\,(1+\cos^2\theta)}{R_\theta}\right]_{c=0}$ vs $\sin^2\theta/2$ is a straight line. In Fig. 5, 6 and 7 the results relating $P(\theta)$, x, D/λ and Z for polydisperse coils were obtained using the scattering factor for this particular distribution. When using the dissymmetry method for unfractionated polymers the correction factors for polydisperse coils should be used in preference to the corresponding ones for monodisperse coils.

VI. Anisotropy and Depolarisation.

The discussion so far has been restricted to light scattering of isotropic particles — that is the polarisabilities are the same in every direction and the induced electric moment is always parallel to the electric field vector. As we have seen earlier for unpolarised incident light, the light scattered at 90° is vertically polarised. If however the scattering particle is anisotropic the polarisabilities are not the same in every direction and can be characterised by a polarisability ellipsoid the three principal axes of which correspond to different values of the polarisability. In such cases there is a slight deviation of the axis of the induced dipole from the plane of polarisation of the incident beam. This results in "depolarisation," i.e., the light scattered at 90° with respect to the incident beam is not fully polarised but will contain a small horizontal component in addition to the vertical one. When depolarisation occurs more light is scattered than in the case of equivalent isotropic molecules and hence corrections for the additional light scattered must be made in molecular weight determination. Anisotropy may be due to two factors — intrinsic anisotropy due to differences of refractive index along different directions in the molecule; form or shape anisotropy due to large anisometric particles.

Consider first anisotropic particles which are small compared to the wave length of light. At $\theta = 90°$ the horizontal component of the scattered intensity does not vanish. If we let H_u and V_u be the horizontally and vertically polarised components, the subscript u denoting that the incident light is unpolarised then the total reduced intensity $R_{\theta,u} = V_{\theta,u} + H_{\theta,u}$. For isotropic particles $R_{90,u} = V_{90,u}$, and for anisotropic particles $R_{90,u} = V_{90,u} + H_{90,u}$. The anisotropy is closely related to the depolarisation ratio, ϱ_u, which is the ratio of the horizontal to the vertical components of the scattered light at 90° when the incident light is unpolarised. CABANNES (1929) has shown that for anisotropic particles the correction factor to be applied to R_{90} is $\frac{6-7\varrho_u}{6+6\varrho_u}$ and to τ is $\frac{3-7\varrho_u}{6+7\varrho_u}$. Different factors are to be used if the incident light is polarised.

The depolarisation values of globular proteins in aqueous solution are generally low, the corrections published (DOTY and EDSALL 1951) being of the order of a few per cent. However serious errors can arise in the measurement of depolarisation if secondary scattering is present and hence special precautions are necessary for this measurement (LOTMAR 1938; GEIDUSCHEK 1954). When these precautions are taken GEIDUSCHEK has concluded that the depolarisation correction is negligible for proteins.

For systems of large anisotropic particles, i.e. when internal interference exists, then not only R_{90} but also the angular dependence of the scattering is affected by the anisotropy. For random coils the effect of anisotropy is low and decreases with increasing molecular weight and is negligible for coils which are large enough to determine the radius of gyration from the limiting slope of $P^{-1}(\theta)$ (HORN 1955). The effect for large rod-like particles has been studied by HORN et al. (1951) who have stated that the maximum error in neglecting anisotropy is of the order of three per cent, in calculating the radius of gyration from the limiting slope of $P^{-1}(\theta)$. However since most rod-like proteins so far studied are negatively birefringent this appears to cancel the effect of shape anisotropy and accounts for the very low values of ϱ_u found for asymmetric molecules like tobacco mosaic virus. To quote the conclusions of GEIDUSCHEK and HOLTZER (1958) — "The optical anisotropy of macromolecules leads to modifications of the relations between molecular weight, shape, size, and scattering power which, in most cases likely to be encountered in investigations of biologically important macromolecules, are only of minor importance. Conversely the depolarisation of light scattering appears to be a somewhat insensitive tool for the measurement of optical anisotropy in macromolecular systems." In the case of very low molecular weights however the depolarisation may be high.

VII. Equations for Polarised Incident Light.

The light scattering equations so far have been for unpolarised incident light. If the incident light is vertically polarised then only the vertical component is observed in the scattered light at any angle to the incident beam and the scattered light is fully polarised at all angles. A beam of unpolarised light of intensity I is equivalent to two polarised beams of equal intensity, $I/2$, polarised in mutually perpendicular directions and incoherent in phase. Therefore for two incident beams, one unpolarised, the other vertically polarised each of unit intensity the scattered intensity at 90° for the polarised beam is twice as great as for an unpolarised beam. This is evident from equations (5) and (6) since for vertically polarised incident light the $(1+\cos^2\theta)$ term disappears and it can be readily shown that $R_{90,v} = 2\,R_{90,u}$. Equation (17) then becomes for vertically polarised light

$$\left(\frac{2\,Kc}{R_{90}}\right)_{c=0} = \frac{1}{M}\,. \tag{34}$$

VIII. Scattering from Large Spherical Particles — Validity of Approximate Solution for Internal Interference.

In deriving light scattering equations where internal interference exists, the assumption was made that the distortion of the field of the incident beam by the particle could be neglected. The effect of the distortion will depend on the difference of refractive index between the particles and the medium and on the size of the particle. A complete theory which takes these factors into account, has been derived by MIE (1908) for spheres of any size and refractive index. The validity of the approximate solution can be assessed by means of MIE's results for spherical particles. The maximum deviation from the approximate theory occurs for spherical particles since a sphere of diameter D has more pairs of scattering elements than a thin rod of length D. The use of equation (28) for the determination of the diameter of spherical protein molecules should be restricted to molecules having a diameter of less than about a quarter of the wave length of the incident light, i.e. approximately 1000 Å (DOTY and EDSALL 1951). For rod-like molecules of refractive index comparable to proteins the error in the magnitude of the scattering will be considerably less and equation (28) will be valid up to far greater dimensions than 1000 Å — for thin rods and random coils the scattering can well be described by the simple theory up to maximum dimensions of at least 10,000 Å (GEIDUSCHEK and HOLTZER 1958). This includes the majority of proteins and nucleic acids.

IX. Multicomponent Systems.

The use of mixed solvents for the determination of the molecular weight of macromolecules is usually unsatisfactory. So far we have assumed that the system consists solely of two components but the use of a mixed solvent involving the addition of a third component complicates the interpretation of light scattering data. The intercept of the light scattering plot [equations (17), (27) and (29)] is not in general equal to the reciprocal of the molecular weight of the macromolecular component but depends also on the mutual interactions and optical properties of individual components. The problem became apparent in DEBYE's study of polystyrene in benzene-methanol mixtures (EWART et al. 1946). The effect was interpreted as due to preferential adsorption of the benzene on the polymer particle. For a mixture of three components (solvent = component 1, macromolecular solute = component 2, second solvent (or solute) = component 3], the following equation was shown to hold at infinite dilution

$$R_{90} = \frac{2\pi^2 n_0^2}{N\lambda^4}\left[\frac{\partial n}{\partial c_2} + \alpha\left(\frac{\partial n}{\partial c_3}\right)\right]^2 c_2 M_2$$

where $\alpha = (\partial c_3/\partial c_2)_{t,p,\mu_3}$ denotes the binding of component 3 by the solute component 2, $\partial n/\partial c_2$ is the solute refractive index increment in the mixed solvent at the composition of the experiment, and $\partial n/\partial c_3$ is the refractive index increment of component 3 at the composition of the experiment. The equation can be rewritten

$$\left(\frac{K c_2}{R_{90}}\right)_{c=o} = \frac{1}{M_2}\,\frac{(\partial n/\partial c_2)^2}{[\partial n/\partial c_2 + \alpha\,\partial n/\partial c_3]^2}\,. \tag{35}$$

This equation shows that the intercept at $c = o$ is only equal to the molecular weight if $\alpha = o$, that is the selective binding of component 3 by the solute component 2 is zero, or if $\partial n/\partial c_3 = o$ in which case the refractive index of the solvent mixture is independent of composition. Methanol-water and butanol-isopropanol mixtures approximate to the latter case. There are many published examples of selective binding of urea by proteins where measurements are made in 8 M urea (DOTY and KATZ 1950; KAY and EDSALL 1956). Due to selective binding of urea by proteins spuriously high molecular weights would be obtained in such systems if the molecular weight were taken as the reciprocal of $\left(\frac{Kc_2}{R_{90}}\right)_{c=0}$. For this case α is positive but it may also have negative values in the case of preferential interaction of component 2 with component 1. Araban fractions in aqueous solutions in the presence of small amounts of dimethylformamide furnish an example of a negative α (TOMIMATSU et al. 1959). The molecular weight of 38,400 determined by sedimentation-viscosity was independent of solvent composition, whereas the light scattering value was 20,700 in the presence of 1 percent dimethyl formamide but 36,300 in its absence. The difference in light scattering intercepts could only be attributed to the existence of repulsive forces between araban and dimethylformamide; that is preferential binding of water.

Fluctuation theory has been applied to multicomponent systems by several authors and gives a more general and exact theory applicable to the light scattering of systems of any number of components at finite concentrations (BRINKMAN and HERMANS 1949; KIRKWOOD and GOLDBERG 1950; and STOCKMAYER 1950). In a two component system the scattered intensity depends only on the solute refractive index increment and the derivative of the excess chemical potential of the solute with respect to its own concentration. In a multicomponent system the scattered intensity depends on all possible thermodynamic interactions, i.e. on the derivatives of the chemical potential of each component with respect to the others, and the refractive index increments of each of the components. The theory will not be reviewed here and the reader is referred to the original publications cited above. The theory has been successfully applied to the analysis of the effects of charge and ionic strength in solutions of bovine serum albumin (EDSALL et al. 1950).

In general the molecular weight of a macromolecular component cannot be determined in a three-component system unless the thermodynamic interaction of the protein with the second solute (or second solvent in a solvent mixture) can be evaluated by some other means or shown to be zero. The interaction term disappears in a mixed solvent whose refractive index is independent of composition. The determination of the molecular weights of charged macromolecules in the presence of salts is discussed in the next section. The size may be determined from the dissymmetry ($Z = i_{45}/i_{135}$) in a mixed solvent since only the ratio of scattered intensities at two symmetrical angles is required for this purpose.

X. Charged Macromolecules: Non-Random Systems.

In solutions of highly charged macromolecules at low ionic strengths i.e. where electrostatic screening is low, very strong repulsive forces are set up between the individual molecules. This leads to a very high value for the second virial coefficient A_2, the Kc/R_{90} plot showing a very steep initial slope with a flattening off at higher concentrations. This effect has been studied by several authors (DOTY and STEINER 1952; MYSELS 1954; KRONMAN and TIMASHEFF 1959; STIGTER 1960). One disadvantage of measurements in the absence of salt is that the turbidity for a one per cent solution of a macromolecule comparable in size to serum albumin may be only one tenth or less of the value obtained in the presence of high salt concentrations. This places a severe limitation on the accuracy of light scattering measurements in salt-free solutions where even at high concentrations the excess turbidity due to the solute is only of the same order as that of the solvent. The strong repulsive forces lead to an ordering effect and a non-random distribution of scattering centres, with a resultant decrease in the scattering intensity due to external interference and a very strong concentration dependence of Kc/R_{90}. Addition of neutral salt produces an electrostatic screening effect and so lowers the slopes of the curves. The flattening off of the curves at higher concentrations in the absence of salt can also be attributed to increased screening. There is no exact quantitative treatment of this behaviour available but the approximate treatment of DOTY and STEINER (1952) accounts qualitatively for the data. The external interference due to electrostatic repulsion causes not only a diminution of the intensity of scattered light but also a change in its angular distribution. There is a greater loss in the forward than in the backward direction and it is predicted from DOTY and STEINER's theory that this should lead to a decrease of dissymmetry with increase of concentration. Thus for serum albumin in acid solutions where the intrinsic dissymmetry is unity, values less than one are predicted from theory and these have been observed experimentally.

For the case of a highly charged macromolecule which shows no depolarisation or dissymmetry STIGTER (1960) has given a treatment based on the exact light scattering equation for multicomponent systems which connects the turbidity with the thermodynamic properties of the solution.

Determination of molecular weights of charged macromolecules. By using multicomponent theory it can be readily shown that extrapolation of Kc/R_{90} to zero concentration for a two component system gives the true value of $1/M_2$ where M_2 is the molecular weight of the macromolecular component. For charged macromolecules in the absence of salt the extrapolation to zero concentration may be difficult or almost impossible unless measurements are carried out at very low concentrations where the experimental accuracy is low. This is overcome by the addition of neutral salt where reliable extrapolations may be made. However in the presence of salt the system is a three-component one and theoretically the value of Kc/R_{90} does not extrapolate to the correct value of $1/M_2$ because the intercept also contains a contribution due to protein-salt interaction. It has been shown by EDSALL et al. (1950) that the molecular weight calculated by means of equation (17) gives the molecular weight of the uncharged macromolecule with an error which is generally less than the experimental error in light scattering. There is an excellent discussion on the extrapolation of light scattering data to zero concentration in a paper by TIMASHEFF and KRONMAN (1959). Macromolecules, in particular proteins, in the presence of salt are generally treated as two component systems.

XI. Interacting Systems.

Light scattering has been found to be one of the most valuable methods for studying the interaction of macromolecules. In fact any reaction which involves changes in molecular weight and size can in principle be studied by the light scattering method. Some typical interactions studied by this method include antigen-antibody reactions, formation and dissociation of mercury mercaptalbumin dimer, scattering and dissolution of gels, complex formation between serum albumin and deoxyribonucleic acid, complex formation between serum albumin and lysozyme, aggregation of dyes, polymers, proteins, polysaccharides and nucleic acids. If the time of reaction is not too fast then the kinetics may be studied. It is not intended to review these interaction studies since these are well treated in the review articles listed. Associating systems which are concentration dependent may give complicated $Kc/R_{90} - c$ curves making extrapolation to zero concentration difficult. Several protein systems such as insulin, lactoglobulin and conarachin belong to this category and a typical example from the authors' laboratory is shown in Fig. 8. Here at high concentrations solutions of α-keratose show a high degree of association and the second virial coefficient is positive but as the solution is diluted Kc/R_{90} passes through a minimum and increases as dissociation proceeds. Cases are also known in which dissociation reaches completion and the curve then shows a maximum at low concentrations and a decrease of Kc/R_{90} with decrease of c, the slope of the plot representing the second virial coefficient of unassociated protein. It may not be possible to make sufficiently accurate measurements at low concentrations to reveal the exact form of the curve in this region.

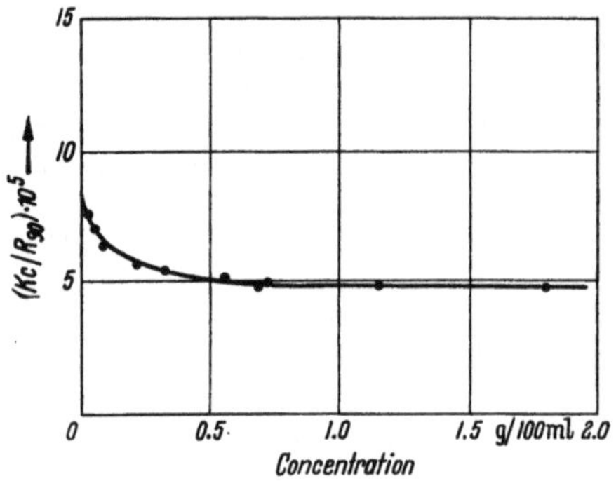

Fig. 8. Variation of $\frac{Kc}{R_{90}}$ v.C. for an oxidized wool keratin fraction (α-keratose) showing dissociation on dilution typical of associating systems which are concentration dependent.

XII. Experimental Methods.

1. Measurement of Reduced Intensity of Scattering.

A large variety of instruments have been described for the measurement of the intensity of scattered light. All depend on the use of sensitive photomultiplier tubes to measure the scattered light since the scattered intensity is so low, for example i_{90} at a distance of a few cm is of the order of $10^{-6} I_0$. The general requirements for a suitable light scattering apparatus are the following:

(1) An intense and stable source of monochromatic light collimated to give a beam of maximum intensity in the light scattering cell and uniform illumination. Although the theory of light scattering assumes parallel incident light, the error due to the use of converging light is not serious if only a narrow beam is used. Mercury arcs are commonly used as light sources for light scattering measurements in combination with a suitable filter to give the required wavelength. Measurements are usually made at 4360 Å or 5460 Å. It is advisable to measure at both wave lengths and if there are no complications such as fluorescence or depolarisation, the molecular weights obtained should be the same.

(2) The instrument should be capable of measurements of the relative intensity of scattering at angles from 30° to 150° — this is achieved by mounting the photocell on a turn-table.

(3) For turbidities of the order of 10^{-4} cm^{-1}, such as with common organic solvents the measurements should be made with a precision of one per cent. The reader is referred to the reviews by Stacey (1956) and Peterlin (1959) for references to the design of light scattering instruments and cells. The more important corrections which must be made to the measurements are listed below. It is very important in the design of instruments and cells to reduce errors due to secondary scattering and minimise the possibility of strong reflections.

The turbidities of most solutions do not vary more than 0.5 per cent per degree so that elaborate temperature control is not required for measurements made at room temperature but they are essential when investigating systems in which an equilibrium reaction (e.g. association or dissociation) is significantly temperature-dependent. It is therefore recommended that light scattering photometers are kept in rooms in which the temperature can be regulated to within 2° C.

Sufficient time must be allowed for the solution to attain thermal equilibrium before starting measurements particularly if it has been clarified in a centrifuge at low temperatures.

2. Cells.

Various types of cells have been used for the measurement of light scattering. Stray light is generally eliminated by coating all possible surfaces including the backs of cells with non-reflecting paint. For 90° scattering any of the cells described in the literature is satisfactory although a square cell is quite adequate. For the dissymmetry method and angular intensity measurements cylindrical cells have been widely used. Cylindrical cells having flat entrance and exit windows as described by WITNAUER and SCHERR (1952) are particularly suitable since multiple reflections from the cell walls are reduced. Conical shaped cells have been used by many workers and have the added advantage that no correction is necessary for back reflection (see later). The symmetry of cells for angular scattering should be ascertained by determining the fluorescence with a dilute solution of fluorescein using a sharp cut-off filter to eliminate the scattered light. The fluorescence should be spherically symmetrical and therefore the observed intensity should be constant for all angles measured. The intensities must be multiplied by $\sin\theta$ since the volume viewed by the phototube varies, decreasing as θ varies from 0 to 90° by an amount inversely proportional to $\sin\theta$.

3. Volume and Refraction Effects.

The value of RAYLEIGH's ratio requires correction for several optical effects and these have been described by CARR and ZIMM (1950), BRICE, HALWER and SPEISER (1950) and STACEY (1956). The first of these, the volume correction, is due to the fact that the photomultiplier measures some light outside the volume defined by the dimensions of the scattered beam. The correction is negligible except when making absolute measurements and can be minimised by adjustment of the dimensions of the photometer. However a correction must always be applied for the reduction in the light "seen" by the photomultiplier due to the spreading of the light rays on leaving the cell containing the medium because of refractive index differences. This correction depends approximately on the square of the refractive index of the scattering medium.

4. Back reflection Correction.

For light which is incident normally on an interface between two transparent media a small fraction of light is reflected. Thus in any apparatus in which the scattering cell is in air and the exit beam is normal to the cell wall there is a small amount of light which is travelling in the reverse direction to the incident beam and hence exactly 180° out of phase with it. The correction to be applied to the scattered intensity can be readily calculated (SHEFFER and HYDE 1952) and for the glass-air interface is

$$i_\theta^{corr} = i_\theta^{obs} - 0.047\, i_{180-\theta}^{obs}\,. \tag{36}$$

If the scattering is symmetrical then only the 90° scattering needs correction, but where there is interference all angular measurements must be corrected. The correction is not necessary when a cell with sloping walls is used since in this case the back reflected beam is deflected to the cell floor provided the angle made by the sides of the cell is sufficient.

5. Calibration of Light Scattering Photometers.

Scattering determinations depend on the measurement of the reduced intensity R_θ, and since the ratio of the scattered intensities and incident beams is of the order of 10^{-6} a direct comparison of the two beams using the same photocell is difficult. This can be achieved by means of two photocells one of which measures the intensity of the incident beam and the other the scattered intensity. The signals from the two photocells after amplification are balanced against each other with a potentiometer. In another method the intensity of the primary beam is diminished by a series of neutral filters, the percentage transmission of each at the wavelength λ having been accurately measured. Comparison of the intensity of light scattered by the solution with the intensity of the incident beam can then be made. In most instruments the intensity of the transmitted beam is measured since this is very little different from the incident beam. Appropriate corrections must be applied for geometrical factors affecting the scattered beam to convert the reduced intensity into absolute units. More usually the measured values of R_θ are compared with those of a well-known standard such as benzene or other liquid or solid of known turbidity. There has been a long controversy about the correct value of the RAYLEIGH ratio of the standard liquids. Owing to difficulties of absolute calibration a reference standard whose turbidity can be readily measured spectrophotometrically is invariably used. The reference standard may be a stable colloidal suspension such as "Ludox" (a colloidal solution of spherical silica particles), a standard preparation of polystyrene or other solution of well-known molecular weight, or solid blocks of plastic or opal glass may be used as working standards which are first calibrated and regularly checked against the primary standard. The precautions necessary in the use of "Ludox" for calibration have been discussed by GORING et al. (1957).

6. Measurements on Coloured Solutions.

The theory of scattering assumes that there is no absorption by the scattering particles. If the particles under examination absorb light at the wavelength used or the solution contains a light absorbing impurity then such absorption, unless corrected for, will lead to low estimates of scattering or molecular weight. The practical limitations in investigating coloured solutions are due to greatly decreased scattering due to the absorption of light. The problem has been discussed by DOTY and EDSALL (1951) and BRICE, NUTTING and HALWER (1953). If the solution is only slightly coloured the extrapolation to zero concentration eliminates the effect of the colour and the correct intercept is obtained but the concentration dependence and hence the second virial coefficient A_2 may be incorrect. Absorption effects in dye solution were avoided by ALEXANDER and STACEY (1952) by working in the near infra-red where the absorption was low even for intensely coloured dyes. BRICE et al. (1953) have applied the analysis of PUTZEYS and DORY (1940) and LAUER (1951) to correct for absorption in light scattering measurements and shown that turbidities of 0.01 cm^{-1} were determined with errors of less than three per cent even when attenuation by absorption was 150 times that by scattering. In dissymmetry measurements for values of Z less than 1.2 the presence of absorption should have no effect, but needs to be considered at high values.

7. Fluorescent Solutions.

The presence of fluorescence is often a complicating factor in light scattering measurements. Fluorescence is due to the optical excitation of the molecules by the absorption of the incident light followed by the emission of the absorbed radiation generally at a longer wavelength. The occurrence of fluorescence will result in an overestimation of the molecular weight. Its occurrence can be detected by interposing in the scattered beam a filter which absorbs the scattered light but transmits the fluorescence which is of a longer wavelength. Errors due to fluorescence can be corrected by choice of a longer wavelength of light which does not excite fluorescence — for example if the solution shows fluorescence at 436 mμ then it may be possible to work at 546 mμ. Since the scattering depends on the inverse fourth power of the wavelength the use of longer wavelengths involves greatly reduced scattering power. A filter which eliminates the longer wavelength component due to fluorescence from the emitted beam may be placed between the scattering solution and phototube (GUPTA and GORING 1960). BRICE et al. (1953) have given another method which depends on the fact that the scattered light is plane polarised while the fluorescent light is largely unpolarised and of longer wavelength than the scattered light. If a polarising disc is placed in the 90° scattered beam, then with the polariser vertical both the scattered and fluorescent radiation are measured, whereas in the horizontal position only the fluorescent radiation is measured since at 90° the scattered beam is vertically polarised. The measurements need to be corrected for depolarisation of the fluorescent light.

8. Measurement of Refractive Index Increment.

The refractive index increment of the solute dn/dc must be accurately known since it appears as its square in the constants K or H which need to be evaluated for the calculation of the molecular weight. For macromolecular solutions below 2 per cent the refractive index increment is almost independent of concentration so measurements are generally made on solutions of 1—2 per cent. Since the values of dn/dc are of the order 0.05—0.2 (c in g/ml), the refractive index difference between solution and solvent is thus of the order of 0.001 and measurements need to be accurate to 5×10^{-6} to reduce the error in the calculated molecular weight to the order of one percent. For measurements of this precision several differential refractometers have been developed by which the refractive index may be measured to about 3×10^{-6}. Measurements should be made at the same temperature at which light scattering measurements are made.

9. Clarification of Solutions.

The success of any light scattering experiment depends on absolute cleanliness of all apparatus used and the optical clarification of the solution. The cells should be cleaned inside and outside and kept clean throughout the measurement. Light scattering cells should be cleaned with detergent, rinsed thoroughly with tap water, then distilled water and finally with freshly distilled dust-free water. All apparatus to be used such as pipettes, flasks, etc. should be cleaned in a similar manner. The solvent used should be cleaned by distillation or by filtering through an ultrafine sintered glass filter under pressure. The apparent dissymmetry of the solvent has little practical significance since it includes extraneous reflected light as well as light scattered from the solvent. It will reach a minimum value as a given solvent is clarified and will vary from cell to cell and with different photometers of the same design. It is of course always subtracted out in light scattering measurements. This is not meant to imply that extraneous reflected light is of no consequence; if it is too high then this reflected light may be rescattered and this is not subtracted out.

The method of clarification of the solution will depend on the solute-solvent system under investigation, some systems being more difficult to clarify than others, e.g. polar solvents, aqueous systems in the absence of added electrolyte, and alkaline solutions. Solutions may be clarified by centrifugation at high speed. This involves transfer from the centrifuge tube with a dust-free pipette although several light scattering cells have been described which can be centrifuged making the transfer unnecessary. Alternatively filtration through ultrafine filters under pressure directly into the light scattering cell may be used, or both centrifugation and filtration. The procedure followed should give reproducible minimum turbidities without appreciable change of solute concentration. Very large molecules (e.g. nucleic acids) generally will not pass through ultra-fine filters and are usually centrifuged. Microgel particles constituting perhaps less than one per cent of the total concentration are often troublesome and pass through filters or resist centrifugation. Prolonged centrifugation at high speeds may remove these. It should be borne in mind that since the molecular weight is a weight average and the radius of gyration a z-average, a small amount of dust or aggregated or incompletely dispersed solute may give a value of the molecular weight which is not characteristic of the bulk of the solution unless special care is taken with the clarification.

10. Concentration of Solute.

It is desirable to work at solute concentrations high enough so that the solution scattering is at least 3—4 times that of the solvent. However since the data must be extrapolated to zero concentration and the concentration dependence of the scattering solution is expressed in terms of a virial expansion, measurements must be made at a concentration low enough to give a reliable extrapolation of Kc/R_{90}. As seen in Section X for charged macromolecules in the absence of salt it is often impossible to make a reliable extrapolation, and measurements in these cases should be made in salt solutions of sufficient ionic strength (> 0.01 M) to prevent ordering of the solute molecules.

11. Measurement of 90° Scattering.

When there is no destructive interference due to the size of the molecules i.e. when the solute molecules are less than 1/20 the wavelength of light the scattering should be independent of angle and 90° scattering measurements are in order. The solvent scattering is generally measured first. Each solution is transferred from the centrifuge tube or filtered directly into the light scattering cell and should be checked for any change of concentration on filtration or centrifugation. Alternatively dilutions are made directly in the light scattering cell either by measuring the most dilute solution first, adding known amounts of a more concentrated solution and stirring by means of a magnetic stirrer; or by measuring the most concentrated solution first and then adding known amounts of solvent. The concentrations are readily determined by weighing the light scattering cell before and after each addition. At least four concentrations should be measured covering a five-fold concentration range. After subtracting the solvent scattering, R_{90} is calculated for the solute using the calibration factor for the instrument (often supplied in a commercial instrument) and applying the corrections for refraction and back reflection. The value of Kc/R_{90} is then plotted against concentration and the molecular weight is given by the reciprocal of the intercept [see equations (12), (13) and (17)].

12. Dissymmetry Method.

When the solute molecule is large enough to cause interference the scattering is not symmetrical about 90° and measurements at several angles are necessary. Dissymmetry measurements can be used provided Z is less than 1.4; at larger values of Z complete information is sacrificed for rapidity of measurement. More accurate dissymmetry measurements are obtained when a narrow slit system is used. If a slit width is used different to that with which the apparatus was calibrated it is necessary to correct the reduced intensities at 0° for change in the width of the primary beam. No correction is needed for the Z values since only the ratio i_{45}/i_{135} is required. Kc/R_{90} is plotted against c and extrapolated to zero concentration; Z is plotted against c or preferably $1/(Z-1)$ is plotted against c to obtain the intrinsic dissymmetry $[Z]$. From the tables of DOTY and STEINER (1950), STACEY (1956), or BEATTIE and BOOTH (1960), the value of $1/P(90)$ corresponding to the observed dissymmetry is found. In principle the shape of the particle should be known but since in the range of $[Z]$ from 1.0 to 1.4 there is little difference in the various models, the error introduced will be negligible. The molecular weight may be calculated from equation (26). It is better to correct each point in the plot of Kc/R_{90} for internal interference and use equation (27) as described in C IV 1. From $[Z]$ thevalue of D/λ can be found corresponding to any particular shape and if the shape is known then the largest dimension is readily calculated.

13. Angular Intensity Measurements.

The angular dependence should be measured if the dissymmetry is greater than 1.4 as information on solute shape and polydispersity can be obtained.

The angular scattering of both solution and solvent are measured at 5° intervals from 30° to 60°, then at 10° intervals up to 140°. The excess scattering is multiplied by $\sin\theta$ to compensate for the varying volume viewed by the phototube, and converted to $R_\theta/(1+\cos^2\theta)$ using the appropriate calibration constants for the instrument and correcting for refraction and back reflection. Tables of $\sin\theta$, $1+\cos^2\theta$, $\sin\theta/(1+\cos^2\theta)$, and $\sin^2(\theta/2)$, are given by STACEY (1956) and facilitate the conversion of the measurements to $R_\theta/(1+\cos^2\theta)$. $Kc(1+\cos^2\theta)/R_\theta$ is plotted against $\sin^2(\theta/2)+kc$, and k is an arbitrary constant to spread out the data. It may be as low as 100 or as great as 10,000. A factor of 1,000 is probably the best to try. All the points for any one concentration lie on either a straight line or smooth curve the initial portion of which is linear. A straight line is drawn through this linear portion and extrapolated to $\theta=0$, i.e. kc. The lines should be parallel. If they are not then this indicates that the concentrations used were too large. Parallel lines are then drawn through the points at constant angle, and these are extrapolated to $c=0$, i.e. $\sin^2(\theta/2)$. Straight lines are drawn through the two sets of extrapolated values and extrapolated to $\sin^2(\theta/2)+kc=0$. The intercept from each extrapolation should be identical and is equal to $1/M$ [equation (29)] for unpolarised incident light. The radius of gyration can be determined directly from the slope of the zero concentration plot by means of equation (33). For larger particles (of the order of 1000 Å) more information may be obtained about their shape and polydispersity from $P^{-1}(\theta)$.

D. Diffusion.

I. Information Available from Diffusion.

Diffusion in liquids has been studied experimentally and theoretically for many years in an attempt to understand and describe the phenomena involved; in aqueous systems molecules from the size of common inorganic salts up to proteins of molecular weights of some millions have been studied. For the types of problems underlying the general theme of this chapter where the emphasis is on macromolecules there are one major and two lesser uses to which measurements of diffusion processes have been put. The major use is in the evaluation of molecular weights by combination of the diffusion and the sedimentation coefficients in the Svedberg equation (see E. V. 1.). The minor uses are testing for solute homogeneity, which is much more highly developed for compounds of low molecular weight e.g. inorganic salts, than for macromolecules, and the evaluation of the frictional coefficient of a molecule.

Since the sedimentation methods, in particular sedimentation velocity, describe and determine heterogeneity of macromolecules much more precisely than diffusion can, it would seem that the main reason for determining the diffusion coefficient of macromolecules having regard to the theme of this chapter is for use in the Svedberg equation for determining molecular weight. This method of determining molecular weight is, from a practical point of view, being superseded by sedimentation equilibrium measurements now that the time taken for these experiments has been reduced by the approach-to-equilibrium method of Archibald (1947) and the use of short solution columns by Van Holde and Baldwin (1958). Also as will be seen there is some slight uncertainty in the value of the diffusion coefficients of macromolecules as measured at the present time (see section D. V).

Weighing these considerations the authors feel there is no need to give examples of actual calculation of diffusion coefficients using the accurate and more complex interferometric methods of following free diffusion. We have given only the calculation for the simple schlieren method of observation of free diffusion, and generally described the other methods. More details of these methods, if required, are given in the review by Gosting (1956). The reader is referred to this and also reviews by Ogston (1956) and Robinson and Stokes (1960).

The symbols and form of equations used in the chapter are those of Gosting (1956).

II. Diffusion and the Laws of Diffusion.

Diffusion of a solute in a liquid is the spontaneous transfer of solute from regions of higher to lower concentration. At the same time solvent molecules move in the opposite direction to the solute and by this process of diffusion concentration differences in a solution finally disappear.

1. Types of Diffusion Measurements Carried out in Practice.

There are four common types of diffusion measurements made and these depend on the manner in which the diffusion process is allowed to proceed.

a) Steady-State Diffusion.

In this method diffusion takes place between two solutions (in practice through a porous disc) each solution being of practically uniform composition. The method has been used a great deal by Stokes and co-workers (see Robinson and Stokes 1959) mainly for the diffusion of compounds of low molecular weight.

b) Free Diffusion.

Free diffusion is the method most used for determining the diffusion coefficients of both small and large molecules and for the determination of homogeneity of a solute. Here an initially sharp boundary between two solutions of different concentrations is allowed to diffuse. The height of the solutions on either side of the boundary are sufficiently great so that there is no change in solute concentration at the two ends of a cell during an experiment. In the most common method for measurement of free diffusion a boundary is formed at the centre of a rectangular cell. The variation of solute concentration with time after formation of the boundary is shown in Fig. 9 together with a plot of concentration gradient versus distance in the cell some time after diffusion has commenced.

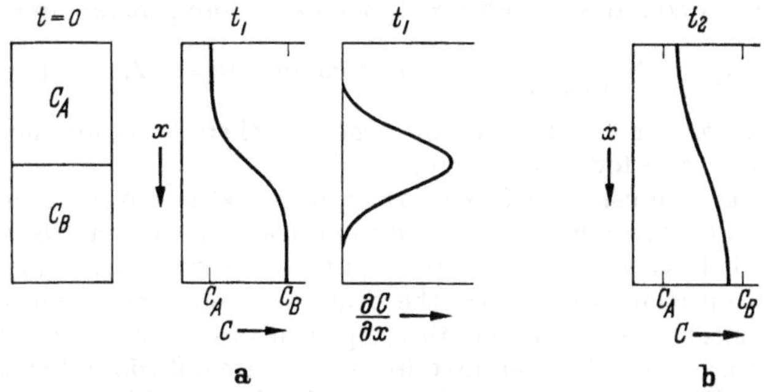

Fig. 9. (a) Free diffusion. The initial arrangement of solutions in the cell ($t = 0$) and illustrative concentration and concentration gradient curves at some later time, t. (b) Restricted diffusion. Illustrative concentration curve for a time t_2 after the concentrations have changed appreciably at the ends of the cell (from GOSTING 1956).

The main emphasis will be placed here on free diffusion which can give two types of information; from the shape of a curve at a given time the homogeneity of a solute can be assessed (if there are no interaction of flows — see D. V. 1); from the shapes of curves at various times the diffusion coefficient of the solute can be calculated.

c) Restricted Diffusion.

If free diffusion is allowed to proceed until there are changes in solute concentration at the top and bottom ends of a cell the process is called restricted diffusion and the curve of concentration versus position in the cell is shown in Fig. 9b. This method has had very little use with large molecules and will not be further discussed here.

d) Diffusion During a Sedimentation Velocity Experiment.

The spreading of a boundary during a sedimentation velocity experiment in an ultracentrifuge is determined partially by diffusion of the sedimenting molecules; if the solute is monodisperse and its dependence of sedimentation coefficient on concentration is linear and known, then the diffusion coefficient of the solute can be calculated (see E. IV. 4).

2. FICK's First Law and Definition of Diffusion Coefficient.

At constant temperature and pressure in a two-component system FICK's law for one-dimensional diffusion can be written as follows:

$$J_1 = -D\left(\frac{\partial c_1}{\partial x}\right)_t \tag{37}$$

(the solvent component is given the subscript o, and the solute components 1, 2, 3 etc.) i.e. the solute flow, J_1, at any point, x, and at any time, t, relative to the cell is directly proportional to the concentration gradient $\partial c_1/\partial x$. J_1 is defined as the amount of solute, i.e. component 1, crossing unit area in unit time perpendicular to the direction of flow and is positive in the same manner as x. If J_1 is in grams of solute crossing a square centimeter per second then c_1 should be g solute/ml and x should be measured in cm; it is seen the dimensions of D in this case are $cm^2\ sec^{-1}$ and these are the standard units used in this review.

D varies at a particular temperature (and pressure) with solute concentration and the name "diffusion coefficient" is used for it.

If the system contains q solute components a more general equation is used.

$$J_i = -D_i \left(\frac{\partial c_i}{\partial x}\right)_t (i = \text{solute components} = 1,2, \ldots q)\,. \tag{38}$$

If volume changes on mixing are not large and there is no interaction of flows this equation is valid for every solute.

Values of the diffusion coefficient vary over a wide range e.g. glycine has a value of $10.6\times10^{-6}\ cm^2\ sec^{-1}$ at zero concentration in water at 25° while bovine plasma albumin has a value of about $0.68\times10^{-6}\ cm^2\ sec^{-1}$ under similar conditions.

Although equation (37) defines the diffusion coefficient in terms of flow of solute and concentration gradient these quantities cannot be readily measured directly (they are indirectly measured in steady state diffusion). For free diffusion and restricted diffusion, integrated forms of FICK's second law are utilized in the practical evaluation of D.

3. Thermodynamic Interpretation of the Diffusion Coefficient.

To express the diffusion coefficient in terms of properties of the solute molecules the thermodynamic approach must be used, the kinetic approach giving an answer which is true only at infinite dilution. Using thermodynamic methods in a two-component system one form of the equation which emerges is

$$D = \frac{RT}{Nf_1}\left[1 + C_1 \frac{d \ln y_1}{dC_1}\right] \tag{39}$$

where C_1 is now in moles per litre, y_1 is the solute activity, f_1 is the frictional coefficient of the solute molecule, and N is AVOGADRO's number. As C_1 approaches zero, D approaches $\frac{RT}{Nf_1}$ which is the result derived from the kinetic approach for an infinitely dilute solution.

4. FICK's Second Law.

FICK's second law is obtained by substituting equation (37) into the continuity equation. The continuity equation, which is an expression of conservation of mass in a system, is applicable to every component in the system and, provided there are no chemical reactions, the equation has the form

$$\frac{\partial c_i}{\partial t_i} = \frac{\partial J_i}{\partial x} (i = 0, 1 \ldots q)\,. \tag{40}$$

Combination of equations (38) and (40) gives FICK's second law (this assumes that equation (38) is valid i.e. there is no interaction of solute flows]

$$\frac{\partial c_i}{\partial t_i} = \frac{\partial}{\partial x}\left(D \frac{\partial c_i}{\partial x}\right) = D \frac{\partial^2 c_i}{\partial x^2} (i = 0, 1 \ldots q) \tag{41}$$

the last equality holding *if* D is independent of c.

Integrated form of FICK's second law for free diffusion. FICK's second law is true for any diffusing system if there is no interaction of solute flows. However for its integration for use in measuring the diffusion coefficient the boundary conditions which must be applied vary with the method used for studying the diffusion and the form of integration is hence different in each case. We will only be concerned with free diffusion here.

If D is independent of c in a two-component system and free diffusion occurs between two solutions of solute concentrations c_A and c_B the integrated form of equation (41) is:

$$c = \bar{c} + \frac{\Delta c}{2}\left(\frac{2}{\sqrt{\pi}}\right)\int\limits_0^{x/2\sqrt{Dt}} e^{-\beta^2}\,d\beta \tag{42}$$

and

$$\frac{\partial c}{\partial x} = \frac{\Delta c}{2\sqrt{\pi D t}}\, e^{-x^2/(4Dt)} \tag{43}$$

where $\bar{c} = \frac{c_A + c_B}{2}$ = mean solute concentration and $\Delta c = c_B - c_A$ = concentration difference across the initial boundary; $\beta = x^2/4Dt$.

Equations (42) and (43) give the concentration and concentration gradient curves in free diffusion.

If D is considerably dependent on c or *if* there is interaction of solute flows or a change in volume on mixing the case is more complicated. However for many systems, including the diffusion of proteins, equations (42) and (43) can be used and yield an accuracy of probably 5 per cent or better.

Mostly in the optical observation of boundaries in free diffusion the refractive index and refractive index gradient are determined rather than the corresponding concentration and concentration gradients. If the variation of refractive index, n, is practically linear with c over the range of solute concentrations in the cell then equations (42) and (43) are re-written with n's substituted for c's. Either one or the other of these equations is applied to the results in free diffusion to calculate the diffusion coefficient, the choice of equation depending on whether our experimental results are used to give refractive index or refractive index gradient plots.

5. Equations for Evaluating Diffusion Coefficients Using Measurements of Free Diffusion Experiments.

Suppose from an experiment using free diffusion a plot of the refractive index gradient curve has been obtained. From this experimental curve certain properties can be measured. They will be written as $\mathfrak{D}$'s (after GOSTING 1956) and are related to the diffusion coefficient. These experimentally determined parameters are the *reduced height-area ratio*

$$\mathfrak{D}_A = \frac{(\Delta n)^2}{4\pi t(\partial n/\partial x)^2_{max}} \tag{44}$$

where $(\partial n/\partial x)_{max}$ is the maximum value of $\partial n/\partial x$, and the *reduced second moment*

$$\mathfrak{D}_{2m} = \frac{1}{2t\Delta n}\int\limits_{-\infty}^{+\infty} x^2\left(\frac{\partial n}{\partial x}\right)dx\,. \tag{45}$$

Higher moments may be defined but experimental accuracy at present is not great enough to utilize moments beyond the fourth. $\mathfrak{D}_A$ is the simplest function to measure and is the one most often used.

To get diffusion coefficients from the reduced height-area ratio or the reduced second moment it can be shown that for a two-component system (i.e. solvent + one solute), with no dependence of D upon c and n a linear function of c that

$$\mathfrak{D}_A = \mathfrak{D}_{2m} = D\,.$$

Average diffusion coefficients. *If* a system contains q solutes and the diffusion coefficient of each solute is independent of c, and *if* n is linear with respect to c and *if* there are no interactions of solute flows then the measured $\mathfrak{D}_A$ and $\mathfrak{D}_{2m}$ are summations of the individual diffusion coefficients. For such a system

$$\mathfrak{D}_A = \sum_{i=1}^{q} [\alpha_i/\sqrt{D_i}]^{-2} \tag{46}$$

$$\mathfrak{D}_{2m} = \sum_{i=1}^{q} \alpha_i D_i \tag{47}$$

where α_i is the fraction of solute i on a refractive index basis. Hence in a multicomponent system the reduced height-area ratio is not a simple average but a weight-average of $1/\sqrt{D}$. The reduced second moment is a weight average of the various diffusion coefficients.

Theoretically then, as in sedimentation equilibrium experiments, one should be able for a heterogeneous system to assume a distribution curve of components and from various $\mathfrak{D}$'s fit the parameters to it. However experiments are just not accurate enough at present to do this for more than two solute components.

6. Correction of Diffusion Coefficients to Standard Conditions.

It is usual for comparative purposes to refer diffusion coefficients of macromolecules measured experimentally to some standard temperature and solvent (for proteins they are commonly referred to 20° or 25° in water), and to extrapolate them to zero solute concentration. So three corrections have to be applied for the variation of D with temperature, viscosity and solute concentration.

a) Correction of D for Viscosity.

The evidence to date indicates that a diffusion coefficient of a solute at infinite dilution, D_0, can be converted from its value in one solvent of viscosity η to another of viscosity η_r, by the STOKES-EINSTEIN relation

$$(D_0)_r = D_0(\eta/\eta_r)\,. \tag{48}$$

b) Correction of D for Temperature.

Experimental results indicate that there is no general formula applicable for correction of diffusion coefficients from one temperature to another. Consequently experiments should be reported at the temperatures at which they were carried out and corrections should only be applied by the following equation for a fraction of a degree

$$(D_0)_T = (D_0)_{T'} \frac{T\eta'}{T'\eta} \tag{49}$$

where T and T' = absolute temperatures and η and η' = the corresponding viscosities of the solvent.

c) Extrapolation of D to Zero Solute Concentration.

It is an experimental fact that D is a function of the solute concentration. By extrapolation of D to zero solute concentration a diffusion coefficient (D_0) is obtained which in a two component system depends only on the temperature. In some cases e.g. with very asymmetric molecules such as deoxyribonucleic acid[1] the variation of D with c will be large and one must perform experiments at low-enough solute concentrations to be certain of the extrapolation to zero concentration. If the RAYLEIGH fringe method (see below) is used information regarding the dependence of D upon c can be obtained from one experiment. The method using polarized light (BRYNGDAHL and LJUNGGREN 1960, TSVETKOV and KLENIN 1958) will allow precise work at much lower concentrations than even the RAYLEIGH and GOUY methods discussed later.

7. Detection of Heterogeneity by Free-Diffusion Experiments.

If in a system there are only two components (solute plus solvent) and *if* D does not depend greatly upon c, and *if* n is linear with c and *if* there is no volume change on mixing then the refractive index gradient curve is GAUSSIAN.

Now if a system of q solute components is subject to the above restrictions and the additional one that the solute flows of the q solutes do not interact the resulting refractive index gradient curve will not in general be GAUSSIAN. In this case the heterogeneity of the solute components can be expressed by the deviation of the curve from the GAUSSIAN form. This deviation can be expressed by $\mathfrak{D}_A$, $\mathfrak{D}_{2m}$ etc., and their non-agreement can be a definition of the extent of the heterogeneity. However now-a-days it would seem general that the two quantities initially measured by interference methods are $\mathfrak{D}_A$ and a "fringe deviation graph." This last-named depicts departures of a set of interference fringes from those which correspond to a GAUSSIAN refractive index gradient curve.

If a solute is homogeneous and subject to the above restrictions the fringe deviation graph will be zero. If there are only two solutes present their contribution to $\mathfrak{D}_A$ can be approximately determined. If the proportion of the two solutes is known from some other measurement such as sedimentation then the precision of D_1 and D_2, the solute diffusion coefficients comprising $\mathfrak{D}_A$, can be increased.

If the system consists of several solute components and it is desired to give expression to the degree of heterogeneity, probably the best method is to calculate the reduced moments, thus giving various averages of the diffusion coefficient.

As far as the restrictions above are concerned, say for proteins in aqueous solution, the diffusion coefficients can be considered as independent of concentration for the concentration increment of protein used in the experiment; n can be taken as linear with c, and interaction of solutes flows seems to be small but this requires further study before making definite statements.

III. Experimental Method for Steady-State and Free-Diffusion Experiments.

1. Steady-State Diffusion — the Diaphragm Cell.

In this method diffusion is allowed to take place between two solutions separated by a porous disc. After a sufficient time has elapsed the concentration

[1] At higher concentrations with deoxyribonucleic acid, solutions are also too viscous to form boundaries and at low concentrations there are not enough fringes for much precision with the present fringe methods.

difference is determined analytically. The type of cell used is known as a diaphragm cell and has been extensively used recently by STOKES and co-workers (see ROBINSON and STOKES 1959 for further details) for inorganic salts; under favourable conditions it yields a high precision (about 0.1 per cent in D). The precision for proteins would be less. Its advantages for studying macro-molecules are its simplicity and the fact that solute concentrations are determined specifically by analysis. One disadvantage is the long time taken in an experiment because the use of analytical methods requires the process to go on until a sufficient concentration change has taken place; a second disadvantage is that it is not an absolute method and must be calibrated with a salt of known diffusion coefficient e.g. sodium chloride.

2. Free Diffusion.

The cell used in free diffusion is commonly of rectangular cross section and is of the type used in moving-boundary electrophoresis. It is shown in Fig. 10. The boundary between the two solutions is formed at the bottom of the cell by shearing and then the boundary is brought up to the centre of the cell by suction through a capillary, the boundary being sharpened in the process (KAHN and POLSON 1947). After several ml of solution have been taken out the capillary is removed; this is taken as zero time but the boundary will not be perfectly sharp and a "starting-time correction" must be made (D, III, 4).

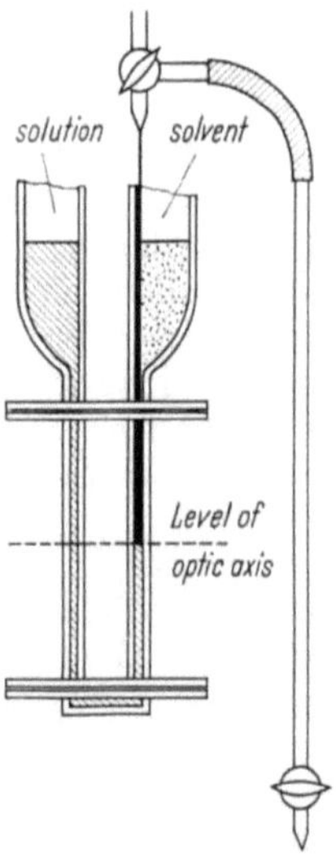

Fig. 10. The KAHN and POLSON boundary sharpening technique as applied to a TISELIUS electrophoresis cell used for diffusion measurements (from GOSTING 1956).

The spreading of the boundary is followed by photography at various times using one of the three optical systems described below. From these photographs $\mathfrak{D}_A$, $\mathfrak{D}_{2m}$ etc. and the fringe deviation graph can be calculated if desired. There are many types of apparatus used for free diffusion. A TISELIUS moving boundary electrophoresis apparatus is often used and most of the commercial instruments combine the functions of diffusion and electrophoresis. An apparatus, the Spinco Model H, suitable for electrophoresis and diffusion has the three optical systems listed below. Furthermore, for an approximate value of the diffusion coefficient the ultracentrifuge can be used. Here the boundary between solution and solvent (or two solutions) is formed in an artificial-boundary cell; the cell is then allowed to run at low speed (about 10,000 r.p.m.) so that sedimentation is negligible and the spreading of the boundary is followed. The answer will not be as precise as when using a special diffusion apparatus but will give answers within 5 per cent of those from the better instruments. Extensive and precise experimental details for using the Spinco Model H for diffusion measurements are given by SCHACHMAN (1957) together with a list of some other commercial instruments available.

3. Optical Methods for Free Diffusion.

a) Schlieren Method.

This optical system which was the one most commonly used until the advent of the interference fringe methods (and is still the one most used in electrophoretic and ultracentrifugal work) gives a graph of dn/dx versus x i.e. the refractive index gradient versus position in the cell at a particular time. The precision of the determination of dn/dx by this method is limited to about one per cent.

Equation (44) can be applied directly to determine the reduced height-area ratio, $\mathfrak{D}_A$, of this curve.

b) GOUY Interference Fringe Method.

With the advent of the GOUY and the RAYLEIGH interference fringe methods it was found that D could be measured with a precision 10 times greater than with the schlieren method, i.e. values of D for two-component systems can be measured to about 0.1 per cent. These interference methods give a series of interference fringes whose spacings can be used to calculate a refractive index gradient curve (GOUY method) or refractive index curve (RAYLEIGH method).

The GOUY interference method has been developed by COULSON et al. (1948), LONGSWORTH (1945) and by GOSTING and co-workers (see GOSTING 1956). In terms of the factors measurable from one GOUY diffusion photograph at a particular time the reduced height-area ratio

$$\mathfrak{D}_A = \frac{(J\lambda b)^2}{4\pi t C_t^2} \tag{50}$$

where

$$J = \text{total number of fringes} = a\Delta n/\lambda \tag{51}$$

where λ = wavelength of light being used, a is the cell thickness along the light path, Δn = the refractive index difference across the initial boundary and t is the time since the start of diffusion. b is the optical lever arm from the cell to the photographic plate and C_t is the maximum deflection of light predicted by ray optics for time t. C_t is proportional to $\partial n/\partial x$. The total integral number of fringes J is obtained from a GOUY photograph but the fractional part is best obtained with the instrument, temporarily modified, to yield RAYLEIGH fringes. The boundary is then photographed during the sharpening process. This method is very well discussed in the review by GOSTING (1956), and the reader is referred to page 478 of that review for evaluation of C_t from the fringe numbers, j, and fringe displacements from the undeviated slit image, Y_j.

c) RAYLEIGH Interference Fringe Method.

In terms of the quantities measurable from one photograph of fringes using the RAYLEIGH interference method, the refractive index gradient curve cannot be accurately determined but rather the fringes give us a refractive index versus x curve and the diffusion coefficient in a two component system must be calculated from this. If there are a number of solutes the reduced height-area ratio can be calculated.

The reader is again referred to GOSTING's reviews for theoretical and practical details and to SCHACHMAN (1957). Each of the interferometric fringe methods has virtues of its own and the two methods are complementary. In analyzing for solute heterogeneity the GOUY method is best since this requires a method which is not affected severely by the dependence of diffusion coefficients on concentration or by a non-linear dependence of n upon c. On the other hand the GOUY method in one experiment will give no information regarding the concentration dependence of a diffusion coefficient whereas the RAYLEIGH method will. The GOUY method is most precise early in a diffusion experiment whereas the RAYLEIGH method is better later in an experiment.

d) Polarized Light Method.

This method which is being developed for use with diffusion by BRYNGDAHL and LJUNGGREN (1960) and TSVETKOV (see TSVETKOV and KLENIN 1958) will allow much lower concentration increments to be precisely dealt with than the GOUY or RAYLEIGH methods; in the polarized-light method it is possible to work with concentration differences corresponding to an optical path difference as small as one wavelength i.e. for a 5 cm thick cell a sucrose solution of 0.01 per cent concentration could be examined precisely. It appears that this method will have a precision 10—20 times greater than the GOUY and RAYLEIGH methods.

4. Zero Time Correction.

Since the starting boundary formed in a diffusion experiment is never ideally sharp the time, t', actually measured is less than the theoretical time, t, at which diffusion effectively commenced. We can write $t = t' + \Delta t$, Δt representing the initial disturbance of the boundary. When measuring the true reduced height-area ratio, $\mathfrak{D}_A$, from those determined experimentally LONGSWORTH (1947) used the relation

$$\mathfrak{D}'_A = \mathfrak{D}_A (1 + \Delta t/t') \,. \qquad (52)$$

If $\mathfrak{D}'_A$ is plotted against $1/t'$ the value of $\mathfrak{D}'_A$ at $1/t' = 0$ equals $\mathfrak{D}_A$. Equation (52) assumes that the boundary was GAUSSIAN when the sharpening was stopped.

5. Method of Expressing Results.

The diffusion data of an experiment should contain

a) what the final figures are $\mathfrak{D}_A$, $\mathfrak{D}_{2m}$ or D_1, etc. and what temperature and medium they refer to,

b) the temperature at which the experiment was actually carried out,

c) the average solute concentration and the concentration units used,

d) buffer compositions and pH if necessary.

6. Example of Calculation of Reduced Height-Area Ratio.

For an example of an actual calculation of reduced height-area ratio, $\mathfrak{D}_A$ by the RAYLEIGH interference fringe method see SCHACHMAN (1957, p. 88) and by the schlieren method see LUNDGREN and WARD (1951). For details of use the GOUY method see GOSTING (1956, p. 478).

From Schlieren Photographs of an Artificial Boundary in the Ultracentrifuge.

Below is given a calculation of reduced height-area ratio, $\mathfrak{D}_A$, using the schlieren method applied to an artificial boundary formed in the ultracentrifuge at low speed.

Example. A solution (approximately 0.8 per cent) of crystalline bovine plasma albumin was made up in buffer containing 0.10 M potassium chloride. 0.01 M acetic acid and 0.01 M sodium acetate at pH = 4.6. The protein solution was dialyzed against buffer for 24 hours on a rocking dialyzer at 4° using 18/32 Visking cellulose tubing. Both buffer and solution were filtered through a Jena G5 sintered glass filter. The refractive increment of the protein solution was 0.00154 (BRICE-PHOENIX Differential Refractometer). This is equivalent to 0.83 per cent protein concentration. A boundary was then formed between the protein solution and dialysate in an artificial boundary cell (HARRINGTON-SCHACHMAN type) in a

Spinco ultracentrifuge. The centrifuge was run at 12,590 r.p.m. and the schlieren pattern photographed every 8 min taking 8—10 photographs altogether. The photographs were taken at a bar angle of 80°. Temperature = 24.6°. Viscosity of buffer relative to water at 25° = 1.005.

The areas (A) of each of the peaks were obtained by measuring with a micro-comparator the height of the curve at 0.025 or 0.050 cm intervals (in the x direction). The maximum height (H) of each peak was also recorded. The calculation of the apparent reduced height-area ratio ($\mathfrak{D}'_A$) is shown in Table 1.

IV. The Use of the Diffusion Coefficient in Determining Molecular Weight.

1. Combination of Sedimentation and Diffusion Coefficients to give Molecular Weight (see section E.V.1).

2. Approximate Method Using STOKES' Law.

With an infinitely dilute solution $D = RT/Nf_s$ where f_s = frictional coefficient of the solute molecule. The frictional coefficient, f_0, of a sphere moving in a medium of viscosity η is

$$f_0 = 6\pi\eta r \quad (\text{STOKES' law}) \tag{53}$$

where r is the radius of the sphere. These two equations can be combined to give

$$D = \frac{RT}{6\pi\eta r N}. \tag{54}$$

This equation is subject to the restrictions that the solution is infinitely dilute, and that the solute molecule is a sphere and is very large compared with the solvent molecules.

Table 1. *Calculation of $\mathfrak{D}_A$ of bovine plasma albumin from schlieren photographs of boundaries* (artificial boundary cell in a Spinco ultracentrifuge at low speed).

Photo	A (sq. cm)	H (cm)	Time = t' (sec)	$\mathfrak{D}_A$*
4	0.152_0	0.748	13×60	8.97×10^{-7}
5	0.148_7	0.625	21×60	7.94×10^{-7}
6	0.147_3	0.540	29×60	7.70×10^{-7}
7	0.146_6	0.489	37×60	7.36×10^{-7}
8	0.148_0	0.450	45×60	7.15×10^{-7}
9	0.149_3	0.416	53×60	7.10×10^{-7}
10	0.148_3	0.391	61×60	6.98×10^{-7}

Average $A = A' = 0.148_6$

* $\mathfrak{D}'_A = \left(\frac{A}{H}\right)^2 \times \frac{1}{4\pi t' F^2}$.

F = vertical magnification factor = 2.12. All other magnification factors cancel out. If an enlarger is used in measurement of the photographs, F should be multiplied by the enlarger magnification factor also.

Value of $\mathfrak{D}'_A$ at $1/t' = 0$ is $\mathfrak{D}_A$ and equals 6.44×10^{-7} cm² sec⁻¹.

$(\mathfrak{D}_A)_{water} = 6.44\times10^{-7}\times1.005 = 6.47\times10^{-7}$ cm^2 sec^{-1}.

$(\mathfrak{D}_A)_{25°,\,water} = 6.47\times10^{-7}\times\frac{298}{297.6}\times\frac{9.034}{8.949} = 6.54\times10^{-7}$ cm^2 sec^{-1}.

$(\bar{c}_{BPA} = 0.41$ g/100 ml $T = 24.6°)$.

If r is now related to the molecular weight and partial specific volume of the sphere it is found

$$D = \frac{RT}{6\,\pi\,\eta\,N\left(\frac{3\,M\bar{v}}{4\,\pi N}\right)^{1/3}}\,. \tag{55}$$

This equation holds subject to the above three restrictions and also the fourth one that the volume of the molecule $= \bar{v}M/N$, $\bar{v}$ being the partial specific volume of the solute. This is not really true because of hydration and electrostriction factors.

Equation (55) can be used to calculate M from D and the partial specific volume of the solute but the values would be uncertain. It has some use for comparing molecular weights of similar materials (see GOSTING 1957, p. 526).

It is emphasized that a molecular weight of a solute can not be obtained from one hydrodynamic measurement alone.

3. Combination of Diffusion Coeffiicient and Intrinsic Viscosity (see Section G).

4. Diffusion Coefficient Relationship to Molecular Shape.

From the value of D for a two-component system the frictional coefficient f of the solute can be calculated.

Many odd shapes and sizes of particles could have the same value for f. The relationship of f to shape is treated in a separate section (see G, I, 1).

V. Factors Contributing to Uncertainty in the Experimental Determination of the Diffusion Coefficients of Macro-Molecules.

1. Interaction of Solute Flows.

So far when systems containing more than one solute component have been discussed it has been assumed that each component would behave as though the others were not present, i.e. each would follow FICK's first law. In actual practice this seems to be not strictly true in many cases and a concentration gradient of one solute component can cause movement of another, i.e. the solute flows interact. With a non-specific method such as the refractometric one of determining concentration changes during diffusion it is seen that these interacting solute flows could cause trouble when the diffusion coefficient of only one component is required. The magnitude of these effects have been determined for some systems involving small molecules (see GOSTING 1956, DUNLOP 1957, O'DONNELL and GOSTING 1959) but not yet for macromolecules; it is reasonable in the case of proteins to expect them to be minor effects since the SVEDBERG equation which uses D_0 as measured at present to determine molecular weights gives the correct answer (probably within 5 per cent) for those cases where there is an independent means for measuring M.

2. Charge Effects with Macromolecules.

With the diffusion of macromolecules (as with their sedimentation) there will be restriction on the thermal movement if they carry a charge (e.g. proteins away from their isoelectric point) and this will give an erroneous value for the diffusion coefficient. So to overcome these charge effects it is customary to add salt; the diffusion experiments on proteins are normally also buffered at a particular pH. With salts present the system is a three-component one (at least) and hence there will be interacting solute flows which could give rise to error in equating $\mathfrak{D}_A$ measured by a non-specific refractometric method with the D of the protein. No work has been done to determine the magnitude of these effects in macromolecular systems yet and the protein-buffer salt-water is at present treated as a two component system — protein being the solute and buffer salt plus water the solvent. Again as in the previous section results using the measured diffusion coefficient of say a protein which have an independent method of confirmation of molecular weight indicate that the error is small.

3. Initial Conditions — Dialysis.

It is present practice to dialyze solutions of charged macromolecules against buffer before doing a diffusion experiment between solution of macromolecule and its dialysate. The procedure seems satisfactory, but is by no means proven.

4. Purity of Solute.

In addition to the three previous factors a serious limitation to the precision of measurement of diffusion coefficients of a homogeneous macromolecular solution is the purity of solute, i.e. it is very difficult to get a homogeneous (with regard to molecular weight and shape) solution of macromolecules. If there is only one macromolecular impurity present then the diffusion coefficient of the main component can be determined, but if there are more, the reduced height-area ratio $\mathfrak{D}_A$ or reduced second moment $\mathfrak{D}_{2m}$ cannot be resolved into the individual contributions of the constituent solutes. In this case only $\mathfrak{D}_A$ or $\mathfrak{D}_{2m}$ can be used in the SVEDBERG equation for determining molecular weight.

E. Ultracentrifugation.

I. Information Available from Sedimentation, and General Aspects of Sedimentation Analysis.

Sedimentation studies on a solution of macromolecules (proteins, polysaccharides, nucleic acids, synthetic polymers) can yield the following information:

a) homogeneity or degree of polydispersity of the macromolecular solute; hence it can be used to follow purification,

b) identification of components,

c) molecular weight of the solute (from 160 up to many millions) which is an average value if the solute is polydisperse,

d) the frictional coefficient of the solute molecules; if a model is arbitrarily chosen e.g. ellipsoid or random coil, then dimensions of this model can be calculated.

e) values of some thermodynamic quantities.

The standard books are "The Ultracentrifuge" (SVEDBERG and PEDERSEN 1940), and "Ultracentrifugation in Biochemistry" (SCHACHMAN 1959); some of the reviews and useful articles are OGSTON (1956), SCHACHMAN (1957), WILLIAMS, VAN HOLDE, BALDWIN and FUJITA (1958), BALDWIN and VAN HOLDE (1960) and WILLIAMS (1960).

Sedimentation analysis is very simple in principle and only requires a small amount of material (of the order of 1—10 mg or less) which can be recovered at the end of an experiment. A solution (0.3—0.7 ml) of the macromolecule is centrifuged in a sector-shaped cell with transparent quartz windows at speeds which produce centrifugal fields up to 250,000 g. Under these conditions the macromolecules sediment towards the bottom of the cell i.e. the periphery of the rotor. (If the molecules are less dense than the solvent e.g. the lipoproteins, then they move in the opposite direction). The observation by optical means of this movement gives data from which the above-listed information can be obtained.

The best commercial instruments available now are electrically driven and the one most frequently used is that made by Beckman Instruments[1]. The temperature of the rotor is controlled in this instrument. There are many types of cells each of which has its own special function. These include cells of varying optical path length for different concentrations of solution, artificial-boundary cells for forming boundaries of substances of low molecular weight or for measuring refractive increments of a solution, double-sector cells for automatic registration of base-line due to solvent sedimentation and for use with interferometric optics, separation cells for isolating sedimented components which normally would become mixed due to convection as the rotor slowed down. There is a dummy counterbalance cell in the rotor; this has two holes and these are used as reference points.

II. The Two Facets of Sedimentation Analysis.

There are two distinct aspects of sedimentation analysis; one, termed the sedimentation velocity method measures the speed at which the macromolecules sediment through the solution. It utilizes high rotor speeds up to 60,000 r.p.m. This, to date, has been the commonest type of usage of the ultracentrifuge and, in addition to yielding information on the frictional coefficient of a solute, it is the simplest and best method for determining and defining heterogeneity of a macromolecular solute. It is a dynamic method, i.e. there is a continuous restraint on the solute. With the aid of subsidiary measurements such as diffusion or viscosity (diffusion coefficients can be measured approximately in the ultracentrifuge) a molecular weight of the sedimenting material can be obtained. For this calculation the partial specific volume ($\bar{v}$) of the sedimenting material must be known. If the solute is heterogeneous it is not easy to obtain a well-defined average of the sedimentation coefficient; even if one did, its combination with a weight-average diffusion coefficient does not give a simple average molecular weight. It seems that this method will give way to the sedimentation equilibrium method for determining molecular weight, the latter giving well-defined average values.

The second aspect of sedimentation analysis is the equilibrium method which utilizes lower rotor speeds. Here the centrifugal force on the solute molecules at the various places in the cell due to rotation of the rotor is balanced exactly by the centripetal force (due to the gradient of chemical potential) causing the molecules to diffuse in the opposite direction. The macromolecular solute becomes distributed in a particular manner throughout the cell and this distribution does not then change with time. The time taken for attainment of equilibrium increases with increasing molecular weight (and decreasing diffusion coefficient) and for molecules with a molecular weight of say 50,000 it may take 1—6 days, depending on the length of the solution in the cell (Van Holde and Baldwin 1958). The long time taken to reach equilibrium has been a big disadvantage in the use of equilibrium ultracentrifugation but two important developments in recent years have shortened the time necessary for this method. These are firstly the Archibald (1947) method of "approach to equilibrium," and secondly the use of short columns of

[1] Spinco Division, Beckman Instruments, Inc., Palo Alto, California.

solution by VAN HOLDE and BALDWIN (1958) to speed-up the attainment of equilibrium.

The sedimentation equilibrium method gives molecular weights directly; no subsidiary measurement is necessary but again the partial specific volume of the macromolecule must be known. If the material is heterogeneous the method readily gives a weight-average and a z-average molecular weight (higher averages can also be obtained). The equilibrium method can decide whether a material is homogeneous or not, but if it is polydisperse an actual distribution of molecular weights can not be obtained. If some form of the distribution curve, e.g. log form, is known or assumed, the parameters for the distribution curve can be calculated from the average molecular weights (M_w, M_z, M_{z+1}, M_n) of the polydisperse system (cf. WILLIAMS et al. 1958, p. 740).

Calculations of distributions of sedimentation coefficients or molecular weights involve much labor and if one wishes to avoid this labor it is possible to obtain with much less work a measure of the extent that the material departs from homogeneity, i.e. such determinations can help to make confident statements that "the material is homogeneous" or "it is fairly homogeneous," or "it is more homogeneous than another particular preparation" or "it is very heterogeneous."

Sedimentation coefficients and sedimentation equilibria vary with the viscosity and temperature of the medium. For sedimentation equilibrium it is essential to have constant temperature; for sedimentation velocity it is desirable to do so as it simplifies the calculations.

III. Non-Ideal Behaviour and Charge Effects.

It should be pointed out that most of the ideas and equations developed below apply only to two-component (i.e. macromolecular solute plus pure solvent) ideal systems. However extrapolation to zero concentration from the concentrations which are measurable have been generally successful for the globular proteins because the non-ideality terms are small. It is for this reason and also because many proteins are nearly homogeneous or only consist of a few discrete components that sedimentation practice has had so much use in the field of protein chemistry.

For long flexible molecules such as synthetic polymers (which exhibit behaviour far removed from ideality) present methods of observation often make it difficult to study these solutions at concentrations low enough so that the behaviour is not too far removed from ideality; it is best to use a theta solvent for these materials if one exists.

For a particular polymer this is a solvent for which there exists a temperature θ, called the "FLORY temperature" where the polymer-solvent system is thermodynamically ideal (see BALDWIN and VAN HOLDE 1960 for the current situation regarding synthetic, uncharged polymers).

For nucleic acids which exhibit large deviations from ideality even at 0.02 per cent concentration the use of quartz optics and the absorption of ultra-violet light as an optical system has allowed much lower concentrations than this (down to 0.001 per cent) to be studied (SHOOTER and BUTLER 1956).

If the proteins (or any other macromolecule) contain charged groups neutral salt, such as potassium or sodium chloride, must be added to minimize charge effects (SVEDBERG and PEDERSEN 1940, PEDERSEN 1958a, b). Usually 0.1—0.2 M salt is sufficient to overcome the charge effects if the protein is not too far removed from its isoelectric point but the only way to tell is to add a higher concentration of salt and repeat the particular experiment to see if there is any effect. If above, say, 0.2 M salt there is an effect of increased salt concentration on either the shape of the sedimentation pattern or the sedimentation coefficient it must be decided whether this is due to charge effect, dissociation or association of the macromolecular solute, or change of molecular shape.

The solvent plus salt is treated as one component. This approximation works satisfactorily for relatively dilute salt solutions (see SCHACHMAN 1959, p. 208, 225 for summary of charge effects).

In general the use of mixed solvents is to be avoided since with three component systems the evaluation of the molecular weight from sedimentation experiments requires terms whose values are not usually known.

IV. Sedimentation Velocity.

1. The Boundary and Optical Means of Observation.

When a macromolecular solution is subjected to a sufficiently high centrifugal field the macromolecules migrate towards the bottom of the cell and pile up, leaving a region of buffer at the top of the cell devoid of macromolecules. The boundary between the "macromolecule-free" and "macromolecule-containing" solvent might be expected to be a sharp line if a single homogeneous substance were sedimenting. In fact, however, diffusion of the macromolecules, due to random thermal movement, causes blurring of this boundary and instead of a "boundary line" (Fig. 11a) there exists a "boundary region" (Fig. 11b) commonly referred to as "the boundary." A plot of concentration of macromolecule in the cell as a function of distance (x) of the position in the cell from the centre of rotation is given in Fig. 11b for different times (i.e. as the sedimentation velocity experiment proceeds); the boundary moves towards the bottom of the cell and in doing so becomes more diffuse due to diffusion.

If there are several macromolecular species there will be several sedimenting boundaries. Whether these actually separate or not, will depend on whether they are sufficiently different to overcome the tendency for them to become intermingled due to diffusion. (If the material is heterogeneous and the boundaries do not separate then the additional spreading due to heterogeneity also increases as the boundary moves towards the bottom of the cell).

A plot of concentration (or refractive index difference) versus radial distance in the cell can be obtained from two of the optical systems in current use; one is the light-absorption (including ultra-violet absorption) system, the sedimenting molecule absorbing more light than the solvent. The plate is photometered and from several photographs a series of curves like those in Fig. 11b is obtained; the second optical system which gives a comparable type of plot is the RAYLEIGH interference system; this is slowly coming into more general use (e.g. see RICHARDS and SCHACHMAN 1959). To use this optical system a double-sector cell fitted with slits is required, one sector containing solvent and the other containing solvent plus protein. As the macromolecule sediments an interference pattern is set up and the shape of these fringes is directly proportional to c versus x in the cell. This method has a higher sensitivity than the schlieren method (see below) and will hence allow lower concentrations to be studied. The measurements obtained during sedimentation experiments (excluding the use of the light absorption system) mostly refer to refractive index increment, but since this is usually nearly linear with respect to concentration we use the word concentration instead of refractive index increment for convenience.

The advantage of the light-absorption method is that if the sedimenting macromolecule has a high absorption at a particular wave length, e.g. nucleic acids and nucleo-proteins at 260 mμ, proteins at 280 mμ or haem proteins at 407 mμ, then sedimentation can be studied at much lower concentrations than with the RAYLEIGH interference method or the schlieren method to be described in the next paragraph, the resolution of which depend on the refractive index difference between the protein solution and the solvent. The limit of usefulness of the schlieren method is about 0.2 g/100 ml and the RAYLEIGH method will be lower than this. It is possible that the use of polarized light (see section D, III, 3, d) may allow much more dilute solutions to be studied in the ultracentrifuge than is possible with the RAYLEIGH or schlieren methods.

If the curves of concentration (or refractive index increment) versus radial distance in the cell (x) are differentiated with respect to x a plot of concentration gradient versus x is obtained and this is shown in Fig. 11c. There are obvious advantages from some points of view for obtaining a record of sedimentation in this manner. Discrete components appear as discrete peaks and measurement of movement of the peak is made easy. Also the concentration gradient is read directly from the photograph and this parameter is used in many of the calculations when measuring heterogeneity. It seems also that this method allows heterogeneity to be gauged more readily visually. However, if the concentration at any point in the cell is required this differential curve has to be integrated whereas the first two optical systems give this figure directly. All three optical systems have their advantages but it is true to say that the gradient observation (or schlieren method) is by far the one most often used to date. This type of gradient record on the photographic plate is obtained by optical differentiation of the refractive index increment versus x curve existing in the centrifuge cell; a cylindrical lens and inclined schlieren diaphragm perform the operation.

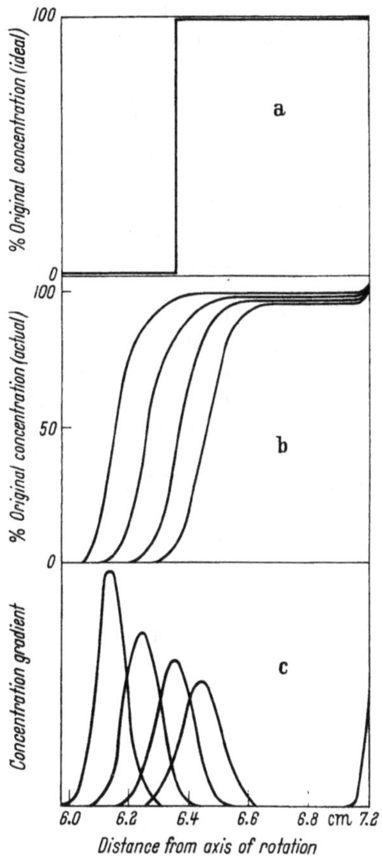

Fig. 11. Plot of concentration (a and b) and concentration gradient (c) versus distance from the axis of rotation during a sedimentation velocity experiment. Sedimentation is to the right. (a) = behaviour if there were no diffusion.

2. The Sedimentation Coefficient and its Experimental Evaluation.

The sedimentation coefficient, (s), of a molecule is defined as the velocity, (dx/dt), at which a molecule will sediment under unit gravitational field. Its units are sec^{+1}. The practical unit of sedimentation velocity is the SVEDBERG (S) and $1\ S = 10^{-13}\ sec^{+1}$. The sedimentation coefficient of a molecule depends on its weight, shape and density. Sucrose (M.W. 342) has a sedimentation coefficient of about 0.2 S, bovine plasma albumin (M.W. 67,000) of about 4.7 S, and a large molecule like tobacco mosaic virus (M.W. approximately 40×10^6) has a value of about 200 S.

The average rate of movement of the sedimenting particles is the same as the rate of movement of the boundary (see Williams et al. 1958, p. 756). The easiest way to determine this is to follow the movement of the maximum of the peak of the gradient curve obtained using schlieren optics (to be strictly correct the point representing the square root of the second moment of the gradient curve should be used (Goldberg 1953) but this is rarely done in practice unless the boundary is very skew). If the molecule is too small to overcome diffusion and form a free boundary (i.e. has a sedimentation coefficient of less than about 1) it is common practice now to use an artificial-boundary cell to form the boundary in the centre of the moving cell by layering solvent over solution; with such a cell the sedimentation coefficient of molecules of molecular weight as low as 150—200 can be measured. Sedimentation coefficients of boundaries incompletely resolved from the air meniscus (using the conventional cell) can also be obtained (Gutfreund and Ogston 1949, Baldwin 1953, Klainer and Kegeles 1955) but the calculation is not as direct or as quick as when using the artificial-boundary cell. However, if one uses the rate of movement of the peak to find the sedimentation coefficient then the accuracy decreases as the molecular weight decreases because the difference between the maximum gradient and the square root of the second moment becomes more important. Hence, using an artificial-boundary cell the error for sucrose would be of the order of 15 per cent. It is probably best to restrict the use of the artificial-boundary cell to molecular weights around 5000, and for molecular weights below this to use the conventional cell and calculate the sedimentation coefficient from the incompletely resolved boundary (see Williams et al. p. 759, 1958).

A sedimentation velocity experiment should be run if possible at such a speed that the sedimenting material takes 1—2 h to travel the length of the cell. As a

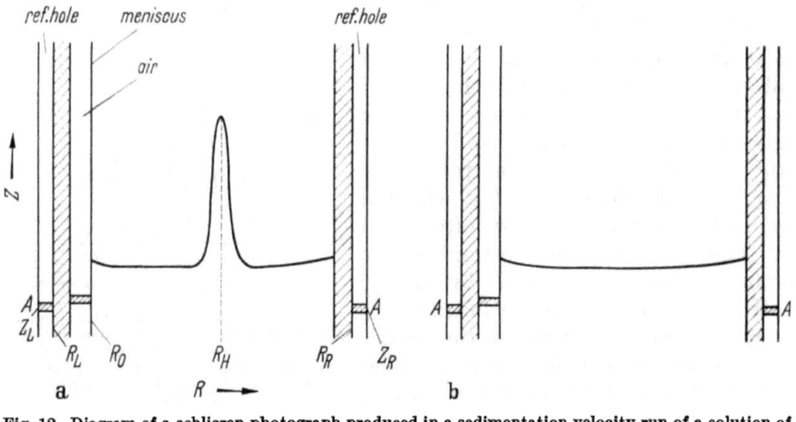

Fig. 12. Diagram of a schlieren photograph produced in a sedimentation velocity run of a solution of macromolecules (a) and the buffer in which it is dissolved (b).

rough guide for globular proteins it might be said that for molecules with a molecular weight less than 70,000 it is satisfactory to use the highest speed available (approximately 60,000 r.p.m.). If the molecular weight is 600,000 then 40,000 r.p.m. is a suitable speed.

By definition the sedimentation coefficient

$$s = (dx/dt)/\omega^2 x \tag{56}$$

where x = distance of the molecule in the cell from the centre of the rotor, t = time in seconds, ω = angular velocity of rotor = $2\pi n$ radians per second (where

n = number of revolutions per second). If s is constant this can be integrated to give

$$\ln(x_H/x_0) = s\,\omega^2 t \tag{57}$$

where x_H and x_0 are the distance of the peak of the boundary and meniscus respectively from the centre of the rotor. If x_0, i.e. the position of the meniscus, is constant throughout the run, i.e. the cell does not leak, then

$$\ln x_H = s\,\omega^2 t + \text{constant}\,. \tag{58}$$

Therefore the slope of a plot of $\log x_H$ against the time gives a figure from which s can be calculated. If the meniscus moves then $\log(x_H/x_0)$ must be plotted against t. If the temperature varies by several degrees during a run the viscosity of the medium and hence the sedimentation coefficient varies. The above equation can still be used under these conditions but must be handled by some method such as that of ONCLEY (1941) who used corrected time intervals to compensate for the temperature variation. However with the most modern ultracentrifuges when they are operating correctly neither does the cell leak nor the temperature vary by more than a few tenths of a degree.

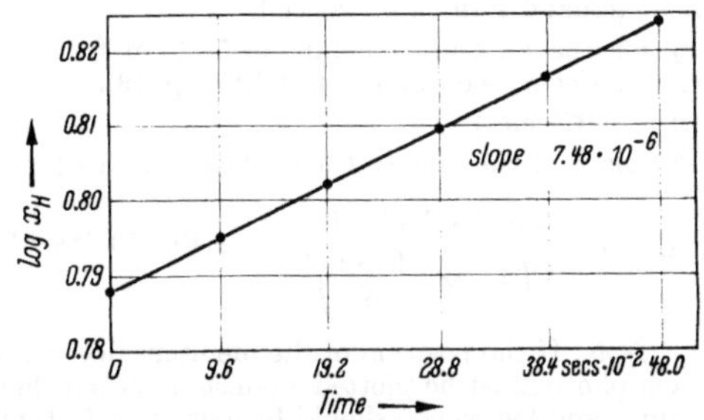

Fig. 13. Plot of $\log x_H$ (distance of peak from centre of rotation) versus time in the sedimentation velocity experiment of bovine plasma albumin.

The measurements of the photographs on the plates are made with a two-way travelling microscope, the air meniscus (see Fig. 12) being lined up with one of the cross wires of the comparator, or by putting the photographs in an enlarger and tracing the images on graph paper. The former method is the more precise. The position of the reference edge, peak and meniscus should be read to the nearest 10 microns. Alternatively when using the comparator the plate can be lined up by making the spaces (A's in Fig. 12) in the reference bars have equal readings within about 0.001 cm when the comparator is racked across. The authors prefer this latter method.

Table 2 and Fig. 12 show the measurements taken from the photographs obtained during a sedimentation velocity run and the plot of $\log x_H$ *vs* t is shown in Fig. 13.

It is usual to refer the sedimentation coefficient to standard conditions e.g. 25° in water. The conversion for temperature differences of a few degrees is

$$s_{25,w} = s_{t,b}(\eta_t/\eta_{25})_w(\eta_b/\eta_w)_{25}(1-\bar{v}\varrho)_{25,w}/1-\bar{v}\varrho)_{t,b} \tag{59}$$

where η, w, t and b stand for viscosity, water, temperature and buffer. It is assumed that $\bar{v}$ is the same in the reference solvent as in the buffer used in the experiment and it is also assumed that $\bar{v}$ does not vary within a small temperature range.

For the buffer used in the above experiment $(\eta_t/\eta_{25})_w = 1$, $(\eta_b/\eta_w)_{25} = 1.037$[1], and

$$(1-\bar{v}\varrho)_{25,w}/(1-\bar{v}\varrho)_{t,b} = \frac{1-0.735\times 0.9971}{1-0.735\times 1.0087^1} = 1.033$$

$$s_{25,w} = 4.40\times 1\times 1.037\times 1.033 = 4.71\ \mathrm{S}\,.$$

[1] Figures taken or calculated from SVEDBERG and PEDERSEN 1940; we have taken ϱ as ϱ_b i.e. the density of buffer; it should actually be density of the solution.

Table 2. *Calculation of sedimentation coefficient of bovine plasma albumin.*

t secs	R_L	R_O	R_H	$R_O - R_L$	$R_H - R_L$	$x_H = x_L + (R_H - R_L)/F$	$\log x_H$
0	*11.016*	*11.520*	*11.850*	0.504	0.834	6.133_3	0.7877
16×60	*14.910*	*15.412*	*15.961*	0.502	1.051	6.235_8	0.7949
32×60	*18.808*	*19.311*	*20.076*	0.503	1.268	6.338_1	0.8019
48×60	*22.722*	*23.223*	*24.214*	0.501	1.492	6.443_8	0.8092
64×60	*26.649*	*27.150*	*28.366*	0.501	1.717	6.549_9	0.8162
80×60	*30.586*	*31.091*	*32.541*	0.505	1.955	6.662_3	0.8236

R's = measurements on photographic plate and italicized figures are those actually read from plate.

x = actual distances in the cell.

x_L for the authors' machine = 5.74 cm from centre of rotation at 59,780 r.p.m. (at rest it = 5.70 cm; see SCHACHMAN 1959, p. 19).

(x_0 = 5.977 cm.)

The slope of the line $= d(\log x_H)/dt = 7.48 \times 10^{-6}$.

$$s_{25,\,buffer} = \frac{2.303 \times 7.48 \times 10^{-6}}{\left(2\,\pi \times \frac{59780}{60}\right)^2} = 4.40 \times 10^{-13}\ \mathrm{sec}^{+1} = 4.40\ S.$$

Note: If the position of the meniscus does vary slightly due to leakage then $\log(x_H/x_0)$ for each photo must be plotted against t; (x_H/x_0) should have five, e.g. (1.0155×10^{-2}) significant figures and $\log(x_H/x_0)$ should be looked up in tables of five-place logarithms.

Concentration of bovine plasma albumin = 0.84 g/100 ml solution. Buffer = 0.15 M NaCl, 0.025 M Na_2HPO_4, 0.025 M NaH_2PO_4, pH = 6.6. 12 mm. 4° cell. Speed 59,780 r.p.m. Temperature of 25°.

$$F = \text{vertical magnification factor} = \frac{\text{distance moved on photographic plate}}{\text{distance moved in cell}} = 2.12.$$

The sedimentation coefficient varies with concentration of the sedimenting substance in a manner which cannot be predicted; practically the form may be $s = s_0(1 - kc)$, $s = s_0(1 - kc \pm 1c^2)$, or $1/s = (1/s_0)(1 + kc)$. None of these equations apply over a large concentration range. The simple linear plot is adequate for many proteins over a narrow concentration range (say 0—1 per cent) but for elongated particles or flexible polymers more complicated forms are necessary unless much lower concentration ranges are considered. Some examples of the dependence of the sedimentation coefficient on concentration are given below: for bovine plasma albumin in the concentration range 0.29—1.76 per cent $s_{25,w} = 5.01\ (1 - 0.054c - 0.004c^2)$ S (BALDWIN 1957b) and for thyroglobulin in the concentration range 0.08—0.8 per cent $s_{25,w} = 21.59\ (1 - 0.147c + 0.006c^2)$ S (O'DONNELL, BALDWIN and WILLIAMS 1958). These are globular proteins; for elongated particles such as tobacco mosaic virus the dependence is much greater, i.e.

$$10^3/s_{20,w} = 5.40\ (1 + 0.278c) \quad \text{(LAUFFER 1944)},$$

and for myosin

$$s_{20,w} = (6.43 - 4.3\ c)\ \mathrm{S} \quad \text{(JOHNSON and ROWE 1960)}.$$

Due to the variation of sedimentation coefficient with concentration it is necessary to refer s to some concentration of the sedimenting solute for each particular experiment. The concentration of solute changes during a sedimentation experiment due to the sector shape of the cell so it is rational to either calculate s in such a manner that it refers to the actual original solute concentration (e.g. BALDWIN 1957b) or to refer it to an average concentration which exists during the time of sedimentation (KEGELES and GUTTER 1951). The first is the more accurate and also the more time consuming; for all ordinary work the second method at least should be used. Using the second method this average concentration would be $\bar{c} = c_0(x_0/x_H)^2$ where x_H is taken from the middle of the experiment (c_0 is the initial concentration).

Looking at the example above it is seen that an average x_H is 6.394 and $x_0 = 5.74 + 0.503/2.12 = 5.977$.

$$\therefore \bar{c} = c_0 \times \left(\frac{5.977}{6.394}\right)^2 = 0.84 \times 0.874 = 0.73 \text{ g/100 ml}.$$

So the experimental result is reported that $s_{25,w} = 4.71$ S with $\bar{c} = 0.73$ g/100ml.

The sedimentation coefficient must be extrapolated to zero concentration for insertion into the SVEDBERG equation (see later).

The sedimentation coefficient of a solute and its variation with concentration under stated conditions of pH, buffer composition and solute concentration gives a good definition of a particular material. It should be reproducible and if care is taken it can be measured to within a few tenths of a per cent (BALDWIN 1957b).

3. Minimum Requirements from Sedimentation Analysis which must be Fulfilled if a Substance is to be Claimed as Homogeneous with Respect to Sedimentation Coefficient.

First of all it is an obvious necessity that for a homogeneous (non-dissociating) solute there must only be one peak in the sedimentation pattern as it passes through the cell in a period of 1—2 h. A more stringent requirement of homogeneity is that all the macromolecular material present in the original solution should sediment under this peak. If some stays in the meniscus the material would not be homogeneous. To make certain that initially at least these two criteria are fulfilled a base line accounting for optical distortion of the cell and sedimentation of the buffer salts must be fitted to the gradient picture of the sedimenting macromolecules. This "fitting of a base line" can be done in one of two ways. The first method involves running the buffer (or dialysate) in a separate run in the same cell and under the same conditions as the protein itself was run. The corresponding photographs for the buffer are substracted from those for the protein by superimposing the air gaps in the reference lines of the two photographs (points A in Fig. 12) and then substracting the ordinates. This can be done by superimposing enlarged tracings of the photographs on graph paper or measuring each curve with a two-way travelling microscope (the latter is the more precise procedure). The second method of fitting the base-line involves the use of a double sector cell. Buffer alone is run in one sector and the protein solution in the other. Here there is automatic registration of the base line. This is the better method but the centre-pieces of the cells are likely to break or leak at speeds higher than 40,000 r.p.m.; there is no doubt that this disadvantage will soon be overcome.

The peak may now be examined more critically than before a base line was fitted — small amounts of heavier or lighter material may be seen or there may be long diffuse trailing or leading edges (see Fig. 14). From this picture the area under the peak (which is directly proportional to the protein concentration and refractive increment) can readily be calculated and it can be checked whether all the initial protein is in this peak. If A is the area of the peak (in sq. cm) on the photographic plate then it should equal

$$A = c_0 a b F G (\Delta n/\Delta c)\, E^2 (x_0/x_H)^2 \cot\theta$$

or

$$A = c_0 \frac{\Delta n}{\Delta c} (x_0/x_H)^2 \cot\theta \cdot \underline{k} \qquad (60)$$

where E = enlarger factor which equals unity if a microcomparator is used to measure the photographic plates;

c_0 = initial protein concentration,

b = optical lever arm which is approximately the focal length of the upper collimating lens of the centrifuge and is usually given by the manufacturer,

a = cell thickness in cm along optical path,

F = vertical magnification factor;

G = horizontal magnification factor,

$\Delta n/\Delta c$ = specific refractive increment; this has a value of about 0.0019 for many proteins,

$(x_0/x_H)^2$ corrects for radial dilution due to sector shaped cell,

$\underline{k} = abFGE^2$.

$c_0(\Delta n/\Delta c)$ may be obtained either from the concentration and the specific refractive increment of the macromolecular solute or by direct measurement in a differential refractometer.

$\underline{k}$ can be determined without a knowledge of the instrument constants by forming a boundary between a solution of known refractive increment e.g. 1 per cent bovine plasma albumin, and its dialysate in an artificial boundary cell. The boundary is allowed to diffuse at a low speed where sedimentation will be negligible i.e. $(x_0/x_H)^2 = 1$. When the peak has spread sufficiently the area of the peak is measured and the angle of the schlieren diaphragm noted. A separate base-line run is not necessary as sedimentation of buffer salts at this low speed will be negligible; it is sufficient to continue the base-line across under the peak. $\underline{k}$ can then be calculated from equation (60). For the authors' Spinco machine $\underline{k}$ was found experimentally to be 542 while the value calculated from the instrument constants was 538. If the area obtained on the photographic plate is less than this calculated value some material has not sedimented away from the meniscus and the solute is heterogeneous.

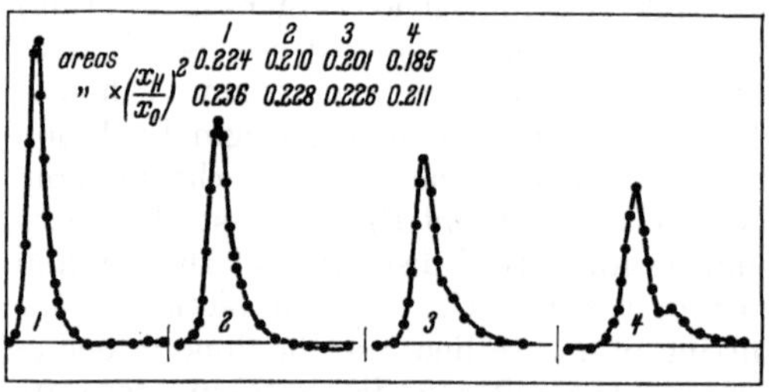

Fig. 14. Diagrams of the sedimentation patterns of a sample of bovine plasma albumin after the base-lines have been fitted and subtracted. Areas of actual patterns on the photographic plate are given as well as those corrected for sectorial dilution.

Solutions of proteins or other macromolecules may be dialyzed against buffer to eliminate material of low molecular weight before measuring their refractive increment either in an artificial-boundary cell or in a separate differential refractometer. The cellulose dialysis tubing of finest pore size which is commercially available is Visking $\frac{18}{32}$ (CRAIG, KING and STRACHER 1957).

The base-line sedimentation patterns corresponding to the bovine plasma albumin run in Table 2 were substracted and the drawings of the curves after fitting the base-lines are shown in Fig. 14; the corresponding areas corrected for sectorial dilution are shown underneath. It is seen that the first three areas vary by about 4 per cent which is to be expected for a broad peak or a peak with a tail. In photograph 4 the area is definitely decreased which means that the leading edge of the boundary has reached the bottom of the cell. With an artificial-boundary cell run at low speed the area constancy from photograph to photograph will be 1 per cent or better.

If for a particular material the dependence of the sedimentation coefficient upon concentration is linear then FUJITA (1956) has shown that for a single component the gradient curve is quite symmetrical except at the edges; marked asymmetry in itself is indicative of heterogeneity.

4. More Stringent Tests which must be Satisfied by a Homogeneous Material.

Suppose that a macromolecular solute on sedimentation through the cell yields one peak and all of the solute sediments under this peak leaving a flat base line on either side after the buffer base line has been subtracted. If it also has a linear dependence of s upon c and the gradient curve is symmetrical it is possible to apply much more stringent tests for homogeneity but these entail laborious calculations. The shape of the single peak must now be considered. At a particular speed this shape depends on a) the heterogeneity of the components composing it (the greater the heterogeneity the greater the degree of spreading) b) the diffusion coefficient (D) of the component (s) present (the higher D is the greater the spread) c) the dependence of the sedimentation coefficient upon concentration of the sedimenting substance (if k is large and positive the boundary will be artifically sharpened). Any one of these three factors may become dominant and over-ride the others; or there could be an extreme case of counter-balancing where a very heterogeneous preparation of low diffusion coefficient and strong dependence of the sedimentation coefficient upon concentration could give a sedimentation pattern which might be mistaken for a reasonably homogeneous material e.g. deoxyribonucleic acids.

So to apply tests based on the departure of the shape of a curve from that expected for a homogeneous sedimenting material the dependence of sedimentation coefficient upon concentration must be determined. If s is linear with c (and it probably will be if experiments are conducted at low enough concentrations) then from this data and from the shape of the sedimentation curve at various times a value for the diffusion coefficient can be calculated from the slope of a line obtained (FUJITA 1956, for application of method see BALDWIN 1957b). If this calculated value of D is constant with time this is a good test of homogeneity with regard to sedimentation coefficient; if the solute is heterogeneous then the calculated D will increase with time. If the line is not straight it can only be said that the material is heterogeneous, i.e. the theory does not give a quantitative description of the heterogeneity.

A second sensitive test of homogeneity which considers the shape of the boundary, taking into account the effects of diffusion and the variation of s with c, is given by BALDWIN (1957a). The standard deviation of the distribution curve of sedimentation coefficients at infinite dilution is calculated. If this is zero then this is a good indication that only a single component is present [see also E, IV, 5 (a)].

If the dependence of s upon c is negligible (and such cases would be rare) the boundary gradient curve should be GAUSSIAN and a plot of $\ln(dc/dx)$ versus $(x - x_H)^2$ should be linear for values of x on either side of the maximum ordinate x_H (BALDWIN 1957a; HALL and OGSTON 1956). However it must be realized that even a small dependence of s upon c has a considerable effect on the shape of the boundary curve.

5. Expressions of the Degree of Heterogeneity of a Material.

From the foregoing it can be said whether or not a material is homogeneous with respect to its sedimentation coefficients. If it is not homogeneous the next question is how is quantitative expression given to its degree of heterogeneity. This can be done in one of two ways: either the actual distribution of the sedimentation coefficients can be obtained or one can obtain a figure indicating its degree of departure from homogeneity.

a) Actual Distribution of Sedimentation Coefficients.

If diffusion is negligible such as with glycogen (BRIDGMAN 1942) or deoxyribonucleic acid (SHOOTER and BUTLER 1956; SCHUMACHER and SCHACHMAN 1957) a curve of actual distribution coefficients is calculated by the method of BRIDGMAN (1942) and SIGNER and GROSS (1934). If diffusion is not negligible, but if the contribution of diffusion to the spreading of the boundary is considerably less than that due to heterogeneity, the distribution curve can be calculated by methods developed by BALDWIN and WILLIAMS (see BALDWIN 1954). If the diffusion contribution is too large an extrapolation procedure can be used (BALDWIN 1959). It should be pointed out that these methods assume a continuous distribution of sedimentation coefficients under any one peak.

b) Method of Indicating Departure of a Sedimentation Curve from that Representative of Homogeneity.

Suppose it is not desired to calculate the exact distribution of sedimentation coefficients in a particular material but the information required is "how far does this material depart from homogeneity." This question can be answered by use of the general method of BALDWIN and WILLIAMS (see BALDWIN 1957a). The method takes into account heterogeneity, diffusion, and the dependence of the sedimentation and diffusion coefficients on concentration. It calculates the standard deviation, p, of the sedimentation coefficient distribution from the second moment of the gradient curve. The standard deviation would equal zero if only a single component were present and departure from zero indicates heterogeneity. The resolving power of the method is sufficient to detect a value of p equal to about 5 per cent of the mean sedimentation coefficient. With this limitation BALDWIN (1957a) found $p = 0$ for a sample of bovine β-lactoglobulin and equal to 1.08 S and 0.69 S for two samples of crystalline bovine plasma albumin.

6. Density Differences as a Test of Heterogeneity. Equilibrium Sedimentation in a Density Gradient.

If molecules differ in density this heterogeneity can be revealed by changing the density of the medium in which the molecules are sedimenting. A polydispersity with respect to density then shows up as an increased spread in the range of sedimentation coefficients as the density of the medium is made to approach that of the sedimenting particles.

MESELSON, STAHL and VINOGRAD (1957) have used solutions of caesium chloride to form a density gradient in the spinning ultracentrifuge cell. If a mixture of macromolecules is placed in this gradient an equilibrium is eventually set up, the molecules being distributed at those places where the density equals the density of the medium. In this way MESELSON and STAHL (1958) were able to separate deoxyribonucleic acid molecules containing N^{15} from those containing N^{14} (a difference in molecular density of 0.014 g/cm^3). A band representative of molecules which are monodisperse with respect to density and molecular weight will be GAUSSIAN with a standard deviation which is a function of the molecular weight of the macromolecules. If there is molecular weight heterogeneity both number- and weight-average molecular weights can be obtained (see SUEOKA 1959 and WILLIAMS et al. 1958).

V. Equations for Determining Molecular Weight from Sedimentation Velocity.

When a molecule of a substance, 1, of partial specific volume $\bar{v}_1$ sediments in a solution of density ϱ in a centrifugal field of strength $\omega^2 x$ the force acting centrifugally on it is *effective mass* × *acceleration* = $(M_1/N)\,(1 - \bar{v}_1\,\varrho)\,\omega^2 x$. If this were the only force acting on the substance it would continually accelerate, but it is opposed by the frictional resistance of the molecule moving through the liquid. This frictional force equals the *molecular frictional coefficient* × *speed* = $f(dx/dt)$. Therefore in sedimentation, where the substance at any particular time and field strength is moving with constant velocity, the resulting force on it must equal zero.

i.e. $$(M_1/N)\,(1 - \bar{v}_1\,\varrho)\,\omega^2 x = f(dx/dt) \quad \text{or} \quad M = N s_1 f/(1 - \bar{v}_1\,\varrho)\,. \tag{61}$$

This is the kinetic approach. Recently these equations have also been derived from the flow equations developed by the methods of thermodynamics of irreversible processes (see WILLIAMS et al. 1958). This latter approach shows that ϱ in the equation is the density of the solution.

1. The SVEDBERG Equation.

To eliminate the frictional coefficient from equation (61) it is known that $D = \frac{RT}{Nf}$ for an *ideal* solution of a homogeneous solute (see equ. 39). The frictional coefficients in sedimentation and diffusion are the same according to the thermodynamic theory of irreversible processes, provided the sedimentation coefficient is independent of rotor speed, a point which can be tested experimentally. Combining these two equations

$$M_1 = \frac{R T s_1}{D(1 - \bar{v}_1\,\varrho)} \tag{62}$$

which is the SVEDBERG sedimentation velocity equation for determination of molecular weight. If D is expressed as $\text{cm}^2\ \text{sec}^{-1}$ and s as sec^{+1}: then $R = 8.315 \times \times 10^7$ ergs/mole degree.

This equation applies strictly only at infinite dilution (i.e. ideal system) so the sedimentation and diffusion coefficients (D) must be extrapolated to zero concentration of solute. D normally has much less dependence upon c than has s. The limiting factors in the accuracy of M for a homogeneous solute from the SVEDBERG equation are the value of D (see section D, V) and the value of $\bar{v}$.

For a *non-ideal* two-component system the corresponding equation is

$$M_1 = \frac{R T s_1 \left(1 + c_1 \frac{d \ln y_1}{d c_1}\right)}{D(1 - \bar{v}_1 \varrho)}\,.$$

y_1 being the thermodynamic activity coeffecient on the g./100 ml concentration scale.

The diffusion coefficient (or rather the reduced height-area ratio $\mathfrak{D}_A$) can be measured more precisely in a separate apparatus but if an approximate value is all that is required (within 5 per cent) it can be obtained from the ultracentrifuge by using the artificial boundary cell at low speed. Here a boundary between solution (e.g. 1 per cent bovine plasma albumin in 0.1 M potassium chloride 0.01 M acetate at pH 4.5) and a dialysate is formed and allowed to diffuse at rotor speeds low enough (10,000 r.p.m. or less) for sedimentation to be negligible. Photographs are taken at various times (16 min intervals) and an example of calculation of $\mathfrak{D}_A$ from the schlieren patterns by the height-area method for this example is given in section D, III, 6.

If a substance is substantially homogeneous then the diffusion coefficient and sedimentation coefficient can be calculated from one and the same velocity run provided that the boundary sharpening effect due to the concentration dependence of the sedimentation coefficient is taken into account (FUJITA 1956; BALDWIN 1957b); i.e. the dependence of s on c must be known.

If a sedimenting material is polydisperse then the molecular weight calculated from the sedimentation and diffusion coefficients at infinite dilution by the SVEDBERG equation gives an average molecular weight which is not related in a simple manner to the weight-average, number average, or z-average molecular weights. When the sedimentation and diffusion coefficients are weight-average values the molecular weight calculated from the SVEDBERG equation is referred to as $M_{w,w}$.

2. Approximate Methods.

Equation (61) can be used to obtain an approximate minimum value of the molecular weight without recourse to a second hydrodynamic measurement just as was done for the diffusion coefficient. The assumptions made are the same (see D, IV, 2). If the molecule is very asymmetric or hydrated the approximate minimum value obtained could be incorrect by a factor of 10 or more. The method is not to be recommended. However the equation $s = K M^a$ is useful for comparing the molecular weights of a series of similar materials using only the sedimentation coefficient e.g. a series of fractions of a polymer preparation.

A practically useful method for determining approximate molecular weights (or limits of molecular weights) from sedimentation coefficients at infinite dilution and the intrinsic viscosity $[\eta]$ is by using the β-function (see section G, I, 3). If the molecule in solution is known to have the form of a randomly-coiled chain and the solute is homogeneous its molecular weight can be determined from s_0 and $[\eta]$ (see G, II).

3. Partial Specific Volume.

This figure appears in the term $(1 - \bar{v}\varrho)$ and needs to be known fairly accurately. For proteins it is of the order of 0.70—0.75 ml/g at 20° in water or dilute buffers and it can be calculated with quite good accuracy from the amino acid composition of a protein (McMEEKIN and MARSHALL 1952). To measure $\bar{v}$ experimentally the details are given in SVEDBERG and PEDERSEN (page 59) and also by SCHACHMAN (1957).

The classical method involves determining the density pycnometrically of the solution containing the protein (ϱ) and also of the solvent alone (ϱ_0). If c is the concentration of the protein in g/ml of solution the apparent partial specific volume of the protein

$$\bar{v}_{app} = 1/\varrho_0 - (\varrho - \varrho_0)/c\,\varrho_0 \,. \qquad (63)$$

It has been found that the apparent partial specific volume for proteins does not vary with concentration and therefore it may be taken as equal to the true partial specific volume. The partial specific volume of a protein does vary a little with the solvent (CHARLWOOD 1957).

The molecular weight derived from the SVEDBERG equation is the anhydrous molecular weight if the partial specific volume refers to dry weight of the protein. If the partial specific volume refers to the hydrated particle then the M from the SVEDBERG equation is the hydrated molecular weight. The hydrated partial specific volume could not be determined unless the solvated solute had a well-defined composition which is not likely.

VI. Sedimentation Equilibrium and Approach-to-Equilibrium.

This technique requires that solutions of macromolecular solutes be run in the ultracentrifuge at low speeds until the centrifugal force (due to field strength) driving the molecules outwards is balanced by the centripetal force (due to the gradient of chemical potential) driving the molecules inwards. When this condition is attained the solute distribution throughout the cell is invariable with time. Fig. 15 shows a plot of concentration and concentration gradient versus distance in the cell at various times in a sedimentation equilibrium experiment.

The equation describing the equilibrium state can be formulated by thermodynamics as follows:

$$M(1 - \bar{v}\varrho)\,\omega^2 x = (\partial\mu/\partial x)_{T,p} \quad (64)$$

where μ is the chemical potential per mole of solute and ϱ is the density of the solution.

Now for an ideal solution

$$\mu = \mu_0 + RT \ln c \quad (65)$$

and therefore

$$RT(dc/dx)/c = M(1 - \bar{v}\varrho)\,\omega^2 x\,. \quad (66)$$

This equation holds at any point in the cell at equilibrium where c is the concentration at that point. Thus the molecular weight in an equilibrium experiment can be obtained without any subsidiary measurement such as the diffusion coefficient. It is obvious that this method is not as good or simple in detecting heterogeneity as is the sedimentation velocity method; it is run at too low a speed to give discrete peaks. Moreover it cannot give an actual distribution of molecular weights (or sedimentation coefficients); the form of the distribution must be assumed and the results found can be fitted to this. However if the molecular weight is constant at various points in the cell once equilibrium has been attained this is a good indication of homogeneity with respect to molecular weight.

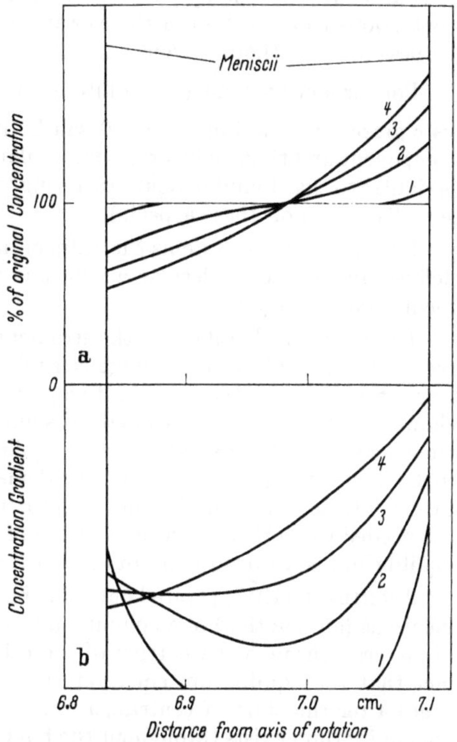

Fig. 15. Plot of concentration (a) and concentration gradient (b) versus distance from the centre of rotation during a sedimentation equilibrium experiment. Curve 4 represents the system at equilibrium.

Equation (66) is strictly valid only for binary ideal systems. For binary non-ideal systems an approach similar to the virial coefficient approach of osmotic pressure and light-scattering methods can be used. An equation can be written

$$\frac{1}{M^{app}} = \frac{1}{M} + 2A_2 c \quad (67)$$

c is expressed here in g/ml as it is in equations (3) and (17). This A_2 is the same as that in the osmotic pressure and light scattering equations. The true value of M can be obtained from a plot of the reciprocal of the apparent molecular weight versus solute concentration.

For the rod-like myosin molecule (M.W. 6.19×10^5) for example the value of A_2 in 0.5 M potassium chloride is approximately 0.9×10^{-4} moles ml/g² while in 5 M guanidine-hydrochloride it is approximately 5×10^{-4} moles ml/g² (KIELLEY and HARRINGTON 1960).

A globular protein in the presence of neutral salt for the suppression of charge effects approximates to a two-component ideal system but synthetic polymer solutions may exhibit marked deviations from ideality. In the latter case it is difficult with the optical systems currently in use to get to low enough concentrations to be certain of the extrapolation to zero concentration and it would be best to use a theta solvent.

For ternary systems (e.g. solute + mixed solvent) both the slope of the $\frac{1}{M^{app}}$ vs c graph and its limit at $c = 0$ will be influenced by the presence of the third component and thus an incorrect molecular weight may be obtained. The situation is similar to that found in light-scattering (see WILLIAMS et al. 1958 for comprehensive discussion on these aspects).

For polydisperse systems the sedimentation equilibrium method can give well-defined averages of molecular weight and this is its big advantage over the sedimentation velocity method.

The main disadvantage of the sedimentation equilibrium method is that until recently a normal ultracentrifuge cell full of a solution of macromolecular solute would take several days to come to equilibrium. VAN HOLDE and BALDWIN (1958) showed that if the length (radial) of solution column was reduced to 3 mm then ribonuclease (M.W. 13,682) reached equilibrium in 14 h and sucrose (M.W. 342) in 3.5 h and they gave equations for calculating the approximate time for a particular solute to reach equilibrium; approximate values for its diffusion coefficient are needed to do this. Solution columns of 1 mm and less have been used and here equilibrium is attained more quickly still (YPHANTIS 1960).

ARCHIBALD (1947) pointed out that there was no transfer of material across the meniscus (air-solution) of a spinning ultracentrifuge cell nor was there any passage of material through the bottom of the cell and that these conditions lead to the fact that the equilibrium equation applies at these two radii of rotation at all times after the start of centrifugation in a sector cell irrespective of whether or not equilibrium exists throughout the bulk of the solution. Thus molecular weights can be calculated at the meniscus and at the bottom of the cell in runs of short duration and before equilibrium is even closely approached. If they are equal and constant with time this is a good criterion of homogeneity. The method requires a knowledge of the concentration and concentration gradient at the top and bottom menisci of the cell and this requires extrapolation from positions close to the menisci as well as accurate location of the menisci. These are the disadvantages of the method. The molecular weight at the bottom meniscus (usually formed by some immiscible transparent liquid so that the correct radius of curvature is obtained) is more difficult to measure than at the top of the cell. The bottom gradient curve initially becomes steeper with time and this makes extrapolation difficult; extrapolation here should be carried out early in the experiment. The gradient curve at the top meniscus becomes shallower as time proceeds and extrapolation here is easy. Commonly the top meniscus only is used. This method has been called the ARCHIBALD "approach-to-equilibrium method" or, simply, "the ARCHIBALD method" and it is a quick method.

For a polydisperse non-ideal system, such as a synthetic polymer in solvent, the differential equation for sedimentation equilibrium still applies at the cell boundaries and can give direct information on the weight-average molecular weight and the second virial coefficient of the polymer (KEGELES, KLAINER and SALEM 1957).

For systems such as proteins in concentrated urea or guanidine hydrochloride theory has not yet been worked out to indicate whether the simple equilibrium equation (66) is applicable. However practically it has been found that it does give the correct answers in some cases investigated (TRAUTMAN and CRAMPTON 1959; KIELLEY and HARRINGTON 1960).

The results from these equilibrium and approach-to-equilibrium experiments can be calculated in several ways four of which are given below.

1. The KLAINER-KEGELES (1955) Calculation of the ARCHIBALD Method when a Plateau Region is Still Present.

The following equations hold at all times after the start of sedimentation.

$$M_{x_0}^{app} = RT(dc/dx)_{x_0}/(1-\bar{v}\varrho)\,\omega^2 x_0 c_{x_0} \tag{68}$$

$$M_{x_B}^{app} = RT(dc/dx)_{x_B}/(1-\bar{v}\varrho)\,\omega^2 x_B c_{x_B} \tag{69}$$

where x_0 = position of the liquid-air meniscus, i.e. the top one, and x_B = position of the liquid-silicone meniscus, i.e. the bottom one. c_{x_0} and c_{x_B} and $(dc/dx)_{x_0}$ and $(dc/dx)_{x_B}$, are the corresponding concentrations and concentration gradients of solute.

If there is a plateau region persisting throughout the experiment, i.e. a portion of the base line remains flat throughout the experiment, then the equations and calculations are relatively simple. The concentrations c_{x_0} and c_{x_B} are then given by the following equations due to KLAINER and KEGELES (1955)

$$c_{x_0} = c_0 - \int_{x_0}^{X} x^2 (dc/dx)\, dx/x_0^2 \tag{70}$$

$$c_{x_B} = c_0 + \int_{X}^{x_b} x^2 (dc/dx)\, dx/x_B^2 \tag{71}$$

X being the plateau region where $(dc/dx) = 0$ and c_0 the original concentration of the solution.

These equations can only be used for the approach-to-equilibrium method. They cannot be applied at equilibrium since under these conditions there would be no constant plateau region.

An example of the calculation of molecular weights of bovine plasma albumin at the top meniscus when there is some plateau region remaining is given in Table 3; the measurements taken from a photograph are shown in Fig. 16.

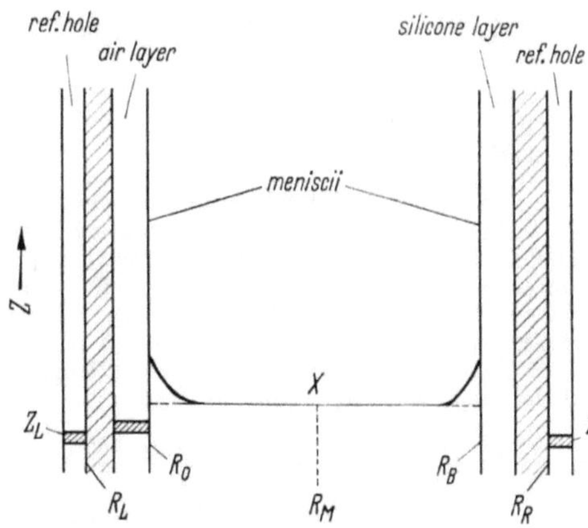

Fig. 16. Diagram of schlieren sedimentation pattern from an "approach-to-equilibrium" experiment using ordinary 12 mm cell.

Notes on the ARCHIBALD method of approach-to-equilibrium:

a) In order to see the bottom of the solution clearly and to have the correct radius of curvature, it is advisable to use approximately 0.1 ml of silicone oil or some such heavy liquid, e.g. Dow Corning 555, Kel F polymer oil No. 1, or carbon tetrachloride (GINSBURG, APPEL and SCHACHMAN 1956).

Some proteins denature at this liquid-liquid interface and this is evidenced by a progressive thickening of the bottom meniscus with time due to the piling up

Table 3 (top meniscus)[1]. *Sedimentation equilibrium of a sample bovine plasma albumin*[2].

KLAINER-KEGELES calculation of molecular weight at the meniscus using ARCHIBALD's approach-to-equilibrium method *(plateau region present)*. Comparator used to measure plates. F = vertical magnification factor = 2.12. Speed = 7928 r.p.m. $\omega^2 = 6.892\times10^5$. Distance from centre of cell to centre of rotor = 6.500 cm = x_m. All measurements are in cm. Area of photo from artificial-boundary cell = 0.1621 sq. cm.

$$Z_L = 6.559,\ R_L = 16.812,\ R_0 = 17.466,\ R_M - R_0 = 1.033.$$

$$Z_B = 6.559,\ R_B = 20.185,\ R_B = 19.644,\ R_M - R_B = -1.145.$$

$$\bar{Z} = 6.559,\ R_M = 18.499,\ x_0 = 6.500 - \frac{1.033}{2.12} = 6.013\ .$$

$$x_B = 6.500 + \frac{1.145}{2.12} = 7.040\ .$$

R_n	$R_n - R_0$	$x_n = 6.013 + (R_n - R_0)/F$	x_n^2	Z_n	$Z_n - \bar{Z}$	$Z_n - \bar{Z} - 0.007 = y \propto (dc/dx)$	$x_n^2 y$
17.466	0	6.013	36.16		(0.290[3])	0.283	10.23
.476	0.01						
.486	0.02						
.496	0.03			*6.807*	0.248		
17.506	0.04			*6.793*	0.234		
.516	0.05	6.036	36.43	*6.780*	0.221	0.214	7.80
.526	0.06			*6.767*	0.208		
.536	0.07			*6.753*	0.194		
.546	0.08			*6.738*	0.179		
.556	0.09			*6.724*	0.165		
.566	0.10	6.060	36.72	*6.708*	0.149	0.142	5.21
17.616	0.15	6.083	37.00	*6.649*	0.090	0.083	3.07
.666	0.20	6.107	37.30	*6.606*	0.047	0.040	1.49
.716	0.25	6.130	37.58	*6.587*	0.028	0.021	0.79
.766	0.30	6.154	37.87	*6.575*	0.016	0.009	0.34
.816	0.35			*6.566*	0.007	0	
.866	0.40			*6.566*	0.007	0	$\Sigma x_n^2 y = 28.93$
.916	0.45			*6.566*	0.007	0	
.966	0.50			*6.566*	0.007	0	

[1] A similar calculation can be carried out at the bottom meniscus.

[2] Protein concentration approximately 1 per cent. Buffer (pH 6.6) = 0.025 M Na_2HPO_4, 0.025 M NaH_2PO_4, 0.15 M NaCl (density approximately 1.0087 at 25°). There was 0.6 ml protein solution and 0.1 ml CCl_4 in a 4° 12 mm Spinco cell. Angle for photographs = 80°. Time of the photograph was 44 min after reaching speed. The protein and its dialysate were also run in an artificial boundary cell at 10,000 r.p.m. and photographed at the same angle.

[3] This figure is obtained by linear extrapolation to $R_n - R_0 = 0$ of the $(Z_n - \bar{Z})$ readings in the range $R_n - R_0 = 0.03$—0.10 cm.

$$c_{x_0} = \text{(by equn. 70) } 0.1621 - \frac{28.93\times0.05}{36.16} = 0.1221$$

$$M_{x_0} = \text{(by equn. 68) } \frac{8.313\times10^7\times298\times0.283\times2.12\,(F)}{(1-0.735\times1.0087)\times6.892\times10^5\times6.013\times0.1221} = 113{,}400\ .$$

1. This figure would be within 1—5% of the true value. It is higher than the molecular weight of bovine plasma albumin (69,000) due to the heavy material present (see Fig. 14).

2. R's and Z's = measurements on the photographic plate, and italicized figures are those actually read from the plate by micro-comparator.

3. x = actual distances in cell; n refers to a point in the cell. (dc/dx) is directly proportional to y, and all magnification factors cancel out except the one (F) shown.

of sedimented material. If this happens with each of the immiscible liquids available then this bottom meniscus cannot be used to make measurements.

b) M_{x_0} and M_{x_B} are weight-average molecular weights. For a homogeneous material they should equal one another and be constant with time. For heterogeneous material M_{x_0} will decrease with time and M_{x_B} will increase with time. However non-identity or otherwise within say about 15—20 per cent for material the size of bovine plasma albumin of these two figures with time should not be relied upon very much to indicate heterogeneity unless one is certain of the accuracy of the results at both menisci — i.e. the optics have been aligned or used according to TRAUTMAN (1958) and trial runs with known materials have indicated what accuracy is being obtained.

To get a weight-average molecular weight of a polydisperse material (i.e. before redistribution due to sedimentation) the values of M at each meniscus should be extrapolated to zero time.

c) For runs at speeds below about 10,000 r.p.m. there is no need to do a separate base line run of buffer only. It is sufficient to project the flat portions of the curve to either meniscus. For runs above this speed a separate run of solvent (buffer) alone should be done under the same conditions as the run on the macromolecular solute. Photographs should be taken at the same times and these base-lines are subtracted from the solute curve. To get identical conditions with the buffer and solute run it is best to weigh the silicone oil and solution into the cell so that the run can be duplicated when buffer is used. Or instead of doing a separate base-line run a double-sector cell can be used with buffer in one side; this is the better method.

d) The accuracy of the method in the hands of the authors varies with the molecular weight of the material being studied. For raffinose, results within 2—3 per cent of the true figure are relatively easily obtainable at both top and bottom menisci. This is because the menisci, particularly the bottom one, have better definition at the high speed (59,780 r.p.m.) required for raffinose than at lower speeds required for say bovine plasma albumin (approximately 8,000 r.p.m.). Also the lower value of s/D for raffinose gives a more gentle slope to the line which must be extrapolated to the menisci and hence gives a greater certainty of extrapolation with raffinose than with much heavier materials. s/D increases with increasing molecular weight. With reasonable care the molecular weight of a protein the size of bovine plasma albumin can be obtained to better than 5 per cent.

EHRENBERG (1957) has used an approach similar to that of KLAINER and KEGELES but has run the centrifuge at a higher speed so that the maximum of the gradient curve starts to separate from the meniscus. This makes extrapolation of the gradient curve to the meniscus more certain — the bottom meniscus is neglected.

e) The ultracentrifuge should be run at such a speed that the small areas at the top and bottom menisci are 10—20 per cent of the total area after an hour or so's running. For raffinose (M.W. 502) 59,780 r.p.m. is satisfactory while for ribonuclease (M.W. 13,682) and bovine plasma albumin (M. W. 69,000) 11,150 r.p.m. and 7928 r.p.m. (or less) respectively are satisfactory.

f) One should aim in a schlieren pattern to have an intercept of 2—3 mm at the menisci on the photographic plate at an angle of 80°. This intercept depends on the concentration of the solution, the speed of centrifugation, the ratio of s/D (or the M.W.) of the material, and the time of centrifugation. As centrifugation proceeds the gradient curve at the top meniscus gets shallower (making extrapolation easier) while that at the bottom meniscus gets steeper (making extrapolation more difficult). So judicious choice of centrifuge speed and time of photographs are necessary for good results.

g) Strictly speaking in the evaluation of the $\int x^2(dc/dx)\,dx$ for the top and bottom of the cell average values of x_n^2 and y should be used in the table so the summation should really be $(\overline{x_n^2})\,\overline{y}$ instead of $x_n^2\,y$. (There is no need to do this for the artificial boundary cell run). The magnitude of the error introduced here will depend on the size of the increments in $(R_n - R_0)$ and the ratio $(c_0 - c_{x_0})/c_0$. If the latter factor is 0.1—0.2 the error will be small.

2. A General Method of Calculation which can be Applied throughout the Cell whether or not a Plateau Region Exists.

The following application of the sedimentation equilibrium equation *does not rely on there being a plateau region*: at equilibrium there will be no plateau region.

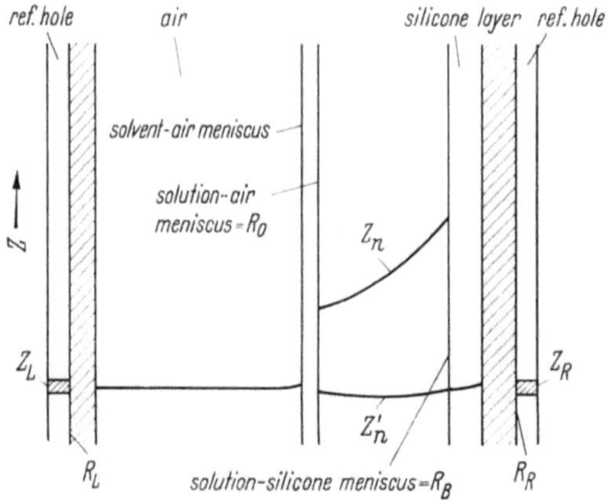

Fig. 17

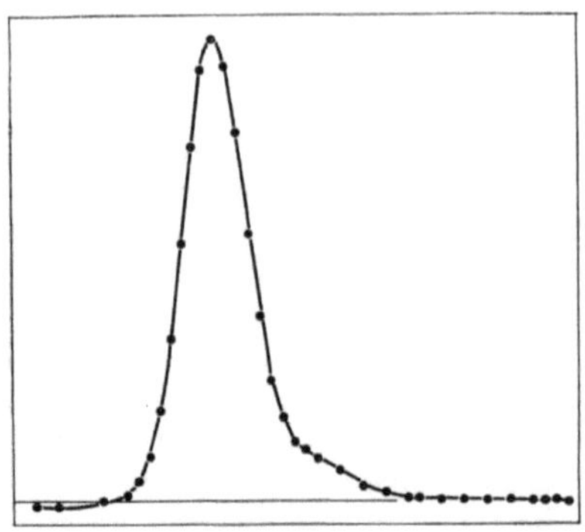

Fig. 18

Fig. 17. Diagram of schlieren pattern of bovine plasma albumin at equilibrium using double sector cell (Table 4).

Fig. 18. Diagram of the sedimentation velocity pattern (59,780 r.p.m.) of the bovine plasma albumin of Table 4. Base line has been fitted and substracted.

If a sedimentation equilibrium experiment is run this application gives molecular weights at the menisci early in the experiment and the approach to complete equilibrium can be followed if desired.

At the upper meniscus the following equation applies (see GINSBURG, APPEL and SCHACHMAN 1956)

$$c_{x_0} = c_0 + \left[\int_{x_0}^{x_B} x^2 (dc/dx)\,dx - x_B^2 \int_{x_0}^{x_B} (dc/dx)\,dx\right] / (x_B^2 - x_0^2) \tag{72}$$

Having determined c_{x_0} for a particular photograph the concentration at any point in the cell, c_x, can be determined by the equation

$$c_x = c_{x_0} + \int_{x_0}^{x} (dc/dx)\,dx\,. \tag{73}$$

The data can then be plotted as suggested by ARCHIBALD (1947) as $(dc/dx)/xc$ vs x, and a figure for M can then be calculated at any point in the cell according to the equation above by multiplying the value of $\frac{dc/dx}{xc}$ by $RT/\omega^2(1-\bar{v}\varrho)$. At equilibrium this method gives the weight-average molecular weight (M_w) at every

Table 4. *Sedimentation equilibrium run on bovine plasma albumin*[1]
(Calculations whether there is a plateau region present or not.)

$$Z_L = 6.829 \quad R_L = 14.143 \quad R_0 = 16.541 \quad R_M - R_0 = -0.711 \quad x_0 = 6.500 - \left(\frac{-0.711}{2.12}\right) = 6.835$$

$$Z_R = 6.829 \quad R_R = 17.518 \quad R_B = 17.102 \quad R_M - R_B = -1.272 \quad x_B = 6.500 - \left(\frac{-1.272}{2.12}\right) = 7.100$$

$$R_M = 15.830 \qquad x_M = 6.500$$

Time 23 h[2] $F = 2.12$ $\qquad \omega^2 = 7.4182 \times 10^5$ (8225 r.p.m.)

R_n	$R_n - R_0$	$x_n = 6.835 + (R_n - R_0)/F$	x_n^2	$\overline{(x_n^2)}$	Z_n	Z'_n	$Z_n - Z'_n$	$\overline{(Z_n - Z'_n)}$ $(\alpha l(dc/dx))$	c_x	$\frac{1}{x}\frac{dc}{dx}$	$\frac{1}{xc}\frac{dc}{dx}$
16.541	0	6.835	46.7172	46.7788	*7.129*	*6.811*	0.318	0.320	0.07621	0.04653	0.610
16.561	0.02	6.844	46.8403	46.9088	*7.134*	*6.811*	0.323	0.330	0.07923	0.04719	0.596
16.581	0.04	6.854	46.9773	47.0390	*7.152*	*6.814*	0.338	0.350	0.08234	0.04931	0.599
16.601	0.06	6.863	47.1008	47.1694	*7.173*	*6.811*	0.362	0.372	0.08564	0.05274	0.616
16.621	0.08	6.873	47.2381	47.3000	*7.190*	*6.808*	0.382	0.388	0.08915	0.05558	0.623
16.641	0.10	6.882	47.3619	47.4308	*7.203*	*6.810*	0.393	0.402	0.09281	0.05710	0.615
16.661	0.12	6.892	47.4997	47.5618	*7.221*	*6.809*	0.412	0.428	0.09660	0.05978	0.619
16.681	0.14	6.901	47.6238	47.6860	*7.239*	*6.811*	0.428	0.438	0.1006	0.06202	0.616
16.701	0.16	6.910	47.7481	47.8172	*7.259*	*6.811*	0.448	0.460	0.1048	0.06483	0.618
16.721	0.18	6.920	47.8864	47.9487	*7.282*	*6.809*	6.473	0.481	0.1091	0.06835	0.626
16.741	0.20	6.929	48.0110	48.0803	*7.300*	*6.811*	0.489	0.496	0.1136	0.07057	0.621
16.761	0.22	6.939	48.1497	48.2122	*7.313*	*6.810*	0.503	0.520	0.1183	0.07249	0.612
16.781	0.24	6.948	48.2747	48.3442	*7.348*	*6.810*	0.538	0.548	0.1232	0.07743	0.628
16.801	0.26	6.958	48.4138	48.4764	*7.369*	*6.810*	0.559	0.572	0.1284	0.08034	0.626
16.821	0.28	6.967	48.5390	48.6038	*7.394*	*6.810*	0.584	0.598	0.1338	0.08382	0.626
16.841	0.30	6.977	48.6785	48.7413	*7.420*	*6.808*	0.612	0.628	0.1394	0.08772	0.629
16.861	0.32	6.986	48.8041	48.8670	*7.453*	*6.810*	0.643	0.665	0.1454	0.09204	0.633
16.881	0.34	6.995	48.9300	49.0000	*7.485*	*6.809*	0.676	0.698	0.1516	0.09664	0.637
16.901	0.36	7.005	49.0700	49.1330	*7.518*	*6.807*	0.711	0.730	0.1582	0.10150	0.641
16.921	0.38	7.014	49.1961	49.2663	*7.560*	*6.810*	0.750	0.765	0.1651	0.10693	0.648
16.941	0.40	7.024	49.3365	49.3998	*7.590*	*6.810*	0.780	0.802	0.1723	0.11104	0.644
16.961	0.42	7.033	49.4631	49.5334	*7.634*	*6.810*	0.824	0.841	0.1799	0.11716	0.651
16.981	0.44	7.043	49.6038	49.6672	*7.668*	*6.810*	0.858	0.884	0.1878	0.12182	0.649
17.001	0.46	7.052	49.7307	49.7942	*7.718*	*6.807*	0.911	0.932	0.1962	0.12918	0.658
17.021	0.48	7.061	49.8577	49.9283	*7.761*	*6.807*	0.954	0.974	0.2050	0.13511	0.659
17.041	0.50	7.071	49.9990	50.0627	*7.800*	*6.807*	0.993	1.018	0.2142	0.14043	0.656
17.061	0.52	7.080	50.1264	50.1972	*7.850*	*6.807*	1.043	1.074	0.2233	0.14732	0.658
17.081	0.54	7.090	50.2681	50.3390	*7.914*	*6.808*	1.106	1.131	0.2339	0.15599	0.667
17.101	0.56	7.100	50.4100		*7.964*	*6.808*	1.156		0.2446	0.16282	0.666

$$\int_{x_0}^{x_B} (dc/dx)\, dx = (0.02)/F^{3} \sum_{R_n - R_0 = 0}^{R_n - R_0 = 0.56} \overline{(Z_n - Z'_n)} = 0.02 \times 17.845/2.12 = 0.1684$$

$$\int_{x_0}^{x_B} x^2 (dc/dx)\, dx = (0.02)/F \sum_{R_n - R_0 = 0}^{R_n - R_0 = 0.56} \overline{(Z_n - Z'_n)} \cdot \overline{(x^2)} = 0.02 \times 873.27/2.12 = 8.238$$

$$c_{x_0} = (\text{equn. } 72)\ \frac{0.3040\,^{4}}{2.12} + \frac{8.238 - 8.486}{50.4100 - 46.7172} = 0.0762$$

$$c_{x_n} = (\text{equn. } 73)\ 0.0762 + (0.02)/F \sum_{R_n - R_0 = 0}^{R_n - R_0 = x_n} \overline{(Z_n - Z'_n)} = 0.0762 + 0.00943 \sum_{R_n - R_0 = 0}^{R_n - R_0 = x_n} \ldots$$

$RT/\omega^2(1 - \bar{v}\varrho) = 1.2687 \times 10^5$. M_w at any point $= 1.2687 \times 10^5 \times (dc/dx)/xc$.

[1] Approximately 1 per cent solution of bovine plasma albumin ($\bar{v} = 0.735$) in buffer containing 0.10 M potassium chloride, 0.01 M sodium acetate and 0.01 M acetic acid pH 4.8 (density = 1.0024 at 25°). The protein solution was dialyzed against buffer overnight and run at 8225 r.p.m. in a 12 mm double-sector Spinco cell at 25°. Height of the column in the solution sector was 2.64 mm (0.100 g solution); also 0.055 gm carbon tetrachloride on this side. Depth of liquid in solvent sector was slightly more (0.148 g solvent); no carbon tetrachloride on the solvent side. Bar angle = 70°.

[2] The system has reached equilibrium by this time.

[3] The magnification factors cancel out except the one (F) shown in the calculation of $(dc/dx)/xc$ below. The figures in the last three columns of the Table have hence not been completely corrected for magnification factors.

[4] This is the area in sq. cm from a separate artificial-boundary cell run at the same angle as this photograph.

point in the cell. During the approach to equilibrium the molecular weights can be calculated from $(dc/dx)/xc$ at the menisci. Equilibrium has been achieved when successive photographs taken say 2 hours apart are identical.

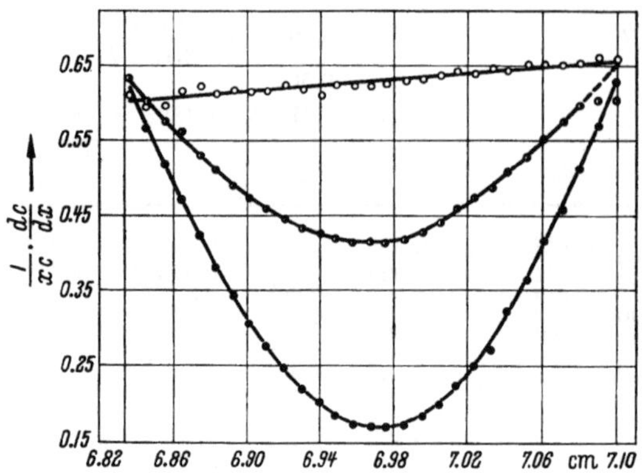

Fig. 19. ARCHIBALD's plot of the data during a sedimentation equilibrium experiment on bovine plasma albumin. ○ = 23 hr ◑ = 4.3 hr ● = 1.7 hr (see Tabele 4).

If a solute is homogeneous then at equilibrium $(dc/dx)/xc$ is constant with x. If the plot has a positive slope as in Fig. 19 the solute is heterogeneous; a negative slope is an indication that the solute is not behaving in an ideal manner and that account should be taken of activities.

A calculation according to this method is shown in Table 4 (Fig. 19) from a photograph similar to the drawings in Fig. 17.

3. Method of Calculation of Molecular Weight which Gives the Weight-Average Molecular Weight of the Whole Solute once Equilibrium has been Attained (VAN HOLDE and BALDWIN 1958).

This uses some of the figures of Table 4. The equation used is

$$2\ln(c_{x_B}/c_{x_0})/(x_B^2 - x_0^2) = \omega^2 M(1 - \bar{v}\varrho)/RT \tag{74}$$

Taking

$$c_{x_B} = 0.2446,\ c_{x_0} = 0.0762,\ x_B^2 = 50.4100,\ x_0^2 = 46.7172$$

it is seen

$$M_w = 80{,}100\ .$$

4. Method of Calculation which Gives the z-Average Molecular Weight of the Whole Solute once Equilibrium has been Attained (VAN HOLDE and BALDWIN 1958).

$(dc/dx)/x$ is plotted against c_x (Fig. 20). This is a straight line once equilibrium has been attained. The slope of this line multiplied by $RT/\omega^2(1 - \bar{v}\varrho)$ (equ. 66) gives the z-average molecular weight. This plot for the bovine plasma albumin experiment has a slope = 0.694 and M_z therefore = 88,050.

If $M_z = M_w$ the material is homogeneous with regard to molecular weight. For the bovine plasma albumin studied here they are not equal nor is the line in Fig. 19 horizontal and therefore this material is not homogeneous which is in agreement with the diagram of the sedimentation velocity pattern (after the base line has been substracted) shown in Fig. 18. Note that this is a different sample of bovine plasma albumin from that shown in Fig. 14.

These methods are given fuller treatment in the paper of VAN HOLDE and BALDWIN (1958).

So the conclusions from this section on sedimentation equilibrium are that it is probably best to use a short solution column (3 mm or less) and a double-sector cell and run the experiment until it is either at equilibrium or almost there, i.e. past the stage where a plateau exists. From rough values of M and D the time taken for a column of solution at a particular length and speed to reach equilibrium can be approximated (VAN HOLDE and BALDWIN 1958). If equilibrium has not been attained the intercepts at the menisci of the plot $(dc/dx)/xc$ versus x (method 2) will give the molecular weights at the top and bottom of the cell and the nearer the system is to equilibrium the more accurate will these values be (cf. SCHACHMAN 1960). This is a similar approach to the KLAINER-KEGELES method (method 1) but it involves more calculation. However the answers are also more precise[1]. If complete equilibrium is reached then the weight-average (method 3) and z-average molecular weights (method 4) can be readily calculated and these can tell us if the solute is homogeneous with respect to molecular weight.

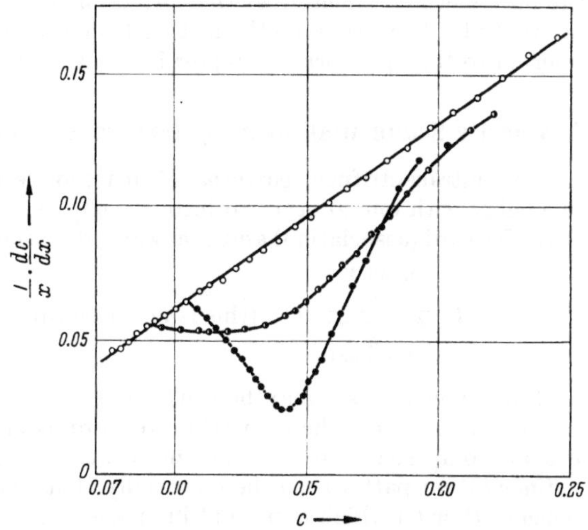

Fig. 20. Plot of $(dc/dx)/x$ versus c_x for the equilibrium run of the example used in Table 4. When the line is straight (at equilibrium) the z-average molecular weight of the whole solute can be calculated from its slope ○ = 23 hr, ◑ = 4.3 hr, ● = 1.7 hr.

VII. The Proportions of Components in a Sedimenting Mixture.

1. The JOHNSTON-OGSTON Effect.

Each component in a mixture will sediment with its own characteristic velocity. However two peaks or more may merge together to give one distorted or even symmetrical peak. This is the reason why emphasis was placed earlier on boundary analysis to test for heterogeneity. If peaks are partly separated the full curves corresponding to each peak can be approximately drawn in and the area of each of these peaks computed. Turning to the proportions of components in a sedimenting mixture it was pointed out by JOHNSTON and OGSTON (1946) that the percentage composition of a mixture calculated from the areas of peaks in a schlieren pattern does not give a true picture of the composition of the original mixture. This is due to differing rates of sedimentation of a slower solute above and below the boundary of a faster solute (due to the dependence of s upon c); this causes, if a non-specific method of observation is used (e.g. schlieren or other method based on refractive index) the areas from the peaks in a schlieren photograph to be not in the same ratio as the actual concentrations of the components in the mixture. The area of the fast component will be too small and the area of the slower component correspondingly too big. The sedimentation coefficients of the boundaries will also be affected and so will the shape of the faster moving boundary. These effects decrease with decreasing concentration and can be eliminated by doing a series of sedimentation runs at different concentrations and

[1] If the material being studied denatures at the bottom solution — silicone meniscus and piles up, thus giving a meniscus which thickens with time, then it is not possible to use the equations (72) and (73) and the complete equilibrium calculations which depend on them. In such circumstances only the KLAINER-KEGELES type of calculation at the top meniscus (with some plateau region remaining) can be used.

extrapolation of the areas and sedimentation coefficients to zero concentration — the correct sedimentation coefficients and composition of the mixture are then obtained. This method may not always be convenient or suitable but it is probably the only method which gives the correct answers (in those systems where it can be applied). For other methods based on theoretical treatments the reader is referred to the appropriate chapter in Schachman's book (1959).

2. Velocity Sedimentation in Systems of Reversibly Interacting Components.

The ultracentrifuge patterns of macromolecular components which react reversibly with one another to form a complex will depend on the relative rates of the forward (association) and backward (dissociation) reactions. If the reaction is

$$A + B \underset{\substack{k_2 \\ \text{backward}}}{\overset{\substack{\text{forward} \\ k_1}}{\rightleftarrows}} C \quad \text{(this is not necessarily a stoichiometric reaction).}$$

the following cases may be considered:

a) If $k_1 \gg k_2$, i.e. the association reaction is fast and the dissociation reaction is slow, compared with the differential rate of migration of the components, the sedimentation pattern will be essentially that of the complex (C) and that component (B or C) which is present in excess.

b) If $k_1 \ll k_2$, i.e. the dissociating reaction is fast and the association is slow then the pattern will be that of components A and B only.

c) If k_1 is of the same order as k_2 and is slow compared with the differential rate of migration of the components then the pattern will be that of A, B and C.

d) If k_1 and k_2 are of the same order as the rate of separation of the components there is no simple explanation of what the pattern should be.

e) If k_1 and k_2 are of the same order and are very large, i.e. the equilibrium is established rapidly compared with the differential rate of migration of the boundaries the sedimentation pattern is not simple (Gilbert 1959, Gilbert and Jenkins 1959; see also Williams et al. 1958 p. 779 and Schachman 1959, p. 160).

f) This is a special case of (e) and is the situation where a monomer is in equilibrium with its polymer and the association-dissociation equilibria are established rapidly, compared with the differential rate of migration.

$$mA \underset{k_2}{\overset{k_1}{\rightleftarrows}} A_m \, .$$

This example is not uncommon in protein systems. Recent developments by Gilbert (1959) have yielded the unexpected result that if $m \gg 3$ there will be two maxima in the boundary region; if $m = 2$ there will be only one. Furthermore for $m > 2$ the gradient curve dc/dx will not fall to zero between the maxima (if it does it is certain that the system is not a rapidly-reversible aggregating one). If $m > 2$ the area and apparent sedimentation coefficient of the faster peak will increase as total concentration of protein increases whereas the apparent sedimentation coefficient and area of the slower peak will remain unaltered as the total concentration increases (once a sufficient minimum concentration to complete the lower peak has been reached).

Townend, Winterbottom and Timasheff (1960) have applied the Gilbert theory to the association of β-lactoglobulin at around pH 4.4—4.65 and this work is a good example of the application of Gilbert's treatment to actual experimental data.

F. Viscosity.

Viscosity may be considered as the resistance by a fluid to any forces tending to induce differential motion between adjacent layers of the fluid. The shearing force (τ) between two parallel planes in a liquid in relative motion is a function of the velocity gradient between them (rate of shear) and the area, A, of the planes. This is NEWTON's well known law and is expressed in the form

$$\tau = \eta \frac{dv}{dx} A \tag{75}$$

where dv/dx is the velocity gradient perpendicular to the planes, and η is the viscosity coefficient or viscosity. Thus the viscosity is the shearing force per unit area and unit velocity gradient and is expressed as gm cm^{-1} sec^{-1}, the absolute c.g.s. unit being called the poise. The viscosity of water at 20° is almost exactly 0.01 poise. The relation between shearing force and velocity gradient has been verified for most pure liquids and many solutions, provided they exhibit stream line or laminar flow. By stream line flow is meant a steady state such that at any point of the system the velocity of the moving liquid does not vary with time in magnitude or direction. Thus in flow through a tube this involves (except at the ends) a liquid velocity parallel to the walls of the tube and of magnitude depending only on the distance from the walls. Laminar flow occurs provided the velocity is not too great compared with the viscosity and dimensions of the system. Above a critical velocity equation (75) does not apply and laminar flow is replaced by turbulent flow. Measurements of viscosities should therefore be made under conditions that the liquid velocities utilised are well below the critical.

The introduction of colloidal materials into a solvent, i.e. particles which are large compared with those of the solvent, always causes an increase in viscosity, the effect varying with the nature of the colloidal particles. All particles will cause a distortion of the stream lines of the flowing liquid and the extra dissipation of energy involved by flow distortion in the solvent around each particle leads to an increase in viscosity. For asymmetric particles extra energy is dissipated by the rotation of the particles by the velocity gradient, and this effect decreases the more nearly spherical the suspended particles.

I. Newtonian and Non-Newtonian Viscosity.

Most pure liquids and many solutions obey equation (75) and the viscosity is independent of the velocity gradient. This is called "NEWTONIAN" viscosity. However for many of the more concentrated colloidal solutions and for dilute solutions of very asymmetric macromolecules the viscosity decreases with increasing velocity gradients. Such fluids are said to show anomalous or "non-NEWTONIAN" viscosity and this may be caused by two main factors:

(i) The occurrence in solution of some type of structure due to specific interactions between particles hindering the rotation of the particles and causing an increase in viscosity. The occurence of flow will tend to break down such interactions and the viscosity will decrease as the velocity gradient increases.

(ii) The orientation of very asymmetric particles caused by velocity gradients in solution. Asymmetric particles will tend to take up positions with their longer dimensions more or less parallel with the stream lines and in this position will therefore make less contribution to the viscosity than when they are randomly oriented. Increasing the velocity gradient will increase the orientation consequently the viscosity of a solution of such particles will fall as the viscosity gradient increases. Also involved in this effect may be the distortion of flexible chain molecules in the velocity gradient.

In the measurement of the viscosities of macromolecules we generally wish to eliminate the causes of anomalous viscosity. Structural viscosity may be eliminated by the use of dilute solutions so that the solute molecules are so far apart that structure formation is impossible. However in some systems, e.g. gelatin solutions, the tendency to structural viscosity may be so great that at the lowest usable concentrations it may still occur. When orientation occurs it is usual to obtain complete randomness by working at low velocity gradients where orientation is negligible or by carrying out the measurements at a series of velocity gradients and extrapolating to zero gradient. One may also study asymmetric particles at high velocity gradients where complete orientation is approached. The degree of orientation under a particular set of conditions (measured by variation of the viscosity with velocity gradient or by streaming birefringence) may be used to derive the rotational diffusion coefficient, from which information on particle size and shape may be derived. This latter aspect will not be discussed in this article.

II. Functions of Viscosity.

In solutions we are generally concerned with the increase in viscosity due to the added solute rather than with absolute values. For this reason different functions of viscosity are used to express the results. The ratio of the viscosity of the solution η, to that of the solvent η_0, is termed the "relative viscosity." The function $(\eta/\eta_0) - 1$ which measures the relative increase in viscosity of the solution over that of the solvent is termed the "specific viscosity," η_{sp}, and η_{sp}/c where c is expressed in g/100 ml is called the "reduced viscosity." The reduced viscosity is dependent on concentration due to the overlapping of the disturbed regions of flow around each particle, and will be more marked for more asymmetric particles. This is apart from structural effects mentioned previously. Thus η_{sp}/c depends on concentration and tends to increase as the concentration increases. The limit of η_{sp}/c at zero concentration is called the intrinsic viscosity $[\eta]$ where c is in g/100 ml. According to the terminology approved by the International Union of Pure and Applied Chemistry (1952) η/η_0 should be termed the "viscosity ratio"; $(\eta/\eta_0) - 1$, the "relative increase in viscosity"; and the function $[(\eta/\eta_0) - 1]/c$, the "viscosity number" if c is expressed in grams per millilitre. The limit of $[(\eta/\eta_0) - 1]/c$ at zero concentration where c is in g/ml has been designated the "limiting viscosity number" or more recently the "STAUDINGER index."

The procedure for extrapolation to infinite dilution is relatively simple since plots of η_{sp}/c against c are approximately linear for relative viscosities less than 2 and the slopes of these plots are proportional to the square of the intrinsic viscosity. The relation, due to HUGGINS (1942), is expressed in the form

$$\eta_{sp}/c = [\eta] + k^1 [\eta]^2 c \,. \tag{76}$$

The coefficient k^1 varies from system to system, but is approximately constant for a given polymer, solvent and temperature. Although derived for randomly coiled chains, equation (76) is found to be generally applicable to all macromolecular solutions. Other relations expressing the variation of η_{sp}/c with concentration have been proposed.

III. Viscosity Relations for Particles of Different Shapes

(See section g).

IV. Experimental Methods.

For a general account of the types of viscometers which can be used the reader is referred to SWINDELLS, ULLMAN and MARK (1960). An investigation of the errors encountered and precautions necessary for the precise measurement of the viscosity of extremely dilute solutions of high polymers is given in the paper by ÖHRN (1958).

For the measurement of viscosities of dilute solutions two chief methods are available — (i) the capillary flow method in which the flow of the liquid through a suitable tube is timed accurately and (ii) the concentric cylinder or COUETTE method in which the liquid contained between two concentric cylinders rotating relatively to one another transmits a torque which is measured to obtain the viscosity of the liquid. The most popular method is the capillary method in which the volume V of liquid flowing in time t secs through a capillary of radius a and length l is described by POISEUILLE's law

$$V = P\pi a^4 t/8\,\eta l \tag{77}$$

where P is the pressure (dyne/cm²), and η is the viscosity in poises. It is usual to measure the relative viscosity which is obtained from the rates of flow of solution and solvent under the same conditions. For viscometers in which the liquid flows under its own head of pressure $P = hg\varrho$ where h = liquid head in cm, g = gravitational constant in cm sec^{-2} and ϱ = density in g/cc, and we may write $\eta = K\varrho t$ where t is the time of flow of a given volume of liquid in the viscometer between two definite points and K is a constant depending on the construction of the viscometer. Thus we can write for the relative viscosity

$$\eta/\eta_0 = (t/t_0)\,(\varrho/\varrho_0) \tag{78}$$

where η_0, t_0, ϱ_0 are the viscosity, outflow time, and density for the reference liquid. The ratio t/t_0 is referred to as the relative kinematic viscosity. One of the most convenient and widely used of capillary viscometers is the OSTWALD viscometer.

Since a stop-watch is generally used for measuring outflow of capillary viscometers it is desirable to use viscometers with an outflow time of the order of 100 sec and for smaller relative viscosities (or when more accurate measurements at low concentrations of solute are required) larger outflow times. Temperature control during viscosity measurements is very important since the absolute viscosity has a high negative temperature coefficient — the viscosity of water changes by 2% per degree at 20°. It is advisable that measurements be made in a water bath controlled to $\pm$ 0.02° or better. A further requirement particularly when fine capillaries are used is the removal of dust, etc. by filtration of liquids before introduction into the viscometer. Centrifugation at low speeds or filtration through sintered glass filters is generally used.

An important correction in capillary viscometry is the kinetic energy correction which can be minimised by careful design of the viscometer. In the derivation of equation (77) it was assumed that the acting pressure was effective only in overcoming viscous resistance whereas, in practice, since the liquid is forced to enter the capillary tube from the drainage bulb and then to leave the capillary into the reservoir bulb, there will be energy losses in establishing the flow pattern of the capillary. The correction, to be substracted from η determined from equation (77), is $v\varrho k/8\,\pi l t$ where k is a constant depending on the apparatus but which is approximately unity. Thus by making V/l small, i.e. slow rates of flow and long capillaries this correction is small. The error due to this factor is less when relative viscosities are measured since kinetic energy corrections of the same order occur for both solvent and solution.

The velocity gradient in a capillary viscometer may be varied by applying additional pressure. The velocity gradient is a mean value varying from zero at the centre to a maximum at the wall; so that for non-NEWTONIAN fluids a mean viscosity is obtained. The effect of velocity gradient may be extrapolated to zero gradient by making measurements over a range of pressures. However for non-NEWTONIAN viscosities the concentric cylinder apparatus is preferable since the velocity gradient is much more nearly uniform and more easily controlled than for a capillary viscometer and measurements can be made rapidly over a wide range of velocity gradient and at much lower gradients than is generally possible with a capillary viscometer. In the usual form of COUETTE viscometer the outer cylinder containing the liquid is rotated and the inner is suspended freely by a torsion wire. The torque transmitted by the liquid from the outer to the inner cylinder is measured by the deflection of the inner cylinder by the use of a mirorr and scale arrangement. As with capillary viscometers the determination of the absolute viscosity is not required and the ratio of the scale deflections for solution and solvent gives the relative viscosities. Although particularly suitable for detecting and measuring non-NEWTONIAN viscosity the precision of the coaxial cylinder viscometer is less than that of the capillary viscometer.

G. Particle Shape from Hydrodynamic Measurements.

It has been customary to describe the hydrodynamic properties of a solute particle in terms of its shape and thus obtain the particle dimensions. Theories can describe the hydrodynamic behaviour of suspensions of several particle shapes such as spheres, ellipsoids, cylindrical rods and randomly-coiled chains, other shapes not being readily amenable to mathematical treatment. The following discussion is limited to the spheroidal model, which has been most successful in interpreting the measurements on protein solutions, and the randomly-coiled chain model which has found wide application in the field of flexible chain molecules such as synthetic polymers.

I. Spheroidal Molecules.

1. Frictional Coefficient.

The frictional resistance of a particle to motion in a fluid is generally described in terms of its frictional coefficient and this will depend on the properties of the solvent, the type of motion (i.e. whether translational or rotational) and the size and shape of the particle. The translational frictional coefficient f of a solute molecule (of any shape) can be obtained from the translational diffusion coefficient at infinite dilution

$$D_0 = RT/Nf \tag{39}$$

or from the sedimentation coefficient at infinite dilution

$$s_0 = M(1 - \bar{v}\varrho)/Nf\,. \tag{61}$$

The frictional coefficient f_0 of a spherical particle is given by STOKES' law (1851)

$$f_0 = 6\,\pi\,\eta_0 r_0 \tag{53}$$

r_0 being the radius of the particle and η_0 the viscosity of the solvent.

The axial ratio b/a of an ellipsoidal particle may be expressed in terms of f/f_0 (called the frictional ratio) where f_0 is the frictional coefficient of a hypothetical sphere having the same volume as the ellipsoid. This relation was given by PERRIN (1936) for prolate and oblate ellipsoids, and values are tabulated in SVEDBERG and PEDERSEN (1940, p. 41) and in COHN and EDSALL (1943, p. 406). In order to utilize this relation it is necessary to know the volume, V, of the particle (or molecule) to determine r_0 and thus f_0.

$$r_0 = (3/4\pi)^{1/3}\, V^{1/3}\,. \tag{79}$$

It has been customary to equate V to $\bar{v}M/N$ plus some arbitrarily assumed or measured degree of solvation, $\bar{v}$ being the partial specific volume of the solute; graphs relating f/f_0 to the axial ratio b/a for various degrees of hydration were derived by ONCLEY (1941) and are also tabulated in COHN and EDSALL (1943, p. 425).

This procedure has in general been found unsatisfactory and a sounder approach is given below by combination of the frictional coefficient with intrinsic viscosity.

2. Intrinsic Viscosity.

EINSTEIN (1906, 1911) has treated theoretically the relative viscosity of a very dilute suspension or solution of uncharged rigid spherical particles which are large compared with the solvent molecules. His quantitative relation is

$$(\eta/\eta_0) - 1 = 2.5\,\phi \tag{80}$$

where ϕ is the volume fraction of the system occupied by the solute molecules and $(\eta/\eta_0) - 1$ the relative viscosity.

It is convenient to define in general the quantity

$$\underset{\Phi\to 0}{\mathrm{Lt}}\,[(\eta/\eta_0) - 1]/\phi = \nu \tag{81}$$

where ν is called the "viscosity increment" of the solute. Thus $\nu = 2.5$ for spheres and it is found that for non-spherical particles ν is always greater than 2.5. SIMHA (1940) has extended this treatment to cover large ellipsoidal particles (provided these are randomly oriented) and has given ν in terms of the axial ratio b/a of prolate and oblate ellipsoids. This function has been computed by MEHL, ONCLEY and SIMHA (1940), and also tabulated in COHN and EDSALL (1943, p. 519) and OGSTON (1953).

ν is related to the intrinsic viscosity $[\eta]$ since $\phi = VNc/100\,M$, if c is the concentration of dry solute in g/100 ml. As in the case of the interpretation of the frictional ratio, V has been equated to $\bar{v}M/N$ plus an assumed degree of hydration; graphs relating b/a to ν for various degrees of hydration are given by ONCLEY (1941) and are in COHN and EDSALL (1943, p. 521).

This procedure often does not give the same answer as the previous one using the frictional ratio. *This procedure is also not satisfactory due to the fact that the extent of solvation of the solute in solution cannot be determined and therefore V is not known*; V would only equal $\bar{v}M/N$ if no solvent moved with the particle and electrostriction effects were absent. The treatment in the next section appears to overcome the problem of defining particle volume.

3. Combination of Hydrodynamic Methods to Give Particle Shape.

The determination of particle shape has been discussed by several authors including OGSTON (1953), SADRON (1953) and SCHERAGA and MANDELKERN (1953). It was realized that no assumption need be made as to the volume factor which should be used. If f/f_0 and ν, together with f from equations (39) or (61) are combined they can be simultaneously solved for both the axial ratio and the effective volume of the particle in solution (which will in general not be equal to $\bar{v}M/N$). Scheraga and Mandelkern stressed the idea that for every particle in solution (no matter what its shape) there is a rigid equivalent hydrodynamic ellipsoid impenetrable to solvent which behaves exactly in all its hydrodynamic properties as the actual particle. They called the effective volume of this equivalent particle V_e.

Scheraga and Mandelkern combined the Perrin function f/f_0 and the Simha function ν into a new function, the β function, and the solution of the above equations using V_e instead of $\bar{v}M/N$

$$\beta = \left(\frac{N}{16200\,\pi^2}\right)^{1/3} \frac{f_0}{f}\,\nu^{1/3} = \frac{N s_0 [\eta]^{1/3} \eta_0}{M^{2/3}(1-\bar{v}\varrho)} = \frac{N D_0 [\eta]^{1/3} M^{1/3} \eta_0}{RT} = \frac{(M[\eta])^{1/3}\eta_0}{f} \tag{82}$$

where η_0 is the viscosity of the solvent. They list tables relating β to the axial ratio of the equivalent hydrodynamic ellipsoid (both flattened and elongated). Thus from M, $[\eta]$ $\bar{v}$ and s_0 or D_0 (or $[\eta]$, $\bar{v}$, s_0 and D_0) β can be determined, then from the axial ratio the shape factor can be obtained from tables of the Simha functions. V_e can then be obtained from equation (81) now expressed in the form $[\eta] = (N\nu/100)\,(V_e/M)$ where the concentration c in the intrinsic viscosity term is the dry weight concentration in g/100 ml solution. (A similar development for a combination of intrinsic viscosity and rotary diffusion coefficient is given by Scheraga and Mandelkern, 1953).

Accurate experimental data are needed to derive axial ratios and V_e from β since β only changes slowly with axial ratio for the elongated ellipsoids — for flattened ellipsoids it is nearly constant.

Although the hydrodynamic theory developed here appears sound, actual particles in solution may be far removed from spheroidal shape and calculated axial ratios would then be at the best semi-quantitative; the closer the actual shape of the particle in solution is to a spheroid (i.e. the assumed model) the closer is the axial ratio determined experimentally to that of the actual particle. If V_e as determined experimentally is much larger than $\bar{v}M/N$ it can be said that the actual molecule is highly swollen, e.g. for proteins in 8 M urea V_e is likely to be large. From changes in b/a and V_e for a molecule in different environments it is sometimes possible to deduce what has happened, e.g. if on changing an environment it was found that b/a decreased and V_e increased one could say the molecule had become less elongated and more swollen.

Since β does not vary over a large range (even for elongated ellipsoids varying in axial ratio from 1 to 300, β only changes from 2.12 to 3.60) it is possible to calculate limits to values of M from $[\eta]$, $\bar{v}$ and s_0 or D_0. This is a useful method.

II. Randomly-Coiled Molecules.

Randomly-coiled chain molecules give viscosity increments ν greater than 2.5 if the partial specific volume, $\bar{v}$, is used to calculate the volume fraction of solute, ϕ. This is because the molecule occupies a domain which is many times (100 or more) greater than that which it would occupy in the compact form and some of the solvent in this domain moves with the polymer, i.e. the molecule has a large hydrodynamic volume.

For randomly-coiled chain molecules in a particular solvent the intrinsic viscosity $[\eta]$ will be related to the mean end-to-end distance of the chains and hence to the molecular weight. The viscosity-molecular weight relationship for these chain molecules can be expressed by the empirical relation

$$[\eta] = KM^a \tag{83}$$

where K and a are dependent on the properties of the solute and solvent molecules. Experimentally a is found to vary from 0.5 to 1.0 for randomly-coiled chain molecules. For free-draining coils $a = 1$ while for a solvent-immobilizing coil $a = 0.5$; for rigid rod-like molecules a should equal 2.

Equation (83) is empirical in origin but because of its convenience has found much use in practice. More complicated theoretical expressions have been developed to describe the conformational and frictional properties of randomly-coiled chain molecules and are outlined below. The field is summarized by FLORY (1953) and BALDWIN and VAN HOLDE (1960).

In terms of the FLORY theory the intrinsic viscosity of a randomly-coiled chain molecule can be expressed as

$$[\eta] = \Phi\,(\overline{r^2})^{3/2}/M \tag{84}$$

and the frictional coefficient (at zero solute concentration) as

$$f = \eta_0 P(\overline{r^2})^{1/2} \tag{85}$$

where $(\overline{r^2})^{1/2}$ is the root-mean-square end-to-end distance of the molecule (this parameter is taken as a convenient measure of size and completely defines a random coil), η_0 is the viscosity of the solvent, and Φ and P are universal constants for randomly-coiled chain molecules. If $\overline{r^2}$ is eliminated from these two equations

$$\Phi^{1/3} P^{-1} = (M[\eta])^{1/3}\,\eta_0/f\,. \tag{86}$$

$\Phi^{1/3} P^{-1}$ is exactly analogous to β in the treatment of spheroids (equation 82) and has the value 2.5×10^{-6} (for randomly-coiled chain molecules).

The root-mean-square end-to-end distance $(\overline{r^2})^{1/2}$ can thus be determined from $[\eta]$ and M (equation 84), or from s_0 and D_0 (or s_0 and M, or D_0 and M) by combination of equations (61) and/or (39) with (85).

Alternatively the molecular weight of a randomly-coiled molecule can be calculated from its intrinsic viscosity and sedimentation or diffusion coefficient at infinite dilution. This procedure should not be used uncritically since $\Phi^{1/3} P^{-1}$ will not equal 2.5×10^{-6} if the molecule departs in shape from a compact random coil. Furthermore the value varies with polydispersity.

It is thus seen that if $(M[\eta])^{1/3}\,\eta_0/f\ (\equiv \beta \equiv \Phi^{1/3} P^{-1})$ was found experimentally to be 2.5×10^{-6} we could not say whether the molecule was a random coil or ellipsoid (of axial ratio 15:1) from hydrodynamic measurements alone.

The authors wish to thank Dr. R. L. BALDWIN, Dr. P. J. DUNLOP and Dr. B. S. HARRAP for helpful criticism of certain sections of this manuscript.

Sectional References.

B. Osmotic Pressure.

ADAIR, G. S.: Proc. Roy. Soc. (Lond.) A **108**, 627 (1925); — Analytical methods of protein chemistry. Vol. 3, p. 24. Eds. ALEXANDER and BLOCK. Pergamon Press 1961. — ALEXANDER, A. E., and P. JOHNSON: Colloid Science Vol. 1, p. 150. Oxford: Clarendon Press 1949. — ALEXANDROWICZ, Z.: J. Polymer Sci. **40**, 91 (1959); **40** 113 (1959).

BURK, N. F., and D. F. GREENBERG: J. biol. Chem. **87**, 197 (1930).

CRAIG, L. C., W. KONIGSBERG, A. STRACHER and T. P. KING: Symposium on protein structure. Ed. NEUBERGER, p. 104. London: Methuen 1958; New York: Wiley and Sons 1958.

DONNET, J. B., R. ROTH and G. MEYERHOFF: J. Polymer Sci. **27**, 591 (1958).

ENOKSSON, B.: J. Polymer Sci. **3**, 314 (1948). — EDSALL, J. T.: The proteins, Vol. 1 Part B. Eds. NEURATH and BAILEY. New York: Academic Press 1953.

FLORY, P. J.: Principles of polymer chemistry. Ithaca, N.Y.: Cornell Univ. Press 1953.
JULLANDER, I.: Arkiv. Kemi Mineral Geol. A 21, No. 8 (1945).
OVERBEEK, J. TH. G.: Progr. in Biophys. and Biophys. Chem. 6, 58 (1956).
RIESEL, E., and A. BERGER: J. Polymer Sci. 37, 337 (1959). — ROWE, D. S., and M. E. ABRAMS: Biochem. J. 67, 431 (1957).
SCATCHARD, G.: J. Am. Chem. Soc. 68, 2315 (1946). — SCATCHARD, G., A. C. BATCHFELDER and A. BROWN: J. Am. Chem. Soc. 68, 2320 (1946). — STABIN, J. V., and E. H. IMMERGUT: J. Polymer Sci. 14, 209 (1954).
WAGNER, R. H., and L. D. MOORE: Physical methods of organic chemistry, Vol. 1, Part 1, 3rd Edition, Ed. A. WEISSBERGER p. 815. New York, London: Interscience 1959.
ZIMM, B. H., and I. MYERSON: J. Am. Chem. Soc. 68, 911 (1946).

C. Light Scattering.

ALEXANDER, P., and K. A. STACEY: Proc. Roy. Soc. (Lond.) A 212, 274 (1952).
BEATTIE, W. H., and C. BOOTH: J. Phys. Chem. 64, 696 (1960); — J. Polymer Sci. 44, 81 (1960). — BENOIT, H.: J. Polymer Sci. 11, 507 (1953). — BENOIT, H., A. M. HOLTZER and P. DOTY: J. Phys. Chem. 58, 635 (1954). — BRINKMAN, H. C., and J. J. HERMANS: J. Chem. Phys. 17, 574 (1949). — BRICE, B. A., M. HALWER and R. SPEISER: J. Opt. Soc. Am. 40, 768 (1950). — BRICE, B. A., G. C. NUTTING and M. HALWER: J. Am. Chem. Soc. 75, 824 (1953).
CABANNES, J.: La diffusion molèculaire de la lumière. Paris: Presses Universitaires 1929. — CARR, C. I., and B. H. ZIMM: J. Chem. Phys. 18, 1616 (1950).
DEBYE, P.: J. appl. Phys. 15, 338 (1944); — J. Phys. Colloid Chem. 51, 18 (1947). — DOTY, P., and J. T. EDSALL: Advanc. Protein Chem. 6, 35 (1951). — DOTY, P., and S. KATZ: Abstr. Am. Chem. Soc. Meeting (1950) quoted by DOTY and EDSALL, Advanc. Protein Chem. 6, 35 (1951). — DOTY, P., and R. F. STEINER: J. Chem. Phys. 18, 1211 (1950); 20, 85 (1952).
EDSALL, J. T., and W. B. DANDLIKER: Fortschr. Chem. Forsch. 2, 1 (1951). — EHRLICH, G., and P. DOTY: J. Am. Chem. Soc. 76, 3764 (1954). — EDSALL, J. T., H. EDELHOCH, R. LONTIE and P. R. MORRISON: J. Am. Chem. Soc. 72, 4641 (1950). — EWART, R. H., C. P. ROE, P. DEBYE and J. R. MCCARTNEY: J. Chem. Phys. 14, 687 (1946).
GEIDUSCHEK, E. P.: J. Polymer Sci. 13, 408 (1954). — GEIDUSCHEK, E. P., and A. M. HOLTZER: Advances in Biol. and Med. Phys. 6, 431 (1958). — GOLDSTEIN, M.: J. Chem. Phys. 21, 1255 (1953). — GORING, D. A. I., M. SENEZ, B. MELANSON and M. M. HUQUE: J. Coll. Sci. 12, 412 (1957). — GUPTA, P. R., and D. A. I. GORING: Can. J. Chem. 38, 270 (1960).
HOLTZER, A. M.: J. Polymer Sci. 17, 432 (1955). — HORN, P.: Ann. Phys. 10, 386 (1955). — HORN, P., H. BENOIT and G. OSTER: J. chim. phys. 48, 530 (1951).
KAY, C. M., and J. T. EDSALL: Arch. Biochem. Biophys. 65, 354 (1956). — KIRKWOOD, J. G., and R. J. GOLDBERG: J. Chem. Phys. 18, 54 (1950). — KRONMAN, M. J., and S. N. TIMASHEFF: J. Phys. Chem. 63, 629 (1959).
LAUER, J. L.: J. Opt. Soc. Am. 41, 482 (1951). — LOTMAR, W.: Helv. Chim. Acta 21, 953 (1938).
MIE, G.: Ann. Physik 25, 377 (1908). — MYSELS, K. J.: J. Phys. Chem. 58, 303 (1954).
NEUGEBAUER, T.: Ann. Physik 42, 509 (1943).
OSTER, G.: Chem. Revs. 43, 319 (1948).
PETERLIN, A.: J. Polymer Sci. 10, 425 (1953); — Progr. in Biophys. and Biophys. Chem. 9, 175 (1959). — PUTZEYS, P., and J. BROSTEAUX: Trans. Faraday Soc. 31, 1314 (1935). — PUTZEYS, P., and E. DORY: Ann. Soc. Sci. Brux. Ser. I 60, 37 (1940).
RAYLEIGH (Lord): Proc. Roy. Soc. A 90, 219 (1914). — RICE, S. A.: J. Polymer Sci. 16, 94 (1955).
SHEFFER, H., and J. C. HYDE: Can. J. Chem. 30, 817 (1952). — STACEY, K. A.: Light scattering in physical chemistry. London: Butterworths 1956. — STIGTER, D.: J. Phys. Chem. 64, 842 (1960). — STOCKMAYER, W. H.: J. Chem. Phys. 18, 58 (1950).
TIMASHEFF, S. N., and M. J. KRONMAN: Arch. Biochem. Biophys. 83, 60 (1959). — TOMIMATSU, Y., K. J. PALMER, A. E. GOODBAN and W. H. WARD: J. Polymer Sci. 36, 129 (1959).
WITNAUER, L. P., and H. J. SCHERR: Rev. Sci. Instr. 23, 99 (1952).
ZIMM, B.: J. Chem. Phys. 16, 1093, 1099 (1948).

D. Diffusion.

ARCHIBALD, W. J.: J. Phys. Colloid Chem. 51, 1204 (1947).
BRYNGDAHL, O., and S. LJUNGGREN: J. Phys. Chem. 64, 1264 (1964).

COULSON, C. A., J. T. COX, A. G. OGSTON and J. ST. L. PHILIPOT: Proc. Roy. Soc. (Lond.) A **192**, 382 (1948).

DUNLOP, P. J.: J. Phys. Chem. **61**, 994 (1957).

GOSTING, L. J.: Advanc. Protein Chem. **11**, 429 (1956).

HOLDE, K. E. VAN, and R. L. BALDWIN: J. Phys. Chem. **62**, 734 (1958).

KAHN, D. S., and A. POLSON: J. Phys. Colloid Chem. **51**, 816 (1947).

LONGSWORTH, L. G.: Ann. N. Y. Acad. Sci. **46**, 211 (1945); — J. Am. Chem. Soc. **69**, 2510 (1947). — LUNDGREN, H. P., and W. H. WARD: Amino acids and proteins. Ed. D. M. GREENBERG. Springfield: Charles C. Thomas 1951.

O'DONNELL, I. J., and L. J. GOSTING: The structure of electrolytic solutions. Ed. W. J. HAMER. New York: John Wiley and Sons; London: Chapman and Hall 1959. — OGSTON, A. G.: Physical techniques in biological research. Vol. II, Eds. G. OSTER and A. W. POLLISTER. New York: Academic Press 1956.

ROBINSON, R. A., and R. H. STOKES: Electrolyte solutions. London: Butterworths 1959.

SCHACHMAN, H. K.: Methods in enzymology. Vol. IV, Eds. S. P. COLOWICK and N. O. KAPLAN. New York: Academic Press 1957.

TSVETKOV, V. N., and S. I. KLENIN: J. Polymer Sci. **30**, 187 (1958).

E. Ultracentrifugation.

ARCHIBALD, W. J.: J. Phys. Colloid Chem. **51**, 1204 (1947).

BALDWIN, R. L.: Biochem. J. **55**, 644 (1953); — J. Phys. Chem. **58**, 1081 (1954); — Biochem. J. **65**, 490 (1957a); **65**, 503 (1957b); — J. Phys. Chem. **63**, 1570 (1959). — BALDWIN, R. L., and K. E. VAN HOLDE: Fortschr. Hochpolym. Forsch. **1**, 451 (1960). — BRIDGMAN, W. B.: J. Am. Chem. Soc. **64**, 2349 (1942).

CHARLWOOD, P. A.: J. Am. Chem. Soc. **79**, 776 (1957). — CRAIG, L. C., T. P. KING and A. STRACHER: J. Am. Chem. Soc. **79**, 3729 (1957).

EHRENBERG, A.: Acta Chem. Scand. **11**, 1257 (1957).

FUJITA, H.: J. Chem. Phys. **24**, 1084 (1956).

GILBERT, G. A.: Proc. Roy. Soc. (Lond.) A **250**, 377 (1959). — GILBERT, G. A., and R. C. LL. JENKINS: Proc. Roy. Soc. (Lond.) A **253**, 420 (1959). — GINSBURG, A., P. APPEL and H. K. SCHACHMAN: Arch. Biochem. Biophys. **65**, 545 (1956). — GOLDBERG, R. J.: J. Phys. Chem. **57**, 194 (1953). — GUTFREUND, H., and A. G. OGSTON: Biochem J. **44**, 163 (1949).

HALL, J. R., and A. G. OGSTON: Biochem. J. **62**, 401 (1956). — HOLDE, K. E. VAN, and R. L. BALDWIN: J. Phys. Chem. **62**, 734 (1958).

JOHNSTON, J. P., and A. G. OGSTON: Trans. Faraday Soc. **42**, 789 (1946). — JOHNSON, P., and A. J. ROWE: Biochem. J. **74**, 432 (1960).

KEGELES, G., and F. J. GUTTER: J. Am. Chem. Soc. **73**, 3770 (1951). — KEGELES, G., S. M. KLAINER and W. J. SALEM: J. Phys. Chem. **61**, 1286 (1957). — KIELLEY, W. W., and W. F. HARRINGTON: Biochim. Biophys. Acta **41**, 401 (1960). — KLAINER, S. M., and G. KEGELES: J. Phys. Chem. **59**, 952 (1955).

LAUFFER, M.: J. Am. Chem. Soc. **66**, 1195 (1944).

MESELSON, M., and F. W. STAHL: Proc. Natl. Acad. Sci. (U.S.) **44**, 671 (1958). — MESELSON, M., F. W. STAHL and J. VINOGRAD: Proc. Natl. Acad. Sci. (U.S.) **43**, 581 (1957). — MCMEEKIN, T. L., and K. MARSHALL: Science **116**, 142 (1952).

O'DONNELL, I. J., R. L. BALDWIN and J. W. WILLIAMS: Biochim. Biophys. Acta **28**, 294 (1958). — OGSTON, A. G.: Physical techniques in biological research. Vol. II, Eds. G. OSTER and A. W. POLLISTER. New York: Academic Press 1956. — ONCLEY, J. L.: Ann. N.Y. Acad. Sci. **41**, 121 (1941).

PEDERSEN, K. O.: J. Phys. Chem. **62**, 1282 (1958a); — J. Franklin Inst. **265**, 503 (1958b).

RICHARDS, E. G., and H. K. SCHACHMAN: J. Phys. Chem. **63**, 1578 (1959).

SCHACHMAN, H. K.: Methods in enzymology, Vol. IV. Eds. S. P. COLOWICK and N. O. KAPLAN. New York: Academic Press 1957; — Ultracentrifugation in biochemistry. London and New York: Academic Press 1959. — SCHUMAKER, V. N., and H. K. SCHACHMAN: Biochim. Biophys. Acta **23**, 628 (1957). — SHOOTER, K. V., and J. A. V. BUTLER: Trans. Faraday Soc. **52**, 734 (1956). — SIGNER, R., and H. GROSS: Helv. Chim. Acta **17**, 726 (1934). — SUEOKA, N.: Proc. Natl. Acad. Sci. U. S. **45**, 1480 (1959). — SVEDBERG, T., and K. O. PEDERSEN: The ultracentrifuge. London and New York: Oxford University Press; New York: Johnson Reprint Corporation 1940.

TOWNEND, R., R. J. WINTERBOTTOM and S. N. TIMASHEFF: J. Am. Chem. Soc. **82**, 3161 (1960). — TRAUTMAN, R.: Biochim. Biophys. Acta **28**, 417 (1958). — TRAUTMAN, R., and C. F. CRAMPTON: J. Am. Chem. Soc. **81**, 4036 (1959).

WILLIAMS, J. W.: Progress in the chemistry of organic natural products. Ed. L. ZECHMEISTER. Vienna: Springer-Verlag 1960. — WILLIAMS, J. W., K. E. VAN HOLDE, R. L. BALDWIN and H. FUJITA: Chem. Revs. **58**, 715 (1958)

YPHANTIS, D. A.: Ann. N.Y. Acad. Sci. (in press).

F. Viscosity.

Huggins, M. L.: J. Am. Chem. Soc. **64**, 2716 (1942).

International Union of Pure and Applied Chemistry, Report on Nomenclature in the Field of Macromolecules: J. Polymer Sci. **8**, 257 (1952).

Öhrn, O. E.: Arkiv Kemi **12**, 397 (1958).

Swindells, J. F., R. Ullman and H. Mark: Physical methods of organic chemistry, Vol. I, Part 1, 3rd Ed., Ed. Weissberger, p. 689. New York, London: Interscience 1959.

G. Particle Shape from Hydrodynamic Measurements.

Baldwin, R. L., and K. E. van Holde: Fortschr. Hochpolym. Forsch. **1**, 451 (1960).

Cohn, E. J., and J. T. Edsall: Proteins, amino acids and peptides. New York: Reinhold 1943.

Einstein, A.: Ann. Physik **19**, 289 (1906); **34**, 591 (1911).

Flory, P. J.: Principles of polymer chemistry. Ithaca (N.Y.): Cornell Univ. Press 1953.

Mehl, J. W., J. L. Oncley and R. Simha: Science **92**, 132 (1940).

Ogston, A. G.: Trans. Faraday Soc. **49**, 1481 (1953). — Oncley, J. L.: Ann. N.Y. Acad. Sci. **41**, 121 (1941).

Perrin, F.: J. Phys. radium **7**, 1 (1936).

Sadron, C.: Progr. in Biophys. and Biophys. Chem. **3**, 237 (1953). — Scheraga, H. A., and L. Mandelkern: J. Am. Chem. Soc. **75**, 179 (1953). — Simha, R.: J. Phys. Chem. **44**, 25 (1940). — Stokes, G. G.: Trans. Cambridge Phil. Soc. **9**, Part II, 8 (1851). — Svedberg, T., and K. O. Pedersen: The ultracentrifuge. London, New York: Oxford University Press; New York: Johnson Reprint Corporation 1940.

Optical Rotatory Dispersion. Its Application to Protein Conformation.

By

E. F. Woods and I. J. O'Donnell.

Optical rotation has been found to be one of the most convenient methods of following the denaturation of proteins. Generally speaking denaturation can be defined as a process or sequence of processes in which the spatial arrangement of the polypeptide chains within the molecule is changed from that typical of the native protein to a more disordered arrangement (KAUZMANN 1959). The terms "*configuration*", "*conformation*" and "*state of folding*" are widely used for spatial arrangement. It is probably best to follow the suggestion of BLOUT (1960) and restrict the use of "configuration" to its original sense, i.e. the spatial arrangement around an asymmetric carbon atom, and to use "conformation" for the shape of the molecule in its entirety. The properties discussed in the previous Chapter i.e., viscosity, diffusion, sedimentation, and light scattering — can all furnish information on the overall shape of proteins or other macromolecules and changes in this shape with environment. Thus DOTY, BRADBURY and HOLTZER (1956) were able to show using these methods, together with streaming birefringence, that poly-γ-benzyl-L-glutamate could exist in two conformations, the α-helix and the solvated randomly coiled form, depending on the solvent. The change from α-helix to random coil was accompanied by marked changes in the optical rotatory properties of the polypeptides. It is to be expected that an α-helical structure should contribute to the rotatory power of a polypeptide since helices are asymmetric and not superimposable on their mirror images. The work on polypeptides has shown that rotatory dispersion is capable of providing information on the folding of the polypeptide chain in proteins and the changes accompanying denaturation.

For the general theory of optical rotation the reader is referred to the following accounts — LOWRY (1935), HELLER (1960), KLYNE and PARKER (1960), KUHN (1958), SCHELLMAN (1958a, 1958b), DJERASSI (1960).

A. Mean Residue Rotation of Polypeptides and Proteins.

The rotatory power of solutions at wave length λ and temperature T is usually reported as the specific optical rotation $[\alpha]_\lambda^T = \frac{100}{lc}$, where c is the concentration of optically active substance in grams per 100 ml of solution and l is the thickness of the solution layer in decimetres. The molar rotation is $[\phi]_\lambda^T = [\alpha]_\lambda^T \frac{M}{100}$ where M is the molecular weight of the solute. For polypeptides derived from optically active monomers $[\phi]$ is not very useful because the molar rotation is primarily a function of the molecular weight rather than of conformation. The most informative measure of the rotatory properties is the rotation per monomer unit or mean residue rotation and this is defined by

$$[m]_\lambda^T = \frac{[\phi]_\lambda^T}{p} = [\alpha]_\lambda^T \frac{M_0}{100} \tag{1}$$

where p is the number of residue units in the molecule and M_0 is the mean residue weight. The reduced residue rotation is given by

$$[m']_\lambda^T = \frac{3}{n^2+2} [m]_\lambda^T \tag{2}$$

where n is the refractive index of the medium; this factor converts all measurements to a medium of refractive index unity (vacuo). Other symbols used for $[m']$ are $[R_{VAC}]$, $[R']$ and $[M]$.

The refractive index correction is important when quantitative comparisons in different solvents are to be made. In comparing the optical rotations of different solutions the rotations are often adjusted to the values in water by multiplying the observed values by

$$\frac{n^2_{water}+2}{n^2_{solvent}+2}.$$

Since the amino end-group of an L-polypeptide has a positive partial rotation and the amino acid residues a negative rotation then for the mean residue rotation to approximate to the rotation of an interior residue the chain must be long. In the completely random conformation the only contribution to the rotation is that of the individual amino acid residues. Thus for many polypeptides the mean residue rotations are of the order $-100°$ to $-120°$. However amino acid residues with two asymmetric centres (e.g. iso-leucine, threonine) or with aromatic residues attached to the β-carbon atom will differ from this value. Tyrosine, phenylalanine, and tryptophan give values more positive than this while proline and cystine give more negative values. The mean residue rotations of proteins in the completely disordered state are generally in the range $-100°$ to $-120°$ unless the protein contains large amounts of an amino acid which makes an abnormal contribution to the rotation. Thus gelatin has a high rotation due to the large content of proline and hydroxyproline, while silk which contains large amounts of the optically inactive glycine would be expected to have a low rotation.

B. Optical Rotatory Dispersion — Drude Equation.

The variation of optical rotation with wave length is described by the Drude dispersion equation

$$[\alpha]_\lambda = \Sigma \frac{K_i}{(\lambda^2 - \lambda_i^2)} \tag{3}$$

where λ_i = the wavelength of the ith absorption band of the molecule (optical activity is always associated with one or more optically active absorption bands) and K_i is a constant characteristic of the ith band and may be positive or negative. However when an optically active absorption band is approached equation (3) does not apply but the contribution to the rotation from this band increases to a maximum (or decreases to a minimum) and changes sign within the region of absorption. This is called the Cotton effect and the dispersion is said to be anomalous. For a single absorption band the following equation applies

$$[\alpha]_\lambda = \frac{K}{\lambda^2 - \lambda_c^2} \tag{4}$$

and this is called a simple Drude equation. λ_c is a dispersion constant.

The development of spectropolarimeters which enable measurements to be readily made down to 250 mμ has enabled the rotatory dispersion of a large number of organic compounds to be measured in the region of the optically active absorption bands (DJERASSI 1960). For many molecules, e.g. polypeptides and proteins, the optically active absorption bands are inaccessible and measurements must be made at higher wavelengths. The optically active absorption bands of proteins are below 230 mμ.

When measurements are made at wavelengths longer than any of the absorption bands of the substance then it is frequently found that the simple DRUDE equation (4) can be used to express the dispersion. This does not mean that a single absorption band is responsible for the optical rotation but only that at long wave lengths equation (3) can be approximated by equation (4). SCHELLMAN (1958a) has shown that if there is more than one absorption band then λ_c is a weighted average of the λ_i's. Ift he absorption cannot be described by a single term DRUDE equation then it is often found that the observations can be expressed by a two-term equation

$$[\alpha]_\lambda = \frac{K_1}{\lambda^2 - \lambda_1^2} + \frac{K_2}{\lambda^2 - \lambda_2^2}\,. \qquad (5)$$

This type of behaviour is also called *anomalous* dispersion or *complex* dispersion. Generally K_1 and K_2 are of opposite sign and as the wavelength is decreased $[\alpha]_\lambda$ will reverse its sign even though λ is not in the region of any absorption band of the molecule. Equation (5) like equations (3) and (4) will only be valid at wavelengths above the region of the optically active absorption bands.

C. Rotatory Properties of Synthetic Polypeptides.

The first experimental evidence for a relation between the optical rotation and chain folding of synthetic polypeptides was obtained by ROBINSON and BOTT (1951). They found for a series of copolymers of γ-methyl-L-glutamate with DL-phenylalanine that the optical rotation depended on whether the sample was dissolved in *m*-cresol or formic acid, and in *m*-cresol it depended on the molecular weight of the polypeptides. Following the work of DOTY et al. (1956) where poly-γ-benzyl-L-glutamate was shown to exist in random coil or α-helical forms depending on the solvent, YANG and DOTY (1957) demonstrated that the randomly coiled forms of the polypeptides showed simple rotatory dispersion whereas the dispersion of the helical forms was complex.

Theories of the rotatory dispersion of helices have been worked out by MOFFITT (1956, 1957) and FITTS and KIRKWOOD (1956, 1957). The theoretical origin of the rotatory dispersion of helices has been satisfactorily established by these workers. The equations of MOFFITT have been the most widely used but since his final equations contain more than two terms it has been found more convenient to treat the data by means of the following equation

$$[m']_\lambda = a_0 \frac{\lambda_0^2}{\lambda^2 - \lambda_0^2} + b_0 \frac{\lambda_0^4}{(\lambda^2 - \lambda_0^2)^2} \qquad (6)$$

where $[m']_\lambda$ is the reduced residue rotation, and a_0, b_0, and λ_0 are constants. The first term in (6) is the DRUDE term and the second term is the complex one, the coefficient b_0 being wholly due to the effects of the helical configuration on the rotatory dispersion and is unaffected by the nature of the solvent. The coefficient a_0 contains both helix and residue contributions and is also dependent on the solvent. Equation (6) is of the same form as that originally proposed by MOFFITT and was found to be applicable to dispersion data on helical polypeptides. MOFFITT's final equations however do not take the same form as equation (6) which is now regarded as an empirical equation. Complex dispersion data on polypeptides can also be fitted with a two term Drude equation [equation (5)], but the Moffitt equation is preferred because of its theoretical basis and the use of fewer constants. For a polypeptide composed of L-amino acid residues Moffitt has calculated that b_0 should be in the region of $-600°$ if it forms a right-handed α-helix and the sign would be reversed for a left-handed helix. Similar deductions were made by Fitts and Kirkwood (1956, 1957) who also predicted that L-polypeptides would exist in a right-handed helical conformation.

In the interpretation of equation (6) it should be remembered that it is not correct to consider the optical rotation of a polypeptide as the sum of the contribution by the asymmetric centres in the chain and that for the helix itself because this neglects the environmental changes that accompany the transition from the helical to the non-helical conformation (Kauzmann 1957). Complex rotatory dispersion is not peculiar to helical conformations and should not be taken as evidence for the presence of helices without additional information from other experimental methods. Any molecule containing a pair of identical weakly interacting groups would give similar rotatory dispersion characteristics to equation (6) but with different constants (Kauzmann 1957).

The work of Doty and coworkers (1956, 1957, 1959), Blout (1960) and Downie, Elliott, Hanby and Malcolm (1957) has shown that the Moffitt equation is successful in describing the rotatory dispersion of a number of polypeptides — poly-γ-benzyl-L-glutamate, poly-L-glutamic acid, poly-ε-carbobenzoxy-L-lysine, poly-L-alanine and poly-L-leucine. In helix-forming solvents, a b_0 value of approximately $-600°$ was obtained for $\lambda_0 = 212\ \text{m}\mu$, while in random-coil solvents $b_0 = 0$. This makes it possible to estimate helical contents of these polypeptides in solvent mixtures since b_0 has been shown to be linearly related to the helix content (Schellman and Schellman 1960).

Recently it has been found that for poly-β-benzyl-L-aspartate in chloroform (Bradbury et al. 1960, Karlson et al. 1960) $b_0 = +630°$ and the evidence suggests that this is due to a left-handed α-helix. The left-handed helix is not as stable as a right-handed one in agreement with the predictions of Huggins (1952).

Certain polypeptides differ markedly in their rotatory properties because of the influence of the nature of the side chain attached to the β-carbon atom of the amino acid, e.g. poly-L-histidine, poly-L-tyrosine, poly-L-serine and poly-L-proline. Proline and hydroxyproline are imino acids and in peptides can only form hydrogen-bonded structures through their C=O groups since the nitrogen of the proline residue contains no hydrogen and the cyclic imino group prevents free rotation between the imino group and the α-carbon atom. Elucidation of the structure of polyproline (Harrington and Sela 1958, Katchalski 1959, Steinberg et al. 1960) has shown that it does not occur in solution as either a random coil or an α-helix but there are two helical forms which are interconvertible both in the solid form and in solution. Certain solvents such as acetic acid, formic acid and trifluoro-acetic acid stretch the peptide molecule into an extended left-handed helical form (polyproline II $[\alpha]_D^{25} = -540°$) while in poor solvents such as n-propanol and n-butanol the molecule was contracted into a right-handed helix (polyproline I $[\alpha]_D^{25} = 40°$). The conformations of poly-L-tyrosine and poly-L-histidine in solution are at present uncertain but neither appears to form an α-helix in solution. (See Coombes et al. 1960) for discussion of conformation of poly-L-tyrosine).

Evidence has also been adduced for the existence of the β-conformation in synthetic polypeptides. Yang and Doty (1957) considered that low molecular weight poly-γ-L-glutamate existed in the β-form at high concentrations in dichloroacetic acid. Fasman and Blout (1960) found that poly-O-acetyl-L-serine could exist in either a β- or random conformation depending on the solvent whereas poly-L-serine was shown to exist only in the random conformation and no evidence has been found for an α-helical conformation. Since β-structures are formed by aggregation reactions in solution which are difficult to control they are the least well-characterised of the basic structures.

D. Optical Rotatory Properties of Proteins.

The majority of proteins have values of $[\alpha]_D$ in the range $-30°$ to $-75°$ and on denaturation this becomes more negative to give rotations in the region $-100°$ to $-120°$. There is also a change in the wavelength dependence on denaturation. It was shown by LINDERSTROM-LANG and SCHELLMAN (1954) that both native and denatured proteins follow the simple DRUDE equation (4), with the exception of the myosin group of proteins which show complex dispersion, and on denaturation the value of λ_c for several proteins decreased from about 250—260 mμ to approximately 220 mμ. COHEN (1955) suggested that the folded conformation of a native protein — identified as an α-helix on the basis of PAULING and COREY's work — makes a direct contribution to the rotation and on denaturation there is a change to a more random conformation. The changes in optical rotation observed on denaturation are in the same direction as the changes which accompany the helix-coil transition in polypeptides.

$[m']_D$ (or $[\alpha]_D$) and λ_c as criteria of conformational changes in proteins. Both these parameters have been used to determine the percentage of helix in proteins on the assumption that only right-handed α-helical and randomly-coiled forms are present. The change in mean residue rotation of a protein when it is unfolded was assumed to be proportional to the fraction of peptide units which are in the helical form provided that the helices are long and of only one sense (SCHELLMAN and SCHELLMAN 1958b). From the limiting values of $[m']_D$ found for synthetic polypeptides in the random coil and helix conformations DOTY and LUNDBERG (1957) estimate that the protein in the completely helical form should have a value of $[m']_D$ which is about 100° above that of the randomly-coiled form. Thus each degree change in $[m']_D$ corresponds to a change of one per cent in the number of residues in the helical conformation.

Measurements by YANG and DOTY (1957) on poly-α-glutamic acid showed that in mixtures of helices and unfolded chains the dispersion becomes complex when the helical content becomes greater than 40 per cent. Thus taking a value of $\lambda_c = 212$ mμ at zero helix content and 255 mμ at 30 per cent helix the helical content can be readily estimated from the observed values of λ_c.

There are several serious objections of the use of $[m']_D$ and/or λ_c to estimate the helical content of proteins: (i) the myosin group of proteins show complex dispersion and the simple DRUDE equation is not satisfactory, (ii) λ_c cannot be used in this way for solvents which induce helical contents greater than 40 per cent because the dispersion becomes complex, (iii) the method is empirical since there is no strict correlation between $[m']_D$ or λ_c and conformation, (iv) both $[m']_D$ and λ_c are not independent of the solvent, (v) although the simple DRUDE equation has been confirmed for several dozen proteins (JIRGENSONS 1958, 1960) the value of λ_c has been found to vary from protein to protein. JIRGENSONS classifies proteins into three groups — (a) proteins which show a decrease in λ_c on denaturation and this includes most of the globular proteins, (b) those proteins whose λ_c varies little on denaturation (e.g. ovalbumin, chymotrypsinogen, lysozyme and ribonuclease), (c) proteins whose λ_c rises on denaturation (e.g. pepsin, γ-globulin, BENCE-JONES protein). In general if the secondary structure of a protein is destroyed λ_c approaches the value 220 mμ and deviations from this value indicate the presence of secondary structure.

b_0 as a criterion of conformational changes in proteins. The parameter b_0 of the MOFFITT equation has been established as a structural property which is independent of solvent for the randomly coiled and α-helical forms of polypeptides. Thus although most native and denatured proteins do not display complex dispersion,

MOFFITT's semi-empirical equation (6) can be applied and seems to be the most reliable method of estimating the helical content. Assuming a value of $b_0 = -630$ for a 100 per cent right-handed α-helical structure (on the basis of the values found for synthetic polypeptides) and $b_0 = 0$ for the completely denatured form, the fraction of residues in the helical form, f_H, can be determined from a plot of the MOFFITT equation $\left(f_H = -\frac{b_0}{630}\right)$. The residues of a protein not existing in the helical conformation need not necessarily be in the randomly coiled form; it is only necessary that they are not in any kind of periodic arrangement since this is the condition that the rotatory dispersion will be of the same kind as that of the randomly coiled form (DOTY 1959).

The usefulness of measuring b_0 for determining the conformation of proteins is exemplified by optical rotation studies on β-lactoglobulin. The native protein has an $[\alpha]_D = -28°$ and it was hence assumed to be highly helical. TANFORD, DE and TAGGART (1960) found b_0 to be $-59°$ to $-76°$ for the native protein indicating an α-helical content no greater than 10 per cent. On denaturation there is a large change in $[\alpha]_D$ but b_0 is scarcely affected. The rotatory properties of β-lactoglobulin were found to be similar to some of the proteins described by JIRGENSONS (1958, 1960). TANFORD considers that the large difference between $[\alpha]_D$ of native and denatured proteins in aqueous media may be partly due to the opening up of the hydrophobic interior which accompanies denaturation and not to a helix-coil transition. The low value of $[\alpha]_D$ for native β-lactoglobulin is attributed to the hydrophobic environment of the peptide groups in the native folded protein.

Existence of Structures in Proteins other than the Right-Handed α-Helix and Random Coil.

So far we have assumed that the right-handed α-helix and random coil are the only structures present in protein molecules and that denaturation is largely a helix-coil transition. However in synthetic polypeptides the random coil, both right-handed and left-handed α-helices, β-structures and the polyproline helix are known to exist. The polyproline helix is probably only important for the collagen group of proteins. The data of TANFORD et al. (1960) on β-lactoglobulin could be explained by the existence of right-handed and left-handed helices in equal proportions but TANFORD et al. give evidence to show that this is probably not so.

The optical rotation data on proteins is consistent with the view that the right-handed α-helix predominates. The only direct evidence on this point comes from the work of KENDREW et al. (1960) on the crystal structure of myoglobin, the helix in this case being exclusively right-handed. The myoglobin molecule also appears to have a region which is helical but not an α-helix.

SCHELLMAN and SCHELLMAN (1960) have considered the rotatory dispersion characteristics of systems containing more than one conformation using the rotatory dispersion constants found for standard polypeptide structures. They give tables of λ_c, b_0, and $[R']_{546} (\equiv [m']_{546})$ for the following systems: left-handed and right-handed α-helices; β-structures and random coils; right-handed α-helices and random coils; left-handed α-helices and random coils. It may be possible to identify the rotatory dispersion constants determined for a particular protein with one of these standard mixtures.

DOTY (1959) further subdivides the constant a_0 in equation (6) into $a_0^H + \Sigma a_0^R$ where a_0^H is due to the effect of the helical conformation and Σa_0^R the sum of the intrinsic residue rotations characteristic of each residue in the protein. Since in the completely denatured state $b_0 = 0$ and $a_0^H = 0$ measurements of the rotatory dispersion in a suitable random coil solvent gives a_0^R. Knowing a_0^R, then a_0^H can be calculated in other solvents from the value of a_0 determined from a MOFFITT plot. a_0^H is positive for a right-handed helix and negative for a left-handed one with an absolute value in the region of 650°. IMAHORI (1960) has used both a_0^H and b_0 to distinguish between the helical, random and β-conformations in bovine serum albumin after denaturation, and to follow the change in the helical conformation in the activation of chymotrypsinogen (IMAHORI, YOSHIDA and HASHIZUME 1960). From the sign of a_0^H and b_0 the conformation can be assigned as follows:

a_0^H pos.,	b_0 neg.	right-handed helix
a_0^H neg.,	b_0 pos.	left-handed helix
a_0^H pos.,	b_0 pos.	β-structure
a_0^H 0 ,	b_0 0	randomly coiled

There is insufficient data at present to evaluate the usefulness of a_0^H as a structural parameter. One serious criticism of this procedure is that the value of a_0^R is very solvent dependent and when going from a random coil to a helical solvent its value probably changes (DOTY 1959).

The effect of amino acid sequence on protein conformation is largely unknown. Since proline cannot fit into an α-helix its occurrence in a polypeptide chain will restrict full development of the helix. Collagen and gelatin which have high proline and hydroxyproline contents do not form an α-helix and the rotatory properties seem to be similar to polyproline I and II.

BLOUT et al. (1960) have determined the conformation of oriented films of polypeptides using infra-red techniques. They found two classes of α-amino acids — those which form an α-helix and those which form either random or β-structures. BLOUT et al. consider that these factors may be operative in protein structures. When a sequence of β- or random-forming amino acids is adjacent to an α-helical section of a protein chain this may be the site where a loop or reversal of direction occurs. They consider that the presence of large amounts of β- or random-forming amino acids may explain the low helix content of many proteins.

The presence of cross-linkages such as disulphide bonds or phosphate ester cross-links between different parts of the chain may affect the conformation of a polypeptide chain and change the optical rotatory properties. Disulphide bonds could either stabilise or restrict helix formation. A discussion of the effect of disulphide bonds on protein conformation is given in the review by LEACH (1960).

The minimum number of residues in a polypeptide which are necessary to form a helical structure may be important in determining the conformation of proteins. The stability of helical conformations will be affected by the nature of the peptide, solvent and temperature (SCHELLMAN and SCHELLMAN 1958b). HARRINGTON and SCHELLMAN (1956) have discussed the stability of polypeptide helices in aqueous solution and from the heat and entropy changes concluded that helix stability is low in aqueous solutions and that the actual stability of folded protein molecules in solution must depend on further structural features, for example, sidechain interactions and cross-linking-S-S-bonds. Helix stability will of course be higher in organic solvents.

GOODMAN, SCHMITT and YPHANTIS (1960) studied the critical range for α-helix formation for poly-γ-methyl-L-glutamates in dimethylformamide and have concluded from the rotatory dispersion results that the critical range appears to be between hepta- and nona-peptides. They concluded that intramolecular hydrogen bonds are formed at chain lengths between the penta and nonapeptides and more stable forms commenced at the nonapeptide since at this chain length residues appear which can intramolecularly hydrogen bond both through the carbonyl and amino group simultaneously.

For a more comprehensive survey of the literature on the optical rotation of proteins the following reviews should be consulted: DOTY (1959), BLOUT (1960), SHOOTER (1960), SCHELLMAN and SCHELLMAN (1960). For a more general treatment of protein conformation, including other techniques, the reviews of LEACH (1960), and KAUZMANN (1957, 1959) are recommended.

E. Temperature Dependence of Optical Rotation.

Increase of temperature will lead to greater freedom of orientation of the groups in all molecules i.e. the magnitude of the rotation is reduced. However if there are restrictions on freedom of orientation, such as in ring compounds which contain asymmetric centres, or due to hydrogen bonding as in proteins, then an increase in temperature cannot increase freedom of orientation about single bonds without rupturing some of the bonds in the ring and the temperature coefficient of the absolute value of the rotation may be positive or negative (see Schellman 1958a). The temperature coefficient of the levorotation of native proteins is found to be small and positive, whereas for denatured proteins i.e. proteins in random coil solvents, the levorotation decreases markedly with increasing temperature as would be expected if freedom of orientation increased with increasing temperature. Hence it should be possible to detect whether a protein is folded or not by observing the manner in which the rotation varies with temperature but some care is needed since the presence of other temperature-dependent effects may give erroneous conclusions (Schellman 1958a). Inverted temperature transitions, i.e. a change from a random to a more helical conformation on increase of temperature have been observed by Doty and Yang (1956) for poly-γ-benzyl-L-glutamate in organic solvent mixtures. Foss and Schellman (1959) and Foss (1960) have also observed similar changes in urea solutions of ribonuclease and lysozyme. On heating these proteins in concentrated urea they first become more helical and on further heating this is followed by a normal temperature transition. According to Schellman (1958c) such inverted transitions can only occur in solvent mixtures.

F. Measurement of Optical Rotation.

The methods of measuring optical rotation have been discussed in numerous reviews on polarimetry (for example Heller 1960, Kortüm and Kortüm-Seiler 1956). For the measurement of optical rotatory dispersion a photoelectric instrument is necessary and it is essential to measure as far into the ultraviolet as possible. The lowest limit with a visual instrument is about 436 mμ. The earlier instruments used for measurement in the ultraviolet region depended on photographic methods and the measurements were laborious. In recent years photoelectric spectropolarimeters have been developed which permit fairly rapid measurements of rotatory dispersion over an appreciable U.V. spectral range. For measurements in the U.V. the polarimeters must be fitted with quartz optics, and the polarimeter tubes must have quartz end-plates.

In the first of the photoelectric instruments described by Rudolph (1955) the "half-shade" device in a conventional type polarimeter is removed leaving a simple system of fixed polariser and rotatable analyser and the telescope is replaced by a phototube connected with suitable amplification to a galvanometer. The sodium lamp is replaced with a monochromator and suitable light source to provide light of different wave lengths. A complete description of the Rudolph instrument is given in the book by Djerassi (1960), and also by Klyne and Parker (1960). It is manufactured by O. C. Rudolph and Sons, Caldwell, New Jersey, U.S.A., and they have also produced a recording spectropolarimeter which enables continuously recorded optical rotation measurements to be made.

Malcolm and Elliott (1957) have described a simple photoelectric polarimeter which is now being manufactured by W. F. Stanley and Co., Ltd., New Eltham, London. In both the Rudolph and the Malcolm-Elliott instruments the measurements are made by a modification of the method of symmetrical angles in which the analyser is rocked (manually or mechanically) between two positions a few degrees apart and on opposite sides of the extinction position and the polariser is rotated until the two positions of the analyser correspond to equal transmission. The transmitted light falls on a phototube whose output is observed by means of a galvanometer. The extinction points are reproducible to 0.001° arc although the absolute accuracy may not be as high as this since it depends on the quality of the optical components and the uniformity of the photomultiplier surface. A photoelectric spectro-polarimeter is also available from Bellingham and Stanley Ltd, 71 Horney Rise, London and covers the range 220 mμ to 600 mμ.

Several accessories have been described to convert ultraviolet spectrophotometers into spectropolarimeters. MITCHELL (1957) has described accessories to enable rotatory dispersion measurements (by the method of symmetrical angles) to be made with the Unicam type SP500 spectrophotometer. The KESTON unit manufactured by the Standard Polarimeter Co., New York, U.S.A. is for attachment to the Beckmann DU spectrophotometer but measurements below 400 mμ are not possible so it cannot be used for ultraviolet spectropolarimetry. A recording spectropolarimeter attachment manufactured by the Perkin-Elmer Corp., Norwalk, Conn. U.S.A., is described by DJERASSI (1960) and can be used with a Perkin-Elmer Model 4000A spectrophotometer or with the Cary recording spectro photometer.

Table 1.

$\lambda_0 = 212$ mμ		$\lambda_0^2 = 0.04494 \times 10^6$	
λ(mμ)	$\lambda^2\ 10^{-6}$	$\frac{\lambda^2 - \lambda_0^2}{\lambda_0^2}$	$\frac{\lambda_0^2}{\lambda^2 - \lambda_0^2}$
700	0.4900	9.903	0.1010
680	0.4624	9.289	0.1077
660	0.4356	8.693	0.1150
640	0.4096	8.114	0.1232
620	0.3844	7.554	0.1324
600	0.3600	7.011	0.1426
589	0.3469	6.719	0.1488
580	0.3364	6.486	0.1542
578	0.3341	6.433	0.1554
577	0.3329	6.408	0.1561
560	0.3136	5.978	0.1673
546	0.2981	5.633	0.1775
540	0.2916	5.489	0.1822
520	0.2704	5.017	0.1993
500	0.2500	4.563	0.2192
493	0.2430	4.407	0.2269
480	0.2304	4.127	0.2423
460	0.2116	3.709	0.2696
440	0.1936	3.308	0.3023
436	0.1901	3.230	0.3096
420	0.1764	2.925	0.3419
405	0.1640	2.649	0.3775
400	0.1600	2.560	0.3906
380	0.1444	2.213	0.4519
365	0.1332	1.964	0.5092
360	0.1296	1.884	0.5308
340	0.1156	1.572	0.6361
334	0.1116	1.483	0.6744
320	0.1024	1.279	0.7819
313	0.09797	1.180	0.8475
302	0.09120	1.029	0.9718
300	0.09000	1.003	0.9970
297	0.08821	0.9628	1.0386

G. Treatment of Data.

Drude equation. For simple dispersion [equation (4)] it has been customary to plot $\frac{1}{[\alpha]_\lambda}\left(\text{or } \frac{1}{[m']_\lambda}\right)$ against λ^2 and determine K from the slope and λ_c from the intercept. YANG and DOTY (1957) consider it more advantageous to plot $\lambda^2[\alpha]_\lambda$ against $[\alpha]_\lambda$ and derive λ_c from the square root of the slope and k from the intercept.

For proteins there is only a small difference between $[\alpha]_\lambda$ and $[m']_\lambda$. This is readily seen from equations (1) and (2) since M_0 for most proteins is in the region 115 and the refractive index correction is about 0.9. However for many polypeptides $[\alpha]_\lambda$ and $[m']_\lambda$ differ appreciably.

MOFFITT equation. Since equation (6) contains three unknowns a value of λ_0 must be found (by a trial and error process) which gives a linear plot when

$[m']_\lambda \frac{\lambda^2 - \lambda_0^2}{\lambda_0^2}$ is plotted against $\frac{\lambda_0^2}{\lambda^2 - \lambda_0^2}$. From this a_0 may be calculated from the intercept and b_0 from the slope. Generally $\lambda_0 = 212$ mμ is tried first and if this does not give a linear plot other values are tried. Alternatively one may plot $[m']_\lambda(\lambda^2 - \lambda_0^2)$ against $\frac{1}{\lambda^2 - \lambda_0^2}$. The accompanying table which is calculated for $\lambda_0 = 212$ mμ is useful for carrying out a MOFFITT plot.

As discussed in this article there is no simple and straight-forward interpretation of the parameters describing optical rotatory dispersion in terms of protein conformation. The parameter b_0 of the MOFFITT equation is preferred to λ_c because of its independence of the solvent. The changes in b_0 and $[m']_\lambda$ as the solvent is changed to a random coil solvent or to a helix-forming solvent assist in determining the conformation of the native protein.

The authors wish to thank Dr. B. S. HARRAP and Dr. S. J. LEACH of these laboratories for helpful criticism during the preparation of this manuscript.

References.

BLOUT, E. R.: In Optical rotatory dispersion, by C. DJERASSI, Chap. 17. New York: McGraw-Hill 1960. — BLOUT, E. R., C. DE LOZÉ, S. M. BLOOM, G. D. FASMAN: J. Am. Chem. Soc. **82**, 3787 (1960). — BRADBURY, E. M., A. R. DOWNIE, A. ELLIOTT and W. E. HANBY: Nature **187**, 321 (1960).

COHEN, C.: Nature **175**, 129 (1955). — COOMBES, J. D., E. KATCHALSKI and P. DOTY: Nature **185**, 534 (1960).

DJERASSI, C.: Optical rotatory dispersion. New York: McGraw-Hill 1960. — DOTY, P., J. H. BRADBURY and A. M. HOLTZER: J. Am. Chem. Soc. **78**, 947 (1956). — DOTY, P., and J. T. YANG: J. Am. Chem. Soc. **78**, 498 (1956). — DOTY, P., and R. D. LUNDBERG: Proc. U.S. Nat. Acad. Sci. **43**, 213 (1957). — DOTY, P.: Proceedings of the IVth International Congress of Biochemistry, Vienna, Vol. 9, p. 9. London: Pergamon Press 1959. — DOWNIE, A. R., A. ELLIOTT, W. E. HANBY and B. R. MALCOLM: Proc. Roy. Soc. A **242**, 325 (1957).

FASMAN, G. D., and E. R. BLOUT: J. Am. Chem. Soc. **82**, 2262 (1960). — FITTS, D. D., and J. G. KIRKWOOD: Proc. U.S. Nat. Acad. Sci. **42**, 33 (1956); **43**, 1046 (1957); — J. Am. Chem. Soc. **78**, 2650 (1956). — FOSS, J. G.: Biochem. Biophys. Acta **43**, 300 (1960). — FOSS, J. G., and J. A. SCHELLMAN: J. Phys. Chem. **63**, 2007 (1960).

GOODMAN, M., E. E. SCHMIDT and D. YPHANTIS: J. Am. Chem. Soc. **82**, 3483 (1960).

HARRINGTON, W. F., and J. A. SCHELLMAN: Compt. rend. trav. lab. Carlsberg Ser Chim. **30**, 21 (1956). — HARRINGTON, W. F., and M. SELA: Biochim. Biophys. Acta **27**, 24 (1958). — HELLER, W.: Polarimetry in A. WEISSBERGER (ed.), Physical methods of organic chemistry, 3rd ed., Vol. 1, Part 3, Chap. 33. New York: Interscience Publishers, Inc. 1949. — HUGGINS, M. L.: J. Am. Chem. Soc. **74**, 3963 (1952).

IMAHORI, K.: Biochim. Biophys. Acta **37**, 336 (1960). — IMAHORI, K., A. YOSHIDA and H. HASHIZUME: Biochim. Biophys. Acta **45**, 380 (1960).

JIRGENSONS, B.: Arch. Biochem. Biophys. **74**, 70 (1958); **78**, 227, 235 (1958); **89**, 48 (1960) and other earlier papers cited in these references.

KATCHALSKI, E.: Proceedings of the IVth International Congress of Biochemistry, Vienna, Vol. 9, p. 21. London: Pergamon Press 1959. — KARLSON, R. H., K. S. NORLAND, G. D. FASMAN and E. R. BLOUT: J. Am. Chem. Soc. **82**, 2268 (1960). — KAUZMANN, W.: Ann. Rev. Phys. Chem. 8, 413 (1957); — Adv. Protein Chem. **XIV**, 1 (1959). — KENDREW, J. C., R. E. DICKERSON, B. E. STRANDBERG, R. G. HART, D. R. DAVIES, D. C. PHILLIPS and V. C. SHORE: Nature **185**, 422 (1960). — KLYNE, W., and A. C. PARKER: Optical rotatory dispersion in A. WEISSBERGER (ed.), Physical methods of organic chemistry, 3rd ed. Vol. 1, Part III, Chapt. 34. New York: Interscience Publishers, Inc. 1960. — KORTÜM, G., and M. KORTÜM-SEILER: Mod. Meth. Plant Analysis **1**, 278 (1956). — KUHN, W.: Ann. Rev. Phys. Chem. **9**, 417 (1958).

LEACH, S. J.: Rev. Pure App. Chem. (Australia) 9, **33** (1959). — LINDERSTROM-LANG, K. U., and J. A. SCHELLMAN: Biochem. Biophys. Acta **15**, 156 (1954). — LOWRY, T. M.: Optical rotatory power. London: Longman, Green & Co. Ltd. 1935.

MALCOLM, B. R., and A. E. ELLIOTT: J. Sci. Instr. **34**, 48 (1957). — MITCHELL, S.: J. Sci. Instr. **34**, 89 (1957). — MOFFITT, W.: J. Chem. Phys. **25**, 467 (1956). — MOFFITT, W., and J. T.

YANG: Proc. U.S. Nat. Acad. Sci. **42**, 596 (1956). — MOFFITT, W., D. D. FITTS and J. G. KIRKWOOD: Proc. U.S. Nat. Acad. Sci. **43**, 723 (1957).

ROBINSON, C., and M. J. BOTT: Nature **168**, 325 (1951). — RUDOLPH, H.: J. Opt. Sci. Am. **45**, 50 (1955).

SCHELLMAN, J. A.: Compt. Rend. trav. lab. Carlsberg, Ser. chim. **30**, 363, 395, 415, 429, 439, 450 (1958a). — SCHELLMAN, C. G., and J. A. SCHELLMAN: Compt. rend. trav. lab. Carlsberg, Ser chim. **30**, 465 (1958b). — SCHELLMAN, J. A.: J. Phys. Chem. **62**, 1485 (1958c). — SCHELLMAN, J. A., and C. G. SCHELLMAN: J. Polymer Sci. **49**, 129 (1961) — SHOOTER, E. M.: Progr. in Biophys. and Biophys. Chem. **10**, 196 (1960). — STEINBERG, I. Z., M. SELA, W. F. HARRINGTON, A. BERGER and E. KATCHALSKI: J. Am. Chem. Soc. **82**, 5263 (1960).

TANFORD, C., P. K. DE, and V. G. TAGGART: J. Am. Chem. Soc. **82**, 6028 (1960).

YANG, J. T., and P. DOTY: J. Am. Chem. Soc. **79**, 761 (1957).

Diffuse Röntgenkleinwinkelstreuung.

Von

O. Kratky.

Mit 26 Abbildungen.

Die Röntgenkleinwinkelmethode ist ein Verfahren, das wertvolle Aussagen über Größe, Gestalt, Anordnung und Gewicht von Kolloidteilchen und Makromolekülen zu machen gestattet. Bei sehr dichten Systemen sind umgekehrt auch Aussagen über die Hohlräume möglich, sofern sie kolloide Dimensionen haben. Es liegt auf der Hand, daß mit Hilfe dieser Methode viele Fragen des submikroskopischen Aufbaues von Strukturelementen der Pflanzen erfolgreich studiert werden können. Im folgenden sollen nach einer kurzen, einfach gehaltenen Darlegung der theoretischen Grundlagen einige Beispiele aus dem Pflanzenreich gebracht werden.

Neuere zusammenfassende Darstellungen des Gesamtgebietes sind erschienen von GUINIER u. FOURNET (1955), BEEMAN, KAESBERG, ANDEREGG u. WEBB (1957), KRATKY (1955, 1956, 1960), POROD (1960), GEROLD (1957), auf welche auch für die theoretischen Grundlagen verwiesen wird.

A. Theorie.

I. Die reine Partikelstreuung.

Wir denken uns von einer Röntgenwelle eine kolloide Partikel bestrahlt, deren Dimension groß im Vergleich mit der Wellenlänge ist (Abb. 1, links). Die Elektronen der Partikel werden dann im Rhythmus der durchlaufenden Welle mitschwingen

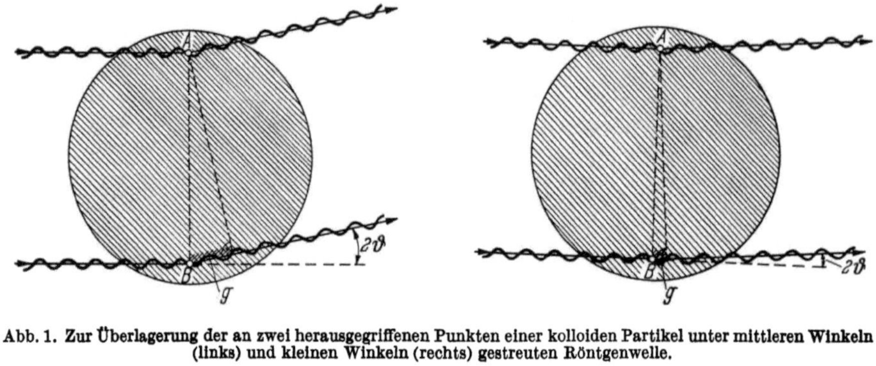

Abb. 1. Zur Überlagerung der an zwei herausgegriffenen Punkten einer kolloiden Partikel unter mittleren Winkeln (links) und kleinen Winkeln (rechts) gestreuten Röntgenwelle.

und ihrerseits nach allen Richtungen Sekundärwellen aussenden. Das interessierende Problem ist nun die Wechselwirkung, d. h. Interferenz dieser Sekundärwellen. Zur Veranschaulichung des Vorganges greifen wir aus dieser Partikel zwei Punkte *A* und *B* heraus und denken uns die an diesen Punkten im eingezeichneten Winkel

2ϑ abgebeugten Wellen überlagert. Wir sehen, daß ein Lichtwegunterschied g auftritt, der von der zufälligen Lage der beiden Punkte abhängt und meist viel größer ist als die Wellenlänge. Verallgemeinern wir die Überlegung, indem wir an *allen* Punkten die Abbeugung in der gleichen Richtung betrachten, so erkennen wir leicht, daß sich die in den mannigfachsten Phasen überlagerten Wellen praktisch vollkommen auslöschen. Fassen wir aber einen sehr viel kleineren Abbeugungswinkel ins Auge (Abb. 1 rechts), so sind die Gangunterschiede kleiner als die Wellenlänge, die Wellen beginnen sich zur verstärken und bei verschwindend kleinen Abbeugungswinkeln überlagern sie sich schließlich exakt in Phase, so daß dort das Maximum der abgebeugten Intensität auftritt. Tragen wir die Intensität gegen den Abbeugungswinkel auf, so werden wir also nur bei sehr kleinen Winkeln endliche Intensitäten erhalten, die „*Streukurve*" des Teilchens wird etwa eine Gestalt haben, wie sie Abb. 2 darstellt. Da die Streuung auf kleine Winkel beschränkt ist, werden wir mit Recht von einer „*Kleinwinkelstreuung*" der kolloiden Partikel sprechen dürfen.

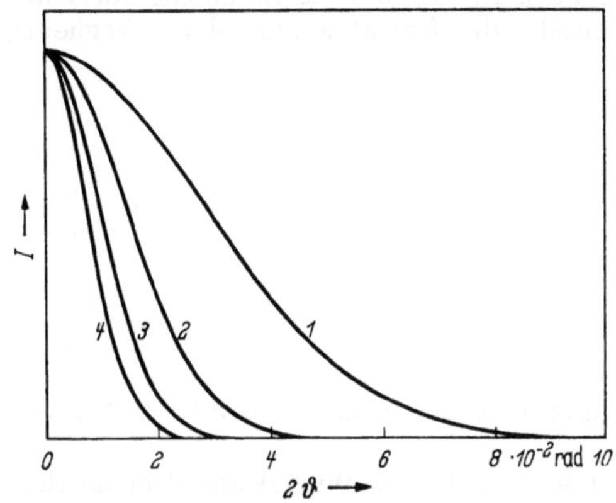

Abb. 2. Streukurven von Teilchen, deren lineare Dimensionen sich wie die bei den Kurven angegebenen Zahlen, also wie 1:2:3:4, verhalten.

Denken wir uns nun die Partikel vergrößert, dann liegt auf der Hand, daß wir erst bei noch kleineren Winkeln eine Verstärkung der abgebeugten Wellen beobachten können. Wie bei allen Beugungsphänomenen gilt auch hier, daß die Dimensionen der beugenden Teilchen und die entsprechenden Beugungswinkel antibat gehen. Das Beugungsbild ist also in einem gewissen Sinne ein reziprokes Abbild der Wirklichkeit. Die Streukurven 1, 2, 3 und 4 in Abb. 2 würden danach Teilchen entsprechen, deren lineare Dimensionen sich wie 1:2:3:4 verhalten.

Nun wäre bei einem realen Experiment der Bruchteil der einfallenden Strahlung, der durch eine *einzelne* Partikel abgebeugt wird, allerdings so klein, daß die Empfindlichkeit der Meßmethoden zum Nachweis der Abbeugung nicht ausreichte. Wir müssen vielmehr dem Röntgenstrahl ein *verdünntes System* von Einzelpartikeln zur Verfügung stellen, also etwa eine verdünnte Lösung, in der die Partikeln unregelmäßige Abstände haben, die groß im Vergleich mit der Dimension der Teilchen sind. In diesem Fall findet einfach eine Addition der von den Einzelteilchen gestreuten Intensitäten statt und es wird sich an der Gestalt der Streukurve nichts ändern. Sind die Partikeln anisotrop, so erfolgt bei dieser Art der Aufnahme bereits eine Überlagerung der Abbeugung für alle Orientierungen der Partikeln relativ zum einfallenden Strahl und die erhaltene Streukurve ist die des makroskopisch isotropen Systems.

Ohne in diesem Rahmen auf die genaue Analyse solcher Streukurven eingehen zu können, sei festgestellt, daß es beim Vorliegen eines homodispersen Systems im allgemeinen möglich ist, aus der Streukurve eine der wirklichen Partikel „*streuungsäquivalente*" Gestalt auszurechnen, die der wirklichen wohl recht nahekommt. Der Weg einer solchen Analyse sei aber durch einige kurze Hinweise charakterisiert.

Gang der Analyse. 1. Wie oben bereits angedeutet, ist die Streukurve um so schmäler, je größer die Partikeln sind und umgekehrt. Bei gleicher Teilchengestalt sind also z. B. Halbwertsbreite der Streukurve und lineare Dimension der Teilchen einander umgekehrt proportional.

2. Im Falle einer der Kugel nahekommenden Gestalt entspricht die Streukurve I etwa der Gaussschen Form (Guinier)

$$I = I_0 \cdot e^{-KR^2\vartheta^2} \tag{1a}$$

darin bedeutet I_0 die Streuintensität beim Winkel null, K ist eine universelle Konstante $\left(=\frac{16\pi^2}{3\lambda^2}\right)$, ϑ ist der halbe Ablenkungswinkel und R der sog. *Streumassenradius*. Er ist eine die räumliche Ausdehnung des Teilchens charakterisierende Größe, die formal analog dem Trägheitsradius der Mechanik definiert wird,

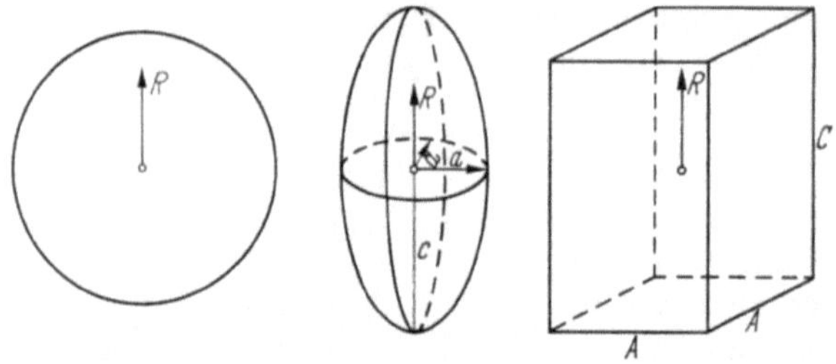

Abb. 3. Größe des Streumassenradius R für eine Kugel, ein Rotationsellipsoid und ein quadratisches Prisma.

d. h. er stellt die Wurzel aus dem mittleren Abstandsquadrat aller streuenden Elektronen vom Schwerpunkt dar. In Abb. 3 sind zur Veranschaulichung in einige Partikeln die entsprechenden Streumassenradien eingezeichnet. Trägt man nach

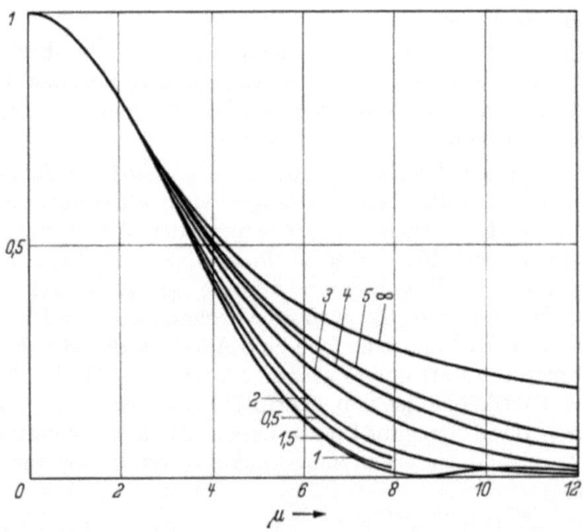

Abb. 4. Streukurven für Rotationsellipsoide. Das Verhältnis zwischen Rotationsachse und Nebenachse ist auf den Kurven angegeben.

dem Vorgang von Guinier den Logarithmus der gestreuten Intensität gegen das Quadrat des Ablenkungswinkels auf, so erhält man, wie die Logarithmierung von (1a) erkennen läßt,

$$\ln I = \ln I_0 - KR^2\vartheta^2 \tag{1b}$$

also eine Gerade, deren Neigungstangente durch $-KR^2$ gegeben ist. Die GUINIERsche Auftragung führt somit unmittelbar auf den Streumassenradius des Teilchens.

3. Sind die Teilchen nicht annähernd kugelförmig, so weicht die Streukurve des verdünnten Systems erheblich von der GAUSSschen Form ab, vor allem im äußeren Teil, wie dies Abb. 4 für Rotationsellipsoide verschiedener Achsenverhältnisse zeigt. Es läßt sich aber beweisen, daß auch in diesem Falle bei GUINIERscher Auftragung die Tangente beim Winkel null auf den wahren Streumassenradius führt. Aus der Abweichung vom GUINIERschen Verlauf ist, wie aus Abb. 4 verständlich wird, eine Gestaltsermittlung möglich. Diese kann oft kontrolliert und wesentlich unterstützt werden durch die Verwertung von Nebenmaximas, wie solche z. B. in Abb. 15 zu sehen sind.

4. Liegen Teilchen vor, deren Länge sehr groß ist im Vergleich mit der Dicke, so kann die Streukurve als Produkt zweier Faktoren dargestellt werden

$$I = I_q \cdot \frac{1}{\vartheta}\,. \tag{2a}$$

Darin bezieht sich I_q auf den Querschnitt (*„Querschnittsfaktor“*) und es zeigt diese Größe eine ähnliche Winkelabhängigkeit wie die Streukurve eines corpuscularen Teilchens. $1/\vartheta$ bezieht sich auf die Länge („LORENTZ-Faktor“). Man braucht also die experimentell erhaltene Streukurve I nur mit ϑ zu multiplizieren:

$$I \cdot \vartheta = I_q \tag{2b}$$

und erhält dadurch den Querschnittsfaktor (KRATKY u. POROD 1948, POROD 1948). Seine Auftragung im Sinne von GUINIER ($\log I_q$ gegen ϑ^2) ergibt eine Kurve, deren Tangente beim Winkel null gleich wie bei den corpuscularen Partikeln den Streumassenradius des Querschnittes zu berechnen gestattet. Die zusätzliche Gestaltsanalyse der Kurve führt auf die Form des Querschnittes.

5. Sind die Partikeln Blättchen, deren Fläche sehr groß ist im Vergleich mit der Dicke, so ist die Streukurve wie folgt darzustellen:

$$I = I_d \cdot \frac{1}{\vartheta^2}\,. \tag{3a}$$

Darin bedeutet I_d den *„Dickenfaktor“*, $\frac{1}{\vartheta^2}$ den „LORENTZ-*Faktor*“ der Fläche (KRATKY u. POROD 1948, POROD 1948). Durch Multiplikation der experimentellen Kurve mit ϑ^2 erhält man danach den Dickenfaktor

$$I\vartheta^2 = I_d\,. \tag{3b}$$

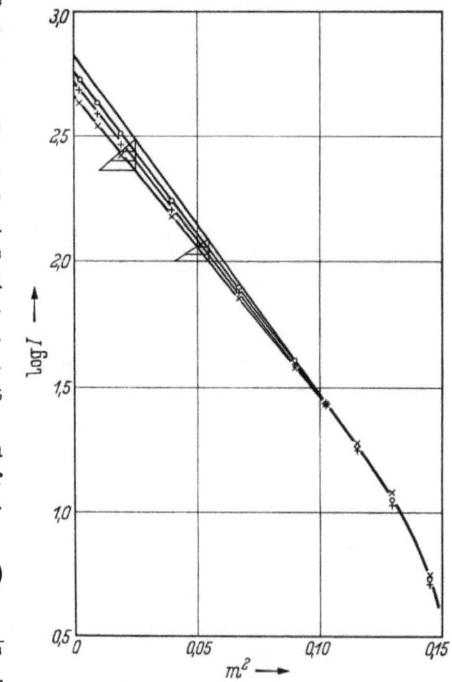

Abb. 5. Extrapolation der Streukurve des Pflanzenglobulins Edestin in GUINIERscher Darstellung auf unendliche Verdünnung. × ... $c = 1.40$ g/100 ml, + ... $c = 0.98$ g/100 ml, ○ ... $c = 0.655$ g/100 ml. Das Ergebnis ist die oberste Kurve (CLEEMANN und KRATKY 1960).

Aus der Tangente beim Winkel null in einer GUINIERschen Auftragung ergibt sich analog den beiden anderen Fällen der Streumassenradius der Dicke, aus dem man durch Multiplikation mit $\sqrt{12}$ die Dicke selbst berechnet.

6. Die dargelegten Zusammenhänge gelten nur, wenn das System tatsächlich sehr verdünnt ist. Bei zu hoher Konzentration kommt es zu einer interferenzmäßigen Wechselwirkung der von den einzelnen Teilchen gestreuten Wellen — wir verzichten auf die Darlegung der theoretischen Grundlagen —, was sich besonders in einer Abflachung im inneren Teil der Streukurve bemerkbar macht. Der übliche Weg zur Überwindung dieser Schwierigkeit besteht darin, Konzentrationsreihen zu untersuchen und das Streuverhalten auf die Konzentration null zu extrapolieren. Abb. 5 zeigt z. B., wie aus Messungen an Lösungen des Pflanzenglobulins Edestin (Cleemann u. Kratky 1960) die für die Konzentration null zu erwartende Kurvengestalt erschlossen wurde.

II. Dichtgepackte isotrope Systeme.

Dem Extremfall des verdünnten Systems ohne jedwede Ordnung kann der Extremfall der völlig regelmäßigen räumlichen Anordnung entgegengestellt werden, wie er im Kristall vorliegt. Uns soll nun eine dazwischenliegende Möglichkeit interessieren, nämlich eine Packung der Partikeln, die so dicht ist, daß schon durch die räumliche Enge eine gewisse Ordnung erzwungen wird. Die Vollkommenheit der kristallinen Ordnung soll aber deshalb nicht erreichbar sein, weil die Partikeln in ihrer Größe statistisch schwanken.

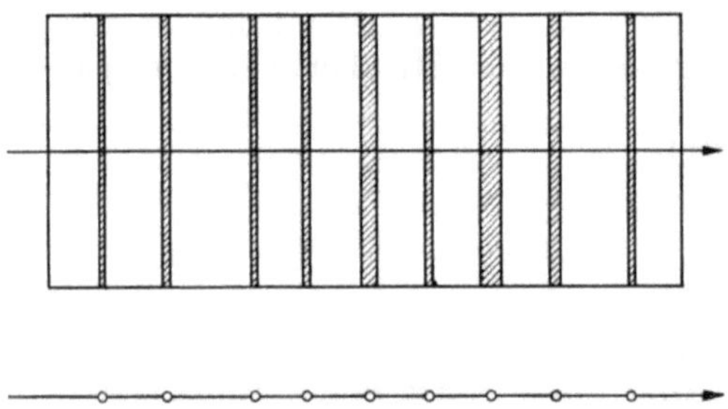

Abb. 6. Schnitt durch ein Lamellenpaket parallel zur Achse. Die gestrichelten Flächen sollen die Hohlräume darstellen. Die Streuung an der im unteren Teil der Abbildung gezeichneten Punktreihe stellt den „Gitterfaktor" dar.

Wir wollen uns damit begnügen, aus der großen Mannigfaltigkeit der hierher zu zählenden Möglichkeiten das *Lamellenpaket* herauszugreifen, das als System einer eindimensionalen Ordnung noch verhältnismäßig einfach ist und zugleich praktische Wichtigkeit besitzt. Die Zwischenräume der Lamellen mögen im Vergleich mit der mittleren Dicke der Einzellamelle sehr schmal sein, wie dies Abb. 6 schematisch andeutet, und sowohl Lamellen als auch Zwischenräume sollen in ihrer Dimension statistisch schwanken. Solche Pakete sollen im untersuchten Objekt in allen räumlichen Orientierungen vorhanden sein.

Bei der Behandlung des Streueffektes gehen wir vom *Reziprozitätsgesetz* der Optik aus. Dieses besagt, daß der Streueffekt eines Objektes unverändert bleibt, wenn wir die streuende Materie durch Hohlräume und die Hohlräume umgekehrt durch Materie ersetzen. In diesem Sinne können wir jedes Lamellenpaket durch ein System von Lamellen ersetzen, die die Dicke der früheren Hohlräume besitzen und voneinander Abstände haben, die den früheren Lamellendicken entsprechen. Der Streueffekt eines Haufwerkes solcher Lamellensysteme ist durch das Produkt zweier Größen darzustellen: der Streuung der Einzellamelle, die etwa $1/\vartheta^2$ entspricht (der Dickenfaktor kann wegen der Kleinheit der Dicke im interessierenden Streubereich als konstant betrachtet werden) und dem „Gitterfaktor". Wir verstehen darunter die Streuung einer Punktreihe, wie sie sich beim Durchstich des Paketes mit einer normal zur Lamellenfläche verlaufenden Geraden ergibt. Ein solches lineares Gitter, in allen räumlichen Orientierungen dem Röntgenstrahl dargeboten, gibt nun Streukurven von der Art der Abb. 7 (Porod 1949). Es tritt stets ein Maximum auf, das etwa beim Braggschen Winkel liegt, der sich aus der mittleren „Gitterkonstanten" ergibt, die ja zugleich der mittlere Schwerpunktsabstand der wirklichen Lamellen ist. Das Maximum ist bei kleiner relativer Schwankung der Gitterkonstanten — mithin auch der Lamellendicke — verhältnismäßig scharf, flacht sich aber mit zunehmender Schwankung rasch ab. Die tatsächliche Streukurve des beschriebenen Objektes ist also durch Multiplikation des in Abb. 7 dargestellten Gitterfaktors mit dem „LORENTZ-Faktor" der Fläche $1/\vartheta^2$ gegeben. Abb. 8 (POROD 1949) zeigt die Streukurven, welche man danach je nach der Größe der statistischen Schwankung der Micelldicken zu erwarten hat.

Wir sehen, daß sich bei einer mittleren Schwankung von 0,5 der Gitterfaktor — vom Innenteil abgesehen — schon beträchtlich der Konstanz nähert, d. h. in diesem Bereich streuen die Blättchen nahezu unabhängig voneinander. Da

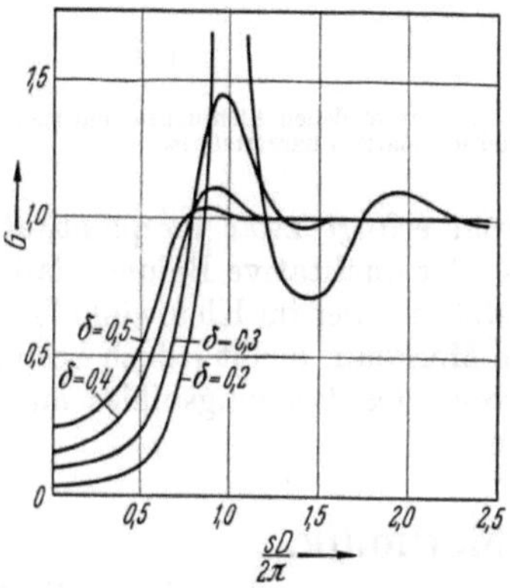

Abb. 7. Gitterfaktoren von Lamellenpaketen mit den mittleren Dickenschwankungen δ (POROD 1949).

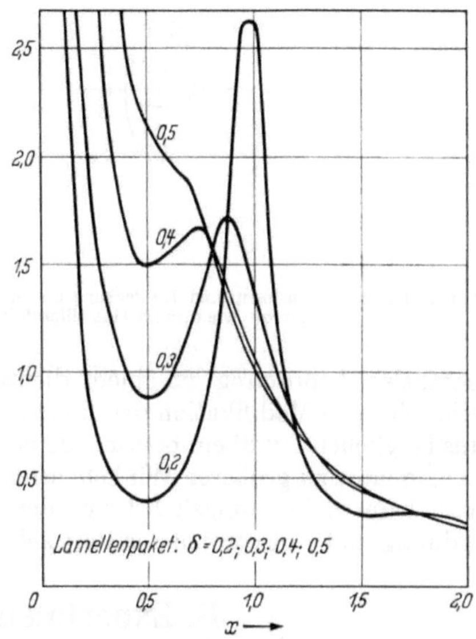

Abb. 8. Gesamt-Streukurven von Lamellenpaketen mit den mittleren Dickenschwankungen δ (POROD 1949).

eine Schwankung der „Gitterkonstanten" nicht nur durch eine verschiedene Dicke der Lamellen, sondern auch durch eine Erniedrigung der Packungsdichte zustande kommt, können wir sagen, daß sowohl Polydispersität als auch verringerte räumliche Dichte den Inteferenzeffekt herabsetzt und eine Annäherung an die Streuung der Einzelpartikel bewirkt. Stark polydisperse Systeme kann man daher auch bei verhältnismäßig dichter Packung in gewisser Annäherung so auswerten, als wären sie verdünnt.

Solange die kolloiden Partikel durch Zwischenräume getrennt sind, erfährt man nach obigem aus der Periodizität, welche die Anordnung der Hohlräume besitzt, die wirklichen Dimensionen der Partikel. Nun liegt allerdings oft der Fall vor, daß benachbarte Partikel so dicht aneinandergelagert sind, daß die Zwischenräume praktisch verschwinden. Die Einzelpartikel treten dann im Röntgenkleinwinkelbereich nicht mehr in Erscheinung. Enthält das gleiche System aber doch gewisse, allenfalls unregelmäßig verteilte Hohlräume, so kann man im Sinne des Reziprozitätsgesetzes diese als die Ursache eines Streueffektes auffassen und die Röntgenkleinwinkelanalyse wird zur Porenanalyse des Systems.

III. Orientierte Objekte.

Wir betonten, daß größere Dimensionen kleineren Beugungswinkeln entsprechen und umgekehrt, mithin das Beugungsbild in einem gewissen Sinne ein reziprokes Abbild des wirklichen Objektes darstellt. Diese Bemerkung gilt auch für die verschiedenen Dimensionen ein und derselben Partikel. Denken wir uns eine einzelne z. B. eiförmig-ellipsoidische Partikel so in den Röntgenstrahl gebracht wie dies Abb. 9 zeigt, so wird ein ellipsoidisches Beugungsbild zustande kommen, das gegenüber dem beugenden Teilchen eine Vertauschung der Achsenverhältnisse zeigt. Durch den Zusammentritt vieler anisotroper Teilchen in

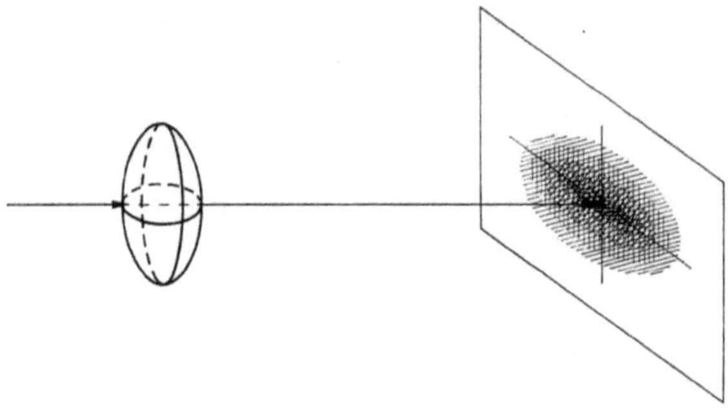

Abb. 9. Partikeldimensionen und Erstreckung des Streubereiches in verschiedenen Richtungen sind zueinander reziprok wie dies für eine ellipsoidische Partikel schematisch dargestellt ist.

paralleler Anordnung zu einem dichten System erfolgt zwar in quantitativer Hinsicht eine Modifikation des Beugungsbildes. Der qualitative Befund, daß eine aus länglichen Partikeln bestehende vertikalgestellte Faser im Kleinwinkelbereich am Äquator zu größeren Winkeln streut als am Meridian, bleibt jedoch bestehen. Man kann daher umgekehrt aus der Anisotropie des Beugungsbildes auf Anordnung und Anisotropie der Partikel schließen.

B. Experimentelle Methodik.

Das Tempo der Röntgenkleinwinkelforschung war vor allem durch die Entwicklung der experimentellen Methodik bestimmt. Heute können die Schwierigkeiten im wesentlichen als überwunden gelten, so daß es möglich ist, Hinweise auf zweckmäßige und leistungsfähige Einrichtungen zu geben. Es gilt vor allem drei Probleme zu meistern.

a) Erzielung einer *ausreichenden Intensität* bei genügend enger Ausblendung des Primärstrahls.

b) Vermeidung bzw. *Ausschaltung der Blendenstreuung*, d. i. der Kleinwinkelstreuung, welche von den den Röntgenstrahl begrenzenden Blenden ausgeht.

c) Exakte *Messung der Streukurve.*

Eine genügend enge Ausblendung des Primärstrahls zur Erreichung eines entsprechend kleinen Streuwinkels ist an sich keine Schwierigkeit; aber es sinkt mit der Verringerung des Primärstrahlquerschnittes einerseits die Intensität der zu messenden Streuung und andererseits steigt mit abnehmendem Streuwinkel die störende „Blendenstreuung" rapid an. Nun variiert wohl das Streuvermögen der Präparate in weiten Grenzen. So hat z. B. eine luftgequollene Cellulose eine um etwa drei Größenordnungen stärkere Streuung als eine so verdünnte Proteinlösung, wie man sie zur Bestimmung von Größe und Gestalt der Proteinmoleküle zu studieren hat. Tatsächlich tritt bei stark streuenden Präparaten die Schwierigkeit der Erzielung einer ausreichenden Streuintensität in den Hintergrund und auch die Blendenstreuung stört wenig neben einer intensiven Präparatstreuung. Bei schwach streuenden Präparaten sind dagegen die unter a) und b) angegebenen Aufgaben ernstzunehmende experimentelle Probleme.

I. Die wichtigsten Kameratypen.

Im folgenden sollen einige Hinweise auf Kleinwinkelkameras gegeben werden, die praktische Bedeutung erlangt haben und die den bei schwach streuenden Präparaten auftretenden Anforderungen gewachsen sind.

1. Kameras vom Lochblendentyp.

Beim Studium der Anisotropie von Objekten verwendet man in der Regel einen Primärstrahl mit kreisförmigem Querschnitt. Durch zwei Lochblenden wird ein schmales Strahlenbündel ausgesondert. Es ist nicht zu vermeiden, daß am Metall der zweiten Blende eine Kleinwinkelstreuung entsteht. Daher wird eine dritte zur Strahlenachse konzentrische Lochblende angebracht, deren Durchmesser so gewählt ist, daß sie vom direkten Strahl nicht getroffen wird, die Streuung der zweiten Blende aber soweit als möglich einengt. Kameras dieser Art sind wiederholt beschrieben worden (Kiessig 1942). Bei dieser Anordnung gelangt, verglichen mit den unten zu besprechenden Spaltkameras, nur eine geringe Primärintensität zur Wirkung. Die Untersuchung der anisotropen Fasern erfordert aber im allgemeinen keine allzu feine Ausblendung, weil die interessierenden Effekte bei verhältnismäßig großen Winkeln liegen, etwa entsprechend Braggschen Werten von 100 bis 200 Å; außerdem haben diese Objekte ein verhältnismäßig hohes Streuvermögen, so daß mit dieser Anordnung zahlreiche Untersuchungen erfolgreich durchgeführt werden konnten. Sie hat aber auch Anwendung beim Studium von Proteinlösungen gefunden, und zwar zur Vermessung der bei verhältnismäßig großen Winkeln gelegenen Nebenmaxima. Naturgemäß ist auch hier keine sehr feine Ausblendung erforderlich, d. h. das primäre Strahlenbündel kann ziemlich dick sein. Beeman u. Mitarb. (Rothwell 1954) bedienen sich bei Untersuchungen dieser Art des sehr wirksamen Kunstgriffes, daß sie die unter einem bestimmten Winkel gestreuten Strahlen durch eine Ringblende aussondern und in ihrer Gesamtheit in einem dahinter geschalteten Zählrohr messen. Durch Verschieben der Blende in Richtung der Strahlenachse kann man die Streuung bei verschiedenen Winkeln erfassen.

2. Kameras vom Spaltblendentyp.

Eine wesentlich größere Intensität ist durch Anwendung eines Primärstrahls von rechteckigem Querschnitt erzielbar, wie er durch Spaltblenden ausgesondert werden kann. Die Streukurve wird durch diese Primärstrahlform allerdings erheblich modifiziert, und es hat ausgedehnter Studien bedurft, um Wege zur Beseitigung dieses „Kollimationseinflusses" aufzufinden, d. h. um auf die eigentliche Streukurve zurückzurechnen, die mit einem Strahl von punktförmigem Querschnitt erhalten würde. Dabei wird im allgemeinen vorausgesetzt, daß die Objekte isotrop sind, was für ruhende Lösungen stets zutrifft. Aber auch bei idealer oder nahezu idealer Faserstruktur kann eine solche Anordnung zur Vermessung der Streuung, besonders in Richtung des Äquators verwendet werden. Die Faserrichtung muß dann mit der Längsrichtung des Primärstrahlquerschnittes zusammenfallen, d. h. die zu studierende Äquatorrichtung muß sich normal dazu erstrecken. Aus Gründen, die im Rahmen dieser kurzen Darlegung nicht näher ausgeführt werden können, ist die so gemessene Streukurve allerdings mit der in der Lochkamera erhältlichen nicht identisch. Die letztere haben wir uns mit dem Streuwinkel 2ϑ durchmultipliziert zu denken, um zur ersteren zu kommen (zusammenfassende Darstellung bei Kratky, Porod u. Skala 1960).

Was nun die Anordnung der Blenden zur Eingrenzung des Primärstrahls betrifft, so wurde bis vor einigen Jahren allgemein eine Anordnung benutzt, wie sie Abb. 10 darstellt (erstmalig beschrieben von HOSEMANN 1939). Die beiden Schneidenpaare S_1 und S_2 verlaufen senkrecht zur Papierebene und geben einen sehr engen Spalt von der Größenordnung 0,1 mm frei. Um die an S_2 auftretende Kleinwinkelstreuung unschädlich zu machen, ist ein drittes Spaltpaar S_3 angebracht, das selbst nicht in den Primärstrahl eintaucht, die schädliche Sekundärstreuung aber auf einen möglichst engen Bereich einengt. Eine Messung des vom Präparat herrührenden Streueffektes im Bereich der von Blendenstreuung erfüllten Zonen Z ist im allgemeinen unmöglich, so daß neben dem Primärstrahl eine für die Messung ungenützte Zone besteht, deren Existenz die Auflösung erheblich herabsetzt.

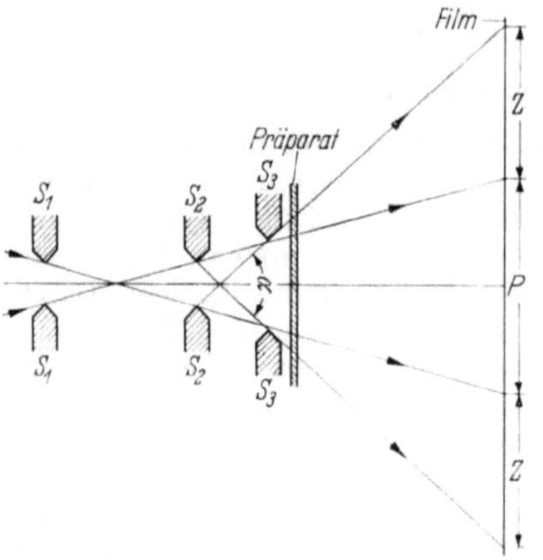

Abb. 10. Blendensystem für Kleinwinkelaufnahmen aus drei Spaltpaaren. Die vertikalen Dimensionen sind sehr stark verkleinert zu denken.

Diese Schwierigkeit wird bei einer Anordnung vermieden, die vom Referenten vor mehreren Jahren angegeben wurde (KRATKY 1954, 1955, 1958; KRATKY u. SKALA 1958). Die in Abb. 11 dargestellten Körper B_2 und B_3 sind Blöcke, deren horizontale Begrenzungsflächen senkrecht zur Zeichenebene verlaufen. Der Röntgenstrahl tritt durch den aus B_1 und B_2 gebildeten Spalt Sp ein und wird

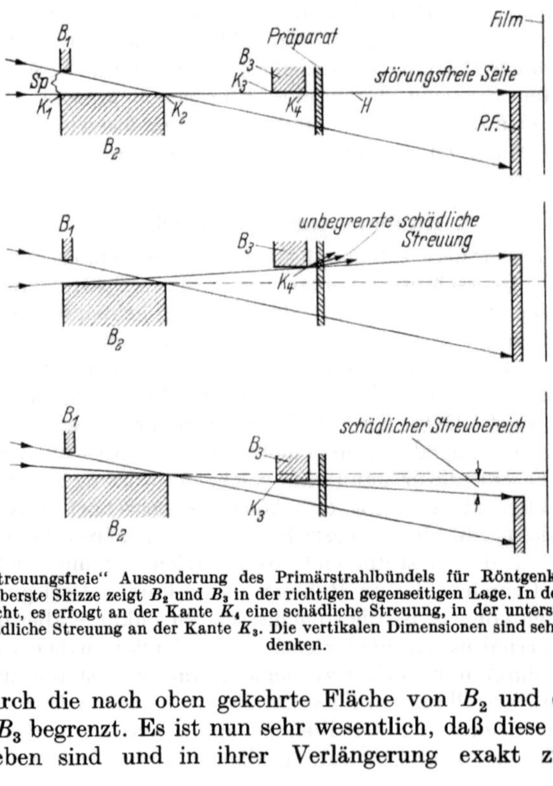

Abb. 11. „Blendenstreuungsfreie" Aussonderung des Primärstrahlbündels für Röntgenkleinwinkelaufnahmen nach KRATKY. Die oberste Skizze zeigt B_2 und B_3 in der richtigen gegenseitigen Lage. In der mittleren Skizze ist B_3 zu hoch angebracht, es erfolgt an der Kante K_4 eine schädliche Streuung, in der untersten Skizze sitzt B_3 zu tief, es entsteht schädliche Streuung an der Kante K_3. Die vertikalen Dimensionen sind sehr stark verkleinert zu denken.

im übrigen durch die nach oben gekehrte Fläche von B_2 und die nach unten gekehrte von B_3 begrenzt. Es ist nun sehr wesentlich, daß diese beiden Flächen vollkommen eben sind und in ihrer Verlängerung exakt zusammenfallen.

Man versteht dann leicht, daß oberhalb des „Hauptschnittes" H keine unerwünschte Streuung auftreten kann, also die Grenze der Streuung mit der Grenze des Primärstrahls zusammenfällt. Die Abbildung deutet weiter an, daß jedoch unerwünschte Streuung oberhalb des Hauptschnittes auftritt, sowie die Flächen von B_2 und B_3 nicht vollkommen zusammenfallen. Wie dieses Kollimationsprinzip realisiert wird, ist in Abb. 12 schematisch dargestellt. Über den

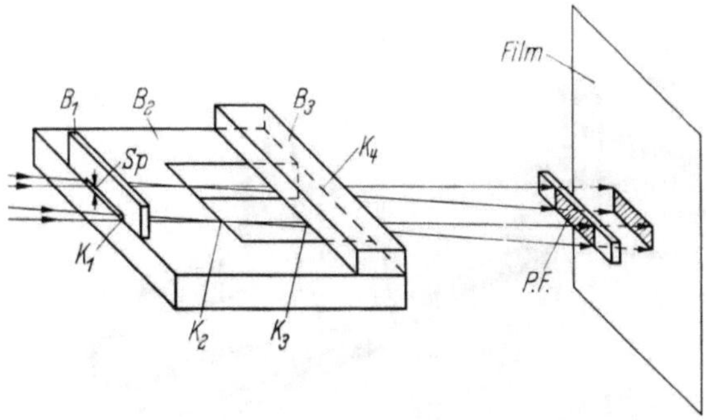

Abb. 12. Realisierung der „blendenstreuungsfreien" Blendenanordnung durch einen U-förmigen Körper mit einer über den Schenkeln liegenden „Brücke".

Schenkeln eines an der Oberfläche exakt geschliffenen U-förmigen Körpers, der dem Block B_2 von Abb. 11 entspricht, liegt eine Brücke, die den Block B_3 von Abb. 11 darstellt. Der beim Spalt *Sp* eintretende Röntgenstrahl geht über die Kante K_2 hinweg und tritt unter der Brücke B_3 durch. In Abb. 13 sehen wir schließlich, wie der U-Körper wirklich aussieht.

Abb. 13. Photo des blendenstreuungsfreien Blendensystems. Am linken Ende des prismatischen Körpers sitzt der mit Mikrometerschraube verstellbare Eintrittsspalt, am rechten Ende die Brücke. Vor dieser ist eine Platte sichtbar, die den Präparatträger darstellt. Der gegen den Hintergrund zu schräg liegende Körper ist der aufgeklappte Deckel, der die Längsbegrenzung des Strahls zu besorgen hat und bei der Aufnahme das Blendensystem nach oben abschließt (vgl. auch Abb. 14).

Die beste Oberflächenbearbeitung ist bei Stahl erzielbar und deshalb wurde dieses Material gewählt trotz der durch die Kupferstrahlung verursachten Sekundärstrahlung des Eisens. Das Prinzip funktioniert so gut, daß diese keinerlei Störung bewirkt. Die Grenze der Vermeßbarkeit liegt unter 1′, entsprechend einem Bragg'schen Wert von ca. 8000 Å.

Die Gesamtansicht einer nach diesem Prinzip arbeitenden Kamera gibt Abb. 14. Der Strahl durchsetzt das Blendensystem *K* und das Präparat *P*. Zwischen Präparat und Registriergerät ist ein evakuiertes Rohr *V* eingeschaltet, um Luftstreuung zu vermeiden. (Diese Anordnung ist einer kompletten Vakuumkamera vorzuziehen, weil man das Präparat besser unter Kontrolle hat und leichter auswechseln kann.) Will man photographisch arbeiten, so ist hinter dem Vakuumrohr

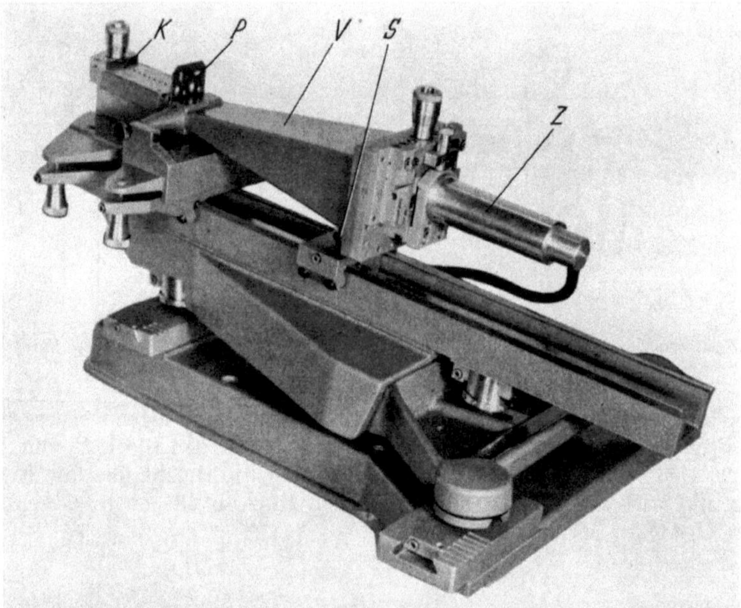

Abb. 14. Photo der gesamten Röntgenkleinwinkelkamera nach der Konstruktion von KRATKY und SKALA (1958).

der Film aufzustellen. Bei den modernen Impulszählungen wird auf das Vakuumrohr ein Zählrohr *Z* aufgesetzt, wie das die Abbildung zeigt. Mittels einer Präzisionsspindel *S* kann das Vakuumrohr durch den Streubereich hindurch bewegt werden. Die gestreute Intensität ist punktweise in genügend engen Intervallen zu messen.

II. Die Monochromatisierung der Strahlung.

Einige Worte sind über die Primärstrahlung zu sagen. In der Röntgenkleinwinkelforschung bedient man sich fast ausschließlich der CuK_α-Linie mit $\lambda = 1{\cdot}54$ Å. Die naheliegende Idee der Verwendung einer Strahlung mit wesentlich größerer Wellenlänge zur Erhöhung der Auflösung scheitert an der mit der dritten Potenz der Wellenlänge zunehmenden Absorption im Präparat und des dadurch bedingten untragbaren Intensitätsverlustes. Zur Monochromatisierung reicht die übliche Nickelfilterung nur in seltenen Fällen aus, allenfalls dann, wenn nur die Bestimmung der Orientierung beabsichtigt ist. In der Regel hat man aber zwischen folgenden Wegen zu wählen.

1. Monochromatisierung durch Kristallreflexion.

Sehr bewährt hat sich ein geschliffener und gebogener Kristall nach JOHANSSON (1933) und GUINIER (1946) für einen strichförmigen Primärstrahl. Vgl. auch JAGODZINSKI u. WOHLLEBEN (1960). Ein interessanter Weg zur Erzielung eines intensiven Primärstrahls mit punktförmigem Querschnitt ist ein neuerdings von Hagström und Siegbahn (1960) angegebenes Verfahren der Verwendung eines doppeltgebogenen Kristalls.

2. Verwendung des Rossschen Filterdifferenzverfahrens.

Bei dem Rossschen Filterdifferenzverfahren (Ross 1928, Hoffmann 1938, 1954) stellt man unter Verwendung eines Nickel- und eines Kobaltfilters, die in ihrer Dicke aufeinander genau abgepaßt sind, je eine Aufnahme her und subtrahiert die damit erhaltenen Streukurven voneinander. Die gebildete Differenz entspricht einer Streukurve, die auf die Wirkung des Spektralbereiches zwischen den Absorptionskonstanten von Nickel (1,483 Å) und Kobalt (1,604 Å) zurückgeht. Damit dominiert die in diesem Intervall gelegene CuK_α-Linie von 1·54 Å so stark, daß das Verfahren praktisch einer Monochromatisierung gleichkommt.

3. Aussonderung der gewünschten Strahlung durch einen Impulshöhendiskriminator in Verbindung mit einem Proportionalzählrohr.

Vgl. den nächsten Abschnitt.

III. Die Registrierung.

1. Photographische Messung.

Hier tritt die Schwierigkeit auf, daß Filme, die eine genügende Empfindlichkeit besitzen, notwendig verhältnismäßig grobkörnig sind. Eine Mehrfachphotometrierung (Kratky, Kahovec u. Werner 1959), d. h. die Übereinanderzeichnung mehrerer Registrierungen ergibt erheblich glattere Kurven, als sie bei einfacher Photometrierung zu erhalten sind. Dieses Verfahren ist allerdings nur bei strichförmigem Primärstrahl anwendbar. Da das photographische Verfahren eine raschere Gesamtübersicht des Streuverhaltens vermittelt als die Impulszählung, wird der photographische Film bei Kleinwinkelmessungen, trotz zunehmender Verbreitung der Impulszähltechnik, eine gewisse Bedeutung beibehalten. Die Orientierungsbestimmung wird bisher ausschließlich auf diesem Wege durchgeführt.

2. Impulszählung.

Wie oben erwähnt, wird das Zählrohr (mit vorgeschalteter schmaler Eintrittsblende) schrittweise durch den Streubereich hindurchbewegt und jedesmal die Streuintensität gemessen. Eine Direktanzeige kommt im allgemeinen wegen der zu geringen Intensität nicht in Betracht. Um den mittleren statistischen Fehler mit 1 % zu begrenzen, müssen an jeder Stelle mindestens 10000 Impulse gemessen werden. Von den Zählrohren hat sich bisher der Geiger-Müller-Zähler und das Proportionalzählrohr mit Impulshöhendiskriminator am besten bewährt. Im ersten Fall kann die Monochromatisierung durch Kristallreflexion oder — mit besserer Energieausbeute — durch Anwendung des Rossschen Filterdifferenzverfahrens erfolgen. Im zweiten Fall unterdrückt der Impulshöhendiskriminator die Wirkung aller Wellenlängen mit Ausnahme der eingestellten CuK_α-Linie und ihrer nächsten Umgebung. Man hat lediglich zur weiteren Schwächung der mit merklicher Intensität durchkommenden CuK_β-Linie ein Nickelfilter anzuwenden. Dieses Verfahren ist insofern das eleganteste, als man die volle Intensität der CuK_α-Linie, wie sie im primären Spektrum enthalten ist, ausnützen kann und — im Gegensatz zum Filterdifferenzverfahren — nur *eine* Messung durchzuführen hat. Es läßt sich allerdings nicht leugnen, daß die Meßanordnung komplizierter ist und daher eine größere Störanfälligkeit besteht als beim einfacheren GEIGER-MÜLLER-Zähler. Wir verzichten auf Einzelliteraturangaben und verweisen auf zusammenfassende Darstellungen (FÜNFER u. NEUERT 1959, PARRISH u. KOHLER 1956).

C. Anwendungsbeispiele.

I. Homodisperse Proteinlösungen.

Für das bereits oben erwähnte Pflanzenglobulin Edestin lieferte die Röntgenkleinwinkelanalyse (CLEEMANN u. KRATKY 1960) die Gestalt eines von der Kugel wenig abweichenden Teilchens. Als die wahrscheinlichsten Werte für die Achsen ergab sich $a = 48 \cdot 5$ Å, $b = 48 \cdot 5$ Å, $c = 87$ Å. Das Volumen von 860000 Å³ ist etwa doppelt so groß wie man nach dem Ultrazentrifugen-Molekulargewicht und dem spezifischen Volumen erwarten würde. Daraus ist zu schließen, daß das Molekül eine erhebliche innere Solvatation zeigt. Daß es sich bei den röntgenographisch untersuchten Teilchen nicht um Agglomerate handelt — es ist immer an die Möglichkeit zu denken, daß die Röntgenbestrahlung derartige Effekte bewirkt —, konnte dadurch erwiesen werden, daß aus der *Absolutintensität*, d. h. dem Verhältnis der gestreuten Energie zur Primärenergie, das Molekulargewicht berechnet wurde. Es liegt mit $M = 324000$ durchaus im Bereich der mit anderen Methoden bestimmten Werte.

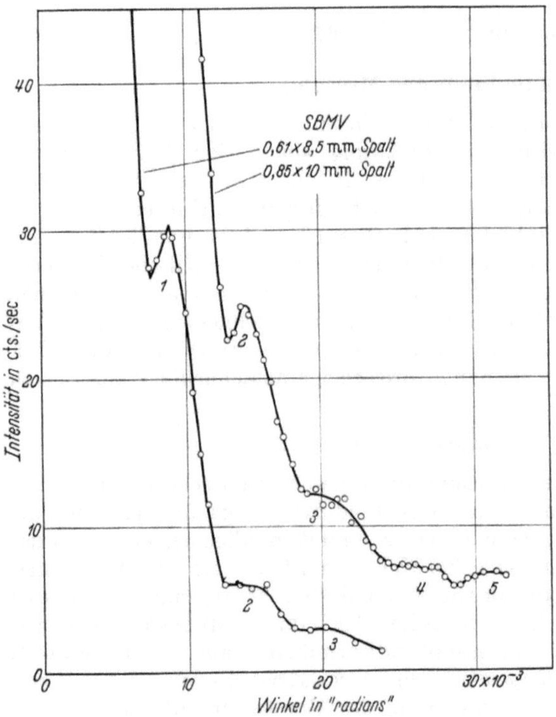

Abb. 15. Streukurven des südlichen Bohnenmosaikvirus (LEONHARD, ANDEREGG, SHULMAN, KAESBERG und BEEMAN 1953).

II. Homodisperse Pflanzenviren.

Ein interessanter Anwendungsbereich der Röntgenkleinwinkelanalyse ist die Untersuchung von Pflanzenviren. Abb. 15 stellt den äußeren Teil der Streukurve von südlichem Bohnenmosaikvirus dar (LEONHARD, ANDEREGG, SHULMAN, KAESBERG u. BEEMAN 1953). Die Interpretation war die eines nahezu *kugelförmigen* Teilchens vom Durchmesser 286 Å und die auf der Abbildung ersichtlichen Nebenmaxima bestätigen dieses Ergebnis aufs beste. Das Molekül ist allerdings wesentlich größer als nach dem Molekulargewicht zu erwarten war, so daß eine innere Hydratation angenommen werden mußte. Ähnliche Ergebnisse konnten beim Tabaknecrosisvirus und beim Tomato-bushy-stunt-Virus erhalten werden. Auf eine weitere interessante Arbeit am Gelben-Rüben-Mosaik-Virus sei verwiesen (SCHMIDT, KAESBERG u. BEEMAN 1954), die zur Vorstellung eines Teilchens von der Gestalt einer Kugelschale geführt hat, die mit Quellungswasser gefüllt ist. Der äußere Radius beträgt 140 Å, das Verhältnis des inneren zum äußeren Radius ist 0,75.

Eine wesentlich andere, nämlich *stäbchenförmige* Gestalt hat bekanntlich das Teilchen des Tabak-Mosaik-Virus. Es mußte danach mittels der Röntgenkleinwinkelstreuung möglich sein, eine Analyse des Querschnittes durchzuführen (KRATKY, PALETTA, POROD u. STROHMAIER 1957). Die Auftragung von $\ln I\vartheta$ gegen ϑ^2 gibt Abb. 16. Die Kurven zeigten allerdings nicht den für unendlich lange

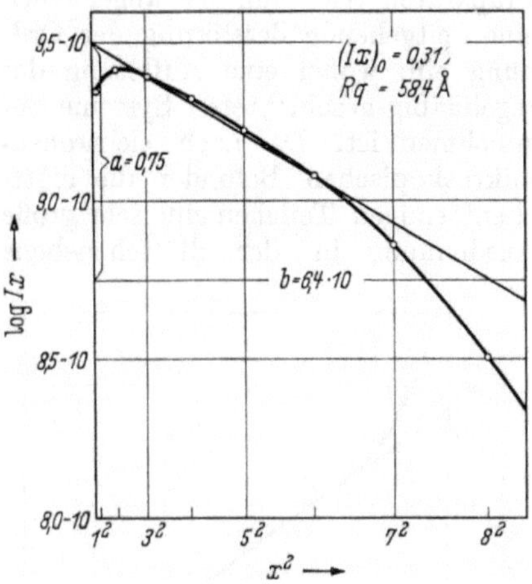

Abb. 16. Querschnittsfaktor von Tabakmosaikvirus in GUINIERscher Darstellung (KRATKY, PALETTA, POROD und STROHMAIER 1957).

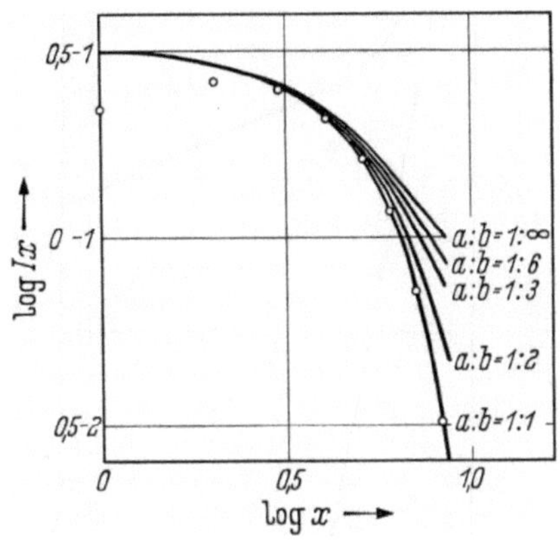

Abb. 17. Querschnittsfaktor von Tabakmosaikvirus in doppeltlogarithmischer Darstellung (KRATKY, PALETTA, POROD und STROHMAIER 1957).

Stäbchen charakteristischen geradlinigen Verlauf im innersten Teil, eben weil sie endliche Länge besitzen. Aus dieser Abweichung ist grundsätzlich eine Berechnung der Stäbchenlänge möglich. Wir wollen uns aber mit der Feststellung begnügen, daß die Tangente im Mittelteil auf einen Streumassenradius des Querschnittes von 58,4 Å führte. Die genaue Gestaltsanalyse der Kurve wird am besten in doppeltlogarithmischer Auftragung vorgenommen (Abb. 17), wo die ausgezogenen Kurven den theoretischen Erwartungen für Querschnitte von verschiedenen Achsenverhältnissen entsprechen. Die aus dem Experiment entnommenen eingezeichneten Punkte passen genau auf den kreisförmigen Querschnitt. Nur im inneren Teil tritt wieder die schon erwähnte auf endliche Länge zurückzuführende Abweichung auf.

III. Dicke der Proteidlamellen von isolierten Chloroplasten.

Der lamellare Bau der Chloroplasten konnte bekanntlich mittels des Elektronenmikroskopes direkt sichtbar gemacht werden (MENKE 1941, FREY-WYSSLING u. MÜHLETHALER 1949, FREY-WYSSLING 1956, WOLKEN u. PALADE 1952, COHN u. BOWLER 1953, LEYON 1953). Die Dickenmessungen der Einzellamellen auf diesem Wege sind aber aus verschiedenen Gründen unsicher, so daß es naheliegt, die Röntgenkleinwinkelmethode zum genaueren Studium heranzuziehen. Zunächst sollte die Dicke der Proteidlamellen an einem lipidfreien, gefriergetrockneten Präparat studiert werden. Zweifellos stellen solche Präparate Haufwerke von Einzellamellen und deren Aggregaten zu Lamellenpaketen dar. Das System ist allerdings keinesweg wirklich „verdünnt“, so daß die Frage auftritt, ob es danach berechtigt sei, die gewonnenen Bilder nach den Prinzipien der Partikelstreuung auszuwerten — noch dazu wo möglicherweise weitgehend Einzellamellen vorliegen und daher auch keine ausreichende Polydispersität besteht. Nun ist aber ganz allgemein gezeigt worden, daß bei extrem anisotropen Partikeln, also bei langen Stäbchen und noch mehr bei großen Blättchen der die Partikelstreuung modifizierende Interferenzeffekt praktisch vernachlässigbar ist, wenn benachbarte Partikel auch nur eine kleine Abweichung von der Parallelität aufweisen (Kratky u. Porod 1956). Bei gefriergetrockneten Präparaten tritt nun im allgemeinen eine weitgehende Zerstörung der Ordnung ein, wobei eine Auflösung der regelmäßig geschichteten Systeme anzunehmen ist. Da nach elektronenmikroskopischen Befunden die blättchenförmigen Teilchen eine sehr große Ausdehnung in der Blättchenebene

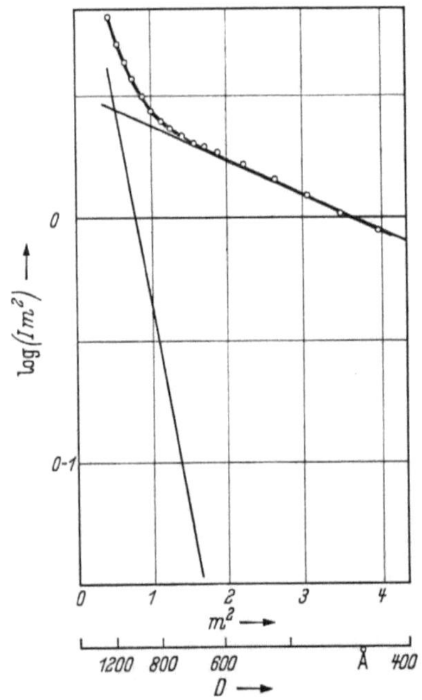

Abb. 18. Dickenfaktor gefriergetrockneter Chloroplasten von Allium porrum in Guinierscher Darstellung (Kratky, Menke, Sekora, Paletta und Bischof 1959).

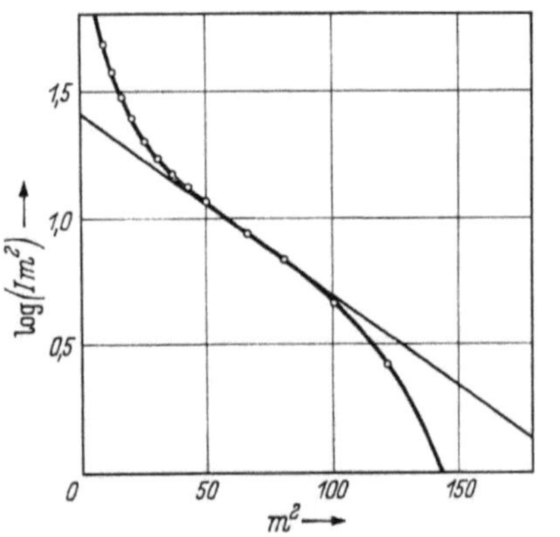

Abb. 19. Dickenfaktor lipidfreier Chloroplasten von Antirrhinum majus (Sippe 50), in Guinierscher Darstellung (Kreutz und Menke 1960).

im Vergleich mit der Dicke haben müssen, war es von vornherein gegeben, den Blättchenfaktor $I\vartheta^2$ darzustellen. Die Guiniersche Auftragung lieferte die geknickte Kurve in Abb. 18, aus der die Anwesenheit von mindestens zwei Teilchendimensionen erkennbar ist (Kratky, Menke, Sekora, Paletta u. Bischof 1959). Die berechneten Dicken lagen allerdings bei 250 Å (für den äußeren Teil) und dem Drei- bis Vierfachen (für den inneren Teil). Da aber an anderen geordneten Präparaten Braggsche Reflexe auftraten, die nur einen Bruchteil des kleineren Wertes ausmachten (63 und 48 Å), lag die Vermutung nahe, daß es sich auch bei den 250 Å-Teilchen nicht um Einzellamellen, sondern bereits um Pakete handle. Kreutz und Menke (1960) haben in Weiterführung dieser Arbeiten nach einem verbesserten Herstellungsverfahren lipidfrei gemachte Präparate — und zwar von *Antirrhinum majus*, Sippe 50 — gewonnen, mit dem Ziel, eine möglichst weitgehende Aufteilung der Proteidlamellen herbeizuführen. Abb. 19 zeigt die Guiniersche Auftragung des Dickenfaktors einer von einem solchen Präparat erhaltenen Streukurve. Sie ist eine typische Mischstreukurve und man kommt zwanglos zur Deutung, daß der mittlere und äußere Teil den einfachen

Teilchen zuzuordnen sei. Aus der Neigung der eingezeichneten Wendetangente erhält man die Lamellendicke zu 50—51 Å. Allerdings war man wegen der Überlagerung mit dem Streueffekt dickerer Teilchen nicht in der Lage, die Tangente im innersten Teil anzulegen, was einen gewissen Fehler verursacht. Aus der zu erwartenden Form des Dickenfaktors für Einzelteilchen ist aber leicht eine diesbezügliche Korrektur durchführbar, die dann eine Dicke von etwa 48 Å ergibt. In ähnlicher Weise wird der Innenteil auf Lamellen von ziemlich genau der doppelten Dicke, nämlich von 99—100 Å zurückgeführt, was die Existenz von Zweier-Aggregaten beweist. Aus einem noch weiter innen gelegenen Bereich finden KREUTZ und MENKE (1960) eine Dicke von 240 Å, was dem kleineren der schon früher gefundenen Werte entspricht, womit der experimentelle Zusammenhang zwischen den älteren und den späteren Messungen hergestellt ist. Allerdings hat man diesen Wert aus ähnlichen Gründen, wie sie bei der Einzellamelle angedeutet wurden, zu korrigieren und man erhält schließlich etwa 200 Å, was einer Viereraggregation entsprechen könnte.

Die Untersuchung von lipidhaltigen, gefriergetrockneten Chloroplasten durch die gleichen Autoren (KREUTZ u. MENKE 1960) ergab überraschenderweise bis etwa 50 Å herunter die gleiche Streukurve wie sie bei lipidfreien Chloroplasten erhalten wurde und es ergibt sich daher die gleiche Dicke für die Einzellamellen. Der Wasserentzug bewirkt also offenbar bereits eine Trennung des Proteids und der Lipidkomponente.

Angetrocknete Schichten aus isolierten Chloroplasten geben eine Reihe von Reflexen, die sämtlich auf Lipide zurückgeführt werden können, mit Ausnahme der beiden typischen Proteinreflexe bei 8—10 bzw. 4,3—5 Å.

IV. Bestimmung der Micelldicke in „luftgequollener“ Cellulose.

Ein wichtiges Anwendungsgebiet der Kleinwinkelforschung ist das Studium der Pflanzenfasern. Am besten untersucht ist die Cellulose, vielfach aber in der Form der aus Lösungen regenerierten Fasern. Da die Ergebnisse einerseits von großem methodischen Interesse sind und sie andererseits gewisse Rückschlüsse auf den Bau der nativen Fasern zulassen, wollen wir die regenerierte Cellulose in unsere Darlegungen einbeziehen.

Bei der *trockenen* Cellulose liegt zweifellos die Vorstellung nahe, daß die übermolekularen Aggregate (Micellen, kristalline Bereiche) zufolge zu dichter Packung ihren Charakter als selbständige Partikel für die Röntgenkleinwinkelstreuung verloren haben; im Sinne des Reziprozitätsgesetzes der Optik werden daher nur Hohlräume erfaßt, von denen nicht ohne weiteres behauptet werden kann, daß sie in einer klaren quantitativen Beziehung zu den Dimensionen der Partikeln stehen. Man wird daher bestrebt sein müssen, durch Auflockerungen irgendwelcher Art die Teilchen in ausreichendem Maße voneinander zu entfernen. Dieser Vorbereitung der Präparate für die Analyse kommt also eine wesentliche Bedeutung zu, in mancher Hinsicht eine ähnliche wie den Maßnahmen vor der elektronenmikroskopischen Untersuchung der Fasern, wo das Studium eines nicht entsprechend zerteilten Objektes praktisch wertlos wäre.

Als aufschlußreich hat sich das Studium der sog. *„luftgequollenen“ regenerierten* Cellulose erwiesen. Wird in einem in Wasser hochgequollenen Faden von regenerierter Cellulose das Wasser durch Alkohol, dieser durch das lyophobe Benzol ersetzt und dann im Vakuum getrocknet, so geht der Zusammenhalt der Micellen, wie HERMANS u. PATZEK (1939) gezeigt haben, in ähnlicher Weise verloren wie der Zusammenhalt der Makromoleküle bei der Gefriertrocknung biologischer

Präparate und man kann einen unorientierten Faden von einem Luftquellungsgrad in der Gegend von 500% erhalten. Da aus vielfachen Gründen, auf welche

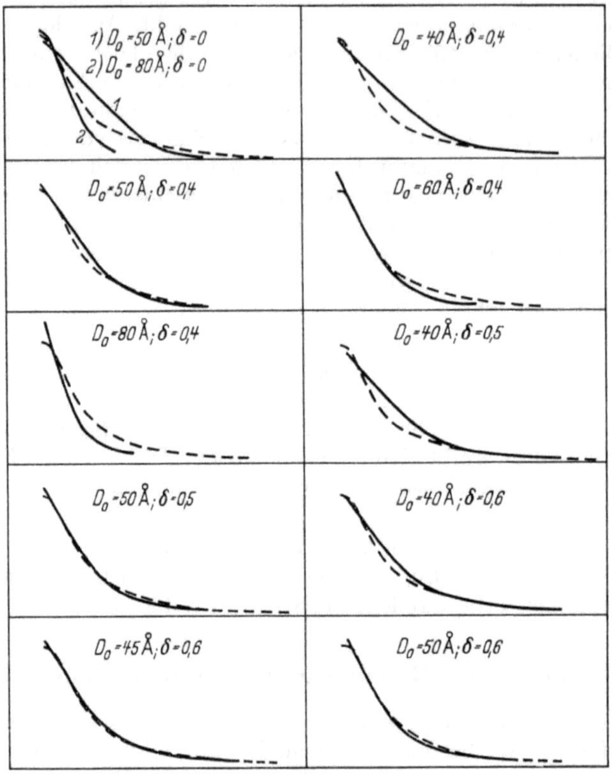

Abb. 20. Dickenfaktor von hochluftgequollener regenerierter Cellulose, verglichen mit theoretischen Kurven für die angegebenen mittleren Dicken D_0 und mittleren Streuungen δ (Janeschitz-Kriegl, Kratky und Porod 1952).

hier nicht eingegangen werden kann, eine bändchenförmige Gestalt der übermolekularen Teilchen anzunehmen war, erschien es sinnvoll, bei der Röntgenkleinwinkelanalyse gleich den „Dickenfaktor"-darzustellen (Janeschitz-Kriegl, Kratky u. Porod 1952). Wie Abb. 20 zeigt, führt er bei der Auswertung nach den Prinzipien der Partikelstreuung auf eine Dicke von etwa 45 Å. Die Nichtberücksichtigung eventueller Interferenzeffekte ist dadurch gerechtfertigt, daß einerseits immerhin eine ziemliche „Verdünnung" vorliegt und andererseits die ausgeprägte Anisotropie in Breite und Länge die Interferenzeffekte sehr weitgehend beseitigt (Kratky u. Porod 1956).

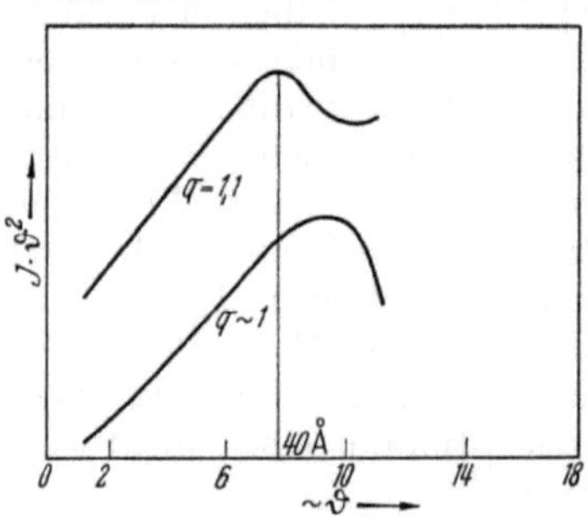

Abb. 21. Gitterfaktor von Luftfäden regenerierter Cellulose niedrigen Quellungsgrades (Janeschitz-Kriegl, Porod und Kratky 1952).

Durch etwas andere Führung der Präparation kann man auch sehr dichte Präparate erhalten (Quellungsgrad 1,1), bei welchen die lamellenförmigen Teilchen aus Gründen der räumlichen Enge nahezu parallel gelagert sein müssen. Es ist dann berechtigt, das Vorliegen von Lamellenpaketen im Sinne von Abschnitt A/2 anzunehmen, d. h. die Multiplikation der Streukurve mit ϑ^2 führt auf den Gitterfaktor. Solche Experimente ergaben Abb. 21. Durch Vergleich mit Abb. 7 schließt man auf eine mittlere Dickenschwankung von etwa 0·4 und aus der Lage des Maximas errechnet man die mittlere Micelldicke zu 45 Å. Somit besteht ausreichende Übereinstimmung mit den aus dem verdünnten System gewonnenen Werten.

V. Bestimmung der Micelldicke in wassergequollener Cellulose.

Im Gegensatz zu den aus lyophobem Milieu getrockneten Präparaten bleibt auch bei ziemlich hoher Wasserquellung der Cellulose deutlich eine Parallellagerung benachbarter Micellen bestehen. Wenn die Micelldicken wenig schwanken,

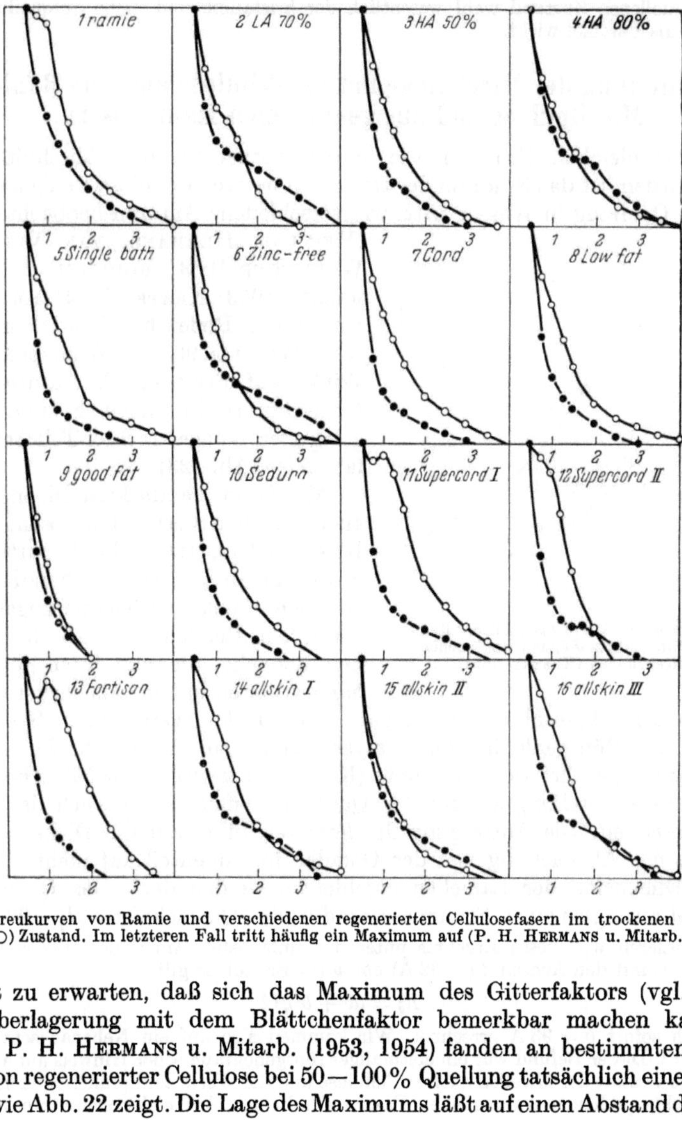

Abb. 22. Streukurven von Ramie und verschiedenen regenerierten Cellulosefasern im trockenen (●) und gequollenen (○) Zustand. Im letzteren Fall tritt häufig ein Maximum auf (P. H. HERMANS u. Mitarb. 1953, 1954).

dann ist zu erwarten, daß sich das Maximum des Gitterfaktors (vgl. Abb. 7) trotz Überlagerung mit dem Blättchenfaktor bemerkbar machen kann (vgl. Abb. 8). P. H. HERMANS u. Mitarb. (1953, 1954) fanden an bestimmten Spezialfasern von regenerierter Cellulose bei 50—100% Quellung tatsächlich einen solchen Effekt, wie Abb. 22 zeigt. Die Lage des Maximums läßt auf einen Abstand der Mitten benachbarter Micellen von 50—85 Å schließen. Er stellt also nicht die Dicke der Micelle dar, sondern entspricht der Summe von Micelldicke und Dicke der Flüssigkeitsschicht. Stellt man den Quellungsgrad in Rechnung, so kommt man annähernd auf den gleichen Wert, der sich aus den luftgequollenen Präparaten ergeben hatte. Das Auftreten des Maximums dokumentiert aber eine noch kleinere Schwankung der Micelldicken (etwa 0,2 vom Mittelwert), als sie aus der diffusen Streuung der luftgequollenen Präparate erschlossen wurde.

Es sei schließlich auf die Untersuchungen an regenerierten Fasern hingewiesen, die einige hundert Prozent gequollen waren (KRATKY u. SEMBACH 1955). Die diffuse Streuung wurde nach den Prinzipien der Partikelstreuung ausgewertet und ergab einen Querschnitt lamellenförmiger Teilchen von etwa 30mal 100 Å. Daß dieser Wert etwas kleiner ist als der an luftgequollenen Fasern erhaltene, dürfte seinen Grund darin haben, daß im letzteren Fall der amorphe Anteil (Rindenschicht) offenbar mit zur Größe der Partikeln beiträgt, während im hochgequollenen Zustand wohl wesentlich der kristalline und nicht aufquellbare Kern der Micelle als Partikel wirkt.

VI. Bestimmung der Micelldicke nativer Cellulose aus dem Rånbyschen Micellpulver und aus gequollenen Ramiefasern.

Aus den gleichen Gründen, die in den vorhergehenden Abschnitten ausgeführt wurden, ist das Studium der trockenen nativen Cellulose wenig aufschlußreich. Bei Quellung in Wasser tritt, wie verschiedene Autoren beobachtet haben (HEIKENS, HERMANS, VAN VELDEN u. WEIDINGER 1953; FOURNET u. ANTZENBERGER 1953; KRATKY 1954), am Äquator ein breiter Reflex bei einem dem BRAGGschen Wert von 90—100 Å entsprechendem Winkel auf, der wohl als Ausdruck eines Gitterfaktors in einem System parallel gelagerter langgestreckter Teilchen aufzufassen ist (Abb. 23).

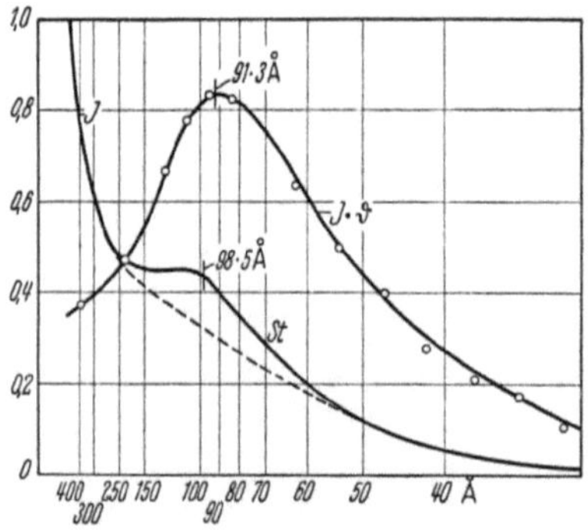

Abb. 23. Äquatorstreukurve J gequollener Ramiefaser und Gitterfaktor $J\,\vartheta$, mit Spaltblende aufgenommen (KRATKY 1954).

Man kann aber die kristallinen Bereiche auch aus dem Verband loslösen, wie das RÅNBY (1949, 1950) durch partielle Hydrolyse getan hat. Diese Behandlung führt zu einem Abbau der leichter zugänglichen amorphen Bereiche, so daß die kristallinen Bereiche, die nun ihren Zusammenhang verloren haben, als lockeres Pulver vorliegen. Die Aufschlämmung ergab ein ideales verdünntes System, an welchem die Röntgenkleinwinkelanalyse nach den Prinzipien der Partikelstreuung durchgeführt werden konnte (KRATKY u. SEMBACH 1955). Abb. 24 läßt die GUINIER-Darstellung sowohl des Querschnittsfaktors als auch des Dickenfaktors erkennen. Die Auftragung des letzteren führt auf die Dicke von 32 Å, wobei aus der Abweichung von der Geraden im Innenteil auf nicht allzugroße seitliche Dimension der Lamellen geschlossen werden darf. Der Querschnittsfaktor gibt zwanglos einen Streumassenradius des Querschnittes $R_q = 28$ Å.

Interpretieren wir dieses Ergebnis unter der einfachsten Annahme eines rechteckigen Querschnittes mit den Achsen d ($= 32$ Å) und b ($=$ Breite), so gilt

$$R_q^2 = (d^2 + b^2)/12$$

woraus sich sofort $b = 93$ Å errechnet. Wir kommen also auf ein Nebenachsenverhältnis von etwa 1:3. Dieser Befund war ein erstmaliger röntgenographischer Hinweis auf eine bändchenförmige Gestalt der Micellen nativer Cellulose und die Werte sind praktisch identisch mit den an hochgequollenen regenerierten Fasern erhaltenen. Auf elektronenmikroskopischem

Wege waren MUKHERJEE, SIKORSKI und WOODS (1951, 1952) zu einem qualitativ gleichen Ergebnis an einem durch Abbau mit starker Säure erhaltenen Micellpulver gelangt. Die Dicke ihrer Teilchen stimmt mit 30 Å quantitativ mit unserem Wert überein. In letzter Zeit ist allerdings K. MÜHLETHALER (1960) zu einem etwas anderen Ergebnis gelangt. Elektronenmikroskopisch mißt er Elementarfibrillen von isotropem Querschnitt von 35 Å Durchmesser (welcher Wert wieder der röntgenographischen Dicke entspricht). Sie können sich nach seinen Feststellungen allerdings zu Mikrofibrillen von 100—300 Å Dicke zusammenlagern. Die beiden Befunde, unser röntgenographischer und der elektronenmikroskopische von MÜHLETHALER, können wohl zwanglos durch die Annahme zur Übereinstimmung gebracht werden, daß im RÅNBYschen Micellpulver im Durchschnitt einige der — nach MÜHLETHALER (1960) — im

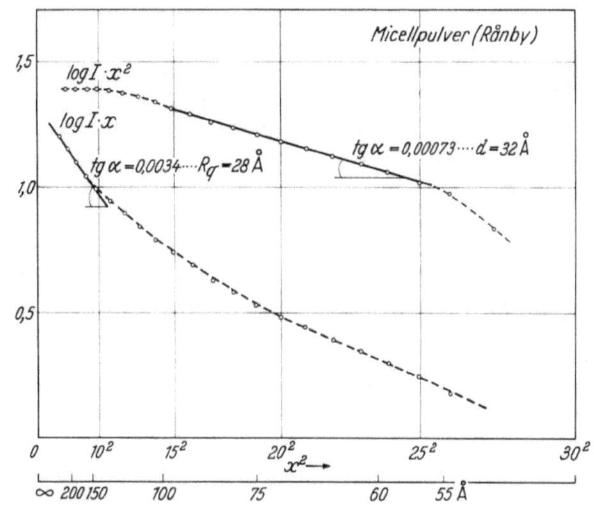

Abb. 24. Querschnitts- und Dickenfaktor von RÅNBYschem Micellpulver in GUINIERscher Darstellung (KRATKY und SEMBACH 1955).

Querschnitt isotropen Elementarfibrillen seitlich anneliert sind. Mit der Breite der mit der Röntgenkleinwinkelmethode im RÅNBYschen Pulver festgestellten Bändchen von etwa 93 Å stimmt übrigens auch die Lage des oben erwähnten Reflexes an schwach gequollener Ramie (90—100 Å) überein, ein Wert, für den das elektronenmikroskopische Ergebnis MÜHLETHALERs keine unmittelbare Evidenz bietet.

Wenn somit die Untersuchungen über die seitlichen Dimensionen der Micellen von nativer und regenerierter Cellulose untereinander einigermaßen übereinstimmende Resultate gegeben und zu einer gewissen Klärung geführt haben, liegt hinsichtlich der Ermittlung der Länge noch sehr wenig Material vor. Die Verwertung der diffusen Streuung in Richtung des Meridians war bisher überhaupt noch nicht in Angriff genommen worden. Nun sind allerdings am Meridian bei gequollenen und in verschiedener Weise vorbehandelten regenerierten Fasern schwache Reflexe bei BRAGGschen Werten zwischen 110 und 220 Å gefunden worden (KIESSIG 1957, STATTON 1956). Da die aus Linienbreitemessungen errechneten Längen mehrfach größer sind, kann es sich bei den Kleinwinkelreflexen nicht um den Ausdruck einer durch die regelmäßige Anordnung von kristallinen Bereichen bedingte Periodizität handeln. Eine überzeugende Deutung der Effekte konnte bisher noch nicht gegeben werden, weshalb wir von einer genaueren Besprechung absehen. An nativer Cellulose sind bisher Effekte dieser Art nicht aufgefunden worden.

VII. Gehaltsbestimmung der Micellen aus der Anisotropie der Kleinwinkelstreuung bei Cellulosefasern.

Durchleuchtet man ein Bündel paralleler Cellulosefasern mit einem dünnen Röntgenstrahlbündel von kreisförmigem Querschnitt, so erhält man eine Kleinwinkelstreuung, die in Richtung des Äquators (d. h. normal zur Längsrichtung der Fasern) zu wesentlich größeren Winkeln ausgedehnt ist, also in der Richtung des Meridians (d. h. in Richtung der Faserachse). Wenngleich in einem so dichtgepackten System nicht die reine Partikelstreuung der Einzelmicelle zum Ausdruck kommt, so kann qualitativ im Sinne des in A/III Gesagten geschlossen werden, daß die Dimension der Micellen — umgekehrt wie die Größe der Streubereiche — in Richtung der Faserachse wesentlich größer ist als quer dazu.

Manche gewachsenen Fasern und Membranen besitzen eine sog. höhere Orientierung, d. h. die offenbar bändchenförmigen Micellen sind nach zwei Achsen ausgerichtet. In manchen Fällen läßt sich eine solche Ausrichtung durch mechanische Einwirkung — Walzung — erzwingen. Kann man eine Faserstruktur mit einem Bündel von Bleistiften vergleichen, deren „Aufschriften" nach beliebigen Richtungen gekehrt sind, so haben wir uns bei höherer Orientierung die Aufschriften alle nach der gleichen Richtung gedreht zu denken.

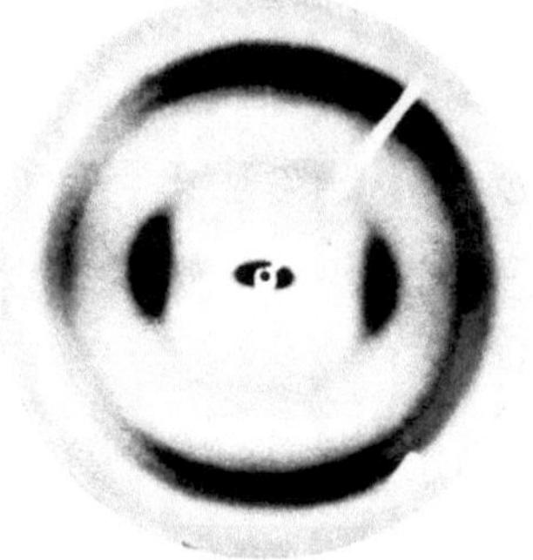

Abb. 25. Röntgenbild, das bei Durchleuchtung höher orientierter regenerierter Cellulose in Faserrichtung erhalten wird. Die eiförmige Gestalt der Kleinwinkelstreuung bildet reziprok die Anisotropie der seitlichen Dimension der Micellen ab (KRATKY, SEKORA und TREER 1942).

Durchleuchtet man ein solches Objekt in der Faserrichtung, also in Richtung der längsten Dimension der Teilchen, so macht sich eine Anisotropie der Kleinwinkelstreuung bemerkbar, welche vom Unterschied der seitlichen Dimensionen der Teilchen — Breite und Dicke — herrührt, wobei natürlich wieder der kleineren Dimension der größere Streubereich entspricht und umgekehrt. Abb. 25 zeigt als Beispiel die Aufnahme einer gewalzten, allerdings regenerierten Cellulose, die in der Walzrichtung durchleuchtet wurde (KRATKY, SEKORA u. TREER 1942). Aus dem Verhältnis der Streubereiche im Kleinwinkelgebiet in horizontaler und vertikaler Richtung kann auf ein Verhältnis der Dicke zur Breite von etwa 1 : 3 geschlossen werden, etwa das gleiche Verhältnis, das aus der Kleinwinkelstreuung hochgequollener Fasern errechnet wurde.

VIII. Orientierungsbestimmung aus der Anisotropie der Kleinwinkelstreuung.

Die soeben besprochene Anisotropie der Kleinwinkelstreuung kann, die Kenntnis der ungefähren Gestalt der Micellen vorausgesetzt, auch als Verfahren zur Orientierungsbestimmung angesehen werden. Der höchste Grad von Anisotropie wird natürlich bei idealer Ordnung erreicht. Denken wir uns, von diesem Grenzfall ausgehend, die Ordnung immer mehr verschlechtert — z. B. im Falle einer Faserstruktur die Ausrichtung der Längsachsen immer unvollkommener gemacht, — so geht die anisotrope Ausbildung der Äquatorinterferenz mehr und mehr in eine isotrope d. h. rotationssymmetrische Streuung über. Die Weitwinkelinterferenzen des Äquators werden gleichzeitig zu immer längeren Sicheln, so daß beide Streueffekte, der des Kleinwinkel- und des Weitwinkelbereiches im Prinzip die gleiche Aussage liefern. Es gibt aber Fälle, wo das Kleinwinkelbild eine Information gibt, die aus der Weitwinkelstreuung nicht entnommen werden kann. Das überzeugendste Beispiel dieser Art ist die Spiralfaserstruktur. Darunter versteht man eine Anordnung, bei welcher die Micellängsachsen einen bestimmten Winkel mit der Längsachse der Fasern einschließen. Während das Weitwinkelbild in diesem Fall eine ganz uncharakteristische Verlängerung der Interferenzsicheln zeigt, wie sie auch bei statistischer Streuung der Richtungen um die Faserachse auftritt, läßt die Kleinwinkelstreuung direkt den Spiralwinkel erkennen, wie dies an den von HEYN (1948, 1949) stammenden Aufnahmen in Abb. 26 überzeugend zum Ausdruck kommt. Die vollständige Darlegung des Zusammenhanges zwischen Orientierung und Anisotropie der Kleinwinkeldiagramme würde den Rahmen der vorliegenden Darstellung weit übersteigen, so daß wir uns mit diesen Bemerkungen begnügen und im übrigen auf eine ausführlichere Darstellung verweisen (KRATKY 1955).

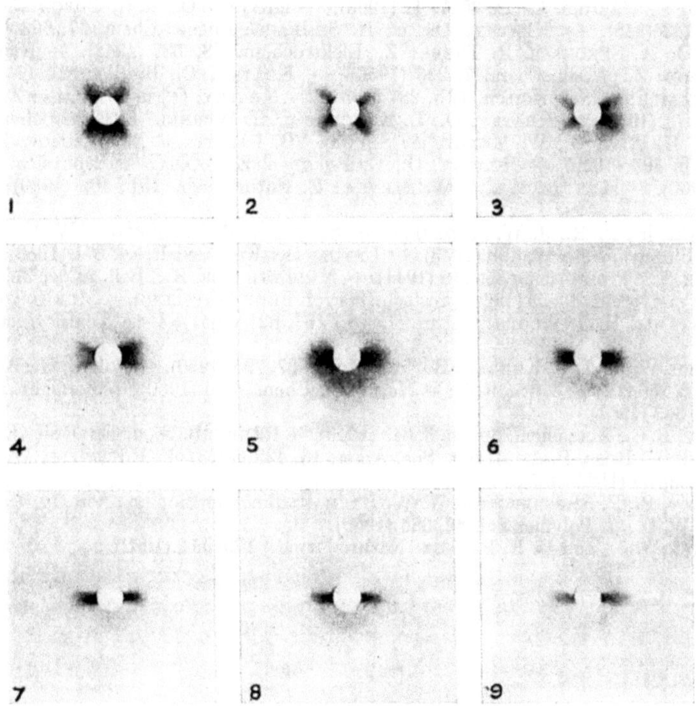

Abb. 26. Kleinwinkeldiagramme von nativen Cellulosefasern (1. Cocos nucifera, 2. Agave lophanta. 3. Agave heteracantha, 4. Agave fourcroydes, 5. Sansevieria guinensis, 6. Agave sisalana. 7. Yucca, 8. Musa textilis, 9. Ananas comosus) mit verschiedenen Spiralwinkeln in der Anordnung der Micellen (HEYN 1948, 1949).

Literatur.

BEEMAN, W. W., P. KAESBERG, J. W. ANDEREGG and M. B. WEBB: Size of particles and lattice defects; in Handbuch der Physik, Band XXXII, herausgeg. von S. FLÜGGE. Berlin-Göttingen-Heidelberg: Springer-Verlag 1957.

CLEEMANN, J. C., u. O. KRATKY: Z. Naturforsch. **15**b, 525 (1960). — COHN, M., u. E. BOWLER: Protoplasma **42**, 414 (1953)

FOURNET, G., et P. ANTZENBERGER: Compt. rend. **236**, 394 (1953). — FREY-WYSSLING, A.: Suomen Kemistilethi A **29**, 147 (1956). — FREY-WYSSLING, A., u. K. MÜHLETHALER: Vjschr. naturforsch. Ges. Zürich **94**, 179 (1949). — FÜNFER, E., u. H. NEUERT: Zählrohre und Szintillationszähler, Karlsruhe: Verlag G. Braun 1959.

GEROLD, V.: Z. angew. Physik **9**, **43** (1957). — GUINIER, A.: Compt. rend. **223**, 31 (1946). — GUINIER, A., and G. FOURNET: Small angle scattering of X-rays. New York: J. Wiley & Sons, Inc. 1955; London: Chapman & Hall, Ltd. 1955.

HAGSTRÖM, S., u. K. SIEGBAHN: J. Ultrastructure Res. **3**, 401 (1960). — HEIKENS, D., P. H. HERMANS, P. F. VAN VELDEN u. A. WEIDINGER: J. Polymer Sci. **11**, 433 (1953). — HERMANS, P. H.,u. P. PLATZEK: Z. physik. Chem. (A) **185**, 260 (1939). — HERMANS, P. H., u. A. WEIDINGER: J. Polymer Sci. **14**, 397 (1954); — Makromol. Chem. **13**, 30 (1954). — HEYN, A. N. J.: Textile Res. **19**, 163 (1949); — J. Amer. chem. Soc. **70**, 3138 (1948). —

Hoffmann, K.: Physik. Z. **39**, 695 (1938); — Angew. Chem. **66**, 711 (1954). — Hosemann, R.: Z. Physik **113**, 751; **114**, 133 (1939).

Jagodzinski, H., u. K. Wohlleben: Z. Elektrochem. **64**, 212 (1960). — Janeschitz-Kriegl, H., O. Kratky u. G. Porod: Z. Elektrochem. **56**, 146 (1952). — Johansson, T.: Z. Physik **82**, 587 (1933).

Kiessig, H.: Kolloid-Z. **98**, 213 (1942); **152**, 62 (1957). — Kratky, O.: Svensk Papperstidning **9**, 1 (1954); — Z. Elektrochem. **58**, 49 (1954); — Kolloid-Z. **144**, 110 (1955); — Die Physik der Hochpolymeren. Herausgeg. von H. A. Stuart, III. Band, S. 305ff. Berlin-Göttingen-Heidelberg: Springer-Verlag 1955; — Z. Elektrochem. **60**, 245 (1956); — Naturwissenschaften **42**, 237 (1955); — Z. Elektrochem. **62**, 66 (1958); — Angew. Chem. **72**, 467 (1960); — Makrom. Chem. **35** A, 12 (1960). — Kratky, O., u. G. Porod: Acta phys. Aust. **2**, 133 (948). — Kratky, O., u. H. Sembach: Angew. Chem. **67**, 603 (1955). — Kratky, O., A. Sekora u. R. Treer: Z. Elektrochem. **48**, 587 (1942). — Kratky, O., u. G. Porod: Z. physik. Chem. **7**, 236 (1956). — Kratky, O., B. Paletta, G. Porod u. K. Strohmaier: Z. Naturforsch. **12**b, 287 (1957). — Kratky, O., u. Z. Skala: Z. Elektrochem. **62**, 73 (1958). — Kratky, O., L. Kahovec u. H. Werner: Z. Elektrochem. **63**, 64 (1959). — Kratky, O., W. Menke, A. Sekora, B. Paletta u. M. Bischof: Z. Naturforsch. **14**b, 307 (1959). — Kratky, O., G. Porod u. Z. Skala: Acta physica Austriaca **13**, 76 (1960). — Kreutz, W., u. W. Menke: Z. Naturforsch. **15**b, 402 (1960); **15**b, 483 (1960).

Leonard jr., B. R., J. W. Anderegg, S. Shulman, P. Kaesberg u. W. W. Beeman: Biochim. Biophys. Acta **12**, 499 (1953). — Leyon, H.: Exp. Cell Res.**4**, 371 (1953).

Menke, W.: Protoplasma **35**, 115 (1941). — Mühlethaler, K.: Beil. zu der Zeitschr. des Schweiz. Forstvereins **30**, 55 (1960) (Festschrift Prof. Frey-Wyssling). — Mukherjee, S. M., S. Sikorski and H. J. Woods: Nature (Lond.) **167**, 821 (1951); — J. Textile Inst. **43**, 196 (1952).

Parrish, W., u. T. R. Kohler: Rev. Sci. Instr. **27**, 795 (1956). — Porod, G.: Acta Phys. Austriaca **2**, 255 (1948); **3**, 66 (1949); — Makromol. Chem. **35**, 1 (1960); — Fortschr. Hochpol.-Forsch. **2**, 363 (1961).

Rånby, B. G.: Acta chem. Scand. **3**, 649 (1949). — Rånby, B. G., u. Ed. Ribi: Experentia **6**, 12 (1950). — Ross, P. A.: J. opt. Soc. Amer. **16**, 433 (1928). — Rothwell, W. S.: Diss. Univ. Wisconsin (USA) 1954.

Schmidt, P., P. Kaesberg and W. W. Beeman: Biochim.Biophys. Acta **14**, 1 (1954). — Statton, W. O.: J. Polymer Sci. **22**, 385 (1956).

Wolken, J. J., and G. E. Palade: Nature (Lond.) **170**, 114 (1952).

Méthodes Calorimétriques pour l'Analyse des Végétaux.

Par

Henri Prat.

Avec 3 Figures.

Depuis les travaux de LAVOISIER au XVIIIème Siècle et de BERTHELOT au XIXème, la mesure des dégagements de chaleur a pris une importance sans cesse croissante dans l'étude des phénomènes chimiques et biologiques. Tout évènement se produisant dans un système physico-chimique ou dans un être vivant se traduit, en dernière analyse, par une production ou une absorption de calories, et la mesure de celles-ci donne aussitôt une évaluation exacte de l'intensité du phénomène. Les variations de la thermogénèse constituent notamment la réponse la plus rapide, la plus fidèle, la plus aisément mesurable et la plus constante d'un organisme vivant à toute action d'un facteur physique ou chimique.

On sait que les plantes, comme tous les autres êtres vivants, dégagent de la chaleur. Ce dégagement passe souvent inaperçu, car il est en général plus faible en valeur relative que celui que l'on observe chez les animaux et dans les cultures de micro-organismes. Il est cependant toujours présent dans les organes végétaux à l'état de vie active. Dans certains cas, par exemple chez les méristèmes en croissance, les graines en germination, les fleurs, il peut atteindre des valeurs tout aussi élevées que celles offertes par les tissus animaux.

Les méthodes calorimétriques offrent donc un grand intérêt pour analyser le fonctionnement des plantes dans diverses conditions de milieu et de nutrition et pour comparer entre elles les espèces végétales au point de vue de leur activité physiologique. Il s'agit là de l'étude de la *thermogénèse biologique*. Ces mêmes méthodes présentent une autre gamme d'applications dans l'étude calorimétrique des produits élaborés par les végétaux. Leur combustion dans une bombe calorimétrique fournit, en effet, la mesure de l'énergie emmagasinée par la plante sous forme chimique, énergie dont l'origine première réside toujours dans le rayonnement solaire. Il s'agit alors de la *calorimétrie biochimique*.

Dans ce chapître nous examinerons successivement ces deux points de vue, en insistant surtout sur le premier, c'est-à-dire sur la mesure des quantités de chaleur dégagées par les plantes à l'état d'activité. Nous trouverons là un certain parallélisme avec le chapître consacré à la mesure du p_H (Vol. I) qui, lui aussi, considérait d'une part les tissus vivants, d'autre part les extraits végétaux.

I. Appareillage.

Les calorimètres employés dans les recherches chimiques et biologiques peuvent être ramenés aux trois types principaux schématisés dans la figure 1. La lettre x represente la portion de la chaleur totale produite qui est conservée dans la cellule calorimétrique A; la lettre y, la portion dispersée à l'extérieur. Le schéma *I* est celui d'un calorimètre *adiabatique* (du grec *αδιαβατως*: impénétrable). Chez lui la cellule *A*, où se produit le phénomène thermique à étudier, est directement entourée d'une enveloppe isolante *B*. Ainsi presque toute la chaleur produite est conservée ($x = 99\%$) et échauffe la cellule. Un thermomètre *T* mesure l'élévation de température Δt. Connaissant la capacité calorifique *M* du calorimètre on déduit aisément le nombre de calories dégagés: $Q = M. \Delta t$. Ce type d'appareil, dit «intégrateur» ou «à échauffement» (Berthelot), convient bien pour la mesure des chaleurs de combustion. Les *bombes calorimétriques* sont géneralement

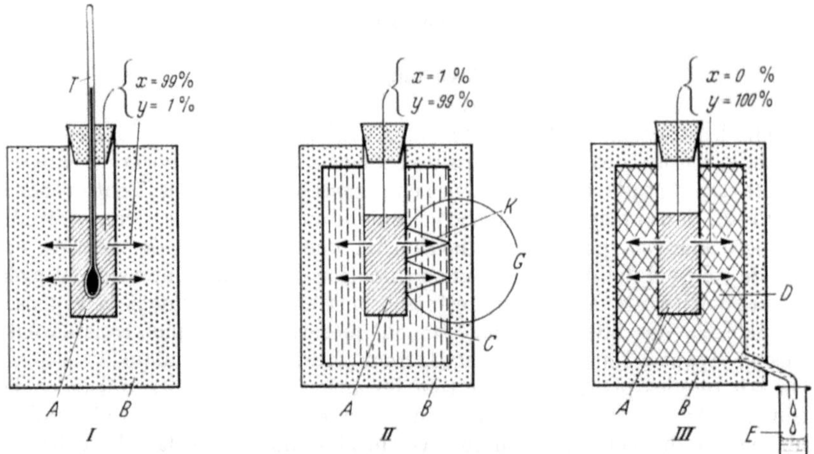

Fig. 1. Principaux types de calorimètres: *I:* adiabatique; *II:* Tian-Calvet; *III:* isothermique. *A* (hachures): zône de production de chaleur dans la «cellule calorimétrique». *B* (pointillé): parties isolantes; *C* (traits verticaux): partie conductrice de la chaleur; *D* (quadrillage): glace, corps fusible ou circulation de liquide ou de gaz. Appareils de mesure; *T:* thermomètre; *K:* thermocouples reliés au galvanomètre enregistreur *G*; *E:* éprouvette graduée (ou compteur de circulation); *X:* portion de la chaleur totale produite qui est conservée dans la cellule calorimétrique *A*; *Y:* portion dispersée à l'extérieur de la cellule. Les flèches indiquent les fuites thermiques.

établies sur ce modèle. Elles sont utilisables pour les mesures dont nous parlerons au paragraphe V; c'est-à-dire celles qui portent sur l'analyse calorimétrique des substances végétales mortes: organes entiers ou produits isolés (observations «in vitro»). Par contre, ce type d'appareil ne convient pas pour l'étude des dégagements de chaleur effectués par des organismes vivants (observations «in vivo»), car il présente l'inconvénient majeur d'opérer à température variable: la température finale est toujours supérieure à la température initiale. Or, on sait que toutes les fonctions d'un organisme ou d'un tissu vivant sont modifiées par la moindre variation de température. La première condition de leur étude est donc d'opérer à une température aussi bien définie et aussi constante que possible.

Les calorimètres du type III, dits *isothermiques*, satisfont à cette condition. La chaleur produite dans la cellule *A* est éliminée au fur et à mesure de son apparition ($y = 100\%$), soit en servant à fondre une certaine quantité d'un corps fusible *D* (p.ex. calorimètre à glace de Lavoisier), soit en étant enlevée par un courant de gaz ou de liquide (calorimètre à circulation). Dans le premier cas le poids de liquide résultant de la fusion de *D* fournit une mesure du nombre de calories dégagées. Dans le second, cette mesure est donnée par la différence des températures d'entrée et de sortie du fluide circulant, multipliée par le débit de celui-ci, le tout ramené à l'unité de temps. Ces appareils conviennent assez bien pour mesurer la thermogénèse d'organismes animaux. Cependant ils présentent l'inconvénient de posséder une grande inertie thermique et de se prêter difficilement à l'enregistrement continu du débit de chaleur. Ils sont peu utilisables pour l'étude des végétaux.

Les appareils du type II: Calorimètres TIAN-CALVET, présentent de grands avantages sur les précédents, tant pour l'étude des organismes vivants: animaux, végétaux ou microorganismes, que pour celle de la cinétique des phénomènes chimiques. En effet, ils opèrent à température pratiquement constante (les variations étant de l'ordre du dix millième de degré), leur faible inertie permet l'analyse détaillée de la cinétique des phénomènes, enfin ils permettent *l'enregistrement continu* du débit de chaleur fourni.

Leur principe est le suivant: un grand nombre de thermocouples K, montés en série, réunissent l'éprouvette calorimétrique A à un bloc métallique C. Ce dernier doit avoir une capacité calorifique très grande par rapport à celle de l'éprouvette et du matériel qu'elle contient; ainsi les calories produites en A sont éliminées aussitôt dans le bloc C, par conduction dans les thermocouples, sans changer notablement sa température.

TIAN et CALVET ont démontré que, dans ces conditions, la force électromotrice produite dans la pile de thermocouples K est proportionnelle non à la quantité totale de chaleur accumulée, mais au *débit thermique*, c'est-à-dire, au nombre de calories produites dans un temps donné. La déviation d'un galvanomètre G relié à la pile thermo-électrique K est donc proportionnelle à la thermogénèse rèalisée en A, exprimée en *calories-heure*. En munissant G d'un dispositif approprié d'enregistrement (tambour photographique, suiveur de spot, microvoltmètre enregistreur) on obtient ainsi un *thermogramme*: tracé indiquant, minute par minute, les fluctuations du débit calorifique produit par l'organisme ou par la réaction chimique en observation.

On peut rendre un calorimètre de ce type complétement isothermique en compensant les dégagements de chaleur par l'effet PELTIER (utilisant pour cela une partie des thermocouples K). D'autre part, CALVET, en réalisant un montage en opposition (différentiel) de deux blocs identiques, a réussi à assurer une indépendance totale de l'appareil vis-à-vis des fluctuations extérieures, donc la stabilité parfaite du zéro; ainsi l'enregistrement du thermogramme peut se poursuivre pendant plusieurs jours ou même plusieurs semaines d'une façon continue. Enfin, en augmentant le nombre des thermocouples, on peut donner à l'instrument une très haute sensibilité atteignant le dix millième de calorie-gramme; il s'agit donc d'un *micro-calorimètre*. La théorie et la construction de ces appareils sont exposées en détail dans l'ouvrage: «Microcalorimétrie» (1956) par CALVET et PRAT. Précisons seulement qu'il s'en construit en ce moment deux types principaux, l'un comportant des cellules de 18 mm. de diamètre et 15 cc de capacité, entourées de 180 à 200 thermocouples; l'autre, des cellules de 35 mm. de diamètre et 100 cc de capacité, entourées de 1800 à 2000 thermocouples. Les premières sont utilisées pour les recherches portant sur des matériels de faibles masse et exigeant une grande sensibilité; les secondes, pour des matériels plus volumineux ou des expériences comportant des dispositifs de «remote-control» fonctionnant dans la cellule même. Les cellules de 35 mm. sont d'un volume suffisant pour qu'il soit possible d'y loger un véritable petit laboratoire «miniaturisé» et commandé à distance au sein même du bloc micro-calorimétrique. Notamment, comme nous le verrons au paragraphe V, il est possible d'y loger une microbombe calorimétrique permettant de mesurer les chaleurs de combustion sur de très petites quantités de matière (analyse calorimétrique «in vitro»), chose précieuse pour l'analyse de produits végétaux rares ne pouvant être extraits que par milligrammes, ou même moins.

II. Thermogenèse de Germinations.

Le dégagement de chaleur produit par des graines que l'on mouille et qui commencent à germer a été remarqué depuis l'antiquité. Dans des silos où se produit une infiltration d'eau il peut même provoquer des accidents graves, d'autant plus qu'il est associé d'un gonflement de ces mêmes graines et à une forte production de gaz carbonique. Mais en général, pour mettre ce phénomène en évidence, on emploie des calorimètres du type I (figure 1), ce qui est défectueux: Par exemple on prend une bouteille «Thermos» (vase de Dewar), on y introduit les grains mouillés et on y plonge un thermomètre. Celui-ci accuse une élévation de température de plusieurs degrés, ce qui enlève toute valeur physiologique à l'observation. En effet, la thermogénèse des graines à la température initiale est très différente de celle qui a lieu à la température finale, et la courbe obtenue en pointant les lectures successives du thermomètre est une combinaison mal définie des courbes correspondant à ces deux extrêmes et aux températures intermédiaires. En outre ce procédé exige l'emploi d'une grande quantité de graines, ce qui fait que celles du fond se trouvent placées dans des conditions asphyxiques.

Seuls des calorimètres du type III, c'est-à-dire opérant à température sensiblement constante, avec une faible inertie thermique, sur de faibles quantités de matériel et en donnant un enregistrement continu, permettent d'analyser avec précision la cinétique du phénomène. La figure 2 montre le thermogramme fourni par un calorimètre Calvet avec des grains de blé en germination à la température de 24° C. Dans cette expérience on prend un gramme de grains secs (32 grains) et, à l'instant initial 0, on fait tomber sur eux un gramme d'eau distillée. On observe aussitôt un fort dégagement de chaleur, atteignant en quelques minutes un débit de l'ordre d'une calorie-heure (OA). Puis le débit thermique décroit; au bout de cinq heures il devient sensiblement nul (parfois même négatif en raison de l'intervention de phénomènes endothermiques tels que le gonflement des colloïdes). Nous avons apppelé cette première phase OAB *thermogénèse physico-chimique*, car aucun phénomène proprement biologique n'y intervient encore: il s'agit seulement du dégagement de chaleur provoqué par l'hydratation du matériel sec. Elle se produit aussi bien avec des graines préalablement tuées qu'avec des graines vivantes. Mais après un *temps mort* BXC, qui peut durer six ou sept heures, un nouveau dégagement de chaleur C D apparaît, qui présente un caractère tout différent. Il s'agit cette fois de la *thermogénèse biologique* résultant de la mise en marche des phénomènes de la germination: digestion des réserves par les diastases, accroissement rapide de l'intensité respiratoire, croissance de la plantule, avec multiplication cellulaire, élongation des organes, etc. Bien entendu elle n'a lieu qu'avec des graines vivantes.

Les ordonnées de la courbe OABCD (thermogramme de la germination) varient en fonction des conditions physiques de l'expérience: température ambiante, quantité d'eau fournie, volume et composition de l'atmosphère de la cellule calorimétrique, et surtout conditions initiales des semences. Lorsqu'avant l'expérience on déshydrate les grains de blé, le résultat en est une augmentation considérable de la chaleur produite pendant la première phase (thermogénèse physico-chimique). Au moment de l'humectation, le maximum AA' du débit thermique peut alors atteindre une valeur de 2,4 ou 3 cal h., pour des grains préalablement desséchés pendant un mois dans le vide en présence d'acide phosphorique (courbe II fig. 2) Dans ces conditions la thermogénèse biologique elle-même (CD') peut se trouver. accrue et le début de la germination peut en être accéléré. Ceci explique en partie la rapidité de germination que l'on observe chez les semences des régions désertiques dès qu'une pluie survient.

Toutes les semences fournissent des thermogrammes présentant les mêmes phases générales: thermogénèse physico-chimique, temps mort, thermogénèse biologique. Mais la forme de la courbe OABCD et les valeurs numériques de ses éléments sont très différentes d'une espèce végétale à l'autre. Lorsqu'on compare, dans des conditions identiques, les thermogrammes de germination de diverses espèces on constate ainsi des profondes différences dans la grandeur des débits thermiques fournis et dans la durée des diverses phases. Par exemple, les graines de poireau *(Allium porrum)* donnent 20 cal./h. par gramme comme maximum de la thermogénèse physico-chimique, soit 20 fois plus que le blé dans les mêmes conditions (24°); chez le céleri ce débit est de 5 cal./h. et le temps mort dure 14 à 15 heures, etc. Ces thermogrammes peuvent donc être employés comme *caractères spécifiques* pour distinguer les espèces végétales ou les variétés. Ce sont des caractères de *physiologie comparée*, tout aussi utilisables que ceux d'anatomie comparée, de morphologie ou d'histologie. En outre ils renseignent sur les *aptitudes écologiques* des espèces: par exemple celles qui possèdent une germination rapide avec un fort débit thermique sont plus aptes à se développer dans des stations sèches n'offrant que de courtes périodes de pluie. Cette analyse des propriétés des plantes par des moyens calorimétriques présente donc un intérêt agricole. Il en va de même pour l'étude, par ces mêmes moyens, des conditions de conservation des semences et de leurs conséquences quant aux aptitudes germinatives, signalées plus haut. L'analyse calorimétrique est donc appelée à rendre de grands services, dans l'avenir, pour les recherches agronomiques.

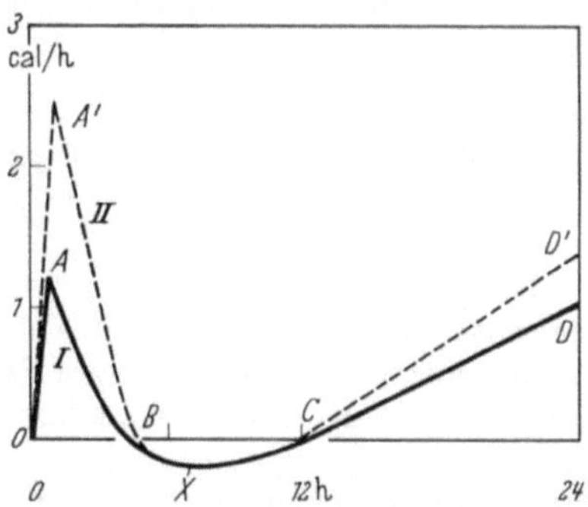

Fig. 2. *I* (courbe en trait plein): Thermogramme fourni par la germination de 32 grains de blé, pesant au total 1 g., à 24° C. *O:* mise en contact des grains avec 1 cc d'eau. *OAB:* thermogénèse physico-chimique; *BXC:* temps mort; *C D:* thermogénèse biologique. *II* (courbe en pointillé): thermogramme fourni par la même quantité de grains de blé, préalablement desséchés pendant un mois; *A':* maximum de la thermogénèse physico-chimique. En ordonnées les débits thermiques, en calories-heures; en abscisse les temps, en heures, à partir du début de l'expérience (humectation des grains).

III. Gradients de Thermogénèse dans les Organes en Croissance.

Une tige ou une racine en cours de croissance dégage de la chaleur et ceci, particulièrement, dans les zones où la division cellulaire est active, c'est-à-dire dans les *méristèmes.* On peut le mettre en évidence, par exemple, en découpant une tige d'asperge en segments d'égale masse et en plaçant ceux-ci successivement dans la cellule d'un micro-calorimètre Calvet. On constate ainsi qu'à 31° C le sommet de la tige dégage par exemple 2,8 cal. h. par gramme, tandis que la base, située 20 cm plus bas, ne dégage que 0,4 cal. h. g. Entre les deux on observe un *gradient* de la thermogénèse régulièrement croissant de la base jusqu'au sommet. Il est parallèle à un gradient d'intensité respiratoire et, plus généralement, de métabolisme des tissus.

Un peu différent est le cas de la thermogénèse de floraison: En plaçant dans un calorimètre des fragments de fleurs au moment de leur épanouissement, on observe que les carpelles, les étamines, les pétales, dégagent plus de chaleur par unité de poids que les pédoncules. Mais ceci ne tient pas tant à la fréquence des divisions cellulaires qu'à une exaltation du métabolisme des tissus sous l'influence des hormones de floraison, avec pour résultat final la mort, par épuisement, de certains des organes floraux: pétales, étamines, styles, etc. A 31° C, par exemple, un capitule de *Tragopogon pratense* fournit un débit thermique de 2 cal. h. g. tandis que le pédoncule qui le porte ne dégage que 0,07 cal. h. g., soit 30 fois moins; une inflorescence de *Dactylis glomerata* fournit 1,8 cal. h. g., contre 0,1 cal. h. g. pour l'entre-noeud qui la supporte, etc.

L'étude calorimétrique des cultures de tissus présente un intérêt spécial, car les cellules se trouvent là soustraites aux contraintes physiques et chimiques exercées par les autres parties de la plante lorsque celle-ci est entière. Ceci permet d'apprécier l'influence de chaque agent chimique (aliment, toxique, hormone, etc.) sur le métabolisme des cellules, ceci sans intervention des intermédiaires habituels les séparant du milieu ambiant (racines, système conducteur). Par exemple une culture de tissus de *Sequoia sempervirens* nous a fourni, à 25°, sur liquide de KNOP additionné de 3% de glucose, un débit de 0,75 cal. h. g., et ce débit varie lorsqu'on modifie la composition du milieu nutritif, notamment quand la nature du sucre change.

IV. Thermogénèse des Cultures Bactériennes et Mycéliennes; Fermentations.

Les bactéries et les champignons faisant partie du règne végétal, l'étude des fermentations est inséparable des études de phytophysiologie. Au point de vue de la calorimétrie leur cas est toutefois un peu différent de celui des plantes supérieures, car il faut généralement faire entrer en ligne de compte non seulement la chaleur produite par l'organisme lui-même, mais encore celle qui prend naissance dans le milieu de culture environnant, ceci sous l'effet des exodiastases diffusées et des réactions chimiques qu'elles provoquent. Cependant, cette réserve faite, il n'en demeure pas moins que ces organismes se comportent bien réellement comme des végétaux en ce qui concerne leur thermogénèse, *celle-ci étant encore strictement liée chez eux à la multiplication cellulaire.* C'est seulement dans les cultures actives des bactéries ou des levures, c'est-à-dire, dans celles qui sont en voie de multiplication, que l'on observe une production importante de chaleur. Signe particulier: cette thermogénèse est toujours légèrement fluctuante. La figure 3 met bien en évidence le tracé sinueux des thermogrammes de micro-organismes, contrastant avec le tracé net et régulier fourni par les plantes supérieures (fig. 2). Celà tient à ce qu'une culture de bactéries ou de levures est une «foule» d'individus incoordonnés ayant chacun son métabolisme propre, non pas un individu cohérent comme l'est un végétal pluricellulaire.

La figure 3 montre le thermogramme d'une culture de *Coli*-bacille *(Escherichia coli)* à 24°, sur 4 cc de bouillon peptoné. Les heures sont indiquées à partir de l'ensemencement. On voit que le débit thermique croît d'abord régulièrement de la 5ème à la 18 ème heure, atteignant alors un maximum D de 0,34 cal. h. au total, soit 0,08 cal. h. par centimètre cube. Puis il décroît progressivement, s'abaissant jusqu'à 0,14 cal. h. au bout de 48 heures. Des numérations successives montrent que la période de thermogénèse croissante correspond à une phase de multiplication rapide et qu'alors le débit thermique est d'environ 5.10^{-10} cal. h. par bactérie. Naturellement ce n'est là qu'une évaluation moyenne, une partie des bacilles étant au repos et ne dégageant que peu de chaleur, alors que le reste, se divisant, en fournit beaucoup plus. Ce chiffre de 5.10^{-10} cal. h. semble faible, en réalité il est considérable: la masse d'une bactérie étant de l'ordre de $5 \cdot 10^{-12}$ g., cette thermogénèse représente 100 cal. h. par gramme, soit 25 fois plus que ce que fournissent, en moyenne, les tissus de notre corps. Et, si l'on pouvait évaluer avec précision le débit propre d'une bactérie au moment où elle se divise, on trouverait des valeurs encore beaucoup plus élevées.

Cette étude calorimétrique des cultures bactériennes présente de nombreux avantages en permettant de comparer les propriétés physiologiques, l'activité de croissance des diverses espèces (ou souches d'une même espèce); d'apprécier les qualités des divers milieux de culture; enfin de mesurer l'efficacité des antibiotiques, et plus généralement l'influence de tous agents chimiques ou physiques. Lorsqu'on laisse tomber (en S, fig. 3) une solution d'antibiotique sur une culture bactérienne en cours de développement on constate que le débit thermique de celle-ci décroît, et ceci d'autant plus rapidement que la solution est plus active (pointillé SF, en II, fig. 3). Ainsi que nous l'avons signalé plus haut, les variations de la thermogénèse constituent la réponse la plus rapide, la plus fidèle, la plus constante d'un organisme à toute action d'un facteur physique ou chimique.

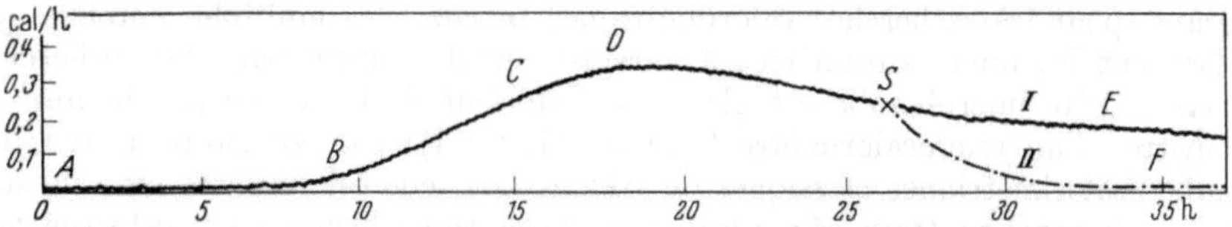

Fig. 3. Thermogramme d'une culture de Coli-Bacille sur 4 cc de bouillon peptoné à 24° (trait pleins). La courbe *II* (pointillée) *S F* montre la dépression de la thermogénèse provoquée par l'addition de 30 micro-grammes de streptomycine, sans agitation, à la 27 ème heure après l'ensemencement.

Il est bien connu qu'un corps en fermentation dégage de la chaleur. Par exemple, un moût de bière, qui est, en fait, une culture de levure (*Saccharomyces cerevisiae*), c'est-à-dire d'un champignon ascomycète microscopique, nous a donné, à 31°, un débit maximum de 0,6 cal. h. par.centimètre cube, ceci 8 heures après l'ensemencement. Ensuite la thermogénèse décroît lentement, le tracé affectant ce même aspect sinueux caractéristique des cultures de microorganismes, pour atteindre 0,3 cal. h. par cc à la 22ème heure. Sur un même milieu chaque espèce mycélienne fournit son thermogramme propre. Inversement, sur des milieux différents, la même espèce donne des tracés différents, révélant l'influence des constituants de ces milieux sur son métabolisme.

V. Chaleurs de Combustion d'Organes et de Produits Végétaux.

Ce qui précède avait trait à l'analyse calorimétrique du métabolisme végétal, c'est-à-dire du fonctionnement de la petite usine solaire constituée par la plante (analyse calorimétrique «*in vivo*»). Restent à examiner les résultats de ce fonctionnement: les produits élaborés par le végétal, et à mesurer l'énergie qui s'y trouve emmagasinée (analyse calorimétrique «in vitro»). Ce problème n'est pas différent de l'étude des chaleurs de combustion de n'importe quelle substance organique ou minérale, de synthèse ou d'origine animale. Il est résolu au moyen des mêmes appareils, qui sont *des bombes calorimétriques*. Celles-ci sont établies en général sur le principe des calorimètres *adiabatiques* (I, fig. 1).

Le corps à étudier, qu'il s'agisse d'une substance à l'état pur ou d'un organe végétal entier, est préalablement desséché. Sa combustion est ensuite assurée dans la bombe en présence d'oxygène, avec une mise à feu électrique. L'élévation de température indiquée par le thermomètre (enregistreur ou à lecture directe), multipliée par la capacité calorifique de l'appareil, donne le nombre de calories produites.

Ainsi on peut constater qu'une fraction importante de l'énergie solaire effectivement captée par la plante se trouve, en définitive, fixée sous forme chimique.

Après l'étude de la cinétique des fonctions végétales, les calorimètres nous permettent donc de poursuivre l'examen du résultat final: l'étude du potentiel énergétique accumulé dans le corps végétal élaboré et dans chacune des substances chimiques qui le constituent. Le *rendement* de l'usine chimique constituée par la pante peut ainsi être mesuré exactement. Or, l'agriculture consiste fondamentalement à choisir les espèces les plus aptes à transformer en substances utiles et à emmagasiner la part la plus grande possible de l'énergie des radiations et des aliments qu'elles reçoivent. Les techniques calorimétriques peuvent donc l'aider considérablement dans cette recherche méthodique, tout comme l'ont fait celles de la génétique, de la biométrie, de l'analyse chimique, de la mesure des échanges gazeux, de la chromatographie, de l'électrophorèse, etc.

Calvet et ses collaborateurs (1960) ont mis au point une «*microbombe*» complétant, pour les recherches calorimétriques *in vitro*, les multiples emplois que présentent les microcalorimètres à enregistrement continu dans les recherches *in vivo*. Cette microbombe est placée à l'intérieur de la cellule (de 35 mm. de diamètre) d'un microcalorimètre Calvet (II, fig. 1); par comparaison avec les bombes calorimétriques classiques de Berthelot, elle est disposée sens dessus-dessous: le bouchon, traversé par le dispositif électrique de mise à feu, est situé vers le bas et le fond hémisphérique de la bombe vers le haut, ceci pour éviter les projections et corrosions. La tige de mise à feu supporte le creuset en platine contenant la substance à brûler. Cette dernière peut être en très petite quantité: de l'ordre du milligramme, en raison de la haute sensibilité de l'appareil. Cette particularité est d'un grand intérêt dans l'analyse calorimétrique des produits végétaux, certains de ceux-ci ne pouvant être extraits qu'en très petite quantité. La mesure de la quantité de chaleur produite peut, dans ces conditions, être ramenée à celle d'une simple longueur: celle de l'ordonnée du maximum de la courbe enregistrée au moment de l'explosion, celle-ci étant alors bien une «courbe balistique» (Calvet et al. 1960).

VI. Conclusion.

Dans ce chapître nous avons exposé très brièvement quelques-uns des services que peuvent rendre les méthodes calorimétriques pour analyser les fonctions des végétaux, leurs réactions au milieu ambiant et les produits résultant de leur activité. En particulier elles permettent, grâce aux nouveaux micro-calorimétres Calvet à enregistrement continu:

1. d'analyser les diverses phases de la thermogénèse au cours de la germination des graines, offrant ainsi, en association avec les méthodes de la génétique, une grande utilité pour la sélection des semences. Le *thermogramme* peut être employé comme caractère spécifique.

2. d'analyser la croissance et le développement d'un végétal par l'observation des variations de sa thermogénèse et de la localisation de celle-ci; d'où leur intérêt dans les recherches sur les herbicides, les engrais, les hormones de croissance, les effets de la vernalisation, du photopériodisme, du thermo-périodisme; plus généralement dans toutes les recherches écologiques, physiologiques et pharmacodynamiques.

3. d'analyser le développement des cultures bactériennes et mycéliennes et leurs fluctuations en fonction des conditions du milieu, offrant ainsi une grande utilité dans les recherches bactériologiques, l'industrie des fermentations, la préparation des antibiotiques, etc.

4. de mesurer *in vitro* la quantité d'énergie fixée dans les produits de l'activité d'une plante ou d'un micro-organisme; d'où découlent de nombreuses applications

pour découvrir les espèces possédant le plus haut rendement et les meilleures aptitudes écologiques dans des conditions données. Les nouvelles «*micro-bombes*» pouvant s'adapter aux modèles de micro-calorimètres CALVET à grosses cellules permettent maintenant des mesures de chaleurs de combustion portant sur un milligramme de matière, circonstance très favorable pour l'analyse de produits végétaux rares, que l'on ne peut obtenir qu'en faible quantité.

Références Bibliographiques.

CALVET, E., et H. PRAT: Microcalorimétrie; applications physico-chimiques et biologiques. Masson 1956. — CALVET, E., et H. PRAT: Récents progrès en micro-calorimétrie. Dunod 1958. — CALVET, E., CHOVIN, MOUREU et TACHOIRE: J. chim. phys. **57**, 593—596 (1960).

PRAT, H.: Rev. can. biol. **4**, 543—693 (1945). — PRAT, H.: Can. J. Botany **30**, 379—394 (1952). — PRAT, H.: Rev. Can. Biol. **12**, 19—34 (1953). — PRAT, H., et E. CALVET: C.R. Soc. Biol. **138**, 662—666 (1944).

TIAN, A.: Bull. Soc. Chim. **33**, 427 (1923).

Surface Factors Affecting the Penetration of Compounds into Plants.

By

A. E. Dimond.

With 2 Figures.

Surface factors affecting the penetration of compounds into plants is a subject presenting a challenge to the investigator. The surface and interfacial properties of solvents and solutions can be measured by well established methods. But to use these measurements meaningfully in relation to plant surfaces is not a simple problem. The plant presents an irregular surface that is chemically heterogeneous and both of these factors introduce complications when measurements are involved. The permeation of compounds into cells beneath the leaf surface is related in an indefinite way to surface properties. There are many hypotheses dealing with one phase or another of the subject but there is no unified theory treating the subject as a whole upon which unequivocal measurements can be based. However, useful predictions can be made concerning the fate of a restricted number of compounds when they are applied to plants.

The treatment applied to plants have widely differing objectives. Protective fungicides and certain insecticides are not designed to enter the plant at all. By contrast, growth regulants, systemic insecticides, chemotherapeutants and foliar nutrients must enter the plant. Obviously, then, one must be able to control not only how rapidly compounds enter plants but whether or not any entry occurs.

In such problems, little can be done to alter the nature of the plant surface. Two courses are open to the investigator. The first is to vary the molecular structure associated with a physiologically interesting grouping so as to modify the permeative properties of the molecule as a whole. The second is to alter the conditions under which an active compound is applied.

A. Effect of Molecular Structure on Permeation.

The leaf surface is more readily wetted by oils than by water. The cuticular surface is composed of waxes and other long-chain fatty substances that give to it a degree of water repellency.

Solutes that reduce the surface tension of water accumulate in the surface film, the hydrophilic portion of the molecule being oriented toward the water and the lipophilic tail being oriented outward. On a sprayed leaf, this orientation brings the lipophilic portion of the molecule into contact with the cuticle.

Many molecules have a center of biochemical activity, the toxophore, which is attached to a fat-soluble group, the lipophore. HORSFALL (1956) and VAN OVERBEEK (1956) have both discussed the importance of a proper balance between toxophore and lipophore functions to achieve a maximum degree of biological activity. A classical example is resorcinol, which is a moderately good bactericide, but hexyl resorcinol, with its lipophilic side chain, is much more effective. Aliphatic chains, aromatic rings, and chlorine are examples of lipophilic substitutions that can give to a toxophore greatly increased permeative ability.

Cell membranes contain water and lipids and these are involved in the permeability of cells to certain types of molecules. The partitioning of a molecule between phases of oil and water is often related to the ability of the molecule to enter cells. For certain fungicides, RICH and HORSFALL (1952) have demonstrated a correlation between toxicity to fungal spores and the partition coefficient between oil and water. In general, neither complete oil solubility nor complete water solubility is desirable. Entry into fungal cells is favored by a higher partitioning in oil than is entry into leaf tissue. An example is the fungicide 2-heptadecyl-2-imidazoline, which is low in phytotoxicity. By contrast, 1-hydroxyethyl-2-undecyl-2-imidazoline is a poor fungicide and is phytotoxic on leaves, according to WELLMAN and McCALLAN (1946). Presumably the former compound, being more lipophilic, enters fungal cells well and leaves poorly, whereas the latter compound does the reverse, owing to differences in the lipids at surfaces of fungal spores and leaves. Favorable partitioning between aqueous and lipid phases is probably also involved in the improved entry into leaves of chelate complexes of nutrient elements as contrasted that of soluble inorganic salts, e.g., the ferric complex of ethylene diamine tetraacetic acid as compared with ferric chloride.

The partition coefficient. A solute will distribute itself between immiscible solvents in accordance with its solubility in each. The ratio of the concentrations in the two solvents is the partition or distribution coefficient. At equilibrium, the partition coefficient,

$$K = c_1/c_2 \tag{1}$$

where c_1 and c_2 are the concentrations of solute in the two solvents. When molecular association occurs in one of the solvents, the partition coefficient will vary with concentration and will probably not be a useful index of permeative properties of the solute molecule into cells.

In practice at least three concentrations of the test compound are prepared in water. A measured volume of aqueous solution is placed in a separatory funnel with an equal volume of oil, and the mixture is shaken frequently over a period of half an hour, during which time it is kept in a thermostatically controlled water bath. The two phases are separated and analyzed separately for the amount of solute present. At each initial concentration, the ratio of the partitioning in the two solvents is established. If K is not a constant, association is involved in one of the solvents.

Choice of the oil phase for partitioning studies will probably be dictated by the physiological system. While ether and petroleum oils have been used in other studies, RICH and HORSFALL (1952) used olive oil as a material of plant origin and the choice of a semi-polar oil is undoubtedly wise. The partition coefficient is not invariably a useful index in predicting the permeative properties of molecules. Within some chemical groupings, it is useful; in others the correlation is poor. Because permeability of cells to solutes is not solely governed by water-lipid relations, this is to be expected.

B. Effect of Surface and Interfacial Factors on Potency of Treatments.

I. Factors Affecting Deposition of Materials on Foliage.

Leaves are surrounded by a thin layer of quiet air, which varies in thickness with the rate of air movement and the roughness of the leaf surface. A spray droplet must have a certain momentum to penetrate this layer (POTTS, 1958). The components of this momentum are mass and velocity. SMITH (1946) has noted the effect of droplet size on the efficiency of deposition and has confirmed this in practice by observing the greater effectiveness of aqueous 2,4-D sprays on plants when spray droplets averaged 300 μ in diameter as contrasted with 30 μ. POTTS (1958) has estimated the carrying power through quiet air of droplets of constant size as related to their velocity. Size of droplets is related to pump pressure, velocity of the carrying stream, surface tension of the spray fluid, and volatility of the carrier.

When foliage is poorly wetted, the deposit of spray is low. Spray droplets rebound from waxy banana leaves that they do not wet. Droplets that land on a water repellent surface roll off, leaving little deposit behind. The addition of a small amount of wetting agent in this situation has several simultaneous effects. It causes the droplets to adhere to waxy leaves, rather than to bounce off (BURCHFIELD and GOENAGA, 1957). By reducing the surface tension, it causes the spray liquid to be atomized as finer droplets, which deposit less efficiently than larger ones do (SMITH, 1946).

When too much wetting agent is present, foliage retains a small volume of liquid at run-off, and the concentration of the spray liquid must be increased so that an effective residue will be retained (AHLGREN, KLINGMAN, and WOLF, 1951; ROBBINS, CRAFTS, and RAYNOR, 1942). Thus a small amount of wetting agent may increase the deposit of spray on water repellent foliage, but a large amount will decrease the deposit retained at run-off (EVANS and MARTIN, 1935; HOSKINS and BEN-AMOTZ, 1938). The deposit and retention of spray on foliage is greatest when wetting agents are used sparingly if at all, and only when necessary for wetting foliage. MARTIN (1940) has found the maximal initial retention of spray by foliage is a function of both wetting and spreading properties and is equal to the quantity $\gamma_1(1 - \cos\theta)$, where γ_1 is the surface tension of the spray liquid and θ is the contact angle between the spray drops and the leaf surface.

The presence of a wetting agent may change the character of the spray deposit on foliage and thus affect the entry of active ingredients into leaves. Absorption of active ingredients by foliage occurs through the cuticle and stomata (CURRIER and DYBING, 1959), and the more perfectly the leaf surface is covered, the greater the surface through which absorption can occur. The beneficial effect of wetting agents on absorption of biologically active materials has been noted by BOYNTON for foliar nutrients (1954), and by VAN OVERBEEK (1956) and CURRIER and DYBING (1959) for growth regulants and herbicides.

When the leaf is wetted poorly, spray droplets remain discrete. The dry deposit remaining is discontinuous. Such a deposit enters underlying tissues poorly, a condition sought in the application of protective fungicides and non-systemic insecticides. Such a deposit also rewets poorly when no wetting agent is present and thus weathers better in rainstorms (MARTIN, 1940).

II. Factors Affecting the Absorption of Materials by Foliage.

1. Nature of the Formulation Applied.

The absorption of materials through the cuticle is relatively rapid while the spray film remains wet (MITCHELL, SMALE, and METCALF, 1960). When the film has dried, absorption either ceases or proceeds very slowly. Thus rate and amount of absorption is affected by the speed with which a spray dries on foliage. A spray deposit consisting of many fine droplets exposes a larger surface for evaporation and dries more rapidly than a deposit consisting of large drops.

Other factors also affect the speed with which a spray dries. A high ambient relative humidity when sprays are applied delays drying and aids entry of materials into the plant, as noted for inorganic salts by YARWOOD (1943), for antibiotics by DYE (1956) and for growth regulants by HOPP and LINDER (1946). To delay the drying of sprays on foliage, humectants are added to sprays in the form of glycerol (RICE, 1948; GRAY, 1956), ethylene glycol, or carbowax (ENNIS and BOYD, 1946; GOODMAN and DOWLER, 1958). Oil emulsions likewise dry more slowly than aqueous sprays and GUSTAFSON (1956) has found the entry of Co^{60} into bean and cucumber leaves is aided by application of cobalt ion in a refined petroleum oil (Sovaspray) emulsion.

Foliar absorption of solutes is aided in at least two ways by maintaining the spray as a moist film for a prolonged period. First, the permeability of the cuticular layer is increased by a film of moisture (VAN OVERBEEK, 1956). Second, the mobility of solutes is greater when in solution than when dry. When a spray film dries and the solutes crystallize out, a sharp decrease occurs in the rate of absorption by leaves (KOONTZ and BIDDULPH, 1957). The addition of cosolvents to spray formulations increases the solubility of biologically active materials in the spray. The grinding of insoluble materials to fine particle size before use in sprays increases their rate of solution. Both of these techniques result in improved performance of active materials (MITCHELL, SMALE, and METCALF, 1960).

Entry of materials into leaves is sometimes aided by applying them as emulsions in spray form. Many biologically active materials partition between water and oil and the amount of material carried in solution in emulsions may be greater than in aqueous solution. The oil phase of the emulsion is likely to be less volatile than an aqueous spray, and when this is true, the spray does not dry rapidly. With emulsions, too, conditions for spreading on the leaf surface may be met when quick-breaking emulsions are used. When the emulsion breaks the active ingredient may be carried through the stomata by the oil. The creeping of oils through stomata has been discussed by KNIGHT, CHAMBERLIN, and SAMUELS (1929), ROHRBAUGH (1934) and by HOSKINS (1933). Absorption of materials entering through stomata is much more rapid than through the cuticle (CURRIER and DYBING, 1959).

The entry of materials into leaves is not favored by dry deposits. When insecticides are applied as dusts to foliage, little absorption into the leaf occurs. Mobility of active ingredients is further restricted by the fact that deposition on all leaf surfaces is encouraged by applying dust particles that bear an opposite electrostatic charge to that of the plant surface (BURCHFIELD, 1959). Deposition is increased in this way, but so long as the electrostatic charge remains, entry of the leaf is unlikely.

Mobility of active ingredients of sprays is restricted when they are adsorbed to the leaf surface strongly. RICH (1954) has noted that the wet leaf surface bears a negative electrokinetic charge and has studied the effect of the charge on a fungicide on its deposition and redeposition during weathering. Positively charged particles of Bordeaux mixture become adsorbed to the leaf and during weathering, loosely bound material redistributes to areas that are not already covered. By contrast, negatively charged particles, such as zinc ethylene bisdithiocarbamate, are not adsorbed, remain mobile on the leaf surface and are readily lost during rainstorms. Significantly, the growth regulants that are most active as foliar sprays and that enter leaves freely are either anionic in character or are neutral.

2. Nature of the Plant Surface.

The behavior of spray drops on a leaf varies according to whether the leaf surface is smooth, rough, or porous. Waxy banana leaves are macroscopically smooth and free of leaf hairs. Wheat leaves are rough with rows of ridges running parallel to the long axis of the leaf. Tomato leaves are hairy. All leaves are porous because of the stomata. In some plants stomata occur on the lower leaf surface only and in others on both leaf surfaces.

Leaves differ widely in their resistance to wetting (LINSKENS, 1950). Banana and crucifer leaves are hard to wet whereas potato leaves are readily wetted. The same leaf varies in water repellency with stage of development and environment (LINSKENS, 1952a, 1952b). STANIFORTH and LOOMIS (1949) have attributed certain aspects of the selective action of herbicides to natural differences in wettability of foliage.

The degree of water repellency of a leaf surface is the resultant of two variables: the amount and kind of wax in the cuticle and the degree of roughness of the surface. Resistance to wetting of the leaf is conferred by the cuticle, which is a mixture of polar compounds overlying the epidermal cells. The epidermal cells have cellulosic walls with plasmodesmatal openings that function in the absorption of molecules (KERLING, 1959). The cuticle itself consists of pectins, cutin, and waxes that are interspersed with one another (VAN OVERBEEK, 1956). Cutin consists of di- and hydroxy-carboxylic acids of varying molecular weight. The outermost layers of cutin are oxidized and polymerized to a hard, varnish-like consistency, as is a drying oil in a hardened paint film (SKOSS, 1955). The waxes occur in two forms, as wax platelets in mixture with cutin and pectins (MUELLER, CARR, and LOOMIS, 1954) and as wax rodlets, which are extruded through the cuticle to the surface. Waxes consist of alcohols, ranging from 24 to 30 carbon atoms in chain length (VAN OVERBEEK, 1956). The wax rodlets vary in abundance and in pattern on the surface of various leaves, giving the surface a microscopic roughness (SCHIEFERSTEIN and LOOMIS, 1956, 1959; JUNIPER, 1959). There is a correlation between the occurrence of wax projections on the leaf surface and the contact angles that water makes with these leaves (LINSKENS, 1950; JUNIPER, 1959).

Cutin itself is reasonably permeable to water, once it is wet (SKOSS, 1955). The cuticle apparently swells as the pectic components take on water. This forces the wax platelets apart so that cuticular layers become more permeable to water and to solutes (VAN OVERBEEK, 1956; MITCHELL, et al., 1960). This behaviour offers a mechanistic explanation for why entry of materials is favored by high humidity, humectants, and by wetting agents, which displace air trapped under droplets standing on wax rodlets.

The rate of entry of compounds into leaf tissue is affected by differences in cuticle composition from leaf to leaf and from plant to plant. It is also affected by the thickness of the cuticular layer. Even spongy mesophyll cells in the interior of the leaf have a very thin cuticular layer where they are in contact with air (SCOTT, SCHROEDER, and TURRELL, 1948). Entry of material into spongy parenchyma cells is rapid when liquids on the leaf surface can enter stomatal openings (SKOSS, 1955; CURRIER and DYBING, 1959). Usually the cuticle of the lower leaf surface is thinner than that of the upper leaf surface, and while compounds penetrate leaves more slowly through the outer leaf surface than when stomatal entry occurs, rate of entry through the cuticle is more rapid through a thin than a thick cuticular layer (CURRIER and DYBING, 1959; GUSTAFSON, 1957).

C. Surface and Interfacial Tension.

Surface tension is a consequence of cohesion, of attraction between like molecules. It is a thermodynamic concept, involving the behavior of masses of molecules, rather than individual ones. In the bulk of a liquid, molecules are surrounded by like molecules that are randomly distributed and the attractive intermolecular forces are balanced. At the surface of a liquid, however, molecules are attracted by many molecules beneath and around, but there are few molecules of vapor above. The result is an inward force on surface molecules so that a liquid exposes a minimum of surface when other forces allow. Surface tension is a force per unit length of film, with dimensions of F/l and is measured in dynes/cm. The surface tension of a liquid decreases as temperature increases and becomes zero at the critical temperature.

When the area of a film of liquid is increased, work must be done; when the film contracts, surface energy is released. Energy is thus associated with the surface of a liquid. Surface energy is dimensionally equivalent to surface tension, having dimensions of Fl/l^2 and is expressed in ergs/cm². Confusion has arisen over the terms surface tension and surface energy because they are sometimes wrongly used interchangeably. Because surface tension and surface energy are difficult to visualize in specific problems, especially where interfaces are involved, the forces at work are sometimes resolved in terms of cohesion and adhesion.

I. Measurement of Surface Tension.

Surface tension can be measured in many ways. The phenomena of surface and interfacial tension and methods of measuring them have been thoroughly treated by ADAM (1941), HARKINS (1946) and by SCHWARTZ and PERRY (1949). For this reason no attempt will be made to treat this subject in detail. However, scrupulous cleanliness is necessary in all measurements of surface tension.

1. Capillary Methods.

The familiar equation for the rise of liquids in capillaries is the basis of a method of measuring the surface tension of liquids. In a capillary of uniform radius, r, a liquid of surface tension, γ, and density, d, will rise to a height, h, above the level of the bulk liquid in accordance with the equation

$$\gamma = \frac{1}{2} r h d g \tag{2}$$

where g is the acceleration due to gravity. This equation is valid only when the liquid readily wets the walls of the capillary, so that the contact angle between the liquid and the walls of the capillary is negligibly small.

The accurate measurement of surface tension by capillary methods presents a few difficulties, which have been summarized by HARKINS and BROWN (1919). Capillaries must be of uniform bore. Accurate readings of the height of the meniscus and of the surface of the bulk liquid are difficult, even with a horizontal telescope. Some liquids do not yield consistent results, but depend upon the grade of glass of which the capillary is made and the time the liquid is in contact with the glass. Liquids of a basic reaction give values as much as 20 per cent low. Viscous liquids may give incorrect results unless drawn up in the capillary past the equilibrium point before determining the height of capillary rise.

An approximate method of measuring surface tension utilizes equation (2) directly. A clean capillary of uniform bore is mounted vertically with its lower end dipping into water contained in a beaker, having an internal diameter of at least 4 cm., which is sufficiently large a surface to avoid curvature of the meniscus of the bulk fluid at its center. At a known temperature, the height of rise of the water is determined. Then the radius of the capillary is calculated from equation (2), the surface tension and density of the water at the prevailing temperature being known. The capillary is cleaned and the procedure is repeated on the liquid of unknown surface tension. Density of this liquid, if unknown, is carefully determined with a pycnometer at the temperature of measurement of the surface tension. The height of capillary rise being determined, surface tension is calculated by means of equation (2).

A differential method of measuring surface tension has been described by KITCHENER (1954), which leads to accuracy of measurement greater than 1 per cent. Two capillaries of differing bore (0.3 and 1.5 mm approximately) are inserted in a rubber stopper together with a blow tube (Fig. 1). The stopper and capillaries are then loosely inserted into a large glass tube, having a diameter of at least 4 cm. The two capillaries are mounted close together, so that the meniscus in each is visible in the field of a horizontal telescope. The horizontal telescope should be capable of detecting differences in height of 0.05 mm. The tube and capillaries can be immersed in a thermostatically controlled water bath.

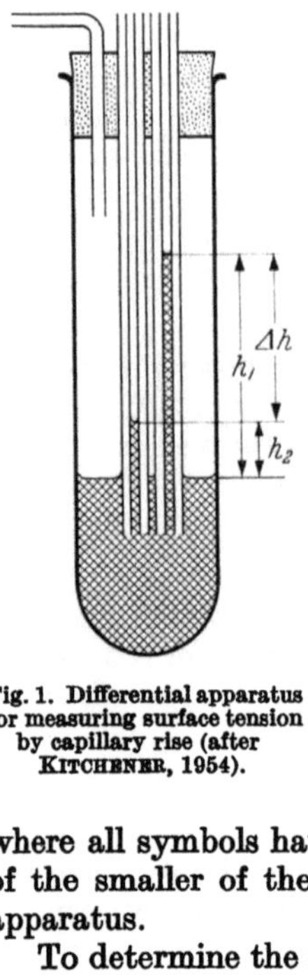

Fig. 1. Differential apparatus for measuring surface tension by capillary rise (after KITCHENER, 1954).

This apparatus is calibrated by using benzene or water as reference liquids for which surface tension is accurately known. By means of the blow tube, slight pressure is exerted on the liquid, so that it rises in the capillaries beyond the equilibrium point. The pressure is released, and, after a few minutes, the height of each meniscus is read with the telescope. If the radii of the two capillaries are r_1 and r_2 and the heights of the menisci in them are h_1 and h_2, then $r_1 h_1 = r_2 h_2$, a relation derivable from equation (2). In this apparatus all other factors in equation (2) are common to the two capillaries and cancel out. Then the difference in height of the menisci in the two capillaries, Δh, is equal to

$$\Delta h = \frac{\gamma}{d}\left[\frac{2}{g}\left(\frac{1}{r_1} - \frac{1}{r_2}\right)\right] \tag{3}$$

where all symbols have the same meaning as in equation (2) and r_1 is the radius of the smaller of the two capillaries. The term in brackets is constant for the apparatus.

To determine the surface tension of an unknown liquid of density d', which in the apparatus has a difference in capillary rise of $\Delta h'$, one uses equation (4)

$$\frac{\gamma'}{\gamma} = \frac{\Delta h' d'}{\Delta h\, d} \tag{4}$$

where values of γ, Δh, and d have been previously determined for the reference liquid and d' has been determined for the unknown liquid. When allowance is made for the weight of the liquid in the curved meniscus, accuracy approaching 0.05 per cent can be attained (KITCHENER, 1954).

2. Falling Drop Method.

Measuring surface tension by means of the weight of a falling drop has become a standard method. Surface tension is measured from freshly formed drops, so that the effect of surface contamination is minimized. Measurements are made under conditions analogous to those existing when sprays are applied to plants.

In principle the drop method depends upon the fact that the size of drops falling from a circular tip with sharp edges having radius r bears a definite relation to surface tension, as given by the expression

$$\gamma = \frac{F m g}{r} \tag{5}$$

where m is the mass of the drop, g is the acceleration due to gravity and F is a factor related to v/r^3, v being the volume of the drop. Values of F can be interpolated from the tables of Young and Harkins (1928).

In practice a standard pipette is used, the pipette having a tip that has been accurately ground to a circular cross section with sharp edges. The radius of the tip is accurately determined with a microscope. The pipette is filled with the test liquid and the delivery of liquid is carefully controlled. Liquid is delivered to form the drop rapidly at first, but the final stage of drop formation must be slow so that drops fall freely, leaving the tip clear of liquid. The value of volume of the current drop depends upon the proper delivery of the previous drop. Several drops, having been formed in this manner, are collected and weighed on an analytical balance. Density of the liquid is determined and the volume of a drop is calculated. Values of v/r^3 obtained in this way are used to obtain the corresponding value of F in the table of Young and Harkins (1928). Surface tension is then computed from equation (5).

3. The Ring Method.

The ring method of measuring surface tension utilizes the DuNuoy tensiometer, which consists of a torsion balance to which is attached a ring of wire. The pull P required to draw the ring through a liquid surface is a measure of surface tension. One would expect surface tension to be given by the pull required, divided by twice the circumference of the ring. However, a correction factor F must be introduced. F varies from 0.75 to 1.11 in value and allows for radius of the ring R, radius of the wire of which the ring is made r, and the volume v of liquid raised, where $v = P/(gd)$. Surface tension is given by the expression

$$\gamma = \frac{P F}{4\pi R} \tag{6}$$

and values of F corresponding to varying values of R/r and R^3/v can be obtained from the tables published by Harkins and Jordan (1930).

The ring method is rapid in practice, and once the torsion balance has been calibrated and dimensions of the ring have been accurately determined, only the density of the liquid need be known for measurements to be made. The method is best suited to liquids of low viscosity.

II. Measurement of Interfacial Tension.

Just as surface tension of a liquid is a characteristic at a given temperature, so is the interfacial tension existing at an interface between immiscible liquids a characteristic of the system.

1. Capillary Method.

Bartell and Miller (1928) have developed a simple method of measuring the interfacial tension between immiscible liquids, based on capillary rise. Two cups, A and B, each about 4 cm in diameter and 8 cm deep are connected through the bottoms by a U tube (Fig. 2). The connection between cup B and the vertical riser leading to cup A is of ordinary glass tubing, whereas the vertical riser itself below cup A is of capillary tubing with an index line etched in the capillary at about the midway point.

The radius of the capillary can be determined by filling cup A with water or benzene, immersing the apparatus in a beaker containing the same reference liquid, inverting the apparatus, and, after fixing it in a stable position so that the capillary is vertical, noting the height of capillary rise. The radius of the capillary can then be calculated, using equation (2) and the known surface tension of the reference liquid.

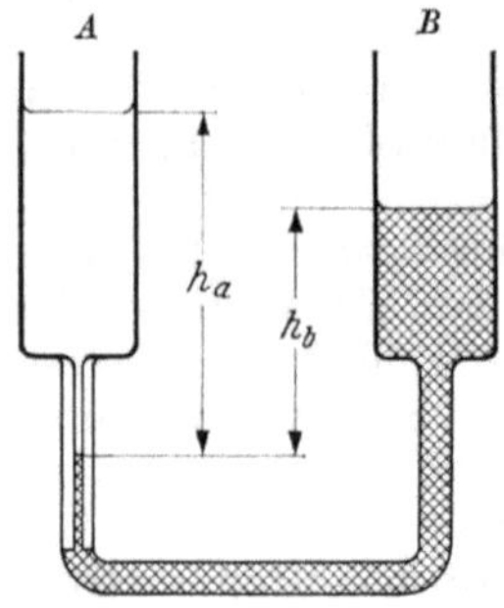

Fig. 2. Apparatus for measuring interfacial tension between immiscible liquids.

In use, the less dense liquid, A, is added to cup A until the meniscus just rises to the bottom of cup B. The denser liquid, b, is then added to cup B until the interface between the liquids is just at the index line in the capillary (Fig. 2). The height of the liquids in each cup above the index line is then determined with a horizontal telescope. Interfacial tension is given by the relation

$$\gamma_{ab} = \frac{rg}{2} (h_a d_a - h_b d_b) \quad (7)$$

where symbols all have the same meaning as in equation (2) and γ_{ab} is interfacial tension. Equation (7) is merely a variant in derivation of equation (2).

2. Falling Drop Method.

This method is more accurate than the capillary method of measuring interfacial tension, but less convenient when the less dense liquid is highly colored. In principle, it is the same as for the falling drop method of measuring surface tension. The denser of the two liquids is allowed to form drops slowly at a standardized tip, which is immersed in the less dense liquid. Because larger drops are formed than when surface tension measurements are made, care must be taken in the choice of a standard tip with a given liquid so that values of v/r^3 will fall in the range of the table of YOUNG and HARKINS (1928). To deliver the denser liquid in a controlled manner and to measure drop volume accurately, a hypodermic syringe is used to which a micrometer screw is attached. The standard tip is placed on the delivery end of the hypodermic syringe.

Each drop of the denser fluid is formed very slowly as it reaches final size. This is necessary because the volume of the drop will be incorrectly determined from the micrometer unless the previously formed drop was properly delivered. The volume of the drop, v, is determined from the initial and final readings of the micrometer and densities of the two liquids are determined pycnometrically. Interfacial tension is then calculated from the relation

$$\gamma_{ab} = \frac{V(d_1 - d_2)g}{r} F \quad (8)$$

where r is the radius of the tip and F is determined from the tables of YOUNG and HARKINS (1928) for the corresponding value of v/r^3.

3. Ring Method.

A special form of the DuNuoy tensiometer is adapted to measuring both interfacial and surface tension. The principle of the method, the precautions, and the general procedure are similar for either kind of measurement. The ring method is simple and rapid, but the drop weight method probably leads to more accurate results.

D. Wetting and Spreading.

I. Wetting and the Contact Angle.

When a drop of liquid stands on a solid surface, it exhibits a contact angle θ that is determined by the degree to which the liquid wets the solid. The surface tensions of the solid γ_s and of the liquid γ_l and the interfacial tension of solid-liquid γ_{sl} are then in equilibrium and define the contact angle θ, according to the equation

$$\gamma_s = \gamma_{sl} + \gamma_l \cos\theta \, . \tag{9}$$

The work required per unit area to separate the liquid from the solid is the work of adhesion and is given by DUPRE's equation

$$W_a = \gamma_s + \gamma_l - \gamma_{sl} \, . \tag{10}$$

By combining equations (9) and (10), one obtains YOUNG's equation

$$W_a = \gamma_l (1 + \cos\theta) \, . \tag{11}$$

This equation is useful because it expresses the attraction between liquid and solid in terms that can be measured: the surface tension of the liquid and the contact angle.

Contact angles have been measured in a variety of ways. O'KANE, WESTGATE, GLOVER, and LOWRY (1930), working with sprays on leaves, photographed drops on leaf surfaces, enlarged the images, and measured them with a protractor. EBELING (1938) magnified the image of the drop by projecting it and then measured the height of the drop, h, above the solid surface and its diameter d. He then computed the contact angle from the equation of MACK (1936)

$$\theta = 2 \tan^{-1} h/x \, . \tag{12}$$

This method is valid only when the lens formed by the drop is a segment of a sphere, a condition fulfilled when the contact angle is less than 90° and the drop is less than 3 mm in diameter or when the contact angle is less than 65° and the drop is no greater than 5 mm in diameter.

The tilting plate method is suitable for measuring contact angles when a flat surface can be coated with the material whose wettability is being studied. The coated plate is mounted on a bracket so that its elevation can be controlled and so that the plate can be rotated freely and slowly, but fixed in any position of rotation. The angle of inclination of the plate to the liquid surface is determined with a protractor. After the plate is set at a given inclination, it is lowered into the liquid. At the proper setting, the meniscus of the liquid has no curvature, but approaches the plate as a plane surface. This is the contact angle. A horizontal microscope is helpful in making accurate determinations of this position. Various forms of tilting plate apparatus have been described by EVANS and MARTIN (1935), ADAM (1941), and by WENZEL (1946).

With the form of apparatus discussed by ADAM (1941), the angle of the plate should be set before the plate is lowered into the liquid. This will require a few trials before a reading is taken. Waves on the liquid must be avoided when readings are taken and the interval between immersion of the plate in the liquid and the observation of the angle of contact should be a constant because the angle changes with time.

Even in controlled physical situations the contact angle will vary as much as 2° or 3° from one part of a surface to another (ADAM, 1941). On biological surfaces variations are even greater because they are seldom either clean or cleanable and never smooth. Variations in the contact angle arise from contamination of the surface and from surface irregularities. Also, as a liquid stands on a solid surface, the contact angle decreases with time as the surface becomes wetted from seepage of the liquid into surface pores. Because of this, a distinction is made between the advancing contact angle, θ_a, presented by a freshly wetted surface, and the receding contact angle, θ_r.

The tilting plate method can be used to measure both the advancing and the receding contact angles. The advancing contact angle is obtained when the plate is lowered vertically, exposing freshly wetted surface. The receding angle is observed when the plate is raised, exposing previously wetted surface.

EVANS and MARTIN (1935) have modified the apparatus to eliminate the necessity of raising and lowering the plate, which rotates about an axis so that protracter readings of the angle of tilt are conveniently made. As the surface rotates into the liquid surface, advancing contact angles are observed, and as it rotates away from the liquid surface, receding angles are observed. When angles are less than 90°, readings are taken on one side of the plate, and when greater than 90°, readings are made on the opposite plate surface.

Advancing and receding contact angles have also been less accurately measured by photographing drops moving across a solid surface and measuring the angles at the advancing and receding edge with a protractor. Although the angles are less accurately determined in this case, they are obtained under conditions approximating those in the field.

EVANS and MARTIN (1935) have found a high correlation between the wetting properties of a spray liquid and the receding contact angle, and MARTIN (1940) refers to the simple test of dipping a waxed card into a spray liquid and noting whether the liquid retained by the card, on drawing it out again, forms droplets or remains as a continuous film. In the latter case, the receding contact angle is 0°.

II. Water Repellency.

When a surface has a high contact angle with water, it is water repellent. As a matter of convenience, a water repellent surface can be considered as having a contact angle to water of more than 90°. ADAM (1958) has discussed the factors making for water repellency. These are the hydrophobic properties of smooth surfaces, arising from chemical constitution, and surface roughness, which is physical in nature. Smooth surfaces seldom present contact angles to water greater than 105°, which is characteristic of paraffin. Most smooth surfaces have considerably lower contact angles than this to water.

A rough surface may have a considerably higher apparent contact angle to water than a smooth surface of the same composition. WENZEL (1946) has investigated the quantitative effect of surface roughness on wettability, water repellency and the contact angle and has shown that the average contact angle, $\bar{\theta}$, on a rough surface in which the ratio of real to apparent surface is r will be

$$\cos\theta = r\cos\bar{\theta} \qquad (13)$$

where θ is the contact angle on a smooth surface of the same composition.

A consideration of the nature of the cosine function will suggest the effect of surface roughness on the average contact angle $\bar{\theta}$. The cosine of 0° is +1, of 90° is 0, and of 180° is −1. As is suggested by equation (13), if the contact angle on a smooth surface is less than 90°, the effect of surface roughness will be to decrease the average contact angle, to make the surface more readily wetted. When the contact angle is greater than 90°, the effect of surface roughness will be just the opposite: the average contact angle will be greater on a rough than on a smooth surface and the surface will be more water repellent.

Surfaces must be roughened to have a permanent water repellency. In this case the contact angle also must be greater than 90°. A drop on a solid water repellent surface makes only partial contact with the surface, air being trapped in the irregularities under the drop. CASSIE (1948) and CASSIE and BAXTER (1944) have considered the nature of such water repellency and found that if the liquid contacts an area of f_1 cm², whereas air gaps on the solid surface under the drop amount to f_2 cm², the average contact angle amounts to

$$\cos\bar{\theta} = f_1 \cos\theta - f_2 , \tag{14}$$

a relation valid for both advancing and receding contact angles.

As ADAM (1958) has noted, the surface projections responsible for roughness must be regularly spaced, closely spaced, and stiff to impart water repellency. These are the very conditions imposed by the wax rodlets on the surface of the cuticle, and both LINSKENS (1950) and JUNIPER (1959) have noted the correlation between high contact angles on leaf surfaces and the presence and pattern of wax deposits. Not all leaves have wax deposits on their surface (SCHIEFERSTEIN and LOOMIS, 1956, 1959; JUNIPER, 1958). Differences in wax deposits on young and old leaves, and on leaves produced in darkness and light apparently account for differences in the contact angle to water and the wettability of these leaves (LINSKENS, 1952a; LINSKENS, 1952b). Contact angles as high as 140° have been observed for pure water on wheat leaves (FOGG, 1948), resulting from surface roughness. In addition to wax rodlets, small leaf hairs also contribute to surface roughness and they also contribute to water repellency if they are sufficiently stiff, when their real, i.e., smooth surface, contact angle is high.

III. Spreading and the Spreading Coefficient.

The spreading of a liquid across a solid surface will occur spontaneously when the work of adhesion between solid and liquid, W_a, is greater than the work of cohesion, W_c. This condition must also be fulfilled when one liquid spreads over a denser liquid. HARKINS and FELDMAN (1922) have measured the spreading tendency of a liquid on another surface in terms of the spreading coefficient, S, which equals $W_a - W_c$. For the spreading of one liquid, A, over another, B, one can derive S simply from DUPRE's equation (10), remembering that $W_c = 2\gamma_A$, whence

$$S = W_a - W_c = \gamma_B - \gamma_A - \gamma_{AB} \tag{15}$$

or from equation (12), whence

$$S = \gamma_l (\cos\theta_a - 1) \tag{16}$$

where θ_a is the advancing contact angle customarily used in spreading studies. Equation (15) applies conveniently to speading of one liquid on another and equation (16) to the spreading of a liquid on a solid surface. WILCOXON and HARTZELL (1931) found the spreading coefficient of spray liquids on insect integument was correlated with toxicity.

EVANS and MARTIN (1935) have found the advancing contact angle is well correlated with spreading properties of sprays on leaves, and CUPPLES (1941) has used the spreading coefficient as given in equation (15) in evaluating spreading properties of spray liquids. Spreading does not occur spontaneously unless $\theta_a = 0°$, but use of equation (15) with suitable liquids leads to positive values of the spreading coefficient, a situation describing spontaneous spreading.

An index related to spreading of sprays on leaf surfaces has been proposed by O'KANE, et al. (1931). Assuming the lens of spray on the leaf forms a segment of a sphere, a condition met when drops are small and the contact angle is less than 90°, they have computed the area of leaf surface covered by a drop of spray from the relation

$$A = \frac{(3V)^{2/3} \pi^{1/3} (1 + \cos\theta)}{[(1 - \cos\theta)(2 + \cos\theta)^2]^{1/3}} . \tag{17}$$

These considerations bring up the contribution of surface dirt to the wettability of leaves. Leaves in the field are rarely clean. Solid dirt contributes to roughness of the surface but is more wettable than the leaf surface itself, and when leaves are very dirty, they are more wettable than when relatively clean. Thus EBELING (1938) has noted the spontaneous spreading of oil sprays on dirty leaves when spreading fails to occur on clean leaves. SHUTTLEWORTH and BAILEY (1948) have discussed the spreading of liquids on rough surfaces from the physical point of view.

IV. Penetration of Porous Surfaces.

The extent to which sprays enter stomata is debated, as also is the importance of absorption of materials via stomata. When conditions favor stomatal entry, absorption of materials is rapid (CURRIER and DYBING, 1959). Special conditions must be met for sprays to penetrate stomata, however. Aqueous sprays can be driven through stomata experimentally, but this is rarely achieved in nature except in driving rainstorms. Conditions for penetration of stomata by means of surface forces are most frequently fulfilled by oil sprays. Surface tension must be low in aqueous sprays to meet the necessary conditions.

The conditions for penetration of porous surfaces have been discussed by ADAM (1941, 1948). Equation (2), describing capillary rise of a liquid, may be more completely written as

$$\gamma_l = \frac{r h d g}{2 \cos\theta_a} . \tag{18}$$

At equilibrium, the pressure P due to the weight of the liquid equals hdg dynes per cm^2 and this equation may be rewritten, solving for P_i, the penetrating pressure that drives the liquid into the capillary at the start, as

$$P_i = \frac{2 \gamma_l \cos\theta_a}{r} . \tag{19}$$

When a liquid passes a constriction and enters an underlying pore, the value of P_i is less than would be predicted from the size of the opening in the surface, but no less than would be predicted from a knowledge of the radius of the underlying pore and the contact angle. For the spray liquid, penetrating properties are proportional to the quantity $\gamma_1 \cos\theta_a$ and will have maximum value when θ_a is zero or nearly so. For the entry through stomata in leaves, the problem is not yet completely solved as it applies to stomata and the substomatal chambers. However, the surfaces of spongy parenchyma cells are wet and must present a negligibly low contact angle to water. On this basis, one would surmise that entry of spray liquids through stomata could occur to a greater extent than is generally appreciated.

When one considers that aqueous sprays evaporate quickly, rate of entry may become important, a matter that has been considered by HOSKINS (1933). Under these conditions, viscosity of the liquid becomes important and by applying POISEUILLE's law to the problem, one deduces that the linear rate of flow, f, through a pore is

$$f = \frac{\gamma_l \cos\theta_a}{4\eta l} \tag{20}$$

where η is the absolute viscosity of the liquid and l is the length of the capillary. Thus rate of entry is proportional to the quantity $\gamma_l \cos\theta_a$ and is inversely proportional to absolute viscosity of the spray fluid. HOSKINS (1933) has demonstrated the applicability of this equation to the penetrating ability of a series of petroleum oils into masses of insects and bark.

Thus the magnitude of the quantity $\gamma_l \cos\theta_a$ is a controlling factor in the penetration of spray fluids through stomata. This quantity is a maximum when the contact angle is zero or nearly so. In general, as γ_l decreases, so does θ_a, but there is no simple relation between them. EBELING (1939) has observed that when θ is low, only surface tension effects need be considered, but when θ has a finite value, the quantity $\gamma_l \cos\theta$ may be increased or decreased after the addition of a wetting agent. The penetrating ability of a spray may be increased or decreased by the addition of a surfactant, and when θ is greater than zero, both γ_1 and θ must be determined to predict the overall effect.

References.

ADAM, N. K.: The physics and chemistry of surfaces, 3rd ed. London: Humphrey Milford 1941; — Disc. Faraday Soc. **3**, 5 (1948); — Endeavour **17**, 37 (1958). — AHLGREN, G. H., G. KLINGMAN and D. E. WOLF: Weed Control, New York: Wiley 1951.

BARTELL, F. E., and F. L. MILLER: J. Amer. Chem. Soc. **50**, 1961 (1928). — BOYNTON, D.: Ann. Rev. Plant Physiol. **5**, 31 (1954). — BURCHFIELD, H. P.: in Plant pathology — problems and progress, 1908—1958, p. 293, Madison, Wis.: U. Wisconsin Press 1959. — BURCHFIELD, H. P., and A GOENAGA: Contrib. Boyce Thompson Inst. **19**, 141 (1957).

CASSIE, A. B. D.: Disc. Faraday Soc. **3**, 11 (1948). — CASSIE, A. B. D., and S. BAXTER: Trans. Faraday Soc. **40**, 546 (1944). — CUPPLES, H. L.: J. Agric. Res. **63**, 681 (1941). — CURRIER, H. B., and C. D. DYBING: Weeds **7**, 195 (1959).

DYE, M. H.: Nature (Lond.) **178**, 551 (1956).

EBELING, W.: VII Intern. Entomol. Congr. **4**, 2966 (1938); — Hilgardia **12**, 665 (1939). — LNNIS, W. B., and F. T. BOYD: Bot. Gaz. **107**, 552 (1946). — EVANS, A. C., and H. MARTIN: J. Pomol. **13**, 261 (1935).

FOGG, G. E.: Disc. Faraday Soc. **3**, 162 (1948).

GOODMAN, R. N., and W. M. DOWLER: Plant Dis. Reptr. **42**, 122 (1958). — GRAY, R. A.: Phytopathology **46**, 105 (1956). — GUSTAFSON, F. G.: Amer. J. Bot. **43**, 157 (1956); — Plant. Physiol. **32**, 141 (1957).

HARKINS, W. D.: in WEISSBERGER's Physical methods of organic chemistry, vol. 1. New York: Interscience 1949. — HARKINS, W. D., and F. E. BROWN: J. Amer. Chem. Soc. **41**, 499 (1919). — HARKINS, W. D., and A. FELDMAN: J. Amer. Chem. Soc. **44**, 2665 (1922). — HARKINS, W. D., and H. F. JORDAN: J. Amer. Chem. Soc. **52**, 1751 (1930). — HOPP, H., and P. J. LINDER: Amer. J. Bot. **33**, 598 (1946). — HORSFALL, J. G.: Principles of fungicidal action. Waltham, Mass.: Chronica Botanica 1956. — HOSKINS, W. M.: Hilgardia **8**, 49 (1933). HOSKINS, W. M., and Y. BEN-AMOTZ: Hilgardia **12**, 83 (1938).

JUNIPER, B. E.: Endeavour **18**, 20 (1959).

KERLING, L. C. P.: Meded. LandbHoogesch. Gent **24**, 577 (1959). — KITCHENER, J. A.: FINDLAY's practical physical chemistry, 8th ed. New York: Longmans Green 1954. — KNIGHT, H., J. C. CHAMBERLIN and C. D. SAMUELS: Plant Physiol. **4**, 299 (1929). — KOONTZ, J., and O. BIDDULPH: Plant Physiol. **32**, 463 (1957).

LINSKENS, H. F.: Planta **38**, 591 (1950); — Naturwissenschaften **39**, 65 (1952a); — Planta **41**, 40 (1952b).

MACK, G. L.: J. Phys. Chem. **40**, 159 (1936). — MARTIN, H.: The scientific principles of plant protection, 3rd ed. London: Edw. Arnold 1940. — MITCHELL, J. W., B. C. SMALE and R. L. METCALF: Advances in Pest Control Research **3**, 359 (1960). — MUELLER, L. E., P. H. CARR and W. E. LOOMIS: Amer. J. Bot. **41**, 593 (1954).

O'KANE, W. C., W. A. WESTGATE, L. C. GLOVER and P. R. LOWRY: N. H. Agric. Exp. Station Tech. Bul. **39** (1930); **46** (1931).

POTTS, S. F.: Concentrated spray equipment. Caldwell, N. J.: Dorland 1958.

RICE, E. L.: Bot. Gaz. **109**, 301 (1948). — RICH, S.: Phytopathology **44**, 203 (1954). — RICH, S., and J. G. HORSFALL: Phytopathology **42**, 457 (1952). — ROBBINS, W. W., A. S. CRAFTS and R. N. RAYNOR: Weed Control. New York: McGraw-Hill 1942. — ROHRBAUGH, P. W.: Plant Physiol. **9**, 699 (1934).

SCHIEFERSTEIN, R. H., and W. E. LOOMIS: Plant Physiol. **31**, 240 (1956); — Amer. J. Bot. **46**, 625 (1959). — SCHWARTZ, A. M., and J. W. PERRY: Surface active agents, their chemistry and technology. New York: Interscience 1949. — SCOTT, FLORA M., MARY R. SCHROEDER and F. M. TURRELL: Bot. Gaz. **109**, 381 (1948). — SHUTTLEWORTH, R., and G. L. J. BAILEY: Disc. Faraday Soc. **3**, 16 (1948). — SKOSS, J. D.: Bot. Gaz. **117**, 55 (1955). — SMITH, H. H.: Bot. Gaz. **107**, 544 (1946). — STANIFORTH, D. W., and W. E. LOOMIS: Science **109**, 628 (1949).

VAN OVERBEEK, J.: Ann. Rev. Plant Physiol. **7**, 355 (1956).

WELLMAN, R. H., and S. E. A. MCCALLAN: Contrib. Boyce Thompson Inst. **14**, 151 (1946). — WENZEL, R. N.: Industr. Engng. Chem. (Industr.) **28**, 988 (1936). — WILCOXON, F., and A. HARTZELL: Contrib. Boyce Thompson Inst. **3**, 1 (1931).

YARWOOD, C. E.: Phytopathology **33**, 1146 (1943). — YOUNG, T. F., and W. D. HARKINS: Int. Crit. Tab. **4**, 435 (1928). New York: McGraw-Hill.

Tissue and Single Cell Cultures of Higher Plants as a Basic Experimental Method[1].

By

A. C. Hildebrandt.

With 26 Figures.

The growth of cells, tissues, and organs of higher plants under sterile conditions on chemically-defined media in vitro has been possible for many plant species. The technique of tissue culture of higher plant materials has, in other words, passed through the phase where cultures were attempted just to see what species of the isolated tissues would grow for unlimited periods in vitro. The method has become well established and many different types of plant materials are now routinely cultured for extended periods. Furthermore, application of the method has been made to studies of a wide range of basic problems related both to normal and diseased growth.

No attempt has been made here to present a detailed review of the literature that would give credit to the many individuals who have contributed to the rapidly expanding field of plant tissue culture. Instead, only certain critical background studies relative to a general range of types of tissue culture work are cited below. General details of the methods and materials available are summarized. The beginner, as well as the more experienced in biological research, may find sufficient help to encourage use of tissue culture to attempt application of the method for the solution of particular basic or applied problems. Special emphasis has been made to callus and single cell cultures, but the procedures employed and media are essentially similar to those for cultures of more differentiated plant materials.

A. History of the Method.

The possible culture in vitro of plant and animal cells, tissues, and organs has stimulated the interest of biologists for many years. The early botanist HABERLANDT (1902) perceived many interesting and practical uses for the method and first attempted growth of plant cells in tissue culture. He was unsuccessful, perhaps because he chose the plant materials we now know are unsuitable or difficult to culture. Furthermore, the nutritional and physical chemical requirements for growth were not clearly understood.

Actual success in growing isolated higher plant materials in vitro came many years later. In the meantime, animal tissue cultures were first prepared by HARRISON (1907) using frog nerve tissue in a hanging drop of frog lymph. Animal tissue culture work is documented in books by PARKER (1950), MURRAY and KOPECH

[1] Published with the approval of the Director of the Wisconsin Agricultural Experiment Station. This work was supported in part by the Rockefeller Foundation, the American Cancer Society, the National Institutes of Health and by the Research Committee of the Graduate School from funds supplied by the Wisconsin Alumni Research Foundation.

(1953), WHITE (1954), and WILLMER (1954). Much of the methodology from successful animal tissue culture was utilized to make possible successful plant tissue cultures.

Details of the background and early history of plant tissue culture are given extended treatment in books by GAUTHERET (1942, 1959), and WHITE (1945) and review papers by MOREL (1948), GAUTHERET (1955), KLEIN (1957), NOBECOURT (1957), SKOOG and MILLER (1957), WHITE (1957), RIKER and HILDEBRANDT (1954, 1958), and STEWARD and SHANTZ (1959).

B. Types of Cultures.

The in vitro culture of higher plant materials has centered for a large part around organ (root and stem tips), embryo, callus, and recently cell cultures. Excised roots were grown for weeks (KOTTE, 1922 and ROBBINS, 1922) in nutrient solutions. Isolated tomato roots were shown to be capable of unlimited growth in vitro (WHITE, 1934). The long history of attempts to grow root tip cultures in vitro provided much necessary information needed for the successful cultures of callus tissue. GAUTHERET (1939) and NOBECOURT (1939) with carrot, and WHITE (1939) with tobacco, each working independently established cultures capable of unlimited growth in nutrient media. The isolation and growth of single cells of higher plants that had been the original hope of HABERLANDT was accomplished by MUIR et al. (1954, 1958).

The early successes with root and callus tissues were followed by cultures of isolated embryos (VAN OVERBEEK, CONKLIN, and BLAKESLEE, 1942; RANDOLPH and COX, 1943; CURTIS and NICHOL, 1947; ZIEBUR and BRINK, 1951). Meristem tips were similarly cultured by LOO, 1945; BALL, 1946; LEE, 1950; WETMORE and WARDLAW, 1951; WETMORE, 1954; MOREL and MARTIN, 1955; KASSANIS, 1957; and QUAK, 1958. Additional types of tissue from various origins were cultured, for example, from monocotyledons (MOREL, 1950a; MOREL and WETMORE, 1951a); from Gymnosperms (BALL, 1950, 1955; LOEWENBERG and SKOOG, 1952; REINERT and WHITE, 1956); from Pteridophytes (MOREL, 1950b, 1956; HUREL-PY, 1950; MOREL and WETMORE, 1951b; STEEVES and SUSSEX, 1951; SUSSEX and STEEVES, 1953, 1958); from pollen (TULECKE, 1953, 1957); from endosperm (LA RUE, 1949; STRAUS and LA RUE, 1954; STRAUS, 1954, 1959, 1960; TAMAOKI and ULLSTRUP, 1958); and from flowers (NITSCH, 1951).

C. Tissue Culture Media.

Many formulae have been developed for media for the culture of isolated tissues of higher plants. Many isolated plant tissues grow on completely synthetic media. Basically, many media contain inorganic salts, nitrate for nitrogen, sucrose for carbon, certain vitamins, minor elements and distilled water. Other tissues require, in addition to the basic medium, supplements of complex extracts. If a solid medium is required, agar may be added sufficient to give the tissue a firm support for aeration at the surface of the medium. Media prepared according to the formulae of NOBECOURT (1939), WHITE (1942), GAUTHERET (1942), HILDEBRANDT, RIKER and DUGGAR (1946), BURKHOLDER and NICKELL (1949), and HELLER (1953) have been used extensively for culturing many types of tissues from many species.

I. Concentrated Stock Solution.

It is convenient, where large volumes of media are required, to prepare concentrated stock solutions of the ingredients of the medium and to use aliquots from the concentrated stock solutions for small quantities of the medium. In this laboratory where WHITES' (1942) medium and modifications of his basic medium (HILDEBRANDT, RIKER and DUGGAR, 1946) have been used extensively, the concentrated mineral salt stock solutions were prepared in the quantities shown in Table 1. Chemically pure chemicals are used.

The three basic concentrated stock solutions of inorganic salts are diluted and supplemented as described below. The tobacco medium was originally developed for cultures of callus that came from isolated stem segments of tobacco (*Nicotiana glauca* Grah. ♀ × *N. Langsdorffii* Weinm. ♂), the sunflower medium from callus of crown gall origin from marigold (*Tagetes erecta* L. var. Sunset Giant) and sunflower (*Helianthus annuus* L. var. Giant Russian), and WHITE's medium has been useful

Table 1. *Concentrated stock solutions for tissue culture media.*

Chemical	Tobacco medium gm/5 liters	Sunflower medium gm/5 liters	WHITES' medium gm/5 liters
Section I			
Na_2SO_4	20.0	2.5	10.0
$Ca(NO_3)_2 \cdot 4H_2O$	10.0	20.0	10.0
$MgSO_4 \cdot 7H_2O$ (dissolve separately)	4.5	18.0	18.0
KNO_3	2.0	4.0	4.0
KCl	1.625	3.25	3.25
$Na_2HPO_4 \cdot H_2O$	0.825	3.3	0.825
$MnSO_4 \cdot 4H_2O$	0.1125	0.1125	0.225
$ZnSO_4 \cdot 7H_2O$	0.15	0.075	0.075
H_3BO_3	0.009375	0.075	0.075
KI	0.075	0.009375	0.0375
	use 200 ml/l	use 200 ml/l	use 100 ml/l
Section II			
$Fe_2(SO_4)_3 \cdot 6H_2O$	—	—	2.5 mg/l
$Fe_2(C_4H_4O_6)_3$	40.0 mg/l	5.0 mg/l	—
Glycine	3.0 mg/l	12.0 mg/l	3.0 mg/l
Thiamine	0.1 mg/l	0.1 mg/l	0.1 mg/l
Pyridoxine	—	0.8 mg/l	—
Sucrose	20.0 gm/l	20.0 gm/l	20.0 gm/l
Bacto agar	6.0 gm/l	6.0 gm/l	6.0 gm/l

Section III Supplemented basic media

D-Medium = tobacco medium + coconut milk (150 ml/l), calcium pantothenate (2.5 mg/l), 2,4-dichlorophenoxyacetic acid (6 mg/l), α-naphthaleneacetic acid (0.1 mg/l).
C-Medium = same as D-medium except without 2,4-D.
NAA-Medium = tobacco medium + α-naphthaleneacetic acid (0.1 mg/l).
Adjust all media to pH 5.8—5.9.

for callus cultures from crown gall of periwinkle (*Vinca rosea* L.) and Paris daisy (*Chrysanthemum frutescens* L.). These basic formulae have since proved adequate, with or without certain supplements as indicated below, for growing many types of callus and single cell cultures.

In preparing the 5 liters of concentrated stock solution care must be exercised to completely dissolve the individual salts. Dissolve the $MgSO_4 \cdot 7H_2O$ separately in 1 liter of distilled water. The remaining chemicals listed in Section I, in Table 1 are dissolved in 4 liters of distilled water. Dissolve each chemical completely before adding the next chemical. When the last chemical is completely dissolved, the 1 liter of $MgSO_4 \cdot 7H_2O$ solution is added, with constant stirring, to the 4 liters containing the remaining chemicals. This concentrated stock solution is then sterilized at 15 pounds pressure for 20 minutes in the autoclave. It is important to mark the level of the solution on the glass container before autoclaving so that any loss of water by evaporation is recognized. Water lost should be restored with distilled water before aliquots are removed to prepare the complete media.

II. Preparation of Basic Media.

The formulae for several useful media are listed in Table 1. To prepare one of the three basic media the indicated aliquot of the concentrated salt mixture is used and to this are added all the chemicals listed under Section II, except the agar. The agar is melted separately in distilled water in a glass container and then added to the solution containing the other items listed in Sections I and II, and the solution then brought to the correct volume. Agar is omitted for liquid media. Before dispensing the media to the culture vessels, the acidity is commonly adjusted to pH 5.5 to 6.0.

The three basic media support growth of many types of excised higher plant tissues. These are especially satisfactory for growth of tissues of crown gall origin. Tissues of non-crown gall origin, however, often require supplements of growth substance or more complex plant extracts. In the latter case, the basic tobacco medium has been used with the three media listed in Section III, and supplemented as indicated. Certain tissues of normal origins require the addition of α-naphthaleneacetic acid (NAA) at 0.1 mg/l. Indole-3-acetic acid (IAA) is commonly used as an alternative. The NAA has similar growth stimulating activity to IAA and has the advantage of being relatively heat and light stable. The NAA is prepared as a concentrated stock solution and aliquots taken to prepare individual batches of the media. In preparing the concentrated stock, the NAA is first dissolved in a few drops of 95% ethyl alcohol, brought to volume with distilled water, and sterilized in the autoclave. The solution must be brought to volume after autoclaving in case water has evaporated.

III. Supplements to the Basic Media.

Many other tissues of normal origin grow best on the NAA medium supplemented with coconut milk (CAPLIN and STEWARD, 1948) and calcium pantothenate. Certain tissues require the further addition of 2,4-dichlorophenoxyacetic acid (2,4-D) (STEWARD and CAPLIN, 1951). Ascorbic acid (HOTSON, 1953) at 300 mg/l has been a further supplement useful to culture certain tissues that commonly turn dark when exposed to the air, or that produce a dark pigmentation in the agar. Many additional chemicals, growth substances, vitamins, amino acids, sugars, purines, or pyrimidines are required or beneficial for specific types of tissue (SKOOG and TSUI, 1948; JABLONSKI and SKOOG, 1954; BLACK, 1958; BRAUN, 1958; HILDEBRANDT, 1958; RIKER and HILDEBRANDT, 1958; and GAUTHERET, 1959).

Coconut milk (liquid endosperm) used for the media is secured from fresh coconuts. Coconuts are commonly secured directly from the wholesaler to insure fresh nuts with abundant milk and then are stored in a cold room at 40° F until needed. The milk is drained through cheesecloth to remove debris. Care is exercised that no fermented milk is used. The coconut milk can be added directly to the media and the mixture can be sterilized in the autoclave without apparent loss of growth-stimulating activity. If fresh nuts with milk are not available, the solid white endosperm may be ground and used as a substitute. The ground material may be added directly to the medium or extracted with distilled water and the extract added. Where storage of the nuts is a problem, the solid endosperm may be extracted and frozen for future use. Similarly, the fresh milk can be evaporated under vacuum at low temperature to a consistency of honey, and refrigerated until needed. The evaporated coconut milk when reconstituted with distilled water and added to the media had growth-stimulating activity equally as good or better than the fresh milk for callus cultures.

IV. Purity of the Water.

The purity of the water available may be a problem for certain types of experiments where minor quantities of chemicals as impurities must be avoided. Common laboratory distilled water or demineralized water is usually satisfactory, but it should be tested for purity. It is convenient and advantageous to use double-distilled water, redistilled in a glass still with added potassium permanganate in the evaporating flask.

V. Types of Culture Vessels.

Various types of culture vessels have been used for tissue cultures. Root tips have been commonly grown in liquid media in 125 ml Erlenmeyer flasks or on agar in Petri plates (White, 1945). Embryo, stem tip, and callus cultures may be conveniently grown on agar in tubes of varying lengths and diameters (Gautheret, 1939 and Nobecourt, 1939). Tubes are especially useful when first isolations of many tissue pieces are being made, since contaminated cultures can be discarded without loss of sterile cultures, as may occur when several tissue pieces are isolated to one culture vessel. Callus cultures of certain species also grow well supported on a filter paper wick over a liquid medium (Heller, 1953) or on agar in Petri plates (Bergmann, 1959). Glass wool or glass beads may also be used to support the tissues at the air surface with liquid media. Tubes of various shapes and sizes have been designed to permit aseptic renewal and exchange of nutrients and gases (De Ropp, 1946; Caplin and Steward, 1949; Sossountzov, 1953, and references therein).

The culture vessels with media are often closed with nonabsorbant cotton plugs and sterilized in the autoclave. The culture vessels, however, with or without cotton plugs may be covered with small beakers (White, 1945), metal caps, wax paper, polyethylene sheets or metal foil (Gautheret, 1939). If screw top caps are used, they should be left loose enough on the culture vessel to permit exchange of air. The various wrapping materials placed on top of cotton plug closures reduce the water evaporation from the medium and permit exchange of air. However, in humid atmospheres the various wrappers may be unsatisfactory over the top of cotton plugs, because the cotton plugs become moist enough so that air-borne spores trapped in the plugs germinate and the resulting mycelium contaminates the tissue culture.

Detailed studies of the nutritional requirements for growth of various tissues require containers made of Pyrex or other hard glass. Soft glass prescription bottles are satisfactory however, for most culture work, since the amounts of impurities in the chemicals used in the medium and in the agar far exceed those available from the glass.

Prescription bottles of 6-ounce capacity (Modern Oval) with flat sides on both the narrow and wide sides are comparatively inexpensive and very convenient. The culture bottles with agar or liquid media can be incubated either standing upright or resting on the narrow or flat side. Bottles placed on the flat side expose a greater surface area of the medium, as compared to the culture bottles placed upright, and permit culturing several pieces of tissue in one bottle. Bottle cultures resting on their sides and plugged with cotton, as compared to upright cultures, lose less water by evaporation and the chances of contamination from outside organisms are greatly reduced, as with the slanted-armed Carrel flasks used for animal tissue cultures. When large numbers of culture bottles are required, it is convenient to store these in heavy gauge, wire screen, metal trays approximately 4″ ×7″ ×25″. These trays with cultures can be stored in a minimum amount of space and can be moved easily without excessive tipping or jarring of the cultures.

VI. Cleaning of Glassware.

Glassware may be cleaned in several ways. Used agar media and old cultures are commonly rinsed or brushed out using hot tap water. Dirty glassware can be cleaned by soaking in potassium dichromate or concentrated sulfuric acid, sodium nitrate cleaning solution. Otherwise, glassware can be soaked overnight in detergent, brushed clean of macroscopic debris, rinsed thoroughly in hot tap water and drained. Finally bottles are filled with distilled water, allowed to soak overnight, drained, rinsed with distilled water, and allowed to drain dry. The plastic caps that come with the bottles are usually discarded and nonabsorbant cotton plugs are used instead.

VII. Aeration of the Cultures.

Liquid medium cultures in tubes, bottles, or flasks may need additional aeration. Test tube cultures were rotated in a roller tube drum aerator (WHITE, 1945). Apparatus that permitted renewal and circulation of the medium was also useful (DE ROPP, 1946). Special tube cultures were aerated in a larger roller drum mechanism (CAPLIN and STEWARD, 1949), and large-mouth culture bottles (2 ×2 ×4 inches) with or without 22 ×50 mm microscope cover glass baffles were rotated in a roller drum aerator (MUIR, HILDEBRANDT, and RIKER, 1958). Rotary shakers have been useful for flask cultures. Large volume cultures (NICKELL and TULECKE, 1960) were agitated by using a shaker or by bubbling sterile air through the medium. Comparisons of various culture procedures for growth of tissues on agar or in liquid cultures (MUIR, HILDEBRANDT, and RIKER, 1958) and unpublished results indicated that excellent growth in liquid cultures developed using the 6-ounce prescription bottles in an upright position with 40 ml of liquid medium on a reciprocating shaker (Fig. 1). The reciprocating shaker used has 6 movable shelves (20″ ×46″) and makes 60 complete cycles per minute with a 4-inch swing.

VIII. Sterilization of Media.

Most media can be sterilized in the autoclave at 15 pounds pressure for 20 minutes without apparently significantly altering the composition of the media. Certain carbohydrates, vitamins, growth substances, and plant or animal extracts may be modified or their growth activity destroyed by heating in the autoclave. Additional disadvantages in autoclaving media have been discussed by BALL, NITSCH and EARLE (see STREET, 1957). Chemicals that are heat labile may be conveniently sterilized by filtration and then added aseptically to the sterile medium after it has been autoclaved and before it has cooled to the temperature of solidification of the agar. It is often convenient to filter concentrated solutions of heat labile materials and then to add aliquots of the filtrate to the basic medium to provide the desired final concentrations.

Seitz filters are useful for sterile filtration, but microbial contaminants commonly pass through if large volumes are filtered. Pressure or suction should be reduced so that the filtrate comes through a drop at a time and the filtrate collected in small samples and tested for sterility. Furthermore, the filtrate may be contaminated by materials from the filter pad, or the acidity of the filtrate may be

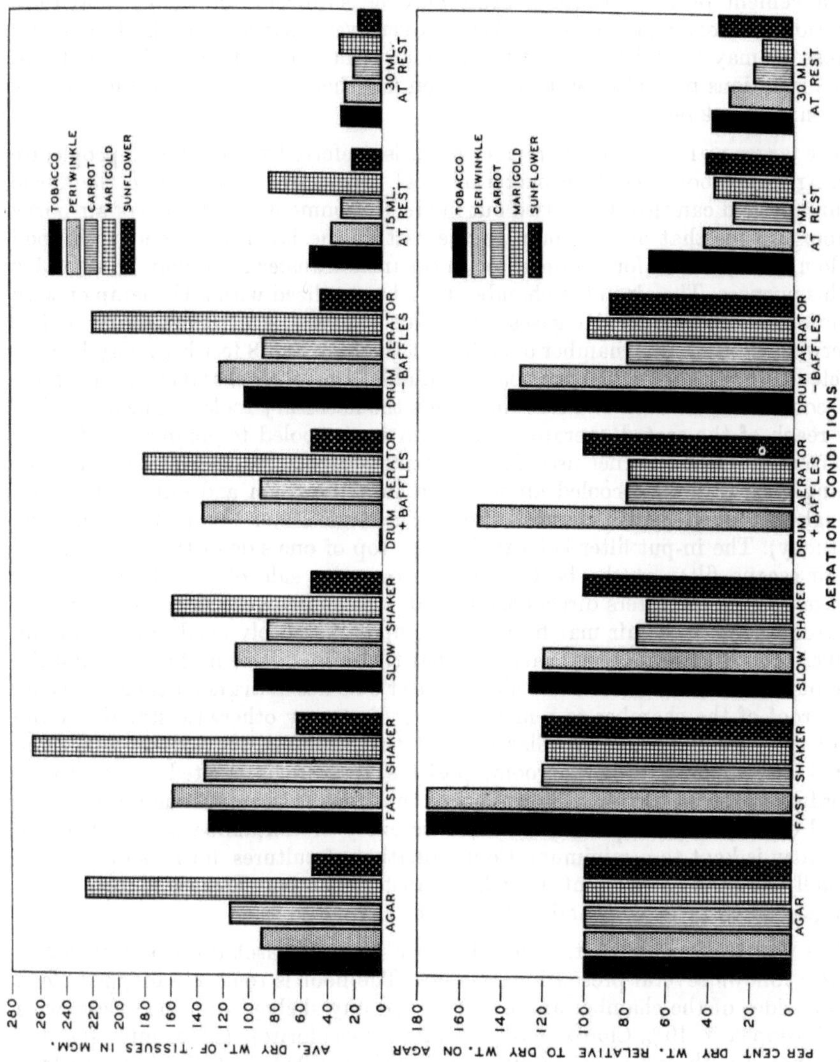

Fig. 1. Growth of callus from marigold, periwinkle, and sunflower crown galls and from normal tobacco and carrot tissue on various aeration conditions as indicated. The volume of medium was 40 ml except where indicated otherwise. The top graph represents the final dry weights. The bottom graph shows dry weights expressed as percents, with the dry weight of each species on agar taken as 100 per cent. [From: W. H. Muir, A. C. Hildebrandt, and A. J. Riker: Am. J. Botany 45, 589—597 (1958)].

modified. The fritted or sintered glass filter of UF porosity and 60 ml capacity has been more satisfactory for sterile filtration. A water pump for suction pressure may be attached to a suction flask holding the glass filter. Since the filters are commonly washed with a concentrated sulfuric acid, sodium nitrate cleaning solution, care must be exercised that enough distilled water is used to rinse out all of the cleaning solution acid before the next filtration.

D. Isolations of Organs, Tissues and Cells.

The various manipulations involved in isolating and establishing plant tissue in vitro are carried out under aseptic conditions. The necessary aseptic conditions are conveniently secured by using a transfer chamber. A small table top transfer chamber with space for the necessary tools, tissues, and media, and space to permit the movement of the operators' arms may be sufficient for small, short-term operations. Occasionally, a sterile plastic cover over a surface-sterilized binocular microscope may be satisfactory to make tissue isolations under aseptic conditions. Tissue isolations may also be made between flat sheets of paper toweling as they come in the package.

The larger walk-in type of transfer room is preferred where several people are working in a laboratory. The bacteriological procedures necessary to the tissue culture method can then be carried out in an environment largely free of air-borne microorganisms that are common to the rest of the laboratory, and with good freedom of movement for the operator. A separate transfer room may be available for this purpose. The air in the chamber may be sterilized with a UV lamp or with bacteriocidal and fungicidal gases or liquids (PECK, 1947; ROBERTSON, 1947). Otherwise, a glassed-in chamber 5×5 feet at the base and 8 feet high may be constructed to fit in the laboratory space available. One such chamber has a 2-foot-wide counter at the back and side that holds the necessary tools and media within easy reach of the seated operator. The room is air-cooled to compensate for the heat from a Bunsen burner used for sterilizing scalpels, transfer needles, and for flaming glassware. The cooled air is forced into the room and out of the room through two oiled paper, air mat filters (American Filter Company, Louisville, Kentucky). The in-put filter is located at the top of one side of the chamber and the air-escape filter at the bottom on the opposite side of the chamber. This arrangement of the filters directs the cooled, sterile air over the counter and the operator's hands. The air mat filters are composed of 6-ply oiled paper, but the number of ply actually used is varied so that there is movement of air through the room but without a draft. A small air mat filter 6 inches in diameter is also inserted in the roof of the chamber to remove hot air that may otherwise stratify at the top of the chamber, and a similar filter at the intake of the blower. Since the filtered air is blown into the room, positive pressure is created in the room. Accordingly, the sterile air is blown out of the transfer room through any cracks around doors or other openings and chance entry of contaminated air from the laboratory is kept to a minimum. Contamination of cultures during subsulturing is usually less than 1 per cent in such a transfer chamber. The room is equipped with gas, fluorescent light, and a steam outlet.

The preparation of the transfer chamber for tissue isolations or transfer of cultures follows several precautionary steps. The floor is routinely mopped clean and the sides of the chamber and counters and any shelves are wiped clean with a cloth dipped in 10% Clorox (6.8% sodium hypochlorite). The counter space is then covered with cotton gauze moistened with distilled or sterile water to trap and hold air-borne contaminants that may touch the cloth surface. Trays of culture bottles with media and other tools needed for transfer are then placed in the room. Instead of using UV light, or gas, or liquid to sterilize the room, live steam is admitted to the room under the counters until the room is filled with condensate but not enough to heat the room. The cooled, filtered air is then blown through the chamber to remove the condensate and any dust or microorganisms that adhere to the water condensate droplets. The room is then ready for operation.

I. Isolation and Transfer Tools.

A great variety of tools are commercially available or can be prepared for plant tissue culture work (WHITE, 1943; GAUTHERET, 1959). The basic necessary tools include: transfer needles of nichrome wire bent to give loops of different sizes to transfer small or large pieces of tissue, or microspatulas can be used; scalpels with interchangeable blades of different sizes; and forceps of several sizes. These tools may be placed upright in alcohol in a glass in the transfer room beside the operator, and, flamed and cooled before use as needed. Some prefer to use a rack to hold the various tools. If one encounters toxicity to the tissues from leaks of natural or manufactured gas in the transfer room, an alcohol flame may be used or the tools may be sterilized in a boiling water bath in the chamber.

II. Organ and Tissue Culture.

Organ and tissue cultures are originally isolated and grown with many similar procedures. Methods for individual types of isolations are described in the references cited. A few general details are given here. In general, for fresh isolations of tissues or organs the less surface sterilization that is applied to plant materials the better. Plant materials that are to be used as sources of isolated tissue culture material should be grown as healthy as possible and kept in areas largely free of air-borne dust, microorganisms and insects.

III. Embryo Cultures.

Sterile cultures of plants have been possible with many species by embryo dissection (see LA RUE, 1936; VAN OVERBEEK, CONKLIN and BLAKESLEE, 1941; BRINK and COOPER, 1947; RAPPAPORT, 1954; and BALL, 1959). Seeds may be washed, treated with 95% ethyl alcohol or detergent, surface-sterilized with 10% sodium hypochlorite or 1:1000 mercuric chloride, and rinsed in several changes of sterile distilled water. In the case of potato, sterile embryos were secured (KUMAR, 1961) when the potato seeds were treated as follows: immerse seeds in 10% Klenzade or sodium hypochlorite solution for 2 minutes, transfer to 95% ethyl alcohol solution for 5 minutes, and rinse in seven changes of sterile distilled water. Individual seeds were treated in individual vials, rather than in groups, since one contaminated seed might contaminate neighboring seeds. Following the surface-sterilization and rinsing in water, the seeds were soaked in sterile water for three days. They were then transferred to sterile Petri plates and the embryos dissected out under aseptic conditions with a needle and small scalpel as the seed was viewed under the binocular miscroscope.

IV. Root Tip Cultures.

Root tips are easily excised aseptically from germinated seeds or taken from stem cuttings. Seeds are placed in 95% ethyl alcohol or other surface tension reductant for a few seconds to several minutes to remove air bubbles that may interfere with penetration of the surface disinfectant. They are then immersed in 10% Klenzade or Clorox solution for 1 to 30 minutes, again depending on the species. Fresh sodium hypochlorite solution prepared from the powdered chemical may be used as an alternative to the Clorox or Klenzade solutions. Seeds are then rinsed in several changes of sterile distilled water and plated out on water agar or on moist, sterile filter paper in Petri plates and allowed to germinate. WHITE (1934) obtained root tips from the adventitious roots infected with virus that grew from cuttings of tomato stems suspended in a jar in a humid atmosphere. Potato and other species will many times produce roots on agar medium from small stem sections. The small stem sections may be secured from aseptically-grown seedlings that have developed from surface-sterilized seeds or embryos. One half inch to one inch sections with a node when laid flat on agar media will commonly produce many roots that can be excised and transferred directly to agar or liquid tissue culture media. Isolated root tips may also be cut from plants grown in vitro from the embryos or seeds (Fig. 2). Isolated roots do not always grow as easily as they are to isolate. With most types of isolated tissues it is advisable to make many isolations in the hopes of securing one or several good growing cultures capable of unlimited growth. Good growing cultures once established must be transferred at regular intervals when they are in a vigorous stage of growth. Neglected cultures may be difficult to rejuvenate to their original vigorous growth.

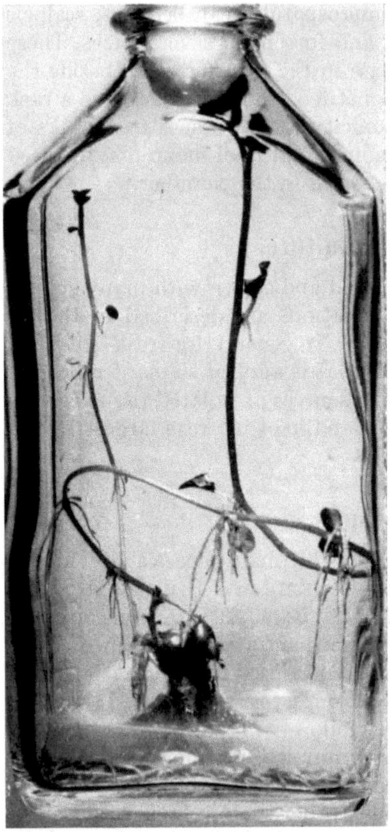

Fig. 2. Roots on potato plant growing *in vitro* on "C" medium in 6-oz. prescription bottles. The plant grew from an isolated embryo. Small tubers have been produced at the base of the plant. (From: S. KUMAR, A. C. HILDEBRANDT, Unpublished photograph.).

V. Stem Tip Cultures.

Stem tips of various lengths may be isolated in sterile culture. The ability of the isolated stem tip to elongate and grow may depend upon the size, the species and the medium. Small apical meristems transferred to sterile nutrient agar often grow into new plants. This has been possible with stem tips of many angiosperms (see LOO, 1946; BALL, 1946; MOREL and MARTIN, 1955; MOREL, 1956; and QUAK, 1958), as well as with fern apices (WETMORE and MOREL, 1949) and apices from *Selaginella*, *Lycopodium*, and *Equisetum* (WETMORE, 1953).

The stem tip cultures are initiated with aseptic techniques similar to those used for isolated root tips. The amount of surface sterilization required may depend upon the species and how free the plants are of surface-borne contaminants. Many stem tips and other tissues are contaminated more or less with microorganisms (TERVET and HOLLIS, 1948; KEENER, 1950; and YOUNG, 1958). For large stem tip cultures of potato, CLAVER et al. (1958) grew potatoes in 6-inch pots and after 45 days cut 15 cm tip cuttings from the plants. The stem tips were sterilized by a 20-minute immersion in 10% Clorox plus a few drops of detergent. The leaves and primordia of leaves were removed from these cuttings, and 2 to 3 mm pieces of apical meristem were removed and transferred to medium in 6-oz. prescription bottles. Similarly, geranium plants to be used as sources of stem tips for stem tip cultures (REINERT, 1960) were grown in a clean greenhouse. The leaves were removed from 3-inch stem tip cuttings and the cuttings immersed in 70% ethyl alcohol for 2 minutes, then in 5% Clorox for 10 minutes, and then rinsed in 3 changes of sterile distilled water and placed in a sterile container to drain. Apical tips from 100 microns to 1 mm long were then removed aseptically with a sharp scalpel. Stem tips less than 1 mm were often difficult to grow into complete plants. MOREL and MARTIN (1955) found it necessary to first isolate the tip of certain species on agar medium and after it had elongated somewhat, to graft this to complete plants to secure sufficient stem tip length so that it could be later propagated as a cutting.

VI. Other Miscellaneous Cultures.

Using aseptic techniques, sterile cultures have also been possible with leaves (STEEVES and SUSSEX, 1957; SUSSEX and STEEVES, 1958), fern prothalli (STEEVES, SUSSEX and PARTANEN, 1955), and other plant parts (LEE, 1950), including flowers (LA RUE, 1942), ovaries (NITSCH, 1951), endosperm (LA RUE, 1949; STRAUS, 1960), and pollen (LA RUE, 1954; TULECKE, 1957, 1960).

VII. Callus Cultures.

Callus cultures have been a major type of tissue culture used for many basic studies. Tissues have been isolated and established from many species and parts of plants. These largely undifferentiated types of cell masses have been established,

Fig. 3. Left: Crown gall on Paris-daisy plant induced by artificial inoculation of plant and used as source of callus tissue. [From: A. J. RIKER and A. C. HILDEBRANDT. Brookhaven Symposium in Biology, No. 6, 79—96 (1954)]. Right: Grape *Phylloxera* galls used as source of callus. (From: FRANCIS PELET, A. C. HILDEBRANDT and A. J. RIKER: Unpublished photograph.)

for example, from embryos (CURTIS and NICHOL, 1948), excised roots (SKOOG, 1944; TORREY and SKIGEMURA, 1957), coniferous seedlings (LOEWENBERG and SKOOG, 1952), redwood burls (BALL, 1950, 1955), crown galls (WHITE and BRAUN, 1942; HILDEBRANDT and RIKER, 1949, 1959) (Fig. 3), white spruce tumors (REINERT and WHITE, 1956), liverworts (ALLSOPP, 1957), virus-induced tumors (BLACK, 1958), mosses (WARD, 1960) and insect galls (PELET, HILDEBRANDT, RIKER, and SKOOG, 1960) (Fig. 3). Stem tissues (WHITE, 1939), or fleshy roots such as carrot (GAUTHERET, 1939; NOBECOURT, 1939; STEWARD and CAPLIN, 1954) or potato (STEWARD and CAPLIN, 1951) have perhaps been used more often as sources of callus cultures.

The callus cultures are secured following strict aseptic techniques. If the callus is desired from stem tissue, stem sections 1 to 5 cm long are cut from the plants, immersed in 95% ethyl alcohol for 1 minute, in 10% Clorox 1 to 15 minutes, and rinsed thoroughly in several changes of sterile distilled water. The stem pieces are transferred to a sterile Petri plate and pieces are cut from both ends of the stem sections to remove portions that may have absorbed the disinfectant. The remaining 1 to 2 cm long section is then either laid flat on the culture medium or inserted, apical end down, half its length into the medium. When the size of the section is critical for certain growth studies, razor blades or other cutting edges may be mounted to cut specific lengths (BOURIQUET, 1952; JABLONSKI and SKOOG, 1953). With potato, carrot or other fleshy root the surface is similarly disinfected and rinsed with sterile distilled water. Small cubes $^1/_8$ to $^1/_2$ inch in thickness are cut from the potato tuber with a sterile scalpel and transferred to culture vessels containing solid medium or to liquid medium in containers on a shaker. Carrot sections from aseptically peeled carrots may be transferred to agar or liquid medium. It is convenient where explants of reproducible weight and size are desired, to remove tubes of carrot tissue with a sterile cannula or cork borer of the desired diameter, to transfer those borings to a sterile Petri plate, and to cut them into sections of desired length with a sterile scalpel or with razor blades mounted on a bolt and separated the desired length with spacers (CAPLIN and STEWARD, 1952). The sections are then transferred to the culture medium with sterile forceps or transfer needle. Callus may also be induced from the cambium of branches of large trees by removing blocks of stem tissue along with the bark (WHITE, 1945). Callus tissues have been similarly derived from crown galls (WHITE and BRAUN, 1941; HILDEBRANDT and RIKER, 1949), virus-induced tumors (BLACK, 1947, 1958), white spruce tumors (REINERT and WHITE, 1956), and insect galls (PELET et al., 1960).

VIII. Some General Precautions.

Many of the crown gall explants whether from primary or secondary galls contain the inciting bacteria so that numerous pieces of tissue must often be isolated in order to obtain pieces that produce good callus and that are free of the crown gall bacteria.

The tender leaf and insect gall tissues of grape (PELET, HILDEBRANDT, RIKER, and SKOOG, 1960) required extra care in washing, following surface sterilization. About 10 grams of leaf material was washed 7 times in 250 ml volumes of sterile distilled water. The galls were excised from still-expanding leaves measuring 3—5 cm in diameter (15 mg). Similar precautions may be essential with fresh isolations from other delicate tissues. Tests for sterility of cultures are described below.

E. Transfer and Maintenance of Cultures.

Freshly isolated tissues may be incubated either in the light or in darkness. Stem tips or embryo cultures may grow best in light; callus and root tip cultures grow well in darkness as well as in light. The optimum temperature varies with the species (HILDEBRANDT, RIKER, DUGGAR, 1945; PELET, HILDEBRANDT, RIKER, and SKOOG, 1960). Most callus cultures grow well in the dark at 26° C and with 55% relative humidity. Stock cultures may also be maintained for long periods under a layer of mineral oil (CAPLIN, 1959).

I. Callus Formation.

Callus usually forms on the top surface of the tissue or stem explant after a few days to several weeks (Fig. 4). This callus is cut from the explant with a scalpel and transferred to fresh medium. Just as it is often necessary to make

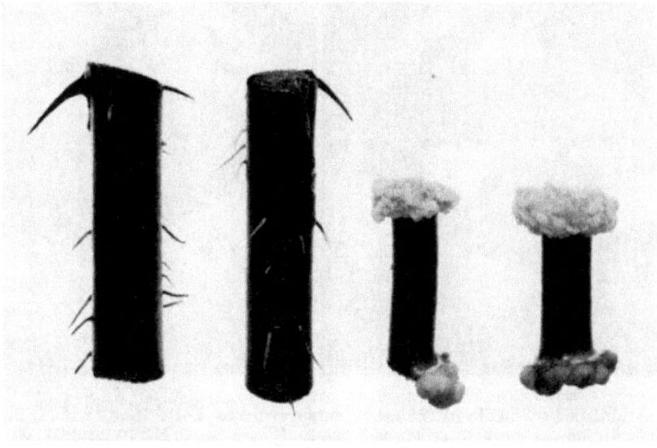

Fig. 4. Stem sections from *Rosa lucida* incubated 6 weeks. Two sections at left isolated on "D" medium failed to produce callus; two sections at right isolated on "C" medium without NAA produced abundant white callus at top and bottom of the sections. Such callus removed with a sterile scalpel and transferred to fresh medium established a strain of tissue capable of indefinite growth. (From: D. COYIER, A. C. HILDEBRANDT, and M. P. BACKUS: Unpublished photograph.)

many isolations to secure tissue capable of unlimited growth, once such an isolate is obtained it should be transferred regularly to fresh medium. Cultures of callus tissue are transferred to fresh medium every 2—6 weeks depending on their

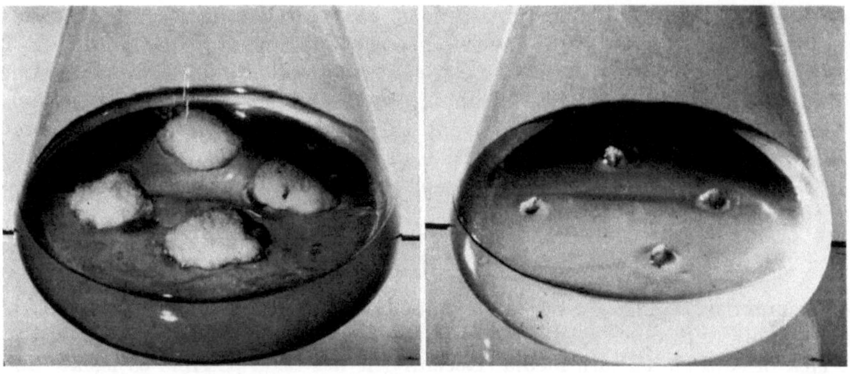

Fig. 5. Left: Established callus tissues of *Nicotiana tabacum*, 6 weeks old and infected with tobacco mosaic virus, as they grew on agar medium in 125 ml Erlenmeyer flask. Right: Four seed pieces of 20 mg weight cut from tissue piece as at left and transferred to fresh medium. No symptoms of virus infection appear. [Left: A. C. HILDEBRANDT: Unpublished photograph; Right: A. C. HILDEBRANDT and A. J. RIKER: Fed. Proc. **17**, 986—993 (1958).]

rate of growth. To make the subcultures of established cultures, the tissue piece is transferred under aseptic conditions with a sterile transfer loop to a sterile Petri plate and cut into pieces 20—50 mg in mass with a sterile scalpel. Four or five such seed pieces may be transferred to fresh medium in each 6-oz. prescription bottle or Erlenmeyer flask (Fig. 5).

Variation in the growth rate and degree of differentiation are often great (NOBECOURT, 1939; GAUTHERET, 1957). Macroscopically, the callus cultures often appear as smooth or lobed masses that may be firm or soft to the touch (Fig. 6). The callus mass consists largely of parenchyma cells, but in cross section may be seen to contain, depending on the strain of tissue, varying amounts of vascular

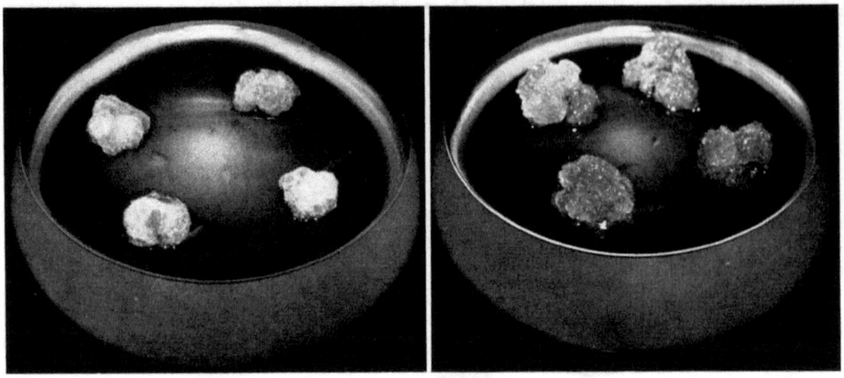

Fig. 6. Callus tissue established on media in 125 ml Erlenmeyer flasks. Left: Stem tissue from *Nicotiana glauca* × *N. Langsdorffii*. Right: Sunflower tissue of crown gall origin. [From: A. C. HILDEBRANDT, A. J. RIKER and B. M. DUGGAR: Am. J. Botany **32**, 357—361 (1945).]

and cambial tissue, and varying degrees of organization of these components (WHITE, 1939; GAUTHERET, 1959). Differences in rates of growth, plastid and pigment formation may also be observed, depending on the isolate.

II. Testing for Sterility of Cultures.

Some cultures contain microorganisms that grow in the tissue and not on the tissue culture medium. Certain precautions may be taken against unknowingly transferring such infected seed pieces. When subculturing, only one mother stock piece should be transferred to each Petri plate for subdividing. After transferring most of the seed pieces to fresh medium, a few representative pieces remaining in the Petri plate may be macerated and the plate poured with a medium such as one containing yeast extract or peptone, or a similar rich medium (ROBBINS and HERVEY, 1951). All cultures from stock pieces that subquently show contamination should be discarded.

F. Single Cell Cultures.

Attempts by HABERLANDT (1902) and others later to grow single cells of higher plants were unsuccessful largely because of the type of cells selected and the lack of suitable culture media. Recently, it has been possible to isolate and grow single cells from higher plants. This means that clones of tissue are possible that grow from a single cell.

Many have observed tissue dissociation in liquid media (see DE ROPP, 1958; MUIR et al., 1954; NOBECOURT, 1957; STEWARD, 1958; TULECKE and NICKELL, 1960). Clones of tissue were first established from handpicked single cells of tobacco and marigold callus cultures by MUIR et al. (1954), using a nurse culture method (Fig. 7). With aseptic methods the single cell was picked up under a binocular microscope with a microspatula from liquid or agar cultures and placed on an 8 × 8 mm square of sterile filter paper that had rested several days on top of an established nurse tissue on agar medium. The nurse tissue piece was either of the same species or from a species different than that of the single cell. Single cells failed to grow if transferred directly to agar medium. However, with the filter paper nurse method, the single cells often divided to produce a mass of cells that grew independently when transferred to the agar medium. The filter paper and single cell could be transferred to a sterile Petri plate and observed at will to follow growth and cell division to insure that the resulting clone of tissue came from the single cell and not from the nurse culture. Single cell clones from many species have been established by this method (MUIR et al., 1958; HILDEBRANDT,

Fig. 7. The filter paper, nurse culture method of growing single isolated cells of higher plants to secure a single cell clone. Left to right: (1) Established culture of *N. tabacum* × *N. glutinosa* host tissue supporting a filter paper. (2) Filter paper with small colony of tobacco cells that grew from a single isolated cell and resting on nurse tissue. Such a colony developed several weeks after the hand-picked single cell was placed on the filter paper. A colony of cells this size usually failed to grow if placed directly on the medium. (3) Such clonal cultures were usually transferred several times to fresh nurse cultures before they reached the size shown. (4) Established clone of tissue that grew from a single cell as it grew independently on the basal medium following the sequences shown at left. [From: A. C. HILDEBRANDT and A. J. RIKER: Fed. Proc. **17**, 986—993 (1958).]

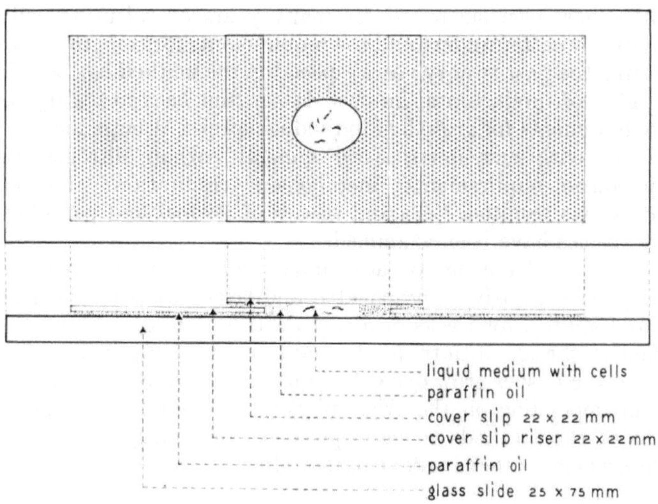

Fig. 8. The microculture chamber for growth and continuous observations of single cells and groups of cells of higher plants in living, unstained condition. [From: L. E. JONES, A. C. HILDEBRANDT, A. J. RIKER, and J. H. WU: Am. J. Botany **47**, 468—475 (1960).]

1958; HILDEBRANDT und RIKER, 1958; BRAUN, 1959; HILDEBRANDT, HOTTINGER and RIKER, 1961). A pea root callus single cell was able to grow to a colony of seven cells on an ager drop an a slide (TORREY, 1957). It has also been shown possible (BERGMANN, 1959) to isolate and grow single cells of tabacco and bean as pure clones by diluting a suspension of these cells (grown in liquid culture) in agar in Petri plates. It has also been possible to isolate single cells and follow growth, maturation, division, and death in a microculture chamber (JONES et al., 1960) (Fig. 8).

G. Growth Measurements.

Growth of callus tissue is commonly expressed as the wet or dry weight after a suitable growth period. However, growth may also be expressed, for example, in terms of colony diameter, increase in cell number, increase in phosphorus, total nitrogen, growth index (ratio of final to original weight) or relative growth rate (CAPLIN, 1947). Tissues can also be aseptically removed from the culture vessel, weighed, and returned to the nutrient at desired intervals.

Callus cultures, depending on the strain of tissue, show degrees of variation between cultures. The choice of tissue for particular studies is important since certain strains grow at relatively uniform rates and certain other strains are exceedingly variable in rate and habit of growth. In many cases 4 tissue pieces distributed in each of 6 culture flasks or bottles provide sufficient replication so that major differences in treatment are readily apparent. When the significance of smaller differences is desirable, additional replications should be added as needed and tests for significance made. Growth of tissues on various media in relation to the initial size of the seed piece was observed by DUHAMET (1955). Variation in growth between tissue pieces in culture has been examined statistically (WHITE, 1943; HILDEBRANDT, RIKER, and DUGGAR, 1946; DE ROPP, 1948; STEWARD and CAPLIN, 1952; HELLER, 1953; GAUTHERET, 1959).

H. Requirements for Growth of Callus Cultures.

The value has been suggested for many years of culturing cell and tissue masses in vitro to better understand the requirements for cell growth and differentiation. During the past twenty years especially, the method has been increasingly used for studies of fundamental interest. It has first been necessary to determine the basic requirements for growth of individual species. A large literature on tissue culture dealing with such items as the mineral, carbon, nitrogen, vitamin and growth substance requirements; metabolism; respiration; and polarity and differentiation has resulted. Cells and tissues of normal and pathological origins from many species have been examined.

Tissue cultures have many advantages for research. Genetically-similar material may be vegetatively propogated and large numbers of cultures may be employed and studied for weeks or months with a minimum of culture work and often without a change of nutrient. In addition, the composition of the medium is defined and the tissues grow depending on what is present in or absent from the medium and depending on the other physical environmental conditions. Some applications of the methods to basic problems in biology are indicated below. These results suggest other fundamental studies.

I. The Acidity of the Medium.

The acidity of the medium is often adjusted to an initial value of pH 5.5—6.0. The optimum acidity, however, depends upon the particular strain of tissue. The acidity of the medium usually drifts from more acid or alkaline conditions toward neutrality during the growth period (HILDEBRANDT, RIKER and DUGGAR, 1945). Studies with the pH indicators in the medium also indicated the tissues change the pH of the medium in the areas of the medium they occupy. Tobacco tissues from the hybrid *(Nicotiana glauca × N. Langsdorffii)* grew best on media with original acidities averaging pH 5.0—5.4 and final acidities from pH 5.5—5.9. Sunflower crown gall tissues grew best on media with original acidities from pH 5.5 to 5.9 and final acidities from pH 6.0—6.4. Little or no growth of cultures appeared

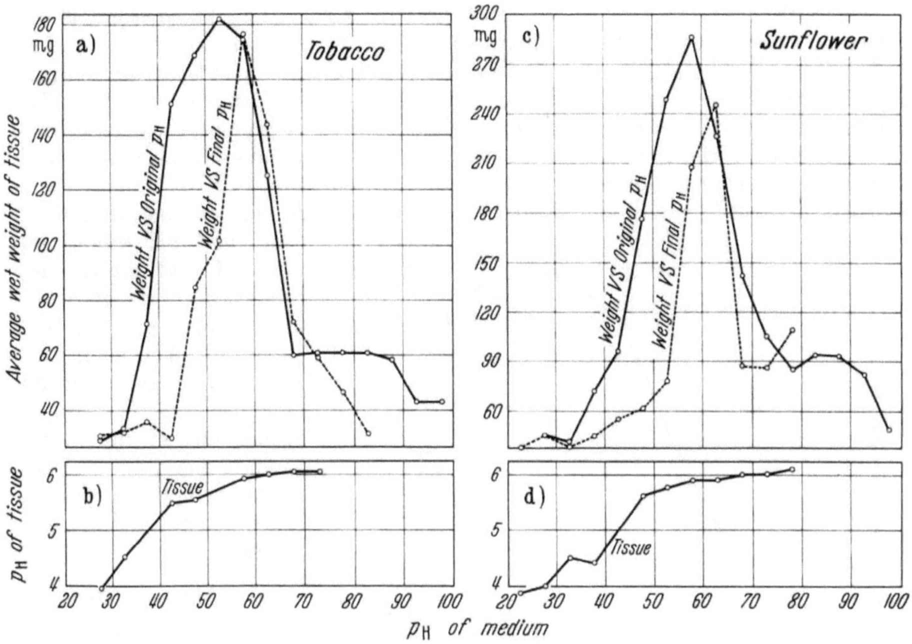

Fig. 9. Effects of acidity of the medium on the weight of tobacco and sunflower tissues growing *in vitro*, and on the pH of the tissues. Weight of the tissues compared with the original and final pH of the media. [From: A. C. Hildebrandt, A. J. Riker, and B. M. Duggar: Am. J. Botany **32**, 357—361 (1945).]

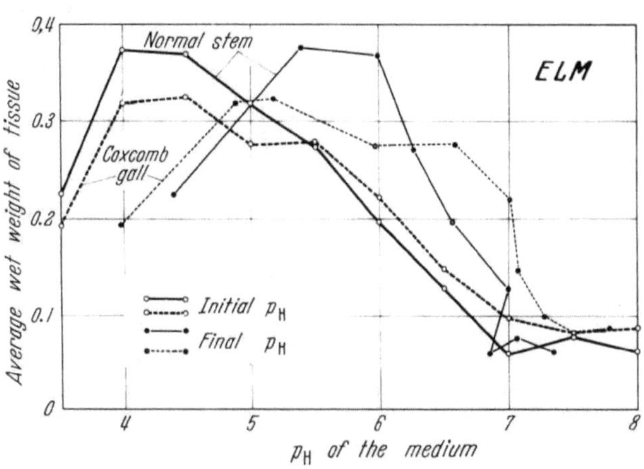

Fig. 10. Growth of callus tissue from normal elm stem and from elm coxcomb gall plotted against the initial and final pH of the medium. [From: F. Pelet, A. C. Hildebrandt, A. J. Riker, and F. Skoog: Am. J. Botany 47, 186—195 (1960).]

on media with values from pH 2,5 to 4,5 or from pH 7.5 to 10.0 (Fig. 9). Corn endosperm (Straus and La Rue, 1954) grew best between pH 6.0 and 7.0. Carrot tissue (GAUTHERET, 1947) was able to grow well over a range from pH 6.0 to 9.0. The optimum for normal sunflower tissue (HENDERSON, 1954) was pH 5.0 to 6.0, while the crown gall tissue grew from pH 4.5 to 6.5. Tobacco tissue infected with tobacco mosaic virus (HILDEBRANDT, 1958) grew optimally on media with final acidities from pH 5.0 to 7.5. The most infective virus was isolated from tissues grown on media at pH 8.0—8.5. There was little or no virus in tissues incubated at pH 9.3 to 9.5. The growth of a variety of callus tissues of insect gall and normal origin was influenced much by the acidity of the medium (PÉLET et al., 1960). The optimum values for individual strains of tissue varied. The importance of knowing the final acidity of the medium at the end of the growth period was again emphasized since regardless of the initial pH, the acidity drifted toward pH 6.0. Thus, where graphs indicated growth of cultures on originally more acid or alkaline media, the final pH values indicated that growth actually occurred at acidities close to pH 6.0 (Fig. 10). The acidity of the medium may exert an influence on a variety of conditions in the media and in the tissues (NIKKELL and BURKHOLDER, 1950; KLEIN and MANOS, 1960).

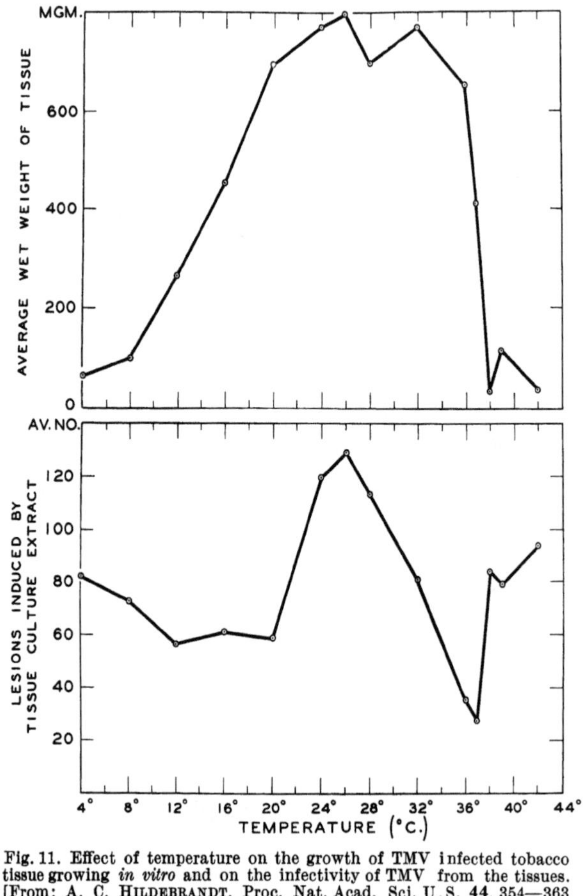

Fig. 11. Effect of temperature on the growth of TMV infected tobacco tissue growing *in vitro* and on the infectivity of TMV from the tissues. [From: A. C. HILDEBRANDT. Proc. Nat. Acad. Sci. U.S. 44, 354—363 (1958).]

II. Temperature.

A growth room maintained at 26° C ± 1° C has been routinely used for cultures of many species (HILDEBRANDT and RIKER, 1958). The optimum temperature, however, for growth of callus cultures varies with the strain of tissue. Sunflower crown gall tissue grew best between 24° C and 28° C, and tobacco hybrid (*N. glauca* × *N. Langsdorffii*) between 28° C and 32° C (HILDEBRANDT, RIKER and DUGGAR, 1945). Similar temperature relationships appeared for tissues from *Rumex* (NICKELL and BURKHOLDER, 1950) and corn endosperm (STRAUS and LA RUE, 1954). The influence of temperature and day length were correlated with growth of normal carrot and grape, and with sunflower crown gall tissues (DE CAPITE, 1955). Temperature influenced the growth substance requirements of sunflower crown gall tissue (RIER and HENDERSON, 1957). It similarly influenced the growth and virus content of TMV infected tobacco tissue (HILDEBRANDT, 1958) (Fig. 11). Various insect gall and normal tissues (PÉLET et al., 1960) grew over a range from 16° C to 36° C (Fig. 12). The optimal temperature was 28° C to 32° C for grape, willow, poplar, and elm tissues and 36° C for normal grape stem tissue.

III. Light.

Tissue cultures are routinely maintained in various laboratories either in the light or dark. GAUTHERET (1959) favored light for carrot and other tissues. Cultures of many species have also been grown for years in a dark culture room (WHITE, 1943; HILDEBRANDT and RIKER, 1949). Where chlorophyll production by the callus in the light may be an undesirable factor, tissues may perhaps best be grown in darkness. Detailed effects of day length or quality of light have been examined with tissues from several species (ENDERLE, 1951; STEWARD, CAPLIN, and MILLER, 1952; BUNNING and WELTE, 1954; DE CAPITE, 1955; and NITSCH and NITSCH, 1956).

IV. Inorganic Nutrition.

The nutrition of callus tissues from many species has been examined in detail by many workers. Some basic media suitable for culture of tissues from many species were described earlier. Many other media for specific tissues have been developed after much precise work (see HELLER, 1953, 1955; NOBECOURT, 1957; STREET, 1957; GAUTHERET, 1959, and references therein) (Fig. 13).

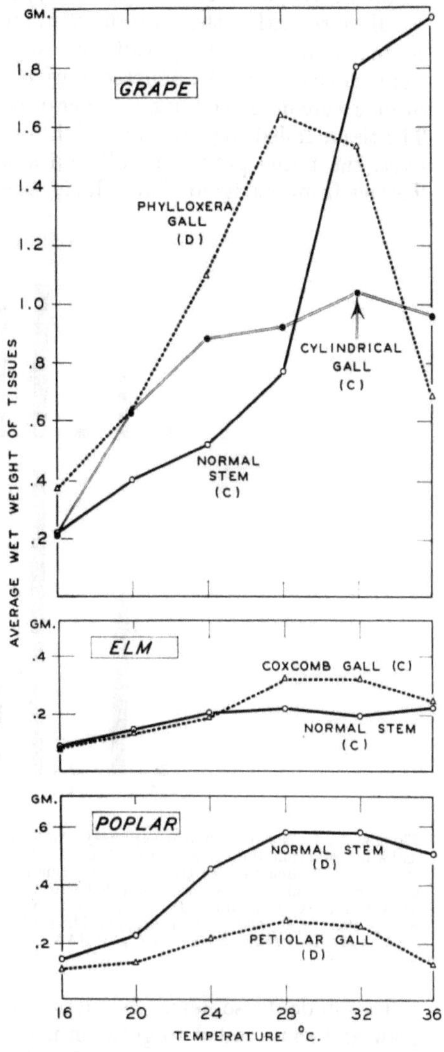

Fig. 12. Effect of temperature on the growth of callus tissue of normal and insect gall origins on "C" or "D" media as indicated. [From: F. PELET, A. C. HILDEBRANDT, A. J. RIKER, and F. SKOOG: Am. J. Botany 47, 186—195 (1960).]

The isolated cell masses many times have tremendous tolerances for a wide range of concentration of the various ingredients used in the media. However, the proper balance of chemicals, and the concentrations of both minor and major elements in the media are often critical for optimum growth of specific tissues (HILDEBRANDT, RIKER and DUGGAR, 1946; NICKELL, 1954; STRAUS, 1960). The purity of the chemicals in the media should be verified for critical analyses of inorganic nutrition.

V. Carbohydrate Nutrition.

The isolated callus tissues require an outside source of carbon for growth in vitro. Freshly isolated tissues may remain alive for months in certain cases but eventually they die if an organic carbon source is not provided. NOBECOURT (1937), and GAUTHERET (1942) used dextrose, and WHITE (1943) and others used sucrose as the source of carbon, respectively, for tissue cultures.

More recently other carbon compounds have been tested for their availability to many species. Many different sugars were tested as sources of carbon for Sequoia (BALL, 1955). Sources of carbon for callus tissue of crown gall and normal origins were described in a number of papers (see HILDEBRANDT and RIKER, 1955). The tissue failed to grow in the light or darkness without a source of carbon. However, the tissue grew generally on a wide range of sugars and polysaccharides. Tissues from marigold, Paris daisy, periwinkle, sunflower and tobacco grew more

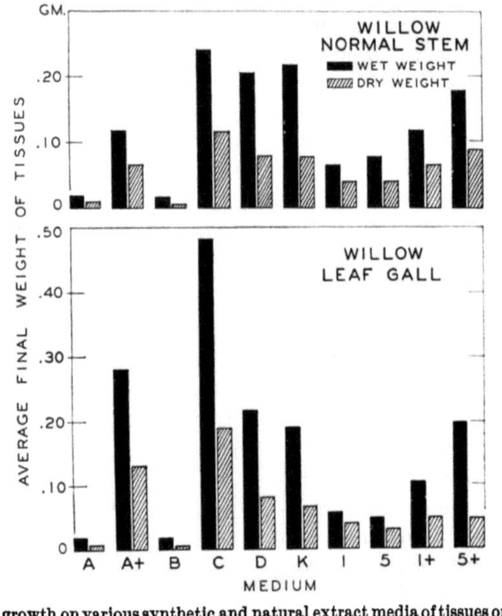

Fig. 13. Comparison of growth on various synthetic and natural extract media of tissues originally isolated from normal willow stem and the *Pontania* insect leaf gall. Wet and dry weights of tissues incubated on the basal "tobacco" medium (A) compared with growth on the same medium but supplemented as follows: A + = A plus coconut milk (CM), 150 ml/l; B = A plus α-naphthaleneacetic acid (NAA), 1×10^{-4} gm/l; C = A plus CM, NAA and calcium pantothenate (CP), 2.5 mg/l; D = A plus CM, NAA, CP and 2,4-dichlorophenoxyacetic acid (2,4-D), 6 mg/l; K = A plus CM, CP and 2,4-D; 1 = A plus kinetin (K), 1×10^{-4} gm/l; 5 = A plus K, 5×10^{-4} gm/l; 1 + = A plus NAA, K, 1×10^{-4} gm/l; and 5 + = A plus NAA, K, 5×10^{-4} gm/l. (From: A. C. HILDEBRANDT and A. J. RIKER: Unpublished.)

or less on dextrose, levulose, sucrose, maltose, cellobiose, dextrin or pectin. Some grew and some failed to grow on mannose, galactose, lactose, raffinose or starch. Arabinose, xylose, rhamnose and inulin were poor sources of carbon (Fig. 14). The concentration of sugars and polysaccharides was important for growth; 0.5 to 2 per cent was often optimum but other concentrations among those tested (8, 4, 2, 1, 0.5, 0.25, 0.125, 0.06, 0.03 and 0.015 per cent) supported growth depending on the carbon source (Fig. 15).

Virus tumor tissue (NICKELL and BURKHOLDER, 1950) grew excellently on 2 per cent sucrose, dextrose, and levulose. Soluble starch was also a good source of carbon. Good sources of carbon were raffinose, melibiose, maltose, and cellobiose. Poor sources of carbon included galactose, xylose, arabinose, lactose and rhamnose. All were tested at 2 per cent as a substitute for sucrose.

Alcohol and organic acids alone were not generally good sources of carbon for cultures of crown gall and normal origins (see HILDEBRANDT and RIKER, 1955) at any of the wide range of concentrations tested. However, in the presence of 2 per cent sucrose all species grew more or less on media with methanol, ethanol, glycerol, erythritol, mannitol or dulcitol. Only periwinkle tissue tolerated butanol and only marigold and sunflower tissue tolerated the phloridzin. Propanol inhibited growth at all concentrations.

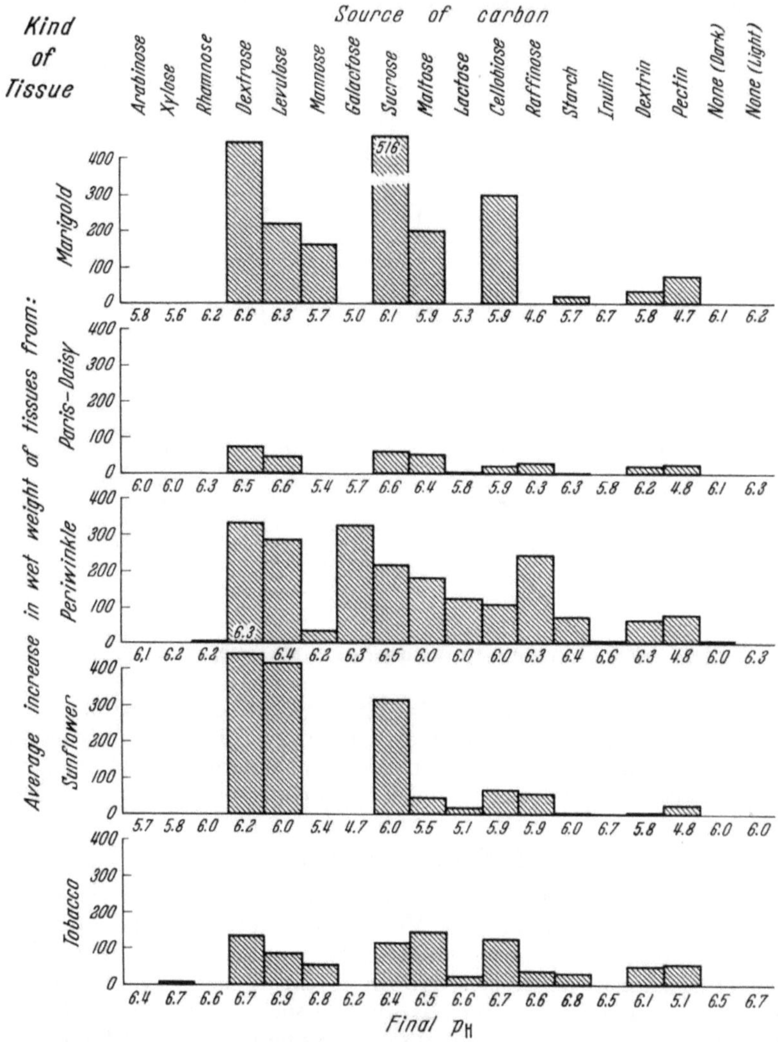

Fig. 14. Growth of callus cultures from marigold, Paris-daisy, periwinkle, and sunflower crown galls and from tobacco stem on synthetic medium with different sources of carbon and the average final acidities of the media. The carbon compounds were tested at 1 per cent as a substitute for sucrose. [From: A. C. HILDEBRANDT and A. J. RIKER: Am. J. Botany **36**, 74—85 (1949).]

Periwinkle tissue, however, grew on a range of concentrations of a number of the alcohols, but the growth was poor compared to that on sucrose controls. Glycerol, methanol, mannitol and dulcitol were used slightly by certain species. In the presence of sucrose many of the alcohols were tolerated but growth was poor. Glycerol was also used for carrot tissue (GAUTHERET, 1948).

The organic acids at pH 6.0 were generally poor sources of carbon for callus tissue (HILDEBRANDT and RIKER, 1949) when used as a substitute for 2 per cent sucrose. All species grew more or less in the presence of sucrose, with stearic, pyruvic, malonic, succinic, fumaric, malic or tartaric acids or calcium gluconate. They grew poorly with glycolic, lactic, oxalic, glutamic or citric acids and none grew in the presence of formic, acetic, or propionic acids. With 2 per cent sucrose

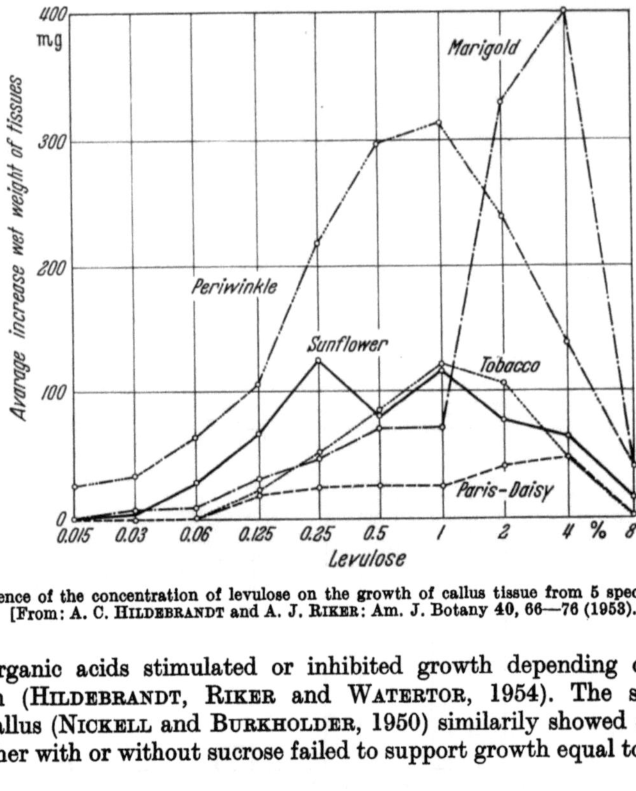

Fig. 15. Influence of the concentration of levulose on the growth of callus tissue from 5 species as indicated. [From: A. C. HILDEBRANDT and A. J. RIKER: Am. J. Botany 40, 66—76 (1953).]

certain organic acids stimulated or inhibited growth depending on the concentration (HILDEBRANDT, RIKER and WATERTOR, 1954). The studies with *Rumex* callus (NICKELL and BURKHOLDER, 1950) similarily showed organic acid media either with or without sucrose failed to support growth equal to the sucrose medium.

VI. Nitrogen Nutrition.

The types and concentrations of nitrogen present in the medium are important for determining growth of the tissues. Many different organic and inorganic compounds have been tested as nitrogen sources for tissues of many species. Nitrate nitrogen often supported excellent growth of the tissues (Fig. 16). Sunflower crown gall tissue grew well on urea as the sole source of nitrogen (RIKER and GUTSCHE, 1948) with a sucrose mineral salts medium (Fig. 17). Most amino acids were not good sources of nitrogen for the callus tissue tested at concentrations from 0.256 to 0.00006 M. Good to poor growth, however, resulted with ammonia, alanine, glycine, arginine, asparagine and aspartic acid. Casein hydrolysate, peptone, and yeast extract also encouraged good growth. Eleven amino acids permitted no growth. When sufficient nitrate was present to provide excellent growth, the growth was inhibited by progressively increasing amounts of ammonium salts and arginine. Growth was inhibited by nitrite, asparagine and all the amino acids at 0.001 M and stronger concentrations. Alanine, aspartic acid and glutamic acid, however, permitted growth at 0.064 M concentration. Transamination studies by EBERTS et al. (1954) suggested the effects of the amino acids resulted from keto acids formed by transamination. Similar nitrogen utilization studies (FRANK et al., 1951) were made with tobacco *N. glauca* × *N. Langsdorffii* callus tissue. Media with ammonium succinate, urea (0.001 M), D-L alanine (0.016 M), or L-glutamic (0.004 M) acid aupported only about half the growth of the nitrate controls. L-leucine did not support growth alone and inhibited growth in the presence of nitrate. The bimodal growth curve with alanine and sunflower tissue (RIKER and GUTSCHE, 1948) (Fig. 17) was not observed with the tobacco tissue.

Growth of *Rumex* virus tumor tissue was better on a medium with potassium and calcium nitrate than on ammonium nitrate or ammonium salts (BURKHOLDER and NICKELL,

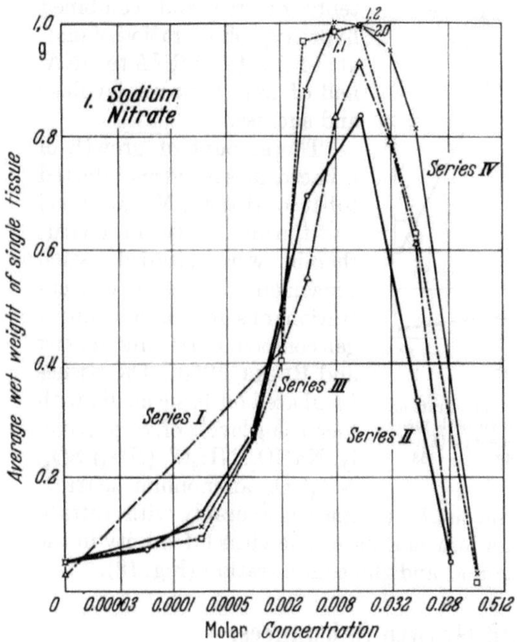

Fig. 16.

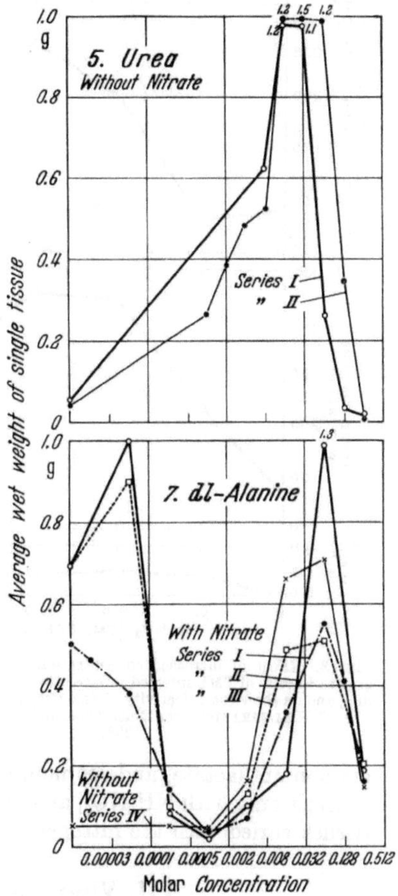

Fig. 17.

Fig. 16. Growth of sunflower crown gall tissue *in vitro* on synthetic media with different concentrations of sodium nitrate. [From: A. J. RIKER and ALICE E. GUTSCHE: Am. J. Botany 35, 227—238 (1948).]

Fig. 17. Growth of sunflower crown gall tissues *in vitro* on synthetic media with different concentrations of urea without nitrate (top) and on different concentrations of DL-alanine with or without nitrate (bottom). [From: A. J. RIKER, and ALICE E. GUTSCHE: Am. J. Botany 35, 227—238 (1948).]

1949). A more complete list of nitrate and ammonium compounds tested later (NICKELL and BURKHOLDER, 1950) showed the optimal concentration of inorganic nitrogen was 8 mM. The pH influenced the utilization of the nitrate and ammonia. At a concentration of 0.002 M, arginine, aspartic acid, glycine, proline and threonine were utilized little and other amino acids not at all. With nitrate present at 0.004 M, and the amino acids at 0.002 M the same amino acids were used slightly. Among the nitrogen compounds tested only aspartic acid and urea were good sources of nitrogen for growth of the tissue.

The utilization of amino acids by Jerusalem artichoke, *Lupinus albus* and willow was similarly studied by DEMETRIADES (1958). Only aspartic acid was utilized by the three species. Asparagine, glutamic acid, glutamine and arginine were also used by certain tissues but not by others.

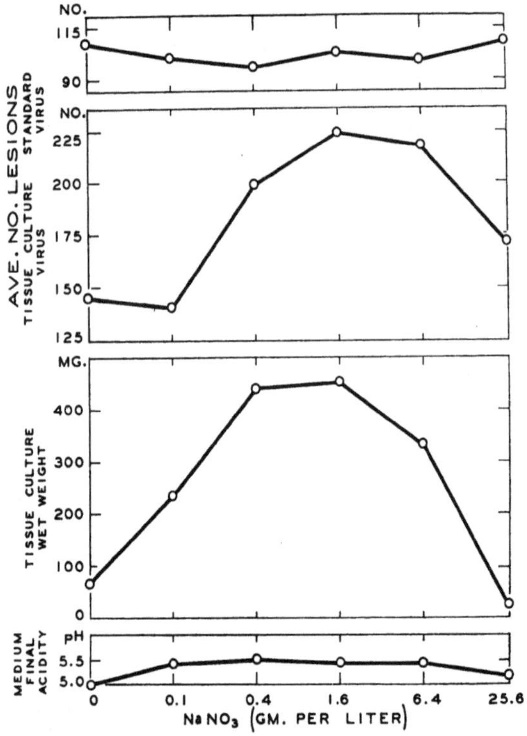

Fig. 18. Effect of increased concentrations of sodium nitrate on growth *in vitro* of TMV infected tobacco tissue on a synthetic medium, and on the virus infectivity of the tissue homogenate. [From: A. C. HILDEBRANDT. Proc. Nat. Acad. Sci. U.S. 44, 354—363 (1958).]

Leaves and tissue cultures from leaves were compared for their contents of amino acids, nonvolatile organic acids, sugars and nucleic acids (WEINSTEIN, NICKELL, LAURENCOT and TULECKE, 1959). Differences appeared in contents of free and combined hydroxyproline, ratios of malate to citrate, of RNA to DNA, and of free dextrose, levulose and sucrose.

The amount of growth of tobacco mosaic virus infected tobacco tissue *(N. tabacum)* and the infectivity of the virus therein were studied when grown on a range of concentrations of some common nitrogen compounds (HILDEBRANDT and RIKER, 1958). The tissues incubated on basic media with increasing levels of, respectively, $NaNO_2$, NH_4Cl, $(NH_4)_2SO_4$, NH_4NO_3, ammonium tartrate, ammonium lactate, and ammonium succinate, compared favorably with nitrate controls only with NH_4NO_3, and ammonium succinate. The virus infectivity in the tissues varied with the nitrogen compound and the concentration (Fig. 18).

VII. Vitamins and Growth Substances.

Many early studies (WHITE, 1943; NOBECOURT, 1957; GAUTHERET, 1959) described the requirement of callus tissue and other tissue for vitamins and growth substances. Many callus tissues evidently synthesize their requirements. Additional requirements are limited to only one or a few vitamins either alone or in combination with other metabolites. Growth of sunflower tissue of normal and crown gall origins with added vitamins was compared by HENDERSON (1955). Vitamin B_1 was favorable for carrot isolates (GAUTHERET, 1939) and for *Rumex* callus (NICKELL, 1952), but NOBECOURT (1940), CZOSNOWSKI (1952), and PARIS (1955) grew carrot isolates without this supplement. Crown gall tissue of *Scorzonera* required vitamin B_1 (PARIS, 1955). In one passage studies with sunflower and tobacco tissue (HILDEBRANDT, RIKER and DUGGAR, 1946) and vitamins B_1, pyridoxine, and nicotinic acid, only pyridoxine produced a little stimulation of growth. Vitamin B_{12} was beneficial for callus from white spruce (REINERT and

WHITE, 1956), pantothenic acid for *Crataegus* (MOREL, 1946), *meso*-inositol for elm (JACQUIOT, 1951), *p*-amino benzoic acid for Jerusalem artichoke (DE CAPITE, 1952) and ascorbic acid for *Juniperus* (HOTSON, 1953).

The growth substance requirements of callus tissue vary with the species and the isolate (Fig. 19). Added growth substance is commonly supplied to the medium as indole-3-acetic acid (IAA). The concentration provided was important in the

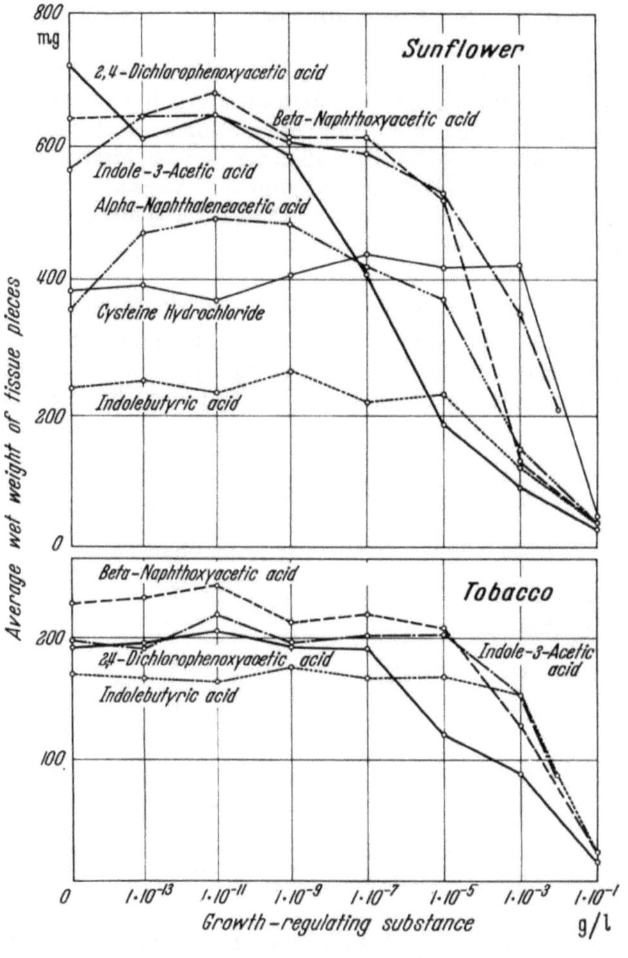

Fig. 19. Influence of the concentration of some growth-regulating substances on growth of sunflower crown gall and normal tobacco tissues on synthetic media. (From: A. C. HILDEBRANDT: Plant growth substances. pp. 391 to 404. University of Wisconsin Press, Madison. 476 p. 1951.)

type and amount of resulting growth (GAUTHERET, 1947; STRUCKMEYER, HILDEBRANDT and RIKER, 1949; SKOOG, 1954). Naphthaleneacetic acid (NAA) is commonly added as a substitute but other growth substances are also active (HILDEBRANDT and RIKER, 1947). The stimulating effect of 2,4-D was noted above.

The varying requirements for IAA of some different isolates of tissue has been detailed (see KULESCHA, 1955; HENDERSON, 1955). GAUTHERET and his associates showed that callus requirements for IAA may be of three types: (1) Certain species grew indefinitely on media without added IAA; (2) other species or strains of tissue required IAA in varying concentrations as a supplement; and (3) so called habituated strains of tissue required added IAA during the first few weeks or months in culture and when subsequently transferred, synthesized their own requirements. Studies of auxin inactivation by normal and crown gall grape callus tissue have been described by Platt (1955).

Forty-nine strains of tissue from 25 species grown on 25 mg/l of gibberellin showed that growth of a few tissues was favored but generally the tissues were inhibited (Nickell and Tulecke, 1959).

The effects of antibiotics on growth of callus tissue have also been observed. Sunflower tissue (De Ropp, 1946) und Jerusalem artichoke (Camus and Lance, 1955) were stimulated by penicillin. Many other compounds (De Ropp, 1950) were inhibiting to the sunflower callus. Penicillin, terramycin, streptomycin, thiolutin and bacitracin (Nickell, 1952) stimulated growth of *Rumex* tissue. Chloramphenicol and polymixin inhibited, and terramycin was toxic to marigold tissue of crown gall origin (Klemmer, Riker and Allen, 1955).

VIII. Nucleic Acids, Purines and Pyrimidines.

The nucleic acids and certain purines and pyrimidines may stimulate or inhibit growth of callus tissue depending on the species and the concentration. The growth of *Rumex* virus tumor tissue was improved by ribose nucleic acid

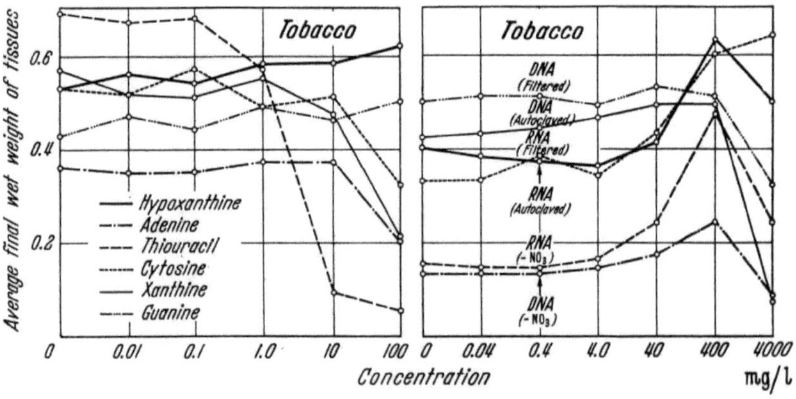

Fig. 20. Growth of tobacco callus on synthetic media containing (left) increasing concentrations of purines and pyrimidines or (right) DNA and RNA with and without nitrate. [From: A. C. Hildebrandt, A. J. Riker and Elizabeth Muir: Plant Physiol. 32, 231—236 (1957).]

(RNA) at concentrations from 0.2 to 0.8 mg and impaired by deoxyribonucleic acid (DNA) (Nickell et al., 1950). Inhibition of the callus growth appeared on media with adenine, adenosine and adenylic acid. Some beneficial effects were observed with guanine, uracil, xanthine or hypoxanthine. Uric acid was toxic. 2,6-diaminopurine at 1 mg/l or more, inhibited growth of callus from *Rumex*, tobacco or sunflower crown gall, pole bean and sweet clover (Nickell, 1955).

Growth of tobacco *(N. glauca × N. Langsdorffii)* callus of normal origin and marigold callus tissue of crown gall origin (Hildebrandt, Riker and Muir, 1957) were strongly stimulated by both glass filtered and autoclaved RNA at high concentrations, while DNA had no effect at low concentrations and was inhibitory at high concentrations. RNA was a good nitrogen source for tobacco tissue but not for marigold tissue. DNA was not a good nitrogen source for either tissue (Fig. 20).

Neither DNA or RNA served as a carbon source. The tobacco tissue was inhibited by adenine, cytosine, xanthine, thymine and uracil, while hypoxanthine, guanine and ribose supported progressively more growth with increasing concentration (Fig. 20). Marigold tissues were inhibited by 400 mg/l of all the purines and pyrimidines tested. Adenine and guanine were the most inhibitory but some stimulation resulted with low levels of cytosine, guanine, uracil and ribose. Thiouracil inhibited growth at 10 or more mg/l.

Nucleic acids, purines, pyrimidines and many analogs influenced growth of virus infected *N. tabacum* tissue and the infectivity of the virus therein depending on the compound and the concentration (see HILDEBRANDT, 1958; HILDEBRANDT and RIKER, 1958; and KURTZMAN et al., 1959). Important differential effects on the host tissue growth and the infectivity of the virus were observed. The importance of the balance of these materials with growth substances and other metabolites has also been emphasized (SKOOG, 1954; SKOOG and MILLER, 1957; GAUTHERET, 1959).

IX. Complex Extracts.

Various plant and animal extracts have been useful and often necessary supplements to the basic mineral salts, sucrose tissue culture medium to induce continued growth in certain tissues. These have included yeast extract, malt extract, tomato and other vegetable juices, casein hydrolysate and various liquid endosperms including those from corn, horse chestnuts and coconuts (see NETIEN and BEAUCHESNE, 1955; GAUTHERET, 1959; STEWARD and SHANTZ, 1959).

Yeast extract was beneficial for some of the original tissue and organ cultures (WHITE, 1943), as well as for more recent isolations (STEEVES and SUSSEX, 1952; TULECKE, 1953; TAMAOKI and ULLSTRUP, 1958). Malt extract has been beneficial for pine tissue cultures (LOEWENBERG and SKOOG, 1952).

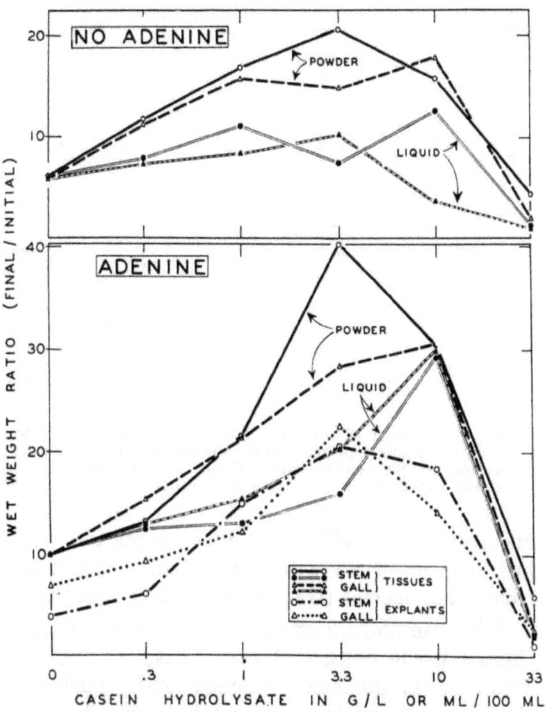

Fig. 21. Growth of recently isolated *Phylloxera* gall and normal grape stem tissues on media with increasing concentrations of casein hydrolysate with or without adenine. [From: F. PELET, A. C. HILDEBRANDT, A. J. RIKER and F. SKOOG: Am. J. Botany 47, 186—195 (1960).]

The growth stimulating activity of tomato juice for embryos was noted by ZEIBUR and BRINK (1951), by DEMETRIADES (1955) for *Parthenocissus* and *Scorzonera*, and by STRAUS (1960) for corn endosperm. Asparagine was an important growth stimulating component of this complex (STRAUS, 1960). Growth of the corn tissue on asparagine alone or on a mixture of aspartic and glutamic acids and glutamine exceeded growth attained with casein hydrolysate, tomato juice or yeast extract. Mixtures of arginine and asparagine supported tissue growth equal to that with tomato juice or yeast extract and better than growth secured with a supplement of casein hydrolysate. Glutamine and aspartic acid alone supported good growth of the endosperm cultures. Glutamic acid was not as good as aspartic acid or glutamine for this tissue strain.

Casein hydrolysate in various concentrations has stimulated growth of many species including normal sunflower stem callus (HENDERSON et al., 1952), carrot callus (CAPLIN, 1956) and normal and insect gall tissues from grape and other species (PELET et al., 1960) (Fig. 21).

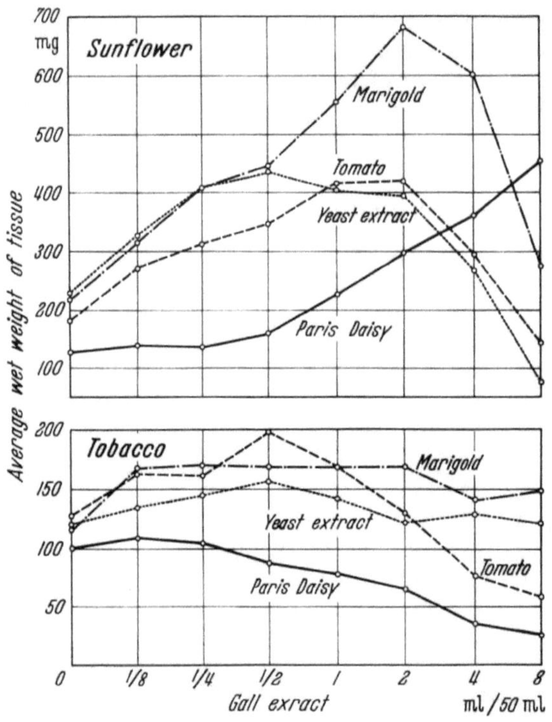

Fig. 22. Growth of sunflower crown gall and normal tobacco tissues on media supplemented with extracts from crown galls on several host plants as indicated. [From: A. C. HILDEBRANDT, A. J. RIKER and B. M. DUGGAR: Cancer Research 6, 368—377 (1946).]

Crown gall tissue extracts stimulated growth of a variety of tissues (see HILDEBRANDT, RIKER and DUGGAR, 1946; KAVOOR, 1955; BRAUN, 1959) (Figure 22).

Striking growth stimulating effects of coconut milk have been observed for callus tissues from many species following the original beneficial effects noted with excised embryos by VAN OVERBEEK, CONKLIN and BLAKESLEE (1941). The favorable effects of coconut milk alone as a supplement to the basic medium (CAPLIN and STEWARD, 1948; DUHAMET, 1950; GAUTHERET and DUHAMET, 1950; NICKELL, 1950; MOREL and WETMORE, 1951; STEWARD and CAPLIN, 1952; MAUNEY et al., 1952) and in combination with 2,4-dichlorophenoxyacetic acid has been observed with carrot callus (STEWARD and CAPLIN, 1951) and callus from other species (GAUTHERET, 1959; PELET et al., 1960).

The stimulating effects of these extracts has been the subject of many continuing tissue culture studies to isolate and identify the beneficial ingredients. Vitamins, amino acids, nucleic acids, purines, pyrimidines, growth substances alone and in various critical balances, have been examined for their growth stimulating activities. STEWARD, CAPLIN and MILLER (1952), and STEWARD and CAPLIN (1952 and 1955); SHANTZ and STEWARD (1952), STEWARD and SHANTZ (1955) reported materials active in coconut milk and endosperm from other species. A concentrate several thousand times more active than fresh coconut milk was reported by MAUNEY et al. (1952). Yeast extract had similar activity. DNA was the source of the highly active kinetin (6-N-furfurylaminopurine), MILLER, SKOOG,

VON SALTZA and STRONG, 1955; MILLER, SKOOG, OKUMURA, VON SALTZA, and STRONG, 1955). The growth stimulating effects of this compound in proper balance with other metabolites has been detailed (DAS et al., 1956; SKOOG and MILLER, 1957; GAUTHERET, 1957).

J. Some Applications of Tissue and Cell Cultures.

The possibilities for basic studies using plant tissue culture methods become evident from this brief account of some types of tissue that have been grown, and their requirements for growth. Special reference has been made to callus tissue, but references cited also refer to studies with more complex organ cultures. Many of the studies indicated above delineate the requirements for growth as well as applications to special problems. In most cases the good degree of physical and chemical environmental control of the cultures by the investigator has attracted attention to the method for particular biological studies. Some of these biological problems of fundamental importance that have been examined with considerable interest and detail, are briefly mentioned below. Further details of particular applications are abundantly available in the references cited.

I. Respiration and Metabolism.

Studies of the respiration of whole pieces of callus tissue, homogenates and cellular fractions provide interesting means of clarifying normal and diseased growth. Enzymatic differences in tissue of crown gall and normal origins for example are retained after growth for extended periods in tissue culture. Reduced respiratory levels were noted in crown gall tissue cultures (PLANTEFOL and GAUTHERET, 1941; WHITE, 1945; KLEIN, 1957; CRAGIE et al., 1958). There was an apparent increase maintained in endogenous auxin synthesis by crown gall tissue culture (HENDERSON et al., 1952; BRAUN and STONIER, 1958; LIPITZ and GALSTON, 1959). C^{14}-labeled fructose, sucrose and glutamine absorption and their incorporation into the alcohol insoluable fraction by crown gall tissues of *Parthenocissus* were lower than in normal tissue cultures (CRAGIE et al., 1958). The respiration of crown gall tissue of *Scorsonere* (LIORET, 1955), that had been subjected to the usual shocks of subculturing, rose appreciably after subculture for about two weeks and reached a maximum after about eight days. The presence of NAA in the medium had only a slight effect on the above observed respiration.

The oxygen uptake of *Rumex* tissue (NICKELL, 1950) was increased by concentrations of IAA, 2,4-D, naphthoxyacetic and 2,3,5-triiodobenzoic acids, and by colchicine at concentrations from 0.001 to 1 mg/l, and decreased at higher concentrations.

The secretion of an α-amylase enzyme by this tissue (BRAKKE and NICKELL, 1955), accounted for the utilization of starch by the tissue, (NICKELL and BURKHOLDER, 1950). Transformation of carbohydrates by carrot tissues was described by GORIS (1954). The metabolism involved in cell enlargement of cultured tobacco pith parenchyma was described by NEWCOMB (1955). Discs of tissue on IAA media as compared to those on media lacking auxin showed a rise in respiratory rate, a large increase in ascorbic acid oxidase activity, an increase (within 72 hours) of intact cells to oxidize SH groups of cysteine and glutathione, increase in invertase and pectin methylesterase activity and a decrease in peroxidase activity. The respiration and ascorbic acid oxidase also increased or decreased with other changes in the composition of the medium. Phenylthiourea at critical concentrations inhibited ascorbic acid oxidase activity and the growth, while permitting a large respiratory increase on auxin, (NEWCOMB 1960). Different phases of metabolism controlling cell enlargement were evidently altered by these changes in the composition of the medium.

The peroxidase, polyphenoloxydase and catalase activity of Jerusalem artichoke was influenced by the concentration of 2,4-D and IAA (MOREL and DEMETRIADES, 1955).

The oxidative and phosphorylative activities of cytoplasmic particles from crown gall and normal tissue cultures of tomato studied with the oxygen electrode (TAMAOKI, HILDEBRANDT, RIKER, BURRIS and HAGIHARA, 1959) indicated they

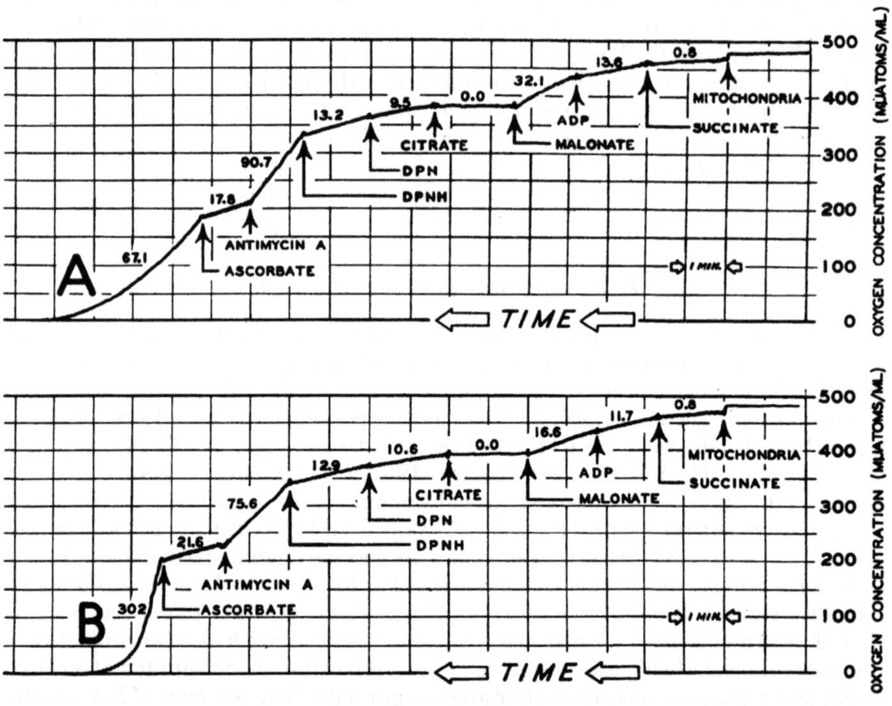

Fig. 23. Respiration curves at 26° C of mitochondria from normal and crown gall tissue cultures of tomato obtained with oxygen electrode. Recordings with VARIAN recorder. Figures in $m\mu$ atoms/min/ml. Total reaction mixture 2.2 ml. Recordings of reactions was from right to left as traced from the original records. A: Respiration curve of normal tissue mitochondria. Mitochondria, 134 μg N in 0.28 ml; succinate, 2.8 μmoles/ml; ADP: 0.2 μmoles per ml; malonate, 14 μmoles/ml; citrate, 2.8 μmoles/ml; DPN: 0.5 μmoles/ml; DPNH: 0.7 μmoles/ml; Antimycin A, 0.4 μg/ml; ascorbate, 2.8 μmoles/ml. B: Respiration curve of crown gall mitochondria. Mitochondria, 146 μg N in 0.28 ml. All others same as in A. [From: T. TAMAOKI, A. C. HILDEBRANDT, R. H. BURRIS, A. J. RIKER and B. HAGIHARA: Plant Physiol. 35, 942—947 (1960).]

were qualitatively similar but quantitatively different. Crown gall and normal tissues (TAMAOKI, 1960) had relatively constant respiratory activity while the tissues were growing actively, but this activity decreased as growth leveled off. Crown gall tissue gave somewhat lower activity than normal tissue on a dry weight basis, but gave similar values on a nitrogen basis. Mitochondria from crown gall tissue oxidized reduced diphosphopyridine nucleotide (DPNH), glutamate, citrate, malate and α-ketoglutarate more slowly, but ascorbic acid more rapidly (4.5 times) than mitochondria from normal tissue cultures (TAMAOKI, HILDEBRANDT, RIKER, BURRIS and HAGIHARA, 1960). (Fig. 23) Normal and crown gall tissue particles (TAMAOKI, HILDEBRANDT, BURRIS and RIKER, 1960) assayed spectrophotometrically for DPNH, DPNH-cytochrome C reductase, cytochrome C oxidase and diaphorase activities always showed normal tissues had higher activity. Pectinase was used to eliminate contaminating polysaccharide substance from the pellet of tissue culture mitochondrial preparation without affecting the enzymatically active sites of these particles (TAMAOKI, HILDEBRANDT and RIKER, 1960).

II. Polarity and Organ Function.

Isolated tissues are especially useful for studies of polarity and organ function (GAUTHERET, 1957). Much information has been learned about these phenomena by GAUTHERET, SKOOG, STEWARD, and WETMORE and their associates. GAUTHERET (1957) has elaborated on three major types of proliferation that develop in isolated tissues. Proliferation may be that of unorganized parenchyma, that of specialized cambial layers or that of organized growing points. These may develop in fresh explants or in established cultures. Variations in these types of proliferation may be observed in a variety of tissues from many species. For example, proliferation from a fresh stem or carrot section commonly occurs as a layer of new growth over the cut surface. Proliferation may also occur from preexisting cambia or from diffuse cambia scattered in the tissue. Also isolated tissue often produces root or bud meristems. The various types of proliferation have been also observed in established strains of tissue. Cultures that were originally parenchymatous may develop vascular bundles with phloem and xylem after several transfers. Eventually root and stem primordia may be developed. Many studies related to these various types of proliferation have been summarized by GAUTHERET (1959).

CAMUS (1949) to clarify polarity and differentiation carried out extensive in vitro grafts.

This interesting process of differentiation of tissues and organs has been further clearly described in a series of papers (STEWARD, 1958; STEWARD, MAPES and SMITH, 1958; STEWARD, MAPES and MEARS, 1958; and WETMORE, 1959). Single cells in mass liquid cultures have been followed through the formation of progressively more complex cell masses, to the production of a carrot plant.

Significant efforts to clarify cell growth, cell division, callus differentiation and organ formation in terms of growth substances and balances of common metabolites have been made by SKOOG and associates (SKOOG, 1954a, 1954b, 1955; DAS et al., 1956; SKOOG and MILLER, 1957). Interaction between IAA and adenine were related to bud formation in tobacco callus. Adenine promoted and IAA prevented bud formation. Concentrations were critical items and the adenine-IAA ratio determined the amount of bud formation. The tobacco callus cells supplied auxin alone (JABLONSKI and SKOOG, 1954) enlarged enormously without dividing. With auxin present cell division resulted by contact with vascular stem tissue, or by adding coconut milk or malt extract. Subsequently kinetin was isolated (MILLER et al., 1956) from DNA that was highly active in promoting cell division in the tobacco callus (DAS et al., 1956). These effects of balance of IAA, kinetin, and other metabolites and extracts, present a picture of much progress in our understanding of growth and differentiation (see SKOOG and MILLER, 1957; GAUTHERET, 1957).

The callus tissue masses may or may not change with time, not only morphologically but also physiologically. Many have noted (see GAUTHERET, 1955; NOBECOURT, 1957) changes that appear in established cultures maintained for extended periods in vitro. Tissue changes have been correlated with auxin requirements (habituation), with pigment formation and with varied sucrose concentration or light conditions (GAUTHERET, 1941). Similarly, modifications appeared in rate of growth, firmness of tissue mass, and requirements for growth substances such as 2,4-D or for coconut milk. Insect gall tissue and normal tissue may have different nutritional requirements depending on the length of time in culture (PELET et al., 1960) (Fig. 24), while crown gall tissue (BRAUN, 1958; TAMAOKI, 1960) are remarkably constant in nutritional and growth substance requirements.

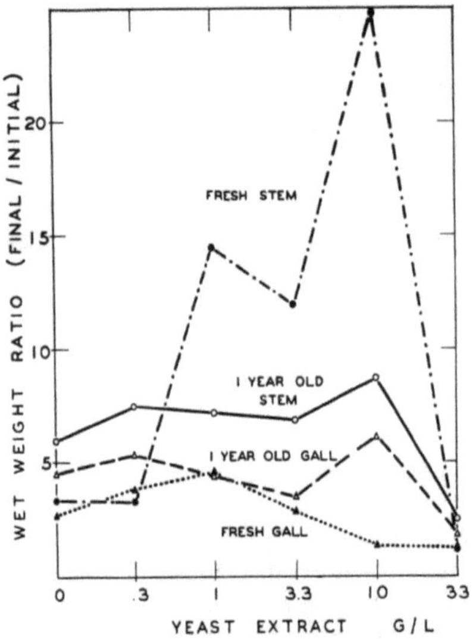

Fig. 24. Effects of various concentrations of yeast extracts on growth of stem and *Phylloxera* gall tissue of grape, related to length of time tissue had been grown *in vitro*. [From: F. PELET, A. C. HILDEBRANDT, A. J. RIKER and F. SKOOG: Am. J. Botany 47, 186—195 (1960).]

III. Studies of Plant Diseases.

Tissue cultures offer a good tool for comparison of normal and diseased tissues of various plants in other respects. The possibility was shown (WHITE and BRAUN, 1942) of growing crown gall tissue in vitro. Many nutritional characteristics of these gall tissues were clarified later (see RIKER and HILDEBRANDT, 1958). The gall tissue had more growth substance activity than normal tissue (KULESCHA and GAUTHERET, 1946). Interactions of gall and normal tissues were studied with in vitro grafts (DE ROPP, 1948). A non-auxinic growth factor was also described from the gall tissue (BRAUN and NAF, 1954). The pathological nature and physiological differences between normal and gall tissue have been clarified through a detailed study (BRAUN, 1957). The change from a normal to a crown gall cell under controlled temperature was related to time, and its ability to synthesize needed metabolites (BRAUN, 1958). The fast growing gall cells synthesized all their growth factor requirements. Less completely altered gall cells required supplements to the medium including glutamine, inositol and NAA, asparagine, cytidylic and guanylic acids, to grow as well as the gall tissues. Normal cells required for growth, kinetin in addition to these compounds. Single cell clones also retained certain crown gall characteristics (MUIR et al., 1954; BRAUN, 1959).

Tissues of galls of genetic origin were among the first grown in vitro (WHITE, 1939). Galls of virus origin (BLACK, 1947) established in tissue cultures have been examined for nutritional requirements and metabolism (see NICKELL, 1955; BRAKKE and NICKELL, 1955). The grape downy mildew fungus was inoculated to tissues growing in vitro by MOREL (1948). Bacterial soft rot bacteria varied in their ability to induce rot in tissue cultures (VOLCANI et al., 1953). Similarly, gall tissue induced by rust fungi (HOTSON, 1953; CUTTER, 1959) and one of unknown origin on white spruce (REINERT and WHITE, 1956) have been isolated and studied.

Callus tissues infected with TMV and other viruses (MOREL, 1948) have been examined to clarify virus infection and multiplication. The amount of virus in the tobacco callus cultures varied depending on the strain of host tissue and the nutritional and physical environments (see MOREL, 1948; KASSANIS, 1957; HILDEBRANDT, 1958; HILDEBRANDT and RIKER, 1958; BERGMANN and MELCHERS, 1959). Growth substances influenced the amount of virus in tobacco tissue cultures

(Kutsky and Rawlins, 1950; Kutsky, 1952). The number of tobacco tissue pieces artifically infected with TMV depended on the type and number of injuries induced at inoculation time (Kassanis, Tinsley and Quak, 1958). No plasmodesmata were found joining cells in the callus cultures but the virus moved between cells at about 1 mm per week, which is comparable to the rate of virus movement in the intact leaf. Nucleic acids, purines, pyrimidines and some analogs, depending on the concentration, influenced the growth of the TMV infected tobacco tissue and the infectivity of the TMV (Hildebrandt, 1958; Hildebrandt and Riker, 1958). Thymine reduced the virus infectivity 20 per cent with 500 mg/l with only slight injury to the tissue. Compounds that inhibited virus commonly also reduced growth or killed the tissue. Other compounds stimulated host growth and virus infectivity. 6-methylpurine (Kurtzman et al., 1960) was especially effective as an antiviral compound. It differentially inhibited the TMV infectivity 40 per cent at 0.1 mg/l without injury to the host tissue (Fig. 25).

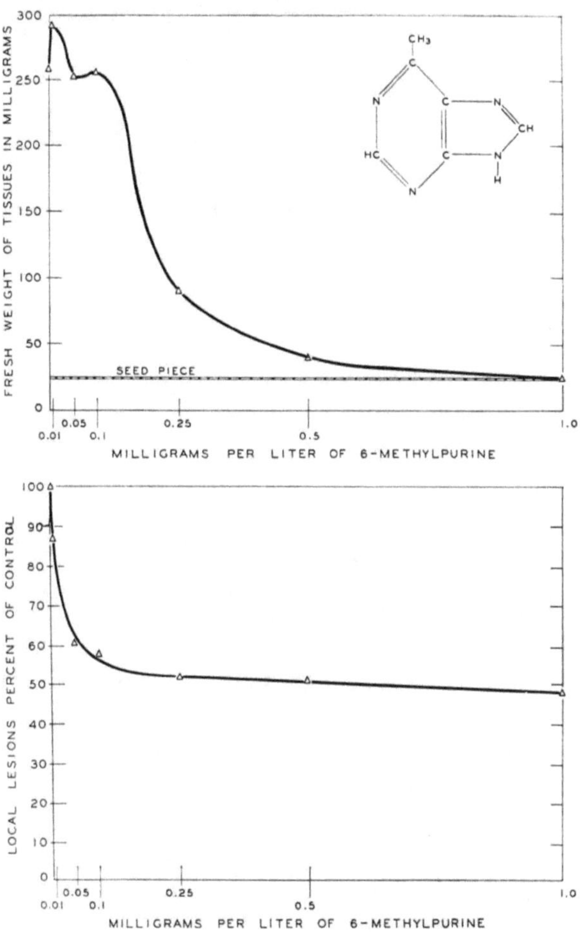

Fig. 25. Growth of TMV infected tobacco tissue on media with various concentrations of 6-methylpurine, and infectivity of the TMV in the tissue homogenates. [From: R. H. Kurtzman jr., A. C. Hildebrandt, R. H. Burris and A. J. Riker: Virology 10, 432—448 (1960).]

Various tissues of insect gall origin were isolated and studied in vitro. Added auxin was necessary to grow isolated gall tissue from *Salvia* (Demetriades, 1953). Extensive studies of the nutritional and physical requirements of insect gall and normal tissues have been tested and described by Pelet et al., 1960. The optimal concentrations and combinations were tested for grape *Phylloxera* gall tissues of vitamins, NAA, IAA, kinetin, casein hydrolysate, yeast extract, adenine and a few amino acids. NAA and kinetin supported moderate continuous growth and excellent growth if casein hydrolysate and adenine were also present (Fig. 26). The NAA requirements for callus cultured one to four years in vitro was much less than for freshly isolated tissue. It was suggested that the galls induced by *Phylloxera* did not represent permanently modified tissue comparable to the condition induced in suitable hosts by crown gall bacteria. The *Phylloxera* insects were also reared through several generations on the sterile gall tissue.

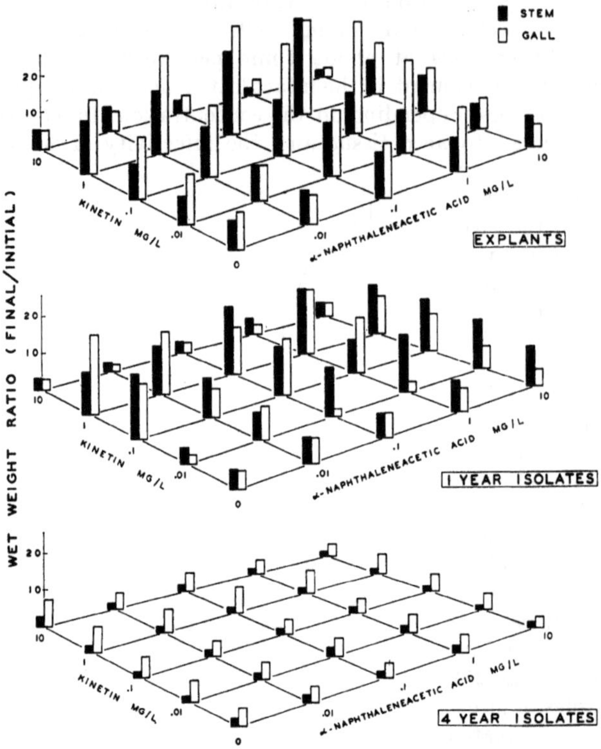

Fig. 26. Growth of stem and *Phylloxera* gall tissues of grape on various combinations of NAA and kinetin. The basic medium contained vitamins, adenine (40 mg/l) and casein hydrolysate (3 g/l). [From: F. PELET, A. C. HILDEBRANDT, A. J. RIKER and F. SKOOG: Am. J. Botany 47, 186—195 (1960).]

IV. Single Cell Cultures of Higher Plant Cells.

Interests in tissue culture methods has expanded over the years from organ cultures to cultures of cells. It is now possible to isolate and grow single cells of higher plants. The question has appeared whether all the cells in a callus tissue mass are similar, or if not, what are the types of variation in the cells that make up the mass. The need to study the growth requirements and potentialities for growth and differentiation was suggested also from variations observed in mass culture of cells from many species (see GAUTHERET, 1955; NOBECOURT, 1957; MUIR et al., 1958).

Single cell clones provide a valuable tool for detailed studies of similarities and differences between normal cells and various kinds of diseased cells. Single cell clones of tissue established by MUIR et al. (1954), from marigold crown gall tissue showed constant differences in growth rate, texture, and color. Some clones with varied media developed shoots or roots in culture. Single cell clones of tissue derived from isolated single cells of tobacco crown gall teratomas (BRAUN, 1959) developed into masses of tissue of morphologically abnormal structures typical of the teratoma cultures from which they were derived. Single cell clones of tobacco tissue established from strains of callus tissue from *Nicotiana tabacum* × *N. glutinosa* and the hybrid from these species studied by HILDEBRANDT and RIKER (1958) showed considerable variation in growth rate, morphology, texture, color and nutrient requirements. Furthermore, the amount of tobacco mosaic virus in respective clones varied greatly after 4 successive monthly passages of inoculated clones in virus-free liquid, shake cultures (Table 2). It was further shown (WU et al., 1960) that young largely meristematic single cell clones of the hybrid tobacco tissue were more resistant to TMV infection and/or subsequent multiplication of TMV than were older cultures composed largely of enlarging or senescent cells. One single cell clone of *N. tabacum* when infected with a mixture of TMV strains, selected a mild strain of TMV which multiplied in the tissue culture while the other virus strains in the mixture were apparently lost. Single cell clones of tissue reisolated from two parent single cell clones of the hybrid tobacco tissue, maintained differences in growth rates which were observed between the two parent clones (SIEVERT, HILDEBRANDT, RIKER and BURRIS, 1961). Differences in color and consistency were observed also among the clones reisolated from a single parent.

Table 2. *Tobacco mosaic virus activity in certain representative single cell clones and host tissue growth.*

Species and clone No.	Local lesions[1] from tissue No.	Local lesions[1] from medium No.	Tissue weight[2] gm
N. tabacum			
M 222	0	6	10.39
M 223	0	0	2.05
M 471	359	184	3.74
N. glutinosa			
G 252	83	4	0.54
G 283	146	5	0.40
N. tabacum × *N. glutinosa*			
H 194	158	7	8.56
H 196	265	2	9.55
H 237	77	6	4.10
H 241	4	2	10.66
H 248	5	5	13.66
H 261	0	0	11.90

[1] 1:1 dilution of tissue homogenate with distilled water assayed after the 4th monthly transfer in liquid "D" medium.
[2] Average weight in grams of tissues in 4 bottles at the end of one month incubation.

Single cell clones were also established from Phylloxera gall and normal grape stem tissues (HILDEBRANDT, HOTTINGER and RIKER, 1961). The clones varied in color, growth rate, texture and the ability to grow on media with various combinations of casein hydrolysate, growth substances and other nitrogen and carbon compounds as substitutes for coconut milk. The type and concentration of sugar influenced the growth of the single cell clones that came from the gall and normal grape stem tissues (ARYA, HILDEBRANDT and RIKER, 1961).

The dilution plate method of BERGMANN (1960) for securing clones of tobacco and bean tissue of single cell origin by plating cells in Petri plates provided a method for getting large numbers of cultures. This method plus the slide methods of DE ROPP (1955) and TORREY (1957) permitted observation of the isolated cells. The microchamber method (JONES et al., 1960) provided a further improvement in that continuous observation of single isolated cells was possible for extended periods. Somatic cells of the *N. tabacum* × *N. glutinosa* were grown and observed for more than four months in a single drop of media on a microscope slide without a change of nutrient. Cultures have since been observed to remain alive over a year under these conditions. Single isolated cells grow well in a medium previously used by a mass of growing cells. The cells divided, enlarged and became senescent. Mitosis was timed in the somatic cells in microculture. Cytoplasmic and nuclear features and content were observed in living unstained cells with bright field and phase contrast optics and recorded with time lapse photography. A few giant parenchyma cells rejuvenated by forming separate, free-floating, endogenous cells. The endogenous cells that escaped from the mother cells grew into apparently normal daughter populations.

The single cell culture methods have suggested many new studies not previously possible. Single cell clones may be further evaluated for additional chemical, morphological, genetical and pathological similarities and differences. The mechanism of virus infection and multiplication, and location of the virus in the host cell may be further clarified. Similarly, mutation of cells, the process of habituation, the action of chemical growth regulators, and of pathogenic agents other than viruses may be further delineated. It may also be possible to develop an entire plant from a single cell as has been demonstrated from cells in mixed cultures. Furthermore the somatic cells growing in microculture provide a simple experimental method for detailed cytological studies of both normal and diseased cells. This material is free of the artifacts so commonly found in killed and stained preparations. The physical and chemical microenvironments can be controlled so that the effects can be observed for extended periods, for example, of such diverse items as metabolites, growth substances, toxic materials and disease incitants.

K. Resume.

Cells from many plant sources may be grown in plant tissue culture to study particular problems. These cultures may be organ cultures of varying degrees of organizations as stem or root tips, or embryos; callus cultures of various normal and diseased origins; or single cell cultures. Callus cultures of varying number of cells and degrees of organization, and organ cultures were first successful. It has been this background information with mass cell cultures and organ cultures that made possible the growth of single cells, and the observation of growth and differentiation of cells and organs. The culture of the individual cells of higher plants has only recently been possible.

The cells in a plant are more or less sheltered from the outside environment and can specialize in the performance of their functions which are determined by a finely balanced set of conditions. In callus culture the cells are largely free from many of the effects of neighboring cells that exist in the whole plant and in organ cultures. Thus the cells growing in vitro on chemically defined media are released from many controls and subjected to other balances in the environment. Complex vascular tissues are not needed as the cells get their nutrients directly from the culture medium.

The culture of higher plant cells and cell masses in tissue culture has opened a new phase of botanical investigations, but the method also has wide application to many other branches of science.

We need not guess further about the importance of the cell as the unit of organization in the tissue or organ. We need now to further clarify the nutritional and physical chemical requirements of the cell for growth and differentiation, and the potentialities of the cell for these phenomena. Isolated cells and tissues that grow for extended periods or fail to grow depending upon what is present in or what is absent from the media and depending upon the other environments, provide a practical tool to experimentally elucidate these basic questions.

References.

ALLSOPP, A.: Nature (Lond.) **179**, 681 (1957). — ARYA, H. C., A. C. HILDEBRANDT and A. J. RIKER: Amer. J. Bot. (In Press) 1962.

BALL, E.: Amer. J. Bot. **33**, 301—318 (1946); — Growth **14**, 295—325 (1950); — Ann. Biol. **31**, 81—105 (1955); — Amer. J. Bot. **46**, 130—139 (1959). — BERGMANN, L.: J. Gen.

Physiol. 43, 841—851 (1959); — Nature (Lond.) 184, 648—649 (1959). — BERGMANN, L., and G. MELCHERS: Z. Naturforschg. 14b, 73—76 (1959). — BLACK, L. M.: Growth, 6th Symposium p. 79—84 (1947); — Proc. Nat. Acid. Sci. (US) 44, 365—367 (1958). — BOUREQUET, R.: C. R. Soc. Biol. 146, 1897—1899 (1952). — BRAKKE, M. K., and L. G. NICKELL: Ann. Biol. 31, 215—226 (1955). — BRAUN, A. C.: J. Nat. Cancer Inst. 19, 753—769 (1957); — Proc. Nat. Acad. Sci. (US) 44, 344—349 (1958); 45, 932—938 (1959). — BRAUN, A. C., and U. NAF: Proc. Soc. Biol. Med. 86, 212—214 (1954). — BRAUN, A. C., and T. STONIER: Protoplasmatologia 10, (5a), 93 p. (1958). — BRINK, R. A., and D. C. COOPER: Bot. Rev. 13, 479—541 (1947). — BUNNING, E., and H. WELTE: Physiol. Plant. 7, 197—203 (1954). — BURKHOLDER, P. R., and L. G. NICKELL: Bot. Gaz. 110, 426—437 (1949).

CAMUS, G.: Rev. Cyto. Biol. Veg. 11, 1—199 (1949). — CAMUS, G., and C. LANCE: Ann. Biol. 31, 63—79 (1955). — CAPLIN, S. M.: Bot. Gaz. 108, 379—393 (1947); — Amer. J. Bot. 43, 749—754 (1956); 46, 324—329 (1959). — CAPLIN, S. M., and F. C. STEWARD: Science 108, 655—657 (1948); — Nature (Lond.) 163, 920—924 (1949); — Ann. Bot. 16, 219—234 (1952). — CLAVER, F. K., A. C. HILDEBRANDT, G. H. REIMAN and D. C. COOPER: Phyton (Argentina) 11, 129—137 (1958). — CURTIS, J. T., and M. A. NICHOL: Bull. Torrey Bot. Club 75, 358—373 (1947). — CAPITE, L. DE: C. R. Soc. Biol. 146, 863—865 (1952); — Amer. J. Bot. 42, 869—873 (1955). — CRAGIE, J., G. KRATKOV and R. H. WETMORE: Amer. J. Bot. 45, 373—377 (1958). — CUTTER, V. M., jr.: Mycologia 51, 248—295 (1959). — CZOSNOWSKI, J.: Prace Kom. Biol. (Poland) 13, 209—246 (1952).

DAS, N. K., K. PATAU and F. SKOOG: Physiol. Plant. 9, 640—651 (1956). — DEMETRIADES, S. D.: Ann. Inst. Phytopath., Benaki 7, 61—67 (1953); — Ann. Biol. 30, 431—436 (1954). — DE ROPP, R. S.: Science 104, 371—373 (1946); — Nature (Lond.) 158, 555 (1946); — Bull. Torrey Bot. Club 75, 45—50 (1948); — Amer. J. Bot. 35, 372—377 (1948); — Science 112, 500—501 (1950); — Proc. Roy. Soc. B. 144, 86—93 (1955). — DUHAMET, L.: C. R. Acad. Sci. (Paris) 230, 770—771 (1950); — Ann. Biol. 31, 123—143 (1955).

EBERTS, F. S., R. H. BURRIS and A. J. RIKER: Plant Physiol. 29, 1—10 (1954). — ENDERLE, W.: Planta 39, 570—588 (1951).

GAUTHERET, R. J.: C. R. Acad. Sci. (Paris) 210, 632—634 (1939); 208, 118 (1939); 135, 875—877 (1941); — Manuel Technique de Culture des Tissus Végétaux 170 p. Paris: Masson ed. Cie. 1942; 6th Growth Symposium p. 21—43 (1947); — Rev. Gen. Bot. 54, 5—30 (1947); — C. R. Soc. Biol. (Paris) 142, 808—810 (1948); — Ann. Biol. 31, 145—171 (1955); — J. Nat. Cancer Inst. 19, 555—590 (1957); — La Culture des Tissus Végétaux. 863 p. Paris: Masson et Cie. 1942; 6th Growth Symposium p. 21—43 (1947); — Rev. Gen. Bot. 54, 5—30 (1947); — GORIS, H.: Ann. Biol. 30, 297—318 (1954).

HABERLANDT, G.: Sitzber. Akad. Wiss. Wien, Math. naturw. Kl. III, 69—92 (1902). — HARRISON, R. G.: Proc. Soc. exp. Biol. (N. Y.) 4, 140—143 (1907). — HELLER, R.: Dr. Sc. Thesis. Univ. of Paris. 223 p. Paris: Masson et Cie, ed. 1953; — Ann. Biol. 30, 261—281 (1954). — HENDERSON, J. H. M.: Rev. Gen. Bot. 54, 5—30 (1954). — HENDERSON, J. H. M., and J. BONNER: Amer. J. Bot. 39, 444—451 (1952). — HENDERSON, J. H. M., MARY E. DURRELL and J. BONNER: Amer. J. Bot. 39, 467—473 (1952). — HILDEBRANDT, A. C.: Proc. Nat. Acad. Sci. (US) 44, 354—364 (1958). — HILDEBRANDT, A. C., H. H. HOTTINGER and A. J. RIKER: Plant Physiol. 36, 29 (1961). — HILDEBRANDT, A. C., and A. J. RIKER: Amer. J. Bot. 34, 421—427 (1947); 36, 74—85 (1949); — J. Cancer Res. 15, 517—522 (1955); — Fed. Proc. 17, 986—993 (1958). — HILDEBRANDT, A. C., A. J. RIKER and B. M. DUGGAR: Amer. J. Bot. 32, 357—361 (1945); — (a) Cancer Res. 6, 368—377 (1946); — (b) Amer. J. Bot. 33, 591—597 (1946). — HILDEBRANDT, A. C., A. J. RIKER and ELIZABETH MUIR: Plant Physiol. 32, 231—236 (1957). — HILDEBRANDT, A. C., A. J. RIKER and JEAN L. WATERTOR: Phytopathology 44, 422—428 (1954). — HOTSON, H. H.: Phytopathology 43, 360—363 (1953). — HUREL-PY, G.: Rev. Gen. Bot. 57, 637—735 (1950).

JACQUIOT, C.: C. R. Acad. Sci. (Paris) 233, 815—817 (1951). — JABLONSKI, J. R., and F. SKOOG: Plant Physiol. 7, 16—24 (1954). — JONES, L. E., A. C. HILDEBRANDT, A. J. RIKER and J. H. WU: Amer. J. Bot. 47, 468—475 (1960).

KASSANIS, B.: (a) Virology 4, 5—13 (1957); — (b) Ann. Appl. Biol. 45, 422—427 (1957). — KASSANIS, B., T. N. TINSLEY and FREDERIKA QUAK: Ann. Appl. Biol. 46, 11—19 (1958). — KEENER, P. D.: Amer. J. Bot. 37, 520—527 (1950). — KLEIN, R. M.: Growth and differentiation of plant tissue cultures. In: Rhythmic and synthetic processes in growth p. 31—58. Princeton Univ. Press. 1957. — KLEIN, R. M., and GEORGIA E. MANOS: Ann. N. Y. Acad. Sci. (US) 88, 416—425 (1960). — KLEMMER, H. W., A. J. RIKER and O. N. ALLEN: Phytopathology 45, 618—625 (1955). — KOTTE, W.: Beitr. Z. Allgem. Bot. 2, 413—434 (1922). — KOVOOR, A.: Ann. Biol. 30, 417—429 (1954). — KULESCHA, Z.: Ann. Biol. 30, 329—348 (1954). — KULESCHA, Z., and R. J. GAUTHERET: C. R. Acad. Sci. (Paris) 227, 292—294 (1948). — KUMAR, S.: A. C. HILDEBRANDT and A. J. RIKER: Plant Physiol. 36, 22 (1961). — KURTZMAN, R. H.,

A. C. HILDEBRANDT, R. H. BURRIS and A. J. RIKER: Virology **10**, 432—448 (1960). — KUTSKY, R.: Science **115**, 19—20 (1952). — KUTSKY, R., and T. E. RAWLINS: J. Bact. **60**, 763—766 (1950).

LA RUE, C. D.: Bull. Torrey Bot. Club **63**, 365—382 (1936); **69**, 332—341 (1942); — Amer. J. Bot. **36**, 798 (1949); — Brookhaven Symp. in Biology No. 6 Brookhaven Nat. Laboratory (US) p. 187—208 (1954). — LEE, E. A.: (a) Amer. J. Bot. **37**, 312—318 (1950); (b) **37**, 528—533 (1950). — LIORET, C.: Ann. Biol. **31**, 185—194 (1955). — LIPITZ, J., and A. W. GALSTON: Amer. J. Bot. **46**, 193—196 (1959). — LOEWENBERG, J. R., and F. SKOOG: Physiol. Plant. **5**, 33—36 (1952). — LOO, S. W.: Amer. J. Bot. **32**, 13—17 (1945); **33**, 295—300 (1946).

MAUNEY, J. R., N. S. HILLMAN, C. O. MILLER, F. SKOOG, R. A. CLAYTON and F. M. STRONG: Physiol. Plant. **5**, 485—497 (1952). — MILLER, C. O., F. SKOOG, M. H. VON SALTZA and F. M. STRONG: J. Amer. chem. Soc. **77**, 1392 (1955). — MILLER, C. O., F. SKOOG, F. S. OKUMURA, M. H. VON SALTZA and F. M. STRONG: J. Amer. chem. Soc. **77**, 2662 (1955). — MOREL, G.: C. R. Acad. Sci. (Paris) **223**, 166—168 (1946); — Ann. Epiphyt. **14**, 123—234 (1948); — (a) C. R. Acad. Sci. (Paris) **230**, 1099 (1950); (b) **230**, 2318—2320 (1950); — (a) Rev. Gen. Bot. (France) **63**, 314—325 (1956); (b) **63**, 325—331 (1956). — MOREL, G., and S. DEMETRIADES: Ann. Biol. **31**, 227—236 (1955). — MOREL, G., and C. MARTIN: Report 14th Internat. Horticulture Congress. p. 303—310. Wageningen (Netherlands): H. Veenman and Sons 1955. — MOREL, G., and R. H. WETMORE: (a) Amer. J. Bot. **38**, 138—140 (1951); (b) **38**, 141—143 (1951). — MUIR, W. H., A. C. HILDEBRANDT and A. J. RIKER: Science **119**, 877—878 (1954); — Amer. J. Bot. **45**, 589—597 (1958). — MURRAY, M. R., and G. KOPECK: A bibliography of the research in tissue culture. 2 vol. New York: Academic Press 1953.

NEWCOMB, E. H.: Ann. Biol. **31**, 195—214 (1955); — Physiol. Plant. **13**, 459—467 (1960). — NETIEN, G., and G. BEAUCHESNE: Ann. Biol. **30**, 437—443 (1954). — NICKELL, L. G.: (a) Bot. Gaz. **112**, 225—228 (1950); — (b) Amer. J. Bot. **37**, 829—835 (1950); — (a) Bull. Torrey Bot. Club **79**, 427—430 (1952); — Proc. Soc. exp. Biol. (N. Y.) **80**, 615—617 (1952); — Brookhaven Symposium in Biol. No. 6 p. 174—186 (1954); — Ann. Biol. **31**, 107—121 (1955); — Antimetabolites and cancer. A.A.A.S. p. 129—151 (1955). — NICKELL, L. G., and P. R. BURKHOLDER: Amer. J. Bot. **37**, 538—547 (1950). — NICKELL, L. G., and W. TULECKE: Bot. Gaz. **120**, 245—250 (1959); — J. Biochem. Microbial Tech. Eng. **2**, 287—297 (1960). — NICKELL, L. G., PHYLLIS GREENFIELD and P. R. BURKHOLDER: Bot. Gaz. **112**, 42—52 (1950). — NITSCH, J. P.: Amer. J. Bot. **38**, 566—577 (1951). — NITSCH, J. P., and Colette NITSCH: Amer. J. Bot. **43**, 839—851 (1956). — NOBECOURT, P.: C. R. Acad. Sci. (Paris) **205**, 521—523 (1937); **130**, 270 (1939); **133**, 530—532 (1940); — Bull. Soc. Bot. (France) **104**, 623—665 (1957).

PARIS, D.: Ann. Biol. **31**, 15—29 (1955). — PARKER, R. C.: Methods of tissue culture. 294 p. 2nd ed. New York: Paul B. Hoeber, Inc. 1950. — PECK, T. T.: J. exp. Med. **85**, 729 bis 757 (1947). — PÉLET, F., A. C. HILDEBRANDT, A. J. RIKER and F. SKOOG: Amer. J. Bot. **47**, 186—195 (1960). — PLANTIFOL, L., and R. J. GAUTHERET: C. R. Acad. Sci. (Paris) **213**, 627—629 (1941). — PLATT, R. S. jr.: Ann. Biol. **30**, 349—359 (1954).

QUAK, FREDERICA: Proc. 14th Internat. Hort. Congress, Nice (1958).

RANDOLPH, L. F., and L. G. COX: Amer. Soc. Hort. Sci. Proc. **43**, 284—300 (1943). — RAPPAPORT, J.: Bot. Rev. **20**, 201—225 (1954). — REINERT, J., and P. R. WHITE: Physiol. Plant. **9**, 177—189 (1956). — REINERT, R. A.: Personal Communication (1960). — RIKER, A. J., and A. C. HILDEBRANDT: Ann. Biol. **30**, [283]—296 (1954); — Ann. Rev. Microbiol. **12**, 469—490 (1958). — RIKER, A. J., and ALICE E. GUTSCHE: Amer. J. Bot. **35**, 227—238 (1948). — ROBBINS, W. J.: Bot. Gaz. **73**, 376—390 (1922). — ROBBINS, W. J., and ANNETTE HERVEY: Science **114**, 322—323 (1951). — ROBERTSON, G. H.: Amer. Med. **46**, 311—347 (1947).

SIEVERT, R. C., A. C. HILDEBRANDT, A. J. RIKER and R. H. BURRIS: Plant Physiol. **36**, 28 (1961). — SOSSOUNTZOV, I.: C. R. Soc. Biol. (Paris) **147**, 1001—1007 (1953). — SHANTZ, E. M., and F. C. STEWARD: J. Amer. chem. Soc. **74**, 6133 (1952). — SKOOG, F.: Amer. J. Bot. **31**, 19—24 (1944); — (a) Brookhaven Symposium in Biology No. 6, 1—21 (1954); — (b) Chapt. 8. Dynamics of growth processes. p. 148—183. Princeton Univ. Press. 1954. — SKOOG, F., and C. O. MILLER: Symposia Soc. Exptl. Biol. No. 11, 118—131 (1957). — SKOOG, F., and C. TSUI: Amer. J. Bot. **35**, 782—787 (1948). — STEEVES, T. A., and I. M. SUSSEX: Nature (Lond.) **170**, 672 (1951); — Amer. J. Bot. **44**, 665—673 (1957). — STEEVES, T. A., I. M. SUSSEX and C. R. PARTANEN: Amer. J. Bot. **42**, 232—245 (1955). — STEWARD, F. C.: Amer. J. Bot. **45**, 709—713 (1958). — STEWARD, F. C., and S. M. CAPLIN: Science **113**, 518—520 (1951); — Amer. J. Bot. **16.** 491—504 (1952); Ann. Biol. **30**, 385—398 (1954). — STEWARD, F. C., C. M. CAPLIN and F. K. MILLER: Ann. Bot. **16**, 57—77 (1952). — STEWARD, F. C., and F. M. SHANTZ: Ann. Biol. **30**, 399—415 (1954); — Plant. Physiol. **34**, 57 (1959). — STEWARD, F. C., MARION O. MAPES and JOAN SMITH: Amer. J. Bot. **45**, 693—703 (1958). — STEWARD, F. C., MARION O. MAPES and KATHRYN MEARS: Amer. J. Bot. **45**, 705—708 (1958). — STRAUS, J.: Amer. J. Bot. **41**, 833—839 (1954); — PLANT. Physiol. **34**, 536—541

(1959); — Amer. J. Bot. **47**, 641—647 (1960). — STRAUS, J., and C. D. LA RUE: Amer. J. Bot. **41**, 687—694 (1954). — STREET, H. E.: J. Nat. Cancer Inst. **19**, 467—494 (1957). — STRUCKMEYER, B. S., A. C. HILDEBRANDT and A. J. RIKER: Amer. J. Bot. **36**, 491—495 (1947). — SUSSEX, I. M., and T. A. STEEVES: Ann. Bot. **17**, 395—401 (1953); — Bot. Gaz. **119**, 203—208 (1958).

TAMAOKI, T.: Ph. D. Thesis University of Wisconsin, 49 p. Madison 1960. — TAMAOKI, T., and A. J. ULLSTRUP: Bull. Torrey Bot. Club **85**, 260—272 (1958). — TAMAOKI, T., A. C. HILDEBRANDT, A. J. RIKER, R. H. BURRIS and B. HAGIHARA: Nature (Lond.) **184**, 1491—1492 (1959); — Plant Physiol. **35**, 942—947 (1960). — TAMAOKI, T., A. C. HILDEBRANDT and A. J. RIKER: (a) Science **132**, 1766—1767 (1960). — TAMAOKI, T., A. C. HILDEBRANDT, R. H. BURRIS and A. J. RIKER: Plant Physiol. **36**, 347—351 (1961). — TORREY, J. G.: Proc. Nat. Acad. Sci. (Wash.) **43**, 887—891 (1957); — Cell, organism and milieu. p. 189—222. Ronald Press 1959. — TORREY, J. G., and Y. SHIGEMURA: Amer. J. Bot. **44**, 334—344 (1957). — TULECKE, W.: Science **117**, 599—600 (1953); — Amer. J. Bot. **44**, 602—608 (1957); — Plant Physiol. **35**, 19—24 (1960). — TULECKE, W., and L. G. NICKELL: Science **117**, 599—600 (1953); — Trans. N. Y. Acad. Sci. **22**, 196—206 (1960). — TURVET, I. N., and J. P. HOLLIS: Phytopathology **38**, 960—967 (1948).

VAN OVERBEEK, J., M. E. CONKLIN and A. F. BLAKESLEE: Science **94**, 350—351 (1941); — Amer. J. Bot. **29**, 472—477 (1942). — VOLCANI, ZAFRIRA, A. J. RIKER and A. C. HILDEBRANDT: Phytopathology **43**, 92—94 (1953).

WEINSTEIN, L. H., L. G. NICKELL, H. J. LAURENCOT and W. TULECKE: Cont. Boyce Thomp. Inst. **20**, 239—250 (1959). — WETMORE, R. H.: Brookhaven Symposia in Biology No. 6 Brookhaven National Laboratory (USA) p. 22—40 (1954); — Amer. Sci. **47**, 326—340 (1959). — WETMORE, R. H., and G. MOREL: Amer. J. Bot. **36**, 805—806 (1949). — WETMORE, R. H., and C. W. WARDLAW: Ann. Rev. Plant Physiol. **2**, 269—292 (1951). — WHITE, P. R.: Plant Physiol. **9**, 585—600 (1934); — Amer. J. Bot. **26**, 59—64 (1939); — Ann. Rev. Biochem. **11**, 615—628 (1942); — A handbook of plant tissue culture. 227 p. Lancaster, Pa.: Jacques Catell Press 1943; — Cancer Res. **5**, 302—311 (1945); — The cultivation of animal and plant cells. 239 p. New York: Ronald Press 1954; — Ed., J. Nat. Cancer Inst. **19**, 467—843 (1957). — WHITE, P. R., and A. C. BRAUN: Science **94**, 239—241 (1941); — Cancer Res. **2**, 597—617 (1942). — WILLMER, E. N.: Tissue culture. 175 p. New York: John Wiley and Sons 1954. — WARD, M.: Science **132**, 1401—1402 (1960). — WU, J. R., A. C. HILDEBRANDT and A. J. RIKER: Phytopathology **50**, 587—594 (1960)..

YOUNG, D. J.: Amer. Potato J. **35**, 960—967 (1958).

ZIEBUR, NANCY K., and R. A. BRINK: Amer. J. Bot. **38**, 253—256 (1951).

Immunological Methods.

By

J. A. van der Veken, D. H. M. van Slogteren and J. P. H. van der Want.

With 12 Figures.

The elucidation of physical and chemical differences between molecules of certain plant products often presents considerable difficulty, since such differences may be too small to be capable of detection by the physico-chemical and other analytical methods at present available. This limitation is particularly felt in the analysis of certain macromolecules, such as proteins, polysaccharides or nucleoproteins.

Although such techniques as ultracentrifugation, diffusion and electrophoresis are widely applied in this field of research, they give little information regarding molecule configuration. In recent years the application of the techniques of column chromatography and column electrophoresis have made possible the demonstration of smaller differences between related substances. By these methods, the heterogeneity of protein preparations hitherto considered as homogenous has been demonstrated.

The antigenic properties of the macromolecules may be made the basis of a still more sensitive analysis, which is capable of reflecting even more subtle variations in internal macromolecular structure. This has led to the use of antibodies, produced in animals according to immunological principles, as highly specific reagents for qualitative and quantitative determinations.

A. The General Principles of Immunological Methods.

Immunological methods are based on the reaction between antigens and their specific antibodies. The latter are modified serum globulins formed in the animal, either in response to a stimulation arising from the environment (e. g. hay-fever), by natural infection with a pathogenic agent or artificially by injection with antigenic substances (ISLIKER, 1957). Thus the first step in applying these methods is the production of antibodies in an animal.

A prerequisite for the success of these methods is the antigenicity of the substances involved. Typical substances known for their antigenicity are proteins, nucleoproteins and polysaccharides, but even among these there are some which stimulate antibody production insufficiently, if at all. On the other hand, certain types of macromolecules exist which provoke a series of antibodies directed against different parts of the molecule, the so-called antigenic determinants. In addition to true antigens there are also substances incapable of stimulating antibody production on their own but doing so if they are first bound to a protein. These substances are called *haptens*.

The antigen-antibody reaction can be observed *in vitro* as well as *in vivo*. In the latter case sensitized animals are used *(anaphylaxis)*. Although the *in vivo* methods are more sensitive, those carried out *in vitro* have received more attention. All work related to the behaviour of antigens and antibodies *in vitro* belongs to the discipline of serology.

For detailed information on this subject the reader is referred to the monographs of BOYD (1956), CARPENTER (1956), CUSHING and CAMPBELL (1957), KABAT and MAYER (1948), and WILSON and MILES (1955).

B. Purification of Antigens, Immunization of Animals and Treatment of Antisera.

I. Purification of Antigens.

The particular aim of any serological investigation will determine whether or not the antigens must be purified. For instance, in studying serological relationships in the plant kingdom it is generally desirable to keep as many antigens as possible in the injection fluid. The virologist, however, will try to purify the antigen as much as possible in order to prevent normal antigenic plant constituents from playing a rôle in antibody production.

Furthermore, purification of the antigen may result in the removal of substances toxic to the animal involved. Often the desired aim (antibodies against only one antigen) cannot be readily achieved because of the difficulty of eliminating admixtures with stronger antigenic properties. In such cases, by adding the same impurities to the antiserum, the unwanted antibodies can be removed (absorption of the antiserum, see page 429).

It is difficult to give general rules for the purification and concentration of antigens. Precipitating reagents such as ammonium sulphate and alcohol can be used. Centrifugation techniques are nowadays of great importance, especially those making use of density gradients e.g. in virus research (BRAKKE, 1960). Other useful techniques include free electrophoresis (see BIER, 1960), zone electrophoresis (PORATH, 1956) and density gradient electrophoresis (BRAKKE, 1955). Purification can also be carried out by means of column chromatography with ion exchange resins, as applied to ribonuclease (CINADER and PEARCE, 1955).

Enzymes can be separated from lipoproteins by extraction with butanol-chloroform (MORTON, 1955). This method also gives good results with certain plant viruses as has been mentioned by STEERE (1959) in his survey of virus purification methods. WETTER (1960) described an analogous method whereby sap of infected plants was treated successively with ether and carbon tetrachloride. A common feature of the methods just mentioned is the coagulation of lipophilic constituents such as chlorophyll. The coagulum can then be separated from the aqueous phase containing the desired substances e.g. virus or enzymes, by subsequent low speed centrifugation.

ROZENDAAL and VAN SLOGTEREN (1958) in a study of the serological relationship between the potato viruses S and M, succeeded in removing lipophilic constituents from freeze-dried virus-infected material by extracting it with chloroform, alcohol, acetone and ether respectively.

Among other methods for concentrating the antigens are evaporation in a dialysing tube, dialysis against concentrated gum arabic or glycerol (KOHN, 1959), ultrafiltration and freeze-drying.

In investigations of relationships between higher plants, seeds are usually first extracted with organic solvents and afterwards with buffer solutions. By regulating the salt concentration the proteins can be fractionated (CHESTER, 1937a; MORITZ, 1958).

It must be kept in mind that during any preliminary purification process conditions such as pH, temperature or the presence of enzymes may affect the antigenic properties of the substance to be studied. For example, McPherson and Heidelberger (1940) found a decrease of antigenic power in ovalbumin after heating.

Finally we wish to mention the possibility that the antigenic properties of the substance may be lost by enzymatic processes in the animal following injection. Pretreatment of such antigens, e.g. papain, with formaldehyde (Ramon, 1945; Zanco and Pratesi, 1947) or applying special injection methods may obviate this difficulty. A general method for the formaldehyde treatment is given by Holt (1950).

II. Production of Antibodies.

1. Choice of Animals.

The production of antibodies can be undertaken in various animal species. In many cases specific requirements dictate which species is to be preferred. For anaphylaxis experiments *in vivo*, guinea pigs are generally used. Moreover guinea pig serum is the main source of the substance called complement which is used in the complement fixation test (see page 446). For most other *in vitro* serological purposes rabbits are considered most convenient to work with. Horses have been used for antiserum production against polysaccharides, bacterial toxins (see review Grabar, 1950) and the potato viruses X and S (van Slogteren and van Slogteren, 1957). Antibodies to the enzyme urease have been produced in chickens (Sumner, 1937).

The choice of an animal species cannot be based solely on the quantity of antiserum it produces, owing to the fact that sera from animals other than rabbits often possess properties complicating their use. It has been found, for instance, that antibodies from chicken serum show maximum activity at high salt concentrations, viz. 8% NaCl (Banovitz et al., 1959). On the other hand fowls, in contrast to guinea pigs, are good producers of precipitating antibodies. No definite rules regarding this aspect are known, however. Some antigens display antigenic activity in one animal species but not in another. Pneumococcus polysaccharide is antigenic in mice but not in rabbits (Cohn, 1952). Antigenicity in rabbits does arise, however, if the polysaccharide isolated from the bacterial cell wall is coupled to a protein.

2. Immunization of Animals.

It is preferable to use fairly large rabbits that are between 9 and 24 months old and weigh at least 4 lbs. Animals having large ears with prominent veins are particularly convenient to work with. It is advisable to inject at least three rabbits for each type of antigen as not every rabbit may be a good antibody producer. Moreover it may happen that an animal dies during the injection period. No definite proof is available so far that differences in facility of antibody production among individuals of an animal species are governed by hereditary characters, although it is conceivable that this may be the case (Sang and Sobey, 1954).

Generally, small quantities of antigen are administered several times to the animal. One obtains better results if a given amount of antigen is injected in several small doses rather than in one large dose (Cohn, 1952).

The antigen injection may be made in a number of different ways e.g. *intravenous, subcutaneous, intramuscular* and *intraperitoneal* (or *intra-abdominal*). Most commonly applied are intravenous injections of 1 to 2 ml of an antigen solution or suspension into the external marginal vein of the ear of the rabbit. This vein runs along the upper surface of the ear, parallel to the hind edge and about 4 mm from it. Fig. 1 shows this vein into which the antigen suspension is injected with a 2 ml or 5 ml syringe and a hypodermic needle of 25 gauge. The needle should be inserted pointing towards the base of the ear to coincide with the direction of blood circulation in this vein. If the vein is difficult to see or if it does not come up sufficiently for easy insertion of the needle, the ear may be gently rubbed with a piece of cotton dipped in 70% alcohol or to which a small quantity of xylene has been applied. In the latter case the xylene should be washed off again with alcohol to prevent irritation. Preferably one ear should be used for injections and the other reserved for taking blood samples later on. The initial injection of a series is made near the tip of the ear and later injections are made progressively nearer the base. In the case of bleeding the opposite applies, the initial bleeding being made near the base of the ear and the later ones progressively towards the tip.

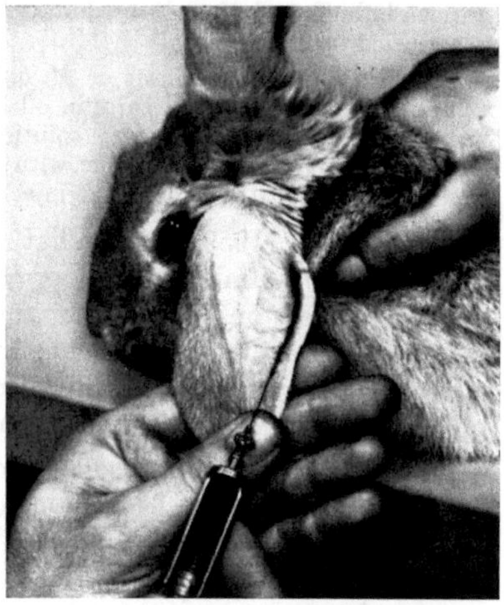

Fig. 1. Intravenous injection into the ear of a rabbit. The vein to be used is outlined in Indian ink.

A certain amount of skill is required to insert the needle properly into the vein and to avoid injection into the perivascular tissue which may spoil the usefulness of the vein for later injections. During the operation the ear is held in the left hand, with the thumb supporting the needle at the point of entry during the injection. The needle is held, bevel upwards, at an angle of less than 45° to the surface of the ear directly above the vein, to be brought parallel to it as soon as the tip has entered. Moving the tip a little further, the plunger of the syringe is now brought gently forward, which movement should proceed freely without any resistance being felt. As the liquid is slowly injected the vein will be seen to dilate whereas any resistance or distention of the perivascular tissue indicates that the needle is not properly in the vein. If in the latter situation pressure is used to depress the plunger, the injection fluid will be forced into the perivascular tissue.

No hard and fast rules can be given about the number of injections needed nor about the time intervals necessary between them. Generally speaking, good results are usually obtained if from 12 to 16 injections are administered over a period of 3 or 4 weeks, at a total dose of from 10 to 100 mg protein per animal.

Boyd (1956) recommends that where a few injections have already been given, any subsequent intravenous injections coming after a lapse of more than one day should be preceded by an intraperitoneal injection with the same antigen to avoid possible anaphylactic shock. In other words, administering the antigen intraperitoneally desensitizes the animal. Subcutaneous injections are used relatively seldom; only in the case of very toxic preparations is this method recommended. In contrast to rabbits, horses do not respond as well if the antigen is administered subcutaneously. The antibodies formed against potato virus X (van Slogteren and van Slogteren, 1957; van der Veken, 1958) for example, and various other proteins such as egg albumin and toxins react only with high antigen concentrations. It is assumed that these antibodies are incompletely adapted to the antigen (Treffers et al., 1947).

Nowadays, intravenous injections are often combined with intramuscular injections. Freund (1951) showed that intramuscular administration of the antigen mixed with "adjuvants" stimulates antibody production.

Adjuvant Method.

Ingredients: 8.5 volume Bayol F light paraffin oil are mixed with killed mycobacteria at a concentration of 10 mg/100 ml; 1.5 volume Arlacel A, a mannide mono-oleate; 10 volumes of antigen in buffer solution. Bayol F and Arlacel A are sterilized by autoclaving and the antigen is sterilized by filtration.

Mycobacterium tuberculosis or *M. butyricum* are killed and dried at 70° C for 30 minutes and then suspended in the paraffin oil by grinding in a mortar. Bayol F and Arlacel A are mixed and the aqueous antigen solution added. These liquids are emulsified in a Waring blendor or by means of a syringe with which the mixture is sucked up and ejected several times. 1 to 2 ml of the emulsion is injected in each hind leg of a rabbit.

Moorhead (1959) found that one intravenous injection one month after the intramuscular administration of virus antigen and adjuvants had a beneficial "booster" effect on antibody production. Similarly, one or more initial intravenous injections of plant virus antigens followed after one month by the intramuscular administration of the same antigens with adjuvants gave antisera with titres as high as $1:10^6$ (Wetter, 1960). For a definition of the term titre see page 430 or 451.

Antiserum production is also promoted by the application of antigen adsorbed on alumina or carbon suspensions (Strauch et al., 1959), quartz particles (Antweiler, 1959) and in the form of antigen-antibody complexes (Adler, 1957).

Intraperitoneal injections are used in chickens and guinea pigs, but subcutaneous or intramuscular administrations are also effective with these animals.

The minimal dose of antigen to be injected depends upon the aim one has in mind. If one is especially interested in producing antibodies against minor components in a preparation, it is advisable to administer large doses during short periods, e.g. 50 to 100 mg antigen per rabbit and to bleed the animal after three weeks.

For quantitative determinations, small doses are injected: 0.5 to 5 mg of antigen per rabbit; bleeding after 6 to 8 weeks. Cohn (1952) succeeded in getting good antisera after one injection of 0.1 to 0.5 mg diphtheria toxoid, egg albumin and β-galactosidase. According to Holt (1950), if diphtheria toxoid adsorbed to aluminium hydroxide is used, 20 to 30 mg of antigen per 3 kg of rabbit have to be injected. The doses applied to horses are proportionally larger, about 2 to 5 grams of diphtheria toxoid per 500—700 kg of bodyweight (Cohn, 1952).

The *life history of the animal* is an important factor. It is known for instance, that the titre of an antiserum can be boosted by renewed injection with an antigen injected earlier (Boyd, 1956). The reverse may also happen. If new-born animals are injected with small quantities of an antigen, antibody formation may not occur after injecting the same antigen at a later age (Felton et al., 1955). It is evident that the application of this phenomenon deserves testing, especially with a view to excluding antibody production against unwanted antigens. According to Moritz (1960), taking about 25 ml of blood from the animal before injecting the antigen may stimulate antibody formation.

During the cycle of injections, small quantities of blood are regularly taken to ascertain antibody concentration (see page 451). As soon as the desired titre is attained, greater quantities of blood are taken off, whilst renewed injections will stimulate further antibody formation.

III. The Preparation of Antisera.

1. Bleeding.

Before bleeding operations are commenced the animals should be starved for a period of 12 to 18 hours in order to reduce the lipid content of the blood, since lipids can cause cloudiness of the antiserum upon standing.

Blood samples as well as fairly large amounts of blood (from 40 to 50 ml) can be taken from the rabbit's ear using the dorsal marginal vein. As already mentioned above, one ear is always reserved for injection and the other for bleeding. To facilitate operations the rabbit may be secured in a special rabbit box (MATTHEWS, 1957) or be held by an assistant.

The site where a small cut is made in the vein for a first bleeding of a series should be near the base of the ear, succeeding bleedings being made progressively more distant from the base. In order to promote a strong blood flow the ear is vigorously rubbed and if necessary, a drop of xylene placed on the tip of the ear. This must be washed off again with liberal amounts of alcohol immediately after the bleeding operation. It may be helpful also to clamp the vein by means of an artery clamp or a clothes peg padded with cotton and placed over the vein between the base of the ear and the place to be punctured. This area is shaved and washed with alcohol after which a small cut is made using a sharp razor blade. Some operators, however, prefer the sharp edge of a freshly broken microscope slide. During the flow of the blood the ear is held by the operator with the punctured vein downwards and with a glass tube for collecting the blood placed under it. When the desired amount of blood has been taken (if a clamp has been used, this is now moved to the other side of the bleeding area), bleeding may be stopped by firmly pressing a wad of cotton wool over the puncture. Persistent bleeding can be stopped by sealing the wound with a small amount of collodion dissolved in alcohol-ether. Three bleedings, each yielding about 40 to 50 ml of blood, can be performed on successive days and may be followed by one further bleeding after an interval of one week. If more serum is required, a further series of bleedings preceded by another course of injections may be made four weeks later.

One series of bleedings will yield more serum than can be obtained by any of the methods in which the animal is sacrificed.

Bleeding by puncturing the heart is to be recommended where it is essential to obtain sterile sera. Blood is drawn through a 3 inch long, 19 gauge needle connected by flexible rubber tubing to a bottle fitted with a second tube for suction. For details of the procedure, reference should be made to SUSSMAN and PRETSCHOLD (1948) and KABAT and MAYER (1948). Cardiac puncture can be repeated several times on the same animal, 50 ml of blood being taken each time. A certain amount of skill by the operator is required as by a wrong manipulation the rabbit may be killed before any blood has been taken. If the animal is sacrificed, from 80 to 120 ml can be collected at once.

Cardiac bleeding should always proceed under complete anaesthesia, which, in the case of rabbits, is most conveniently achieved by an intravenous injection of nembutal (about $1^1/_2$ ml for a 7 pound rabbit).

In the case of guinea pigs, bleeding from the heart is recommended above all other methods, anaesthesia being induced by holding an ether cone over the animal's nose (CARPENTER, 1956). Blood is drawn by means of a 10 ml syringe with a 22 gauge needle. The animal is either secured horizontally to a board or held by an assistant in a vertical position. Just above the sternum and slightly to the right a patch is shaved and washed with alcohol and the needle gently inserted between two ribs at the point where the heart beat is most distinct, the exact site being located by palpation. The moment that the point touches the heart can be felt by the operator if the needle, as it is pushed deeper, is held between thumb and index finger of the left hand. With a short thrust the needle is made to enter the heart and at the same time the plunger is withdrawn slowly so that blood appears immediately in the syringe. Depending on the size of the animal 10 to 15 ml can be taken.

2. Separation of Serum.

When blood has been collected it is left for several hours to clot, either at room temperature or in an incubator at 37° C. When the clot has been formed it is gently loosened from the wall of the tube with the aid of a thin glass rod. The tube is then placed in a refrigerator at a temperature slightly above 0° C for a period of about 24 hours which procedure allows the clot to contract. The clear serum can be periodically taken off with a pipette, but the final aliquots removed may require centrifugation to separate red blood cells.

3. Storage of Serum.

Serum is best stored frozen at a temperature ranging between −15° C to −25° C in small glass bottles. If serum is required, the full content of a bottle should be thawed and thoroughly mixed before the desired amount is withdrawn. As it may deteriorate by being subjected to frequent freezing and thawing, it is best stored in lots not exceeding a quintuple of the average daily consumption.

As most sera are not sterile, the addition of Merthiolate at a final concentration of 1:10,000 is recommended.

Sera can be efficiently preserved also by freeze-drying in sealed ampullas (van der Veken, 1960), preferably in lots not exceeding the amount needed for future immediate use. Freeze-dried sera can be kept at room temperature conditions for prolonged periods without deteriorating. This method of preservation is especially convenient if sera have to be transported over long distances by ordinary mail. For permanent storage they are best kept at a cool temperature.

IV. Pretreatment of the Antisera.

In general the antisera can be used without elaborate treatment. There are, however, some points of interest with which we must deal, viz. isolation and concentration of antibodies, absorption of antisera and the preparation of labelled antisera.

1. Isolation and Concentration of Antibodies.

It may be that the antibody concentration is too low to demonstrate the presence of the antigen involved. By the addition of alcohol to a final concentration of 20% (*v/v*) or ammonium sulphate at the rate of 30 g/100 ml at 0° C the globulin fraction containing the antibodies can be precipitated (Cohn et al., 1946; Deutsch, 1952).

If specific antibodies are wanted, serological precipitates can be used for their isolation. According to Ramon (1923), the addition of acetic acid to decrease the pH to 3 suffices for the dissociation of toxin-antitoxin complexes. Antigens of high molecular weight, such as viruses, can be removed subsequently by centrifuging at high speed, but a considerable loss of antibodies is observed (van der Veken, 1961). Another method is based on denaturing the antigen by means of mercury compounds (Singer et al., 1959).

Isolation of the globulin fraction can be helpful in avoiding non-specific reactions occurring with the use of the complete antiserum. Various factors may be responsible for such non-specific reactions which are not due to interaction of antigen and antibodies. For instance, the isolated globulins, in contrast to the complete antiserum, do not produce non-specific reactions with extracts of tomato plants (ROHRINGER and STAHMANN, 1958).

2. Absorption of Antisera.

Many antisera produced with a view to use in determining only one particular antigen nevertheless contain antibodies formed against a mixture of antigens.

Antisera against viruses, for instance, often also contain antibodies against normal plant constituents, due to the fact that the virus preparation used for the injections was not pure. Adding to the antiserum the antigens corresponding to the undesired antibodies leads to their removal. This process is called the absorption of the antiserum. Such an absorbed antiserum thus finally produces positive reactions with only one of the injected antigens or with a serologically related group of antigens.

Antisera against potato viruses are most commonly absorbed with 10 volumes of healthy plant sap (VAN SLOGTEREN and VAN SLOGTEREN, 1957).

Good results were obtained in absorbing antisera against *Narcissus grey disease virus* with purified antigen from healthy narcissus. If it is difficult to remove normal, antigenically-active sap components from plant viruses, it may be advisable to isolate the virus from host species other than the one with which the serological tests are to be performed. Various potato viruses are for this reason propagated in tobacco.

3. Labelled Antisera.

In certain investigations it may be important to identify antigens within the cell. Good results have been obtained in plant virus research with γ-globulin fractions from antisera against tobacco mosaic virus (SCHRAMM and RÖTTGER, 1959) and narcissus grey disease virus (CREMER et al., 1960) coupled with fluorescein. Instead of fluorescein *iso*cyanate as used by COONS (1954), the more stable fluorescein *iso*thiocyanate is nowadays applied (RIGGS et al., 1958).

a) Production of Fluorescein Labelled Antiserum.

Antiserum is diluted with three volumes of water and the pH adjusted to 7.6—7.8 with acetic acid or $NaHCO_3$. After cooling to 0° C the γ-globulins are precipitated with alcohol at a final concentration of 20% (*v/v*) or ammonium sulphate at a final concentration of 30% (*w/v*). After stirring for 20 minutes, the precipitate is centrifuged down (5 minutes at 14,000 r.p.m.) and resuspended in 0.8% NaCl solution + 0.01 M phosphate buffer, pH 7.0. To remove the alcohol or ammonium sulphate the liquid is dialysed overnight against saline phosphate buffer. The solution must now contain a minimum of 10 mg protein/ml.For conjugation, 3 ml of a 0.5 M carbonate-bicarbonate buffer, pH 9.0 is mixed with 2 ml dry acetone (A.R.) and 500 mg of the γ-globulins, after which 0.8% NaCl solution is added to give a final volume of 25 ml. Finally, 25 mg of fluorescein isothiocyanate is added. The whole procedure is carried out at 0° C. The suspension is shaken for 18 hours and then dialysed in the cold against 0.15 M NaCl – 0.01 M phosphate buffer, pH 7.0, for 5 to 6 days. The dialysis liquid is renewed every day[1]. As soon as this liquid no longer shows any fluorescence the dialysis is terminated. Any precipitates that have formed are removed by centrifuging, and Merthiolate is added until a final concentration of 0.005%. The

[1] Free fluorescein can be removed very quickly by gelfiltration through Sephadex G-25 (GEORGE and WALTON, 1961).

end-product is then kept at $-20°$ C or in freeze-dried condition. Before use, the solution of the fluorescein-labelled antibodies is shaken with freeze-dried material from healthy plants to avoid non-specific staining (see page 446).

b) Production of Ferritin Labelled Antibodies.

Recently Singer (1959, 1960) demonstrated the possibility of marking antibodies with ferritin and of using this complex as cytochemical reagent for electron-microscopy. The only application so far achieved has been to give tobacco mosaic virus particles more contrast in electron microscope preparations. The procedure is as follows:

The globulin fraction is isolated as described above. To 5 ml of a 1.6% ferritin solution in sodium borate buffer of pH 9.5 and ionic strength 0.1 at 0° C, is added 0.1 ml m-xylylene *di-iso*cyanate. After stirring for 45 minutes the mixture is centrifuged and the precipitate, containing the unreacted *di-iso*cyanate, is discarded. After 1 hour 5 ml of a 1.6% γ-globulin in the same borate buffer but containing 1 M NaCl, is added. The solution is now stirred for two days at 6° C and afterwards dialysed against 0.1 M ammonium carbonate to remove the unreacted *iso*cyanate. Finally, the whole is dialysed against phosphate buffer of pH 7.5 and ionic strength 0.1. A precipitate may possibly form during dialysis and must be removed by centrifugation (Singer, 1959).

C. Serological Methods.

The reaction between antigen and antibody is made perceptible in various ways. Widely used are the *precipitin* and *agglutinin* tests. In both, the mixing of antigen and antibodies is followed by flocculation of the antigen-antibody compound. The term agglutination is particularly used for the flocculation of larger bodies, such as bacteria, blood corpuscles and plastids, on the surfaces of which the antigens are present either naturally or artificially.

In *complement fixation* — another type of serological test — use is made of the phenomenon that a thermolabile substance known as complement is fixed by antigen-antibody complexes. Red sheep blood corpuscles showing lysis in the presence of their antibodies and complement, are used as indicator for this substance.

If the concentration of one of the reactants is too low no directly visible reaction will occur. Small aggregates of antigen and antibody may in such a case be detected with the aid of the electron microscope. The antibodies may also affect certain biological properties of their antigens, for example infectivity of viruses or activity of enzymes. In this connection, decrease in infectivity or activity respectively, may be used as a criterion of affinity between antigen and antibody.

In order to confirm the specificity of the observed reactions parallel control tests are performed with normal serum, i.e. serum of an animal not injected with the antigen involved. As in every analytical method, there is a minimum limit to the quantities of the two reacting substances required for a positive reaction. For antisera this minimum quantity is termed the *titre*, indicating at what final dilution the antiserum still produces a visible reaction with the antigen. In the case of the antigen the final dilution is called the dilution end-point.

An antiserum is termed homologous to its own antigen and heterologous to other antigens. Some incomplete antigens (simple or inhibiting haptens) combine with antibodies without the formation of a visible precipitate. This combination is evidenced by inhibition of flocculation when the precipitating antigen is subsequently added to the system.

For detailed information of the *theoretical background* to all these phenomena reference should be made to: Boyd (1956); Carpenter (1956); Cushing and Campbell (1957); Kabat and Mayer (1948); Wilson and Miles (1955).

I. Precipitin and Agglutinin Reactions.

These types of reaction have found most general application in serology because of their relative simplicity. *Precipitin* reactions are characterized by the fact that the bond between antigen and the corresponding antibodies results in the formation of a precipitate, consisting of the antigen-antibody complex.

The term *agglutinin* implies the production of floccules the bulk of which is composed of constituents (chloroplasts, bacteria, collodion particles or red blood cells) which carry the antigen in one way or another. On the other hand antigens may be artificially adsorbed on to collodion particles or red blood cells and to such a suspension the corresponding antiserum may be added. In both cases serological reaction results in agglutination of the coarse particles. However, there is not a distinct borderline between precipitin and agglutinin reactions.

Depending upon the kind of antigen two kinds of floccules are distinguished. Bacteria with flagella (H-antigen) produce open fluffy floccules with their antiserum. The same type occurs with rod- and thread-shaped plant viruses. Non-motile bacteria (O-antigen) produce a dense granular precipitate as do spherical viruses.

1. Considerations of the Mechanism of the Precipitin Reaction.

The formation of insoluble antigen-antibody complexes might be explained in various ways. In terms of one theory, the mutual binding of antigen and antibody is thought to be due to a reaction of polar groups on each of the reagents. The number of polar groups determines the degree of hydration of a substance and

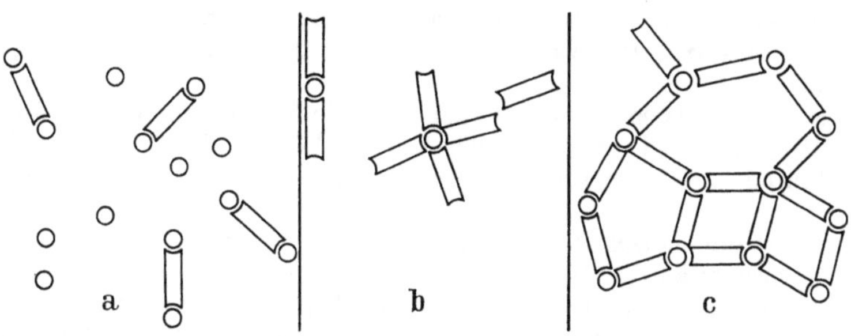

Fig. 2. A diagrammatic representation of antigen-antibody reactions. Antigen ○, antibody ⊏⊐. a) Soluble complex in which antigen is in excess. b) Soluble complex in which antibody is in excess. c) Insoluble antigen-antibody complex [based on Cushing and Campbell (1956)].

therefore its solubility. If the number of polar groups decreases — and this takes place if antigen reacts with antibody — the complex formed will precipitate. This is not, however, the only determining factor since the magnitude of the complex in relation to each reactant is also important. Fig. 2 represents schematically various possible configurations in relation to the amount of the substances involved.

On this basis it is understandable that the *proportion of antigen to antibody* can vary considerably between precipitates. For tobacco mosaic virus it was found that in serological precipitates the ratio of antibody molecules bound to one antigen particle varies between only two in antigen excess and even more than 70 in antibody excess (Kleczkowski, 1941). If the amounts of precipitate are plotted against different concentrations of antigen, a curve such as is shown in Fig. 3a is obtained. In this curve one can distinguish three zones, viz. a zone of antigen excess, a zone of antibody excess and, in between, a zone of antigen and antibody equivalence. It is to be expected that under conditions of appreciable antigen excess no precipitation will occur. Most precipitate forms in the zone of equivalence and in the neighbouring part of the zone of antigen excess. As a consequence one should aim at using proportions of antigen and antibody corresponding to these zones in practical work.

The *equivalence zone* is characterized by the minimum time necessary for the formation of the precipitate. In mixtures of different concentrations of antigen

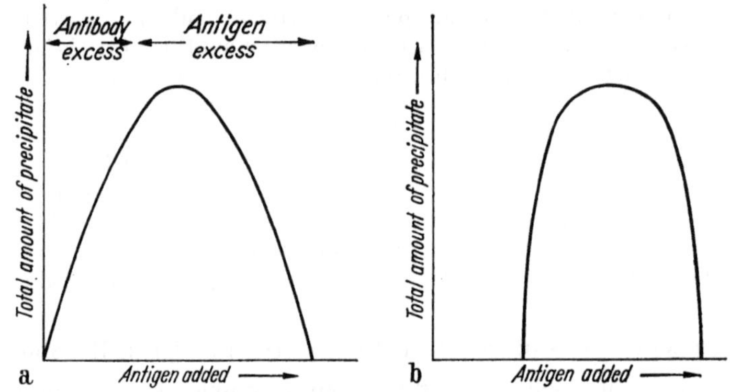

Fig. 3. a) Precipitin curve obtained with a constant amount of rabbit antiserum and varying amounts of antigen. b) Precipitin curve with a constant amount of horse antiserum (containing incomplete antibodies) and varying amounts of antigen (based on BOYD 1956).

reacting with a constant amount of antiserum, the dilution of the antigen giving rise to the quickest precipitation indicates the *α optimum.* The determination of the α optimum provides thus a means to establish the different zones in the precipitin reaction.

This *α optimum* is constant and characteristic for a given batch of antiserum. If an antiserum used at a dilution 1:5 would flocculate most rapidly with antigen diluted 1:100, the antiserum diluted 1:10 will do so with antigen diluted 1:200.

Antisera from horses sometimes present a completely different picture. Neither a precipitate is formed in antigen excess nor in antibody excess. These phenomena have been observed with diphtheria toxin (PAPPENHEIMER and ROBINSON, 1937), chicken ovalbumin (PAPPENHEIMER, 1940), tetanus toxin (MALONEY and HENNESY, 1942), chicken conalbumin (COHN et al., 1949) and human serumalbumin (GITLIN et al., 1951). This can be represented schematically as shown in Fig. 3b. Characteristic of many of these reactions is the fact that precipitation occurs in special zones. A similar phenomenon has been found in connection with antibodies formed against potato virus X after subcutaneous injections (VAN DER VEKEN, 1958). According to TREFFERS et al. (1947) these phenomena can be ascribed to the nature of the antibodies involved, which are considered to be incomplete. In rabbit antisera such incomplete antibodies are generally not present in a high proportion, although there are a few exceptional cases, e.g. ribonuclease of animal origin may elicit as much as 50% incomplete antibodies. Probably the proportion of complete antibodies carrying more reactive groups than the incomplete ones increases in the course of time after prolonged injections.

It is evident therefore that in view of the phenomena described above it is necessary in all precipitin tests to work with various *concentrations of antigen and antibody*.

Apart from the effect of antibody and antigen concentrations on the result of the precipitin reaction, certain other factors are also important, viz. the presence of *electrolytes*, *pH* and *temperature*. Normally, NaCl solutions of 0.9% are used to make up the dilutions but in work with antisera from chickens a NaCl concentration of 8% is recommended (BANOVITZ et al., 1959). Generally speaking, however, the precipitate dissolves in high salt concentration as a result of dissociation of the antigen-antibody complex (HEIDELBERGER et al., 1936). In plant virus serology low salt concentrations tend to promote non-specific flocculations (MATTHEWS, 1957).

Within a pH range of 5 to 8 the amount of precipitate is approximately constant (MARRACK, 1958). Exceptions arise, however, in cases where the antigen is altered by small changes within this range. The sub-unit of tobacco mosaic virus (the so-called X protein) shows, for instance, an antigen dilution end-point of e.g. 1:50 at pH 7, and 1:400 at pH 5.4 (TAKAHASHI and GOLD, 1960). This is a result of aggregation of X protein at pH 5.4, which facilitates observation of the precipitate.

Regarding the effect of temperature, generally between 0° and 37° C the same amount of precipitate is formed. Higher temperatures, up to the critical limit of the proteins involved, may be used to accelerate precipitation (BOYD, 1956). For the determination of the α optimum therefore, a constant temperature must be maintained.

2. Reactions in Free Liquids.

A great many forms of application can be distinguished in which use is made of tubes or capillaries of different diameter, or in drops, corresponding to macro- and micromethods respectively.

a) In Tubes.

Aliquots of antigen and antisera are mixed in tubes, e.g. 0.5 ml of each. These tubes are then placed in a water bath at constant temperature (e.g. 37° C or 50° C) and at certain intervals the appearance of precipitates is observed. Preferably the tubes should be set in such a manner that the meniscus in them is higher than the water level in the bath. Convection then occurs in the tubes, resulting in quicker growth of the floccules which are then more readily observable. The time interval between the mixing of antigen and antibody and the appearance of the precipitate is noted as well as the degree of precipitation as evidenced by the density of the precipitate. The latter is commonly denoted by symbols such as +, ++, +++ etc. To exclude the possibility of no precipitate being formed as a consequence of antigen or antibody excess, the experiments are performed with different concentrations of both reagents (MATTHEWS, 1957).

In another method the antigen-containing liquid is introduced on top of a layer of antibody containing-liquid in a capillary tube (SLATMANN, 1928). Antigen and antibody react after mutual diffusion and consequently a precipitate is formed at or near the border line between both liquids. This type of reaction is incorrectly called the *ring test*. As a further modification, wider tubes may be used and mixing of the reagents avoided by adding 10% glycerol or sucrose to the lower (generally the antibody-containing) liquid (WHITCOMBE and BLACK, 1961). In this case the precipitate is formed at or near the border between the two liquids. Wetting the outside of the tubes with paraffin oil facilitates observation of the precipitate.

b) In Droplets.

Both precipitin and agglutinin reactions in droplets have found wide application in plant virus research.

Bartels (1957) described a precipitin method for indexing potato plants for the presence of potato virus Y. Aliquots each of about five drops of diluted antiserum are mixed on microscope slides with equal volumes of centrifuged sap from plants to be tested. The mixtures are then placed in an incubator and kept at a temperature of 25° C for 40 minutes during which period a high humidity is essential in order to prevent evaporation of the mixture. Reading of the reactions is done microscopically using dark field illumination at 60 to 80 times magnification. Positive reactions are indicated by a white flocculation standing out against the dark background.

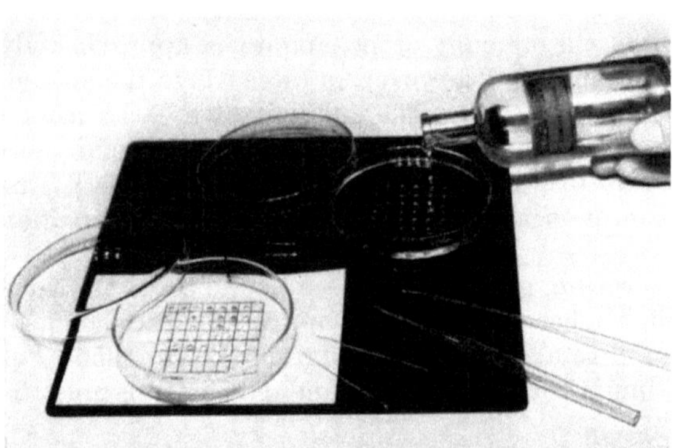

Fig. 4. Aids and appliances for serological micro-reactions under paraffin oil (van Slogteren, 1955).

In the slide agglutinin test, used for indexing potatoes for the viruses X and S, five drops of the respective diluted antisera and two drops of crude sap expressed from the potato leaves are mixed together on a microscope slide (van Slogteren, 1945). Positive reactions are indicated by a typical aggregation of the chloroplasts and can be observed either with the aid of a hand lens or microscopically at low magnification (about 60 X) using direct illumination.

Both the methods mentioned above have been modified into microtechniques in order to economize on sera (van Slogteren, 1955a). In that the microreactions take place under paraffin oil (paraffinum liquidum) they have an advantage over analogous methods because evaporation of the droplets is avoided completely. This is particularly important where a long incubation period is involved. In the micro-precipitin test small droplets of clarified plant sap are added to similar droplets of diluted antiserum previously placed on the bottom of a Formvar-coated Petri dish. The coating of the inner glass surface of the Petri dish with Formvar is done to prevent spreading of the mixed droplets and is achieved by pouring a 1% solution in chloroform of polyvinylformaldehyde[1] into a dry, clean Petri dish and then pouring it out again immediately afterwards. The thin film remaining is left to dry for a period of about 8 hours. A number of dishes can be coated by pouring the solution from one dish into the next. Petri dishes selected for this use should have a flat bottom and the texture of the glass should be free from any flaws.

The introduction into the Petri dishes of the small droplets of antiserum and plant sap is done by means of Pasteur pipettes with finely drawn out points. By placing the Petri dish on sectioned paper with squares of about 8×8 mm, it is possible to put the droplets neatly in rows, 8 mm apart from each other (Fig. 4). Reactions are best read with the aid of a dark field microscope, which operation is greatly facilitated if the drops have been placed in straight rows for then the

[1] Commercially available as Mowital F 40, Farbwerke Hoechst.

Petri dish requires only systematic movement to bring the successive rows of drops in view. The number of reactions per dish is determined by the routine adopted by the operator owing to the restriction that paraffin oil must be poured in before the droplets first introduced begin to dry up. Depending on the particular circumstances of a test the number may vary from 36 to 64. Sometimes it is preferable to cover one row of droplets with oil before the next row is introduced. This can be conveniently done by placing the Petri dish on a tilting stand and adding the first row of droplets with the dish in a tilted position so that the oil remains localized. The oil can then be made to advance to cover this row merely by decreasing the slope of the stand after which the next row can be commenced, and so on (Fig. 5). The micro-precipitin method has been extensively used in testing plants for the presence of virus and is also suitable for determining relative concentrations of either antigen or antibody. (For quantitative estimations see page 449).

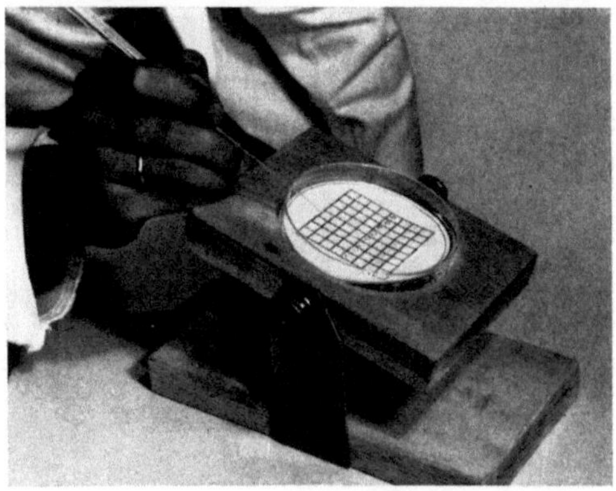

Fig. 5. Precipitin reactions on a revolving stand (VAN SLOGTEREN, 1955).

The micro-agglutinin method has found wide application in the indexing of potato plants for the viruses X, S, M and Y. The procedure involves mixing drops of the respective diluted antisera pipetted into a Petri dish as described above, with tiny amounts of crude plant sap and then applying a covering of paraffin oil. The amount of antiserum should in each case be about three to five times as much as the amount of sap. The latter can therefore be most conveniently added by dipping the rounded end of a thin glass rod once or twice in the sample of sap and stirring it thoroughly in the droplet of serum.

c) The Agglutinin Technique.

As mentioned earlier, the agglutinin method is limited to those reactions in which the antigen forms part of the surface of an organism or of parts thereof, e.g. flagella of bacteria, cell walls and virus adsorbed to chloroplasts.

It is also possible to attach the antigen artificially to certain substances, especially collodion particles or red blood cells. These adsorption techniques are very useful in work with incomplete antibodies where the antiserum otherwise fails to produce a visible precipitate. Applications have been mainly limited to studies with egg albumin adsorbed to collodion particles and with other small proteins adsorbed to red blood cells.

a) Adsorption to collodion particles. Collodion particles are produced by stirring a 5% collodion solution with water. After washing, particles are coated with antigen and finally mixed with the antiserum (KABAT and MAYER, 1948).

b) Adsorption to red blood cells. This procedure is known as BOYDEN's hemagglutination technique (BOYDEN, 1951). Since it has proved of great value in medical research, we feel that it deserves attention in the botanical field. A somewhat detailed description is therefore given here (BOYDEN, 1951; STAVITSKY, 1958). One of the principle features is the tannic acid pretreatment of the red blood cells, since proteins do not adhere directly to their surface.

Method. Sheep's red blood cells, from blood collected in Alsevers solution (BUKANTZ et al., 1946), are washed three times with 0.8% NaCl. (If the supernatant is not clear after this treatment, the cells cannot be used). 1 ml of these cells are diluted with 40 ml buffered saline, pH 7.2. The buffer is prepared by mixing 1 part of 0.8% NaCl and 1 part 0.15 M phosphate buffer.

1 ml of this cell suspension is mixed with an equal volume of 1/20,000 tannic acid solution and incubated for 10 minutes at 37° C. The cells are then centrifuged and washed with 1 ml of buffered saline pH 7.2 and resuspended in 1 ml of saline.

4 ml of buffered saline, pH 6.4, + 1 ml of protein in saline + 1 ml cells treated with tannic acid are mixed in this order at room temperature and kept there for 10 minutes. The cells are centrifuged down and washed once with 2 ml 0.01 normal rabbit serum. The cells must be used within 24 hours. A cell suspension not treated with protein is used as control.

To prevent hemolysis all sera are heated at 56° C for half an hour.

3. Reactions in Gels.

During the last decade various forms of serological reactions in gels have been developed. The advantage which all these methods have in common over precipitin reactions in liquid media is the direct approach to differentiation achieved in mixed systems of multiple antigens and their corresponding antibodies. In such systems multiple zones or lines of precipitation are formed, the number of which gives direct information about the minimum number of antigens present.

Most of the work done so far has been carried out using agar media (OUDIN, 1946; OUCHTERLONY, 1948) but more recently gelatin (POPE et al., 1951), pectins (GRABAR et al., 1956) and alginates (MORITZ, 1960) have been used to advantage.

In simple diffusion a gelified medium containing as reactant either antigen or antibody is in direct contact with another liquid or likewise gelified medium containing the other reactant.

In double diffusion the media containing the reactants are at the start separated by a gel free of either reactant.

Both simple and double diffusion can be carried out in one dimension by the use of glass tubes as well as in two dimensions using layers of gel either on glass plates or in a Petri dish.

In immunoelectrophoresis the components of a mixture of antigens present in a gel are separated electrophoretically prior to their precipitation by antibodies in the same gelified medium.

a) Simple Diffusion in One Dimension.

Simple diffusion in one dimension was first employed by OUDIN (1946) and reference should be made to his survey (OUDIN, 1952) for details about procedure, variations and materials.

Reactions are performed in glass tubes with an inner diameter of between 1.5 and 3 mm. Dilutions of reactant in agar are made in such proportions that the final concentration of agar is 0.6%. The glass tubes require preliminary lining with a film of dried agar in order to prevent the reactants from flowing between glass and gel.

Above a column of agar containing one of the reactants, the so-called internal reactant, an overlayer of a concentrated solution of the other external reactant is added.

It is preferable to have antibody as the internal reactant, with excess antigen diffusing into it. The precipitation zones will then migrate slowly away from the interface, any precipitate formed near to the interface being dissolved in antigen excess. The leading edge of the migrating zone is sharp, owing to maximum precipitation near this edge.

Where antigen forms the internal reactant, maximum precipitation persists at the interface since the precipitate is insoluble in excess antibody[1]. Precipitates forming farther down the column are progressively less dense, without distinct boundaries.

Migration of the zone of precipitation does not apply to antigens with a high molecular weight such as tobacco mosaic virus particles (JEENER et al., 1953) and the antigens in extracts from plant seeds (MORITZ, 1960), owing to the fact that

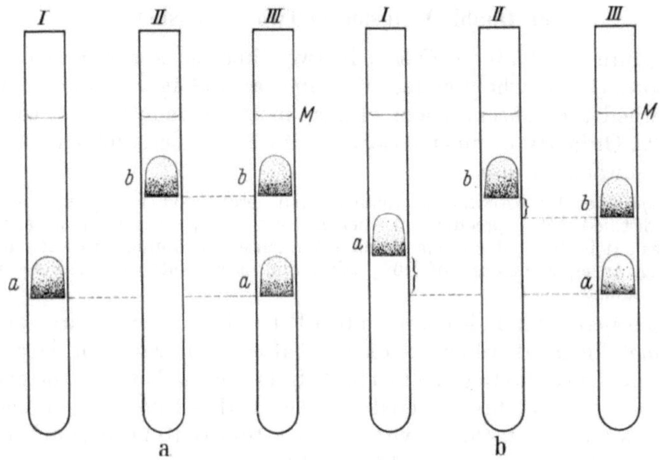

Fig. 6. a) Two unrelated antigens diffusing separately (*I* and *II*) and in a mixture (*III*) into gel containing their antibodies. b) Two related antigens diffusing separately (*I* and *II*) and in a mixture (*III*) into gel containing their antibodies. *M* meniscus of the gel [based on OUDIN (1952)].

precipitates involving these substances are not soluble in excess antigen. On the other hand, the non-infectious, so-called X-proteins, present in infected tobacco in addition to the complete infectious particles, precipitate as a migrating zone (JEENER et al., 1954).

In multiple precipitation systems, antigens that do not cross-react will behave independently. Thus in mixtures of antigens the leading edge associated with each antigen will be found at the same site in the column as would be the case when the respective antigens were diffusing alone, provided that the same concentrations have been used and that the times of diffusion have been synchronized (see Fig. 6).

If, on the other hand, two or more antigens cross-react with the same antibody in the gel, then penetration of both antigens is increased, as has been demonstrated for mixtures of chicken and duck ovalbumin by OUDIN (1952).

b) Double Diffusion in One Dimension.

OAKLEY and FULTHORPE (1953), who studied antigenic analysis by diffusion, describe a method in glass tubes allowing antigens and antibodies to diffuse towards each other through a column of agar. Antiserum diluted in agar is first poured into the tube and when this has solidified a column of agar free of either reactant is poured in. After solidification of the latter, the antigen, which may be either dissolved in liquid medium or likewise diluted in agar, is added. POLSON

[1] With horse antiserum the precipitate may dissolve in excess antibody.

(1958) used this technique for determining both the diffusion constant and concentration of antigens. For this purpose an apparatus composed of polymethacrylate was constructed. Kleczkowski (1957) also made use of this method in an antigenic analysis of tobacco mosaic virus. According to Feinberg (1956) the sensitivity of the reaction can be enhanced by allowing the antibody sufficient time to diffuse into the agar column before the addition of the antigen solution, thus enabling a concentration gradient of antibody to be built up first.

Interpretation of the appearance of precipitation lines in one dimensional double diffusion is in general the same as that discussed below for double diffusion in two dimensions.

c) Double Diffusion in Two Dimensions.

A procedure for double diffusion in two dimensions has been described by Ouchterlony (1948), who used reservoirs in a layer of agar as sources of diffusion. In this method the concentrations of agar most commonly used vary between 1% and 2%. Ordinary commercial agar has to be clarified and freed from pectins before it is suitable for use.

The *receipt* given by Björklund (1952) is cited here: Good quality agar is dissolved in 30 parts of distilled water, precipitated when hot with 0.5% calcium chloride, filtered, and left to congeal. It is then cut into small pieces and rinsed in running tap water for 72 hours. After melting, an equal amount of 1.6% saline and merthiolate to a final concentration of 1:10,000 are added[1].

When the agar medium is poured into a Petri dish, reservoirs are obtained with the aid of metallic molds which are removed after solidification of the agar. To prevent liquid media containing the reactants from flowing between the gel and glass surfaces Ouchterlony first covered the glass with a thin layer of agar. On this he placed his metallic molds, covered with a thin coating of paraffin, and then poured a second agar layer around the molds.

Elek (1949) independently developed a method of using strips of filterpaper as sources of diffusion. Working with the diphtheria toxin-antitoxin system, he immersed one strip soaked in antitoxin in agar warm enough to be still liquid, whilst the other strip containing a concentrated solution of toxin was placed on the surface of the agar after solidification, at a right angle to the immersed strip.

Reservoirs can be made in various shapes, sizes, numbers and arrangements. Ouchterlony used square reservoirs of the order of 10×10 to 15×15 mm in size. Much smaller reservoirs have since been employed to advantage by many workers. Not only are they more economical in their smaller requirements of reagents but they are also more efficient since gel diffusion reactions on a microscale proceed more rapidly. For this purpose thin layers of agar of about 2 mm thickness are poured inside glass rings cemented on to a Petri dish or on to glass slides. The inner diameter of the glass rings can vary within the range 15 to 33 mm (van Slogteren, 1958). Holes with a diameter of 2 to 5 mm can conveniently be punched out from the agar layer with the aid of a piece of brass tubing sharpened like a corkborer at one end. The cylinders of agar thus punched out can be removed by suction with a pipette. In order to prevent liquid suspensions from running between the gel and the glass it is advisable to "*siliconize*" all glass surfaces with "*Desicote Beckman*". Mansi (1958) employed microscope slides covered with a thin layer of agar in which small cylindrical holes are punched out.

To prevent *contamination* by molds or bacteria, substances such as Merthiolate are as a rule included when gels are prepared. Buffers may be included as circumstances require with respect to the antigens studied.

[1] High grade agar free of turbidity when dissolved is nowadays commercially available e.g. Difco Agar "Noble."

In double diffusion, serological reactions become manifest by the *formation of visible bands or lines of precipitation* in the gel that separates the reservoirs containing the reactants. In the case of a number of lines, each line represents a different antigen-antibody system. If we keep in mind the fact that conditions for precipitation, such as ratio and absolute values of concentrations, are not always fulfilled for all possible combinations of antigens with their corresponding antibodies, the number of lines may be regarded as the minimum number of antigen-antibody systems present.

With double diffusion in two dimensions it is possible to place 3 or more reservoirs in the same gel, thus enabling antigens from 2 sources to diffuse towards a third source containing the antibody. This is particularly interesting for the study of relationships between antigens. The most suitable arrangement is composed of 3 square reservoirs with their centres at the corners of an equilateral triangle.

d) Interpretation of Precipitation Patterns in Double Diffusion.

The site of the precipitates in the gel separating the sources of diffusion depends on different factors which have been discussed in detail by OUCHTERLONY (1958) and by OUDIN (1952). A recent review has been given by CROWLE (1960).

The main factors are the diffusion constants of the reactants, and the level of concentration of each reactant and the ratio of these concentrations. In addition, diffusion may be influenced by temperature, composition of the gel medium and pH value.

In general, factors which influence precipitation reactions in liquid media also apply to gel diffusion reactions. On the other hand, size and shape of diffusing antigens with high molecular weights, as in the case of some plant viruses, may be a limiting factor, since there is a direct relationship between these properties and the diffusion constant. HARRISON and NIXON (1959) studied two fractions of tobacco rattle virus, the one comprising "short" non-infectious particles, and the other complete, infectious, rodshaped particles of 185×30 mμ. In diffusion experiments against an antiserum to this virus precipitation lines were obtained only with the "short" fraction.

The rate of diffusion is enhanced by increasing temperatures and as a consequence precipitation lines develop earlier. In double diffusion, precipitates can appear either as sharp, distinct lines or as more or less broad zones. The governing factors are on the one hand the ratio of the concentrations of the reactants and on the other the properties of the participating reactants, antigens and antibodies.

OUDIN (1952), using OUCHTERLONY's technique with square reservoirs facing each other in a parallel arrangement, demonstrated this with ovalbumin at 3 different initial concentrations diffusing against homologous antiserum from the parallel source.

In the case of excess antigen, the precipitate first formed in the gel separating the reservoirs gradually dissolves, which results in the visible zone being displaced away from the source of antigen. As precipitation continues the zone broadens and shows increasing maximum density near the leading edge as this edge moves towards the source of antibody. Precipitation bands formed bv viruses, however, usually do not move.

Where the initial concentrations of antigen and antibody are equivalent, a precipitate is formed which stands out clearly as a sharp immobile line, its density increasing with time.

In most cases of excess antibody, as precipitation progresses towards the antigen source the rear edge facing the antibody source is not dissolved but remains stationary resulting in a broad zone of relatively low density.

Oudin (1952), however, using diphtheria toxoid against excess horse antitoxin showed that in systems with antibody of the so-called flocculating type, in which the precipitate is soluble in excess antibody, the precipitate migrates towards the antigen source as a relatively narrow band.

The formation of broader bands or zones of precipitate may also be expected in gel diffusion reactions between haptens and their corresponding antisera. Oudin (1952), using single diffusion in one dimension, described the precipitation zone arising between O-polysaccharide of *Eberthella typhosa* as external reactant and an antisomatic serum. Precipitation occurs as a broad zone lacking a leading edge at either end. It is reasonable to assume that similar zones may be encountered in double diffusion with reactants possessing haptenic properties. Very likely this type was encountered by van Slogteren (1958) in serological experiments with tobacco infected with cucumber mosaic virus (CMV). Sap of such plants contains an antigenic substance capable of precipitating with a heterologous antiserum prepared against a related strain of tomato aspermy virus (TAV), but which does so only when the gel double diffusion method is employed. If sap from tobacco infected with CMV is mixed with liquid solutions of the serum, no precipitation takes place at all. Fig. 7 shows the gel diffusion precipitation patterns described here. When antiserum TAV is made to diffuse against sap of tobacco infected with CMV a relatively broad band of precipitation is formed about halfway between the sources of diffusion and often nearer to the antiserum reservoir. In the homologous reaction between the same antiserum and sap of plants infected with TAV, the same band is observed, but between it and the sap reservoir a much sharper thin line of precipitate also appears. When the same antiserum is reacted against a third virus strain, "*Price's yellow cucumber mosaic virus*", an even more diffuse band appears in the gel separating the reservoirs.

Fig. 7. Gel diffusion precipitin reactions. Antiserum to tomato aspermy virus (TAV) pipetted in the reservoir in the centre. The surrounding reservoirs contain sap from tobacco plants infected with the following virus strains, respectively: TAV in the top left — "Price's yellow strain" in the top right — CMV type strain in the lower reservoir. For explanation see text.

Precipitation Patterns. The precipitation patterns used to study relationships between antigens deserve special attention. Fig. 8 illustrates, respectively, examples of antigenic identity, partial identity and non-identity of antigens diffusing from parallel reservoirs to a perpendicular source of antiserum.

Where there is identity of the two diffusing antigens their lines of precipitation curve towards each other to coalesce completely at their junction (Fig. 8.1—8.3).

In the case of non-identity the lines formed by the independent antigens cross each other without causing any interference (Fig. 8.8).

Partial identity arises where one antigen source contains a diffusing antigen having at least 2 linked, independent antigenic properties whereas the antigen diffusing from the parallel source has only one of these properties and they are meeting with separate antibodies to both properties coming from the perpendicular source.

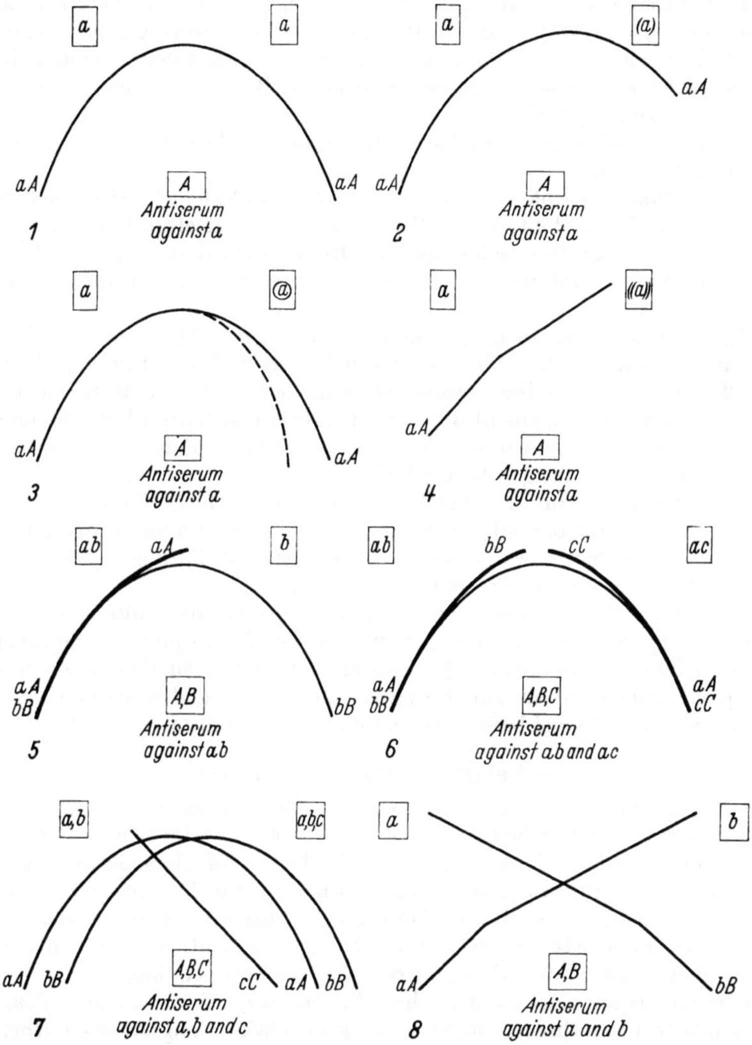

Fig. 8. Patterns of antigen-antibody reactions in gels. Antigenic determinants are represented as *a*, *b* and *c*; the corresponding antibodies as *A*, *B* and *C*. 1. Line of complete identity arising from two identical antigens, *a*, which are present in the same or nearly the same concentration. 2. An asymmetrical arc of interference is formed when the concentration of *a* is lower than (*a*). 3. Line of complete identy. The concentration of the antibody (expressed in equivalent units) is, however, lower than that which corresponds with the concentration of the antigen @ so that two or more lines are formed (WILSON and PRINGLE, 1956). 4. Reaction of two identical antigens but with the concentration of ((*a*)) below the threshold value for the formation of a precipitate. 5. Line of partial identity arising from two cross-reacting antigens, *ab* and *b*, that have one antigenic determinant in common. This causes the precipitation line denoted as *bB*. The substance possessing a second linked antigenic determinant forms a second line, *aA*, which is known as a spur. In many cases the spur is thinner than the line of identity. 6. Lines of partial identity arising from two cross-reacting antigens *ab* and *ac*, that have each one antigenic determinant in common and one that is different. With antiserum containing antibodies to all these determinants, two spurs *bB* and *cC* will be formed in addition to the continuous line *aA*. 7. Comparison of two antigen mixtures having two antigens *a* and *b* in common but at different concentrations. 8. Lines of non-identity. Two unrelated antigens, *a* and *b*, give rise to crossing lines.

In this case the line that forms between the antibody source and the source containing the antigen with 2 properties partially coalesces with the line of the antigen carrying the single common property, the line first mentioned being extended beyond the point of junction in what is known as a spur. As a rule the spur is thinner than the line from which it extends (Fig. 8.5).

When 2 antigens diffuse from parallel sources, the one carrying 2 linked antigenic properties, represented by ab, and the other also carrying linked properties, represented by ac, thus having a in common, a so-called double spur is formed when antibodies to a, b and c, represented by A, B and C, diffuse from the perpendicular source (Fig. 8.6).

The formation of single as well as double spurs has been extensively discussed by KORNGOLD (1956).

The formation of double spurs as a result of partial identity could be confused with a reaction of non-identity, but the former reveals itself in that the spurs are generally thinner than the coalescing lines from which they extend, and furthermore the lines of precipitation bend towards each other just below the point of junction.

In the case of multiple antigens, where 2 or more lines of precipitation should be formed, it may happen that 2 lines representing precipitates of independent antigens with their corresponding antibodies both form at the same site in the gel. This can occur with antigens of different diffusion constants where the one with the lower diffusion constant moves towards its antibody under a more favourable concentration ratio of antigen to antibody.

Varying the concentrations of either the antigen or the antibody mixture may result in a more complete analysis and it is therefore advisable to commence with reactions at more than one dilution of either reactant in order to estimate the extent to which the lines of precipitation overlap.

In cases where the presence of great numbers of antigens results in complicated spectra of many precipitation lines, it may be helpful to prevent the formation of too many lines by specific absorption of antiserum with single components from the complex antigen system. An alternative method is the dispersion of the interfering antigen(s) in the gel before its solidification (BJÖRKLUND, 1952).

e) Immunoelectrophoresis in Gels.

GRABAR and WILLIAMS (1953) have combined electrophoretic separation of antigenic fractions with serological two-dimensional double diffusion within the same layer of agar gel.

The "*macro method*" is carried out with the aid of glass plates (clean and flawless) about 13 × 18 cm in size, covered with a layer of 1% agar gel. In order to prevent the reactants from flowing between the agar layer and the glass surface the plates are previously smeared with hot 2% agar solution. When this coating has dried they must be placed in an exactly horizontal position.

This is for instance achieved in the following way: A rectangular glass tray, large enough to place three or more of the glass plates alongside each other with enough space left at both sides, is permanently fastened on a bench. A horizontal surface is automatically obtained by pouring in a thick layer of a 3% agar solution including 0.1% Merthiolate. The same surface can be used repeatedly if the glass tray is kept covered.

The *medium* in which immunoelectrophoresis is performed consists of a buffered 1% agar solution of ionic strength 0.05. This is prepared by mixing a 2% agar solution with an equal volume of veronal buffer solution of pH 8.2 and ionic strength 0.1. Merthiolate should be included to give a final concentration of 1:10,000 (GRABAR, 1957).

Of this agar solution a layer 2 to 3 mm thick is poured over and around the horizontally placed glass plates. Before pouring the agar, rectangular strips of filter paper drenched in veronal buffer are placed on both ends of each glass plate. They should overlap the glass by about 2 cm and protrude for a further 4 cm, this latter part being supported by placing pieces of glass alongside the glass plates.

When the agar layer has solidified, the glass plates are freed by cutting the agar along their longest sides and around the edges of the paper. Taking care not to damage the underlying paper when the plates are lifted and placed into position for electrophoresis, an extra cut is made over the short sides to enable the paper ends to hang vertically. In so doing the cut widens to an open wedge, which must be filled with hot agar solution. These paper strips serve to connect the agar layer on the glass plates with the two parallel troughs containing buffer solution, in each of which a spiral electrode of thin platinum wire is immersed. The buffer solution should be of twice the ionic strength of the agar and must circulate continually in order to obviate changes in pH.

When the plates are in position, having an end hanging into each trough, rectangular reservoirs (10 × 3 mm in size) are punched using a piece of copper or brass tubing, flattened and sharpened at its end. These reservoirs are sited at right angles to the direction of electrophoresis and as a rule about midway along the length of the plate. This may, however, be altered to suit a particular sample's electrophoretic mobility which is determined not only by electrophoresis but by electro-endosmosis as well. On the same plate there is ample space to analyse 3 samples simultaneously by employing 3 reservoirs having all their sides aligned and parallel (Fig. 9).

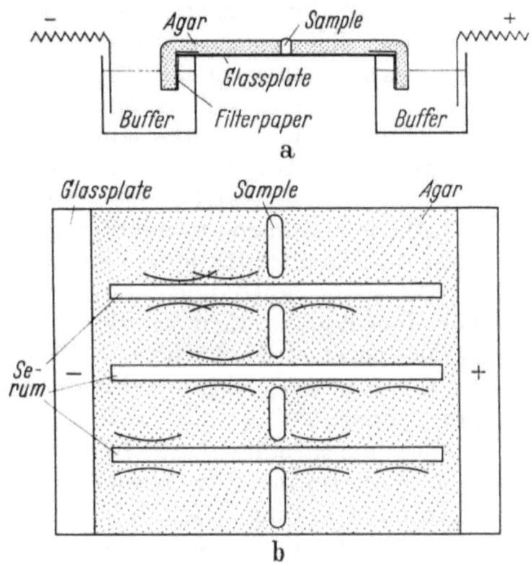

Fig. 9. a) Schematicr epresentation of immunoelectrophoresis technique. b) Schematic illustration of precipitin patterns obtained by immunoelectrophoresis.

Samples for analysis are prepared by mixing in a waterbath at 40° C, equal parts of the suspension to be analysed and a 2% agar solution of ionic strength 0.1, buffered at pH 8.2. Merthiolate should also be included, in the proportion of 1:10,000. Of this mixture 0.2 ml is transferred to the reservoir.

Electrophoresis will commence when the respective electrodes are connected to the cathode and anode of a direct current source. A potential gradient of 3 to 4 V/cm is employed. In the case of glass plates having a size of 13 × 18 cm a potential difference of 65 V is necessary between one end of the plate and the other, corresponding to a current strength of 35 to 45 milli-ampères. Higher voltages can be employed only if the heat generated by the electrophoretic process is dispersed by some kind of cooling system.

The time required for completion of the process depends both on the material to be analysed and the voltage applied. In general it takes from 15 to 18 hours. After this period the current is switched off and elongate, rectangular trenches 140 × 2 mm in size, are cut in the agar layer so that they lie parallel to the direction of electrophoresis. These are then filled with the respective antisera. In the case of a mixture of antigens, electrophoretic separation creates a file of diffusion centres in the gel from which antigens move towards antibodies diffusing from the trench. This results in the development of falcate lines of precipitation which appear after a period of 1 to 10 days. Both the electrophoresis and the diffusion should be carried out within a closed space in which a high humidity can be maintained, such as under a hood of either glass or perspex.

Instead of the macro-method, micro-methods are now used to advantage, not only being more economical on serum as well as antigen but also requiring less time for completion. In analysis of antigens liable to denaturation, the more rapid procedure involves higher sensitivity as well.

In the "*micro-method*" (Scheidegger, 1955) use is made of 76×26 mm microscope slides coated as was described above with a thin film of dried agar. The slides are then covered by pipetting a further agar layer of 1 to 2 mm thickness, for which 2—4 ml of buffered agar solution is required. The composition and ionic strength of the agar are the same as was indicated for the macro-method. After solidification of the agar, two small holes of 2 mm are punched usually about halfway along each slide and about 1 cm apart. Connection of the layer of agar with the cathode and anode terminals is accomplished by means of filter paper strips as was described for the macro-method. In this case, however, it is not necessary that the strips protruding into the jars be covered with agar. Approximately 0.001 ml of liquid antigen solution is then pipetted into each hole and as soon as this has penetrated, the hole is filled with a drop of lukewarm agar solution. Electrophoresis should proceed under a potential gradient of 6 V/cm, lasting from $^1/_2$ to $^3/_4$ hours.

Antigenic substances are often of high molecular weight, associated with slow electrophorectic mobility in which case drying of the agar layer by evaporation becomes the limiting factor. A favourable effect is obtained by covering the slides with liquids having a low boiling point, e.g. *n* pentane or petroleum ether, boiling of which gives rise to a fall in temperature (Wieme, 1959). When the process of electrophoresis is completed a narrow slit is cut in the agar down the middle of the slide and filled with antiserum.

Scheidegger (1955) compared the macro-method and micro-method with regard to amounts of reactants and average lengths of time required for both steps of the procedure, i.e. electrophoresis as well as diffusion (Tab. 1).

Table 1. *Comparison of the macro- and the micro-method of immunoelectrophoresis with regard to requirements of reagent and time* (Scheidegger, 1955).

Method	Volume of antigen in ml	Volume of antiserum in ml	Time required for:	
			electrophoresis	diffusion
macro	0.05—0.2	0.5—1.0	6 hours at 6 V/cm	1 to 10 days
micro	0.001	0.005	$^3/_4$ hour at 6 V/cm	$^1/_2$ to 24 hours

Townsley (1959) described a method using strips of paper drenched in antiserum which are placed on the agar layer immediately above the site to which the antigens have migrated during electrophoresis. This method can be used to advantage where no diffusion of antigens takes place.

An advantage of immunoelectrophoresis over gel diffusion methods is the additional differentiation based on the electrophoretic mobility of the antigens. In gel diffusion experiments it may happen that lines of precipitation, representing independent antigens, overlap. Only sometimes is the separation of these lines possible by changing the antigen-antibody ratio, for example by diluting one of the reactants. Immunoelectrophoresis constitutes a more rapid approach to such problems.

4. Localization of Antigens in Tissues.

The preceding sections all deal with antigens, the separation of which involves disruption of the biological materials in which they occur. However, in recent years it has also been shown to be of great value in histological and cytological research to localize antigens *in situ*. For this purpose the antibodies must be labelled in order to recognize their presence microscopically (COONS, 1954; 1956).

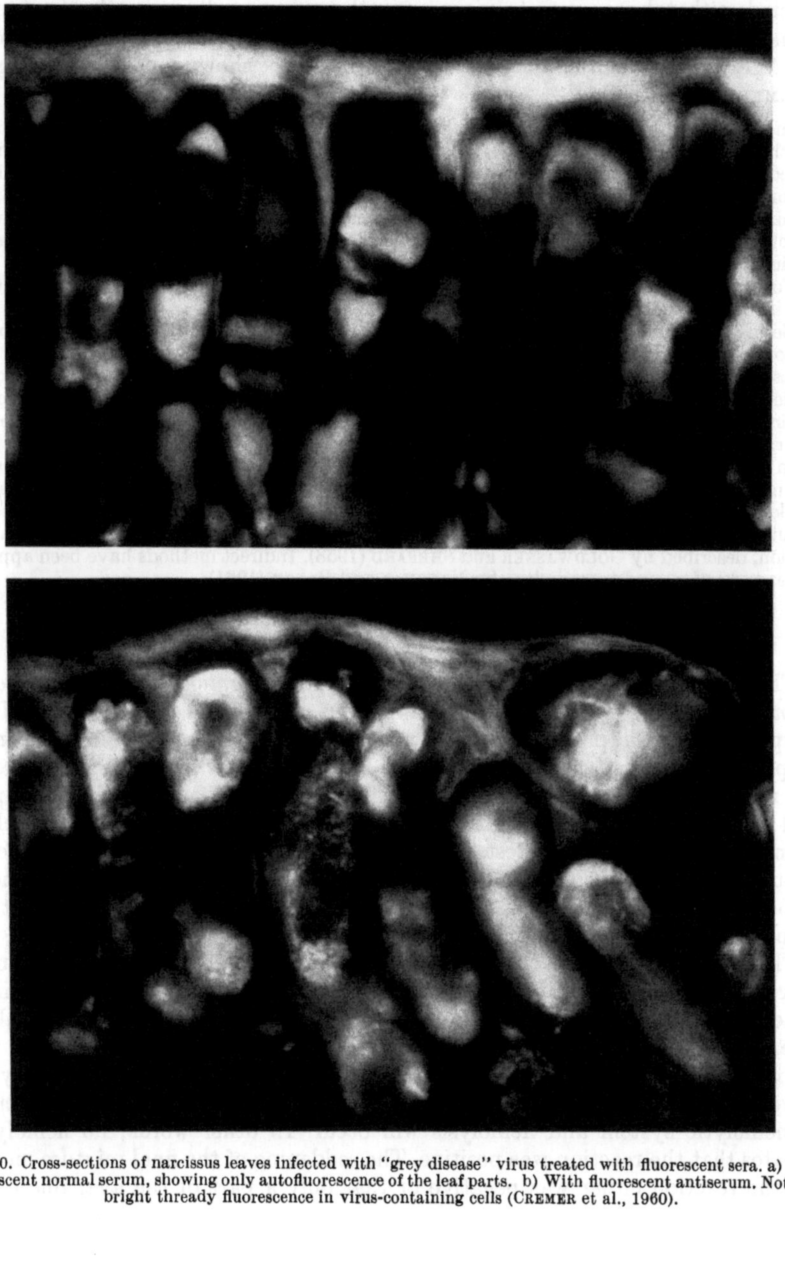

a

b

Fig. 10. Cross-sections of narcissus leaves infected with "grey disease" virus treated with fluorescent sera. a) With fluorescent normal serum, showing only autofluorescence of the leaf parts. b) With fluorescent antiserum. Note the bright thready fluorescence in virus-containing cells (CREMER et al., 1960).

For labelling antibodies, fluorescent compounds have been used (Chadwick et al., 1958; Clayton, 1954). Antibodies labelled with radio-active substances may also be useful but they are less attractive from the point of view of cytochemistry since they cannot be used to localize the antigen inside cells. The latest development is in the use of antibodies labelled with ferritin. Its high iron content possibly facilitates the use of such antibodies as cytochemical reagents for electron microscopy (Singer, 1960).

Fluorescein-labelled antibodies have been used for plant tissue sections of 25 μ thickness cut on a freezing microtome. Carnoy-fixed tobacco leaf tissue infected with tobacco mosaic virus gave the same results as unfixed material (Schramm and Röttger, 1959). In frozen sections of narcissus leaves infected with "grey disease" virus, the presence of virus can be shown quite clearly, Fig. 10 (Cremer et al., 1960).

In addition to the direct staining techniques employing conjugated fluorescent antibodies (for their production see page 429), *indirect methods* are now available. Direct staining is carried out as follows. Sections of material that have been embedded in a glycerol-gelatin mixture and cut with the freezing microtome are kept in fluorescent antiserum for 45 to 60 minutes. Thereafter the excess of antiserum is removed by washing the sections with phosphate buffer of pH 7.2 for 10 to 15 minutes. The sections are then embedded in glycerol-gelatin mixture and viewed with a fluorescence microscope.

For this we used a Zeiss filter combination with an ultraviolet light source of 2500—4000 Å inducing fluorescence in the range of 5500—6000 Å.

Besides the green fluorescence of the specific serological reaction, the operator is often confronted with additional non-specific fluorescence. Since this is due to the presence of free fluorescein a great improvement can usually be obtained by vigorously shaking the fluorescein-labelled antiserum with dried, powdered, healthy plant material before use. Fluorescence of certain normal substances, such as chlorophyll, is most commonly recognized by a different colour.

In the indirect "*sandwich*" *method* of Weller and Coons (1956) the section is first treated with unlabelled antiserum and then with a fluorescein conjugated antiserum directed against the globulins of the animal species used for preparing the first antiserum. Tyrrell (personal communication) has used another indirect method which is based on the complement fixation reaction, described by Goldwasser and Shepard (1958). Indirect methods have been applied to the study of wound-tumor virus by Nagaraj and Black (1961).

II. The Complement Fixation Test.

This is an indirect test to show the *occurrence of antigen-antibody reactions*. "*Complement*" is a thermolabile factor present in all fresh unheated sera. It is indispensable in the process of hemolysis, i.e. the liberation of hemoglobin from sheep erythrocytes after mixing them with their antibodies. Other types of serological reactions, viz. agglutination and precipitation, can proceed without complement: however, if it is present, it becomes fixed to the antigen-antibody complex.

In complement fixation tests we make use of the hemolytic system, i.e. a mixture containing sheep red blood cells and antiserum directed against sheep red blood cells. This antiserum is heated to 56° C beforehand in order to destroy complement in it.

The antiserum against the antigen is similarly heated to 56° C before use. After mixing the antiserum and antigen, fresh complement-containing guinea pig serum is added to this mixture. In cases where a positive reaction has taken place, the complement from the guinea pig serum becomes fixed so that when the hemolytic system is added no complement will be available to cause hemolysis. Conversely, if the reaction is negative, the complement will still be free to act in the hemolytic system and hemolysis will occur. In other words, no hemolysis indicates that the reaction was positive. The reddening of the final mixture means that the reaction either was negative or that only a part of the available complement was fixed. Varying degrees of reddening are possible depending upon the amount of complement fixed.

A prerequisite for the satisfactory progress of the reaction is to make exact determinations beforehand of both the minimum amount of complement necessary to redden the hemolytic system, and the zone of optimal binding of complement. It must also be investigated whether or not complement is fixed non-specifically to substances present in the plant extract, without adding antiserum. On the other hand it is known that substances able to lyse red blood cells occur frequently in plant extracts (TIGGELMAN-VAN KRUGTEN et al., 1956). This must, of course, be taken into account where such non-specific reactions can occur.

It must be emphasized that many precautions must be taken when the complement fixation technique is used. Moreover a series of dilutions of both antigen and antiserum must be tested.

Experience has shown that best results are obtained with purified antigens and that complications are more apt to occur where analysis of raw plant extracts is attempted.

For examples of applications of this type of reaction in botany reference should be made to studies of MORITZ (1932, 1934) and for plant virus work to TALL et al. (1949), LIMASSET (1952) and MOORHEAD (1956, 1959).

III. Serological Methods Based on Electron Microscopy.

The sensitivity of the precipitin test can be increased considerably by using the electron microscope. A drop of the antigen-antibody mixture is dried on the specimen holder and viewed in the electron microscope after shadow casting. In cases of serological reactions with plant viruses, even at dilutions exceeding endpoints in precipitin reactions, the virus particles will be seen adhering together in aggregates (Fig. 11).

This fact may serve as additional evidence for the specificity of the virus antibody reactions.

IV. Serological Methods Based on Specific Activities of Antigens.

When antigens possess some specific activities, the reaction of such antigens with antibodies can sometimes be tested by observing or measuring its effect on these specific activities.

With enzymes one simply centrifuges the liquid after adding the antiserum in order to remove the precipitate and determines the enzyme activity of the supernatant. This method offers important possibilities especially if two unrelated enzymes have to be separated.

Often the enzyme-antibody reaction does not result in a visible precipitate. Then the reaction of the enzyme with its antibodies may be detected by the inhibition of the specific enzyme activity. The same applies to neutralization of toxins by antitoxins (SEVAG, 1945). Inhibition of an enzyme by antiserum has been used to study a particular reaction mechanism (KREBS et al., 1951).

In plant virus literature opinions regarding the process reponsible for the reduction of infectivity by specific antibodies have for a long time been contradictory (BAWDEN, 1954). Recent investigations (FRAENKEL-CONRAT and SINGER, 1957) have revealed that the isolated γ-globulin fraction of the antiserum displays a specific inhibitory effect on infectivity. This contrasts with the non-specific effect of whole homologous as well as heterologous and normal sera (KASSANIS, 1943).

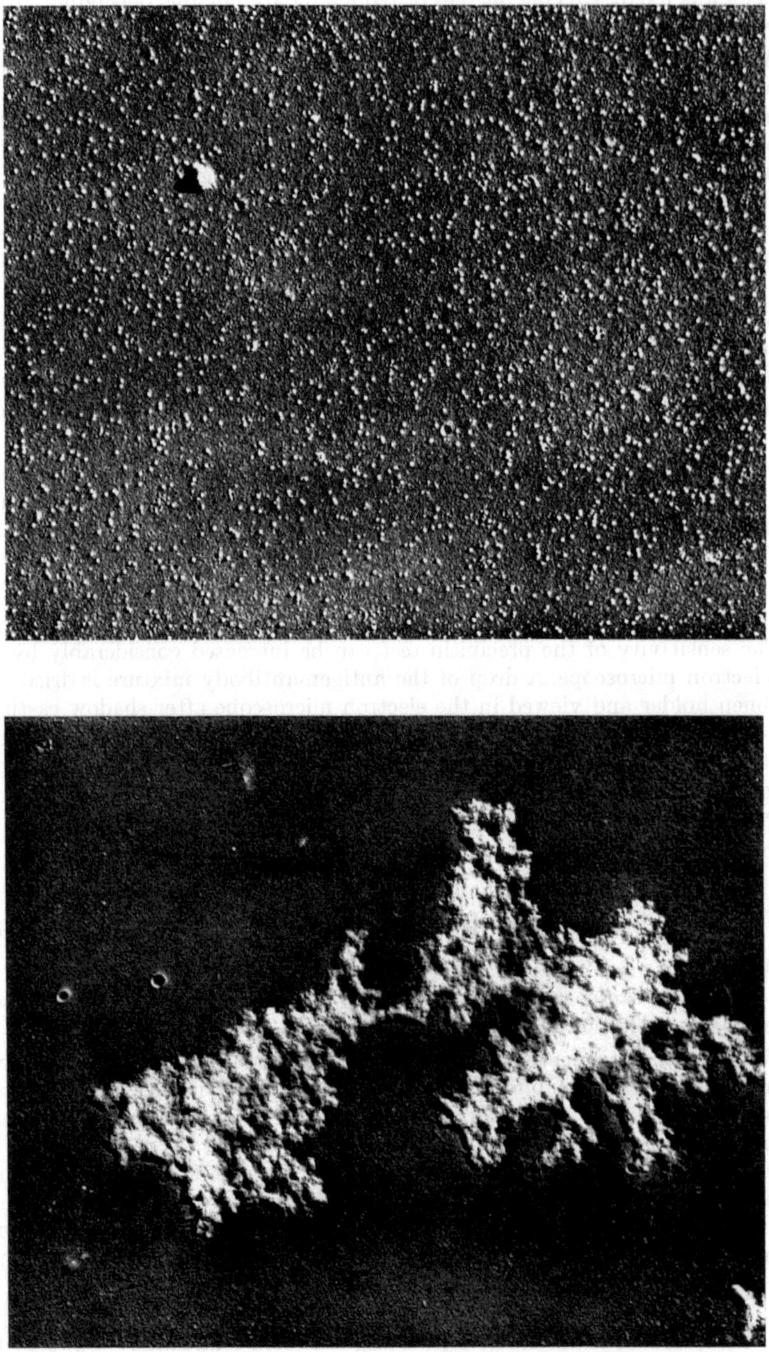

a

b

Fig. 11. Electron micrograph of (a) tobacco necrosis virus, (b) the same virus incubated with an antiserum prepared against this virus (DE BRUYN OUBOTER et al., 1952).

D. Antigen-Antibody Reactions in Sensitized Animals or their Organs.

In serological methods use is made of an antiserum to show the presence of an antigen. We wish now to describe some methods in which the antigen-antibody reaction is made apparent by using animals or their organs. Most important amongst these are *anaphylaxis* and its related reactions.

Anaphylaxis is that phenomenon whereby an animal injected with an antigen reacts severely with shock upon a repeated injection with the same antigen. This is a type of hypersensitivity of the animal which manifests itself even if very small quantities are administered. The shock is the result of a specific antigen-antibody reaction giving rise to contraction of the smooth muscles, e.g. of the bronchiae, the intestine, the bladder and the uterus.

As a rule animals are not used to demonstrate the reaction. The method most frequently used is to inject virgin guinea pig or mouse females with the antigen involved. After some time the animal is killed and the uterus with its horns is taken from the body and placed in saline kept at body temperature. The horns are fixed into a mechanism which can record contractions of the muscles. If the antigen is added to the saline it will react with antibodies present in the muscle tissue, resulting in contraction of the muscles which is recorded. Apparently this can occur only once with any particular antigen. However, if the animal has been sensitized with a mixture of unrelated antigens, a number of contractions can be evoked by adding the respective antigens successively.

This reaction has been used on several occasions for plant proteins, references for which are given by MORITZ (1958).

The so-called *skin-test* is closely related to the phenomenon described above. In this case the antigen to be studied is introduced into the skin of the sensitized animal. When carried out on human beings, e.g. the tuberculin reaction of PIRQUET, the hypersensitive response is considered a form of allergy. In contrast to its widespread use in medical practice, the skin-test has not found application in botanical research.

E. Quantitative Determinations of Antigens.

In quantitative determinations of antigens serological precipitin techniques are, for practical reasons, most commonly used.

Depending upon the purity of the antigen different methods are appropriate. A rough estimate of antigen concentration can be obtained simply by mixing a series of antigen dilutions with a constant amount of antiserum; the dilution still giving a positive result is taken as a measure of the antigen concentration (determination of the dilution end-point). Alternatively one can measure the lapse of time between the moment of mixing antigen and antiserum and the moment the first floccules of precipitation appear.

A much higher degree of accuracy is obtained by weighing the dried antigen-antibody complex, by measuring the amount of precipitate nephelometrically or by analysing the complex chemically (BOYDEN, 1954; KABAT and MAYER, 1948).

Finally we wish to mention the quantitative chromatographic procedure used by MATTHEWS (1954) for analysing the antigen-antibody complex in tobacco mosaic virus studies. The complex is hydrolysed and the hydrolysate chromatographed to reveal the amount of nucleotides present. From the nucleotide content the tobacco mosaic virus concentration can be calculated.

I. Dilution-End-Point Test.

In this test a series of antigen dilutions is mixed with a constant amount of antiserum. There are a number of modifications of the method viz. the mixture can be made in tubes or in droplets, or the ring test can be applied (see page 433). There are not much data available regarding the sensitivity of these modifications of the method. According to GOLD et al. (1960) the ring test is considered to be the most sensitive, at least with certain plant viruses.

As was mentioned above, the highest antigen dilution still giving rise to a visible precipitate is a measure of the antigen concentration in the original solution. Thus a solution having a dilution end-point of 1/64 contains eight times as much antigen as a solution having a dilution end-point of 1/8. In this way data are obtained on relative amounts of antigen. It is clear that the degree of objectivity of the observer must play an important role in all such experiments.

Wright (1959) made use of the gel diffusion method for quantitative determinations of antigens from plant material. This method is of particular interest in analysing mixtures of antigens, as in the case of virus in the sap of infected plants. Different dilutions of antigen are put in holes in the agar located around a central hole filled with antiserum. In this case the dilution end-point corresponds with that dilution of antigen producing a falcate line of precipitation just next to the hole containing the diluted antigen. Higher dilutions will not produce any sign of a precipitation line. It must be mentioned that the state of aggregation of certain antigens which tend to form aggregates may strongly affect the results of this test.

Finally it is worth mentioning that more reliable data are obtained if antigen dilutions to be tested can be compared with dilutions of purified preparations of known concentration.

II. α Optimum Test.

In C/II. the α optimum has been mentioned. This is the antigen dilution producing the quickest precipitation with a given antiserum dilution. As described for the dilution end-point test, a series of antigen dilutions is mixed with a constant dilution of antibody, starting by adding antiserum to the highest dilution of antigen. Then the tubes are shaken (placed in a waterbath) and the moment at which precipitation first occurs is recorded.

In practice when two samples of the same antigen are compared, the α optimum is determined for both. If for sample no. 1 the α optimum lies at the dilution 1/4 and for sample no. 2 at the dilution 1/8, sample no. 2 has twice the antigen-concentration of sample no. 1. If a purified antigen of known concentration can be used as a standard, more accurate estimations can be obtained.

The maximum error in the determination of the α optimum is as high as 50% because the true optimum may be on either side of the tube showing fastest precipitation.

In order to ensure accurate observations too rapid precipitation must be avoided.

III. The Determination of the Amount of Specific Precipitate.

This concerns the amount of precipitate formed after mixing a quantity of antigen with a known volume of antiserum. The two most important problems we have to deal with are, avoiding non-specific precipitation and complete separation of precipitate and supernatant liquid.

The first problem is solved if purified antigens are used and all other factors causing non-specific precipitates are controlled. Moreover we have to keep in mind that antigen-antibody complexes are soluble in excess antigen and in the case of horse sera also in excess antibody.

First a *standard curve* is made in the following way. A series of solutions of an antigen (of a known concentration) at different dilutions are mixed with a constant amount of antiserum. The precipitates formed after incubation for 2 hours at 37° C and kept overnight at 4° C are centrifuged down (10 min 3000 r.p.m.). The precipitates are washed three to four times with saline in order to remove non-specific substances. After drying at 105° C the precipitates are weighed or their nitrogen content is determined according to the method described by PIRIE in volume II, p. 23 of "Mod. Meth. Plant Anal.".

In order to determine the antigen content of a sample, a given volume is mixed with a known amount of antiserum and the amount of precipitate formed is determined. This procedure is repeated at another dilution of the antigen sample. As illustrated by Fig. 3 at least two determinations at two different dilutions are needed in order to see whether the antigen concentration corresponds to either the zone of excess antigen or that of excess antibody. Instead of weighing the precipitate, the turbidity of the antigen-antibody mixture can be measured in a nephelometer. A comparison of the two methods by BOYDEN (1954) showed that they give similar results. In nephelometry a standard curve must first be plotted for a given antiserum.

GLENN (1958) introduced a method for measuring the turbidity caused by antigen-antibody complexes in OUDIN columns. Essentially his method is based on the principle of nephelometry. An advantage of his nephelometric method is that multiple antigens can be determined simultaneously.

F. Measurements of the Amounts of Antibodies in Sera.

A reasonably accurate determination of amounts of antibodies in antisera is possible only if certain conditions are fulfilled, the main condition being a high degree of purity of the respective antigens. Furthermore, precipitation must be carried out at an antigen-antibody ratio previously found to correspond to that of equivalence.

When a known weight of antigen is mixed with a known dilution of antiserum, the nitrogen content of the resulting precipitate as well as that of the corresponding weight of antigen alone can be determined by the micro-KJELDAHL nitrogen determination method. Substraction of antigen nitrogen from total nitrogen gives the amount of antibody nitrogen in the precipitate. Then the amount of antibody nitrogen per ml of undiluted serum can be calculated, for the dilution and the volume of the initial antiserum are known, whereupon the total weight of antibodies per ml can be determined as the amount of nitrogen is usually 15% of it. The presence of small amounts of nitrogen containing compounds other than antigen or antibody in the precipitate may, however, cause an overestimation of the amount of antibody.

For detailed information about quantitative analytical methods for measuring antibodies reference must be made to KABAT and MAYER (1948).

Usually, in plant-serological work estimations of relative amounts of antibodies suffice in order to compare the antibody content of different sera. The two methods most commonly used for this purpose will be described here.

I. The Titre of the Serum.

This can be determined by adding serially increasing dilutions of an antiserum to equal amounts of antigen at constant dilution. The highest dilution showing a positive serological reaction is called the titre of the antiserum. Thus a reasonably good estimate of the relative amounts of antibody present in sera is obtained. It is important, however, to use antigen at sufficiently high dilution to avoid inhibition of the reaction by antigen excess, otherwise the titre will be underestimated.

II. From the Determination of the α Optimum.

The precipitation time for a system of an antigen with its corresponding antiserum at varying concentrations will be different. If the ratio of the concentrations of antigen and antibody at which the time for precipitation is shortest is recorded for every antiserum dilution, it is found that this ratio or α optimum is constant: e.g. if at antiserum dilution of 1/8 the most rapid precipitation occurs with an antigen dilution of 1/32, the α optimum thus becomes $\frac{1/32}{1/8} = 1/4$. As for every antiserum dilution this ratio has to be the same, it follows by mere substitution that with undiluted antiserum (at 1/1), the antigen dilution for the α optimum is 1/4. Thus from a test with a known dilution of antiserum reacted against a dilution series of the antigen one can calculate the antigen concentration for the α optimum with antiserum at 1/1. This value can be taken as a measure of the relative amounts of antibody in different sera. Thus an antiserum for which the α-optimum-antigen dilution at 1/1 is 1/2 has approximately four times as much antibody as an antiserum for which this antigen dilution lies at 1/8.

For further methods of estimating relative amounts of antibody in sera reference is made to MATTHEWS (1957).

G. Methods for the Determination of Relationships between Antigens.

In addition to problems involving the demonstration and quantitative determination of antigens there often arise also questions regarding relationships between substances. As indicated in the introduction, an antiserum reacts not only with its homologous antigen but also with heterologous, related, antigens. Generally one is not content with a mere "yes or no answer" concerning relationship, but one seeks also to know what is the degree of relationship. This can be done immunochemically. Identity established in this way, however, relates only to the antigenically active groups in the molecules and not to the complete structures. Information about this must be obtained by other means of investigation.

1. Relationships between two unpurified antigens can be quite simply determined as follows. Antisera are produced against both antigens a and b and reactions are carried out by mixing the antisera A and B with the antigens a and b in various dilutions. Precautions must be taken to exclude positive reactions due possibly to contaminating antigens which the preparations studied have in common. Subject to this, positive precipitin reactions in every combination of antiserum and antigen indicate without doubt, that the antigens a and b are related. This can be extended to include a great number of antigens. To prove that two antigens are not related, the experiments have to be performed with antisera against each of them. It is possible that two antigens are related but that one of them does not produce a positive reaction with the antisera involved, due to the fact that the concentration of that antigen is too low. An illustrative example is found in the mild and severe etch viruses of tobacco; the former is present in plant sap in too low a concentration to give a visible reaction with either the homologous or the heterologous antisera (BAWDEN and KASSANIS, 1941). The severe does produce a visible precipitate with both antisera indicating a relationship between the viruses. Where the precipitin test fails, complement fixation tests may give a positive result. With haptens especially, complement fixation is preferable. Anaphylaxis is another approach to the study of such problems and has been used by MORITZ (1931) with different plant proteins and by CHESTER (1937 b) to show the presence of minute quantities of normal proteins in preparations of purified tobacco mosaic virus. If the substance carries typical biological properties, such as infectivity or enzymatic activity, relationship can be shown by means of appropriate determinations (page 447). JOHNSON (1954) has based a relationship of antigenic material on turbidity measurements.

2. In many cases relationships between antigens can be determined quickly with *gel diffusion methods* (page 436). We wish here simply to refer to its application in the study of normal plant proteins (HEITEFUSS et al., 1960; WRIGHT, 1960), enzymes (CINADER and PEARCE, 1958) and plant viruses (VAN SLOGTEREN, 1955b, 1955c, 1958).

3. With purified antigens an indication of the degree of relationship can be obtained with the *α-optimum test*. If the unknown purified antigen has the same α optimum as a known substance then the conclusion can be drawn that it is closely related to, or identical with, that substance. A lower α optimum on the part of the unknown antigen would mean a lower degree of relationship, whilst a higher α optimum would point to a mixture of a related antigen with an unrelated one or with an impurity.

4. Better results can be achieved with the *cross-absorption test*. Antiserum against antigen a is absorbed with antigen b and then the amount of remaining antibodies is determined with antigen a itself. The same procedure is followed with an antiserum prepared against antigen b which is absorbed with antigen a. In making the absorption somewhat more antigen is used than agrees with the estimated proportional amount of undiluted antigen at the α optimum. The mixture is incubated for 1 to 2 hours at 22 to 50° C, kept overnight at 4° C and the precipitate centrifuged down. After ensuring that all antibodies against the heterologous antigen have been removed, the antiserum is tested with its homologous antigen. In these tests the antigen should be at approximately dilution end-point concentration in order to exclude inhibition of precipitation by antigen excess. The amounts of antibodies reacting with their heterologous antigens can then be estimated by any of the following determinations.

a) The titres of the antiserum before and after absorption. The drawback of this method is the relatively low sensitivity of the high dilutions used so that small differences do not appear. Its advantage lies in the small amount of reagents necessary.

b) Instead of the titre, the α optimum of the serum can be determined before and after absorption. If the dilution at the α optimum has increased from the value of 1/2 to 1/32, about 1/16 of the amount of antibodies is specific for the homologous antigen.

c) From the amount of the precipitate at the α optimum the amount of specific antibodies remaining in the serum can be estimated. BAWDEN and PIRIE (1937) give an example of this in their work on strains of tobacco mosaic virus.

The determination of the amount of antibodies in the precipitate gives most reliable results, but it is laborious and large amounts of antigen and antibody are required.

The question of relationships among three or more antigens may arise. In that case the antisera involved are, after absorption, tested with each of the heterologous antigens separately and with a mixture of both heterologous antigens (MATTHEWS, 1948).

Complete identity of antigens can only be proven by performing quantitatively the reciprocal test.

Some factors which may hamper the methods described in this section require mentioning. The sensitivity of the reactions is adversely influenced by the dilution of antiserum with the addition of the heterologous antigen. This difficulty can be overcome by concentrating the antigen before, or the antiserum after, absorption.

Furthermore, the titre of the antiserum has an effect on the specificity. According to AUGUSTIN (1959), increasing the number and the period of injections leads to an increase in the amount of antibodies responsible for cross-reactions.

In studies on close serological relationships, e.g. amongst strains of potato virus X (MATTHEWS, 1948), it is therefore desirable to use antisera of fairly low titre.

On the other hand if one wants to trace distant serological relationships, very high titred antisera produced over long periods of injections must be used. A good illustration of this is presented by Bercks (1960) concerning serological relationships between beet mosaic virus, potato virus Y and bean yellow mosaic virus.

H. Immuno-Chemical Criteria of Purity of Antigens.

The immunological procedures described above provide methods for examining the homogeneity of supposedly purified material. The particular circumstances may be such that the substance which has been purified may have antigenic properties, whilst any impurities remaining may or may not be antigenic. On the other hand one may have to deal with material which is not itself antigenic but contains antigenic impurities.

The criteria to be considered here are based on the immunological behaviour of either the desired substance, the contaminants, or both.

The qualitative reactions described in C are relatively sensitive and demonstrate the presence of impurities with antigenic properties. Using the precipitin reaction Kunkel and Ward (1950) could show the presence of 3.0 μg protein (human albumin) with 0.2 ml of antiserum.

As indicated on page 449, anaphylaxis is a very sensitive tool in detecting minute quantities of antigenic impurities.

The quantitative precipitin reaction may also give important data regarding the purity of preparations. The form of the curve obtained by plotting the total amount of precipitated nitrogen as a function of the amount of added antigen serves as a very sensitive index of relative purity (Cohn, 1952). The presence of impurities is shown by a shift of the maximum of the curve to higher antigen concentrations (Fig. 12). Similar information can be obtained by measuring the quantity of antigen leading to maximum precipitation of antibody (Cinader and Pearce, 1956).

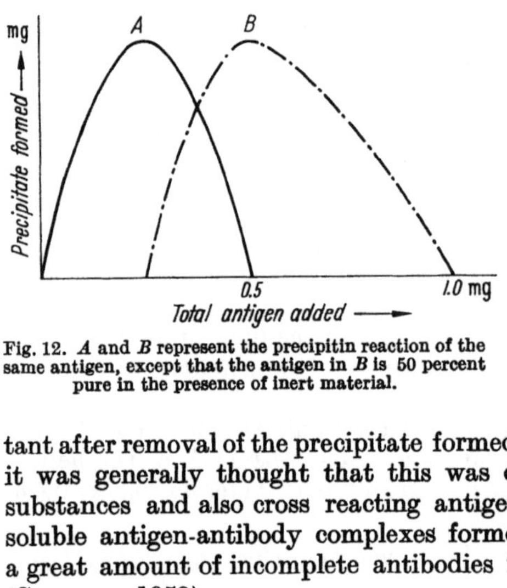

Fig. 12. *A* and *B* represent the precipitin reaction of the same antigen, except that the antigen in *B* is 50 percent pure in the presence of inert material.

Certain conclusions can also be drawn from the simultaneous presence of antigen and antibody in the supernatant after removal of the precipitate formed in the serological reaction. Formerly it was generally thought that this was due to the presence of other antigenic substances and also cross reacting antigens, but this is not always true since soluble antigen-antibody complexes formed as a consequence of the presence of a great amount of incomplete antibodies may produce the same phenomenon (Cinader, 1958).

A completely different procedure for showing the presence of impurities is that based on the decrease of activity of an antiserum against these admixtures by adding the preparation to be tested for homogeneity. Boyden's hemagglutinin technique (see page 435) is particularly suitable in this respect (Boyden, 1951). This technique also gives good results with certain non-precipitating antigens (haptens).

With regard to all these methods, it must be emphasized that any one may fail for one reason or another to show the presence of a *contaminant*. It is advisable therefore, always to employ a combination of various techniques when testing the homogeneity of a substance.

J. Applications.

Rather than being a complete survey of all the literature on serological applications, this section is intended to deal only with certain selected examples, of a general nature, in order to serve as a guide in approaching similar problems in other fields.

I. Serology in Relation to Taxonomy of Higher Plants, Fungi and Bacteria.

As early as 1926 MEZ and ZIEGENSPECK introduced their so-called "*Königsberger serodiagnostic system*". This system is based on the recognition of common antigenic characters within different species of plants. On account of the criticism levelled against this system, MORITZ (1931) undertook elaborate relationship studies with the aid of the SCHULZ-DALE technique of *anaphylaxis*. In so doing it was his intention to eliminate, as possible sources of error, substances with haptenic properties such as lipoids and polysaccharides. Although this standpoint has in the meantime come to be considered as unjustified, the results which MORITZ (1958) obtained remain unchallenged. On the whole they are in agreement with the results of MEZ and ZIEGENSPECK (1926).

MASCHMEIER (1927) obtained negative results in an attempt to classify a number of potato varieties. The reason for this failure was probably the presence of potato virus X in all the varieties (CHESTER, 1937b). On the other hand, GELL et al. (1956), applying the gel diffusion method, succeeded serologically in subdividing 14 Mexican species belonging to the genus *Solanum* into 5 groups.

TEMPEL (1958) made an attempt to differentiate some of the physiologic races of the fungus *Fusarium oxysporum* by the application of serology.

A number of studies have been specially concerned with *immuno-genetical* investigations.

LEWIS (1953) carried out immuno-genetical experiments with pollen originating from seven clones of *Oenothera organensis*. From the data obtained he has tried to draw up a scheme in which there is a direct relationship between single genes and antigenic properties. Similar investigations into antigenic differences caused by genes have been described by MINTZ and LEWIN (1954) with respect to mobile and immobile mutants of *Chlamydomonas moewusii*. In the flagella of these organisms no antigenic components could be demonstrated on which to base a serological differentiation. According to MARKERT and OWEN (1954) who worked serologically with mutants of *Glomerella cingulata* differing in tyrosinase activity, only quantitative and not qualitative antigenic differences could be established.

MORITZ and VOM BERG (1931) have studied hybrids between species as well as between genera. These investigations have contributed to a serological analysis of *Trifolium hybridum*, *Berberis stenophylla* and *Vicia leganyi*, not only with respect to the antigenic properties of the hybrids but also to those of the parents. For a more complete survey of investigations into the field of immunological properties of hybrids reference may be made to CHESTER (1937b) and MORITZ (1958).

II. Identification of Antigenic Substances from Plants.

The serological differences mentioned in the preceding paragraph are related to the presence of distinct antigens. In addition to proteins and nucleoproteins, polysaccharides and lipoids may play a role. Much attention has been given to the antigenic properties of a large number of cell constituents and to the variability of these properties.

1. Proteins in General.

Some plant proteins such as prolamine from wheat, exhibit a low antigenic capacity (CHESTER, 1937b), whereas globulins derived from seeds can be considered as being strongly antigenic. This has been demonstrated serologically by a number of workers e.g. for phasine by Szprel-Seyfriedowa (1950), for edestine by Munoz and Becker (1952), for globulins from *Magnoliaceae* by Johnson (1952), for those from *Cucurbitaceae* and *Leguminosae* by Baum (1954) and finally for globulins from *Linum usitatissimum* and *Cannabis sativa* by Moritz (1934). Using varieties of cabbage, Heitefuss et al. (1960) demonstrated the presence of 4 antigenic components stable in vitro, which he assumed to be either proteins or nucleoproteins. In varieties showing resistance against *Fusarium oxysporum*, no distinct antigenic properties could be established which were not also present in susceptible varieties.

2. Toxins and Allergens.

Polysaccharides and proteins combining antigenic with toxic properties have been isolated from many bacteria. In this respect toxins associated with pathogenic bacteria e.g. *Corynebacterium diphtheriae* and *Clostridium tetani* are a matter of common knowledge. For further details on this subject reference may be made to textbooks on bacteriology (Wilson and Miles, 1958). Investigations by Kabat (1956) into immunological properties of dextran have disclosed that its antigenic capacity seems to be dependent upon the size of the molecule, the degree of antigenicity diminishing as the molecular units become smaller.

Very little is known about the chemical structure of many toxins. It is of importance therefore that something at least has been revealed about the relationships between them by immunological studies. As examples of toxins from higher plants can be mentioned ricin, robin and soyin, derived respectively from *Ricinus communis*, *Robinia pseudacacia* and *Glycine soya*. According to Eisler (1933) there is evidence supporting the assumption that ricin and robin are serologically related. Liener and Rose (1953) studying the immunological properties of soyin have concluded that this substance is antigenically heterogeneous.

Of considerable practical value have been the investigations made by Ramon (1937) demonstrating that adsorption of toxins by means of adjuvants, e.g. tapiocaflour and lanolin can greatly reduce toxicity without affecting the titres of antisera prepared by injecting such mixtures of toxin and adjuvant.

Allergens include some substances which are not antigenic in the ordinary sense (Boyd, 1956). They are excitants of an allergic reaction in individuals, predisposed for this reaction by a previous exposure to the allergen. They are taken up either by inhalation, ingestion, injection or by direct contact, and at the first exposure do not induce any toxic effect.

Allergens from pollen are responsible for hay fever, but little is known of their chemical structure. Rockwell (1943) isolated a substance from *Ambrosia* pollen composed of protein and polysaccharides, having a molecular weight of 4496. Similar substances from pollen of *Phleum* have molecular weights varying between 651 and 4496.

The antigenic properties of plant toxins and allergens are essentially dependent upon taxonomic relationships. Pollen of *Compositae* can be serologically differentiated from that of *Gramineae*, but it is not possible on the basis of antigenic characteristics of the pollen, to differentiate species within the *Gramineae*. Similarly in serological comparisons of *Artemisia trifida* with *A. artemisiaefolia*; *Solidago canadensis* with *S. nemoralis*; and *Festuca rubra* with *F. pratensis* no species specificity could be found in the pollen (Doerr, 1951).

3. Substances with Enzymatic Properties.

Antibodies to enzymes occur in many antisera prepared by immunization with vegetable extracts. Enzymes, being mainly of a protein nature, stimulate production of specific antibodies. Of all enzymes so far studied, Warburg's yellow ferment from yeast has been the only one lacking antigenic properties (VARTERESZ and KESZTYÜS, 1941).

It should be mentioned, however, that a similar enzymatic function shared by different substances by no means indicates a serological relationship as well. According to recent investigations (CINADER, 1957) this can be explained by the fact that a certain enzymatic action need not be dependent upon a definite plant protein. It has been revealed that rather the taxonomic relationship of the plant from which an enzyme is derived, determines its serological specificity.

There exists no serological relationship either between any of the tyrosinases produced by fungi, such as the genera *Neurospora* and *Psalliota*, or between any of those and a tyrosinase from a higher plant such as *Solanum* (MARKERT and OWEN, 1954). ADAMS (1942) had already found the same absence of serological relationship between tyrosinases of the fungi *Psalliota campestris* and *Lactarius piperitas*. Thus it seems reasonable to assume that, as a rule, prosthetic groups on the protein molecule, determining enzymatic action, are not identical to the determinant groups responsible for antigenic properties. Similarly, according to KIRK and SUMNER (1932) there exists a serological difference between the respective ureases of *Glycine soya* and *Canavalia ensiformis*. On the other hand alcohol dehydrogenases derived from closely related forms of *Saccharomyces cerevisiae* are identical. CINADER (1957) has investigated the effect of antibodies on the activity of enzymes.

Finally an investigation into the *homogeneity of enzymes* is worth mentioning. With the aid of immunoelectrophoresis CINADER and PEARCE (1958) have demonstrated that the main component of crystallised ribonuclease is heterogeneous.

4. Plant Viruses.

Serological methods have nowadays become indispensable tools in plant virus research. On account of the relatively high degree of antigenicity exhibited by an important proportion of the known plant viruses, many attempts to prepare specific antisera have been successful. Serological methods for the identification of plant viruses have since been introduced into practice and various techniques have been developed for the rapid diagnosis of plant virus diseases.

In agriculture and horticulture the selection of virus free plants as foundation stock has become practice, and of this work a considerable part is based on serological methods. In this connection the propagation of potatoes from selected, certified material free from the potato viruses X, S, M and Y can be mentioned. For this purpose in the Netherlands great numbers of plants are indexed by the application of the so-called micro-agglutinin technique (page 434). In the case of flowerbulbs such as *Freesia* (VAN KOOT et al., 1954), *Tulipa*, *Hyacinthus*, *Narcissus*, and certain other ornamentals, e.g. *Chrysanthemum* (OERTEL, 1956), diagnostic serological tests have become common practice. For further data in this field, including the theoretical background, reference may be made to MATTHEWS (1957) and to the survey by VAN SLOGTEREN and VAN SLOGTEREN (1957).

Interrelationships between virus strains have been studied by means of the serological cross-absorption method (page 453). BAWDEN and PIRIE (1937) have used this procedure in a study of relationships among strains of tobacco mosaic virus. In a similar study TALL et al. (1949) preferred the more sensitive complement fixation method. In this connection it should be mentioned, however, that difficulties were encountered in attempts to obtain a better interpretation of the results by introducing the concept "antigenic unit" (MOORHEAD, 1959).

A procedure for eliminating the infectivity of virus particles from nucleic acid preparations with the aid of specific antisera has been applied by Schramm (1956) and by Fraenkel-Conrat and Singer (1957).

The establishment of more distant relationships between viruses, such as could not be demonstrated by the methods mentioned above, has now become possible through applying antisera with a very high titre.

The relationship between potato virus Y, bean yellow mosaic virus and beet mosaic virus has been determined with the aid of a "high titred" antiserum to potato virus Y. A serological relationship of the same order has been established between potato virus X, white clover mosaic virus and *Hydrangea* ringspot virus (Bercks, 1960; Wetter, 1960). Conformity in shape and length of virus particles determined by electron microscopy (Brandes and Wetter, 1959) seems ingeneral to be in agreement with such remote serological relationships.

Gel diffusion methods (page 436) offer a new approach in the elucidation of relationships between viruses as well as in the determination of the degree of homogeneity or multiple antigenicity of the components of a single virus.

In extracts from tobacco infected with tobacco mosaic virus, Jeener et al. (1954), van Slogteren (1955b) and Kleczkowski (1957) presented evidence for the occurrence of multiple antigens. In addition to complete virus particles, diffusing into agar gel at a relatively slow rate, antigenic components with a much higher diffusion rate were also found, the latter moreover giving rise to 2 or more lines of precipitation close to each other.

From suspensions of tobacco rattle virus, Harrison and Nixon (1959) separated 2 fractions with particle sizes of 185×30 mμ and 75×30 mμ respectively. Of the two fractions only the one with shorts rods showed a positive serological reaction in the gel diffusion method.

In contrast to the success achieved with at least some of the rigid, rod-shaped plant viruses as mentioned above, less satisfactory results have been obtained with elongated flexible viruses in gel diffusion reactions. According to van Slogteren (unpublished) a typical precipitin reaction in gels readily occurs when potato virus X is made to diffuse against its corresponding antiserum but no results have been obtained with potato viruses S or Y diffusing against their respective antisera. On the other hand promising results have been obtained with plant viruses of small size and especially with those of spherical shape. As examples of these may be mentioned strains of cucumber mosaic virus (van Slogteren, 1958) and infective preparations of some fruit viruses, such as cherry yellows virus and necrotic ringspot virus (Willison et al., 1959).

A serological approach to the structure of tobacco mosaic virus has been attempted by Aach (1957), Kleczkowski (1957) and by Takahashi and Gold (1960). Protein sub-units with a molecular weight of about 20,000, designated by some workers as X-protein, are helically arranged in the tobacco mosaic virus particles (Franklin and Holmes, 1956). Antisera prepared against either complete tobacco mosaic virus or against reconstituted particles of identical shape obtained by reaggregation of X-proteins, contain specific antibodies which cannot be removed by absorption with X-protein suspensions. Consequently X-protein and complete tobacco mosaic virus protein are serologically distinct. It should be mentioned that according to Kleczkowski (1957) X-protein by itself is not in fact homogeneous but has been shown with the gel diffusion method to contain at least two antigenic components. Furthermore, X-proteins are not the only sub-units present in preparations obtained in various ways from infected plants. A protein component of molecular weight 90,000, obtained by Schramm and ZILLIG (1955) by alkali treatment of tobacco mosaic virus and designated as "*A-protein*", was found by KLECZKOWSKI (1957) using the gel diffusion method, to be a mixture of X-protein and Y-proteins, the latter fraction consisting of at least five antigenically different components.

III. Serological Studies of Cell Products.

Although considerable progress has been achieved in our knowledge of structures present inside plant cells, mainly with the aid of electron-microscopy, little is known of the antigenic properties of these various structures and organelles. Far more data are available regarding the antigenicity of animal nuclei and mitochondria, for example, than about their plant counterparts (VOGT, 1960).

DORNER et al. (1958) isolated a fraction from chloroplasts with a sedimentation constant of 19 S, and prepared an antiserum to the substances it contained. They could demonstrate that these components not only exhibited a strong antigenic capacity but also that they occur in widely differing plant families.

In recent years a new interest has developed in microsomes and also in the endoplasmatic reticulum, mainly on account of their function in protein synthesis. In this connection, PIRIE (1950) pointed out that nucleoproteins isolated from microsomes show a relatively poor antigenicity. When plant viruses are purified by alternate low- and high speed centrifugation, the resulting suspensions still contain a fair amount of nucleoprotein material from the microsomes. In spite of this, however, upon injection of these suspensions no antibodies against these nucleoproteins are present in the resulting antisera.

Recently, however, WRIGHT (1960) has succeeded in preparing specific antisera to microsomes derived from the coleoptiles of *Triticum vulgare*. He employed FREUND's adjuvant technique of injection (page 426) and the serological procedure of OUCHTERLONY's gel diffusion method. This antiserum contained in addition to antibodies reacting with a variety of substances present also in other parts of the plant, a fraction of antibodies reacting exclusively with a substance from coleoptiles. According to the author this latter substance may be connected with cell differentiation in the coleoptiles.

In attempts to isolate constituents of plants, denaturation due to chemical transformation following disruption of the cells often cannot be avoided. In the interpretation of the results, however, this factor is frequently left out of consideration.

IV. Serological Applications in the Analysis of Biosynthetic Processes.

1. Biosynthesis During Multiplication of Plant Viruses.

Virus infection is frequently followed by marked changes in the physiology of the host plant. In this respect, reference should be made to BAWDEN and PIRIE (1952) and PORTER (1959). Serological approaches to the complicated processes involved in virus synthesis are not restricted to complete virus proteins but are also concerned with other anomalous proteins found in infected plants. The X-proteins present in tobacco mosaic virus infected tobacco have already been mentioned. Their existence was first demonstrated by TAKAHASHI and ISHII (1952, 1953). COMMONER et al. (1953), employing different procedures of preparation, have isolated 3 non-infectious proteins cross-reacting serologically with tobacco mosaic virus but lacking nucleic acid. JEENER et al. (1954), by serological analysis of preparations, applying OAKLEY's double diffusion in one dimension in tubes, isolated two soluble antigens free of nucleic acid and with distinct electrophoretic mobilities. KLECZKOWSKI (1957), using OUCHTERLONY's double diffusion technique with purified preparations of tobacco mosaic virus, obtained two lines of precipitation indicating soluble antigens (X-proteins) and a third nearer to the diffusion centre representing complete virus antigen. Van Slogteren (1958) obtained the same 3-line pattern when antigens were made to diffuse from crude sap of infected tobacco.

Recent research has revealed the occurrence of anomalous non-infective proteins in addition to complete virus particles in plants infected with viruses other than tobacco mosaic virus. According to Matthews (1960), the biosynthesis of turnip yellow mosaic virus is accompanied by the formation of two non-infectious proteins.

2. Localization of Virus Synthesis within Intact Plant Cells.

Since in most procedures designed to examine constituents of plant cells destruction of the cells cannot be avoided, little is yet known about the localization of biosynthetic processes involved in virus multiplication inside the intact cells. Recently Schramm and Röttger (1959) have employed fluorescent antibodies of tobacco mosaic virus for staining thin sections of tobacco leaves infected with this virus. The results of these investigations have shown that on infection of tobacco with tobacco mosaic virus the formation of antigenic material commences in the vicinity of the nucleus. Cremer et al. (1960), employing the same technique, have similarly demonstrated the presence of antigenic material in intact cells of *Narcissus* infected with "grey disease" virus.

3. Antigens Associated with Fungus Infection.

Heitefuss et al. (1960) have demonstrated at least three distinct antigens in homogenates from a susceptible variety of cabbage infected with *Fusarium oxysporum*. These antigens could not be found in homogenates of pure cultures of the fungus, neither were they present in plants free of the disease. In relation to this problem, no antigenic difference could be established between non-infected plants of susceptible and resistant varieties of cabbage.

4. Biosynthesis of Enzymes.

The biosynthesis of β-galactosidase by *Escherichia coli* can be induced by cultivating the bacteria on a medium which contains β-galactose. Bacteria grown on a medium devoid of this sugar do not synthesize the enzyme. With the aid of specific antisera, Cohn and Torriani (1952) have demonstrated that bacteria without β-galactosidase contain a substance antigenically identical with β-galactosidase but lacking enzymatic activity. Further investigations have revealed that both the enzyme and the serologically related inactive compound are probably derived from the same precursor.

5. The Study of the Process of Vernalisation.

Attempts to differentiate serologically between cytoplasmatic proteins from vernalised and from non-vernalised *Kalanchoë* plants (Aach and Melchers, 1957) have yielded quantitative results only and give no further information about the identity of the antigens involved.

Concluding Remarks.

Serological studies of vegetable organisms are restricted mainly to antigenic constituents such as proteins and polysaccharides. A serious handicap is the inaccessibility of nucleic acids for direct immunological treatment which is all the more regrettable since these constituents play such an important rôle in various phases of biosynthetic and metabolic processes.

Acknowledgements. The authors are indebted to Mr. M. Martin for correcting the English text and to Dr. A. Kleczkowski for many helpful comments.

References.

AACH, H. G.: Z. Naturforsch. **12b**, 614 (1957); — Biochim. biophys. Acta **32**, 140 (1959). — AACH, H. G., u. G. MELCHERS: Biol. Zentrbl. **76**, 466 (1957). — ADAMS, M. H.: J. exp. Med. **76**, 175 (1942). — ADLER, F. L.: J. Immunol. **76**, 217 (1955). — ANTWEILER, H.: Naturwissenschaften **46**, 360 (1959). — AUGUSTIN, R.: Int. Arch. Allergy **11**, 153 (1957); — Immunol. **2**, 148 (1959).

BANOVITZ, J., S. J. SINGER and H. R. WOLFE: J. Immunol. **82**, 481 (1959). — BARTELS, R.: Phytopath. Z. **30**, 1 (1957). — BAUM, W.: Serol. Museum Bull. **13**, 5 (1954). — BAWDEN, F. C.: Advanc. Virus Res. **2**, pg. 32. New York: Acad. Press Inc. Publ. 1954. — BAWDEN, F. C., and B. KASSANIS: Ann. appl. Biol. **28**, 107 (1941). — BAWDEN, F. C., and N. W. PIRIE: Brit. J. exp. Path. **18**, 275 (1937); — The nature of virus multiplication, 2nd. Symp. Soc. Gen. Microbiol. pg. 21. Cambridge Univ. Press 1952. — BERCKS, R.: Virology **12**, 311 (1960). — BIER, M.: Electrophoresis. London: Acad. Books 1959. — BJÖRKLUND, B.: Int. Arch. Allergy **4**, 340 (1953). — BOYD, W. C.: Fundamentals of Immunology, London: Intersci. Publ. (1956). — BOYDEN, A. A.: in Serological approaches to studies of protein structure and metabolism. Rutgers Univ. Press 1954. — BOYDEN, S. V.: J. exp. Med. **93**, 107 (1951). — BRAKKE, M. K.: Arch. Biochem. Biophys. **55**, 175 (1955). — Advanc. Virus Res. **7**, 193 (1960). — BRANDES, J., and C. WETTER: Virology **8**, 99 (1959). — BRUYN OUBOTER, M. P. DE, J. J. BEIJER and E. VAN SLOGTEREN: Antonie v. Leeuwenhoek **17**, 189 (1951). — BUKANTZ, S. C., C. R. REIN and J. F. KENT: J. Lab. clin. Med. **31**, 394 (1946).

CARPENTER, P. L.: Immunology and serology. London: Saunders Comp. 1956. — CHADWICK, C. S., M. G. MCENTEGART and R. C. NAIRN: Immunology **1**, 315 (1958). — CHESTER, K. S.: Phytopath. **27**, 722 (1937a); — Quart. Rev. Biol. **12**, 294 (1937b). — CINADER, B.: Ann. Rev. Biochem. **11**, 371 (1957). — CINADER, B., and J. H. PEARCE: Brit. J. exp. Path. **36**, 515 (1955); **37**, 541 (1956); — Symp. Protein Structure by A. NEUBERGER. London: Methuen & Co. Ltd. 1958. — CLAYTON, R. M.: Nature (Lond.) **174**, 1059 (1954). — COHN, E. J. et al.: J. Amer. chem. Soc. **58**, 459 (1946). — COHN, M.: in Methods in Med. Res. Vol. **5**, 270. Chicago: Year Book Publ. 1952; — Serological approaches to studies of protein structure and metabolism. Rutgers Univ. Press 1954. — COHN, M., L. R. WETTER and H. F. DEUTSCH: J. Immunol. **61**, 283 (1949). — COHN, M., and A. M. TORRIANI: J. Immunol. **69**, 471 (1952). — COMMONER, B., N. YAMADA, S. RODENBERG, T. WANG and E. BASLER: Science **118**, 529 (1953). — COONS, A. H.: Ann. Rev. Microbiol. **8**, 333 (1954); — Int. Rev. Cyt. **5**, 1 (1956). — CREMER, M. C., D. H. M. VAN SLOGTEREN and J. A. VAN DER VEKEN: Proc. Eur. Reg. Conf. Electron Microscopy Vol. II, 974 (1960). — CROWLE, A. J.: Ann. Rev. Microbiol. **14**, 161 (1960). — CUSHING, J. E., and D. H. CAMPBELL: Principles of Immunology. New York: McGraw Hill Book Comp. 1957.

DEUTSCH, H. F.: In Methods in Med. Res. Vol. **5**, 284. Chicago: Year Book Publ. 1952. — DOERR, R.: Die Immunitätsforsch. Bd. VIII, Allergie. Wien: Springer 1951. — DORNER, R. W., A. KAHN and S. G. WILDMAN: Biochim. biophys. Acta **29**, 240 (1958).

EISLER, M.: Kleines Handbuch der Pflanzenanalyse **3**, 987 (1933). — ELEK, S. D.: Brit. J. exp. Path. **30**, 484 (1949).

FEINBERG, J. G.: Nature (Lond.) **177**, 530 (1956); **185**, 565 (1960). — FELTON, L. D., G. KAUFFMANN, B. PRESCOTT and B. OTTINGER: J. Immunol. **74**, 17 (1955). — FRAENKEL-CONRAT, H., and B. SINGER: Biochim. biophys. Acta **24**, 540 (1957). — FRANKLIN, R. E., and K. C. HOLMES: Biochim. biophys. Acta **24**, 540 (1956). — FREUND, J.: Am. J. clin. Pathol. **21**, 645 (1951). — FULTON, R. W.: Virology **9**, 522 (1959).

GELL, P. H. G., S. T. C. WRIGHT and J. G. HAWKES: Nature (Lond.) **177**, 573 (1956). — GEORGE, W., and K. W. WALTON: Nature (Lond.) **192**, 1188 (1961). — GITLIN, D., C. S. DAVIDSON and L. H. WETTERLOW: J. Immunol. **63**, 415 (1949). — GLENN, W. G.: Serological and biochemical comparisons of proteins by W. H. COLE. Rutgers Univ. Press 1958. — GRABAR, P.: Ann. Rev. Biochem. **19**, 453 (1950); — Proc. 3rd. Int. Congr. Bioch., pg. 37, Acad. Press 1955; — Advanc. Protein Chem. **18**, 1 (1957). — GRABAR, P., W. W. NOWINSKI and B. P. GENERAUX: Nature (Lond.) **178**, 430 (1956). — GRABAR, P., and C. A. WILLIAMS: Biochim. biophys. Acta **17**, 67 (1953). — GOLD, A. H. et al.: Phytopath. **50**, 428 (1960). — GOLDWASSER, R. A., and C. C. SHEPARD: J. Immunol. **80**, 122 (1958).

HARRISON, B. D., and H. L. NIXON: J. gen. Microbiol. **21**, 569 (1959). — HEIDELBERGER, M., F. E. KENDALL and T. TEORELL: J. exp. Med. **63**, 819 (1936). — HEITEFUSS, R., D. J. BUCHANAN-DAVIDSON and M. A. STAHMANN: Phytopath. **50**, 198 (1960). — HOLT, L. B.: Development in Diphtheria Prophylaxis. London: W. Heinemann. 1950.

ISLIKER, H. C.: Ann. Rev. Proteinchem. **11**, 388 (1957).

JEENER, R., P. LEMOINE and C. LAVAND'HOMME: Biochim. biophys. Acta **14**, 321 (1954). — JOHNSON, M. A.: Serol. Museum Bull. **13**, 1 (1954).

KABAT, E. A.: J. Immunol. **77**, 377 (1956). — KABAT, E. A., and M. M. MAYER: Experimental immunochemistry. Springfield (USA): C. C. Thomas Publ. 1948. — KASSANIS, B.:

Brit. J. exp. Path. 24, 152 (1943); — Nature (Lond.) 188, 73 (1960). — Kirk, J. S., and J. B. Sumner: Proc. Soc. exp. Biol. (N.Y.) 29, 712 (1932). — Kleczkowski, A.: Brit. J. exp. Path. 22, 44 (1941); — J. gen. Microbiol. 16, 405 (1957). — Kohn, J.: Nature (Lond.) 183, 1055 (1959). — Korngold, L.: J. Immunol. 77, 119 (1956). — Krebs, E. G., R. E. Wright and D. E. Cleveland: J. Biol. Chem. 192, 555 (1951). — Kunkel, H. G., and S. M. Ward: J. biol. Chem. 182, 597 (1950). — Koot, Y. van, D. H. M. van Slogteren and M. C. Cremer: Tijdschr. Plantenziekten 60, 157 (1954).

Lewis, D.: Proc. roy. Soc. (Lond.) 140, 127 (1953). — Liener, I. E., and J. E. Rose: Proc. Soc. exp. Biol. (N.Y.) 83, 539 (1953). — Limasset, P.: Ann. Inst. Nat. Rech. Agron. 1, 1 (1952).

Maloney, P. S., and J. N. Hennesy: Bioch. J. 36, 544 (1942). — Mansi, W.: Nature (Lond.) 181, 1289 (1958). — Markert, C. D., and R. D. Owen: Genetics 39, 818 (1954). — Marrack, W.: Immunology 1, 251 (1958). — Maschmeyer, W.: Diss. Berlin 1927. — Matthews, R. E. F.: Ann. appl. Biol. 36, 460 (1948); — J. gen. Microbiol. 10, 521 (1954); — Plant virus serology. Cambridge: Univ. Press 1957; — Virology 12, 521 (1960). — Mez, C., and H. Ziegenspeck: Bot. Arch. 13, 483 (1926). — McPherson, C. F. C., and M. Heidelberger: Proc. Soc. exp. Biol. (N.Y.) 43, 646 (1940). — Mintz, R. M., and R. A. Lewin: Canad. J. Microbiol. 15, 6 (1954). — Moorhead, E.: Phytopath. 46, 498 (1956); 49, 151 (1959). — Moritz, O.: Ber. dtsch. bot. Ges. 49, 76 (1931); — Beitr. Biol. Pflanz. 22, 51 (1934); — Handb. Pfl.-Physiol. VIII, pg. 356. Springer Verlag: Berlin-Göttingen-Heidelberg 1958; — Serol. Museum Bull. 24, 1 (1960). — Moritz, O., and H. vom Berg: Biol. Zbl. 51, 290 (1931). — Morton, R. K.: Methods in Enzymology I, pg. 25, by S. P. Colewick and N. O. Kaplan. Acad. Press 1955. — Munoz, J., and E. L. Becker: J. Immunol. 68, 405 (1952).

Nagaraj, A. N., and L. M. Black: Virology 15, 289 (1961).

Oakley, C. L., and A. J. Fulthorpe: J. Path. Bact. 65, 49 (1953). — Oertel, C.: Nachr.-Bl. Dtsch. Pflanzsch. 18, 46 (1956). — Ouchterlony, O.: Acta Path. Microbiol. Scand. 25, 186 (1948). — Progr. Allergy 5, 1 (1958). — Oudin, J.: Compt. Rend. 222, 115 (1946); — in Methods in Med. Res. Vol. 5, 353. Chicago: Year Book Publ. 1952. — Owen, R. D., and C. L. Markert: J. Immunol. 71, 238 (1954).

Pappenheimer, A. M. jr.: J. exp. Med. 71, 263 (1940). — Pappenheimer, A. M. jr., and E. S. Robinson: J. Immunol. 32, 291 (1937). — Pirie, N. W.: Bioch. J. 47, 614 (1950). — Polson, A.: Sci. Tools 5, 17 (1958). — Pope, C. G., M. F. Stevens, E. A. Caspary and E. L. Fenton: Brit. J. exp. Path. 32, 246 (1951). — Porath, J.: Biochim. biophys. Acta 22, 151 (1956). — Porter, C. A.: Advanc. Virus Res. 6, 75 (1959). — Pressman, D., Y. Yagi and R. Hiramoto: Int. Arch. Allergy 12, 125 (1958).

Ramon, G.: Compt. Rend. 176, 276 (1923); — Rev. Immunol. 9, 134 (1945). — Ramon, G., L. Lemetayer, R. Richez and L. Nichol: Rev. Immunol. 3, 285 (1937). — Riggs, J. L., R. J. Seiwald, J. H. Burckhalter, C. M. Downs and T. G. Metcall: Amer. J. Path. 34, 1081 (1958). — Rockwell, G. E.: Ohio State Med. J. 39, 128 (1943). — Rohringer, R., and M. A. Stahmann: Science 127, 1336 (1958). — Rozendaal, A., and D. H. M. van Slogteren: Proc. 3rd. Conf. Potato Virus Diseases, pg. 20. Lisse-Wageningen 1957 (1958).

Sang, J. H., and W. R. Sobey: J. Immunol. 72, 52 (1954). — Scheidegger, J. J.: Int. Arch. Allergy 7, 103 (1955). — Schneider, I. R.: Science 117, 30 (1953). — Schramm, G., and B. Röttger: Z. Naturforsch. 14b, 510 (1959). — Schramm, G., and W. Zillig: Z. Naturforsch. 10b, 493 (1955). — Sevag, M. G.: Immunocatalysis. Springfield (USA): C. C. Thomas. 1945. — Singer, S. J.: Nature (Lond.) 183, 1523 (1959); — Chem. Eng. News 38, 55 (1960). — Singer, S. J., J. E. Fothergill and J. R. Shainoff: J. Amer. chem. Soc. 81, 2277 (1959). — Slatmann, W.: Planta (Berl.) 6, 277 (1928). — Slogteren, D. H. M. van: Proc. Sec. Conf. Potato Virus Diseases, pg. 45. Lisse-Wageningen 1954 (1955a); — Proc. Sec. Conf. Potato Virus Diseases, pg. 51. Lisse-Wageningen 1954 (1955b); — Acta Bot. Neerl. 4, 474 (1955c). — Proc. Int. Scient. Tobacco Congr., pg. 144. Bruxelles 1958. — Slogteren, E. van: Meded. N.A.K. 2, 6 (1945). — Slogteren, E. van, and D. H. M. van Slogteren: Ann. Rev. Microbiol. 11, 149 (1957). — Sumner, J. B.: Ergebn. Enzymforsch. 6, 201 (1937). — Sussman, J., and H. Pretschold: J. Amer. clin. Path. 23, 168 (1948). — Stavitsky, A. B.: J. Immunol. 80, 360 (1958). — Steere, R. L.: Advanc. Virus Res. 6, 3 (1959). — Strauch, D., M. S. Stender and H. Winter: J. Immunol. 82, 298 (1958). — Szperl-Seyfriedowa, H.: Poznansk Towarzystwo przyjaciol Nauk 8, 177 (1950).

Takahashi, W. N., and A. H. Gold: Virology 10, 449 (1960). — Takahashi, W. N., and M. Ishii: Nature (Lond.) 169, 419 (1952); — Am. J. Bot. 40, 85 (1953). — Tall, M. G., W. C. Price and K. Wertman: Phytopath. 39, 288 (1949). — Tempel, A.: Med. Landbouwhogeschool Wageningen 59, 1 (1959). — Tiggelman-van Krugten, V. A. H., C. M. Ostendorf-Doyer and W. A. Collier: Antonie v. Leeuwenhoek 22, 289 (1956). — Townsley, P. M.: Canad. J. Biochem. Physiol. 37, 119 (1959). — Treffers, H. P., M. Heidelberger and J. Freund: J. exp. Med. 86, 83 (1947).

VARTERESZ, W., and L. KESZTYÜS: Z. Immunforsch. **99**, 211 (1941). — VEKEN, J. A. VAN DER: Proc. Third. Conf. Potato Virus Diseases, pg. 37. Lisse-Wageningen 1957 (1958); — Tijdschr. Plantenziekten **66**, 1 (1960); — Proc. Fourth Conf. Potato Virus Diseases, pg. 153. Braunschweig 1960 (1961). — VOGT, P. K.: Z. Naturforsch. **15b**, 213 (1960).

WELLER, T. H., and A. H. COONS: Proc. Soc. exp. Biol. (N.Y.) **86**, 789 (1954). — WETTER, C.: Arch. Mikrobiol. **37**, 278 (1960). — WHITCOMB, R. F., and L. M. BLACK: Virology **15**, 507 (1961). — WIEME, R. J.: Studies on agar gel electrophoresis Thesis. Arscia N. V. Brussel, 1959. — WILLISON, S., M. WEINTRAUB and J. H. TREMAINE: Prot. IXth Int. Bot. Congr. Montreal (1959). — WILSON, G. S.. and A. A. MILES: Topley and Wilson's principles of bacteriology and immunity, 2 vols, 4th ed. London: Edward Arnold Publ. 1955. — WILSON, M. W., and B. H. PRINGLE: J. Immunol. **73**, 232 (1954); **77**, 324 (1956). — WRIGHT, S. T. C.: Nature (Lond.) **183**, 1282 (1959); **185**, 82 (1960).

ZANCO, M., and G. PRATESI: Bull. Soc. Ital. Biol. sper. **23**, 419 (1947).

Polarography and Tensammetry.

By

B. Breyer.

With 28 Figures.

It is the object of this article to acquaint the reader with the practice of polarographic analysis; it deals with an indispensable minimum only of theoretical information and its main aim is to enable those interested in its use, to carry out the procedure with a minimum of expense and disappointment. In keeping with this aim, the main stress lies on such matters as the construction and maintenance of simple, but completely adequate polarographs, the preparation and treatment of analytical samples, the taking of polarograms and their interpretations, common faults etc., whilst no attempt has been made to discuss the polarographic analysis of single elements or compounds. A number of monographs on the subject have been published which, together with the regularly published lists of polarographic papers, make it easy to find the detailed procedure to be followed in the analysis of a given substance or group of substances. These monographs and bibliographies are listed at the beginning of the references (page 492).

The techniques to be discussed in this article are: 1. Classical polarography, or polarography with direct potentials, or *d.c. polarography* for short; 2. Polarography with combined direct and alternating potentials, or so called *a.c. polarography*; 3. *Tensammetry*, an off-shoot of the a.c. method, permitting specifically the analysis of surface active compounds.

A. Principles of Polarography.

I. (A) D. C. Polarography.

1. General.

It has been mentioned in vol. I, p. 476 of "Mod. Meth. Plant Anal.", that polarography consists in recording current/voltage curves by means of a dropping mercury electrode (d.m.e.), using an arrangement shown schematically in Fig. 1; in it B represents a source of EMF (usually a storage battery or dry cells); P a potentiometer, consisting of a resistance wire and a sliding contact S; G a galvanometer; the electrolysis cell *Ce*, contains the sample to be analyzed. The cell embodies the d.m.e., i.e. a capillary tube C of very fine bore (about 0.05 mm), connected by a tube to the mercury reservoir R. Small droplets of mercury issue at regular intervals (3 to 6 seconds) from the orifice of the capillary. The mercury reservoir, and hence the capillary, is connected to one pole (most commonly the negative) of the battery, whereas the other pole is connected to a mercury pool M at the bottom of the electrolysis cell, or else to any type of commonly used reference electrode, such as the saturated calomel electrode (S.C.E.).

By moving the sliding contact S along the potentiometer wire the potential difference (E) applied to the cell is gradually increased and at the same time the current (I) flowing through the cell noted from the deflection of the galvanometer; by plotting the magnitude of the current against the value of the applied potential, current/voltage curves such as that shown in Fig. 2 are obtained in those cases where substances are present which can be either oxidized or reduced at the d.m.e. Instruments recording such types of curves automatically are known under the name of "polarographs", while the curve itself is called a "polarogram".

The polarographic method depends essentially on the use of the d.m.e., introduced in 1922 by J. HEYROVSKY, the originator of polarography. The main advantages of the d.m.e. are:

1. Owing to its small surface compared with that of the reference electrode, practically the whole potential drop appears at the mercury drop; in other words, the current/voltage curve shows only the processes occurring at the d.m.e and is independent of the processes occurring at the large reference electrode.

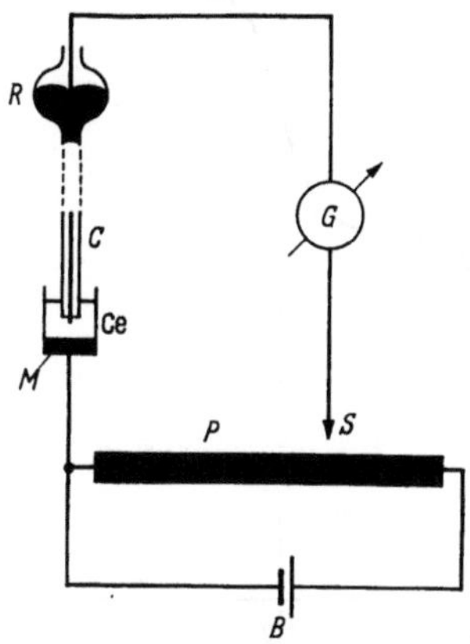

Fig. 1. Schematic representation of the polarograph.

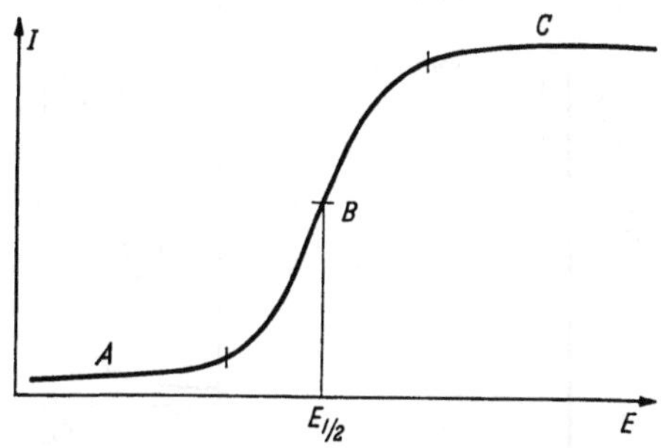

Fig. 2. Typical polarogram. A: Residual current; B: Decomposition current; C: Diffusion current.

2. The d.m.e. is self-renewing, i.e. contamination by products of electrolysis, or by adventitious impurities such as surfactants, is largely obviated.

3. In view of the small surface of the mercury drop, the currents passing through the cell are very small (10^{-5} to 10^{-7} A); thus, the amount of substance electrolyzed in a single measurement is neglegible and the procedure can be repeated as often as desired on the same sample with perfect reproducibility.

4. As distinct from solid electrodes, the surface of the mercury drop is always in the same physical state, always perfectly smooth, a further factor contributing to the excellent reproducibility of polarographic results.

5. Finally, of all the known electrodes, the mercury electrode has the highest hydrogen overvoltage, that is, hydrogen ions are discharged last of all (excepting only in frankly acid solutions); this means that analysis can be carried out in aqueous media, without the water being decomposed in the process.

Returning to Fig. 2, it can be seen that every polarogram shows at least one so called "step" which consists of three sections. During the first section (A) there is only a very slight increase of current with increasing applied potential; this part of the polarogram is known as "residual current". Then follows a section (B) when the current rises steeply over a relatively narrow range of potential, a section known as "depolarization current" or "decomposition current". Finally, in the third section (C), the current reaches a limiting value and does not increase further with increasing potential; this is the so-called "diffusion current" or "limiting current". A brief analysis of the electrode processes corresponding to the different sections of the polarographic step, will make it possible to understand the fundamental principles underlying the polarographic technique.

Any electrochemical reaction consists in the transfer of electrons from the electrode (cathodic reactions) or to the electrode (anodic reactions). Thus, in the simplest case, the cathodic reaction can be represented by

$$M^{n+} + ne \rightarrow M \tag{1}$$

that is, the cations of the electrolyte undergo reduction by accepting electrons from the electrode. This process, however, is not spontaneous, that is, the "electron pressure" (in other words, the applied voltage) must reach a certain minimum value before the electrons are forced onto the cation and reduction ensues. If the applied voltage is less than this critical value, all that happens is that the ions

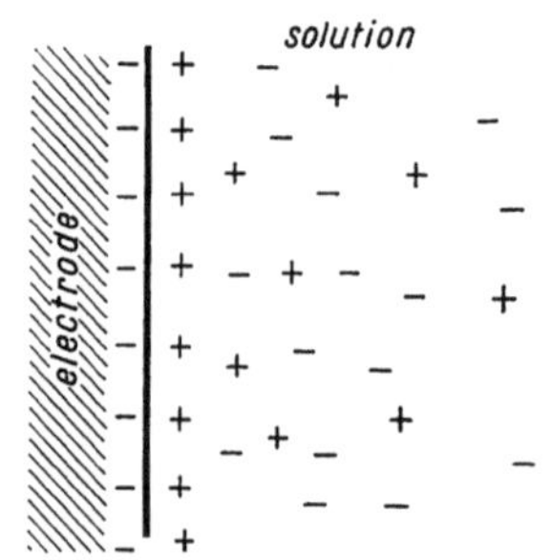

Fig. 3. The polarize de lectrode (cathode).

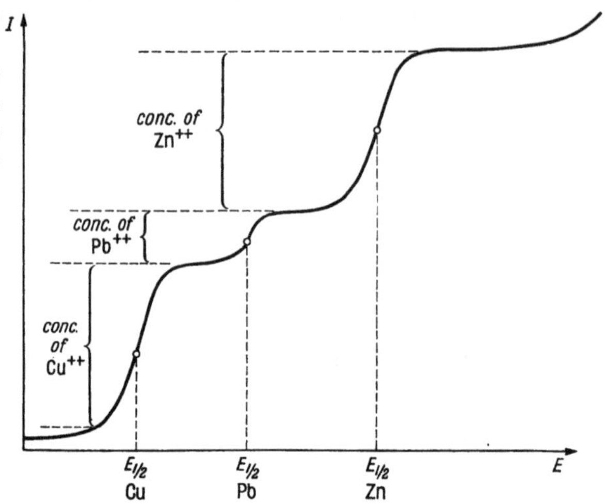

Fig. 4. Polarogram of a mixture of Cu^{++}, Pb^{++}, and Zn^{++}.

accumulate round the electrode by virtue of electrostatic attraction, forming as it were a "screen" of ions, without undergoing discharge (cf. Fig. 3). In other words, the applied potential is used to maintain the screen of oppositely charged ions round the electrode, but virtually no current flows through the cell. This state of affairs is known as "electrical polarization" and is the reason why the current remains essentially constant in section (A) of the polarographic step. The slight increase in residual current with increasing potential is due to the fact that the electrode together with the screen of ions surrounding it, forms an electrical condenser which is recharged every time a new mercury drop forms at the capillary, the number of charges increasing with increasing magnitude of the applied potential.

If the voltage is sufficiently raised, the ions start accepting electrons from the electrode and current flows through the cell continuously. In other words, the electrode is depolarized and the electrolyte decomposed by electrolysis; hence the name depolarization or decomposition current for section (B) of the polarographic step; the voltage corresponding to the decomposition current is known as "decomposition voltage". The value of the decomposition voltage depends on the ion species undergoing the electrochemical change: nobler ions (such as those of silver, mercury, copper etc.) are reduced at relatively small values of applied potential, whereas those of strongly electropositive elements (such as the alkalis and alkaline earths) are discharged last of all, i.e. at very negative potentials (about -2 volts vs. the S.C.E.). In polarography the value of the decomposition potential is measured at the point corresponding to half the step height (cf. Fig. 2); it is known as the "half-step potential" and given the symbol $E_{\frac{1}{2}}$. Since the half-step potential depends on the nature of the depolarizer, its measurement can be used to determine the species of depolarizer present in the solution (cf. Fig. 4). It should, however, be pointed out from the outset that polarography is much more commonly used for quantitative estimations than for qualitative analysis.

In reversible electrode processes, section (B) of the polarographic step is S-shaped, since it follows the NERNST equation

$$E = E^0 + \frac{RT}{nF} \ln \frac{[\mathrm{Ox}]}{[\mathrm{Red}]} \tag{2}$$

where E is the electrode potential, E^0 the standard potential of the same electrode, R the universal gas constant, T the absolute temperature, n the number of electrons transferred in the electrode reaction, F the Faraday, and [Ox] and [Red] are the concentrations of oxidant and reductant respectively; thus, the NERNST equation correlates the ratio of the concentrations of oxidant and reductant with the electrode potential.

In polarography the corresponding equation is that derived by HEYROVSKY and ILKOVIC (1935):

$$E = E_{\frac{1}{2}} + \frac{RT}{nF} \ln \frac{i_d - i}{i} \tag{3}$$

where i_d is the diffusion current and i is the current at that point of the polarographic step which corresponds to the applied potential E; $E_{\frac{1}{2}}$ ist the half-step potential. Hence, by plotting $\log(i_d - i)/i$ against E (the "log-plot" for short), a straight line is obtained, always provided that the electrode process is reversible; in these cases the slope of the line indicates the value of n, i.e. the number of electrons transferred in the electrode reaction.

Quantitative polarographic estimations depend on the measurement of the diffusion current (C, Fig. 2) and, hence, the understanding of its origin is paramount for the understanding of polarography. The constancy of current in this potential region of the polarogram means that the number of particles reacting at the electrode surface per unit time remains constant and is independent of the magnitude of the applied voltage. In the case of a cathode, for instance, all particles reducible in this region of applied potentials, immediately react on reaching the surface of the electrode; but new reducible particles will reach the electrode surface from the bulk of the solution, essentially by two mechanisms: (1) by electrostatic attraction between the electrode and the oppositely charged ions, and (2) by diffusion, because of the difference in concentration of reducible particles in the bulk of the solution and at the surface of the electrode. In the bulk of the solution the concentration of reducible particles has a finite value whereas at the surface of the electrode it equals zero, since no unreduced particle of that given kind, can subsist at the electrode in this potential region. Now, the electrode potential is being constantly increased during the polarogram and, hence, the electrostatic attraction between the electrode and the ions keeps also increasing, that is, the experimental conditions keep changing all the time, a most undesirable fact. In order to eliminate or minimize this effect, an excess of electrolyte whose ions are discharged only at very negative potentials, a so-called "supporting electrolyte", is added to the solution. Thus, the ion screen round the electrode consists mainly of the ions of the supporting electrolyte, i.e. of ions which are not discharged in the potential region corresponding to the polarographic step. The presence of these ions, however, screens the electrostatic attraction of the electrode from the bulk of the solution, so that in presence of the supporting electrolyte, the dischargeable ions reach the electrode practically exclusively in virtue of diffusion forces. Inspection of Fig. 5 should make the position clear.

The magnitude of the diffusion current will obviously depend on the number of particles reaching the electrode surface per unit time and all factors influencing this number will affect the current value. Thus, if the bulk concentration of the substance undergoing the electrochemical change, the so-called "depolarizer", is, say, doubled, then also its concentration gradient between the bulk of the solution

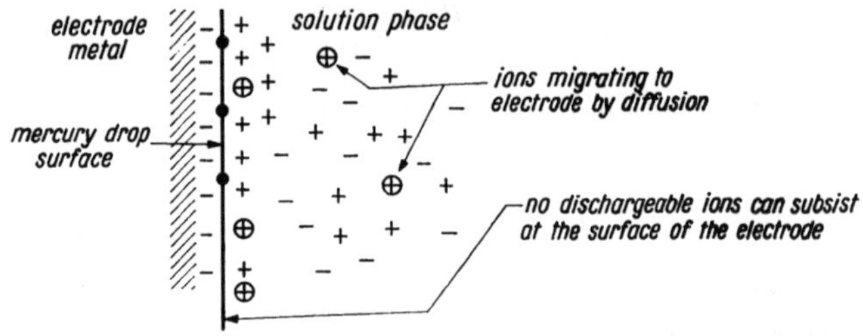

Fig. 5. Schematic representation of the d.m.e. with the surrounding "screen" of ions; +, — represent cations and anions of supporting electrolyte, ⊕ reducible cations, • metal atoms (discharged cations).

and the electrode surface increases twofold; that is, twice as many particles will now reach the surface of the electrode per unit time. But twice the number of particles accepting electrons from the electrode, means that the current flowing through the cell will be twice as large. In words: *the polarographic current is a linear function of the bulk concentration of depolarizer*. On this fact is based the use of polarography for quantitative analysis. For analytical purposes, a so-called "calibration curve" is taken (Fig. 6), that is, a plot of current against the concentration of depolarizer in a number of solutions of known composition. This calibration curve is then used in the analysis of samples having the same composition with regard of supporting electrolyte, but containing the depolarizer in unknown concentration.

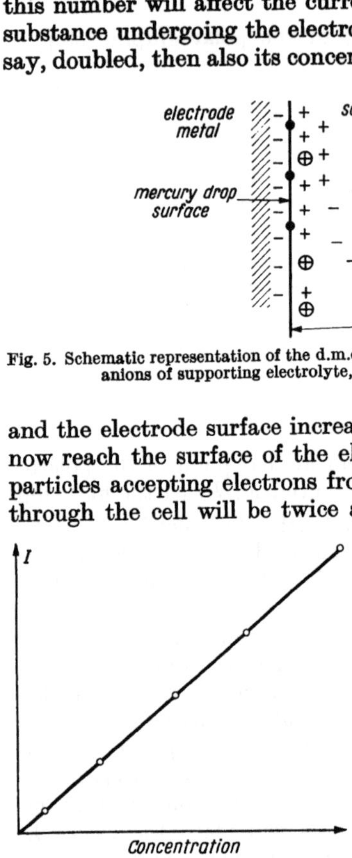

Fig. 6. Calibration curve.

Another factor which will influence the number of particles reaching the electrode per unit time, is obviously the diffusion coefficient of the particles themselves. It is obvious that the larger the diffusion coefficient of the depolarizer, the more particles will reach the electrode per unit time and the larger will be the polarographic current. For similar reasons, the polarographic current increases with temperature (about 2%/° C), with decreasing viscosity of the medium, and it is clear that stirring also will have a marked (but generally inconstant) effect.

The factors governing the diffusion current are quantitatively expressed by the ILKOVIC (1934) equation:

$$i_d = 0.627\, n F D^{\frac{1}{2}} m^{2/3} t^{1/6} C \tag{4}$$

where n and F have the same meaning as in eq. (2), D is the diffusion coefficient of the depolarizer, m the mass of mercury flowing from the capillary per second,

t the drop time of the capillary in seconds, and C the concentration of depolarizer. It follows, therefore, that a calibration curve is valid only for a given capillary, connected to a mercury reservoir to be kept at the same height, since any variation in the reservoir height will affect the so-called "capillary characteristics", i.e. the values of m and t in the ILKOVIC equation.

2. Types of Polarographic Processes.

Anodic currents. Depolarizers wich can be oxidized at the d.m.e. yield anodic currents, i.e. the steps appear below the galvanometer zero line (Fig. 7). Examples of anodic currents are e.g. that due to the oxidation of Fe^{++} to Fe^{+++}, or that of the oxidation of ascorbic acid to the corresponding di-ketone. A number of anions also yield anodic polarographic steps; here, the condition for obtaining a defined limiting current is that the anion "captures" mercurous ions (freed by the d.m.e. in the region of positive polarizing potentials), by forming either insoluble salts or stable complexes with mercury. Examples of this type of reaction are the anodic steps obtained e.g. with Cl^-, Br^-, I^-, OH^-, S^{--}, $S_2O_3^{--}$, CN^-, CNS^-, etc.

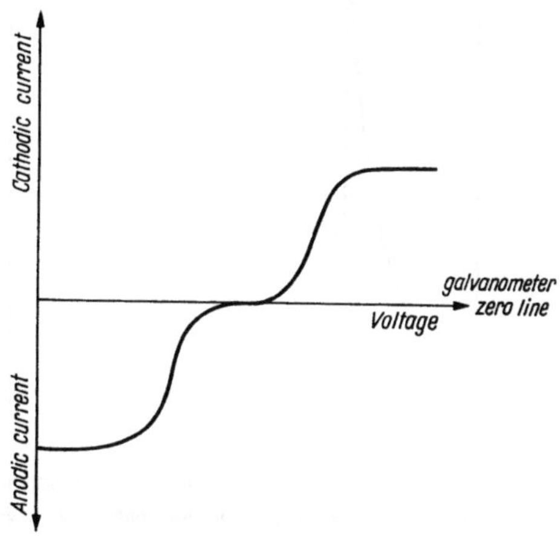

Fig. 7. Polarogram showing an anodic and a cathodic step.

Kinetic currents. It has been pointed out above that the height of a polarographic step is normally determined by the diffusion rate of the depolarizer; in other words, diffusion commonly is the slowest (i.e. the rate determining) process of a polarographic change. In 1947, however, WIESNER discovered polarographic steps whose height depends on the rate at which a depolarizer, in a secondary reaction, is changed chemically at the electrode surface. If now the rate of this change is slower than the rate of diffusion, the chemical change becomes the rate determining process, and the magnitude of the polarographic current will depend on the rate of this secondary change and not on the diffusion rate; such currents are known as "kinetic currents". An example of a kinetic current is that obtained with glucose; here the current depends on the relatively slow change of the cyclic form of glucose into the aldehydo-form at the surface of the electrode, and the height of the polarographic step is governed entirely by the rate at which this change takes place. Hence, polarography can be used to establish the rate of mutarotation processes and to calculate the concentrations of aldehydo-form present under various experimental conditions (LOS and WIESNER, 1953).

Kinetic currents can be readily distinguished from those controlled by diffusion; altering the height of the mercury reservoir changes the magnitude of the diffusion-controlled current according to:

$$i_d = k h^{\frac{1}{2}} \tag{5}$$

(where k is a proportionality constant and h the height of the mercury level above the tip of the capillary), whilst kinetic currents are independent of the height of the mercury vessel, or nearly so.

Catalytic currents. The presence of certain substances catalyzes reactions occurring at the d.m.e. Thus, e.g. a variety of substances produces so-called "catalytic hydrogen steps", i.e. catalyzes the discharge of H^+, sometimes at exceedingly low concentrations; examples are platinum salts, quinine and quinidine compounds, numerous alkaloids, etc. which give rise to catalytic hydrogen steps when present in concentrations as low as 10^{-6} to 10^{-7} M. One of the best known examples of catalytic currents is Brdicka's (1933) reaction for thiol groups: the step height due to proteins containing thiol groups, of cystine etc., increases over 500 times in the presence of Co^{++} ions as a catalyst. This reaction forms the

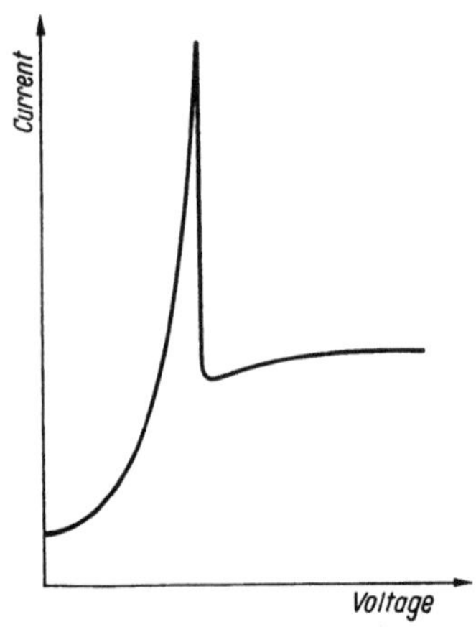

Fig. 8.

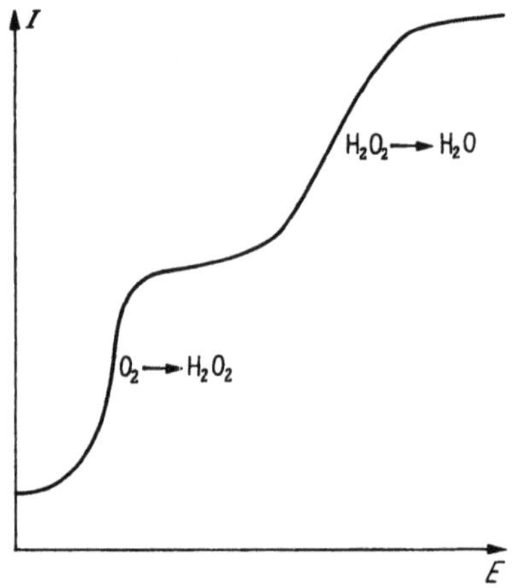

Fig. 9.

Fig. 8. A polarographic maximum.

Fig. 9. Polarogram of a solution containing oxygen (in presence of a maximum suppressor).

basis of the so-called "polarographic cancer test" (Brdicka, 1947). There are numerous examples of catalytic currents to be found in polarographic literature.

Current maxima. A number of depolarizers yield, under defined conditions, polarograms of a shape shown in Fig. 8, i.e. the current reaches a maximum, to fall abruptly to its normal value when a certain potential is reached; this type of curve is known as a "polarographic maximum". The origin of maxima is still widely discussed and will not be entered upon here.

The best known example of a polarographic maximum is that obtained with solutions containing dissolved atmospheric oxygen. Oxygen is reduced at the d.m.e. in two steps (cf. Fig. 9), the first leading to the formation of H_2O_2, while the second is due to the change of H_2O_2 to H_2O; the first of these two reduction steps frequently appears in the form of a maximum. Incidentally, the presence of these steps makes it imperative to remove oxygen from the sample prior to polarographic analysis; details will be found in the section dealing with the practice of polarographic analysis.

From a practical point of view, it is an important fact that maxima are easily suppressed by the presence of traces of surface active substances. This fact has a twofold consequence: firstly, the addition of surfactants makes it possible to suppress unwanted maxima in cases where their presence interferes with the proper evaluation of polarograms; secondly, and more importantly, the suppression of maxima by the addition of surfactants can be used for quantitative estimation of these substances. The method is extremely sensitive and has been successfully applied to the study of a great variety of substances, such as vitamins and hormones (VELDSTRA, 1949), of certain body fluids, such as lymph (GAWALOWSKI, 1941), and many more. Unwanted maxima are most commonly eliminated by the addition of surfactants, such as gelatin or methylcellulose.

3. Applicability of D.C. Polarography.

From the foregoing it will be clear that polarography can be applied to the study and analysis of a wide array of substances. Thus, the great majority of metal-ions yield polarographic steps; furthermore, as already mentioned, polarography can also be applied to the analysis of a variety of anions, whether by anodic polarography, or by cathodic reduction of such anions as ClO_3^-, BrO_3^-, IO_3^-, etc. At times, indirect polarographic analysis will yield useful results, when the direct procedure cannot be applied satisfactorily. Examples of indirect procedures are the estimation of sulphate ion by measuring the lowering of the lead step height after precipitation of some of the Pb^{++} by the addition of a solution whose SO_4^{--} ion content is to be estimated (HOHN, 1937); or, the indirect determination of calcium ion (the latter does not yield satisfactory polarographic steps) by measurement of the lowering of the step height of chloranilic acid (BREYER and MCPHILLIPS, 1953), and many more. Certain non-metals, such as oxygen and sulphur, can also be estimated by polarography.

Organic substances are susceptible to polarography provided they contain groups which are either reducible or oxidizable at the d.m.e. Examples of organic compounds reducible under polarographic conditions are: quinones, halogeno-compounds, unsaturated hydrocarbons and acids (provided they contain conjugated or cummulated double-bond systems), aldehydes, ketones, sugars (glucose, fructose), nitro- and nitroso-compounds, –S–S– and –SH compounds, a great number of heterocyclic compounds, and so on. Hence, a very large number of plant products is amenable to polarographic analysis. Here are some examples: tropone and tropolone derivatives, various alkaloids (e.g. quinine, morphine, narcein, cotarnine and hydrastinine, lobelin, etc.), certain vitamins (such as carotene, vitamins B_1, B_2, B_{12}, folic acid, vitamins C, E, K, etc.), some glycosides (e.g. cardiac glycosides and certain sapogenins), thiols and thiol derivatives (such as cysteine, cystine, glutathione, etc.), and many others. Detailed procedures for the analysis of these compounds will be found in the excellent monograph by BREZINA and ZUMAN (1958); the same monograph also contains copious tables of $E_{\frac{1}{2}}$ values of both inorganic and organic depolarizers.

The sensitivity of the polarographic method is high. The useful concentration range commonly extends from 10^{-3} to about $5 \cdot 10^{-6}$ molar; in certain cases, such as compounds giving rise to catalytic currents, it is possible to carry out polarographic estimations at concentrations as low as 10^{-6} molar, or even lower. Modern developments of the polarographic technique, such as a.c. polarography, and its off-shoots square-wave polarography and tensammetry, methods discussed in a subsequent section of this article, permit analysis in a concentration range as low as 10^{-8} molar.

D.C. polarography has an accuracy of about $\pm$ 2% in the concentration range down to about 10^{-4} molar; in the concentration range between 10^{-4} and 10^{-5} molar the accuracy is of the order of $\pm$ 5%. A still better accuracy can be achieved by the use of a.c. methods.

The high sensitivity of polarography makes it particularly suited to trace analysis. By the use of special electrolysis vessels (cf. Fig. 20), it is possible to analyze as little as $5 \cdot 10^{-3}$ mls. of solution; in other words, classical polarography permits analysis of depolarizer amounts as small as $5 \cdot 10^{-11}$ grams; a.c. methods extend the analytical range to amounts of about 10^{-13} grams and less.

4. Specialized Techniques.

Polarographic titrations. In 1929 Heyrovsky (Heyrovsky and Berezicky, 1929) introduced a new method of volumetric analysis, consisting in measuring the change in diffusion current of a depolarizer when titrated with a suitable reagent. Examples of this type of procedure are e.g. the titration of lead solutions with chromate (Kolthoff and Pan, 1939), of nickel with dimethylglyoxime

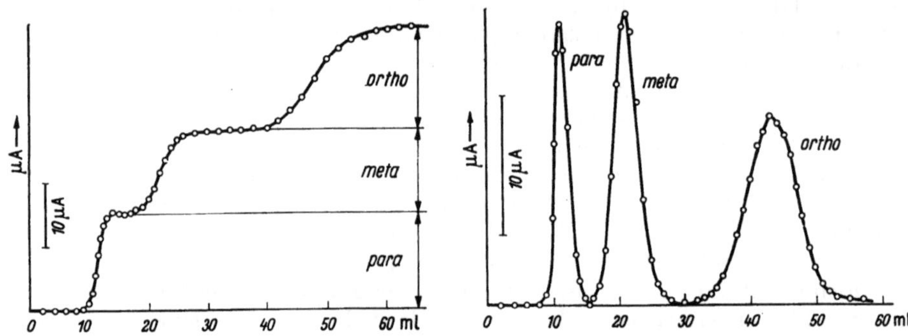

Fig. 10. (a) Chromatopolarogram by frontal elution. (b) Chromatopolarogram by elution of a column of separated o-, m-, p-nitranilines.

(Kolthoff and Langer, 1940), of copper and of silver with potassium ferrocyanide (Kalvoda and Zyka, 1950), and a host of others. In polarographic titrations, the polarizing potential is kept at a constant value (within the region of the diffusion current of the polarographic step) and the current reading plotted against the volume of titrant added. Further developments of this most useful technique are mainly due to Majer (1936) and to Kolthoff; the latter introduced the name "amperometric titrations" for this type of analysis.

The apparatus used for polarographic titrations will be found in the section dealing with the practice of polarography. The method is most useful for the estimation of small concentrations of metal ions; its accuracy is particularly high, of the order of $\pm$ 0.5%.

Chromato-polarography. As the name implies this technique, first introduced by Kemula (1954,1955), combines polarography with chromatography. In chromato-polarography, the eluate from a chromatographic column flows past a d.m.e. and the current is plotted against the volume of eluate. Typical curves are shown in Fig. 10. The method permits analysis of mixtures containing depolarizers of similar or identical reduction potentials, such as a mixture of ortho-, meta-, and para-nitraniline, etc.

Solid electrodes. In the region of positive polarizations the d.m.e is limited to a relatively small potential range (about +0.4 V *vs.* the S.C.E.). In order to extend the useful potential range, platinum micro-electrodes (Glasstone, 1931; Laitinen and Kolthoff, 1941) were introduced into polarography. Since, however, the attainment of diffusion equilibrium with stationary electrodes is rather slow, such variations as the rotating platinum-electrode (Laitinen and Kolthoff, 1941), the vibrating (Lindsey, 1952), and the "bubbling" micro-electrode

(GROZZI and DESIDERI, 1960) have been developed. Platinum micro-electrodes extend the useful range of positive polarizations up to about +1.2 V (*vs.* the S.C.E.); it should, however, be stressed that the reproducibility of results is generally less than that obtained with the d.m.e.

II. A. C. Polarography.

The a.c. polarographic method was first described by C. MACALEAVY in a Belgian patent in 1941. Independently, a similar procedure was developed by BREYER et al. (1945), who introduced the term "a.c. polarography" to designate the new technique. A. C. polarography consists in superposing onto the direct

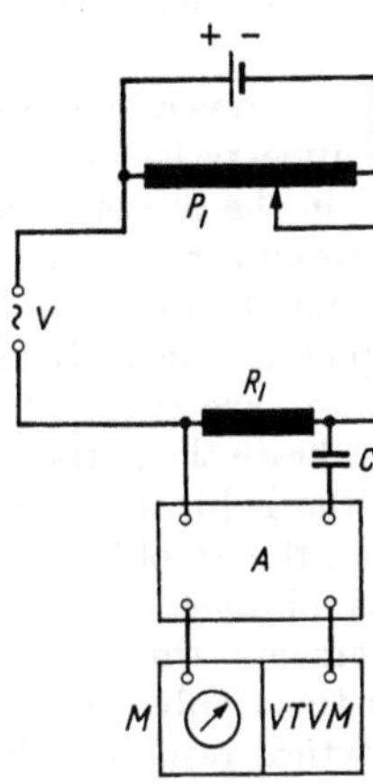

Fig. 11.

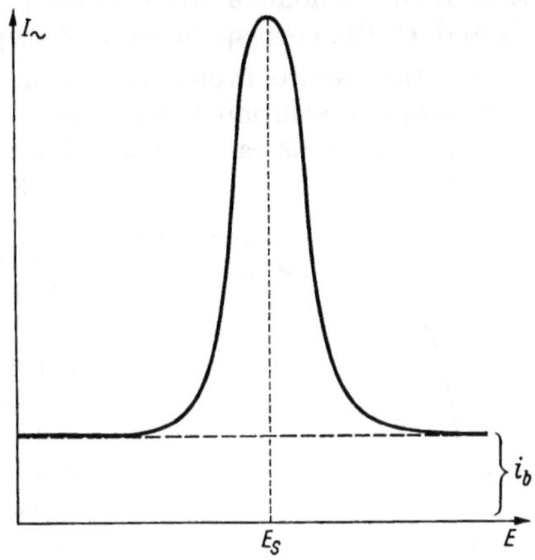

Fig. 12.

Fig. 11. Schematic representation of an a.c. polarograph. P_1, potentiometer; V, sinusoidal e.m.f. 1—30 mV; R_1, voltage-dropping resistor; C, condenser; C_x, polarographic cell; A, amplifier; VTVM, vacuum-tube voltmeter; M, microammeter.

Fig. 12. *A.C.* polarographic wave. E_s, summit potential; i_b, base current.

potential applied to the d.m.e. a small, sinusoidal, low-frequency alternating voltage (about 50—60 c.p.s. and 1—30 millivolts r.m.s.). The apparatus used is represented schematically in Fig. 11. It consists of the potentiometer P_1, from which the applied direct potential is derived, and the usual polarographic cell C_x; at V the alternating voltage is injected, so that both d.c. and a.c. current flow when the cell is depolarized. Instead of the galvanometer a voltage dropping resistor R_1 is used. The magnitude of the voltage drop across this resistor is, of course, a linear function of the current flowing through the cell. After filtering out the d.c. component (e.g. by means of the condenser C in Fig. 11), the a.c. component is amplified with a suitable amplifier A, and its magnitude read from the vacuum-tube voltmeter M.

A.C. polarograms are obtained by plotting the alternating current ($I_\sim$) flowing through the cell against the applied direct potential (E). An example of an a.c. polarogram is shown in Fig. 12. To describe the characteristic features of an a.c. polarogram, the following terminology was introduced: "summit potential", E_s, the direct potential corresponding to the maximum alternating current; "base current", i_b, the alternating current obtained in the absence of a depolarizer; "a.c. polarographic wave", the additional alternating current produced by the presence of a depolarizer. To avoid ambiguity, it is proposed to speak of "a.c. polarographic *waves*" and of "d.c. polarographic *steps*".

BREYER et al. (1945) soon found that in simple cases, such as the reversible reduction of many inorganic ions, the summit potential E_s coincides with the d.c. polarographic half-step potential $E_{\frac{1}{2}}$, and that, exactly as in conventional polarography, the alternating current is a linear function of the concentration of depolarizer. Moreover, it soon became evident that only reversible electrode processes yield a.c. polarographic waves.

The main advantages of a.c. polarography can be briefly summarized as follows:

1. Tedious curve plotting or recording can be obviated, if desired. All that is needed is to adjust the potentiometer P_1 until the meter M shows maximum deflection. If this is done M will indicate the concentration of the substance under analysis and P_1 the corresponding half-step potential.

2. Since the polarographic reduction of oxygen is an irreversible process (except in alkaline solutions), no oxygen-wave normally appears in a.c. polarograms, that is, analysis can be carried out in open vessels in the presence of air.

3. The analytical results are substantially more accurate and reproducible than those obtained with d.c. polarography. If results of the highest accuracy are required, it is good practice to recalibrate the outfit before every analysis with the help of a standard solution and to adjust the amplifier so that a fiducial reading is obtained, just as is done in pH measurements with the glass electrode; the ad justmentonly takes a few seconds and the analytical results so obtained are accurate to at least $\pm 0.5\%$.

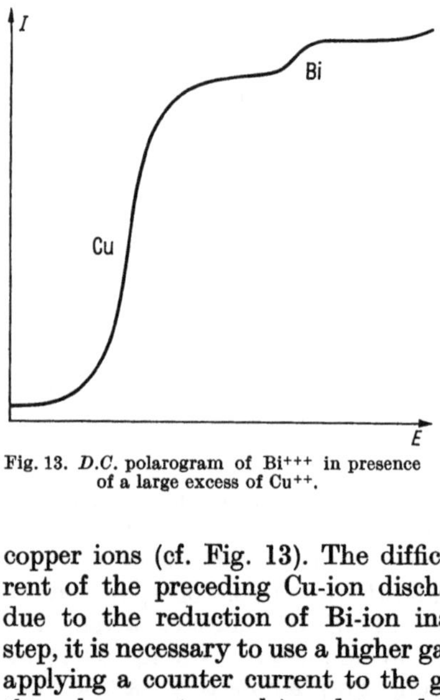

Fig. 13. *D.C.* polarogram of Bi^{+++} in presence of a large excess of Cu^{++}.

4. It is a notorious difficulty in ordinary polarography to analyze mixtures containing a great excess of nobler ions in the presence of traces of less noble ions. An example is the polarographic analysis of traces of bismuth in presence of a large excess of copper ions (cf. Fig. 13). The difficulty arises from the fact that the large current of the preceding Cu-ion discharge makes recording of the small current due to the reduction of Bi-ion inaccurate; in order to obtain a measurable step, it is necessary to use a higher galvanometer sensitivity which in turn involves applying a counter current to the galvanometer (so that the step occurs within the galvanometer scale) and to reduce excessive oscillations of the galvanometer (due to the dropping electrode) with the help of suitable condensers applied to the galvanometer terminals. It is obvious how all these complications can be avoided by the use of a.c. polarography, where the current returns to the base value after the peak potential is exceeded and the difficulties encountered in d.c. polarography do not arise.

5. In classical polarography it is difficult, or outright impossible, to obtain satisfactory analytical results when the half-step potentials of two substances are separated by less than 0.1 volts; the steps are either insufficiently distinct for accurate individual measurement, or else even fuse into one single step, making the evaluation of the single components impossible. In contradistinction, a.c. polarograms show separate peaks even in cases where the respective E_s values are only 0.045 volts apart. Inspection of Fig. 14 shows that a mixture containing

Pb, Tl, In, Cd, and Zn ions yields only three separate d.c. steps, whereas the a.c. polarogram shows every single component as a separate peak.

6. The height of a.c. waves is independent of the height of the mercury level in the reservoir.

7. The temperature coefficient of a.c. waves is smaller than that of d.c. steps; hence, thermostatic control is less critical.

The sensitivity of the a.c. polarographic method is limited by the magnitude of the base current (i_b, Fig. 12); the relatively high base current makes a.c. polarographic waves rather indistinct in a concentration range smaller than about $2 \cdot 10^{-5}$ molar; hence, a number of instruments have been designed to minimize the influence of the base current, thereby increasing the sensitivity of the a.c. method about 1000 fold. Examples of such instruments are Barker's square-wave polarograph (Barker and Jenkins, 1952), Jessop's "Univector" polarograph (1948), and the instruments developed by Niki (1955) and by Ishibashi et al. (1959). The principles underlying these ad aptations of the a.c. method, as well as their practical applications, will be found in the quoted literature.

A further shortcoming of a.c. polarography can be seen in the fact that only reversible electrode processes yield a.c. waves. However, the reversibility of a polarographic process often depends on the nature of the supporting electrolyte, and the judicious choice of the latter is, therefore, an important consideration in a.c. polarography. Examples how the supporting electrolyte might affect the reversibility will be found in a paper by Breyer, Gutmann and Hacobian (1951).

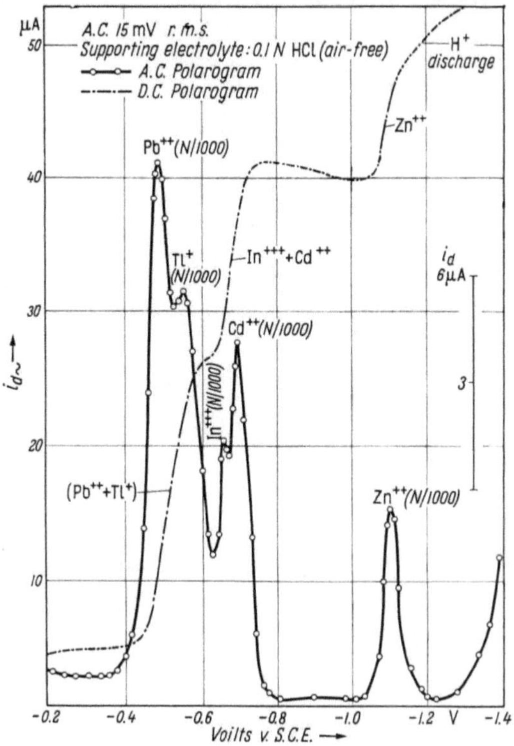

Fig. 14. *A.C.* and *D.C.* polarograms of a mixture containing Pb^{++}, Tl^{+}, In^{+++}, Cd^{++}, and Zn^{++}.

Although a.c. polarography is limited in its applications to reversible electrode processes, the fact that adsorption processes, not necessarily involving electron transfer, give rise to a.c. waves, greatly extends its scope. The importance of this fact and the new fields of investigation opened up, will be discussed in the following section.

III. Tensammetry.

In 1951 Breyer and Breyer and Hacobian (1952) found that solutions containing surface active substances give rise to large a.c. waves, whilst there is no corresponding increase in the d.c. polarographic current. The absence of a corresponding d.c. current means that no electrons are transferred in the electrode process, that is, the reaction occurring is not a polarographic one. It was soon realized that the a.c. waves in these cases are due to reversible adsorption/desorption processes taking place over a narrow potential range. To distinguish this type of wave from those due to a polarographic process, the term "tensammetry" (contracted from surface *tens*ion and *ammetry*, i.e. current measurement) was introduced. Tensammetric waves have been found independently also by Doss and Kalyanasundaram (1952).

Fig. 15 is a schematic representation of a tensammetric curve. It will be seen that a typical tensammetric graph consists of two waves and a considerably depressed current between them. This is not the place to discuss the theoretical

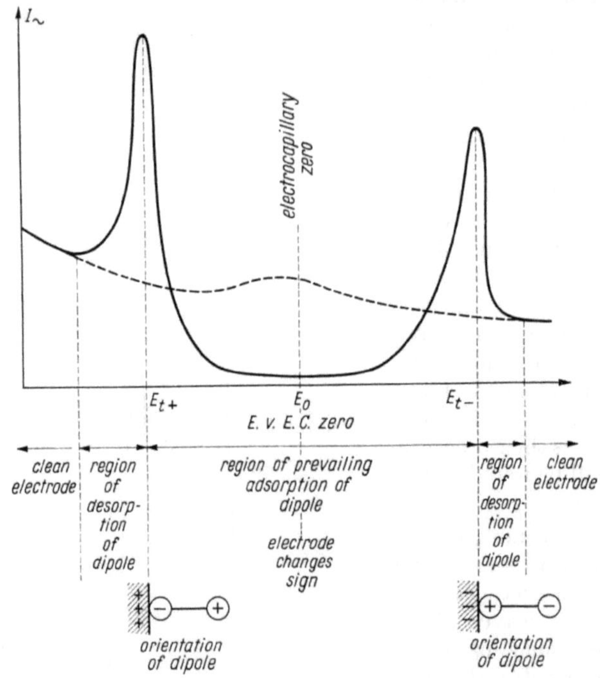

Fig. 15. Schematic representation of tensammetric waves. —— Tensametric curve; ---- polarogram of supporting electrolyte; E_{t+}, E_{t-} peak potential of positive and negative tensammetric waves respectively.

foundations of the tensammetric process, but as Fig. 15 shows, the mechanism can be visualized as one of dipole orientation and competing adsorption/desorption processes.

The height of tensammetric waves is not a linear function of the concentration of the surface active agent; graphs of the tensammetric wave-height against concentration of surfactant show the limiting form of adsorption isotherms. Nevertheless, it is quite possible to use tensammetry for analytical purposes, especially in the low concentration range where the calibration curves are practically linear.

Tensammetry constitutes an extremely sensitive procedure; thus, many substances classified as "insoluble" in water, frequently yield high tensammetric waves. Instances are e.g. the tensammetric waves obtained with anisol, octylalcohol, aniline, and many others.

It is a disadvantage of the tensammetric method that mixtures of surface active compounds cannot be analyzed; here, only the tensammetric wave of the most strongly adsorbed component appears. The possibility, however, exists of separating the surface-active constituents of a mixture chromatographically and of analyzing the eluate by tensammetry, a procedure which has been successfully adopted by KEMULA (1957). A further disadvantage is the fact that, at least so far, no tensammetric waves have been obtained with proteins; the reason is that proteins are so firmly bonded by the mercury of the d.m.e. that they remain adsorbed throughout the whole region of polarizing potentials.

Although numerous compounds have been examined by tensammetry (BREYER and HACOBIAN, 1952; OVERTON, 1957), the method is still in its infancy and much further work remains to be done. So far, tensammetric procedures have been developed for the analysis of essential oils (OVERTON, 1957), of mineral oils (AYCOCK, HOUGH and CRAWFORD, 1957), of tungstate ion (BREYER and HACOBIAN, 1957), etc. It can be foreseen that tensammetry should be useful in the analysis of many surface-active plant products, especially when combined with gas chromatography.

Combined A.C. Polarographic-Tensammetric Processes.

The majority of organic depolarizers undergo adsorption at the d.m.e. When the adsorption/desorption potentials and reduction potentials are sufficiently far apart, graphs result which contain both tensammetric and a.c. polarographic waves; these types of curves are of interest in the elucidation of electrode processes but do not concern us here.

Most commonly, however, the adsorption/desorption potentials and reduction potentials lie close together, so that only a single a.c. wave is evident; thus, this wave is the outcome of two processes occurring simultaneously, viz. adsorption/desorption and electron transfer, each contributing to a varying extent to the total alternating current observed. The total alternating current in these cases has been termed "rearrangement current" (BREYER, 1953). The result is that the alternating current obtained with organic depolarizers frequently is unusually large, i.e. the a.c. method generally is very sensitive for the detection and analysis of organic compounds. Although the calibration curves show a limiting form in the higher concentration ranges, the method is a sensitive and precise analytical tool at low concentrations, where wave height and concentration are linearly related. Substances such as chloranilic acid (BREYER and BAUER, 1955) or methylene blue (BARNES, 1953) can easily be detected in the concentration range of 10^{-6} and 10^{-7} molar, even by the simple a.c. polarograph without base-current suppression. The same instrument will serve very well for the estimation of riboflavin in the concentration range of $2 \cdot 10^{-7}$ to 10^{-5} molar.

Another promising application of tensammetry is the recent finding (BREYER, 1961; BREYER and HAYES, 1961) that the majority of complexogens used in analytical chemistry, yield tensammetric and/or a.c. polarographic waves in the region of positive polarizations and at very low concentrations. Work is in progress to make use of these waves for a.c. titrations; preliminary results show that the method permits estimation of a large number of inorganic ions in a concentration range well below 10^{-5} molar.

B. The Practice of Polarography.

Polarography is essentially a very simple technique. Provided that a few basic rules and precautions are observed, it is possible to obtain highly reproducible, accurate and reliable results. A further and most notable advantage of polarography is that instruments satisfying even the most rigorous requirements can be speedily and easily assembled from parts commonly available in every physico-chemical laboratory. These types of instruments, commonly known as "manual polarographs", will be described in detail in the following.

A great number of commercial firms produce D.C. and A.C. polarographs incorporating automatic registration. These instruments are of greatest use there where a great number of routine analyses are to be carried out. For ordinary laboratory work, however, manual instruments will be found convenient and entirely satisfactory.

Every polarograph essentially consists of two distinct parts: the polarographic (or electrolysis) cell and the measuring (or polarizing) outfit.

I. Measuring Outfit.

1. D.C. Polarography.

The essential parts of the measuring outfit are the galvanometer together with the shunt resistances, the potentiometer, and the source of e.m.f.

Galvanometer. A mirror galvanometer with lamp and scale is commonly used for polarographic measurements. A sensitivity of 10^{-8} to 10^{-9} A/mm. is desirable; the period should be of the order of 2 seconds and its inner resistance of a few hundred ohms. So called taut string galvanometers are particularly recommended because of their ruggedness; they are also relatively insensitive to vibration and consequently need no special mounting.

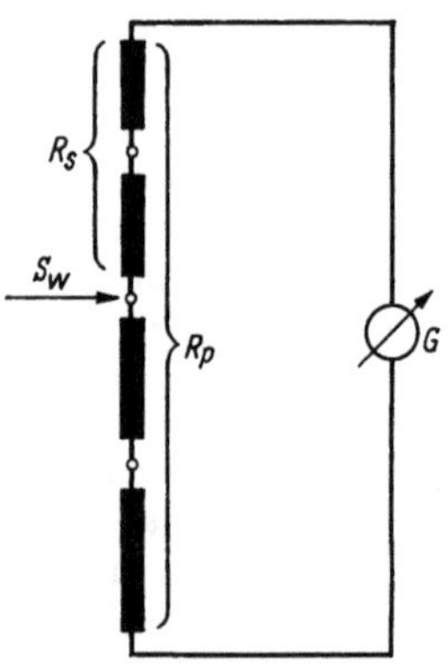

Fig. 16. Ayrton shunt. G, galvanometer; S, switch; R_s and R_p resistances.

The Ayrton shunt. The principle of the shunt is shown in Fig. 16. It consists of a number of resistances in parallel to the galvanometer G. The suitable value of the shunt is chosen with the help of the switch Sw.

The shunt is constructed from a single pole, multiple contact, rotary switch (known in the radio trade under the designation "two-deck" switch) by connecting suitable wire-wound resistors in series between the fixed contacts. If R_s indicates the value of the switched in shunt resistance and R_p the total resistance (cf. Fig. 16) then the relative sensitivity of the galvanometer is given by $S = \frac{R_s}{R_p}$. The value of R_p should be about the same as the critical damping resistance of the galvanometer used. Thus if the critical damping resistance of the galvanometer is, say, 1,000 ohms, a set of resistances of 10, 10, 30, 50, 100, 300 and 500 ohms when switched in successively, will yield relative sensitivities of $^1/_{100}$, $^1/_{50}$, $^1/_{20}$, $^1/_5$, $^1/_2$ and 1. For ordinary work the absolute value of these resistances is in no way critical. All that matters is that the calibration curves be taken with the same setting of the switch, i.e. at the same sensitivity as that used for analysis. Care should be taken, however, to choose good quality wire wound resistors whose resistance values will remain constant over an indefinite time. It goes without saying that resistance boxes to be found in every laboratory are eminently suitable as galvanometer shunts.

The potentiometer. If the outfit is to be used exclusively for quantitative analysis, that is, when no accurate potential measurements are required, the quality of the potentiometer is of secondary importance. Thus, we found that in ordinary routine d.c. work, cheap, wire wound radio potentiometers of about 100 ohms proved quite satisfactory. On the whole, however, helical potentiometers of the KOHLRAUSCH type are to be preferred. These should have 10 to 20 turns of wire and preferably a low resistance (about 20 ohms.); the latter is an important consideration in a.c. polarography (see later).

Source of e.m.f. Commonly storage batteries of 4 volts are used as a source of e.m.f. We find it very convenient, however, to use two "Size F" dry cells of 1.5 volt each, as a source of e.m.f.; with ordinary use these cells can be used for several months.

a) A Simple, Cheap Polarograph.

Keeping in mind what has just been said enables us to construct a simple polarograph at nominal cost.

The wiring is shown schematically in Fig. 17. P_1 is an ordinary radio potentiometer of 50—100 ohms, G is a galvanometer, P_2 and P_3 the shunt resistances, and V a high resistance voltmeter. The latter has 50 divisions and full deflection equals 2.5 volts. Polarograms are taken by moving the sliding contact along the resistance wire P_1 and noting the galvanometer deflection for each increase in potential of 25 millivolts (i.e. half a division on the scale). Since in quantitative polarographic work, all that is of interest is the step-height, it is sufficient to measure a few points before and after the step.

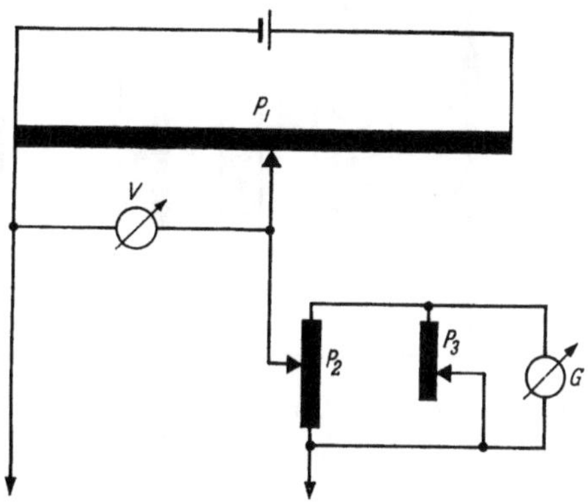

Fig. 17. A simple D.C. polarograph.

b) Useful Additions to the Simple Polarograph.

It has already been mentioned that for accurate potential measurements it is necessary to use a helical bridge of the KOHLRAUSCH type, instead of the cheap potentiometer. At the same time it will be necessary to incorporate a WESTON standard cell and to calibrate the bridge in the ordinary way. It should be stressed again, however, that this kind of arrangement is necessary only then when it is desired to measure half-step potentials accurately whereas it is superfluous for ordinary routine estimations.

There are, however, a few additions which are very useful in ordinary quantitative work. They will be discussed now.

Suppression of excessive galvanometer oscillations. At high galvanometer sensitivities the oscillations due to the periodic growth and fall of the mercury drops from the dropping electrode might prove troublesome. To damp these oscillations, LINGANE and KERLINGER (1940) have introduced electrolytic condensers of high capacity connected in the form of a RC net-work with a high time constant. The great merit of this arrangement is that it increases the effective period of the galvanometer without decreasing its sensitivity.

Galvanometer-zero adjustment and suppression of unwanted steps. A suitable arrangement of resistances enables part of the applied e.m.f. to be used for adjustment of the galvanometer zero (Fig. 18). The same arrangement can also be used for suppression of unwanted steps, as for instance in the case of preceding discharge of nobler ion-species at high concentration in the presence of small concentrations of less noble ions. This method provides useful results up to a roughly tenfold excess of nobler ions.

Combined anodic-cathodic polarograms. At times it is convenient to start the polarogram in the region of positive polarizations and to continue it through to negative polarizations, so that a graph is obtained containing both anodic and cathodic steps (cf. Fig. 7). Such a device incorporates a second potentiometer in parallel to the main one as shown in Fig. 18. The range of positive polarizations is regulated by the setting of the potentiometer CD (Fig. 18).

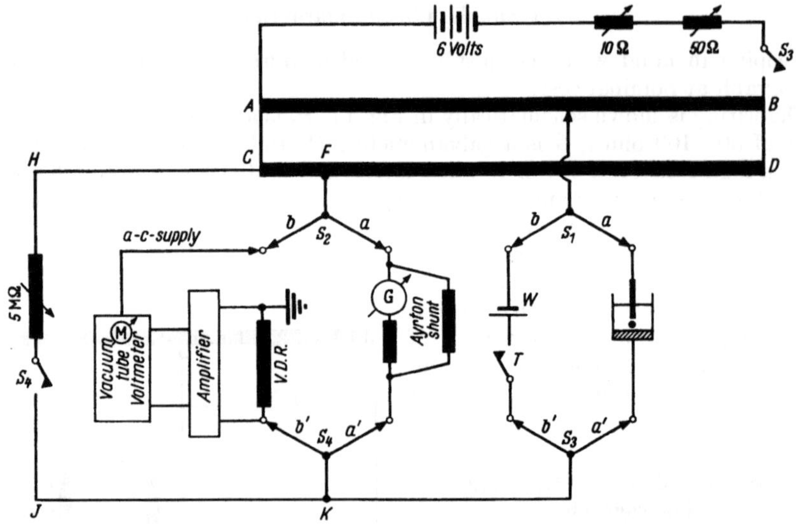

Fig. 18. Complete A.C. and D.C. polarograph. *AB*, variable potentiometer (about 20 Ω); *CD*, potentiometer, divided at fixed point *F* — total resistance about 20 Ω; S_1, S_2, S_3, S_4, switches; *T*, tapping key; *CHJK*, counter-current circuit; *W*, Weston standard cell; *G*, galvanometer; *VDR* voltage-dropping resistor; *M*, micro-ammeter.

2. A.C. Polarography.

Inspection of Fig. 11 shows that the main additions needed in order to convert an ordinary polarograph into an a.c. instrument are a source of alternating voltage, a preamplifier and a vacuum-tube voltmeter.

The alternating voltage supply. Most commonly the alternating voltage is derived from the mains supply via a step-down transformer and a voltage divider as shown in Fig. 19. Mains frequencies are entirely satisfactory for purposes of a.c. polarography; the amplitude of the superposed voltage is about 15 millivolts. It is an advantage to use some kind of stabilizing device between the mains and the step-down transformer; the device most commonly used is a constant-voltage transformer. It should be noted that the alternating voltage and the current show a linear relationship; hence, the alternating voltage should be kept constant to better than 0.5%, a condition easily satisfied by the use of any of the common constant-voltage devices.

Voltage-dropping resistors are most commonly used as input to the amplifier. Since the a.c. impedance of the cell is rather low (about 100 ohms) the value of the voltage dropping resistor must also be kept suitably low. The highest permissible value is about 100 ohms; resistances of about 4 to 25 ohms are the most commonly used. The resistances are connected between the terminals of a double-deck switch in an identical way to that used for the galvanometer shunt in d.c. polarography.

Suitable transformers can also be used with advantage as input to the amplifier. The primary side of such a transformer is inserted into the cell-circuit instead of the voltage dropping resistor while the input voltage is derived from the secondary side. When working with depolarizer concentrations of the order of 10^{-3} molar, the use of an input transformer might obviate the necessity for an pre-amplifier.

The pre-amplifier. The pre-amplifiers suitable for a.c. polarography are of the low frequency type. A gain of about 100 is required. Such amplifiers are commonly found in the radio trade; care should be taken, however, to choose a reliable type, incorporating high-quality component parts, which assure constancy of performance and output.

Vacuum-tube voltmeter. Any good quality, low-frequency, vacuum-tube millivoltmeter will be found satisfactory.

A self-built A.C. polarograph. For those who desire to build their own instrument a circuit diagram with details of component parts etc., will be found in a paper by Breyer, Gutmann and Hacobian (1953). However the use of ready-made amplifiers and voltmeters permits one to carry out a.c. polarography without any further preparations. Our own experience is limited to the Philips A.C. Measuring Amplifier GM 4574 and their low frequency voltmeter GM 6015; many other commercial firms produce similar instruments. When using these instruments, all that is needed is an alternating voltage supply (see above) which can also be easily derived from the power-pack transformer of the millivoltmeter used, with the help of suitable tappings.

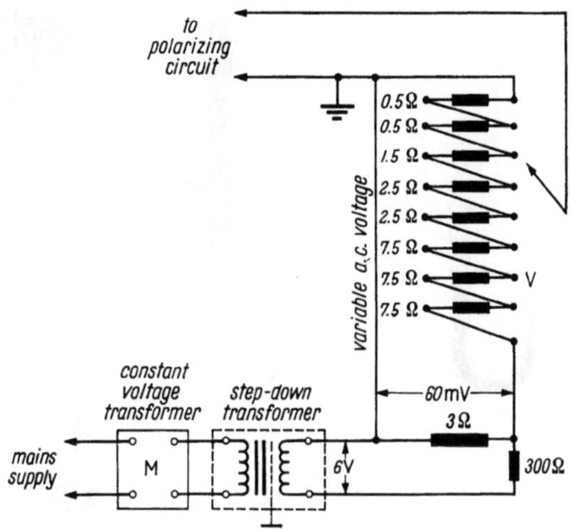

Fig. 19. A.C. voltage supply.

Fig. 18 shows a diagram of a complete a.c. and d.c. polarograph as currently used in these laboratories.

Proper shielding and earthing are most essential. No generally valid rules can be given; the best procedure will vary with the local circumstances.

It is readily possible to incorporate chart recorders with an a.c. polarograph. In order to do so provision is made for the potentiometer to be driven by a synchronous motor, using a suitable de-multiplication gear. It is most important, however, that the voltage increase should not exceed 2 millivolts per second, to make sure that the peak value of the a.c. current be recorded.

II. The Polarographic Cell.

Numerous types of polarographic cells have been proposed. Some of the most commonly used are shown in Fig. 20. Ordinary 50 ml. beakers without lip, fitted with a rubber stopper (cf. Fig. 21) are most conveniently used. Cell contact is usually made with the help of Pt-wires (sealed through glass tubing) making external contact with mercury. When connecting the cell to the mercury reservoir, care should be taken to avoid stranded wire; this type of wire acts as a wick for the mercury which, in time, will creep along the wire and do irreparable damage to the equipment, such as resistance boxes, etc.

Suitable capillary tubing is produced by a number of firms. The type most commonly used is that produced under the name of "marine barometer tubing" made by the Corning Glass Works, Corning, N.Y. A suitable length of this tubing (about 10 cms) is cut by filing a round groove with a glass file and making a clean

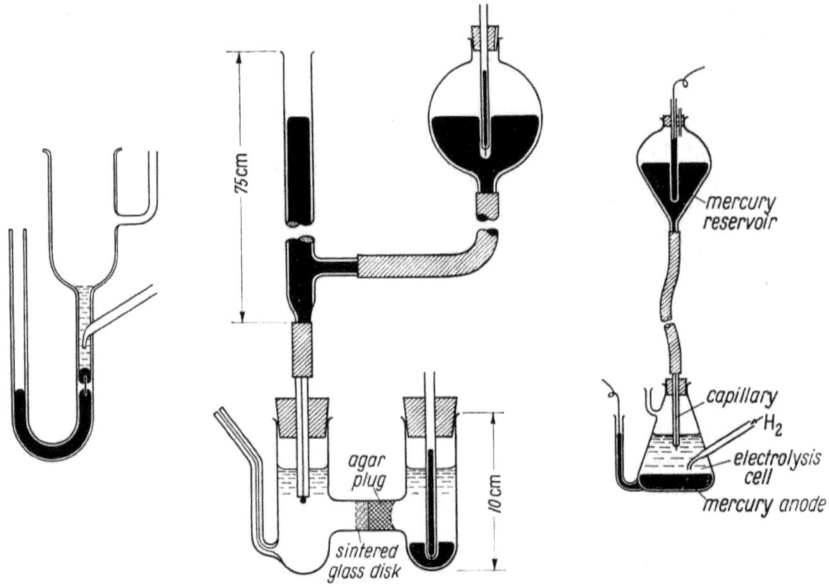

Fig. 20. Types of polarographic cells (from Kolthoff and Lingane "Polarography").

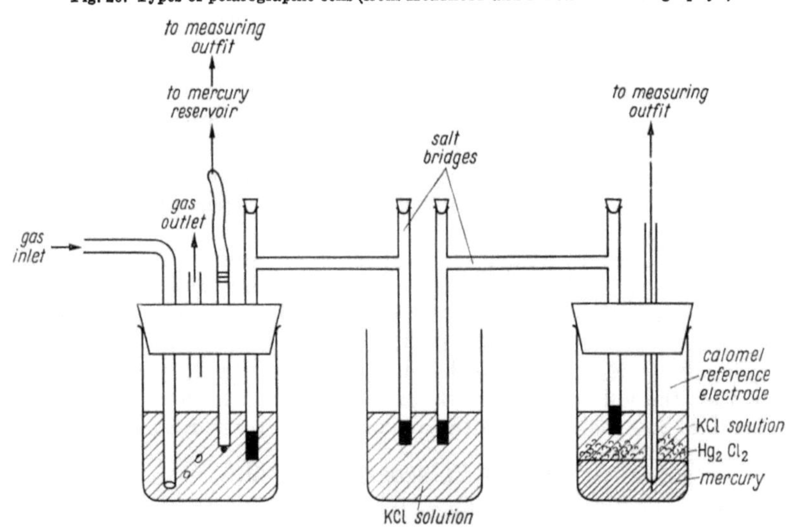

Fig. 21. Polarographic cell complete with reference electrode.

break. When it is desired to shorten the capillary by a small amount only, the breaking off is carried out with the help of a pair of pliers after a suitable groove has been made. In cases where it should prove difficult to procure the type of tubing described, it is possible to obtain a satisfactory capillary by cutting off the ends of an ordinary clinical thermometer and using it as a capillary.

To connect the capillary to the mercury reservoir, it is best to use tubing made from synthetic materials such as nylex, polythene, etc. In case it is desired to use rubber, this should be of the thick-walled pressure tubing type. It is necessary, however, to remove first of all the sulphur compounds contained in it, by boiling it for about 20 minutes with a 10% solution of caustic potash, washing and drying it carefully before use.

Regarding the mercury reservoir, it should be noted that traces of fat on its walls might in time seriously affect the performance of the capillary; hence, the reservoir should be carefully cleaned by the use of chrome-sulphuric acid, washed and thoroughly oven-dried before use. Wire loops are used to connect the tubing firmly, both to the capillary and to the mercury reservoir. Great care should be exercised that these connections be properly made, to avoid accidental slipping of the tubing and the most unpleasant experience of great quantities of spilled mercury. For the same reason the tubing of the mercury reservoir should have the usual ridges preventing accidental slipping of the tubing.

The reservoir is filled by pouring the mercury carefully and slowly along the wall of the vessel; great care must be taken to avoid the formation of air bubbles in the tubing by slapping and pinching the tubing until all the air is expelled. The use of semi-transparent plastic tubing makes it easier to detect air bubbles. The level in the reservoir should be adjusted in such a way that it is at a height of 40 to 60 cms above the tip of the capillary. It is good practice to use a rather wide mercury reservoir so that the level of the mercury changes but little over a period of time; an average capillary over an 8 hour working day consumes about 4 mls of mercury.

Next, the dropping rate is adjusted. For manual instruments a drop-rate of about 3 to 6 seconds is suitable; longer drop-times slow down the polarographic procedure unnecessarily. The drop-rate is checked by immersing the capillary in a salt solution (such as 0.1 molar KCl). The drop-rate can be, of course, adjusted by lowering or raising the mercury reservoir; if, however, the drop rate is too slow even at the maximum height of the mercury reservoir, the capillary must be shortened by cutting off lengths of a few millimeters at a time (using pliers to break off the short pieces, as described above) until the desired drop-time is attained.

A good capillary is the very heart and soul of polarography and thus great care must be taken to extend its useful lifetime as much as possible. One of the main precautions to be observed is that *a capillary must never be immersed into liquid unless the mercury is dropping from it.* It should be remembered that mercury has a negative capillarity whereas water has a positive capillarity; the consequence is that when an aqueous solution has penetrated into the capillary, the mercury will not be able to displace it and the capillary will stop dropping. To ensure the well-being of the capillary, therefore, the following points should be carefully observed:

1. The dropping of the capillary should never be stopped while the capillary is immersed into the liquid. The correct procedure consists in removing it from the solution while the mercury still issues, washing it with a washbottle, drying it thoroughly (not forgetting the tip) with a clean filter paper, and raising it to the height of the mercury reservoir, in order to stop the dropping. When not in use the capillary is kept, tip uppermost, near the mercury reservoir and covered by a small test tube.

2. Before use the capillary is lowered to the level of the cell, but immersed only after mercury is seen to issue from the orifice. If water should inadvertently penetrate into the capillary, there are a number of remedies open. First one should

attempt to force the mercury through the capillary orifice by squeezing the tubing near the top of the capillary thus forcing the mercury out; after the dropping has started again it is necessary to let the dropping continue for some time until it becomes completely regular. Frequently, however, such capillaries do not yield satisfactory polarograms; the effects to be expected are shown in Fig. 22. In cases where this procedure does not give the desired result it might become necessary to disconnect the capillary from the tubing, attach it to a suction pump and suck through it successively water, alcohol and ether, and finally drying it very carefully in a drying oven. If the capillary is beyond repair a new one must be mounted and re-calibrated with the help of a standard solution and by re-setting the potentiometer P_2 in Fig. 17 so that the step height is the same as that obtained with the previous capillary.

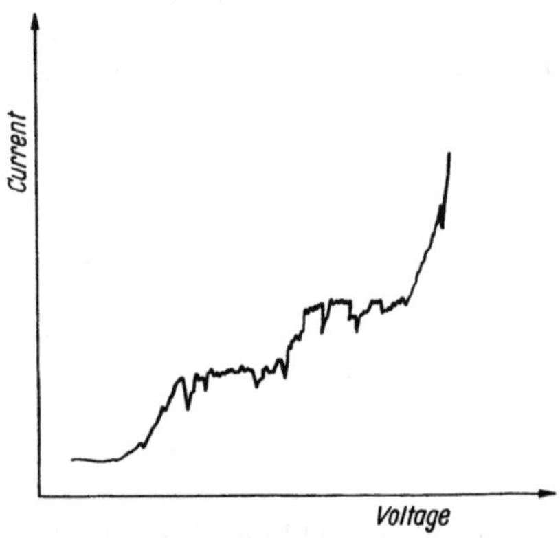

Fig. 22. Irregular tracings obtained with a faulty capillary.

In order to obtain reproducible results the capillary must be always clean; it is good practice to immerse it into concentrated nitric acid and wash it with distilled water after every run making sure, of course, that the dropping of the mercury continues during immersion.

A number of surface-active substances affect the regular dropping of the capillary. In these cases, siliconizing will frequently overcome the difficulty; this is carried out by immersing the dropping capillary for a few seconds into an ethereal solution of chloro-silicone, then into water, drying it and repeating the whole procedure once more; finally, the capillary is left dropping into pure water or salt solution for at least two hours. The film of silicone is invisible but is highly resistant to most chemical agents and it will be found that the effect of one siliconization lasts for about six months.

Although pool electrodes will be found satisfactory in most cases, an external reference electrode is usually preferable because the pool has to be renewed for every analysis, requiring large amounts of purified mercury. Most commonly a saturated calomel reference electrode is used; contact is made with the help of salt bridges in the usual way. Two bridges have to be used always as shown in Fig. 21. The bridge dipping into the analytical sample should contain the same electrolyte as that used in preparing the sample solution. It is most essential that reference electrodes and bridges should have a low resistance, a fact which is best ensured by having a large mercury surface in the reference cell and by using tightly rolled filter paper as plugs for the bridges. High resistances anywhere in the circuit will cause a flattening-out and diminished height of the polarographic steps.

In a.c. polarography, where the impedence has to be kept particularly low, it is necessary to shunt the salt bridges with the help of an electrolytic condenser of about 100 microfarads as shown in Fig. 23, an arrangement known as "pool-condenser electrode" (Breyer, Gutmann and Hacobian, 1951).

The used mercury is collected into a bottle, covered with dilute nitric acid solution (5—10% HNO_3) and frequently agitated. After a sufficient amount of mercury has been collected, it is separated from the acid in a separating funnel and washed successively with 10% nitric acid, distilled water, a 10% solution of caustic potash in alcohol and with water again; the mercury is then separated in a separating funnel, and, finally, filtered into the storage bottle through clean filter paper with a pin hole in it. The storage bottle must be carefully cleaned with chrome-sulphuric acid, washed and thoroughly dried in an oven. On rare occasions, such as when solutions of noble metals like gold, platinum, palladium, etc. have been electrolyzed, it might be necessary to clean the mercury by vacuum distillation.

With regard to solid electrodes the reader is referred to the text-books on polarography. It should be pointed out, however, that great experience is required in their use.

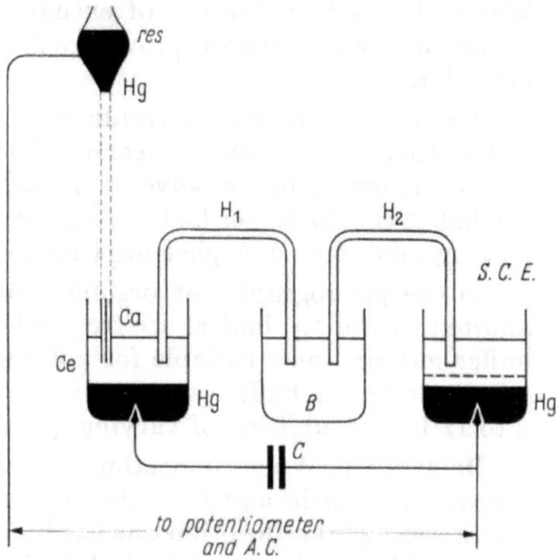

Fig. 23. The pool-condenser electrode.

III. Testing of the Outfit.

Before proceeding to polarographic analysis it is essential that the electrical part of the outfit be tested as to satisfactory performance. In order to do so, a carbon resistor of about 0.1 megohm is connected in place of the polarographic cell. Setting the shunt to different values, regularly increasing voltages (about 0.2 V at a time) are applied from the potentiometer and the resulting galvanometer deflections plotted against bridge readings; in a properly working outfit the plot should represent a straight line. If no straight line is obtained the fault might lie either in the quality of the potentiometer or in the galvanometer itself. In order to decide this issue, it is necessary to use a better quality bridge instead of the cheap ordinary potentiometer and to carry out the same operation again; in case a linear plot is now obtained, the potentiometer should be substituted by another one. If it is found that the fault lies in the galvanometer itself, one should make sure first of all that the scale of a lamp and scale model galvanometer be at least one meter distant from the galvanometer mirror, to avoid errors due to parallax.

The next step consists in establishing the sensitivity of the galvanometer; in purely analytical work of course it is unnecessary to know its exact sensitivity, since no actual current values but galvanometer deflections are plotted against the applied voltage. However, whenever it is required to convert galvanometer deflections into actual current values, the sensitivity of the galvanometer at the different shunt settings must be, of course, known; to do so, a known increase of potential is applied from the potentiometer and the galvanometer deflection noted for different settings of the Ayrton shunt, while a resistance whose value is accurately known is inserted in place of the cell. The sensitivity of the galvanometer is then simply calculated from Ohm's Law.

A.C. polarographs are tested in a completely analogous way. Here, different input (voltage dropping) resistors are successively chosen and the linearity of the amplifier response ascertained.

IV. The Taking of Polarograms.

Supporting electrolyte. Generally speaking, the reader will rarely be confronted with the necessity of having to choose the best suited supporting electrolyte, since, most commonly, all he has to do is to apply wellknown procedures, described in the literature. However, a few general remarks might be useful.

The solvent most commonly used is, of course, water; this should have undergone distillation from glass vessels, since the quantities of metal dissolved from metal stills are quite sufficient to produce adventitious steps in polarograms. Water obtained by the use of exchange resins is apt to contain surface active compounds which might prove troublesome, especially so in the case of a.c. procedures.

A number of organic solvents have also been used in polarography, such as water-ethanol mixtures, dioxane, dimethyl-formamide, and others. It is, of course, necessary to use solvents of high purity in all cases. In the case of ethyl alcohol, it should be verified that it does not contain acetaldehyde which yields a polarographic step at higher negative polarization (about -1.6 to -2.0 V).

In the polarography of organic substances, whose reduction potentials are affected by the p_H, buffers are frequently used as the supporting electrolyte. The buffer mixture most suitable for polarographic work is the PRIDAUX and WARD (1924) universal buffer which permits variation of the p_H range from about 2 to 12 by the addition of varying quantities of 0.2 molar NaOH.

De-aeration. When connecting the cell, care must be taken that good electrical contacts are made and that the capillary be immersed only after the mercury starts issuing from the orifice, as has been previously described; the rubber stopper on top of the beaker must be tightly fitted, and one must ascertain that the salt bridges are properly filled and make good contact with the solutions.

Dissolved oxygen must be removed from all solutions in d.c. polarography. The most common way of removing oxygen consists in passing an inert gas such as hydrogen or nitrogen through the solution to be analyzed. These gases as ordinarily supplied in cylinders are entirely suitable for polarographic purposes; washing is usually unnecessary, certainly in the case of hydrogen, but even the small amount of oxygen normally contained in commercial nitrogen-gas does not affect the results of polarography unless work must be carried out in the highest dilution range, i.e. 10^{-5} molar and below. It is, however, good practice to pass the gas through a washbottle containing the same solvent or solvent mixture as that in the polarographic cell, in order to prevent evaporation during the passing of gas. The gas should enter the solution near the bottom of the vessel; its flow is regulated at such a rate that it is still just possible to count the bubbles as they emerge from the inlet tubing. The time needed for gassing varies, of course, with the quantity and type of the electrolyte used; it is easy to remove oxygen from aqueous solutions, especially when they contain relatively high concentrations of supporting electrolyte, while the removal of oxygen from alcoholic solutions takes an appreciable time. The progress of de-aeration can be easily tested by setting the potentiometer to a value of about -0.4 V, passing the gas and checking the galvanometer deflection at regular time intervals, until there is no appreciable difference between two successive readings; it is necessary to stop the gas flow while the readings are taken. On the whole it will be found that a de-aeration time of 10 to 20 minutes is necessary.

In neutral or alkaline solutions it is possible to use sodium sulphite for the removal of oxygen. The addition of about 1% of solid sodium sulphite to the analytical sample (using, of course, a closed cell), will remove the oxygen in about ten minutes.

Before proceeding with the actual taking of polarograms it is necessary to stop the gas flow, since stirring of the solution while measurements are made must be strictly avoided. It is best, however, not to turn off the gas supply but to raise the gas-inlet tubing above the solution; the gas passing through the cell prevents air leaking into it.

Temperature control in polarography, although desirable, is not critical. It is usually sufficient to keep the temperature constant to ± 1° C; hence, any large container containing water of the required temperature will serve well as a thermostat. In cases where electrical thermostat controls are used, they should be switched off during the actual run; closing and opening of the temperature-control switch might cause irregularities in the polarographic curve, particularly in the case of a.c. polarography.

Before proceeding to the actual taking of a polarogram, the suitable setting of the shunt resistance must be decided; to do so, the potentiometer is adjusted to a voltage corresponding to the last plateau in the polarogram (i.e. before the final rise, due to the discharge of the ions of the supporting electrolyte) and the sensitivity of the meter adjusted in such a way that the reading is within the galvanometer scale.

We are now ready to proceed to the recording of the polarogram. The potentiometer is set to a potential value, about 100 millivolts below the beginning of the step whose height is to be measured, and then increased at regular intervals of about 20 millivolts or so, until the plateau after the step is reached.

V. Evaluation of Polarograms.

It has been repeatedly mentioned that the meter readings during the run of the polarogram show regular oscillations due to the dropping of the electrode. There has been much discussion about the moment during the swing of the meter

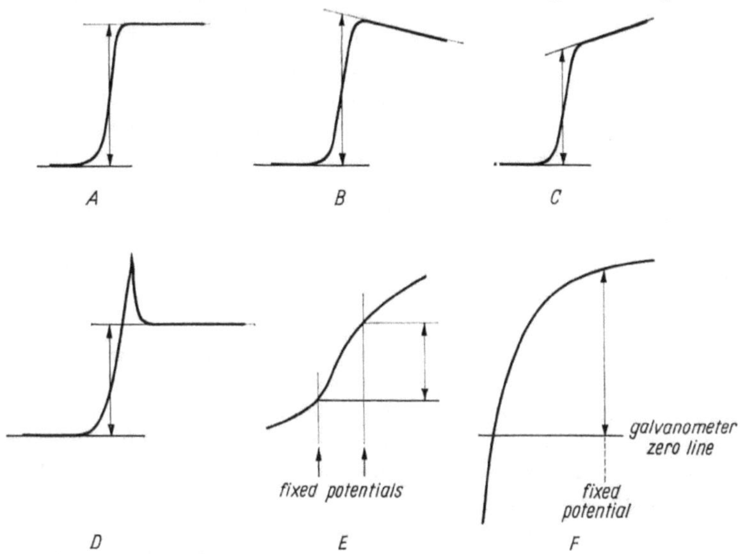

Fig. 24. Evaluation of D.C. polarograms.

at which the readings should be taken. This question is of very much less importance than it seems because all that is necessary is to use the same procedure in actual analysis as that employed during the calibration of the outfit. Although it is frequently stated that the average swing of the galvanometer reading should be measured, it is our experience that it is just as satisfactory (and, indicentally, much simpler) to take the peak values of the galvanometer readings.

Fig. 24 shows the different types of polarographic steps which might be obtained. To measure the step height, it is first of all necessary to correct for the residual current; this is usually done by simple extrapolation as shown in Fig. 24. For very accurate measurements at high meter sensitivities, it might be necessary to take the actual polarogram of the supporting electrolyte alone, in the same potential region as that where the step appears. Graph *D* in Fig. 24 shows that step heights can be evaluated even in the case where maxima are present, although it is good practice to eliminate unwanted maxima whenever possible. For this purpose maxima suppressors (such as 0.005% gelatin, methylcellulose, methyl red, and so on) are added to the analytical sample. In cases of very indistinct steps (cf. *E*, Fig. 24) satisfactory evaluation can often be obtained by measuring the currents at two fixed potential values, one before and one after the step.

It has been shown by VON STACKELBERG (1950) that even in cases where the beginning of the step cannot be seen (cf. *F*, Fig. 24) the step height can be satisfactorily evaluated by measuring the height from the galvanometer zero line.

It has been mentioned repeatedly that each dropping mercury electrode must be calibrated with the help of solutions containing known concentrations of the substance to be analyzed, i.e. a calibration curve as that shown in Fig. 6 must be taken. These graphs are used to read off directly the concentration of depolarizer in the unknown solution.

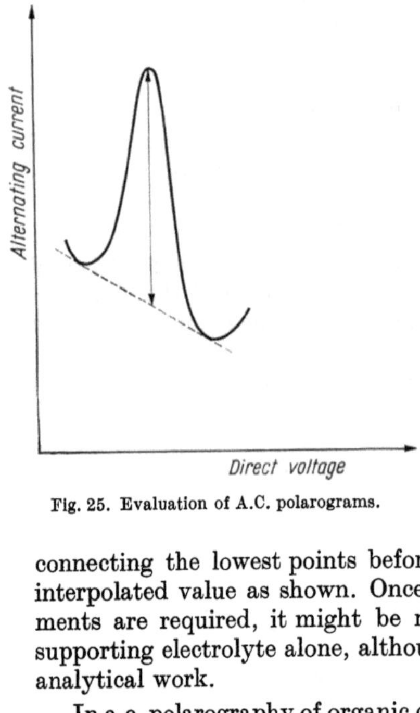

Fig. 25. Evaluation of A.C. polarograms.

The evaluation of a.c. polarograms is shown in Fig. 25; the base current before and after the wave frequently has a different value and the procedure for evaluating the wave height consists in drawing a straight line connecting the lowest points before and after the wave and measuring from the interpolated value as shown. Once again, where exceedingly accurate measurements are required, it might be necessary to take the a.c. polarogram of the supporting electrolyte alone, although this contingency will rarely arise in purely analytical work.

In a.c. polarography of organic compounds, the calibration curves are frequently non-linear in the higher concentration range. It is still possible, however, to analyze successfully at low concentrations of depolarizer, that is, in a concentration region where the calibration curves are practically linear; a.c. polarography of organic substances, although covering only a limited concentration range, shows an extremely high sensitivity.

In tensammetry, calibration curves also show a non-linear form, but once again it is quite possible to carry out successful estimations at low concentrations.

An alternative method consists in measuring the lowering of the base current on addition of surface-active substances (DOSS, GUPTA and RAO, 1952). This procedure, however, seems less accurate and less sensitive than measurement of peak heights, although it is very useful in those cases where the effect of the surfactant consists entirely in the lowering of the base current and no proper tensammetric peaks are produced (e.g. sugars, proteins, etc.).

VI. Polarographic Titrations.

In polarographic titrations, both titrant and analytical sample must be de-aerated. Different apparatus for this purpose has been proposed; a simple outfit satisfying all the requirements is shown in Fig. 26. Care should be taken that de-aeration be carried out as thoroughly as possible. It is good practice to continue passing inert gas for about 20—30 minutes before titration is started; as usual, the gas flow is stopped before each current measurement, to be continued again for about $^1/_2$ minute or so, after each addition of titrant.

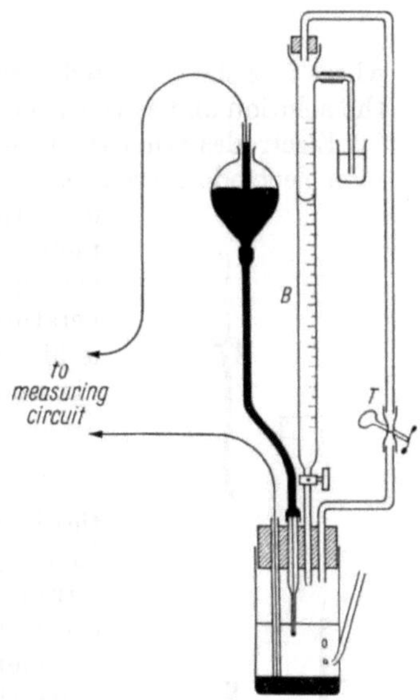

Fig. 26. Apparatus for polarographic titration.

Before beginning the titration itself, the potential is adjusted to a value corresponding to the plateau after the step of the substance to be titrated; the titrant is then run into the solution and the galvanometer reading plotted against the volume of the titrant added. Different forms of the graphs obtained are shown in Fig. 27. In cases where both the titrant and the solution produce a polarographic current, V-shaped graphs as that shown in (*c*), Fig. 27, are obtained; graphs *b* and *c* correspond to the cases where either the titrant only (*a*) or the titrated solution only (*b*) yield a polarographic current; graph (*d*) is obtained when an electro-oxidizable substance (yielding an anodic step) is titrated with a titrant which gives rise to a cathodic polarographic step; an example of such a titration is that of stannous ion with cupric ion (LINGANE, 1943).

Polarographic titrations are very speedy, since it is quite sufficient to measure the current at only two points before and two points after the end point as shown

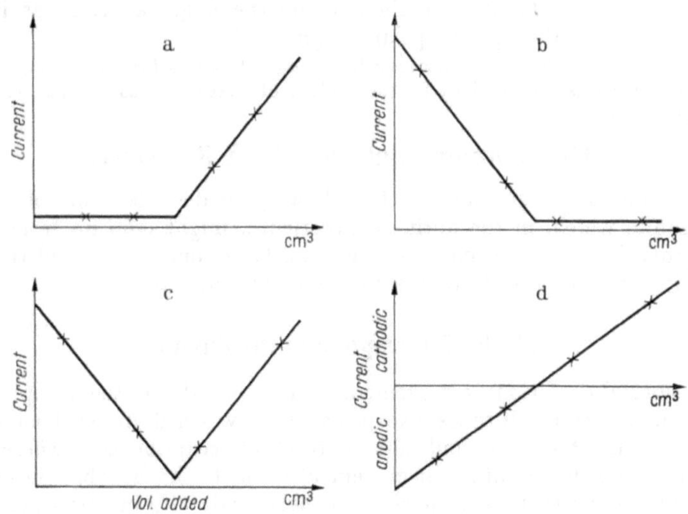

Fig. 27. Types of graphs obtained in polarographic titrations.

in Fig. 27. It must, however, be pointed out that straight lines as shown in Fig. 27 are obtained only then when the titrant is used at a concentration 10 to 20 times higher than that of the solution to be titrated; if the concentrations of titrant and solution are of comparable magnitude, curved lines are obtained, unless the current is corrected for dilution at each measured value using the expression

$$i_c = i_m \,(v_s + v_t)/v_s \tag{6}$$

where i_c is the corrected current value, i_m the measured current, v_s the volume of the solution and v_t the volume of the titrant added.

Electrodes other than the d.m.e., such as the rotating and the vibrating platinum electrode, have been frequently used for purposes of polarographic titrations; when the d.m.e. is used, it is unnecessary to know its characteristics or to recalibrate if the capillary should need changing, since only relative currents are being read; temperature control is also superfluous. The accuracy of results is higher than that obtained with ordinary polarography, viz. about ±0.4%.

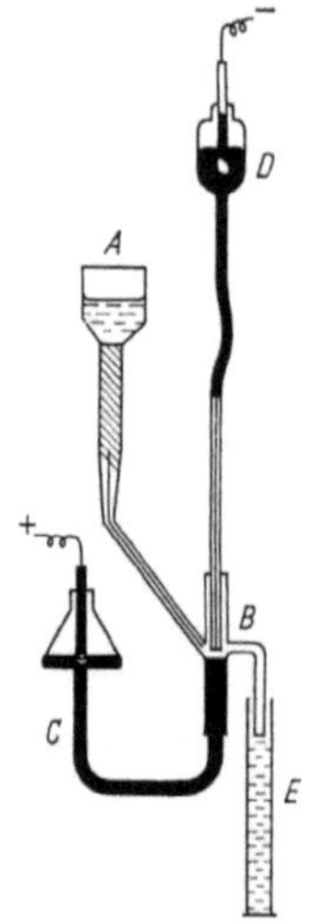

Fig. 28. Apparatus for chromatopolarography.

VII. Chromato-Polarography.

The apparatus used in chromato-polarography is schematically represented in Fig. 28. *A* is the chromatographic column, *B* the polarographic cell, *C* the mercury anode (arranged in such a way that the mercury level in the cell can be adjusted at will), *D* the mercury reservoir, and *E* a cylinder, measuring the volume of the solution which has flown through the cell. By plotting the polarographic current against the volume read from the cylinder *E*, graphs such as those shown in Fig. 10 are obtained; graph *A* results when the solution containing the mixture to be separated is passed through the chromatographic column continuously and the effluent analysed, whereas a graph as that shown in *B* results when the mixture is first separated in the column, then eluated and the eluate analysed as it passes through the polarographic cell.

Chromato-polarography has as yet been little used; it can be foreseen, however, that it will prove extremely useful especially for the analysis of mixtures of compounds with similar reduction potentials.

VIII. Common Faults and their Remedies.

An attempt has been made in the following to describe some of the more common faults which, in the author's experience, might crop up from time to time. A properly kept instrument will develop faults only rarely and the reader will have to make recourse to this section only exceptionally.

1. Total Interruption in the Circuit.

Under normal conditions, a movement of the meter due to the dropping of the mercury can be seen at all times, except at very low sensitivities; the absence of meter movement, therefore, indicates a total interruption somewhere in the circuit, unless the absence of meter movement is due to the fact that the dropping of mercury has ceased; this contingency has been dealt with in a previous section. In order to locate the spot where the interruption has occurred, a resistance of about 0.1 megohms is inserted in place of the polarographic cell; if now, on applying a voltage from the potentiometer galvanometer movement ensues, the fault lies in the cell circuit and the latter has to be tested as described below; if no movement ensues the fault might lie either in the cell circuit and/or in the measuring outfit.

To test the measuring outfit, the continuity of the leads is checked section by section with the help of a continuity meter, until the interruption is found. The points most commonly at fault are the connections of the battery to the potentiometer and the potentiometer contact itself. To test for faults in the galvanometer, the 0.1 megohm resistance is removed and the cell leads short circuited, while at the same time, an ordinary multimeter is connected in place of the galvanometer; if now, by applying a potential, the needle of the test-meter shows movement, the fault lies in the galvanometer. Galvanometer repairs are, of course, a job for the expert.

To search for faults in the cell circuit, the cell is disconnected from the measuring outfit. If the dropping of mercury from the capillary should have ceased one proceeds as described in the section dealing with the treatment of capillaries. Next, the sealed-through platinum contacts are inspected to see that they have not been inadvertently broken. After that, one has to make sure that there is no hidden air bubble in the salt bridges. Finally, the mercury thread in the capillary is checked for continuity with the help of a magnifying glass and the tubing connecting the capillary to the mercury reservoir checked for a hidden air bubble; to test this part of the circuit, it is easiest to lower the capillary right into the mercury pool and test for continuity with the help of a test-meter.

2. Irreproducibility of Results.

Inconstancy of results is most commonly due to bad contacts. Points to be particularly watched are loose connections, especially to and in the potentiometer; to check it, it is best to connect another potentiometer (a cheap radio-potentiometer will do) and see whether reproducible results are now obtained. If the potentiometer is not of the sealed-in type, it is most advisable to clean it at regular intervals; this is done with the help of a rag moistened with alcohol. One of the most common causes of irreproducible results (and very frequently overlooked) are bad battery contacts; it is essential that they should be regularly inspected and kept clean. In order to test the measuring outfit, once again it is best to connect a 0.1 megohm resistor instead of the cell, and to proceed in the way it has been described above; this procedure will also reveal bad contacts in the shunt switch which, on rare occasions, might be the reason for irreproducible results.

If testing of the measuring outfit with the help of the 0.1 megohm resistance yields linear response, the fault lies in the cell circuit. Some of these faults become evident immediately on inspecting the polarograms obtained; thus, drawn-out steps of reduced height are seen when a high resistance is present in the cell circuit. The most common cause for this type of fault is, once again, the presence of an air bubble in the salt bridges; some of the potentiometers have a large resistance in series with the bridge wire, in order to protect the galvanometer during potentiometric measurements and one must make sure that this high resistance is short-circuited during polarography. A trouble-spot frequently overlooked is the state of the mercury platinum-wire contacts of the cell. The tubing is apt to develop minute cracks which can hardly be discerned with the naked eye; if now electrolyte solution penetrates the cracks by capillarity, an additional e.m.f. is set up between the mercury and the copper wire immersed into it. The e.m.f. so produced might be as high as 0.3 V, causing a corresponding shift in all polarographic steps, as well as irreproducible step and wave-heights.

The most frequent source of trouble, however, is due to the capillary. The trouble most commonly found is irregularity of drop formation; it is often impossible to detect these irregularities by mere inspection of the dropping capillary but they are most evident from irregularities seen in the polarograms, such as those shown in Fig. 22. The most common reason for irregularities of drop formation is the adsorption of adventitious surface-active substances at the orifice of the capillary; the drop-rate usually increases and in extreme cases the mercury pours from the electrode, in a continuous stream. The remedy in these cases is proper cleaning of the capillary with concentrated nitric acid as described above.

A rather common source for irreproducible results is the slow penetration of air into the cell during polarography; the best way to avoid this contingency consists in raising the gas-inlet tube above the surface of the electrolyte and continue to pass the gas above the solution during analysis, as already mentioned. Inconsistent results might also be due to a diffusion of electrolyte from the salt bridge in to the cell; hence, it ought to be stressed again that the electrolyte used in the salt bridge connected to the cell should always consist of the supporting electrolyte used in the particular case. This consideration is particularly important in a.c. polarography with regard to supporting electrolytes containing chloride ions; this ion gives a very well pronounced a.c. polarographic wave round about zero volts vs. S.C.E. and it is essential that salt bridges containing chloride solutions should be marked as such and kept apart from those containing other supporting electrolytes.

Monographs and Bibliographies

Bibliography of polarographic literature 1922—1955. Chicago: E. H. Sargent & Co. 1956. — BREYER, B., and H. H. BAUER: A.C. Polarography and Tensammetry. New York: Interscience Publishers, in print. — BREZINA, M., and P. ZUMAN: Polarography in Medicine, Biochemistry, and Pharmacy. New York: Interscience Publishers 1958.

DELAHAY, P.: New Instrumental Methods in Electrochemistry. New York: Interscience Publishers 1952.

HEYROVSKY, J.: Polarographie. Vienna: Springer 1941.

KOLTHOFF, I. M., and J. J. LINGANE: Polarography. New York: Interscience Publishers 1952.

MILNER, G. W. C.: Principles and Applications of Polarography etc. London: Longmans, Green and Co. 1957.

Proceedings 1st International Polarogr. Congr. Prague, part II, Polarogr. Bibliography 1922—1950; Collection Czechosl. Chem. Commun., Supplement I, Polarogr. Bibliography, published annually.

STACKELBERG, M. v.: Polarographische Arbeitsmethoden. Berlin: W. de Gruyter 1950.

References

AYCOCK, W. P., E. W. HOUGH and G. W. CRAWFORD: J. Petrol. Technol. **9**, 53 (1957).

BARKER, G. C., and I. L. JENKINS: Analyst **77**, 685 (1952). — BARNES, G. T.: M. Sc. Thesis, University of Sydney, 1953. — BREYER, B.: Report of the 28th Meeting of A.N.Z.A.A.S., p. 192. Brisbane 1951. — BREYER, B.: Australian J. Sci. **16**, 109 (1953). — BREYER, B.: Australian J. Sci. **23**, 225 (1961). — BREYER, B., and H. H. BAUER: Australian J. Chem. 8, 472 (1955). — BREYER, B., and H. H. BAUER: Australian J. Chem. 8, 472 (1955). — BREYER, B., F. GUTMANN and S. HACOBIAN: Australian J. Sci. Res. A **4**, 595 (1951). — BREYER, B., F. GUTMANN and S. HACOBIAN: Australian J. Chem. **6**, 188 (1953). — BREYER, B., and S. HACOBIAN: Australian J. Sci. Res. A **5**, 500 (1952). —BREYER, B., and S. HACOBIAN: Anal. Chim. Acta **16**, 497 (1957). — BREYER, B., and J. W. HAYES: in preparation. — BREYER, B., and J. MCPHILLIPS: Analyst **78**, 666 (1953). — BRDIČKA, R.: Collection Czechoslov. Chem. Commun. **5**, 112, 148, 238 (1933). — BRDIČKA, R., cf. e.g.: Research **1**, 25 (1947). — BREZINA, M., and P. ZUMAN: Polarography in medicine, biochemistry, and pharmacy. New York: Interscience Publishers 1958.

DOSS, K. S. G., K. K. GUPTA and D. L. N. RAO: 21st Proc. Sugar Technol. Assoc. India, Pt. I, 40. (1952). — DOSS, K. S. G., and A. KALYANASUNDARAM: Proc. Indiana Acad. Sci. A **35**, 27, 173 (1952).

GAWALOWSKI, K.: Misc. Dermatol. Prague, 457 (1941). — GLASSTONE, S.: Trans. Am. Electrochem. Soc. **59**, 277 (1931). — GOZZI, D., and P. G. DESIDERI: J. Electroan. Chem. **1**, 301 (1960).

HEYROVSKY, J., and S. BEREZICKY: Collection Czechoslov. Chem. Commun. **1**, 19 (1929). — HEYROVSKY, J., and D. ILKOVIČ: Collection Czechoslov. Chem. Commun. **7**, 198 (1935). — HOHN, H.: Chemische Analysen mit dem Polarographen. Berlin: Springer 1937.

ILKOVIČ, D.: Collection Czechoslov. Chem. Commun. **6**, 498 (1934). — ISHIBASHI, M., T. FUJINAGA and A. SAITO: Polarography (Japan) **7**, 41 (1959).

JESSOP, G.: British Pat. 640, 768 (1948).

KALVODA, R., and J. ZYKA: Collection Czechoslov. Chem. Commun. **15**, 630 (1950). — KEMULA, W.: personal communication (1957). — KEMULA, W.: Chem. Zvesti **8**, 740 (1954); Rocznicki Chem. **29**, 643 (1955). — KOLTHOFF, I. M., and A. LANGER: J. Am. Chem. Soc. **62**, 211 (1940). — KOLTHOFF, I. M., and Y. D. PAN: J. Am. Chem. Soc. **61**, 3402 (1939).

LAITINEN, H. A., and I. M. KOLTHOFF: J. Phys. Chem. **45**, 1061 (1941). — LINGANE, J. J.: J. Am. Chem. Soc. **65**, 866 (1943). — LINGANE, J. J., and H. KERLINGER: Ind. Eng. Chem., Anal. Ed. **12**, 750 (1940). — LINDSEY, A. J.: J. Phys. Chem. **56**, 439 (1952). — LOS, J. M., and K. WIESNER: J. Am. Chem. Soc. **75**, 6346 (1953).

MACALEAVY, C.: Belgian Pat. 443,003 (1941). — MAJER, V.: Z. Elektrochem. **42**, 120, 122 (1936).

NIKI, E.: Polarography (Japan) **3**, 75 (1955).

OVERTON, RUTH: M. Sc. Thesis, University of Sydney (1957).

PRIDEAUX, E. B. R., and A. T. WARD: J. Chem. Soc. **125**, 426 (1924).

STACKELBERG, M. v.: Polarographische Arbeitsmethoden. Berlin: W. de Gruyter 1950.

VELDSTRA, H.: Bull. Soc. Chim. Biol. **31**, 594 (1949).

WIESNER, K.: Collection Czechoslov. Chem. Commun. **12**, 64 (1947).

Fallout Contamination in Plants.

By

J. V. Possingham and P. S. Davis.

With 4 Figures.

Fallout is a term which has come into widespread use since the advent of atomic fission and is used to describe the airborne debris that results from nuclear explosions.

This debris which consists of radioactive particles eventually settles to the earth's surface contaminating exposed surfaces such as soil, water and plants. From the initial deposition sites of the fallout, individual radioactive nuclides may pass through the food chains in significant quantities and ultimately create a hazard to the health of man.

Plants occupy a central position in the food chains. They may be consumed directly and they also provide food for the animals eaten by man. In this way the radioactive substances present in plants are passed to man. Plants are of particular importance in fallout studies because they can be contaminated both by retaining airborne fallout particles on their leaf surfaces and by absorbing radioactive compounds from contaminated soils through their root systems. However it should be noted that the levels of fallout activity required to limit plant growth by direct radiation damage are many orders of magnitude higher than the fallout levels significant in human food chains, and only occur in the immediate neighbourhood of a nuclear explosion.

The aim of this article is to discuss some of the factors that influence the fallout contamination of plants and to describe methods for the estimation of radioactive contamination at levels commonly found in plants at the present time. Methods are given for the assessment of gross contamination and also for fractionating the total burden into the individual components of greatest significance in human food chains.

A. Factors which Influence the Fallout Contamination of Plants.

I. Properties and Characteristics of Fallout.

The physical and chemical properties of fallout itself are important determinants of the degree and extent of plant contamination. Accordingly some consideration of these properties and of the types of fallout which occur is necessary.

Three main types of radioactive fallout can be distinguished (Libby 1956). The first is known as local fallout. This consists of large particles that contaminate areas adjacent to the site of an atomic explosion and which settle within a few hours. The pattern of deposition is in general an elliptically shaped area, downwind from the explosion site, whose magnitude depends on the type and size of device exploded and the weather conditions at the time of detonation. Fallout of this local type is extremely hazardous because of the intensity of its early radiation which stems largely from short lived radionuclides. LIBBY (1957) has stated that from land surface weapon tests approximately 80% of the total fallout is deposited as local contamination while in surface water tests this fraction is approximately 20%. The local fallout from weapons exploded high in the air is negligible, but contamination from a nuclear reactor accident for example, would be primarily of the local type.

The second type of fallout that can be defined is known as troposheric fallout. This consists of particles that are carried up no higher than the tropopause at the time of the explosion. These then travel many times round the earth in the troposphere and are deposited in a belt around the earth at latitudes fairly close to that of the explosion. Particles of this type are deposited mainly through the scavenging action of rain and snow and their mean residence time in the troposphere is approximately four weeks (MARLEY 1956). The major source of tropospheric fallout is from kiloton size weapons (usually A-bombs) but a small percent age of this type of fallout is produced by megaton bombs of the H-bomb type.

Table 1. *Principle Radioactive Nuclides Formed in the Slow Neutron-induced Fission of Uranium*-235 (STEINBERG and GLENDENIN 1955).

Nuclide	Mass number	Half life	Fission yield %
Kr	85	10.27 y	1.5
Sr	89	53 d	4.8
Sr	90	28 y	5.8
Y	91	61 d	5.9
Zr	95	63 d	6.0
Mo	99	67 h	6.1
Ru	103	40 d	2.9
Ru	106	1.0 y	.38
Cd	115	53 h	.011
Sb	123	136 d	.001
Te	127	90 d	.056
Te	129	33 d	1.00
I	131	8.1 d	3.0
I	133	20.8 h	6.6
Xe	133	5.27 d	6.6
Cs	137	33 y	5.9
Ba	140	12.8 d	6.3
Ce	141	33.1 d	5.7
Ce	144	282 d	5.8
Pm	147	2.6 y	2.6
Eu	156	15.4 d	.013

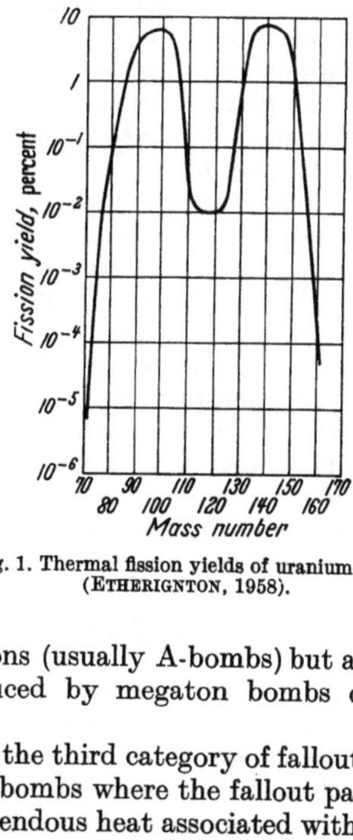

Fig. 1. Thermal fission yields of uranium-235 (ETHERIGNTON, 1958).

Stratospheric fallout is the name assigned to the third category of fallout. This results mainly from the detonation of megaton bombs where the fallout particles are carried up into the stratosphere by the tremendous heat associated with these weapons. Stratospheric fallout can also result from kiloton size weapons exploded at extremely high altitudes. From the stratospheric reservoir these particles fall to the earth's surface over a period of many years. LIBBY (1959) and LOPER (1959) have calculated a mean residence time in the stratosphere of two to four years. As a consequence of this long residence time stratospheric fallout contains predominantly long lived nuclides whereas local and troposheric fallout contain many short lived isotopes. There is a considerable lateral spreading of stratospheric fallout during deposition leading to the contamination of extensive areas of the globe.

All three types of radioactive fallout are similar in that the greater part of their radioactivity is due to fission products which are derived from the splitting of high atomic weight elements such as plutonium and uranium. Some idea of the range and relative abundance of radioactive elements produced by atomic fission is given in Table 1 (Steinberg and Glendenin, 1955), and shown graphically in Fig. 1. The values presented are the fission yields of uranium-235 split by slow neutrons and form only a rough guide to the chemical composition of fallout. Considerable quantitative variation can occur in the composition of individual fallout samples. A major source of variation is the age of the particular sample examined. Fallout consists of a complex mixture of radioactive nuclides, the individual components having half-lives varying from a fraction of a second to many years. Accordingly the chemical composition of a fallout sample will continually change with time, the long-lived radionuclides progressively occupying a larger percentage of the residual activity.

The physical properties of fallout derived from stratospheric and tropospheric sources are similar. Both types consist of small particles of colloidal dimensions which remain suspended in the air until moved down by the action of rain or snowfall. Little work has been carried out on the solubility of fallout samples of this type but tests have been made on the solubility of individual nuclides. Libby has stated that the strontium-90 in stratospheric fallout is fully soluble and freely available to plants. Local fallout, on the other hand, exhibits a considerable variation in particle size and solubility, dependent on the type of weapon exploded and the conditions of detonation. There is also some quantitative chemical variation in fallout dependent on the particular type of nuclear device exploded, but in general the range of nuclides produced is similar so that shown above.

In addition to fission products a number of neutron induced isotopes are found in fallout. Their nature and quantity depend on the materials surrounding the atomic weapon itself, the materials in the supporting tower if one is used and the nature of the ground surface. The proximity of the detonation to ground surface also has an important influence on the quantities of neutron induced isotopes produced. A further contribution to the radioactivity of fallout can come from residual fissile material such as uranium or plutonium. The quantities of these are usually low in fallout from nuclear weapons where the aim is to achieve complete fission. However in the case of contamination caused by a reactor accident the contribution of these alpha emitters could be high.

II. Plant Factors which affect Contamination.

Plants may be contaminated by radioactive fallout in two distinct ways. The most obvious way is by the retention on leaf surfaces of fallout particles that settle from the air. Initially the plant acts only as an inert support for deposited fallout. This fallout material however is freely available to grazing animals pastured on contaminated swards and represents a potent source of biologically available fission products (Marston 1958). With time, some of the individual radioactive compounds in the fallout particles trapped on leaf surfaces are absorbed into the plant itself. Once absorbed, these compounds can be translocated and accumulated preferentially in specific parts of the plant such as the growing points and developing grain and fruit (Middleton 1959).

The second mechanism whereby plants become contaminated with fallout is by the plant roots absorbing individual radioactive nuclides from contaminated soils. Over recent years during a period of virtually continuous generation of new fallout the contamination of plants by this route has been of a much lower order than that caused by direct foliar deposition. BURTON, MILBOURNE and RUSSEL (1960) have calculated that the direct absorption of the important isotope strontium-90 from the accumulated soil reservoir is responsible only for a small percentage of the total strontium-90 found in grass in the United Kingdom. Their calculations take into account many variables but indicate that plant contamination due to root absorption is, at the present time low. Many factors are already known which influence the uptake of particular nuclides from the soil. RUSSEL (1958) has shown that soil calcium level markedly influences strontium-90 uptake by plants; while FOWLER and CHRISTENSON (1959) have shown that soil potassium level is important in determining caesium-137 uptake by plants.

Many factors are also known which influence the amount of fallout material retained on the aerial portions of plants. The first of these is the particle size of the deposited fallout. From experiments carried out in areas of local contamination NISHITA and LARSON (1957) have reported that only fallout particles with an average diameter less than forty four microns are retained to any significant extent by plant leaves. Particles larger than this pass through the plant canopy to the underlying soil. Fallout particles derived from stratospheric and tropospheric sources are virtually all smaller than this critical limit and are substantially retained by a dense plant canopy.

ELLIS and MERCER (1960) have recently shown that a very high percentage of the individual nuclides in fallout can be retained by pastures. These workers have found that up to eighty five percent of the total strontium-90 deposited in Welsh hillside pastures can be retained by the herbage and underlying root mat. This value is an extreme one but indicated that plant foliage can very effectively retain deposited fallout.

A further factor that is important in plant contamination is plant species. In plants contaminated by the direct deposition of world wide fallout no large differences between species in their capacities to retain fallout have been substantiated. Generally, hairly leaved species are regarded as more efficient traps for fallout particles than smooth leaved species (LARSON, 1959); while perennial species because of their long growing periods can be more heavily contaminated than annual species. However RUSSELL and POSSINGHAM (1961) have shown that in an area of local fallout contamination in a mixed pasture, the amounts retained by grass *(Lolium perenne)* are approximately twice those retained by clover *(Trifolium repens)* at a series of heavily contaminated sites.

The plant species has been shown by ROMNEY et al. (1960) and by other workers to be an important determinant of the uptake of individual radionuclides from contaminated soils. These workers found that root uptake of strontium-90 from experimentally contaminated soils varied by a factor as high as ten, on the basis of dry weight with plant species. In the case of strontium-90 which is retained selectively in the upper soil layers shallow rooted plants are more heavily contaminated than deep rooted species.

B. Methods of Estimating Gross Fallout Contamination in Plant Material.

As a first indication of a sudden increase in the level of environmental radioactivity the determination of gross α and β-γ activity may be made. Although such measurements cannot be made quantitative they can provide useful information especially if the baseline level of activity is known either from samples collected previously or, if this has not been done, from samples taken from similar locations which have not been exposed to the particular contamination being investigated. In this way the radioactivity due to naturally occurring isotopes such as potassium-40 can be allowed for when estimating plant contamination levels.

The determination of unspecified gross activity is readily made with simple laboratory equipment. Although such measurements are of limited value when examining samples contaminated by world wide fallout, useful information can be obtained when examining samples contaminated by heavy depositions of local fallout. It must be stressed, however, that when a plant sample contains a mixture of unidentified nuclides there are major limitations to interpretation of gross radioactivity measurements. Self absorption losses are impossible to measure accurately and the counting efficiency of any given assembly for a mixed fission product sample can only be inferred from its performance with known isotopes (W.H.O. Report 1959). In practice, an arbitrary determination of the overall efficiency of the counter is made using a natural uranium source as an α standard and a potassium source as a β standard, and the sample ash is counted as an "infinitely thick" sample to minimise errors from self absorption[1].

Gross γ activity is not normally measured as such in plant material samples. γ activity is most efficiently measured by scintillation counters and even when the counting assembly is shielded by two inches of lead, natural γ backgrounds and photomultiplier noise are together much greater than the fission product activity normally encountered in biological samples. Furthermore, most γ active radionuclides are also β emitters and as such contribute to the gross β count. However, γ spectrometry using multichannel pulse height analysis is becoming widely used to estimate the individual γ emitters in plant samples. This technique will be discussed below when considering isotopes of iodine and caesium.

I. Counting.

Sample preparation (BANKS 1956). When collected, the plant samples are placed in polythene bags so as to avoid the loss of fallout particles in transit to the laboratory. If the samples include roots these must be washed to remove adhering soil. The samples are oven dried at 110° C for about 8 hours and are then ashed in a muffle furnace at 450—500° C until a greyish white ash is obtained. The ash is let cool, ground and passed through a B.S. 60 sieve (which has a sieve opening of about 0.25 mm). It is placed in a glass stoppered bottle and shaken for one hour. The sample is then ready for α and β counting.

Gross α counting. The most generally used method for determining the α activity of plant ashes is the ZnS scintillation counter. It can measure α disintegrations selectively in the presence of β and γ activities, and using simple equipment, counting efficiencies of 40% with background of less than 0.1 c.p.m. are attainable. Because α measurement is greatly influenced by self absorption and back-scattering it is usual to use "infinitely thick" sources of not less than 10 mg $\cdot$ cm^{-2} and to take steps to keep the moisture content of the ash as low as possible. The apparatus and methods most suited to estimating total α activity are described in detail by TURNER et al. (1958a).

It should be noted that there are naturally occurring α emitters of the thorium and uranium series that occur in plant material samples in significant quantities.

[1] It should also be noted that where weak pure beta emitters such as tritium, carbon-14 or nickel-63 are of interest gross β counting with a GEIGER tube is of little use because of sample self absorption and low counting efficiencies. In these cases, special methods such as gas flow counting or scintillation counting techniques must be resorted to (W.H.O. Report, 1959).

The most extreme case is that reported by TURNER et al. (1958b) who found quantities as high as 1690 $\mu\mu c/100$ gms of total α activity in brazil nuts due to naturally occurring α emitters. In general, however, the quantities of these naturally occurring isotopes are not high but precautions must be taken to assess the background level of α emitters in any plant contamination studies.

Gross β counting. Gross β counting is most commonly and most simply carried out using an end-window GEIGER counter. As "infinitely thick" samples are used for the determination of gross activity it is advantageous to use a tube with as large a window area as possible. The sample tube and sample holding assembly are placed in a two inch thick lead castle to reduce the background count. Lower backgrounds achieved by the use of anti-coincidence assemblies are not always essential for gross counts of plant ashes.

The plant ash is counted by mounting the sample in a cardboard container or perspex sample holder to a depth not less than 1.5 cm, the upper surface of the sample being level with the lip of the container. The area of the sample contained depends on the size of the GEIGER tube available.

As a comparative standard, a sample of A.R. potassium chloride similarly ground and passed through B.S. 60 sieve is placed in a cardboard or perspex sample holder to a depth of not less than 1.5 cm. It is necessary that both sample and standard fill the container used so that in each case the top of the source is the same distance from the end-window of the GEIGER tube.

Natural potassium, which contains 0.0119% of K^{40} gives 32 β and 3.4 γ d.p.s./gK.

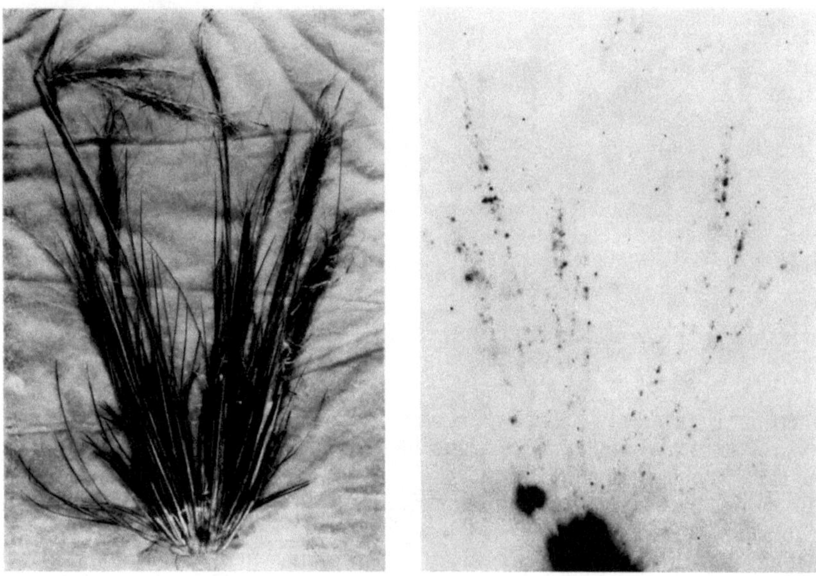

Fig. 2. Mounted specimen of *Stipa* sp on right, and autoradiograph on the left showing heavy accumulations of fallout in the basal cone area.

II. Autoradiography of Contaminated Plant Specimens.

This technique is used primarily to reveal the distribution of fallout contamination in individual heavily contaminated specimens. It is ideally suited for indicating in which plant parts absorbed fission products accumulate in greatest concentration. With care the sites of fallout particle retention can be demonstrated in freshly contaminated plants. Comparative levels of contamination are discerned using this technique rather than total quantities. In autoradiography the sensitivity of detection of α and β emitting nuclides is very much higher than it is for γ emitting nuclides.

The autoradiograph technique is only satisfactory for fallout detection in areas receiving local depositions of the order of 1 mc·m^{-2}. In areas where the contamination is primarily from world wide fallout sources the excessively long exposure times required and contribution of radioactivity from natural K^{40} make this technique unsuitable.

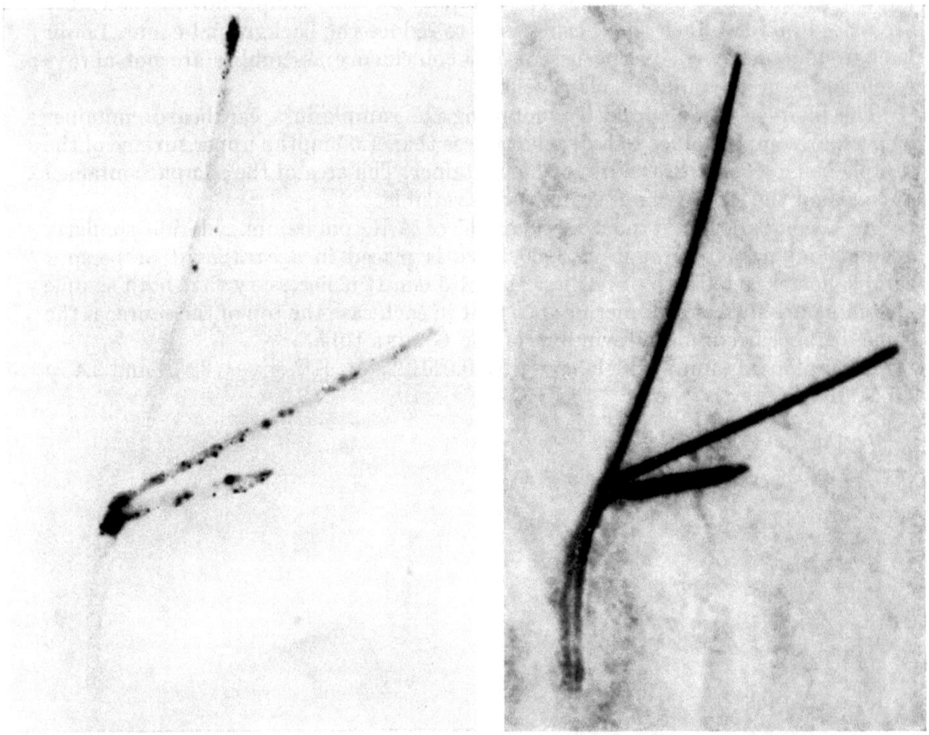

Fig. 3. Mounted specimen of *Lolium perenne* on left and autoradiograph on the right showing fallout particle contamination and accumulation of contamination in leaf sheath and in growing plant.

Boyd (1955) has given a comprehensive review of the methods to be adopted for obtaining satisfactory autoradiographs, while Russell and Possingham (1961) have outlined a simple procedure for examining plants contaminated with local fallout. The plant specimens are mounted on wadding pads backed by-plywood sheets and "Cellophane" is drawn tightly around the boards to retain the specimens in position. The mounted specimens are then exposed to commercial X-ray film in a deep freeze. Some idea of the approximate exposure time to allow before developing the film can be obtained by monitoring the mounted specimens with a portable ratemeter. Figures (2) and (3) are examples of the type of information that can be obtained from autoradiographs.

C. Methods of Estimating the Individual Radionuclides in Contaminated Plant Materials.

As discussed above, gross α and β measurements have only a limited value and it is often necessary to determine the individual nuclides which present the greatest hazard to man. In terrestrial plants these are Sr^{89-90}, Cs^{137}, I^{131}, Ba^{140} and Ra^{226}. In the case of aquatic plants, Co^{60}, Ru^{106}, Zn^{65}, $Fe^{55,59}$ and $Ce^{141,144}$ may also have to be considered as specific concentration mechanisms for these elements are known to occur. Analytical methods will be given below for the determination of Sr^{90}, I^{131} and Cs^{137}. In some instances analysis of samples for individual nuclides can be achieved simply by direct γ-spectrometry; care must be taken, however, with samples containing more than 4—5 gamma components. Where direct γ-spectrometry is not possible, radiochemical analysis must be used.

I. Radioactive Strontium.

Strontium-90 is the most hazardous to man of all the fission products for the following reasons:

It is produced in high yields in fission reactions, and is widely dispersed over the earth's surface because of its formation from an inert gas precursor krypton-90. The radioactive half-life of strontium-90 is long (twenty-eight years) and because of the chemical similarity to calcium it follows the latter from the soil, through vegetation and dairy products, and is eventually incorporated into human bone, where it is retained for a long time. Because of its importance, we are describing in detail the methods for estimating strontium-90.

Strontium-89 because of its shorter half-life of fifty one days is less important biologically than strontium-90, but its determination is of interest because a knowledge of the ratio Sr^{89}/Sr^{90} gives an indication of the age of the fallout.

Most analytical methods for the estimation of strontium-90 in plant material follow the same general pattern. These involve a detailed series of radiochemical steps followed by relatively complex counting procedures to achieve accurate and reliable results. Initially the plant material sample is ashed and a known amount of stable strontium carrier is added. The unwanted components of the sample are removed leaving only the strontium fraction which consists of stable strontium and radioactive strontium.

Stable strontium and yttrium are added as "holdback carriers" to minimise adsorption losses and to enable the percentage recovery of the strontium and yttrium activities to be assessed. For this recovery to be determined the stable strontium content of the original sample must be known. There are a large number of methods available for the determination of strontium in plant ashes. These include emission spectroscopy (THURBER, 1958), flame spectrophotometry, (HARRISON, 1958), X-ray fluorescence (ROBERTS, 1959), atomic absorption spectroscopy (DAVID, 1960), neutron activation analysis (SOWDEN and STITCH, 1957) and stable isotope dilution analysis with the mass spectrometer (LOVERIDGE et al. 1959).

The stable strontium consists partly of added carrier and partly of stable strontium originally present in the plant. The radioactive fraction contains strontium-90 and strontium-89. Strontium-89, a pure β emitter with a half-life of fifty-one days, decomposes to stable yttrium-89. Strontium-90, also a pure β emitter, has a half-life of twenty-eight years and decomposes to yttrium-90 which is also radioactive.

Yttrium-90 which is a pure β emitter with a half-life of 64 hours affords a very convenient means of distinguishing between the strontium-90 and strontium-89 components of the original sample. If yttrium carrier is added to the separated strontium fraction and the solution is stored for fourteen days the yttrium-90 will reach equilibrium with the parent strontium-90. After this storage period the yttrium may be removed from the strontium and its activity determined, thus giving a measure of the strontium-90 content of the sample.

The advantages of this method are that yttrium gives a more energetic β-particle (2.27 MeV) than strontium-90 (0.54 MeV) so that it may be counted with higher efficiency, and furthermore because of its short half-life its decay may be followed so as to ensure that the activity measured is entirely due to yttrium-90 uncontaminated with other nuclides of different half-lives. If necessary further yttrium carrier can be added to the parent strontium solution and after a further fourteen days the yttrium-90 may again be separated and counted.

Many methods have been published for the separation of strontium-90 from biological materials. Most of these fall into one of two classes based either on nitric acid precipitation or on an ion-exchange separation. Ion-exchange methods (MILTON and GRUMMITT, 1957; DAVIS, 1957) although they have not found wide acceptance, offer the promise of greatly simplifying the radio-chemical separation of strontium and calcium. However, the method most widely used at the present time is the nitrate precipitation method described by BRYANT, MORGAN and SPICER (1959). The method described below is essentially the same as that of BRYANT et al.

1. Sample Preparation.

The weight of fresh sample material should be noted. In the case of grass, the areas cut should also be recorded; it is usual to sample an area of about ten square metres and to cut the grass at a height of about 2.5 cm from the ground. From this large bulk of material small representative subsamples can be taken of the ground dry material. This is ashed and the analysis is proceeded with. Alternatively a large bulk of plant material can be ashed, ground and a subsample taken of this. Collection and chemical estimation when carried out on small plant samples leads to variable results and sub-sampling from an initially large sample is preferred.

The ashing of large samples can be difficult because often kilogram amounts of plant material have to be handled. The ashing of the sample can often be facilitated by pre-ashing as follows:

1. The sample is saturated with a 5% alcoholic solution of glycerine and set on fire. When the burning has ceased the resulting mass is porous and ready for final ashing in a muffle furnace; or

2. The dry sample is set on fire by a Bunsen burner from above and the rate of burning is controlled by the spacing of burner and sample.

Pre-ashing is considered complete when the sample volume has been reduced to about one fifth.

The final ashing of the sample should be carried out in a muffle furnace at a temperature of 450° C. Higher temperatures are undesirable because losses of volatile components such as caesium can occur, and because at elevated temperatures corrosion of the sample holding vessels is increased.

After ashing, the sample is ground if necessary and lightly brushed through a sieve (0.125 mm mesh) when any soil particles are retained and can be removed.

2. General Method of Bryant, Morgan and Spicer (1959).

Reagents required.

Strontium carrier solution: Sr^{++} 5 mg/ml, as nitrate.
Barium carrier solution: Ba^{++} 10 mg/ml, as nitrate.
Iron carrier solution: Fe^{+++} 5 mg/ml, as nitrate.
Yttrium carrier solution: Y^{+++} 10 mg/ml, as nitrate.
Nitric acid, 95—96% $^{w}/_{w}$.
Nitric acid, dilute, 5 M.
Hydrochloric acid, 10 M.
Hydrochloric acid, 5 M.
Phosphoric acid, 15 M.
Hydrofluoric acid, 40% $^{w}/_{w}$.
Perchloric acid, 60% $^{w}/_{w}$.
Oxalic acid solution, 8% $^{w}/_{w}$.
Acetic acid, dilute, 6 M.
Ammonium hydroxide, 17 M.
Ammonium hydroxide, carbonate-free, 5 M.
Ammonium carbonate, solid.
Ammonium acetate solution, 6 M.
Sodium chromate, solution, 1.5 M.
Methyl red indicator solution.
Hydrogen peroxide 100 vol.
Methanol.
Wash solution: Add 3 ml of 15 M phosphoric acid and 20 ml of 17 M ammonium hydroxide to 2 l of water.

Notes on carrier solutions. Prepare the strontium carrier solution by dissolving 4.2123 g of $SrCO_3$ (A.R.) in the minimum volume of 5 M.HNO_3 and making up to a volume of 500 ml in a volumetric flask. The yttrium carrier solution is best prepared from Y_2O_3. 3.175 gms of the oxide is dissolved in a few ml of 5 M.HNO_3 and made up to a volume of 250 ml. It is necessary to ensure that the yttrium carrier used is free of radioactivity due to contamination by rare earths. This is conveniently done by taking an aliquot of the carrier solution, precipitating the yttrium as oxalate, and counting the oxalate precipitate.

a) Preparative Procedure.

1. Weigh 10 gm of ash into a 10 cm platinum dish. Add 10 ml of strontium carrier solution, 1 ml of barium carrier solution, 10 ml of water and then cautiously about 40 ml of 40% hydrofluoric acid. Add 30 ml of 60% perchloric acid and evaporate almost to dryness. Cool, add a further 30 ml of perchloric acid and evaporate as before. Dissolve the residue in water with the addition of 5 ml of 10 M hydrochloric acid and transfer to a 600 ml beaker, adjust the volume to 250—300 ml and boil. Cool and filter through a Whatman 15 cm No. 40 filter paper into a second 600 ml beaker.

2. Add 5 ml of syrupy (15 M) phosphoric acid to the filtrate, followed by 17 M ammonium hydroxide while stirring, until an excess of ammonia is detected.

3. Transfer the contents of the beaker to a 200 ml pyrex centrifuge tube, centrifuging and rejecting the supernate, until all the precipitate has been collected in the tube. Wash the precipitate by stirring with 100 ml of wash solution, centrifuge and reject the supernate.

4. Dissolve the precipitate by the addition of 20 ml of fuming nitric acid and transfer to a measuring cylinder. Note the volume (V/ml) and return the solution to the centrifuge. Add 2.5 V—70 ml of fuming nitric acid and cool in running water for at least 30 minutes, stirring occasionally.

5. Centrifuge and reject the supernate. Dissolve the residue in 40 ml of distilled water, add 90 ml of fuming nitric acid and cool with stirring for at least 30 minutes. Centrifuge and reject the supernate. Transfer the residue to a 40 ml centrifuge tube with 20—30 ml of water.

6. Centrifuge and transfer the supernate to a second 40 ml centrifuge tube. Wash the precipitate with 10 ml of distilled water, centrifuge and add the supernate to that in the second tube. Reject the residue.

7. Make the solution in the centrifuge tube alkaline with ammonium hydroxide, add solid ammonium carbonate and heat in a boiling water bath to coagulate the carbonate precipitate. Centrifuge and reject the supernate.

8. Add 10.0 ml of water to the residue in the centrifuge tube and then 22.5 ml of fuming nitric acid cautiously. Cool with stirring for at least 30 minutes, centrifuge and reject the supernate.

9. Repeat paragraph 8. Dissolve the residue in 10—15 ml of water, add 1 ml of barium carrier solution and 1 drop of methyl red indicator. Neutralise excess acid with 5 M, carbonate-free ammonium hydroxide and add 1 ml of 6 M acetic acid and 2 ml of 6 M ammonium acetate. Dilute the solution to 30 ml and heat in a boiling water bath. Add 1 ml of 1.5 M sodium chromate and continue heating for 5 minutes. Centrifuge and transfer supernate to second centrifuge tube. Reject residue.

10. Make the solution alkaline with 17 M ammonium hydroxide, add solid ammonium carbonate and heat in a boiling water bath to coagulate the carbonate precipitate. Centrifuge and reject the supernate.

11. Dissolve the residue in dilute nitric acid, add 1 drop of 100-vol. hydrogen peroxide (see note) and add 1 ml of iron carrier. Heat and stir to remove carbon dioxide. Dilute to 15—20 ml and make alkaline with 5 M, carbonate-free, ammonium hydroxide, heating for 2—3 minutes to complete precipitation. Centrifuge and transfer the supernate to another 40 ml centrifuge tube. Reject the precipitate.

Note: The carbonate precipitate contains occluded chromate which is removed by reducing the chromium to the trivalent state and precipitating with the iron as hydroxide.

12. Precipitate as carbonate from the supernate by adding solid ammonium carbonate and heating in a water bath. Cool and filter the solution on a tared number 4 porosity sintered glass crucible. Wash with water and methanol successively and dry at 105° C for one hour. Cool in a desiccator and weigh as strontium carbonate. Calculate the % recovery of strontium, making allowance, where necessary, for the stable strontium content of the sample.

13. Dissolve the strontium carbonate off the sintered glass crucible in a minimum volume of 5 M nitric acid and transfer to a 40 ml centrifuge tube into which has been pipetted 1 ml of yttrium carrier solution. Store for at least 14 days.

14. Make the solution alkaline with carbonate-free ammonium hydroxide. Centrifuge and transfer the supernate to a second tube. Redissolve the precipitate in 6 M nitric acid, dilute to 10—15 ml and reprecipitate with ammonium hydroxide. Centrifuge and add the supernate to that from the previous precipitation (see note). Note the time of the first precipitation.

Note: The combined supernates should be retained in case counting of a further lot of yttrium is required.

15. Redissolve the precipitate in a *minimum* of 6 M nitric acid and add 20 ml of 8% oxalic acid solution. Heat in a water-bath for 10—15 minutes to produce

a granular precipitate. Cool and filter through a Whatman No. 42 filter paper. Wash three times with methanol and mount on a planchette. Begin counting immediately, noting the time.

16. When counting of the yttrium source is completed remove the paper with the source from the counting tray and transfer to a tared platinum dish, burn off the paper, ignite at 800—900° C for about one hour, cool in a desiccator and weigh as Y_2O_3.

17. Calculate the % recovery of yttrium.

b) Counting Procedure.

The activities obtained from most plant material samples are small and special low background β counters are usually needed. A counter commonly used consists

a

b

Fig. 4a and b. Low-background anticoincidence β-counter. The arrangement of the anticoincidence umbrella around the central end-window Geiger tube is shown in the (a) side-on, and (b) end-on positions.

of a central thin end-window Geiger tube (for example G.E.C. Types EHM 2S or GM 4 or equivalent) surrounded by an umbrella of 13 G24 or GM5 guard tubes through 180° and connected in anti-coincidence. This device enables the background count due to cosmic rays to be greatly reduced. The whole assembly is shielded by 5 cm of lead. Such an arrangement without the lead shield is shown in Fig. 4.

With such a counter a background count of less than 2 c.p.m. and an overall counting efficiency for yttrium-90 of 25—35% can normally be attained. Alternatively a central flow counter may be used in place of a thin end-window GEIGER counter (viz. Tracerlab CE-14 Low Background Beta Counter or equivalent). Such an arrangement, though usually somewhat more expensive, has a higher counting efficiency for weak beta emitters such as carbon-14, tritium and nickel-63.

In order to obtain positive identification of the activity measured, i.e. to ensure that none of it is due to substances other than yttrium-90, it is necessary to follow the decay of the yttrium-90. A number of procedures may be used. The following is that given by BRYANT et al. (1959):

Method. (1) Count the yttrium source for three periods spaced at approximately 3 days intervals, i.e. the whole counting time should occupy about a week. Each count should be 3000 or more if possible. If the total count does not reach this figure in 2 days, the count is recorded and the source left in the counter, readings being taken at intervals of 2 days.

(2) Calculate the mean count rate for each counting period, add the correction for paralysis losses and substract the background of the counter. Evaluate the Y^{90} decay factor f for each counting period from the expression:

$$f = \text{Antilog}_{10} \frac{0.3010\, t}{2.69}$$

where t is the time in days from the yttrium separation to half way through the particular counting periods.

(3) Calculate the count rate at the time of separation from the three count rates.

A series of values $f_1 C_1$, $f_2 C_2$, $f_3 C_3$ will be obtained where f_1, f_2, f_3 = the decay factors for counting periods 1, 2 and 3 and C_1, C_2, C_3 = the mean count rates for counting periods 1, 2 and 3.

The three fC values should agree, showing that the yttrium-90 is radiochemically pure. The presence of a long lived activity gives values of fC which increase steadily. In this case subtract a correction

$$\frac{f_3 C_3 - f_1 C_1}{f_3 - f_1}$$

from each mean count and calculate the corrected count rate at the time of yttrium separation.

Calculation of result. Strontium-90 content of sample

$$= \frac{C \cdot 10^8}{2.22 \cdot E \cdot (\text{Sr-}R)\,(\text{Y-}R) \cdot \text{Ca} \cdot w}\ \mu\mu c\, \text{Sr}^{90}/\text{g Ca}$$

where:

w = weight of plant sample ash taken (g).
C = activity of yttrium oxalate source, extrapolated to time of first yttrium hydroxide precipitation (c.p.m.).
E = % overall counting efficiency for Y^{90} of counter used.
Ca = % $^{w}/_{w}$ of calcium in plant ash sample.
(Sr-R) = % recovery of strontium
(Y-R) = % recovery of yttrium.

c) Addenda.

1. Calibration of counter for Y^{90}. To tubes containing 10 mg each of strontium and yttrium carriers, add known aliquots of standardized Sr^{90} solution. Store for a minimum of 18 days and then proceed as in paragraphs 14, 15, 16 and 17 of "Procedure". Correct the yttrium source count rate at zero time for yttrium recovery and calculate counter efficiency (E) from the known disintegration rate of the aliquot.

2. Determination of calcium content. Weigh accurately about 1 g of ash into a 9 cm platinum dish. Add 20 ml of distilled water and then cautiously add 20 ml of 60% perchloric acid and about 20 ml of 40% hydrofluoric acid. Warm on a hot-plate in the fume cupboard and evaporate until dense white fumes of perchloric acid appear. Cool, add a further 20 ml of hydrofluoric acid and evaporate as before. Continue the evaporation almost to dryness and dissolve the residue in 40 ml of water. Transfer to a 250 ml beaker and add 10 ml 10 M hydrochloric acid. Dilute to 150 ml and add 1—2 drops of methyl orange indicator. Add 20 ml saturated ammonium oxalate solution and 10 g urea. Cover with a watch glass and transfer to a hot plate. The solution should remain clear until boiling has commenced. If cloudy, add the minimum of 10 M hydrochloric acid necessary to clear the solution. The pH is now raised gradually by hydrolysis of the urea to form ammonia in situ. Boil gently until the indicator changes colour (pH 4). Filter off the calcium oxalate precipitate through a tared sintered-glass crucible, porosity 3, transferring and washing the precipitate with 0.1% ammonium oxalate rolution. A final wash with a minimum amount of distilled water is necessary to semove the ammonium oxalate. The precipitate is washed finally with methanol and then dried in the oven at 105° C for 2 hours. Calculate the calcium content of the vegetable ash, the calcium being weighed as calcium oxalate monohydrate, conversion factor 0.27430.

II. Iodine-131.

1. γ-Spectrometry.

For the determination of the iodine-131 content of biological samples, particularly plant material, the γ spectrometer is the preferred instrument. When using γ spectrometry no sample preparation is needed and even bulky samples as hay can be measured directly. As a result, many samples can be analysed in a short time (the exact time will depend upon, among other things, the amount of activity in the sample and the degree of precision required). The method is extremely sensitive, as little as 2 $\mu\mu$c I^{131} per ml can readily be detected using a sample of volume one litre.

Another great advantage of γ-spectrometry is that a number of different nuclides can be determined on the one sample without any chemical separation. Among the more important γ emitters may be mentioned I^{131}, Nb^{95}, Zr^{95}, Cs^{137}, Ru^{106}, Zn^{65} and Co^{60}.

Methods have been given by PERKINS (1958), and WOODMAN et al. (1958).

Essentially, the γ-spectrometer consists of a scintillator, usually a thallium activated sodium iodide crystal linked to a photomultiplier. These are shielded from light and moisture by a thin aluminium can. Where low backgrounds are required, the crystal, photomultiplier and sample may all be contained in a lead castle. γ-rays from the sample easily penetrate the aluminium can and are absorbed in the crystal giving rise to scintillations which can be detected by the photomultiplier. The output pulses from the photomultiplier, after suitable amplification, are passed to a single or multi-channel pulse height analyser and resolved according to the energies of the γ-rays emitted by the sample. The instrument may be calibrated using definite amounts of known nuclides. The principal disadvantage of the instrument is its high initial cost.

2. Radiochemical Method.

Where facilities for γ-spectroscopy are not available or the mixture of nuclides is complex, a radiochemical method must be used. The most generally used method is that due to BERGH (1958). In this method stable iodine carrier is added to the sample solution, the iodide is extracted into carbon tetrachloride, back-extracted into water and precipitated as silver iodide, mounted and counted using an end-window GEIGER-MULLER counter. A somewhat simpler method than BERGH's, developed by LEBOEF (1950) specifically for the determination of radioiodine deposited on vegetation will be given below in detail.

It is interesting to note that as an integrated measure of fallout contamination of pastures in Australia after the atomic weapons tests at Monte Bello and Maralinga in 1956, MARSTON (1958) measured the iodine-131 present in the thyroid glands of sheep and cattle. In his procedure the gland was fused with KOH at 550° C, the aqueous solution of the melt was acidified with H_2SO_4 in the presence of excess $NaNO_3$ and exhaustively extracted with $CHCl_3$, in which solvent, after washing with $NaHCO_3$, the iodine was titrated with 0.01 N.$Na_2S_2O_3$ in the presence of $NaHCO_3$. A one inch diameter well-type scintillation counter was used to estimate the iodine-131 content of the gland. There was no evidence of any radioactive substance other than iodine-131 in the glands examined.

Iodine-131 in vegetation (LEBOEF, 1950).

Reagents.

Iodide carrier solution, 10 mg I^-/ml.
Sodium bisulphite solution, 0.5 M.
Sodium hydroxide solution, 0.2 M.
Aluminium nitrate solution, 0.3 M.
Nitric acid, 8 M.
Silver nitrate solution, 0.1 M.

Procedure.

1. Add 5 g of the sample of vegetation to 97 ml of 0.2 M sodium hydroxide, 1 ml of 0.5 M sodium bisulphite, and 2 ml of iodide carrier solution.
2. Heat the mixture to boiling (about 3 minutes) and filter through an 8.0 cm *asbestos* filter disc using a few milligrammes of filter aid.
3. Boil the filtrate for 20 minutes.
4. Add 20 ml of 0.3 M aluminium nitrate solution with stirring.
5. After cooling, determine the pH of the mixture and adjust to pH 6.0, adding dilute nitric acid or sodium hydroxide as required.
6. Centrifuge the sample, using 90 ml plastic tubes, and discard the aluminium hydroxide precipitate.
7. Adjust the pH of the supernate to 2.0 by adding dilute nitric acid.
8. Add 10.0 ml of 0.1 M silver nitrate solution and heat nearly to boiling.
9. Filter off the silver iodide precipitate using Whatman No. 5 paper and wash with a small amount of alcohol.
10. Dry at room temperature, mount the filter paper on a sample card, and count.

III. Caesium-137.

The preferred method for the analysis of Cs^{137} in plant materials is γ-spectrometry. Radiochemical analysis of caesium in plant materials is usually undertaken together with Sr^{90} analysis and provided the ashing temperature is not allowed to exceed 450° C and stable caesium carrier has been added Cs^{137} can be analysed on the same sample as was used for the Sr^{90} analysis. In this case the supernatant fluid from the nitrate precipitate of strontium will contain the caesium.

The radiochemical separation of caesium is based on two separations involving the relative solubilities of the alkali alums and the alkali chloroplatinates. Separation of caesium from the bulk of the alkali elements and the mixed fission products is accompanied by co-crystallisation of caesium with ammonium aluminium sulphate. The ammonium salts are then decomposed by heating to 500° C for at least 24 hours. The caesium is separated from the aluminium sulphate by precipitation as the chloroplatinate, in which form it is mounted and counted (WHITNEY, 1957).

References.

BANKS, C. J.: U.K.A.E.A. Repl: IGO-AM/W. 35 (1956). — BERGH, H.: Proceedings of Second United Nations International Conference on the Peaceful Uses of Atomic Energy, **18**, 508 (1958). — BOYD, G. A.: Autoradiography in biology and medicine. New York: Academic Press Inc. 1955. — BRYANT, F. J., A. MORGAN and G. S. SPICER: U.K.A.E.A. Rept. AERE-R 3030 (1959). — BURTON, J. D., G. M. MILBOURNE and S. R. RUSSELL: Nature (Lond.) **185**, 498 (1960).

DAVID, D. J.: Nature (Lond.) **187**, 1109 (1960). — DAVIS, P. S.: Nature (Lond.) **183**, 674 (1959).

ELLIS, F. B., and E. R. MERCER: Agricultural Research Council (U.K.) Radiological Laboratory Report No. 2 p. 42 (1960). — ETHERINGTON, H.: Nuclear engineering handbook p. 11—100. New York: McGraw Hill 1958.

FOWLER, E. B., and C. W. CHRISTENSON: Science **130**, 1689 (1959).

HARRISON, G. E.: Nature (Lond.) **182**, 792 (1958).

LARSON, K. H.: Congress of the U.S. Joint Committee on Atomic Energy, Vol. III, p. 2006 (1959). — LEBOEF, M. B.: U.A.S.E.C. Rept. HW-15743 (1950). — LIBBY, W. F.: Science **123**, 657 (1956); — Proc. Nat. Acad. Sci. **43**, 758 (1957); — Congress of the U.S. Joint Committee on Atomic Energy Vol. III p. 2227 (1959); — Science **129**, 866 (1959). — LOPER, H. B.: Science **129**, 884 (1959). — LOVERIDGE, B. A., R. K. WEBSTER, J. W. MORGAN, A. M. THOMAS and A. A. SMALES: U.K.A.E.A. Rept. AERE C/R2657 (1959).

MARLEY, W. G.: M.R.C. (U.K.) Hazards to man of nuclear and allied radiations London: H. M.S.O. (1956). — MARSTON, H. R.: Aust. J. Biol. Sci. **11**, 382 (1958). — MIDDLETON, L. J.: Int.J. Rad. Biol. **1**, 387 (1959). — MILTON, G. W., and W. E. GRUMMITT: Canad. J. Chem. **35**, 541 (1957).

NISHITA, H., and K. H. LARSON: U.S.A.E.C. Rept. U.C.L.A. 401 (1957).

PERKINS, R. W.: Proceedings of Second United Nations International Conference on the Peaceful Uses of Atomic Energy **28**, 445 (1958).

ROBERTS, W. M.: Nature (Lond.) **183**, 887 (1959). — ROMNEY, E. M., W. L. EHRLER, A. H. LANGE and K. H. LARSON: Plant and Soil **12**, 41 (1960). — RUSSELL, R. S.: Nature (Lond.) **182**, 834 (1958). — RUSSELL, R. S., and J. V. POSSINGHAM: Progress in Nuclear Energy, Series 6, Vol. III, Biological Sciences: The entry of fission products into food chains p. 1. London: Pergamon Press 1961.

SOWDEN, E. M., and S. R. STITCH: Biochem. J. **67**, 104 (1957). — STEINBERG, E. P., and L. E. GLENDENIN: Proceedings of United Nations International Conference on the Peaceful Uses of Atomic Energy **7**, 3 (1955).

THURBER, D. L.: Science **128**, 256 (1958). — TURNER, R. C., J. M. RADLEY and W. V. MAYNEORD: Brit. J. Radiol. **31**, 397 (1958a); — Health Physics I, 268 (1958b).

WHITNEY, I. B.: ed. U.S.A.E.C. Rept. NYO-4700 (1957). — *World Health Organization:* Methods of Radiochemical Analysis Tech. Rept. Ser. No. 173 (1959). — WOODMAN, F. J., T. G. CLINTON, W. FLETCHER and C. A. WELCH: Proceedings of Second United Nations International Conference on the Peaceful Uses of Atomic Energy **28**, 423 (1958).

Sachverzeichnis.

(Deutsch — Englisch).

Wegen allgemeiner Stichworte wie Extraktion, Abtrennung, Reinigung usw. der einzelnen Stoffgruppen vergleiche man auch das Inhaltsverzeichnis am Anfang dieses Bandes.

cis-, trans-, n-, D-, L- und ähnliche Isomere sind unter dem Anfangsbuchstaben der Verbindung und nicht unter dem Präfix eingeordnet.

Alle iso-Verbindungen finden sich unter Iso-.

Ä, Ö, Ü sind wie Ae, Oe, Ue eingereiht.

Bei gleicher Schreibweise in beiden Sprachen sind die Verbindungen jeweils einfach aufgeführt.

Subject Index.

(English — German)

For general terms such as extraction, isolation, purification etc. of various groups of compounds cf. also the table of contents at the beginning of the volume.

cis-, trans-, n-, D-, L- and similar isomers are listed according to the first letter of the following word.

All iso-compounds are to be found under Iso-.

Ä, Ö, Ü are taken as Ae, Oe, Ue.

Where English and German spelling of a word is identical, the italicised (German) entry is omitted.

Table des Matières.

Pour la contribution écrite en français:

Henry Prat, Méthodes Calorimétriques pour l'Analyse des Végétaux

(français-allemand)

Sachverzeichnis

Zu dem französischen Beitrag

Henry Prat, Méthodes calorimétriques pour l'Analyse des Végétaux

(deutsch-französisch)

MIX
Papier aus verantwortungsvollen Quellen
Paper from responsible sources
FSC® C105338

If you have any concerns about our products,
you can contact us on
ProductSafety@springernature.com

In case Publisher is established outside the EU,
the EU authorized representative is:
Springer Nature Customer Service Center GmbH
Europaplatz 3, 69115 Heidelberg, Germany

Printed by Libri Plureos GmbH
in Hamburg, Germany